MINE WASTE AND TAILINGS CONFERENCE 2023

Exploring all aspects of mine waste and tailings management

13–14 JULY 2023
BRISBANE, AUSTRALIA

The Australasian Institute of Mining and Metallurgy
Publication Series No 3/2023

AusIMM

Published by:
The Australasian Institute of Mining and Metallurgy
Ground Floor, 204 Lygon Street, Carlton Victoria 3053, Australia

ISBN 978-1-922395-22-1

ORGANISING COMMITTEE AND REVIEWERS

Prof David Williams
FAusIMM
Conference Organising Committee Chair

Angélica Amanda Andrade

Lis Boczek
AAusIMM

David Brett

Dr Hernan Cifuentes
MAusIMM

Prof Andy Fourie
MAusIMM

Theo Gerritsen

Eve Greenbury

Symon Jackson
FAusIMM

Robert Jordan

Dr Marcelo Llano-Serna

Fernanda Maluly Kemeid
MAusIMM

Allan McConnell
MAusIMM

Arun Muhunthan

Heidi Reynders

Kathy Tehrani

Dr Chenming Zhang

AUSIMM

Julie Allen
Head of Events

Fiona Geoghegan
Manager, Events

Samara Brown
Conference Program Manager

FOREWORD

On behalf of the Organising Committee, we are delighted to welcome you to the Mine Waste and Tailings Conference 2023, the fourth in the highly successful series first launched in 2015.

This conference, co-hosted by AusIMM and The University of Queensland, sets the benchmark for sharing knowledge and experience on mine waste and tailings management, sustainable practice, and mine closure.

As the demand for minerals increases exponentially in the face of diminishing ore grades, the mining industry faces sustained threats to its financial and social licences to operate in the face of recent tailings dam failures and the longer-term threat of acid and metalliferous drainage.

The mining industry accepts that the majority of the world's future minerals will come from low-grade, high-tonnage, ultra-mechanised operations. As a result, a higher production of waste rock and tailings is expected. Mining systems will require re-engineering, based on the paradigm shift that the success of mining organisations has become fundamentally dependent upon waste management. As sustainable development principles are increasingly applied worldwide, this conference will cover all aspects of life cycle waste rock and tailings management, from site selection and design to post-closure care.

With four preeminent keynote speakers and four expert panel discussions I am confident this conference will generate valuable information and knowledge sharing that is critical to future safe operations and ensure their success.

I would like to thank the Organising Committee, authors, paper reviewers, attending delegates and all our sponsors and exhibitors.

Yours faithfully,

Prof David Williams FAusIMM
Mine Waste and Tailings Conference Organising Committee Chair

SPONSORS

Major Conference Sponsor

BHP

Platinum Sponsors

Klohn Crippen Berger

red earth engineering
A Geosyntec Company

Innovation Sponsor

GHD

Gold Sponsors

INSITU GEOTECH SERVICES

SGM environmental

TENEO

WSP

Silver Sponsors

ATC WILLIAMS

CONETEC

RioTinto

SCIDEV

Conference Proceedings Sponsor

sixense

Name Badge and Lanyard Sponsor

AngloAmerican

Technical Session Sponsors

Ausenco

WEIR Minerals

Coffee Cart and Break Sponsor

BLACK INSITU TESTING

CONTENTS

Commingling of tailings and coarse wastes

Emergency response

Geochemical and geotechnical testing of mine wastes

Governance and standards

Importance of site climatic, topographic, seismic and social settings

Innovative mine waste management

Mine closure

Monitoring and surveillance

Operational aspects and case histories

Selection of parameters for TSF and WRD design

Tailings dam breach and runout assessment

Tailings dewatering and placement

Tailings liquefaction and seismic response

Water and AMD management

Commingling of tailings and coarse wastes

Co-disposed coal rejects and tailings strengths and existing co-disposal strategies review

J Li[1], D Payne[2], R Rosales[3] and M Stimpfl[4]

1. FAusIMM(CP), Principal Geotechnical Engineer, BHP, Moranbah Qld 4744.
   Email: jianping.li@bhp.com
2. Manager Geotechnical Services, BHP, Moranbah Qld 4744. Email: dan.payne@bhp.com
3. Manager Mine Scheduling, BHP, Moranbah Qld 4744. Email: roy.rosales@bhp.com
4. Principal Environmental Geochemist, BHP, Perth WA 6000. Email: marilena.stimpfl@bhp.com

ABSTRACT

The mine discussed in this paper is an open cut coalmine located in Bowen Basin, Queensland. It has been co-disposing mixed plant rejects (MPR) and dewatered tailings (DT) within spoil dumps for over nine years since commencement of operations. MPR is loaded onto trucks at the Coal Handling and Processing Plant (CHPP), transported to prestrip waste dumps and then co-disposed on dump bench through either paddock dumping or at active tipheads along with waste truck tipping. DT is disposed in tailings cells constructed in prestrip truck dumps due to wet weather events and other operational constraints, and then capped with spoil. Physical and mechanical properties, including total moisture contents, densities and shear strengths of the co-disposed MPR and DT, were unknown until the Closure Planning team, Principal Environmental Geochemist, organised a study program to assess geochemical properties on *in situ* waste. As part of the program, the geotechnical team collected MPR and DT samples from sonic core drilling, which were then sent to the University of Newcastle for laboratory testing. This paper presents the findings from sonic core drilling and laboratory geotechnical testing, and validating and optimising outcomes from reviewing of the existing co-disposal strategies at the mine through dump stability modelling and by using the shear strength results from triaxial compression testing conducted on the samples collected during the fieldwork.

INTRODUCTION

The subject mine discussed in this paper has never considered disposing tailings in tailings storage facilities or tailing dams at all design and operational stages for environmental protection. Instead, all coal washery waste materials from the Coal Handling and Processing Plant (CHPP), ie rejects and tailings, are co-disposed within the mining spoil dumps. Prior to co-disposing, the tailings is mechanically dewatered from slurry to paste or cake through Belt Press Filters. Majority of dewatered tailings (DT) is combined with the rejects to form mixed plant rejects (MPR) and co-disposed within prestrip spoils. Due to wet weather conditions or other operational constraints, only 5–10 percent of DT is separately disposed in wet cells constructed in spoil dumps. The cells are then capped with spoil.

The subject mine implements four main co-disposal methods (Li, Payne and Hooi, 2018; Li *et al*, 2021), which are paddock dumping, active tiphead co-disposing and reject curtain placement for MPR as shown in Figure 1; and DT cell disposing as shown in Figure 2. Initial total moisture (TM) of MPR by weight from CHPP reject bin is up to 22 per cent, while for DT it is approximately 37 per cent. Among all four co-disposal strategies, TM content of MPR dumped at active tiphead and of MPR curtains reduces soon after deposition and it is subject to minimal impact from weather conditions. This translates to a significant increase in MPR strength. Moisture content of the paddock dumped MPR is more variable depending on weather conditions, the extent of desiccation and the time lag before it is over-dumped by the dump lift above. In addition, establishment of drainage prior to paddock dumping prevents water ponding at low points during heavy rainfall events; therefore, ensuring that the paddock dumped MPR remains unsaturated enhances tiphead stability.

Figure 2 also shows consistency variations of DT in cells prior to capping depending on the degree of desiccation. Field observation found that if it is wet (refer to Figure 2, left), DT is more easily mingled with capping Categories 2 and 3 spoils, comprising weathered and fresh Permian, and the DT in cell is also squeezed up during capping. This mingling and blending process results in mixed DT having much higher strength than DT only material due to reduced moisture content. On the other hand, the desiccated DT in cells as shown in Figure 2 (right) experience minimal blending with capping spoils, therefore, a consistent tailing layer is present in the cell after capping. This consistent DT layer has the most impact on dump stability.

FIG 1 – MPR co-disposal by paddock dumping, tiphead co-disposing and MPR curtain placement.

FIG 2 – Dewatered tailings (DT) disposed in cells with different desiccation periods.

MPR and DT gain strength and become significantly more competent than those directly loaded out from the Reject Bin for co-disposing in spoil dumps, due to moisture content reduction. However, the appearance of the co-disposed MPR and DT was unknown until the BHP Closure Team commenced a spoil oxidisation studying project in early 2022. Geotechnical and mine planning teams were involved in the project, provided assistance to determine the most appropriate sonic drilling locations, and collected MPR and DT samples in the field for geotechnical laboratory testing.

This paper presents field observations of the co-disposed MPR and DT. It also deep dives into the sonic drilling and sampling for geotechnical testing, laboratory testing results from the University of Newcastle. Finally, the paper discusses the validation and optimisation of the current MPR co-disposal methods through dump stability modelling supported by the laboratory data.

SONIC DRILLING PROGRAM

The spoil oxidisation project led by the BHP Closure Team was to investigate acid metalliferous drainage (AMD) of the MPR and DT in spoil dumps over time. The intention of the program is to monitor AMD and oxidisation history of MPR and DT materials in the future with all co-disposal methods by geochemical instruments installed in the drill holes. A further consideration for location selection of the drill holes is minimal operational interaction, to allow the geochemical monitoring instruments easy accessibility for data login or acquisition for at least five years. Figures 3 and 4 show plan and section views of the holes related a DT cell and co-disposed MPR, where we can see that:

- Six holes are planned, which are located with 50 m × 55 m area. Hole collars are located on 300RL dump bench.

- Holes H1 and H5 intersect 1.5 to 3.1 m DT in cell and around 3.7 m paddock dumped MPR placed at 260RL before the cell was constructed.

- H2 hole intersects 7.7 m thick DT above the cell embankment batter based on survey data, and one truckload of MPR through tiphead co-disposed MPR.

- H3 hole is similar to H2 and located on the cell embankment batter, intersects a 7.6 m thick DT if not disturbed by capping and around a 2 m thick MPR at base of the cell.

- H4 hole intersects DT by 9.3 m; an approximately 2.5 m deep water pooled in the area prior to DT emplacement in the cell. This water would likely impact on the moisture content of the DT, and particularly, the portion of DT submerged in water would likely have a higher moisture content than the DT at H2 and directly sourced from the CHPP (~37 per cent).

- H2A hole intersects tiphead co-disposed MPR at 18–21 m below hole collar.

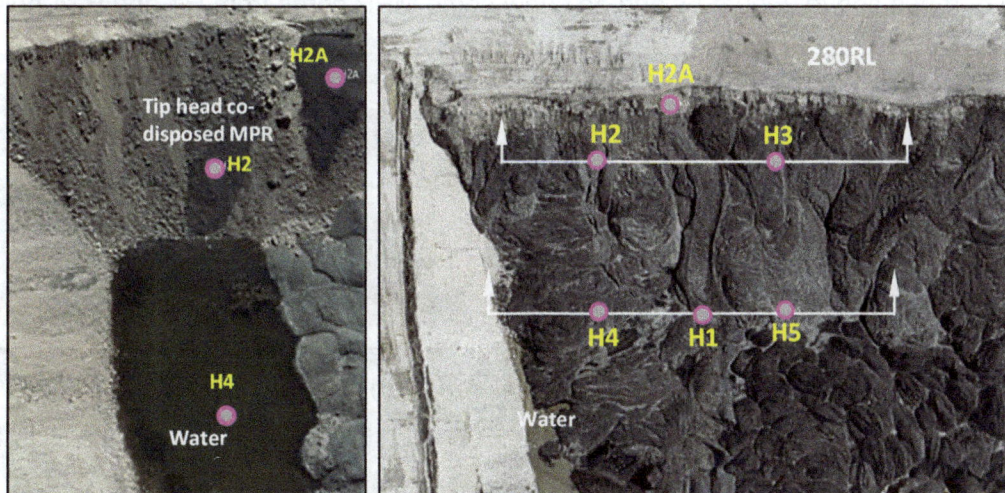

FIG 3 – Sonic hole locations related to a cell before and after DT placement.

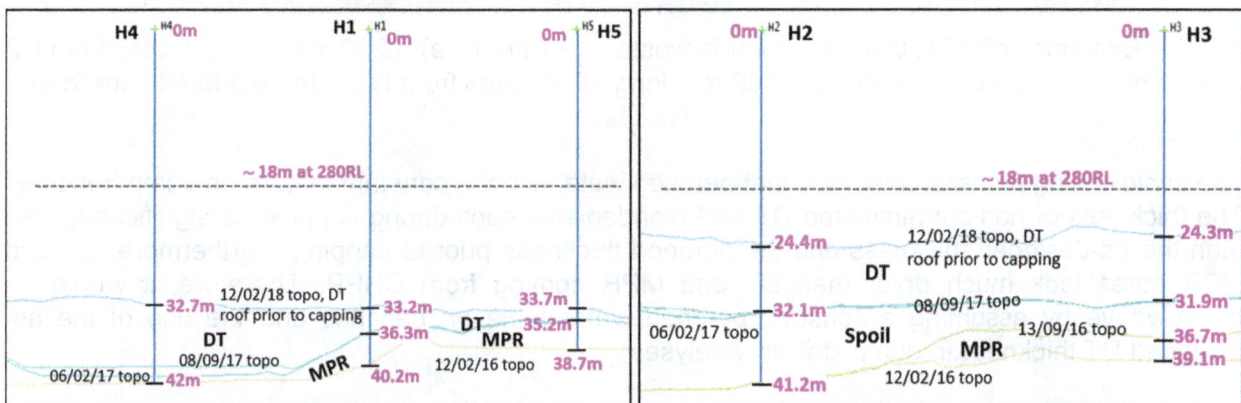

FIG 4 – Section views for relation of the sonic drill holes to materials in the cell.

Sonic drilling does not introduce water and any other liquid into the hole, thus it does not change the moisture contents of the *in situ* waste materials. Therefore, the moisture content of the core reflects the true moisture contents of the spoil, DT and MPR in the dump. The cell capping commenced in May 2018, therefore, sonic drilling took place four years after DT capping, or six years after the MPR placement at the cell base before cell construction. Sonic drill core is 100 mm in diameter and 3 m long for each drilling run. Therefore, the drilling depth is counted in multiples of 3 m. From sonic drilling, it is found that:

- No perched or ponded water on both 260RL and 280RL benches, either inside or outside of the cell is intersected in all holes except H3. Water from 280RL bench in H3 hole caused hole collapse and then drilling ceased.

- Eight discontinuous section of DT cores (the weakest portion of cores having minimal contamination of spoil) are uncovered from H2 hole with a total length of approximately 3 m, representing 40 per cent of total length as shown in Figure 4. The reminder 60 per cent of DT was either mixed with or replaced by capping spoil. Figure 5a shows a short DT core and spoil column replacing DT during capping.

- Only four individual samples up to 0.3 m long were collected from H4 hole (Figure 5b), as the saturated 9.3 m thick DT in H4 hole (as measured before capping, Figures 3 and 4) was well blended with the capping spoil. Total length of the weakest DT core in H4 is approximately 15 per cent of the DT thickness prior to capping. In addition, the DT cores from H4 hole looked more solid and competent than the H2 DT cores, as the DT moisture is diffused into the blended capping spoil. This finding confirmed the visual observations during cell capping as shown in Figure 2 (left).

- MPR from H1, H5 and H2A holes were solid core pieces due to compression exerted by the spoil dumped above. Figure 5c and 5d show some MPR cores from H1 and H2A holes.

FIG 5 – Examples of MPR placement within waste rock dumps: a) a 200 mm long DT core from H2 hole; b) a DT core from H4 hole; c) 100 mm long MPR cores from H1 hole; d) a MPR core from H2A hole.

The sonic drilling process confirmed that perched water is not a constant feature on dump benches. The thickness of non-contaminated DT (not blended with spoil during capping) is significantly less than the as-designed thickness and the dumped thickness prior to capping. Furthermore, DT and MPR cores look much dryer than DT and MPR coming from CHPP. Therefore, it would be conservative by assuming a constant perched water table on benches and the use of the as-designed DT thickness in dump stability analyses.

LABORATORY TESTING AND RESULTS

Laboratory geotechnical testing on the DT and MPR samples were conducted at the University of Newcastle. The testing scope included measurement of physical properties of the samples as supplied to reflect their *in situ* conditions, particle size analysis (AS 1289.3.6.1-2009, 2009) and undrained and unconsolidated (UU) triaxial compressive shear strength testing. The undrained testing is to reflect the actual moisture conditions of the co-disposed DT and MPR in dumps. The moisture contents of DT and MPR from sonic drilling are the results of moisture equilibrium with surrounding spoils over the years, and remain unchanged with time. Five single stage testing for the samples from each hole was conducted with confining stresses of 50 kPa, 200 kPa, 350 kPa, 500 kPa and 1000 kPa. The physical properties include total moisture content, bulk density, dry density and specific gravity for as-cored samples and tested specimens. Table 1 presents the average values of all core samples (as-cored) from each hole and the average total moisture and densities and saturation degree values.

TABLE 1

Moisture, densities, SG and saturation degree for as-cored samples.

Hole ID and material	Moisture content (%)	Bulk density (g/cm³)	Dry density (g/cm³)	Specific gravity	Saturation degree
H1 (MPR)	12.04	1.70	1.46	2.25	0.60
H2 (DT)	19.62	1.74	1.45	2.21	0.92
H4 (DT)	15.82	1.66	1.44	2.29	0.74
H5 (MPR)	12.87	1.54	1.66	1.99	0.51
H2A (MPR)	10.03	2.12	1.93	2.36	0.83

As expected during sonic drilling that the moisture content of DT as-cored (up to 20 per cent) is much lower than the DT when placed (around 37 per cent). Similarly, the moisture content for the co-disposed MPR, 10–13 per cent, is lower than the MPR when they are disposed in dumps (around 22 per cent). The saturation degree values suggest that the co-disposed DT and MPR within spoil dumps are unsaturated in general, but close to saturation for the weakest DT samples from H2 hole.

Because the drilling objective was to assess geochemical properties of *in situ* wastes, only half of each DT core piece and small MPR solid core samples were collected for geotechnical laboratory testing. Therefore, after measurements of the densities and moisture content of the as-cored samples, all material from each borehole was crushed to particles size <2.36 mm, and homogenised prior to mechanical testing (Buzzi *et al*, 2023), due to size constraints on 50 mm diameter and 100 mm long triaxial specimens. As a result, the MPR samples from H1, H5 and H2A have very similar particle size distributions to DT from H2 and H4 (Figure 6). Figure 6 also compared the particle size distributions of the homogenised samples with DT and MPR samples from CHPP (Li, Payne and Hooi, 2018; Li *et al*, 2021). The difference between H2 and H4 DT and CHPP DT suggests that some contamination exists from spoil during capping.

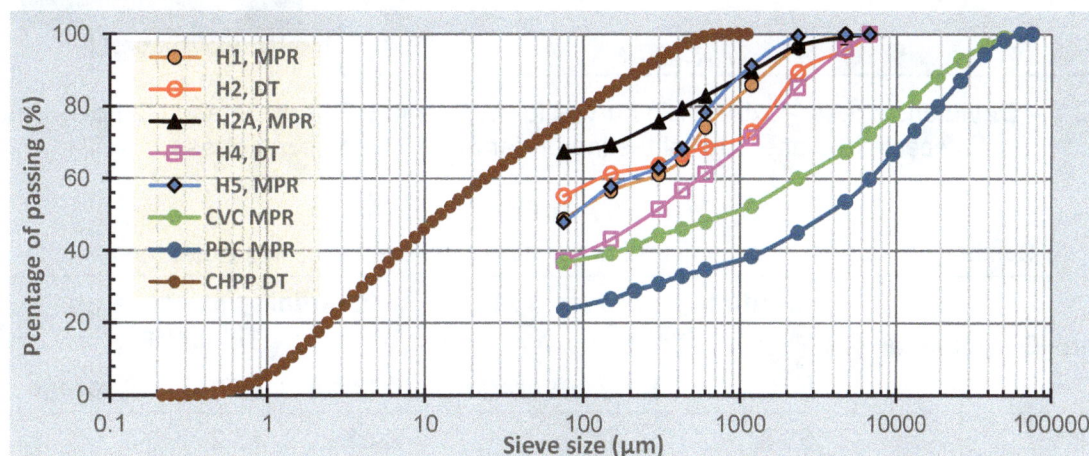

FIG 6 – Particle size distributions for homogenised testing samples and compare with DT and MPR prior to disposing (Li, Payne and Hooi, 2018; Li *et al*, 2021).

Figure 7 gives the deviatoric stress versus axial strain curves for MPR and DT specimens from H1 and H2 holes. The MPR from H1 hole can take much higher deviatoric (or axial) stress than the DT from H2 under the same confining stresses. The higher deviatoric stresses result in higher shear strength. Table 2 summarises the undrained shear strengths (cohesion and friction angle) of DT and MPR samples from each hole and their average values. The strength values for the two confining stress ranges, up to 500 kPa and up to 1000 kPa, could be applicable to shallow and high overall dumps, respectively. Figures 8 and 9 present the Mohr circles and Mohr–Coulomb equations for all MPR and DT samples tested to obtain their average shear strengths. Shear strength from previous tests for CHPP DT and MPR (Tucker, Li and Todd, 2015; Li *et al*, 2021) are also shown in the figures for comparison (confining stresses up to 400 kPa). The co-disposed DT in cell and MPR through paddock dumping are significantly more competent than the DT and MPR when placed in dumps.

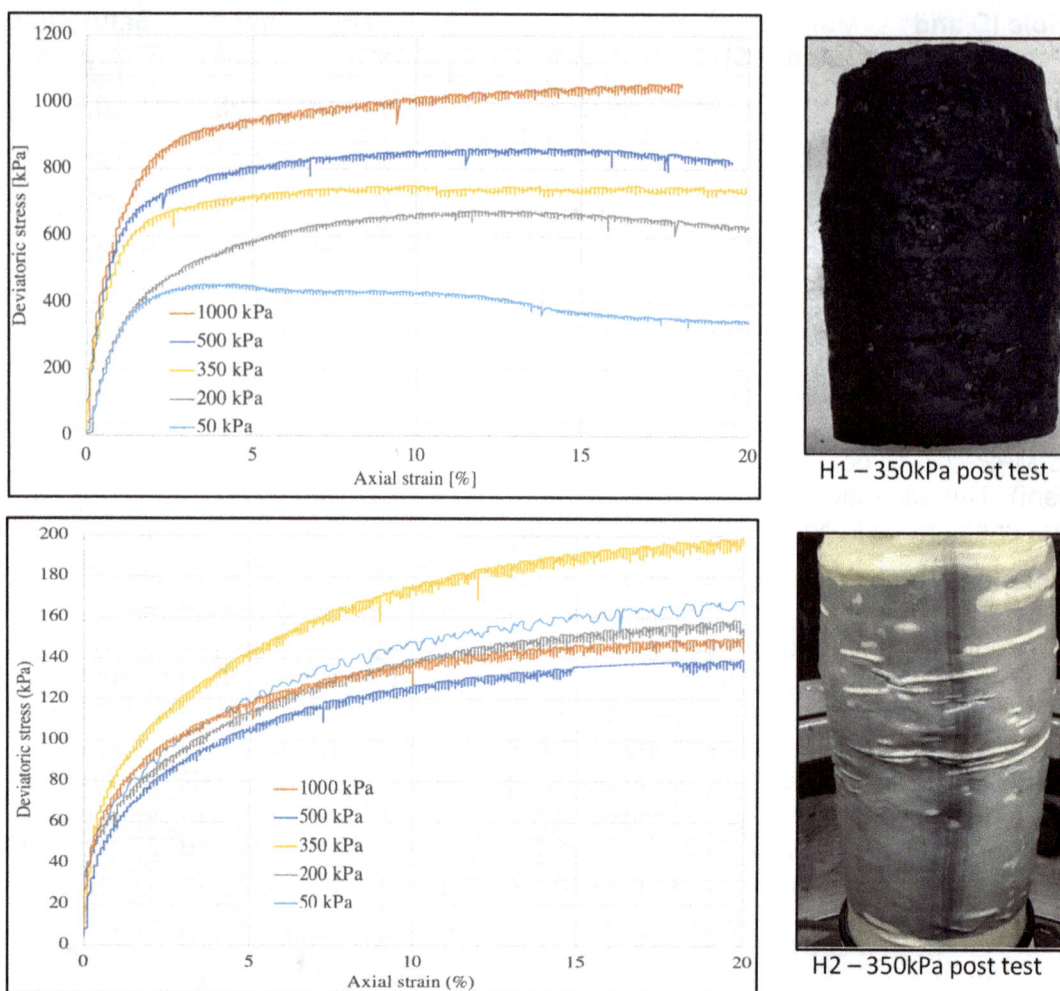

H1 – 350kPa post test

H2 – 350kPa post test

FIG 7 – Deviatoric stress – axial strain curves for MPR and DT from H1 and H2 holes, and specimens after triaxial testing under 350 kPa confining stress.

TABLE 2

Summary of laboratory strength testing results with different confining stress ranges.

Hole ID	Material	Confining stresses up to 500 kPa			Confining stresses up to 1000 kPa		
		C (kPa)	φ (°)	Average	C (kPa)	φ (°)	Average
H1	MPR	163.8	17.6	C = 135 kPa φ = 20°	203.2	13.2	C = 170 kPa φ = 16°
H5		96.4	25.1		110.9	25.9	
H2A		159.2	14.9		209.3	9.0	
H2	DT	66.1	1.8	C = 70 kPa φ = 7°	66.4	1.8	C = 80 kPa φ = 5°
H4		71.3	11.4		96.9	7.4	

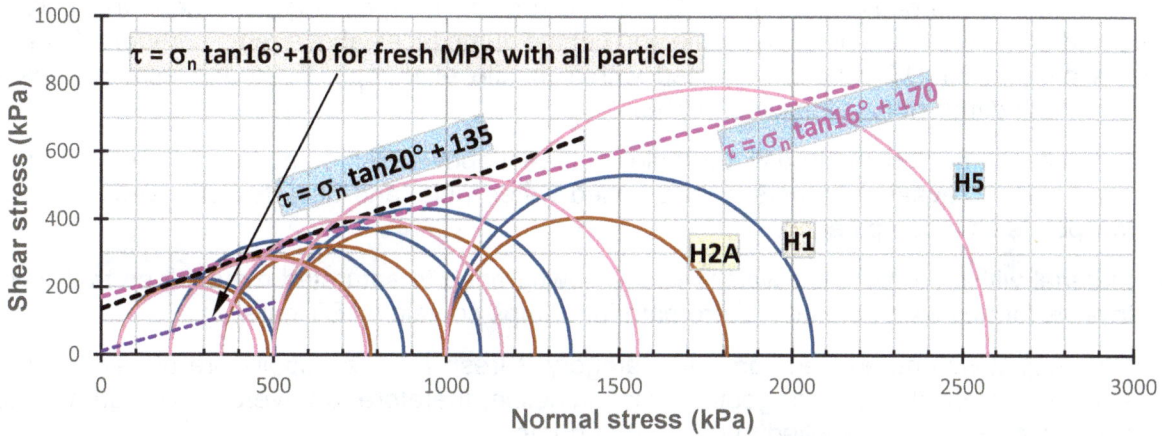

FIG 8 – Average shear strengths of MPR with confinement ranges up to 500 kPa and 1000 kPa, and shear strength for fresh CHPP MPR with all particles (Li *et al*, 2021).

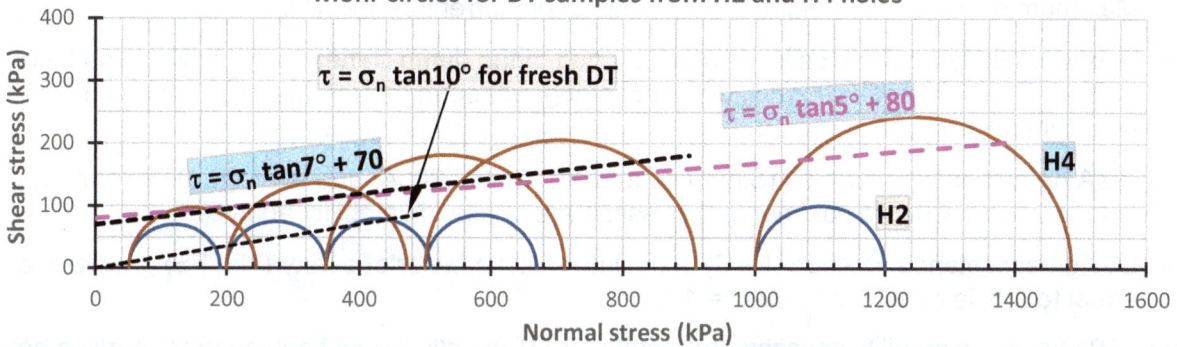

FIG 9 – Average shear strengths of DT with confinement ranges up to 500 kPa and 1000 kPa, and shear strength for fresh CHPP DT (Tucker, Li and Todd, 2015).

The high cohesion and low friction angle for DT would be the true values for the DT with moisture contents ranging from 15 per cent to 20 per cent. On the other hand, the low friction angle for MPR could be related to the lack of large particles. However, it is more conservative to apply the shear strengths with low friction angle to dump stability assessments, as the Mohr–Coulomb linear line for H5 MPR samples would stay above the linear lines (refer to Figure 8), if plotted in the same figure.

Nevertheless, original MPR (with no crushing or scalping) should be tested in the future to obtain actual shear strength values at moisture contents from 5 per cent to 15 per cent, which are similar to the tested sonic drilling core samples. The 5–10 per cent moisture content range reflects the tiphead co-disposed MPR and those placed as MPR curtain. For DT, laboratory testing should also be carried out for Atterberg limits, and relationships of cohesion and friction angle with moisture contents, to support a dump stability assessment with different DT conditions.

VALIDATING AND OPTIMISING EXISTING CO-DISPOSAL STRATEGIES

It is important to validate and optimise the current MPR and DT co-disposing strategies using the laboratory testing results. The focus of this work includes disposing DT in cells, together with MPR tiphead co-disposal, through stability analysis using Spencer and GLE/Morgenstern-Price methods in Rocscience Slide2® software. Although DT placed in cells comprises only about 5–10 per cent of total CHPP washery wastes and the cell placement is implemented in wet weather conditions or due to unforeseeable operational constraints. DT in cells is the weakest material among all mining and CHPP wastes. MPR co-disposed at active tiphead, together with MPR curtain, represents the main mode of disposal for the CHPP wastes, followed by paddock dumping.

DT cell disposing

There are two concerned scenarios of DT cell disposing in spoil dumps, where the cells are constructed under final rehabilitation profile and behind dump faces above ramp corridors. No safety concerns present for DT cells constructed within middle of dumps. Following assumptions are applied in the 2-dimensional stability analyses:

- A constant 1 m high perched water table is present above each dump bench, even though the sonic drilling revealed no persistent perched water table, and detected only associated with low points on a dump bench.

- Material within the perched water table is downgraded to saturated material, so that a weak layer is formed as a potential failure path for dumps containing DT cells.

- Category 2 weathered Permian and Category 3 fresh Permian spoils are placed in dumps along ramp corridors as per geotechnical guideline, therefore, an average strength for these two category spoils is applied to this part of dump.

- For both dump face along ramp corridors and below the final rehabilitation landscape, no offset of cells in adjacent dump lifts are analysed as the most conservative scenario. Any offset of cells in design and operational stages will make the dump more stable.

- Maximum overall dump height is 120 m based on current 5-Year Plan.

- Maximum depth of DT in cells is 10 m, or a half of the dump lift height. No mixing and blending of the DT is assumed with spoil during capping, although the DT blending with capping spoil was confirmed by sonic drilling.

- BMA default spoil properties (Simmons and McManus, 2004) and average DT shear strength values up to 1000 kPa confinement stresses are applied in stability analyses.

- Design acceptance criterion (DAC) is minimum static factor of safety (Min FS) greater than or equal to 1.5, ie Min FS ≥ FSDAC = 1.5.

Figure 10 shows the stability assessment results for DT in cells below final landform profiles before and after rehabilitation is completed. The minimum factor of safety values are 2.19 >> FS_{DAC} = 1.5. This outcome confirms current geotechnical guideline that no dimension and location restriction exists for DT cells design and construction below rehabilitation landform. The internal cells away from the final landform can be constructed with any one of the three category of spoils.

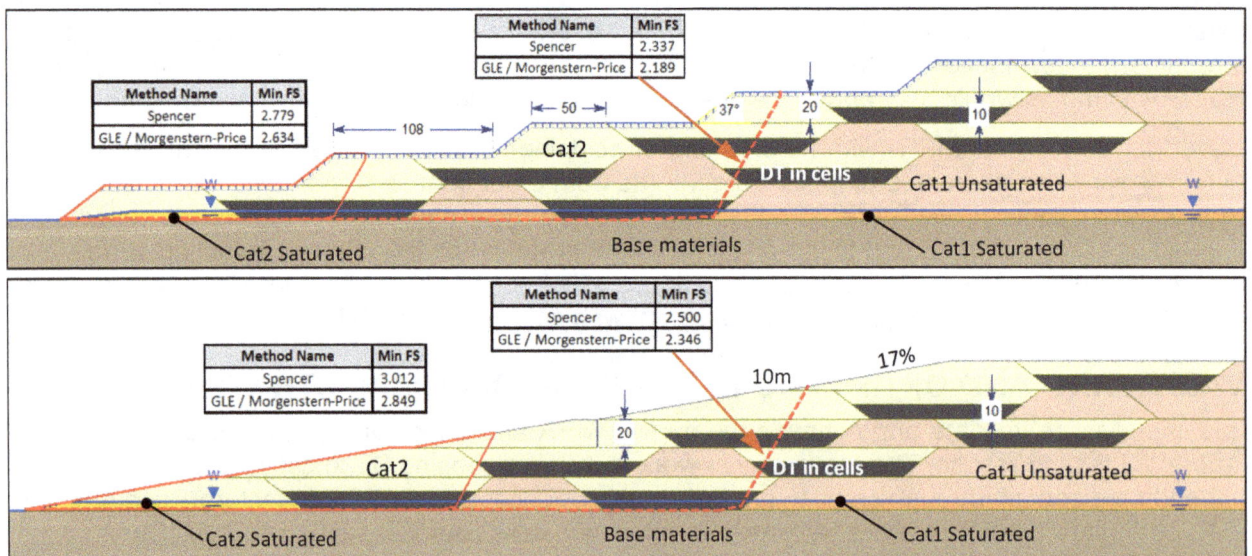

FIG 10 – Stability results for dumps with DT cells below overall 15 per cent landscape profile.

For dumps along ramp corridors, the most conservative case is placing DT cells 50 m from the dump face on DT tipping benches and on all levels with no offset. As shown in Figure 11 (left) the dump is unstable with Min FS = 1.04 << FS_{DAC} = 1.5. One of approaches to improving the dump stability is to offset the DT cells in two adjacent lifts. However, the stability assessment results shown in Figure 11 (right) indicates that the offset only results in minimal Min FS increase, and dump is still unstable and with a Min FS = 1.1. This suggests that offset of cells between dump levels is not an effective approach for improving dump stability. Instead, an improvement in dump stability should achieved by increasing the DT tipping embankment width along the ramp corridor.

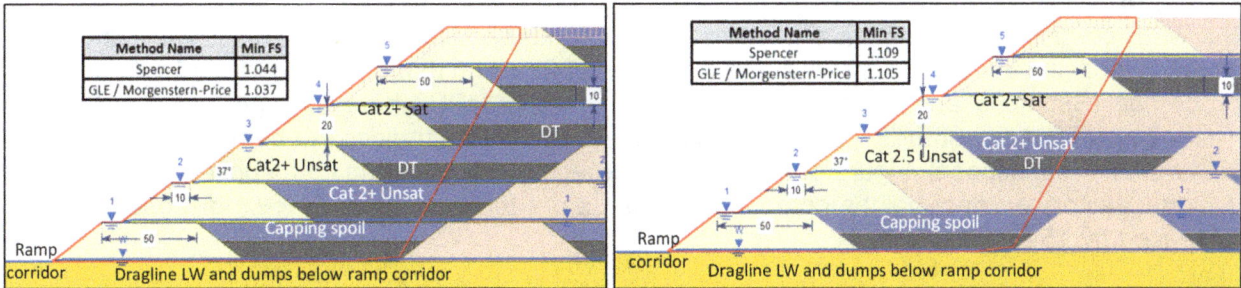

FIG 11 – Stability results for 120 m high dump along a ramp corridor – Unstable if the cells located 50 m from dump face on DT tipping level, either with or without offset in adjacent lifts.

Different embankment width is expected for different dump lifts due to different height from the maximum dump level. Figure 12 presents stability analysing results for all levels up to fourth lift above ramp corridor, as no further stability analysis is required for top two lifts based on the all the stability assessments presented. We can see from Figure 12 that to ensure dump stability the embankment for the first lift above ramp corridor needs a minimum width of 120 m on DT tipping level, then decreased to 95 m for the second lift above the ramp corroder. The standard operational embankment width of 50 m is sufficient to maintain dump stability for the fourth lift and above. In reality, the DT cells in adjacent lifts are unlikely to be constructed on top each other, therefore, the dump stability is further improved with the DT cells offset in adjacent lifts.

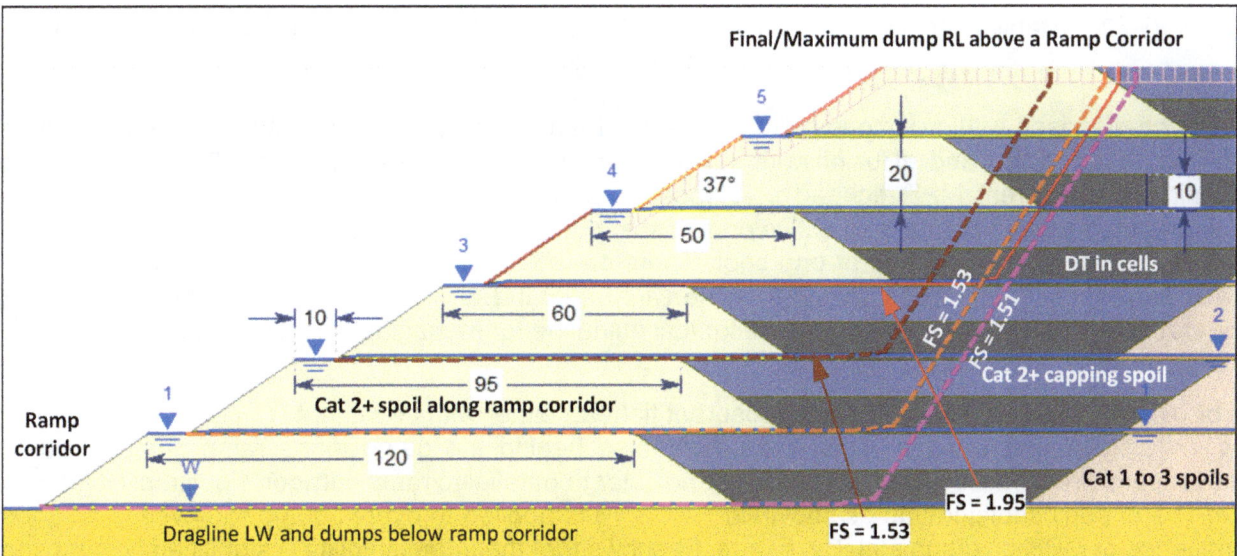

FIG 12 – Stability assessment results for all levels of 120 m high dump along a ramp corridor with modified embankment widths – Stable.

Existing DT cell design recommendations are proposed for maximum 100 m high dumps above ramp corridors. The current recommendation is that no DT cell offset is required between two adjacent lifts if the cells are constructed 100 m away from dump face along ramp corridor. On one hand, this statement ensured a stable overall dump, but makes DT cell design more conservative compared with the stability analysing results shown in Figures 12. It is not necessary for the upper lifts ≤ 80 m in height to have DT cells constructed 100 m away from the dump face along the ramp corridor. On the other hand, the current design recommendation does not provide the exact distances between DT cells and the dump face along the ramp corridor when DT cells in adjacent lifts are offset. A risk of dump instability could exist if the DT cells with offset in adjacent lifts are constructed 50 m away from the dump face along the ramp corridor by referring to Figure 11, which indicates that cell offset provides minimal stability improvement. Therefore, the stability analyses conducted in this review by using the laboratory testing results of sonic drilled DT samples has validated and optimised the DT cells design, in addition to ensuring stability of the dumps in all conditions.

MPR tiphead co-disposing

Site MPR co-disposal geotechnical guideline recommends spoil to MPR ratios of 4:1 with Category 1 Tertiary spoil and 3:1 with Category 2 and 3 spoils for 20 m high dumps. Using the BMA default spoil properties (Simmons and McManus, 2004) and average MPR shear strength values shown in Table 2 and Figure 8, stability of dumps is improved with MPR co-disposed compared to the dump with no MPR co-disposal. Figure 13 illustrates a Category 2 spoil dump co-disposed with MPR has a higher factor of safety values compared to the same dump with no MPR co-disposed, Min FS increased by 15 per cent.

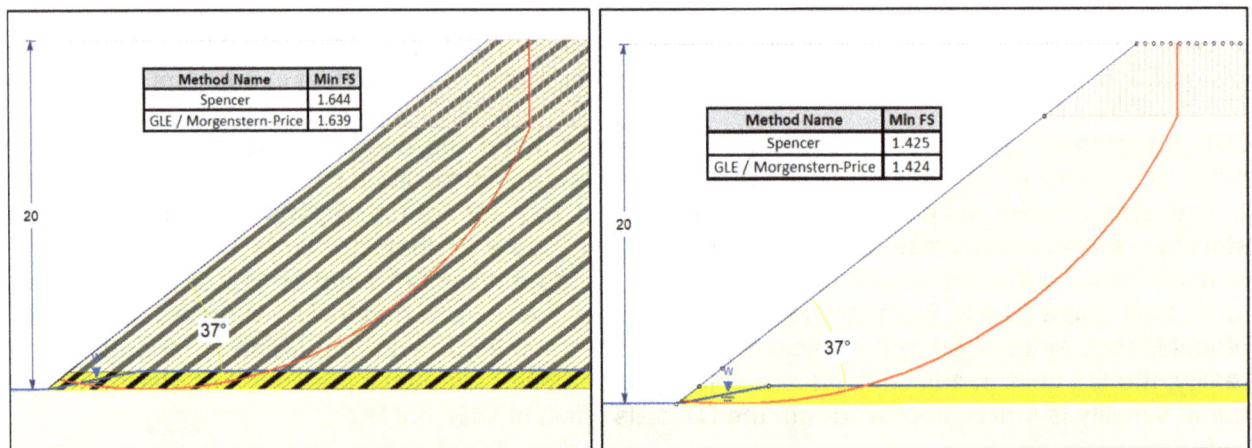

Method Name	Min FS
Spencer	1.644
GLE / Morgenstern-Price	1.639

Method Name	Min FS
Spencer	1.425
GLE / Morgenstern-Price	1.424

FIG 13 – Stability results for 20 m high dump with and without tiphead co-disposed MPR in Category 2 spoil.

MPR co-disposed within dump is more competent than the spoil due to moisture content reduction through evaporation and diffusion with surrounding spoil materials (Koosmen, 2017). This finding confirms the current MPR tiphead co-disposing operational experiences that no instability has experienced around 5 m behind active dump face, but the active tiphead may experience instability when MPR loads are built up at one spot. Therefore, MPR can be co-disposed as close to dump face as practical as long as satisfied with environmental and operational constraints. In addition, operations only need to follow the geotechnical guideline for managing the tiphead instability during co-disposal.

For overall dump MPR is to be co-disposed at tiphead either with or without DT cells constructed. It is understood that when a DT cell is constructed behind the dump along ramp corridor, dump advancing direction may change from perpendicular to parallel to ramp corridor. For simplicity, same MPR and spoil configuration as shown in Figure 13 is applied to overall dump for investigating the influence of MPR on dump stability. Figure 14 shows that the dump stability is significantly improved with MPR co-disposed in dumps through active tiphead, compared to the dump without MPR co-disposed. Within the perched water table the MPR shear strength is downgraded to half of the MPR strength parameters above the water table in the stability analysis.

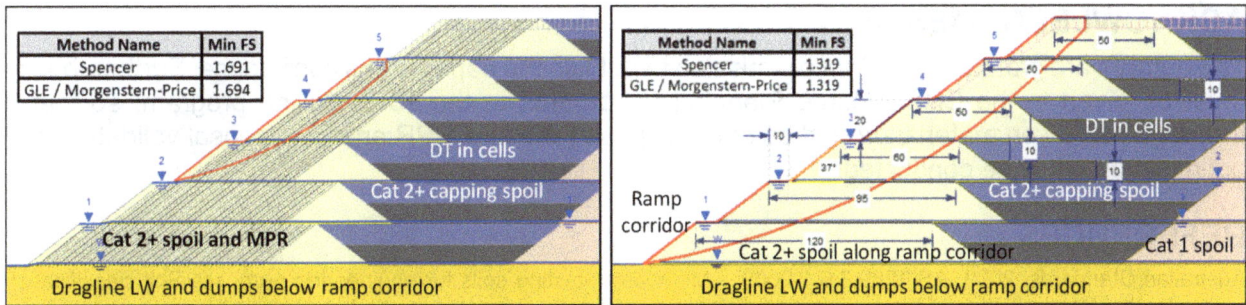

FIG 14 – Stability for 120 m high overall dumps with and without MPR tiphead co-disposal.

The stability analysing results shown in Figures 13 and 14 suggests that MPR co-disposal improves long-term stability of both individual dump lift and overall dumps. Considering the MPR shear strength values were from samples with particle sizes smaller than 2.36 mm, the actual co-disposed MPR having particles up to 70 mm would have much higher strengths, therefore, actual Min FS for both lift and overall dumps would be much higher. Knowing the actual strengths of the MPR in dumps is important to confirm the real stability conditions of the dumps.

CONCLUSIONS

It is rare to core spoils in dumps due to high cost and time penalties. Furthermore, the coring of the co-disposed MPR and DT in dumps is even rarer as MPR co-disposing is not widely implemented in the coal mining industry. Instead, most mines are still storing tailings in TSFs. The sonic drilling revealed no persistent perched water table presence on dump benches, except localised ponded water at low points of the dump. The sonic drilling revealed MPR and tailings physical properties in dumps, particularly the moisture content and densities. Sonic drilling revealed that DT is either blended with or replaced by the capping spoils, resulting in lower moisture content and higher shear strength values. The sonic drilling revealed the co-disposed MPR consolidated to a total moisture content of around 12 per cent. For tiphead co-disposed MPR the moisture content (up to 22 per cent when disposed) would be lower than the paddock dumped MPR due to evaporation and moisture diffusing with spoil of around 3–5 per cent moisture. The moisture contents of DT and MPR in dumps would stay unchanged as it is in the future, therefore, undrained unconsolidated triaxial compression testing should be carried out to obtain shear strength of the DT and MPR.

Shear strength of the DT samples with limited spoil contamination would have reflected the real, but the most conservative conditions of the overall tailings disposed in cells. The shear strength of DT without spoil contamination would still display variability depending on desiccation degrees prior to capping (Figure 2).

Due to limited samples from sonic drill holes, the MPR shear strength was obtained from specimens having particle sizes smaller than 2.36 mm, which are not a true reflection of actual co-disposed MPR particle sizes of up to 70 mm. Therefore, future investigations should consider undrained triaxial compression testing on the MPR with all particles, including 5–15 per cent total moisture content to obtain the true material properties.

Stability analyses using the laboratory testing data validated current geotechnical guideline on DT cell and MPR co-disposal strategies, including placing cells under rehabilitation landforms and active tiphead MPR co-disposal. The stability analyses using the laboratory testing data optimised DT cell placed above ramp corridors. The stability analyses confirmed that the co-disposed MPR makes individual dump lift and overall dump more stable compared with the dump without co-disposed MPR.

In the future, we need to conduct further laboratory testing on DT for Atterberg limits and undrained shear strength with different total moisture contents between 5 per cent and 30 per cent. In addition, large scale undrained triaxial testing on MPR with all or most of the particles (up to 50 mm) under a total moisture range between 5 per cent and 15 per cent, followed by further validating and optimising the DT and MPR co-disposal strategies.

ACKNOWLEDGEMENTS

The authors are grateful to BHP for permission to publish the paper at this conference. Further thanks must be given to the BHP Closure Planning for organising the sonic drilling program, so that a thorough review on material properties of co-disposed DT and MPR and co-disposal validation and optimisation could be conducted.

REFERENCES

Australian Standards, 2009. AS 1289.3.6.1-2009 – Methods of testing soils for engineering purposes Soil classification tests – Determination of the particle size distribution of a soil – Standard method of analysis by sieving.

Buzzi, O, Li, J, Pineda, J, Payne, D, Ouyang, K and Wu, J, 2023. Stress-strain behaviour of unsaturated compacted coal rejects and tailings, in *Proceedings of Eighth International Conference on Unsaturated Soils (UNSAT 2023)*, 6 p (Hellenic Society for Soil Mechanics and Geotechnical Engineering). https://www.e3s-conferences.org/articles/e3sconf/abs/2023/19/e3sconf_unsat2023_16001/e3sconf_unsat2023_16001.html

Koosmen, K, 2017. Considerations for the geotechnical stability of coal mine spoil piles containing co-disposed tailings, PhD Thesis, University of Newcastle.

Li, J, Fityus, D, Payne, Vaughan, S and Burton, G, 2021. Full PSD rejects shear strength testing for co-disposal and assisted tailings dewatering technique selection, in *Proceedings of the Mine Waste and Tailings Conference*, pp 471–484 (The Australasian Institute of Mining and Metallurgy: Melbourne).

Li, J, Payne, D and Hooi, H, 2018. Plant rejects disposal strategies applied at Caval Ridge open cut coal mine, in *Proceedings of the Mine Waste and Tailings Conference*, pp 50–60 (The Australasian Institute of Mining and Metallurgy: Melbourne).

Simmons, J V and McManus, D A, 2004. Shear strength framework for design of dumped spoil slopes for open pit coal mines, in *Proceedings Advances in Geotechnical Engineering* (eds: R J Jardine, D M Potts and K G Higgins), vol 2, pp 981–991 (Thomas Telford Limited: London).

Tucker, N, Li, J and Todd, T K, 2015. Application of laboratory testing to approximate the behaviour of mixed plant rejects, in *Proceedings of the Bowen Basin Symposium 2015* (ed: J W Beeston), pp 453–461 (The Geological Society of Australia Incorporated Coal Geology Group: Brisbane).

Feasibility study of co-disposal of tailings and mine waste rock

A Wanninayake[1] and T Dixon[2]

1. Technical Director – Tailings, GHD, Perth WA 6000. Email: ajitha.wanninayake@ghd.com
2. Senior Tailings and Dams Engineer, GHD, Vancouver BC V5T 1M6, Canada.
   Email: tyler.dixon@ghd.com

ABSTRACT

Alternative tailings disposal methods have gained much attention due to their advantages over conventional slurry tailings storage facilities. Co-disposal of tailings and mine waste rock is one of the leading methods of alternative disposal methods. However, application of this technology has been mostly limited to small-scale and research projects until recently. This paper discusses the use of co-disposal technology in a large-scale mining project in Western Australia.

An options assessment was performed to evaluate alternative tailings disposal methods against conventional slurry deposition, and co-disposal of tailings and mine waste rock was selected as the preferred option for tailings management. A feasibility study was completed for the selected co-disposal option that involved co-mingling of tailings with mine waste rock and construction of alternating layers of waste rock and tailings. Geotechnical and geochemical characteristics of tailings and waste rock, site topography and subsurface conditions, climatic conditions, construction and operational benefits, life cycle costs and safety in design were evaluated during the study.

Suitable mixing ratios of waste rock and tailings were assessed relative to the mine waste production schedule to maintain rock to rock contact ensuring improved static and dynamic stability of the co-disposal facility. Containment of potentially acid forming waste rock was incorporated in the design by including tailings in the waste rock dump.

Preliminary designs were completed for the co-disposal waste rock dump facility and slope stability and deformation analyses were performed for the design. The co-disposal facility will be developed to a detailed design during the next phase of the project.

INTRODUCTION

Project overview

A large-scale mining project, located in Western Australia, is envisaged to be an open pit mine with a concentrate production rate of 5 Mt/a and a 28-year mine life. The project Pre-Feasibility Study (PFS) identified a reduced-capital development option. The pilot process testing showed favourable results for the manufactured tailings, specifically greater than expected quantities of the dry and coarse tailings stream (referred to as grits) from the vertical roller mill (VRM). For this reason, the project evaluated different tailings storage options prior to proceeding with a definitive feasibility study. The evaluation comprised a thorough review of waste disposal techniques and completing a trade-off study by way of multicriteria analysis (MCA).

Site overview

The project is located in an area where the climate can generally be described as temperate with distinctly dry and warm summers. The monthly average temperature ranges from a low of 6°C in August to a high of 25°C in January. Seasonal fluctuations show a noticeable difference between summer and winter, with the warmest months being December to February and the coolest months being June to August. Typically, the wetter months occur during the winter while the dry months occur during the summer, however, some high rainfall events do occur as summer storms. The mean annual evaporation is approximately 1200 mm.

The project site is characterised by undulating topography comprising sand dunes, ridges, and natural depressions. The depressions are generally structurally controlled by faults while prevailing wind patterns control the sand dunes. No significant drainage features are present across the site.

The project site is also recognised to be in a low to moderate seismic activity region.

Summary of waste streams

Waste for the project will be comprised of waste rock and tailings. Waste rock is divided into two streams, non-acid forming (NAF) rock particles and soils and potentially acid forming (PAF) rock particles. The tailings generated from the process plant are divided into three streams as described below:

1. Dry non-magnetic grits are the coarse by-product of the VRM and rougher magnetic separator (RMS) components of the process plant. This stream produces a dry, benign, and uniformly graded sand material.

2. Intermediate magnetic separator (IMS)+Cleaner magnetic separator (CMS) tailings underflow is the finer by-product of the VRM that goes through additional processing. The by-product is a slurry that is pumped through IMS and CMS to produce non-magnetic IMS and CMS tailings, respectively. The non-magnetic IMS+CMS tailings are then re-combined at the tailings thickener and the underflow produces a stream of filtered/paste-like, benign, fine tailings.

3. Sulfide tailings are the by-product of all waste streams that include sulfide containing waste. These tailings will be managed separately.

The dry non-magnetic grits and IMS+CMS tailings are combined into an agglomerated tailings stream that will be co-disposed with mine waste in the Waste Rock Dumps (WRDs).

Sulfide tailings will be stored in a separate lined conventional impoundment Tailings Storage Facility (TSF). The overall process, with tailings streams, is presented in Figure 1.

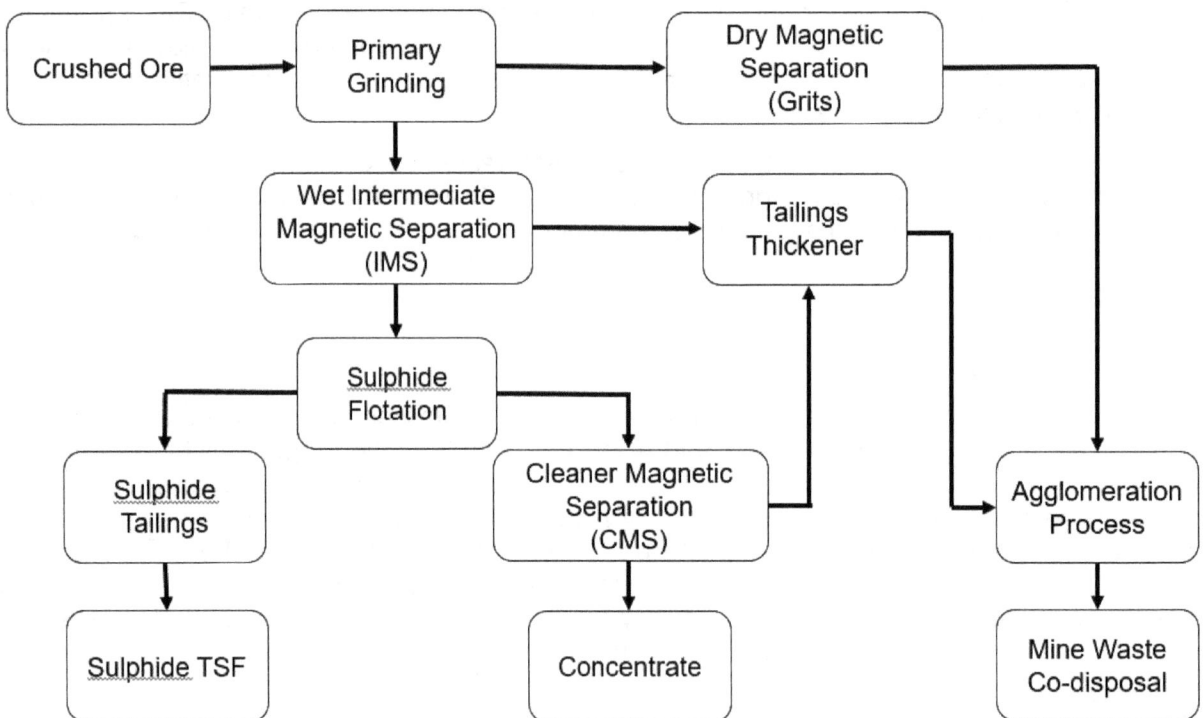

FIG 1 – Simplified tailings processing block flow diagram.

Summary of waste production rates and tailings volumetrics

The volumes of expected NAF waste rock and agglomerated tailings are provided in Figure 2. The ratio of NAF waste rock to agglomerated tailings, by volume, is expected to be approximately 7:1 during initial start-up construction activities and averaging 2.3:1 thereafter. Tailings production inputs are provided in Table 1.

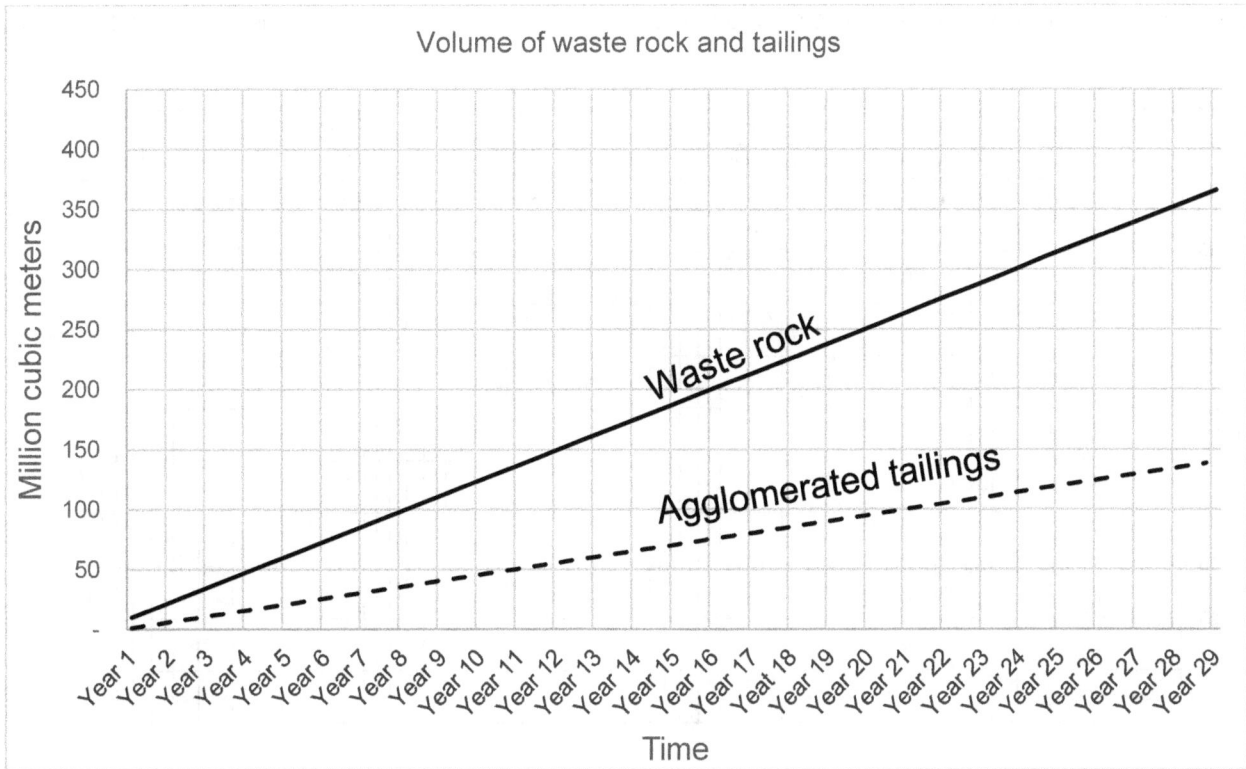

FIG 2 – Waste rock and agglomerated tailings production volumes.

TABLE 1

Tailings streams production data.

	Dry non-magnetic tailings (grits)	IMS+CMS tailings underflow	Agglomerated tailings
Solids mass flow (t/h)	769	458	1227
Water mass flow (t/h)	16	206	222
Total mass flow (t/h)	785	664	1449
Total mass flow (Mt/annum) (Based on 8000 operation hrs/annum)	6.3	5.3	11.6
Solids content (%)	98	69	85
Moisture content (%)	2	45	18
Solids specific gravity	3.01	3.25	3.10
Slurry specific gravity	2.89	1.92	2.35
Volume (m³/h)	271	346	617
Volume (Mm³/annum) (Based on 8000 operation hrs/annum)	2.2	2.8	4.9

Summary of tailings properties

Based on laboratory testing, in accordance with Australian Standard 1726, the agglomerated tailings have been classified as SAND-SILT MIXTURE with non-plastic fines (SM), and the grits are classified as poorly graded SAND with little to no fines (SP). The tested Standard Maximum Dry Density (SMDD) and Optimum Moisture Content (OMC) of agglomerated tailings are 1.97 t/m^3 and 13 per cent, respectively; while the grits achieved a SMDD of 1.93 t/m^3 and an OMC of 15.5 per cent, respectively. The Particle Size Distribution (PSD) of agglomerated tailings and grits are shown in Figure 3.

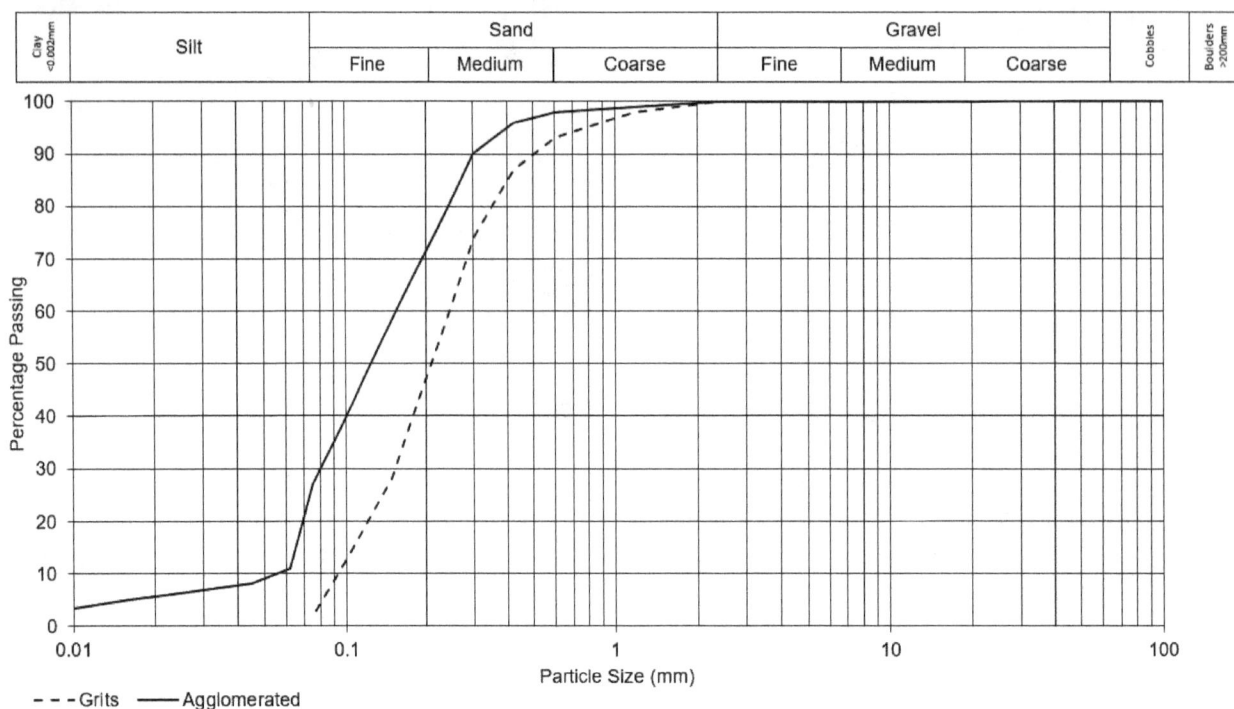

FIG 3 – Tailings particle size distribution.

Triaxial, Oedometer and California Bearing Ratio (CBR) tests were conducted on agglomerated tailings to inform the feasibility design of the co-disposal WRD.

OPTIONS ASSESSMENT

Objective

The objective of the options assessment was to evaluate different storage options for waste rock and tailings, accounting for the relative proportions and properties of different tailings streams, and to assess potential optimisations by combining waste streams for improved geotechnical performance. Conceptualised options were developed and then assessed using multicriteria analyses.

Overview and approach

The options assessment commenced with a period of information gathering and sharing among the owner, the tailings consultant, the mining consultants, the processing consultants, and other stakeholders. The owner provided guidance and offered position statements consistent with best available practices and encouraged the adoption of emerging technologies. Through this consultation process three conceptualised options were proposed for further study which were:

1. A conventional wet TSF.
2. Dry stacking tailings.
3. Combinations of tailings and waste rock co-disposal methods.

The options assessment was then carried out through a two-step MCA, described below.

Key criteria and constraints

The key criteria driving the options assessment were:

- Storage capacity: will containment and storage of the tailings be feasible?

- Ease of production, material handling, and deposition: does the option provide simpler production, material handling, and deposition?

- Geotechnical engineering: does the option face any geotechnical risks?

- Water and environment: does the option face any water/environmental risks?

- Operations and processing: are there any operation and processing risks associated with the option?

- Upfront costs: relative to other options based on a comparative order of magnitude.

- Ongoing costs: relative to other options based on a comparative order of magnitude.

- Permitting and approvals risk: will there be any complexity for approvals for this option?

The key constraints were:

- Storage capacity

- Footprint constraints

- Visual impacts and height restrictions

- Start-up costs and initial start-up risks

- Closure planning and long-term risks.

Options study

Summary of storage concepts

Based on the pre-feasibility studies the approximate general arrangement and footprints available for waste storages were known. Then, through the consultation process, it became clear that spatial constraints and volumetrics of waste materials were critical inputs for the key criteria. Therefore, any storage concept relied on co-disposal to manage at least a proportion of the tailings waste streams. Three combinations of storage options were studied as depicted in Figure 4. The combinations were:

Option 1: Conventional TSF and wet storage of IMC+CMS tailings and co-disposal of grits and excess IMS+CMS tailings at WRD1 considering: a) blended and b) layered co-disposal methods.

Option 2: Dry stack of agglomerated tailings and co-disposal of agglomerated tailings at WRD1, considering: a) blended and b) layered co-disposal methods.

Option 3: Co-disposal of agglomerated tailings only, considering: a) blended co-disposal at the WRD only, and b) layered co-disposal methods at the WRD only, c) blended co-disposal at both locations, and d) layered co-disposal at both locations.

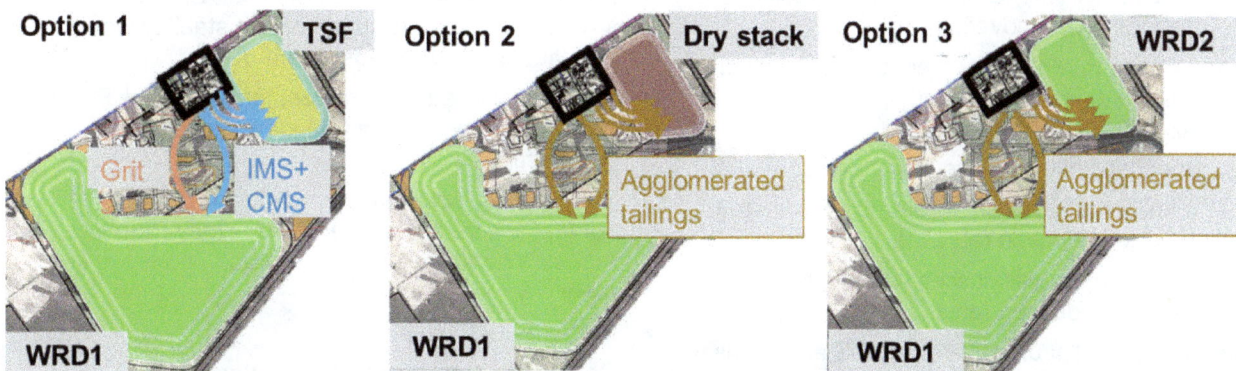

FIG 4 – Waste storage options.

Co-disposal

In this study co-disposal is a generic term used to describe the handling and disposal of different waste materials together at the same location. This study then considered further subclassifications identified as blended co-disposal and layered co-disposal, which are techniques sometimes referred to as co-mingling, co-placement, or co-deposition.

Blended co-disposal is the process when waste rock and tailings are transported separately and blended together during placement in such a way that the tailings ingress into the waste rock void space to generate a composite material. Layered co-disposal is the process when waste rock and tailings are transported separately then placed and compacted in alternating layers.

A necessary requirement of the blended co-disposal technique is to have the shear strength performance of the WRD governed by the properties of the waste rock. To achieve this intent the waste rock particles shall remain in contact whilst tailings fill the void space. The waste rock to tailings mix ratio for the blended co-disposal technique was assumed to be the product of a 60 per cent void filling efficiency and a waste rock porosity of 0.3, which is equivalent to 18 per cent of the waste rock unit volume and a 5.55:1 waste rock to tailings ratio (by volume). The assumed void filling efficiency is consistent with ideal mix ratios of greater than 4:1, preferably 5:1, as supported in findings by Gowan, Lee and Williams (2010), Wickland (2006), Khalili, Wijewickreme and Wilson (2010), Wijewickreme, Khalili and Wilson (2010), and Bainbridge et al (2022).

A comparison of either co-disposal method is given in Table 2.

TABLE 2

Comparison of blended and layered co-disposal methods.

Advantages	Disadvantages
1. Blended co-disposal	
• Clast supported (at mix ratios of 5:1 or greater)	• Requires test fills to determine an appropriate mechanical blending methodology
• Increased physical stability performance as the rock particles will be in contact	• Added equipment costs to achieve blending
• Reduction in potential ARD of mine waste rock due to low ingress of oxygen into tailings filled voids	• Tailings ratio is limited by waste rock porosity and 'rock-to-rock' contact requirements
• Reduces overall waste storage footprint	• Success of blending is highly dependent on waste rock PSD and porosity
• Lowers overall environmental and long-term risks	• Efficiency is susceptible to different waste production rates
• Dust suppression efficiency	
2. Layered co-disposal	
• Reduction of ARD of mine waste rock due to tailings layers covering rock	• Includes connected zones or layers of lower shear strength material (tailings), hence stability would be influenced by tailings layers
• Less requirement for mechanical blending of tailings and waste rock	• Requires civil equipment fleet for additional construction work front (placement and compaction)
• More constructable as opposed to mechanically blending the material	• Tailings as-placed and compacted moisture content will be a critical element – may require moisture conditioning or evaporation ponds or other drying methods
• Flexible design can accommodate different waste production rates	
• Efficiency is less susceptible to different waste production rates	• Requires CQC and CQA
• Independent of waste rock PSD and porosity	• Decreases available storage of waste rock

Advantages and disadvantages of primary storage options

Throughout the options assessment, in consultation with the owner, the owner's other consultants, and other stakeholders, a number of advantages and disadvantages were identified for each of the options and summarised in Table 3.

TABLE 3

A comparison of advantages and disadvantages for primary waste storage options.

Advantages	Disadvantages
1. Conventional wet TSF as the primary storage option with co-disposal at the designated WRD	
• Confined storage area provides a contingency for upset conditions (eg, tailings with high water contents) and space for evaporation	• Total volume of IMS + CMS tailings is expected to exceed the maximum available capacity of TSF storage; therefore, alternative storage would be required later in the mine life
	• Additional handling methods would be required for co-disposal of the grits
	• Greater long-term risks due to difficulties closing wet storage facilities
	• Greater short-term risks associated with breach of the TSF embankments
	• Greater CAPEX costs
	• May require dust suppression measures for grits
2. Dry stack TSF as the primary storage option with co-disposal at the designated WRD	
• Potential lower OPEX	• May be vulnerable to upset conditions (eg tailings with high water contents)
• Simplified tailings storage area	
• Lower short-term risk as embankment breach risks would be mitigated	• Total volume of agglomerated tailings is expected to exceed the storage capacity for the dry stack area, therefore, alternative storage would be required later in the mine life
• Lower long-term risk (than a wet TSF) as closure would be less costly	
• Reduces wet TSF storage to sulfide tailings only, lower overall project risk	• Requires a minimum density for geotechnical stability
	• Low erosion resistance without rock cover
	• Reprofiling for final landform design will be required
	• Susceptible to liquefaction
	• Potentially high Capex
	• Limited in height, therefore greater storage area required and longer conveyor distances
	• Lower process plant annual operating hours due to TSF mechanical equipment availabilities
	• Dust suppression measures would be required
3. Co-disposal only, utilising both storage areas available	
• Simplicity in combining waste rock and tailings into one facility	• May be more susceptible to upset conditions
• Good final environmental outcome limiting long-term environmental and owner risks through a high-density compacted structure	• Blending techniques are not developed and proven
	• Tailings streams require mixing to average moisture content prior to co-disposal. This requires extra handling
• Truck haulage providing flexibility to create zones within the WRD that meet the required characteristics for geotechnical stability, drainage, seepage, and final landform for mine closure commitments	• The height of the storage facility will increase if excess tailings cannot be stored elsewhere (eg in-pit backfill)
• Volume efficiency reducing the project disturbed footprint as grits and tailings will fill	• High OPEX option, due to low Capex expenditure and diesel prices, until electrification for emission

a portion of the voids in the fresh rock and WRD has potentially less height constraints

- Operating procedure would be flexible through truck, dumping and bulldozer combinations
- Dust suppression efficiency through combining dry grits and wet tailings. Wetter products may better penetrate the fresh rock voids
- Low CAPEX option
- No impact on process plant availability due to flexibility in truck fleets
- Reduces wet TSF storage to sulfide tailings only, lower overall project risk
- Option to include blended and layered co-disposal methods

reduction. Trolley assist or conveying extensions will reduce OPEX

- Truckload capacity will depend on material characteristics (eg slumping)

Multicriteria analyses

Methodology

A two-step analysis was performed by completing two MCAs in sequence for the co-disposal and tailings storage options discussed. A multistep process was undertaken to allow for ease of comparison between co-disposal methods and primary storage options as one may have influence over the other. For example, the first step was an MCA to evaluate which co-disposal method was preferred, which was given a weighting factor that was then input into the second MCA for primary storage considerations.

For both MCAs options were assessed against elements that together made up a single key criterion. Each option was then given a score between 1 to 5 for each element and the average score across all elements made up the overall score for that particular key criterion. For example, the key criterion of 'ease of production, material handling, and deposition' was the combined average scores for the elements: tailings delivery/deposition methodology, blending/mixing/agglomeration process, and variable production rates. Following scoring, the key criteria were given a weighting based on a consensus of the level of importance to the overall outcome.

MCA results

The hierarchy for the MCA scoring was as follows:

1. Very negative outcome
2. Negative outcome
3. Neutral outcome
4. Favourable outcome
5. Very favourable outcome.

The results of the MCA favoured Option 3 as the primary storage option and the layered co-disposal method. Then extra advantages were realised when considering utilising both blended and layered co-disposals methods, as will be discussed next.

INTEGRATED WASTE LANDFORM DEFINITIVE FEASIBILITY DESIGN

General

The envisaged design concept emerged from co-disposal waste rock dumps to two integrated waste landforms. The concept is comprised of strategically placed agglomerated tailings and waste rock zones that take advantage of qualities of each material. By doing so the concept optimises the overall site-wide waste storage capacity, provides operational flexibility, lowers short-term and long-term foreseeable risks, and more easily facilitates closure objectives.

Design concept

The selected design concept is integrated waste landforms that include the following elements:

- A perimeter shell comprised of non-acid forming, unblended, waste rock.

- Alternating layers of compacted agglomerated tailings and blended co-disposal tailings/waste rock.

- Internal waste rock columns.

- Internal potentially acid-forming waste rock encapsulation cells.

The design concept is shown in Figure 5.

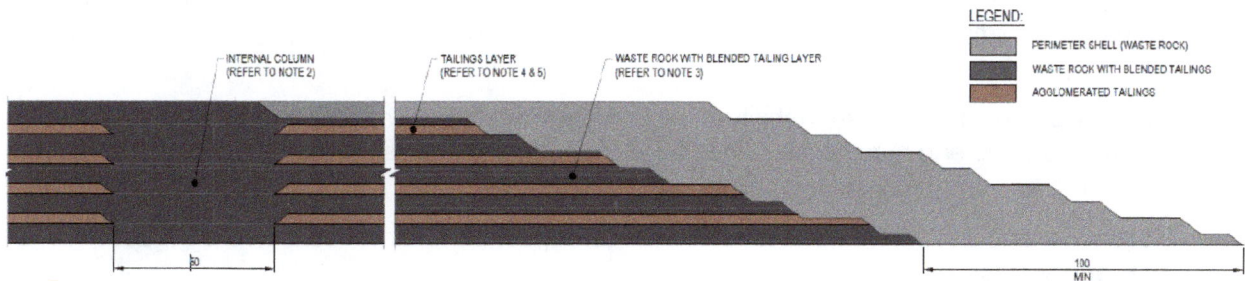

FIG 5 –Typical schematic cross-section (PAF cells not included).

Envisaged advantages of the co-disposal concept

Optimised waste storage and operational flexibility

Within the layered and blended co-disposal regions, the blended tailings/waste rock layers are between 5–6 m placed in 1 m thick layers, and the agglomerated tailings layers are between 3–5 m placed in 0.5 m thick layers considering compatibility. Increase in density of agglomerated tailings due to compaction and dissipation of pore water pressure through internal waste rock columns will help reducing liquefaction risk. Each zone between rock columns is in a grid pattern with spacing between 100–200 m. This arrangement maximises operational flexibility and provides optimised storage capacity of the tailings waste stream.

PAF management

Tailings are lower permeability than waste rock and have a higher degree of entrained moisture. By including tailings in layers and in the void spaces of the waste rock the landform and composite material will have a reduced vertical transmissivity, thus reducing diffusion and advection and the ingress of oxygen throughout the landform. This will result in a reduced flux of both air and water, which will reduce the propensity for PAF waste rock to generate acid.

Short-term and long-term foreseeable risks

Many of the risks associated with a 'conventional' tailings storage would be reduced or mitigated by implementing this design concept. At a high level, the risks and their mitigating factors are summarised below:

- Embankment failure and dam breach: Eliminates the hazard from release of tailings due to a potential embankment breach. Embankment failure is design dependent.

- Geotechnical slope instability: The perimeter shells provide buttressing and can be adapted (batter slopes and dimensions) to suit variable conditions, as necessary, reducing the hazard. Waste rock particle contact, within the blended co-disposal layers, provide additional internal shear resistance improving overall geotechnical stability.

- Geotechnical instability of the foundation: Construction can be staged to allow for consolidation of foundation soils and dissipation of pore pressures and the perimeter shell can be adapted to provide restoring forces, reducing the hazard.

- Liquefaction: Internal agglomerated tailings will be compacted to a suitable relative density to achieve a non-liquefiable mass. The perimeter shell and blended layers are not expected to be susceptible to liquefaction. The landform is not expected to impound water and the waste rock columns provide shortened drainage paths to mitigate against perched water tables.

- Earthquake induced deformation: Deformations as a result of an earthquake may be possible, but large deformations would be tolerable, reducing the overall hazard.

- Variable rate of waste streams: Construction may be staged throughout the landform to provide for optimum operational flexibility. Tolerances of layering thickness provides additional flexibility.

- PAF cell acid generation: Tailings layers will reduce the conditions that typically lead to acid formation by limiting the ingress of oxygen and reducing seepage flux, in addition to the PAF cell design.

- Closure planning and sustainability: Closure planning and rehabilitation at lower levels can begin before the end of mine life. Slopes can be flattened and revegetated during construction, improving the long-term sustainability.

- Post mine land-use: More options are viable for a post mine land-use than a conventional wet TSF or dry stack.

CONCLUSIONS

By incorporating tailings with waste rock into an integrated waste landform, the favourable properties of each material type were used to their advantage for an optimal waste storage design. The agglomeration process combines dry grits with wet IMS+CMS tailings to produce a tailings mixture which is close to the optimum moisture content facilitating handling and compaction. Adding agglomerated tailings to fill rock voids reduces oxygen ingress resulting in less potential for acid rock drainage.

The results of the options assessment and multicriteria analyses indicated that co-disposal of agglomerated tailings and waste rock would provide an effective integrated mine waste landform whilst addressing potential risks associated with other disposal methods and provide advantages during operations and closure.

The concept design was developed to a definitive feasibility level however further study is required during detailed design and during construction. This will include confirmation of the geotechnical properties of the waste materials, a comprehensive geotechnical investigation of the foundation, field trials for optimal blending techniques and compaction methods, development of material handling procedures and a strategy for coordination between mine waste production and tailings production streams.

ACKNOWLEDGEMENTS

The authors would like to express their gratitude towards the owner of the project and the project team for their support throughout the study and this paper. The authors would also like to thank Daniel Maher, Milan Thiaga, John Phillips, Sina Kazemian, Caroline Holmes, Louise Thomas, and Andrew White for their significant contributions, guidance, and review.

REFERENCES

Bainbridge, S, Steele, J, Samlets, J, and Walker, D, 2022. Design approach for a co-managed tailings and waste rock facility in British Columbia, Canada. Proceedings Tailings and Mine Waste 2022, pp 48 – 58. (Tailings and Mine Waste: Denver)

Gowan, M, Lee, M, and Williams, D J, 2010. Co-disposal techniques that may mitigate risks associated with storage and management of potentially acid generating wastes, in Proceedings Mine Waste 2010 – A.B. Fourie and R.J. Jewell (eds), (The Australian Centre for Geomechanics, Perth)

Khalili, A, Wijewickreme, D, and Wilson, G W, 2010. Mechanical response of highly gap-graded mixtures of waste rock and tailings. Part I: monotonic shear response. Canadian Geotechnical Journal, 47: 552–565.

Wickland, B E, 2006. Volume change and permeability of mixtures of waste rock and fine tailings, PhD thesis, University of British Columbia, Vancouver.

Wijewickreme, D, Khalili, A, and Wilson, G W, 2010. Mechanical response of highly gap graded mixtures of waste rock and tailings. Part II: undrained cyclic and post-cyclic shear response.

Emergency response

Response to an emergency caused by a TSF failure

O Wennstrom[1]

1. Technical Director, Crisis and Emergency Management, GHD, Perth WA 6000.
   Email: olle.wennstrom@ghd.com

ABSTRACT

The emergency operation, in response to a failure of a Tailings Storage Facility (TSF) is often a multi-disciplinary, multi-organisational operation, impacting multiple jurisdictions. Capacity of organisations expected to contribute to the operation may vary, eg due to background flooding caused by wet weather.

Within the mining industry it is common to treat a response to a major emergency, such as TSF failure as an isolated short-term operation with a technical focus, and as a solution develop Trigger Action Response Plans (TARP). In contradiction to this practice, the Global Industry Standard on Tailings Management (GISTM) clearly focuses on a 'shared state of readiness', 'immediate response to save lives, supply humanitarian aid and minimise environmental harm', and the ability to execute 'long-term recovery'.

The paper elaborates on the concept of 'principal duties and timescales' applicable to an emergency, and the importance of defining and communicating a Strategic Intent (SI). A model is presented for an advantageous SI and how to apply it.

It is the combination of understanding the Concept of Operation, and the ability to develop and deliver a clear SI, which via the Plan Inventory is turned into executable plans, that forms one of the corner stones for the emergency preparedness.

Finally, the interconnections between organisations participating in a response to an emergency caused by TSF failure, the concept of coordination, and the application of the SI within that concept, is described.

Preparedness to deliver a response to an emergency is the product of Plans, Organisation, and Competency. To benefit this Emergency Preparedness, the plans need to give support on how to execute the entire emergency operation over time. If the plans are incomprehensible and overly complicated, the organisation will never gain the Competency.

'Plans are nothing; planning is everything', the famous quote by Dwight D Eisenhower is just as valid today as it was when he uttered it in 1957 (Eisenhower, 1957).

INTRODUCTION

The Global Industry Standard on Tailings Management (GISTM; International Council on Mining and Metals (ICMM), 2020) came as an industry response to some of the infamous tailings facility failures during 2015–2020. It strives to achieve the ultimate goal of zero harm to people and the environment with zero tolerance for human fatality. The standard requires an owner/operator of a Tailings Storage Facility (TSF) to take responsibility and prioritise the safety of tailings facilities, through all phases of a facility's life cycle, including closure and post-closure. It also requires the disclosure of relevant information to support public accountability.

This paper touches upon the concept of 'shared readiness' and explains in detail how an emergency response operation is planned and executed, and which parameters are important for a successful response. It elaborates on the concept of Principal Duties and Timescales, further described and detailed below in the section Principal Duties and Timescales applicable to an emergency, and the importance of defining and communicating a Strategic Intent (SI). A model for an advantageous SI and how to apply it is presented.

'Shared state of readiness' is in the GISTM, Requirement 13.3 described as: 'Considering community-focused measures and public sector capacity, the Operator shall take all reasonable steps to maintain a shared state of readiness for tailings facility credible flow failure scenarios by securing resources and carrying out annual training and exercises' (ICMM, 2020).

The standard places a responsibility of the operator to take the lead in creating the shared state of readiness.

This paper describes how a response to a TSF failure would look. Initially by describing the nature of a TSF failure, the Concept of Operation, and the nature of a response to an emergency caused by a TSF failure. Thereinafter how an emergency operation is developed, based on the What, When and Where. And finally, the hierarchy of strategy, tactics and technique is described, and a model for a Strategic Intent delivered.

In this paper, the term Emergency Response Plan (ERP) will be used in the context as defined by Emergency Management Australia: *'A plan which sets out the roles and responsibilities of agencies in emergency response and the coordination arrangements which are to be utilised'* (Emergency Management Australia (EMA), 1998). For the purpose of this paper the term is interchangeable with the term Emergency Preparedness and Response Plan (EPRP), which is used by the GISTM.

RESPONDING TO AN EMERGENCY CAUSED BY A TSF FAILURE

The nature of a TSF failure

The failure scenarios for a TSF can be categorised broadly into 'rainy day' and the 'sunny day' scenarios, whilst each scenario can involve one or more of many causes of failure, including earthquake, slope instability, erosion, seepage, overtopping, or foundation failure. The extent of the impact following a TSF failure may vary widely depending on to which level the dam is filled, and its location in relation to people, environment and property.

Depending on the nature of the cause of the failure, the surveillance system in place, and the owner/operator's ability to interpret, react and respond to the failure, the impact from a failure can be near instant, a drawn-out situation, or totally avoided.

Some of the more recent and more pronounced TSF failures in the world, have presented themselves as 'actual or imminent', with little or no warning. This is however far from always the case. Many failures do give warning, such as overtopping or piping failure. Depending on the available time from detection to breach, the operational conditions may vary widely.

Taking into consideration the above, and that many TSFs have potential to fail in multiple directions, it is apparent that there may be an array of potential failure scenarios, both in terms of the impacted area, and the timeline for a failure event to occur.

Concept of operations

The GISTM expresses a clear focus on *'shared state of readiness for tailings facility credible flow failure scenarios'* (ICMM, 2020). The intent of this is to *'In the case of a catastrophic tailings facility failure, provide immediate response to save lives, supply humanitarian aid and minimise environmental harm'* (ICMM, 2020).

In the unlikely event of a TSF failure, the emergency response will most likely be a joint operation, as the impact will probably affect areas outside of the owner's/operator's footprint.

Looking at history, the emergency response operation will be prolonged, and the extent of the impact can follow one of many plausible scenarios. Add to this the long-term recovery operation, which ideally should be commenced before the emergency response is terminated, and a total time span of many months or more is involved.

It is worth noting that if public authorities and emergency services are planned to be part of the emergency response, they may already have their resources exhausted at the onset of an TSF failure. This is especially the case if the scenario is a rainy-day event, as the rain most likely has been falling for some time and inundated the surrounding areas and the community.

The nature of a response to an emergency caused by a TSF failure

The GISTM states that an owner/operator shall: *'In the case of a catastrophic tailings facility failure, provide immediate response to save lives, supply humanitarian aid and minimise environmental harm'* (ICMM, 2020).

In most countries, with some minor variations in wording, an emergency is defined as *'an event, actual or imminent, which endangers or threatens to endanger life, property or the environment, and which requires a significant coordinated response'* (EMA, 1998).

Hence the objective of an emergency response to a TSF failure is to mitigate the threat to life, environment, and property, caused by a TSF failure. That operation may comprise maintenance and repair of the TSF, but as one of the tools to reduce the threat to life, environment, and property.

For many TSFs, a failure would impact areas outside of the mine lease, and impact both the owner/operator's personnel and assets, as well as environmental values, cultural values, neighbouring companies, private property, and damage both the public and public assets.

Developing an emergency response operation

To be able to plan and execute a proactive emergency response, it is imperative to understand the 'Three Ws', namely: *What* has happened? (or will happen), *When?* and *Where*? In the following sections, the 'Three Ws' are briefly explained.

What?

It is important to know what hazard will be impacting the operational area and causing the threat to life, environment, and property, in the unlikely event of a TSF failure, tailings will release and inundate an area. The inundation area will be impacted by tailings to certain depth. The danger of this can be estimated from Figure 1 (The Australian Institute for Disaster Resilience, 2017).

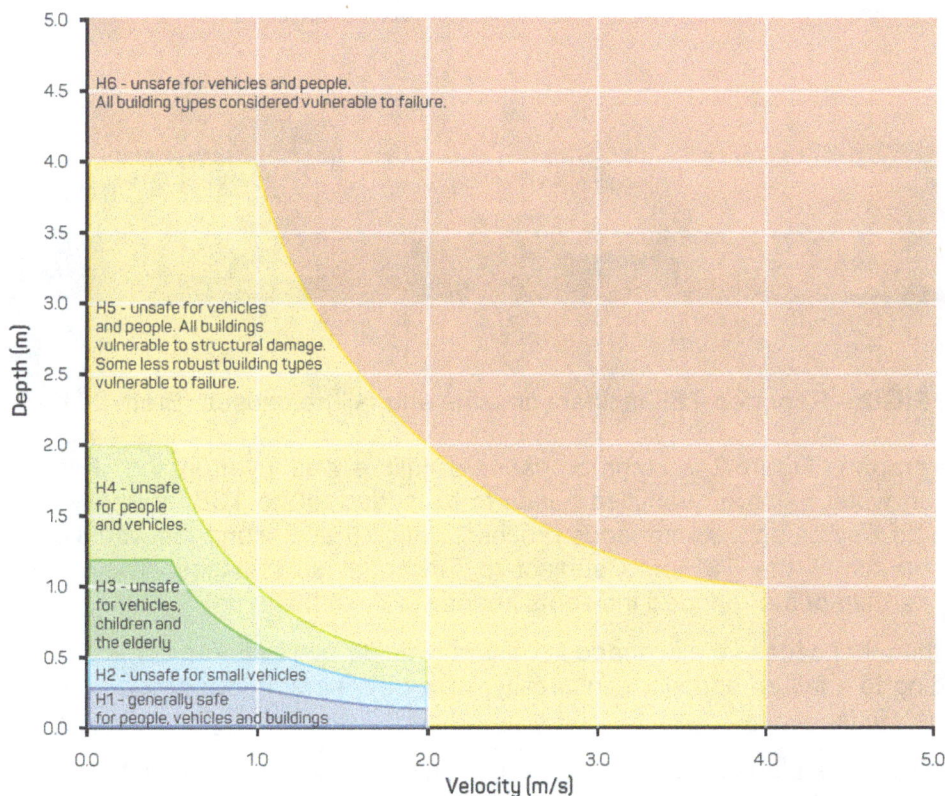

FIG 1 – General flood hazard vulnerability curve (source: The Australian Institute for Disaster Resilience, 2017).

Note that the graph above shows the flood hazard vulnerability pertaining to water. Tailings can be expected to be somewhat more dangerous, as the density is higher. But a good rule of thumb is that water or tailings deeper than 25 cm, if moving faster than 1 m/s, is a threat to life for anyone being caught in it. An average member of an emergency response operation will not have much use for detailed velocity predictions, as determining the actual velocity is difficult. As a rule of thumb, if it is moving and is deeper than knee height – do not try to enter.

Hence, it is the depth of the inundation that forms the 'What'.

When?

A timeline of the TSF failure should not only illustrate the time from breach to impact, but the estimated time from detection, via breach, and to impact. The difference can for many scenarios be substantial.

Looking at a graph of reported TSF incidents over time, in Figure 2, and their failure causes, it is possible to approximate how many TSF failures were *'actual or imminent'*, ie they could occur without any warning signs or undistinguishable signs, and how many had a *'warning time'*, often referred to as *'estimated time from detection to breach'*.

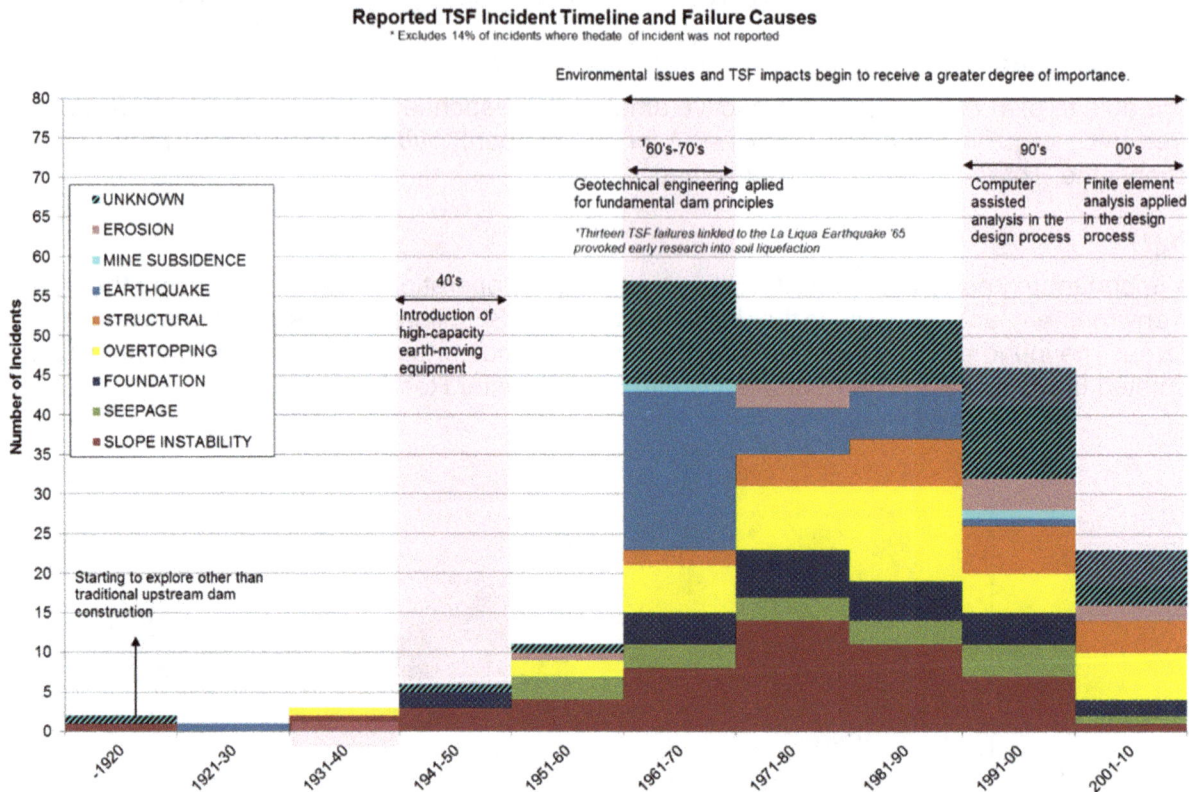

Reported TSF Incident Timeline and Failure Causes
* Excludes 14% of incidents where the date of incident was not reported

FIG 2 – Reported TSF incident timeline and failure causes (Brett, 2019).

Studying the graph in Figure 2, a common cause of failure was 'overtopping'. For an overtopping failure to occur, water in large quantities needs to be added to the TSF, normally as precipitation. This would in no way be an instantaneous process, hence these scenarios would offer a time from detection to breach, if the TSF was subject to functional surveillance. The graph shows that approximately a third of the reported incidents appear to be of this type.

For some of the other causes the scenario in general would have required some time to develop, eg cracking leading to a failure and slope instability, which appears, in the graph, to be at a declining trend since the 1980s.

To better support the planning and execution of the emergency operation the 'When' can be classified into three categories:

- A failure is *actual or imminent*, impact has hit or is likely to hit the surroundings within the next few hours.

- A failure is imminent and the impact to the surroundings is expected to occur *within the next 48 hours*.

- There is an *elevated risk* of a failure, which if it occurs, is likely to happen beyond 48 hours, and could impact the surroundings.

An estimate based on the data presented in Figure 2, would indicate that approximately 50 per cent of failures would have had sufficient warning to qualify for the category *Elevated Risk*, ie a potential failure is more than 48 hours away.

Another 25 per cent would be found in the category *within the next 48 hours*, and the remainder in *actual or imminent*.

That leaves approximately 25 per cent in the category *actual or imminent*.

It is worth mentioning that ignoring warning signs from the operation and surveillance, that indicate the TSF is at risk of failure, will sooner or later put the emergency in the category of *actual or imminent*.

For an emergency operation, when time is of the essence and often scarce, it is essential to supplement the inundation maps (see next section) with an estimated timeline showing both the estimated *time from detection to breach* as well as *the time from breach to impact*. On that timeline, the current time of the planning process, the 'now' should be presented for planning purposes. As time moves, the 'now' have to be moved along that timeline to make sure that planning is focusing on future.

As an example, digital modelling of a near identical condition to the Brumadinho TSF failure (Lumbroso *et al*, 2020), concluded that *'a warning received 15 minutes before the failure could have reduced the number of deaths to zero'*. This was *'if the evacuees knew what to do'*. The competency level for 'knowing what to do' is comparable to the one gained by a passenger from an inflight safety video aboard a plane. The Brumadinho TSF failure killed approximately 300 people.

Where?

Each modelled scenario used in the Emergency Response Plan (ERP), needs to be presented as a map, intended for the emergency response to the potential failure. These maps will mostly be used by individuals with other professions and expertise than dam engineers.

An overview map, as illustrated in Figure 3, showing the modelled alternative scenarios is a good start and should be included in the ERP.

FIG 3 – Overview map showing the modelled scenarios. Fictitious example.

The ERP should comprise individual inundation maps for each scenario, as illustrated in Figure 4.

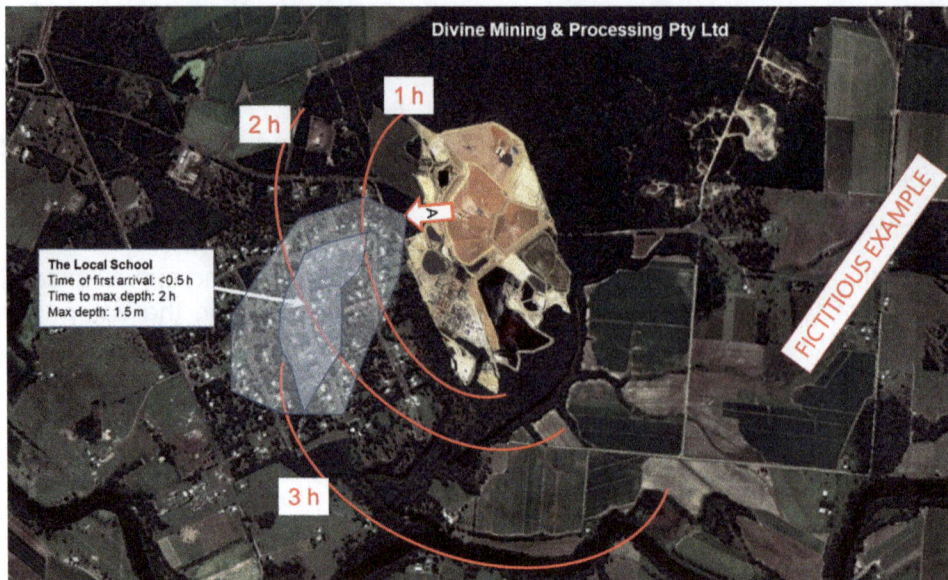

FIG 4 – Individual inundation map with points of interest and isochrones, showing estimated time of arrival. Fictitious example.

It is important not to clutter the maps with too many details. Estimated arrival time for the inundation can be illustrated with isochrones, lines showing the same arrival time from the breach point. It is typical to use half or one-hour intervals.

Landmarks and Points of Interest may be indicated to visualise the potential extent of the inundation and support orientation. Coloured tones may be used to show the expected depth.

The inundation maps should ideally deliver a representation as realistic as possible of the *What*, *When* and *Where*, for each scenario, noting that the *When* will be counted from the time of breach. Hence the timeline needs to supplement the maps.

Presenting the What, When and Where

As an Incident Management Team (IMT) Leader in the owner/operator's organisation, in charge of an emergency response to the impact of a TSF failure, it is important to select the applicable 'When' category. The choice should be between prepared distinct alternatives; *actual or imminent, within 48 hours*, or somewhere e*levated risk beyond 48 hours*, and which of the inundation maps is to be basis for the planning and execution.

Strategy

Strategy is a plan of action designed to achieve a long-term or overall aim (Oxford Languages, 2023).

A good strategy must precisely diagnose the problem being solved, set a guiding policy that will address that problem, and propose a set of coherent actions which will deliver that policy (Rumelt, 2011).

Often strategy is described as visualisation of the 'end game', with a few main steps to reach there. The longer and more complex an emergency operation becomes, the more important a clear strategy becomes.

Strategic intent

It falls upon the IMT Leader to formulate a Strategic Intent for the operation. A good Strategic Intent should comprise two main parts, the objective of the operation, and the implementation steps, no more than three, to reach the objective. These steps are often formed along a timeline, as 'initially', 'thereafter' and 'finally'. The Strategic Intent may also deliver limitations or directives over time. Well formulated, and communicated, the Strategic Intent should be the guidance for the entire operation. For the strategy to become a useful input into the response to, and the long-term recovery from, an emergency caused by a TSF failure, the strategy must be made clear, understandable, and visual to the organisation, especially the emergency management function. Compare emergency management in the figure in the Principal Duties and Timescale section.

This is in line with one of the basic principles of the Australasian Inter-Service Incident Management System (AIIMS), namely 'management by objectives'.

Strategic Intent is in many organisations referred to as an Incident Objective (AIIMS, 2004). However, an Incident Objective does not comprise the elements of crisis management, such as business continuity, finance, reputation and/or legal standing, which the IMT Leader must consider. Hence, there is a clear difference between Incident Objective and Strategic Intent.

A good example is the Strategic Intent formulated, by the Incident Controller, in the initial stages of the Piny Point emergency response, Florida US, April 2021:

> *Objective*
>
> *To prevent a catastrophic failure of the dam. Prepare for a collapse of the wall and ensure no people are harmed.*
>
> *Implementation*
>
> *Initially, remove all people that may come in harm's way, should the dam fail.*
>
> *Thereafter, attempt to rectify the piping failure, to avoid a total wall collapse.*
>
> *Finally, secure the dam and return evacuees.*
>
> *Over time, all evacuated areas shall be protected from intrusion or looting.*

This text has been reconstructed from multiple open sources. The actual text has not been made publicly available.

This simple Strategic Intent stayed valid over the entire operation, which lasted for over a week, and gave a clear directive for two tactical prongs: one repairing the dam, and one keeping the residents out of harm's way. The respective persons in charge of each tactical prong were managed by objectives, and free to develop the best tactical and technical solution.

It is an art to develop a Strategic Intent with substance enough to guide and steer the emergency operation over its duration, without becoming a set of meaningless value statements on one hand, or a set of detailed instructions and tasks on the other. The road to mastering this is training.

A well formulated Strategic Intent would take a competent IMT Leader perhaps a fraction of the time that the emergency operation will run but is a crucial investment for successful operations.

A generic template for a Strategic Intent is included in Appendix 1.

Tactics and technique

Following the choice of a strategy as per Figure 5, expressed in a Strategic Intent as described in the section above, is the choice of tactics. What tactics are available to achieve the Strategic Intent? The choice of tactics requires detailed knowledge on the subject, such as emergency response, TSF maintenance, etc.

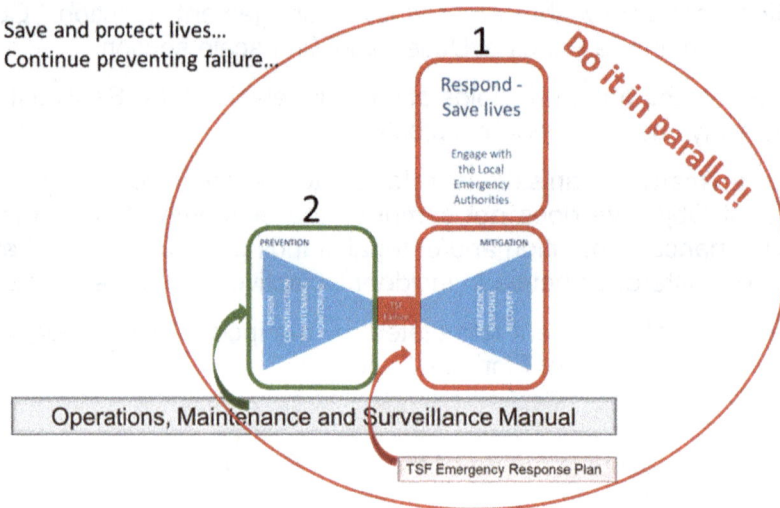

FIG 5 – Generic Strategy example for an emergency response operation to the impact of a TSF failure.

Based on the choice of tactics made, the technique for executing this can be selected. For example, as illustrated in Figure 6, if the tactical choice for the first part of the strategy is 'Evacuation' very different techniques are applied compared to if the tactic of 'Self-Rescue' is selected.

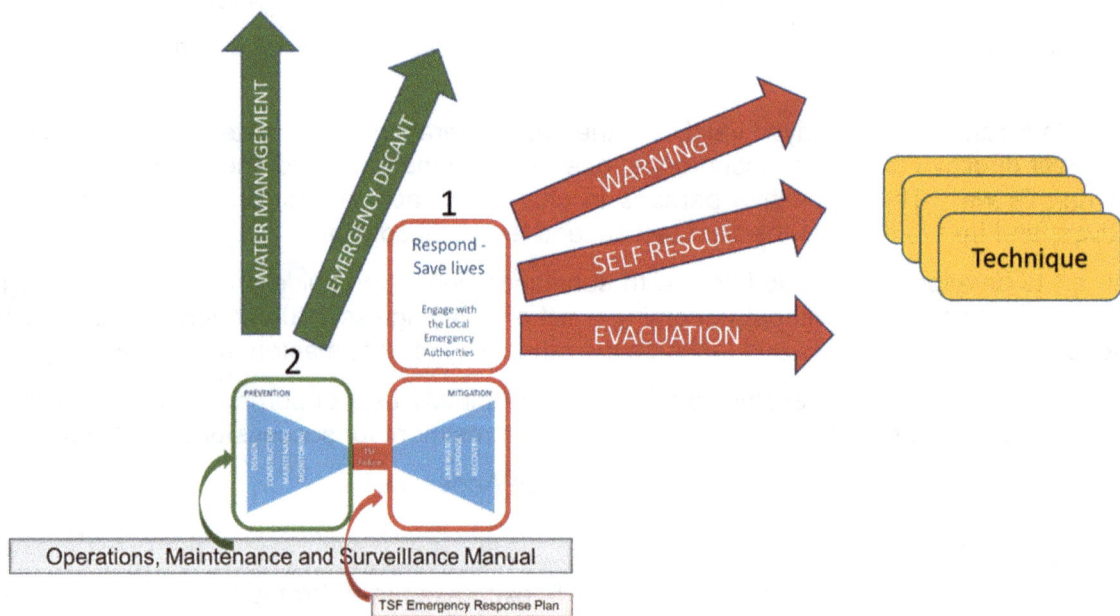

FIG 6 – Choice of tactics and technique.

Starting the process of developing and executing an emergency response at the technique end of the flow, will generate an infinite number of tasks, without any coherence. This only leads to success if the organisation has infinite resources and the strategic and tactical alternatives are limited or none, or as it is referred to in fire and rescue tactics: an over-strong response in a closed system. The proper planning of an emergency response is formulation of Strategic Intent followed by development of tactics to match strategy and the application of specific plans to achieve the tactical objectives.

Planning process

Whilst an operation with few alternatives and short duration may be suited for a simpler pre-set step-by-step plan, sometimes referred to as a Trigger Action Response Plan (TARP) or Trigger Response Plan (TRP) where very little adjustment or adaptation is required, a long-term operation where many steps will require decisions and judgement, requires a clear steering through an understandable strategy. This strategy is best expressed and communicated, in the section above, as a Strategic Intent.

The Strategic Intent is the base for the Plan Inventory, which is a simple list of the plans required or activities that will need planning, to achieve the objective stated in the Strategic Intent. The Plan Inventory is a live document, or rather a list of the plans and planning activities that the IMT needs to achieve.

There might initially be a limited number of actions that are to be executed directly and without further orders or planning. These are referred to as 'Immediate Actions'. However, one must be careful not to build the entire operation on such, as this will limit the operational alternatives and exhaust the organisation in reactive mode.

Some plans might be pre-existing, and only require launching, whilst some, such as evacuation plans may exist, but need to be adjusted for the current situation. The plan inventory is normally managed by the Planning Officer in the IMT. Each plan or planning task is allocated to one of the IMT members, potentially supported by other capacities internal or external to the organisation.

Following on the Planning Inventory, the plans are adjusted, developed, and launched. The planning process, as a sequence, from the initiation of the emergency response, via Immediate Actions, Strategic Intent, Plan Inventory and Planning, and to the What-if Planning is illustrated in Figure 7.

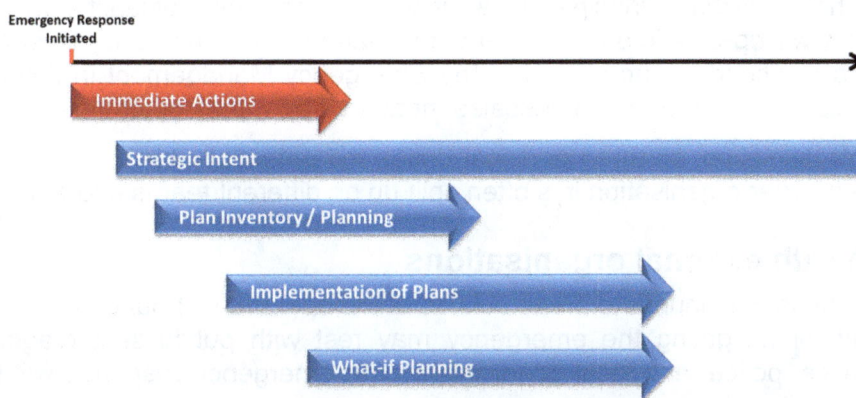

FIG 7 – Graphic representation of the Planning Process in the Incident Management Team.

'No plan survives first contact with the enemy' (von Motke, 1880), which today in civilian context has been transformed into: *'No plan survives first contact with reality'*. One must keep a preparedness for what is referred to as 'What-if Scenarios'.

Not to overwhelm the planning work What-if Plans are normally developed for:

- The most **likely** change of the situation calling for a new plan.
- The most **dangerous** change of the situation or conditions.
- The most **different** development realistically imaginable.

Principal duties and timescale

In the unlikely event of a catastrophic TSF failure, the response to the emergency will most likely draw a high level of public and media attention, the emergency may well develop a crisis in parallel. A crisis being an: *'Abnormal and unstable situation that threatens the organisation's strategic objectives, reputation or viability'* (BS 11200:2014, 2014).

To succeed in mitigating a larger scale emergency, potentially escalating to a crisis, one should move the operation forward with three *'Principal Duties'*. Emergency Response, Emergency Management and Crisis Management (Wennstrom and White, 2021), as illustrated in Figure 8.

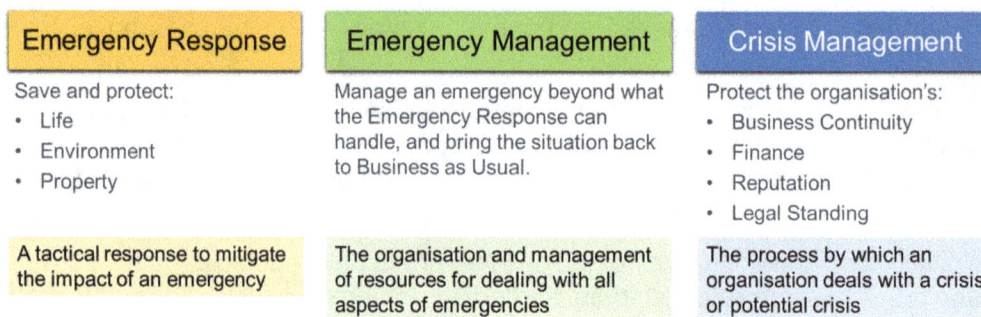

Emergency Response	Emergency Management	Crisis Management
Save and protect: • Life • Environment • Property	Manage an emergency beyond what the Emergency Response can handle, and bring the situation back to Business as Usual.	Protect the organisation's: • Business Continuity • Finance • Reputation • Legal Standing
A tactical response to mitigate the impact of an emergency	The organisation and management of resources for dealing with all aspects of emergencies	The process by which an organisation deals with a crisis or potential crisis

FIG 8 – Principal Duties (Wennstrom and White, 2021).

Emergency Response is the *Tactical response to the impact of an emergency*.

Emergency Management is *the organisation and management of resources for dealing with all aspects of emergencies* (EMA, 1998), hence arranging for a continuity when the initial emergency response has exhausted its capacity.

Crisis Management is *the process by which an organisation deals with a crisis or potential crisis* (Bundy *et al,* 2017). All three principal duties should be initiated early and function in parallel.

It is important to initiate all three Principal Duties, if not simultaneously, at least early in the operation. The timescale they will operate in are however very different. The Emergency Response till operate in a timescale from seconds to hours, whilst the Emergency Management in days and the Crisis Management can sometimes work in timescales measured by weeks or longer.

In a very small organisation, all three principal duties will potentially have to be managed by one person, whilst in a larger organisation it is often split up on different teams and team leaders.

Coordination with external organisations

If a TSF failure threatens to inundate areas outside of the operational boundaries ie the mine lease, the responsibility for mitigating the emergency may rest with public sector agencies, such as emergency services, police, and local government. The emergency response will become a joint operation.

When conducting a joint operation, where different organisations participate, with geographic and legislative differences, and difference in area of responsibility and operation, it is essential to establish a well-functioning mechanism for coordination.

Coordination is *'The bringing together of organisations and elements to ensure an effective response'* (EMA, 1998).

If a catastrophic TSF failure is expected to impact areas outside of site, which is the case for many locations, a concept for coordination, as illustrated in Figure 9, with the organisation/s responsible for these external areas needs to be in place. The responsibility for the different organisations does not always follow geographic boundaries but can be based on the function of the organisation.

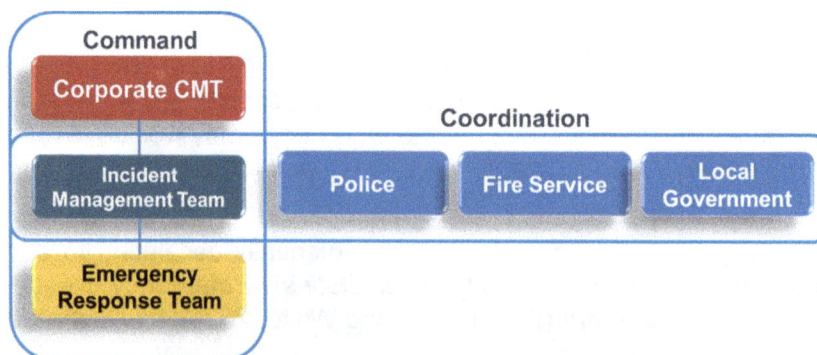

FIG 9 – Concept of coordination.

The coordination should take place, for the whole of operation, on the site/IMT level. Routines and mechanisms are to be developed, described, and trained for. For a jurisdiction where many sites are located, a local government body is in place for this purpose. These are often referred to as the Local Disaster Management Group (LDMG), or Local Emergency Management Committee (LEMC). These bodies are developed for coordination of different organisations engaged in an emergency operation.

The mechanism for establishing coordination through these pre-arranged bodies, is for the IMT Leader to send a Liaison Officer (LO) to the group. The LO remains with the coordination group for as long as the operation is ongoing and is to be the only link to the IMT. Individual points of coordination can be advised such as an Emergency Response Team Captain to coordinate directly with a Foreman from the Fire Service on a site. This should not be confused with the coordination of the operation.

Successful coordination relies upon understanding the other organisations' objectives, area of responsibility, legislative framework, capacity, and limitations. This understanding is developed as an ongoing process and is part of developing the emergency preparedness. It cannot be developed in the moment when an emergency is occurring.

Each organisation participating in the joint operation requires its own plan, as a plan is written for a specific target group, with its own organisation, competency, professional language, and needs. Sharing plans with other organisations has limited value, but keeping the coordination partners informed and up to date on your plan is beneficial and in line with the GISTM concept of shared state of readiness.

Emergency preparedness

'The objective of emergency preparedness is to ensure that the strategic direction and required building blocks for an eventual emergency response are in place' (United Nations Human Rights Council, 2007).

Preparedness to deliver a response to an emergency, or Emergency Preparedness, is the product of Organisation, Plans, and Competency (Wennstrom and White, 2021).

- Organisation – A suitable Emergency Management and Emergency Response Organisation.

- Plan – An understandable plan to guide the operation.

- Competency – Capability to execute the emergency operation.

Organisation

The role of an Emergency Response Organisation is to execute the emergency response, based on strategic and tactical decisions by the Emergency Management Organisation. To deliver a functional emergency response, the Emergency Management Organisation needs to understand the capabilities and limitations of the Emergency Response Organisation.

Most major mining companies have an Existing Emergency Management Organisation, with emergency management or incident management teams. In the following these teams, headed up by General Manager or Site Manager, will be referred to as an IMT irrespective of the company/site nomenclature.

This Emergency Management organisation should be dimensioned for management of all conceivable emergencies at site, hence also management of a response to a potential catastrophic TSF failure. The organisation should be built on a reputable Incident Management System, eg Australasian Inter-Service Incident Management System (AIIMS), National Incident Management System (NIMS; FEMA, 2017; Homeland Security, 2008), Incident Command System (ICS), or similar.

Plan

The GISTM requires the emergency preparedness to be *'based on credible flow failure scenarios and the assessment of potential consequences'*. Hence, one or more credible flow failure scenarios need to be selected for modelling.

Sunny-day and rainy-day scenarios will most likely behave differently, hence both need to be considered. In addition, many TSFs have multiple directions where a breach may occur. It is not uncommon to end up with several modelled scenarios for the same TSF.

If the scenarios comprise a 'rainy-day' scenario, it is beneficial to illustrate the effects of the modelled rain depth on the surrounds, as the rain will not fall solely on the TSF. This is commonly referred to as a 'base case'.

The Emergency Response Plan (ERP) for a TSF failure should be written for and be given ample support from the owner/operator to the organisation that is to lead the emergency operation. In most cases this is the IMT, headed up by site manager or General Manager (GM).

The TSF team, tasked with the mission to operate, maintain, and survey the TSF should find support and guidance from the Operations Maintenance and Surveillance (OMS) Manual. Content and instructions given in the OMS for the purpose of the TSF team and their work should as far as possible not be duplicated in the ERP, as the ERP and the OMS are written for different target groups.

There are however parts that needs to be correlated between the two plans. Such items are the 'operational conditions'; and the 'trigger points' moving from one operational condition to another. These should come from the OMS and be used in the ERP. A generic scale of Operational Conditions is illustrated in Figure 10. Trigger points are the criteria defining the transition from one operational condition and to the next.

FIG 10 – Generic example of operational conditions for a TSF.

Competency

Competency for the personnel that are to lead the operation, should be gained through training and exercise. It is especially the decision-making, command and control and the coordination that must be well functioning. This requires realistic exercise scenarios, where the different stakeholders are working from their respective command posts and centres. Round table meetings and discussions may be beneficial to the relationships but does very little for the competency of running an operation.

Knowing the plan, your role and the objective of the operation is of the essence. This is created by good training, delivered by instructors who are experienced in planning and executing large scale emergency operations.

To support the IMT and the IMT Leader, in planning and executing an emergency response to the impact of a TSF failure, the ERP should comprise simple and understandable checklists. These checklists should be written for the IMT and the IMT Leader and be available for each of the timely alternatives described in Section *When* above as it is the estimated time to impact that is one of the key factors in the operation.

To benefit the Emergency Preparedness, the ERP need to give support on how to execute the entire emergency operation over time, not just how to initiate it, through eg a call-tree. If the plans are incomprehensible, overly complicated, and cluttered with details, the organisation will never gain competency in handling TSF failures.

CONCLUSIONS

The emergency response to mitigate the impact of a TSF failure is often a multi-disciplinary, multi-organisational operation, impacting multiple jurisdictions.

As the scenario can vary widely, the plan must have several alternatives, both in terms of inundation extent and the time to impact.

Imperative from any emergency management perspective, to be able to plan and execute a proactive emergency response, is the *'Three Ws'*, namely: *What* has/will happen, *When*, and *Where*.

The *What* can be modelled to the inundation areas and the depth of the inundation. The *When* is a function of the monitoring on the TSF and the type of failure mechanism. The *Where* is also possible to model to some degree of certainty.

Many TSFs have several failure scenarios, and the time at hand may vary, from a somewhat predictable overfilling and overflowing scenario to a sudden earthquake. Hence a good plan needs to have a set of simple *What* and *Where*, presented in a manageable number of modelled and inundation mapped scenarios.

The *When* alternatives should be approximate and not more than three, for example 'actual or imminent', 'within the next 48 hours' and 'elevated risk – beyond 48 hours'.

As the person in charge of an emergency response to a TSF failure, normally the GM or the Site Manager, in the role as IMT Leader, one does not need technical superiority in the field of tailings dams but need to be able to get many people to do different things, that they are good at, in one direction. An instrument to gain this clear focus for the broader operation, is the early delivery of a understandable and clear Strategic Intent.

The most valuable tool to achieve such a coherent operation, is to develop and communicate a Strategic Intent. Starting the process of developing and executing an emergency response at the technique end of the flow, is likely to generate an infinite number of tasks, without any coherence. This only leads to success if the organisation has infinite resources, and the strategic and tactical alternatives are limited or none, or as it is referred to: an over-strong response in a closed system.

The use of a strategic intent is in line with the principle of management by objective, which is the first of the founding principles in AIIMS and many other international incident management systems.

A multi organisational operation requires good coordination with external stakeholders participating in the operation. In many local government areas initiatives and arrangements are in place for coordination in case of major emergencies or disaster. In line with the GISTM the operator/owner of a TSF should reach out and initiate coordination when and where required.

If public authorities and emergency services are to be part of the operation, they may already have their resource exhausted at the onset of an TSF failure. If the scenario is a rainy-day event, the rain most likely has been falling for some time and inundated the surrounding areas and the community. The emergency services might be otherwise engaged.

The key function to understand and adjust the emergency operation is good communication between the stakeholders. This is achieved through a common concept for coordination and a suitable number of trained Liaison Officer.

The emergency preparedness, which the GISTM requires, is a product of the organisation, plans and competency, and the responsibility rests heavy on the site's IMT Leader. Emergency preparedness requires good training, delivered by instructors who are experienced in planning and executing large scale emergency operations.

ACKNOWLEDGEMENTS

Andrew White, Manger Dams and Tailings, GHD, for valuable information and guidance.

David Brett, Senior Technical Director, GHD, for invaluable insight to the causes of TSF failures.

John Phillips, Senior Technical Director – Tailings, GHD, for valuable inputs to the development of the concept of ERP for TSF.

REFERENCES

Australasian Inter-Service Incident Management System (AIIMS), 2004. A Management System for any Emergency, third edition, version 1.

Australian Institute for Disaster Resilience, 2017. Guideline 7–3; Flood Hazard, on behalf of the Australian Government Attorney-General's Department Australian Disaster Resilience Handbook Collection.

Brett, D, 2019. 'Tailings Dam Guidelines – Adopting International Direction', NZSOLD/ANCOLD Combined Conference, Auckland, New Zealand, October.

BS 11200:2014, 2014. Crisis Management – Guidance and good practice.

Bundy, J, Pfarrer, M D, Short, C E and Coombs, W T, 2017. Crises and Crisis Management: Integration, Interpretation, and Research Development, *Journal of Management*, 43(6):1661–1692.

Eisenhower, D D, 1957. Speech to the National Defence Executive Reserve Conference in Washington, D.C., November 14.

Emergency Management Australia (EMA), 1998. Australian Emergency Manual Series, Part 1, The Fundamentals, Manual 3, Australian Emergency Management Glossary, Emergency Management Australia.

Homeland Security, 2008. National Incident Management System (NIMS), December.

International Council on Mining and Metals (ICMM), 2020. The Global Industry Standard on Tailings Management (GISTM), March.

Lumbroso, D, Davison, M, Body, R, Petkovšek, G, 2020. 1HR Wallingford; Modelling the Brumadinho tailings dam failure, the subsequent loss of life and how it could have been reduced, 4 June.

Oxford Languages, 2023. *Online dictionary* (Oxford University Press).

Rumelt, R, 2011. *Good Strategy/Bad Strategy: The Difference and Why it Matters* (Crown Business).

The Federal Emergency Management Agency (FEMA), 2017. National Incident Management System (NIMS), third edition, October.

United Nations Human Rights Council (UNHRC), 2007. *UNHCR Emergency Handbook*, Section 2, https://emergency.unhcr.org

von Motke, H, 1880. Kriegsgechichtliche Einzelschriften.

Wennstrom, O and White, A, 2021. Emergency Preparedness for Tailings Dams, ANCOLD Proceedings of Technical Groups – Dams: Evolving with Modern Technology; December.

APPENDIX 1

Strategic Intent template for Incident Management Team Leader

A Strategic Intent (SI) is to be formulated by the IMT Leader. The SI is a comprehensive yet compact way of explaining the IMT Leader's intent and overall plan for the combat of the incident and return to Business as Usual.

Before developing the SI, create a Problem Picture to understand the situation in broad.

The SI issued by the IMT Leader shall cover the entire area of responsibility for the IMT, ie the entire site/operation.

An SI comprises of two parts the 'Objective' and the 'Implementation'. Ideally an SI is formulated to be valid over the entire period of the incident and the recovery. It is always written and expressed in first person.

- Objective.
  (What is the overall objective of the operation?)

- Implementation:
  (State what you want to see happen in a few timely stages)

  - Initially.

  - Thereinafter.

  - Finally.

- Restrictions or directives over time.
  (Is there something you want to be happening all the time or something that you don't want to see in the operation/operations?)

Note!

The SI shall cover the entire 'area of responsibility' for the IMT Leader, not just the ongoing incident.

Fire Service and Emergency Services uses a similar term called Incident Objective, but this does not comprise the crisis management elements, the business continuity, or the recovery phase, hence it is different from the SI.

Geochemical and geotechnical testing of mine wastes

Geotechnical characterisation and dump design of historic clay-rich coalmine spoil

J Ahn[1] and K Young[2]

1. Geotechnical Engineer, BHP, Brisbane Qld 4000. Email: jiwoo.ahn@bhp.com
2. Superintendent Geotechnical, BHP, Brisbane Qld 4000. Email: kathryn.young@bhp.com

ABSTRACT

At a BMA site in the Bowen Basin, historic spoil believed to be over 20 years old is required to be re-dug and re-dumped from an historic offset pit to permit mining progression. The spoil had been initially dug and rehandled by dragline in a predominantly tertiary and weathered horizons. Therefore, the spoil is highly variable. Poor tip head performance has been experienced with this spoil and was hypothesised to be due to degradation from long-term exposure to the environment, as well as disturbance caused by re-handling. Bulk samples of the spoil were collected from freshly paddock dumped loads to represent the *in situ* condition at tip heads. A suite of laboratory testing has been undertaken to understand the index characteristics as well as the geotechnical behaviour of this spoil such as Atterberg limits, particle size distribution, triaxial and large-scale direct shear. Shear strength testing was undertaken in both unsaturated and saturated conditions to respectively investigate the spoil's short-term performance at the tip head and long-term performance under prolonged exposure to surface and groundwater. This paper presents the interpretation of the test results, as well as presenting a case study on the operational controls for dumping that have been implemented at the site. It is expected that this study will also inform the long-term stability analyses of weak spoils as part of mine closure studies.

INTRODUCTION

At a BMA site in the Bowen Basin, historic spoil believed to be at least 20 years old is currently being re-handled and re-dumped to allow for pit progression. There is no historic testing data available for this spoil. However, poor tip head performance and back analyses of observed instabilities had indicated that this spoil was of lower strength than the other spoils typically dumped at site. This may be due to the spoil remoulding over time, long-term exposure to the environment as well as experiencing further degradation caused by re-handling when the spoil is dug and hauled to the western dumps. To better understand the characteristics and shear strengths of this spoil, laboratory testing was undertaken and reconciled with actual dump performance.

LABORATORY CHARACTERISATION TESTING RESULTS

Sampling

A range of bulk spoil samples were collected from truckloads of freshly re-dug spoil. From visual and tactile inspections, it is also geomechanically variable. The spoil varies from a moist, very soft to soft, high plasticity clay to a sandy gravel with some low to medium plasticity fines. Sample A was sampled to be representative of the overall spoil mass, while Sample B was selectively sampled for the portions appearing most clay-rich.

Classification testing

The particle size distributions are presented in Figure 1. The results indicate that the spoil is a Sandy CLAY with variable gravel content. Generally, Sample A has a higher clay content than the Sample B. Samples B1 and B3 have a significantly higher fines content than Sample A. The moisture content is typically around 10 per cent, with Sample B1 showing a higher moisture content of 14 per cent. Based on the Atterberg limits summarised in Table 1, the samples are low/medium to medium plasticity.

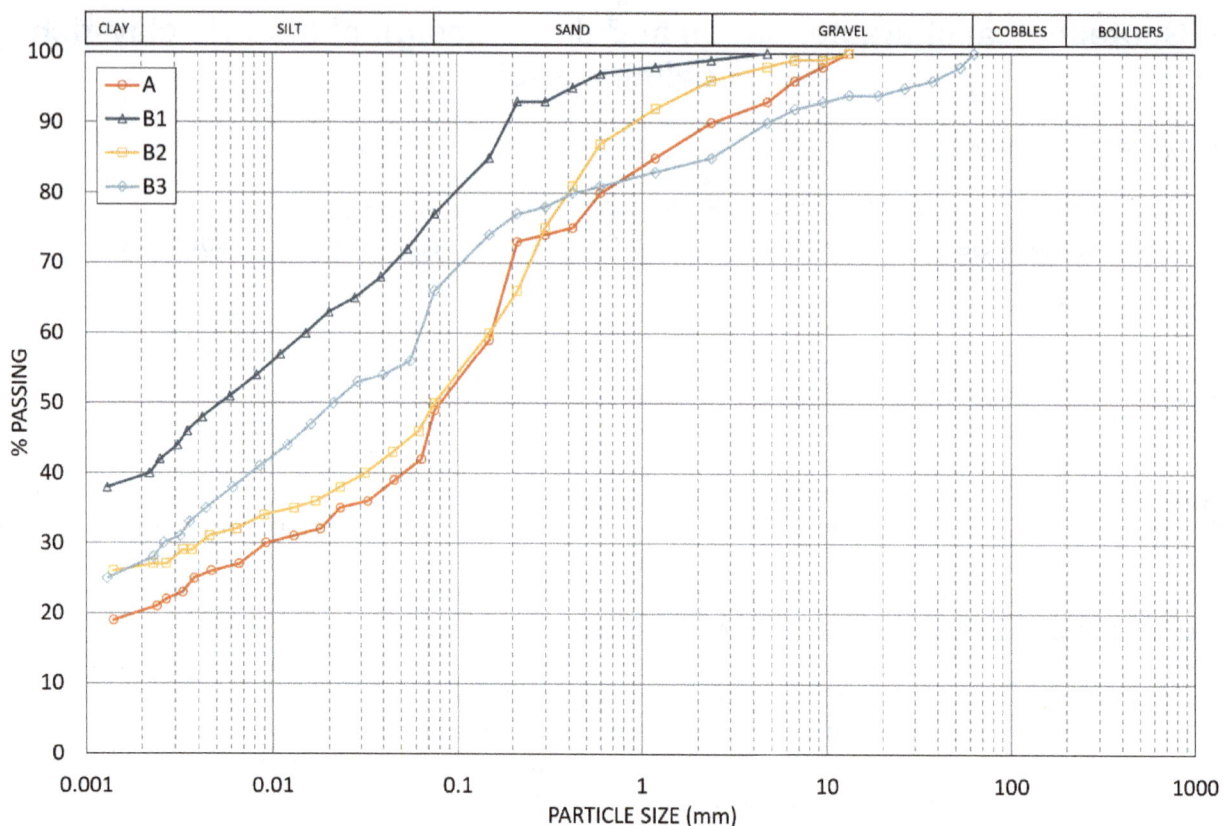

FIG 1 – Particle size distribution.

TABLE 1

Atterberg limits.

Sample	Colour	Moisture Content	Plastic Limit	Liquid Limit	Plasticity Index	Classification
A	Brown	9.4%	13%	35%	22%	CL / CI
B1	Orange / White	13.8%	15%	46%	31%	CI
B2	Orange / White	9.5%	14%	35%	21%	CL / CI
B3	Brown / White	9.6%	15%	44%	29%	CI

Shear strength testing

Consolidated undrained (CU) triaxial tests were undertaken for each of the samples. Photos of the sheared samples are shown in Figure 2. In addition, a 300 mm × 300 mm direct shear box test was undertaken with sample A to assess the benefits to frictional strength with the inclusion of the gravel component. Apparent cohesion and friction angle for both the effective and total stresses were calculated to represent the drained and undrained conditions respectively. These values are summarised together with the direct shear box result in Table 2.

The spoil category system adopted across BMA mines and as well as many other open cut coalmines throughout the Bowen Basin are founded in the work presented by Simmons and McManus (2004). Spoils of parent rock from the Tertiary period are typically classified as 'Category 1' under the framework, which is described as a fine-grained, clay-rich and high plasticity.

The effective cohesion for all samples is relatively low, which was expected as the testing was undertaken on disturbed and remoulded samples to represent field conditions of re-digging and dumping. Therefore, the spoil is expected to be normally consolidated.

Sample A

Sample B1

Sample B2

Sample B3

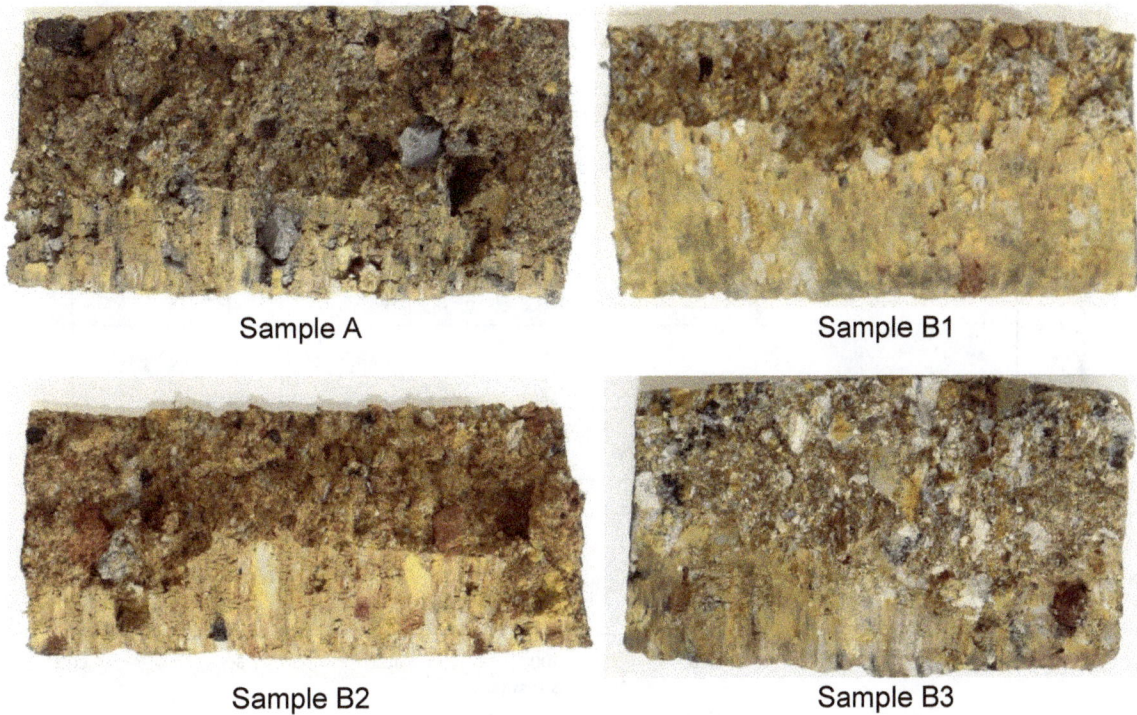

FIG 2 – Photos of the CU triaxial samples after testing.

TABLE 2

Triaxial compression and direct shear testing results.

Sample	Test type	Effective (drained)		Total (undrained)	
		c' (kPa)	Φ' (°)	c (kPa)	Φ (°)
A	Direct shear	0	33	N/A	
A		0	29	0	18
B1		5	20	3	15
B2	Triaxial	0	26	0	12
B3		12	24	5	18

The drained friction angle of sample A is well above the mean for 'unsaturated Category 1', while the undrained friction angle is equal to the mean. On the other hand, both drained and undrained friction angles of the two samples with higher fines content (B1 and B3) are well below the mean for Category 1 and are as low as the second percentile. This confirms that some fines-rich portions of the historic spoil do not fit within the BMA framework.

Effective Mohr–Coulomb envelopes have been fitted for samples with ~50 per cent fines (Samples A and B2) and ~70 per cent fines (Samples B1 and B3) respectively, Figure 3.

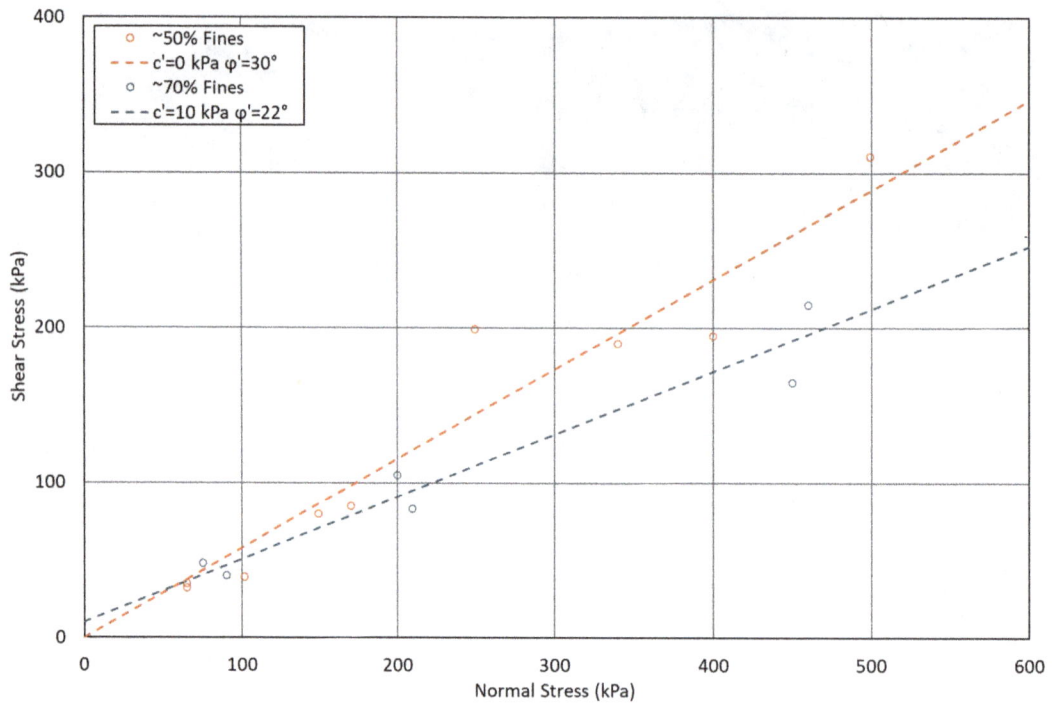

FIG 3 – Mohr–Coulomb envelopes for ~50 per cent and ~70 per cent fines content.

RECONCILIATION WITH OBSERVED TIP HEAD PERFORMANCE

Previously, this spoil was previously classified under the BMA spoil framework (Simmons and McManus, 2004) as Category 1. This historic spoil was previously dumped at 20 m high tip heads, as per the site's typical procedures for dumping Category 1 spoils. However, poor tip head performance has suggested that the material may be of poorer quality than anticipated. An example is presented in Figure 4. Back analysis of the presented failure suggested that the shear strength of the spoil is c' = 15 kPa and φ' = 20°, similar to the tested strengths for the 70 per cent fines samples.

Photo from below

Photo from above

FIG 4 – Photos of failure in 20 m high tip head.

Subsequently, the tip head heights were reduced to 10 m. This resulted in a significant improvement in tip head performance, with a near elimination of geotechnical hazards and instabilities, Figure 5.

FIG 5 – Successful implementation of 10 m high lifts.

Using the design shear strengths presented in Figure 3, slope stability analyses of 20 m and 10 m high tip heads were undertaken using the 2D limit equilibrium slope stability software *Slide2 (Rocscience)*. A repose angle of 37° was adopted based on observations of dumped slopes on-site. The factor of safety was calculated using the GLE method of slices for circular slip surfaces. The results are presented in Figures 6 and 7 respectively. The calculated factor of safety (FOS) for the 20 m high case is below unity for both cases, matching the poor performance observed at site. On the other hand, the FOS for the 10 m high case is above unity for both cases, and for more deep-seated slip surfaces where the backscarp is more than 5 m from the crest (ie the typical width of tip head bund), the FOS is above the acceptance criteria.

It should be noted that the presented stability analyses are based on shear strength testing of samples at the typical *in situ* moisture content of about 10 per cent. At higher moisture contents, the observed spoil strength and associated tip head stability is typically reduced. Therefore when the moisture content is elevated (eg following large rain event), the tip head may require dozing down to a shallower angle and additional controls required to progress the dump (eg dumping short of the crest and dozing). Such controls are commonly implemented on-site.

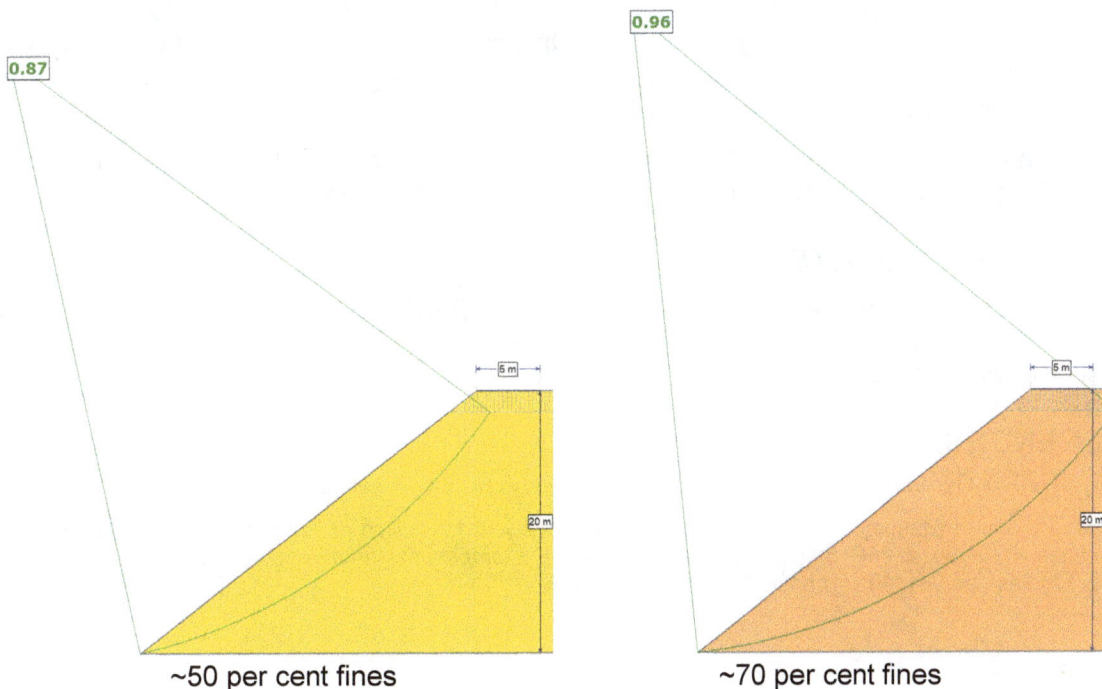

FIG 6 – Slope stability analysis of 20 m high tip heads.

FIG 7 – Slope stability analysis of 10 m high tip heads.

MINE CLOSURE

In the time frame of mine closure, the moisture content may increase to sufficiently high levels such that the spoil becomes fully saturated. The factor of safety (FoS) for an infinite cohesionless slope (Chowdry, 2010) is presented in Equation 1, where α is equal to the slope angle.

$$FoS = \frac{\tan \phi'}{\tan \alpha} \tag{1}$$

For the range of tested undrained friction angles, the maximum slope angle ranges from 8 to 12° to achieve a FoS of 1.5.

CONCLUSIONS

At this site, spoil from the Tertiary horizons is typically classified as Category 1 under the BMA Spoil Framework and is dumped at 20 m high tip heads. However, back analysis of observed tip head performance and laboratory testing suggests that this spoil is weaker than the equivalent freshly dug spoil. This suggests that the spoil degrades, perhaps due to long-term exposure to the environment and/or during re-handling.

This has implications for re-handling and re-dumping historic spoil for either productivity or rehabilitation purposes and should be considered in design and operations.

ACKNOWLEDGEMENTS

The authors would like to thank BHP Mitsubishi Alliance (BMA) for permission to publish this paper. The authors are also grateful to Jianping Li for reviewing, and to Ross Branch and Hannah Elvery for assisting with sample collection.

REFERENCES

Chowdry, R, 2010. *Geotechnical Slope Analysis*, 2nd ed (Taylor & Francis Group: London).

Simmons, J V and McManus, D A, 2004. Shear strength framework for design of dumped spoil slopes for open pit coal mines, in *Advances in Geotechnical Engineering – The Skempton Conference* (eds: R J Jardine, D M Potts and K G Higgins), pp 981–991 (Thomas Telford Ltd: London).

Estimating *in situ* state of tailings using Panda Dynamic Cone Penetrometer

O Dudley[1] and M Llano-Serna[2]

1. Senior Geotechnical Engineer, Red Earth Engineering – a Geosyntec Company, Brisbane Qld 4000. Email: oliver.dudley@redearthengineering.com.au
2. Principal Geotechnical Engineer, Red Earth Engineering – a Geosyntec Company, Brisbane Qld 4000. Email: marcelo.llano@redearthengineering.com.au

ABSTRACT

Identifying tailings layers susceptible to static liquefaction, particularly for contractive soils, is a fundamental question considered by geotechnical engineers involved with the design, construction and operations of tailings storage facilities (TSFs). To determine the likelihood of static liquefaction for any given TSF, usually, geotechnical engineers aim to understand the *in situ* state of the tailings using empirical correlations and/or critical state soil mechanics. The state parameter approach has been recognised as an efficient tool to assess the liquefaction potential of tailings. Cone penetration testing (CPT) has been demonstrated to play an essential role in determining the *in situ* state parameter. However, the size and weight of the CPT rigs may limit their access to soft tailings surfaces. Furthermore, the logistical requirements to mobilise a CPT rig can be cumbersome in remote areas. The Panda dynamic cone penetrometer (Panda DCP) is a lightweight variable energy tool that has been used throughout Europe, particularly in France, for compaction control of engineered fills. Panda DCP applications have also been reported in Chile for compaction control of upstream construction of TSFs. The Panda DCP can rapidly be deployed for testing, is quick to carry out a test, can be undertaken by one person and can access areas inaccessible by low-pressure CPT rigs. The paper presents the results of a trial program that aimed to estimate state characteristics of tailings using the Panda DCP. The trial program was undertaken at a TSF in Queensland, Australia. Firstly, this study compared Panda DCP dynamic cone resistance, q_d against CPT cone tip resistance, q_c. Secondly, this study estimated *in situ* state of the tailings with the Panda DCP. This was done using two methods referred to in this paper as Method-1 and Method-2, with the results compared with CPT-based approach for determining *in situ* state of tailings. The results in this paper indicate that the Panda DCP could become an interesting alternative to screen and estimate the state of contractive and dense tailings.

INTRODUCTION

Mine tailings are hydraulically placed material of predominantly sand, silt or clay-sized particles resulting from mining and mineral processing after the mineral has been extracted or the unwanted material separated, eg shale and clays from coal, clayey fines from iron ore or bauxite (Fell *et al*, 2018). For coal and bauxite, the ore will not be ground, and the process is one of separation of the other rock particles from coal or finer clays and silts from bauxite gravel by washing and separation (Fell *et al*, 2018). Most tailings storage facilities (TSF) consist of tailings of sandy, silty, and clay grain-sized soil particles. There are many forms of tailings disposal, including tailings water slurry, thickened discharge, co-disposal and filtered paste technology. Conventional tailings are thickened in high water content slurry and either pumped or gravity feed to a TSF where tailings are discharged from single, several discharge points or spigots. Deposition of tailings slurry from the spigot encourages the segregation of material particles. Tailings can be discharged into TSF via subaqueous (deposited underwater) or subaerial deposition (deposited above water over the beach) methods. Subaqueous deposition tailings densities are dictated by particle density and settlement of fines which usually results in low densities and low shear strength. Subaerial depositions allow for drying, bleeding and desiccation during deposition cycling, which increases *in situ* densities.

Particle size is important to geotechnical engineers as sand, silt and clay size particles exhibit different material behaviours at different void ratios under the same effective stress conditions. During the deposition of tailings slurry, the sand and finer silt and clay-sized particles are segregated. Experience on sites with subaerial deposition of high-water content tailings slurry (30 per cent to 40 per cent solids content) has shown that heavier sand and coarse silt particles segregate and drop out closer to the spigot/embankment, and the smaller silt and clay size fractions travel further down the tailings beach away from the spigot.

When a TSF is nearing maximum design storage capacity, plans are usually underway to increase storage levels within the dam by raising the dam walls. Three standard methods can extend the storage capacity: upstream raise, centreline raise or downstream raise. Upstream raise embankment construction begins with starter embankment, and the free draining tailings beach close to the embankment becomes the foundation for embankment wall raise.

A common challenge for geotechnical engineers involved in upstream TSF dam raise design and construction is characterising the *in situ* states of the tailings foundation and determining whether the tailings are susceptible to static or seismic liquefaction. Static liquefaction usually occurs when contractive materials are loaded to a state of shear stress ratio such that contractive undrained shearing may occur with a small trigger. Been (2016) states that static liquefaction has been identified as one of the most damaging failure modes for tailings facilities, including TVA Kingston tailings dam, Mt Polley tailings dam and Fundao tailings dam. The TVA Kingston failure was attributed to the undrained static failure of fine-grained fly ash and caused significant environmental contamination (Been and Jefferies, 2016); Mt Polley tailings dam failure was attributed to the rapid foundation yielding resulting in a dam breach and static liquefaction event, causing one of the biggest environmental disasters in modern Canadian history (Independent Expert Engineering Investigation and Review Panel (IEEIRP), 2015); and Fundao tailings dam failure was attributed to static liquefaction with at least 17 dead and estimated damages of billions of dollars (Morgenstern, Vick and Watts, 2016).

Characterising mine tailings and their foundation and understanding strength behaviour is at the forefront of geotechnical engineers, mainly when undertaking stability assessment of upstream TSF dam raise design in accordance with governing guidelines and standards like the Global Industry Standard on Tailings Management (GISTM, Global Tailings Review, 2020) and ANCOLD Guidelines on Tailings Dams (ANCOLD, 2019).

This paper studied a TSF in Queensland, Australia, which has been designed and constructed using the upstream raise construction method. During the geotechnical investigation phase of the feasibility design, Panda dynamic cone penetrometer (Panda DCP) was undertaken adjacent to Cone penetrometer test (CPTu) locations on the existing tailings surface. The objective was to determine whether the Panda DCP dynamic cone tip resistance, q_d, correlated with CPTu static cone tip resistance, q_c, and whether the Panda DCP correlations may be used to determine whether the tailings foundation is in a dilative or contractive state. Estimating the *in situ* state of tailings was done using two alternative methods referred to in this paper as Method-1 and Method-2, with the results compared with the CPT-based approach for determining *in situ* state of tailings.

The Panda DCP is a lightweight device (20 kg), portable and easily transported by one person. The Panda DCP can rapidly be deployed for testing, is quick to perform a test, can be undertaken by one person, and access areas inaccessible by low-pressure CPTu rigs, making Panda DCP a favourable tool for screening *in situ* state of tailings. The Panda DCP with a larger cone tip size (4 cm^2 and 10 cm^2) is capable of testing to depths of up to 6.0 m but limited with possible inclination and damping of hammer energy transfer to the cone tip due to rod flex. The Panda DCP, with a smaller cone tip size (2 cm^2), has a maximum recommended testing depth of 1.5 m as skin friction of the rods may overestimate cone tip resistance. This is seen as a limitation by some practitioners. The Panda DCP is a variable energy tool used throughout Europe for compaction control of engineered fills and in Chile for compaction control of upstream construction of TSFs. A drawback of the Panda DCP is that there are no local Australian standards for Panda DCP equipment and that the user will need to adopt foreign (eg French or Chilean) standards and industry best practices. This study aims to address a data gap existing in the tailings industry with a particular application to the Australian market. The Panda DCP may also serve as suitable testing equipment for tailings operation quality control to ensure tailings foundations is progressively in a dilative state in preparation for upstream raise construction.

In Australia, ground improvement of slurry deposited tailings material by low-ground pressure swamp dozers is usually called mud farming. Mud farming promotes dilative tailings state by compaction and then drying thanks to environmental action. Mud farming also provides surface drainage paths towards the decant. A long beach and central decant structure usually characterise upstream raises. Adequate water management mitigates the risk of static liquefaction of foundation materials. Historically, mud farming quality control for the studied dam comprises field *in situ* dry density (ie nuclear density guage) and conventional Dynamic Cone Penetration (DCP) test. Nuclear density guage are only limited to testing a maximum depth of 0.3 m and DCP test results with low blow count in soft tailings are difficult (and not recommended) to correlate to *in situ* state. Some correlations with relative density exist. The Panda DCP is a useful tool for quality control of mud-farmed tailings and provides useful information to designers and operators to plan and prepare for future upstream raise loads. As the tailings rate of rise is only limited to a maximum of 0.5 m per deposition cycle, the depth limitation of the Panda DCP is often not an issue.

TAILINGS *IN SITU* DENSITY THROUGH STATE PARAMETER

Relative density is a measure of the state of the soil (or tailings). The critical state and state parameter concept, ψ, is often used as a quantitative measure to assess liquefaction potential. State parameter, ψ, is the difference between the current void ratio, e and the void ratio at the critical state, e_{cs}, at the same mean effective stress (Robertson, 2010).

$$\psi = e - e_{cs} \tag{1}$$

The critical state locus (CSL), varies with mean effective stress, and in this study, has been observed to vary with the fines content. ψ is introduced as a measure of deviation from the critical state. The use of the critical state and state parameters is argued to be more useful for understanding soil behaviour and properties than simply using relative density or void ratio (Robertson, 2010).

Currently, in the mining industry, the *in situ* state of the tailings is commonly assessed by *in situ* testing, eg CPTu and triaxial testing correlation to ψ (Jefferies and Been, 2016). Triaxial testing is a laboratory technique that allows determining the location of the CSL, and it is a cost-effective alternative; however, it does not inform the *in situ* state of the tailings on-site unless the tailings samples were in a true undisturbed state at the time of testing. The disadvantage of CPTu, is the cost and availability of low-track pressure rigs in Australia capable of accessing mine tailings (Fourie *et al*, 2013). The high demand for CPTu rigs and triaxial testing is currently experienced throughout the industry since the implementation of GISTM requirements. While CPTu testing can be more onerous than a laboratory campaign, it indirectly measures the state of tailings on its *in situ* condition. A combination of both techniques, laboratory testing and CPTu testing, is considered a recommended option for tailings characterisation.

Despite the advantages of laboratory and field techniques, each presents drawbacks. As mentioned, CPTu testing can cost considerably more than a laboratory testing program. When dealing with soft surfaces some site investigation, alternatives include the construction of fingerpads to ease access. A finger pad often results in tailings disturbance underneath the testing pad making the CPTu results lose representativity. Amphibious CPTu rigs have become common in recent years. Securing an available amphibious CPTu rig can be problematic due to the significant backlog of work seen during the last three years due to the implementation of GISTM requirements. Furthermore, undertaking a CPTu survey during the deposition of tailings can conflict with operations and pose risks to operating personnel.

CPTu test results allow evaluation of existing site conditions. In contrast, a laboratory program is useful to assess present and future stress conditions. However, laboratory testing also presents some limitations. For example, the most common undisturbed sampling method in Australia for soft soils is the open sample (Shelby tube 50 mm to 75 mm). It has been demonstrated that mechanical parameters obtained from laboratory tests on samples recovered from Shelby tubes do not always represent *in situ* conditions (Pineda *et al*, 2018). Another common undisturbed sampling practice in Australia is the fixed-piston sampler. Pineda *et al* (2018) have reported that the quality of the retrieved fixed-piston tube sample is highly influenced by the skills of the driller. Research has shown undisturbed samples of soft clays, sand, or silt undergo disturbance and void ratio change during sampling, waxing, transportation, handling, extrusion, trimming and consolidation in the laboratory. Despite the limitations, laboratory testing allows for assessing the behaviour under various stress paths, including current and future conditions.

Due to the likelihood of sample disturbance during sample transportation resulting in changes to its *in situ* state, geotechnical practitioners are more inclined to gather data via *in situ* testing methods to provide *in situ* state information. The information is used at a later stage to inform reconstituting sample procedures in the laboratory. Laboratory testing is necessary for determining the CSL of particular tailings. Key questions for geotechnical practitioners designing TSF upstream wall raises are:

- What is the state parameter of tailings? Do the tailings being studied exist in a contractive state?

- At what state do the tailings exhibit contractive behaviours, and how would tailings perform under future loads?

By understanding these questions, geotechnical practitioners may develop solutions to future potential problems, especially if it means a high risk of impact on people and the environment.

One challenge to operators and designers of a mud-farmed TSF is achieving design tailings density during deposition before the next cycle of tailings deposition. Quality control of mud-farmed tailings usually includes the measurement of field dry density testing with a nuclear density guage, which is limited to a maximum 300 mm test depth. Information regarding the *in situ* state of deeper tailings is unknown without the use of CPTu equipment. The Panda DCP is proposed here to provide a quick source of information regarding the *in situ* state of the tailings during operating conditions, especially in areas not accessible by CPT rigs. It also facilitates taking corrective actions through quality control procedures when *in situ* state does not meet design conditions.

SOILS CHARACTERISATION THROUGH CONE INDENTERS

Pieter Barentsen developed the Dutch cone penetrometer around 1930 in the Netherlands to measure the resistance of soil resistance on the conical tip. The original purpose of penetration testing was to characterise hydraulic fills' thickness and bearing capacity. The first Dutch cone tip had an area of 10 cm^2 (diameter 37.5 mm) with a 60° apex angle.

The conventional Dynamic cone penetrometer (DCP) was invented in Australia (Scala, 1956). The DCP is a simple and rapid device which was first used in Australia for the characterisation of subgrade soils for pavements. The make-up of the DCP consisted of a 9 kg mass dropping 508 mm onto an anvil and rod with a 30° apex cone tip. The South African variant of the DCP was introduced in 1970 with an 8 kg mass dropping 575 mm onto an anvil and rod with a 60° apex cone tip.

Sol Solutions developed the Panda DCP in France as a lightweight DCP capable of soil strength testing and compaction control. The Panda continually measures a dynamic cone resistance, q_d (MPa) with depth as the Panda DCP hammer drives the probe into the ground. q_d is derived using the modified Dutch formula (Cassan, 1988) given in Equation 2.

$$q_d = \frac{1}{A} \cdot \frac{\frac{1}{2} \times mv^2}{1+\frac{P}{m}} \cdot \frac{1}{x_{90°}} \tag{2}$$

where:

A	Area of Cone
V	Speed of Impact (of the hammer)
m	the weight of the striking mass
P	the weight of the struck mass
$X_{90°}$	Penetration due to one blow (90° cone apex).

The work described herein used cone tips with the following cross-sectional areas: 2 cm^2, 4 cm^2, and 10 cm^2. All the cone tips had a cone apex angle of 90°. The Panda DCP equipment is compliant with French Standard NF P 94–105-April 2012 (French Standards Association (AFNOR), 2012). The testing described in this paper was undertaken in accordance with Chilean Standard NCh3261–2012 (Instituto Nacional de Narmalizacion (INN), 2012).

The Panda DCP testing for this geotechnical investigation was carried out by striking a 5 kg weight onto an accelerometer on the head of a tool which measures the velocity of the impact, and the control panel microprocessor records the penetration depth with each blow on the tool. Based on laboratory calibration chamber experiments for different materials at different *in situ* dry densities and water contents, Laurent (1999) demonstrated the relationship between q_d, dry density, γ_d and water content, *w,* was supported by Equation 3. Regression coefficients A, B and C are introduced.

$$\gamma_d = A(w) + B ln(q_d) + C \tag{3}$$

Benz-Navarrete *et al* (2019) simplified the dry density estimation correlation based on tip resistance to be:

$$\gamma_d = 1.06 \cdot ln(q_d) + 15.82 \tag{4}$$

Benz-Navarrete *et al* (2019) established that bulk density may be estimated from (Robertson, 2022):

$$\gamma/\gamma_w = 0.36 \log\left(\frac{q_d}{p_a}\right) + 1.43 \tag{5}$$

Where γ_w is the unit weight of water and p_a is the atmospheric pressure. Using Equation 6 and the existing empirical Panda correlation in Equation 4, the void ratio, *e* can be estimated:

$$e = \left(\frac{G_s \gamma_w}{\gamma_d}\right) - 1 \tag{6}$$

Where G_s refer to the specific gravity of a given soil particle.

For each void ratio estimate, the mean effective stress, p'$_0$ can be estimated if we assume that the vertical stress, σ_{vo} acting at depth, z is γ'.z, where bulk density can be estimated using Equation 5. The approach requires knowledge of the piezometric surface. Mean effective stress, p'$_0$ can be estimated using Equation 7:

$$p'_0 = \gamma' z \, (1 + 2K_0)/3 \tag{7}$$

where the earth pressure coefficient at rest, K_0 may be derived from measured values using self-boring pressuremeter testing or estimated based on existing soil correlations.

The piezometric water level at the studied site is known to be significantly lower than the tailings beach surface level (>5.0 mbgl). The piezometric surface was established from VWP readings in the tailings and decant water levels. Therefore in this paper, using this information, soil unit weight, γ', used in Equation 7 is equal to the bulk density calculated in Equation 5.

TAILINGS CHARACTERISATION

To understand and characterise the tailings, a geotechnical investigation campaign was undertaken. Tailings surface samples were taken at 10 m, 100 m, 150 m, 200 m and 250 m away from the spigot at two locations on the TSF being studied. Twenty-three Specific Gravity (G_s) tests found the tailings material to range between 2.51 and 2.71, with 2.51 being the closest to the spigot and 2.71 being the furthest away from the spigot. Ten Particle size distribution (PSD) tests were undertaken on tailings samples gathered at set distances from the spigot. The testing aimed to classify tailings particle size with distance from the spigot location down the tailings beach and determine whether particle segregation had occurred. G_s of the tailings material were found to be dependent on fine content percentage. The results of surface sampling at the locations studied are depicted in Figure 1. The figure shows a decrease in sand content percentage and an increase in fines content percentage with distance away from the spigot location.

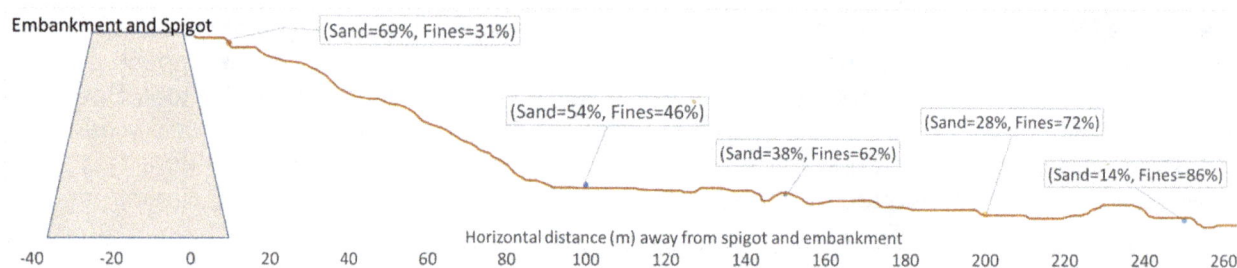

FIG 1 – Typical segregation of tailings particles at TSF study.

Based on the PSD analysis in Figure 1, the 10 m and 100 m tailings samples are of 70 per cent to 55 per cent sand size particles which supports the notion that the heavier sand particles drop out closer to the spigot and the lighter, smaller silt and clay size fractions travel further down the tailings beach away from the spigot. Based on on-site experience, the tailings particle size content at a given location away from the embankment and spigot is subject to change depending on the following:

- Spigot location and tailing beach angle.

- Deposition strategy.

- Mud farming process.

- Washing out of fines during rain events or pipe flushing, which washes the lighter fines further down the tailings beach profile.

A recent wall raise quality control construction data was assessed whereby 436 Atterberg limits and compaction tests were undertaken on tailings borrow material gathered between the spigot location and 150 m distance from the spigot location. The tested tailings borrow material was found to be predominantly of low to intermediate plasticity fines. The results were observed to plot above the A-line in the Casagrande plasticity chart with some outliers below the A-line. A review of the quality control standard compaction results showed that when field dry density is between 1.55 t/m^3 and 1.72 t/m^3, the standard density ratio was found to be greater than 99 per cent resulting in a dilative embankment.

In situ testing of tailings comprised nine CPTu locations utilising 100 MPa Geomil compression cones. To evaluate the Panda DCP suitability for mine tailings characterisation, ten Panda DCP tests were undertaken adjacent to the existing CPTu, approximately within 1.0 m distance from the CPTu location. CPTu testing was undertaken between 20 m and 100 m distance from the spigot location. PANDA testing was driven to a depth between 1.0 m and 4.7 m below existing tailings surface levels. The Panda DCP location and tip sizes used for this study are summarised in Table 1.

Disturbed samples were recovered and taken for laboratory testing to determine the CSL. The location of the CSL is necessary because the methods proposed herein rely on determining the state of tailings, not the state parameter directly. Testing was undertaken on the tailings for various fine contents. This was done to represent the segregation of tailings down the tailings beach and G_s being affected by fines content. The CSL results obtained from the laboratory study are presented in this work because the information is required to demonstrate the method. More details about the CSL test results will be provided in subsequent publications. Samples were prepared at state parameters of between 0.09 and 0.11 (loose of critical). The summary of CSL parameters is shown in Table 2. A remarkably similar CSL slope in the compressibility plane was noted. This was attributed to very similar mineralogy. Fines content was observed to influence the CSL intercept in the compressibility plot.

TABLE 1

Regression analysis of q_d, and q_c at the same depth.

Location	Distance from spigot (m)	Cone tip area (cm²)	Test depth (m)	No. of readings	q_d range (MPa)	q_c range (MPa)	q_c/q_d	R^2
02	100	2	4.62	150	0.16–13.31	0.34–8.16	0.62	0.82
02	100	4	4.34	200	0.25–12.23	0.25–8.16	0.83	0.88
02	100	10	1.01	88	1.06–8.24	1.95–8.16	1.1	0.84
04	30	2	3.01	126	2.53–20.76	0.49–6.6	0.48	0.78[1]
04	30	4	2.41	140	1.31–6.68	0.49–6.60	1.13	0.89
05	20	4	3.6	455	1.49–18.34	2.69–14.67	1.26	0.94
08	20	2	4.69	125	0.5–11.71	0.38–12.21	1.17	0.89
09	20	2	3.11	86	0.97–11.93	1.3–15.58	1.27	0.82
11	100	2	3.04	125	0.65–14.73	0.82–21.18	0.82	0.82[2]
11	100	4	4.66	204	0.05–13.17	0.08–21.18	0.97	0.83[2]

Note   (1): natural material was encountered at 1.4 m and therefore, regression analysis was undertaken to 1.4 m depth.
         (2): Regression analysis of the dry crust between 0.0 m and 0.7 m ignored.

TABLE 2

Summary of Critical state testing.

CSL parameter	Fines content			
	6%	20%	40%	80%
λ_{10}	0.161	0.161	0.161	0.161
λ_e	0.069	0.069	0.069	0.069
Γ	1.049	0.999	0.990	1.065
G_s	2.60	2.63	2.67	2.72
LL	32	36	35	50
PL	25	23	24	21
IP	7	13	11	29
LL*G_s	0.83	0.95	0.93	1.36

REGRESSION ANALYSIS OF CPT AND PANDA DCP SOUNDING

To determine whether a correlation existed between the Panda q_d and CPTu q_c, linear regression analysis was undertaken for pair data of q_d and q_c at the same depths. The q_d readings were staggered and non-consistent when compared with the CPTu, which records q_c reading every 10 mm. The data was filtered to compare and correlate the measured q_d and q_c at the same depths. The result of the linear regression analysis, R^2, is provided in Table 1, with q_c and q_d plots detailed in Figure 2. Based on the plots in Figure 2 and other Panda DCP data not shown in this paper for brevity, it was determined that the Panda DCP cone tip resistance shows a good correlation with CPTu cone tip resistance. Three Panda cone tips with cross-sectional areas of 2 cm², 4 cm² and 10 cm² were trialled at location 2, showing good correlation with each other and with CPTu results, although the 10 cm² was seen to penetrate down to approximately 1.0 m depth. Panda cone tips with cross-sectional areas of 2 cm² and 4 cm² were trialled at locations 4 and 11. A Panda cone tip with a cross-section area of 4 cm² was trialled at location 5. A Panda cone tip with a cross-section area of 2 cm² was trialled at locations 8 and 9. The 4 cm² cone tip was observed to have the highest R^2 on average – 0.89. The 2 cm² cone tip was observed to have the lowest R^2 on average – 0.83. A single 10 cm² cone R^2 calculated for location 2 resulted in 0.83.

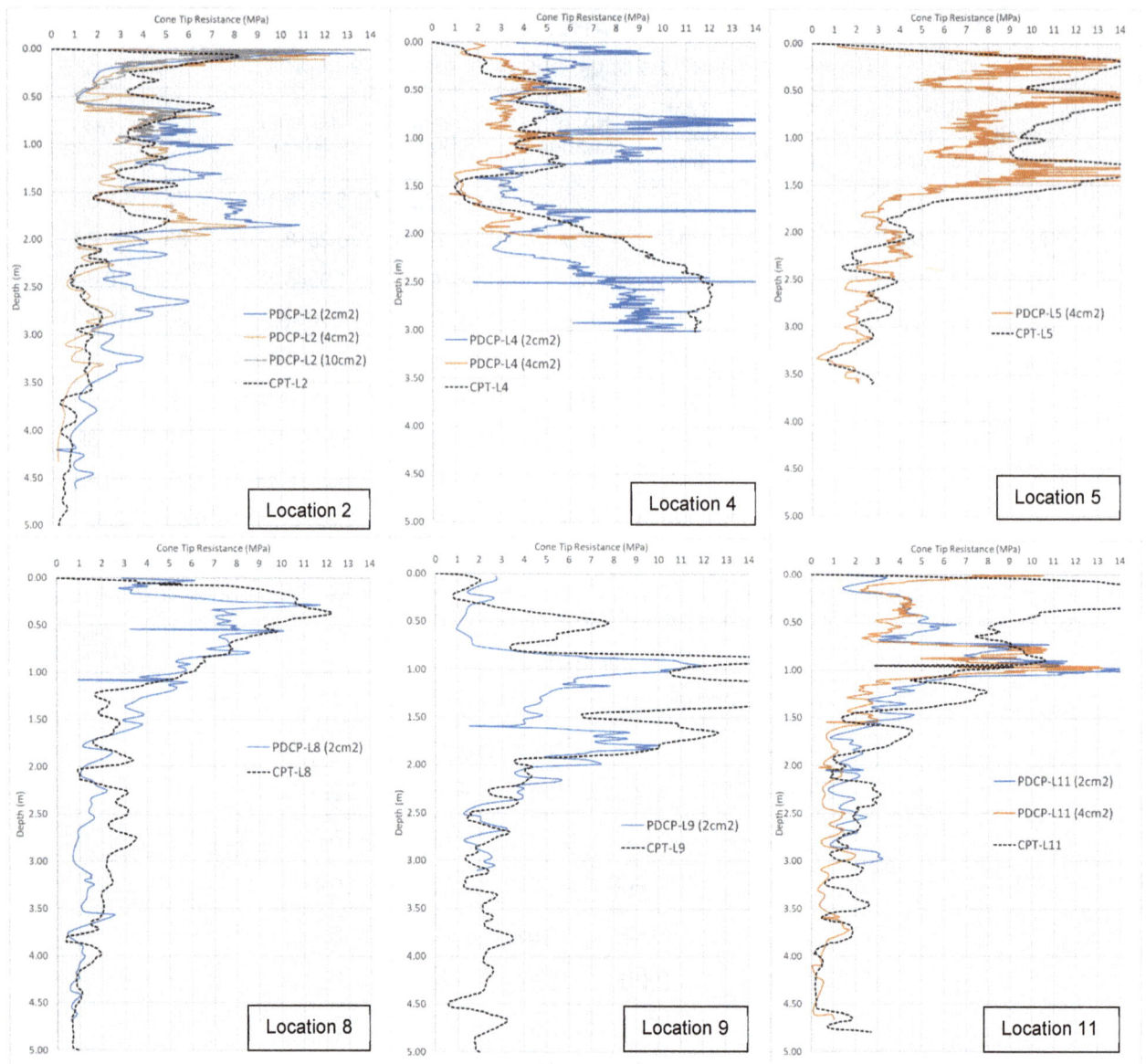

FIG 2 – CPTu, q_c and Panda DCP, q_d cone tip plots.

Experience using the different cone tip sizes on-site has shown that the bigger the cone tip area, the higher the resistance, resulting in higher operator hammer force and repetition to advance the Panda DCP tip to depth. The 2 cm² and 4 cm² were favoured by the Panda DCP operators as they required the least force and hammering repetition in mud-farmed tailings when compared to 10 cm². The 10 cm² tip size was the preferred cone tip size for soft tailings (<1 MPa) if encountered near the surface. This was why the 10 cm² was only trialled in one location as higher operator hammer force and higher repetition were required to advance Panda DCP to depth.

The following points below are a summary of linear regression analysis, R^2 in Table 1, for each cone size:

- 2 cm² cone tip q_d vs q_c relationship range between q_c=0.48 to 1.27 × q_d (R^2=0.78 to 0.89), for locations 2, 4, 9 and 11.

- 4 cm² cone tip q_d vs q_c relationship range between q_c=0.83 to 1.26 × q_d (R^2=0.83 to 0.94), for locations 2, 4, 5 and 11.

- 10 cm² cone tip q_d vs q_c relationship range between q_c=1.1 × q_d (R^2=0.83), for location 2.

Location 4, 2 cm² cone result (refer Figure 2) showed large spikes in tip resistance, which resulted in R^2=0.78 but 4 cm² cone faired better. Based on the summary above, q_d for the cone sizes can be within ±20 per cent of q_c results with similar trend. The difference in cone tip resistance is explained due to moisture and dry density conditions at the time of test, energy transfer loss and driving mechanism.

EVALUATION OF CRITICAL STATE LOCUS CPTU AND PANDA PLOTS

To estimate the *in situ* state of the tailings using the Panda data, two methods are referred to in this section which are Method-1 (M1) and Method-2 (M2). M1 and M2 are described in Tables 3 and 4, respectively.

TABLE 3
State parameter assessment using Panda DCP Method-1 (M1).

Parameter	Panda DCP (M1)
γ_d – dry density	Equation 4 – $\gamma_d = 1.06 \cdot ln(q_d) + 15.82$
e – Void Ratio	Equation 6 – $e = \left(\frac{G_s\gamma_w}{\gamma_d}\right) - 1$
	where G$_s$ estimated from laboratory testing, see Table 2 based on distance from the spigot and estimate fines content
p'_0 – Mean effective stress at depth	Equation 6 – $p'_0 = \gamma'z\,(1 + 2K_0)/3$
	where bulk unit, γ weight can be estimated using Equation 5 and K$_0$ for demonstration of this paper we have adopted a value of 0.3

TABLE 4
State parameter assessment using Panda DCP Method-2 (M2) and CPTu (Been and Jefferies, 2016).

Parameter	Panda DCP (M2)	CPTu (Been and Jefferies, 2016)
Q – normalised cone resistance	$Q_p = \frac{q_d - p}{p'}$ (Been *et al*, 1986) where $p' = \gamma'z\,(1 + 2K_0)/3$ and q$_d$ replacing q$_c$ from Been *et al*, 1986)	$Q_p = \frac{q_t - p}{p'}$ (Been *et al*, 2016) where $p' = \gamma'z\,(1 + 2K_0)/3$
ψ – state parameter	$\psi = \dfrac{-\ln[Q_p\,/k']}{m'}$ (Been *et al*, 1986, 2016) where k' and m' estimated based on the conventional CPTu interpretation	$\psi = \dfrac{[-\ln[Q_p(1 - B_q)\,/k']}{m'}$ (Been *et al*, 1986) where k' and m' estimated based on the conventional CPTu interpretation and B$_q$ is pore pressure ratio
e$_c$ – critical void ratio	e$_c$=Γ-λ_{10}log (p'$_0$) (Been and Jefferies, 2016) where Γ and λ_{10} are provided in Table 1	e$_c$=Γ-λ_{10}log (p'$_0$) (Been and Jefferies, 2016) where Γ and λ_{10} are provided in Table 1
e – Void Ratio	Equation 1 – e= ψ + e$_c$	Equation 1 – e= ψ + e$_c$

A few extrapolations were considered for the study as follows:

- Equation 4 (M1) was originally derived for natural soils by the Panda manufacturers. Conventional site dry density testing at the surface and to depths of max 0.3 m was undertaken and compared against the estimates presented herein. Results compared remarkably well and are detailed in Figure 3.

- m' and k' (M2) were estimated based on the conventional CPTu interpretation. Improved m' and k' estimates will be considered in future studies.

- K$_0$=0.3 was adopted for demonstration of this paper.

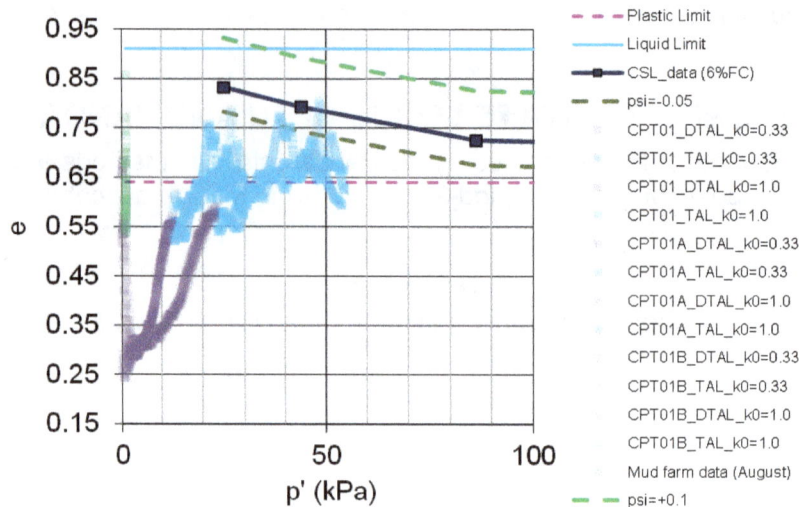

FIG 3 – Comparison between field dry density data (green data points at p'~0 at 0.3 m) measured using a nuclear densitometer and state parameter and CSL estimates.

The Panda method was compared to CPTu data critical state analysis using Been and Jefferies (2016) method to determine whether there is a basis for the method proposed herein. Been *et al* (1986) first demonstrated the relationship between ψ and q_c in a calibration chamber for sand in fully drained conditions. This was also demonstrated by Abedin and Hettiaratchi (2002). The CPTu state parameter assessment methodology (Been and Jefferies, 2016) is summarised in Table 4. Figure 3 shows that Been and Jefferies (2016) state parameter estimates at very low confinement stress agree remarkably well with the void ratio calculated from nuclear density guage field data in the vicinity of the CPT. Future studies will include the measurement of tailings density at depth using a mini-block sampler.

Using the Panda DCP and CPTu data in Table 1, critical state assessment may be derived from the Method-1 and Method-2 mentioned in Tables 3 and 4. The results were plotted on critical state plots, detailed in Figure 4. The CPTu-derived results are presented in broken black lines. Panda-interpreted results are plotted in blue lines (designated 2 cm² cone area), orange lines (designted 4 cm² cone area) and grey lines (designted 10 cm² cone area). Critical state lines corresponding to the fines content at each location are also included in each plot. The results show a generally good agreement when comparing the CPTu-borne results with Panda-borne results. Regions of largely dilative mud-farmed tailings were successfully characterised using both methods. Furthermore, a contractive layer underlying the mud-farmed tailings was identified in locations 2, 5, 8, 9 and 11, which are related to a region below about 3 m having q_c and q_d values below 1 MPa. Locations 2, 5, 9, and 11 show that both methods indicate the presence of potentially contractive tailings. The aforementioned contractive layer was found to lie at mean effective stresses of 30 kPa and above. Based on the data plotted in Figure 4, the CPTu and Panda methods show general agreement with similar trends in estimating *in situ* state of tailings.

The last two steps shown in Table 4 effectively derive the void ratio to facilitate plotting the CSL in a compressibility plot, see Figure 4. The authors recommend the procedure because the horizontal distance between each data point and the CSL can be used to estimate the stress required for that point to transition from dilative to a contractive state.

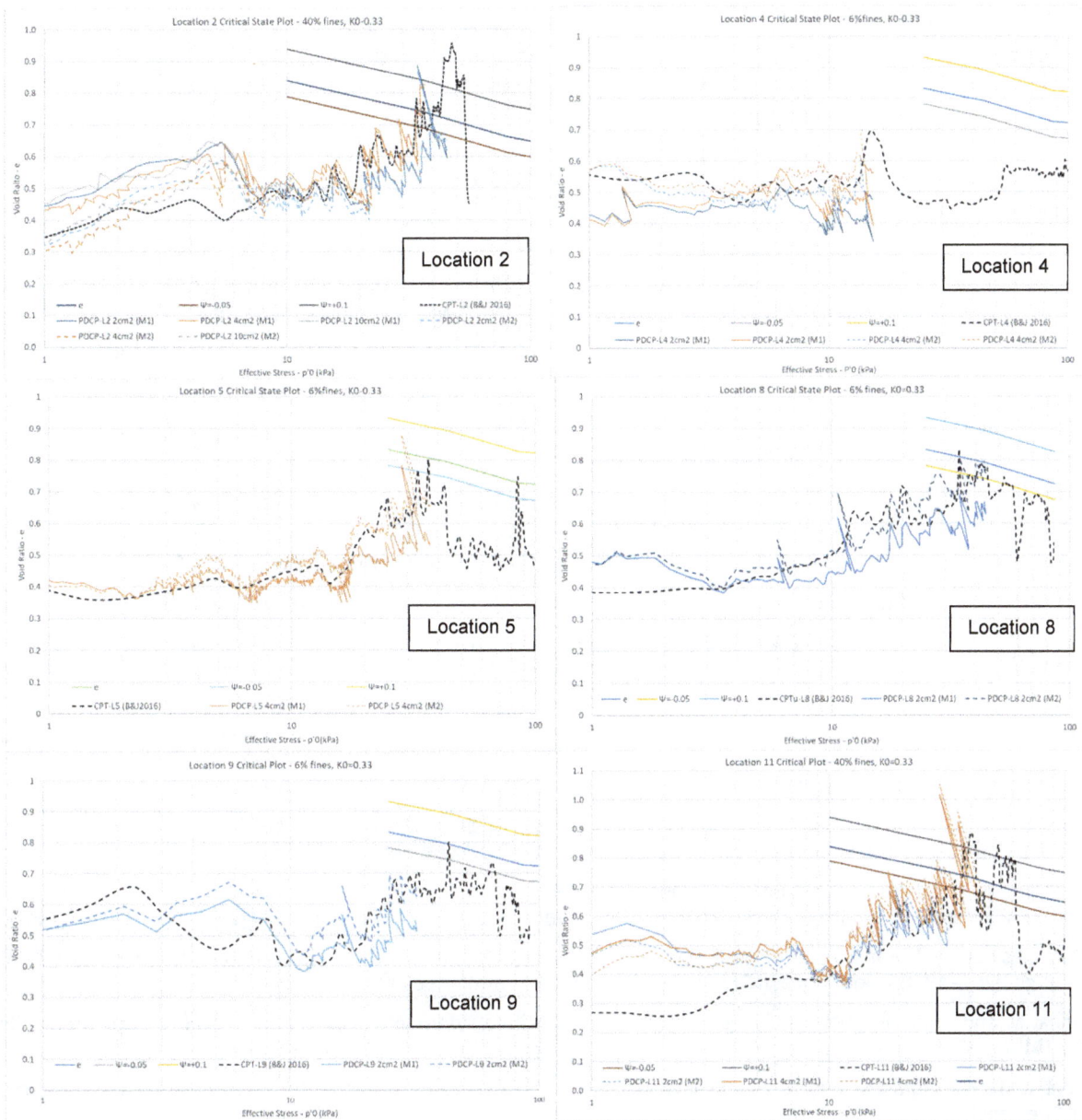

FIG 4 – Critical state plots (Effective stress *versus* void ratio) for Panda DCP Method-1 (M1) and Method-2 (M2) and CPTu (Been and Jefferies, 2016) data.

CONCLUSION

This paper has demonstrated how q_d shows good correlations with the widely accepted q_c for a TSF in Queensland, Australia. The method was trialled in sandy tailings and clayey tailings material. The Panda DCP-derived results for estimating *in situ* state are in general agreement and show a remarkably similar trend to Been and Jefferies (2016) results for determining *in situ* state. The following comments regarding the findings in this paper are as follows:

- Further studies need to be conducted on TSF of different mine ore origins and climates. The study should be replicated in controlled laboratory conditions using a calibration chamber. Tailings type and fines content dictate soil critical state behaviour.

- Measurement of K_0 parameter will provide a better estimation of *in situ* state. A sensitivity analysis of this parameter is recommended in the absence of such results. A sensitivity analysis is underway and will be presented in the future.

- Although useful, the method presented herein requires some extrapolations and is recommended when CSL laboratory data is available.

- The methods described herein are useful techniques for Quality Control during mud farming and targeting potential CPTu locations for more advanced geotechnical investigation. Because the method is recommended for Quality Control, the maximum depth achievable is not seen as a significant drawback.

REFERENCES

Abedin, M Z and Hettiaratchi, D R P, 2002. State Parameter Interpretation of Cone Penetration in Agricultural Soils, *Biosystem Engineering,* 83(4):469–479.

Australian National Committee on Large Dams (ANCOLD), 2019. *Guidelines on Tailings Dam: Planning, Design, Construction, Operation and Closure,* s.l.: Australian National Committee on Large Dams.

Been, K and Jefferies, M, 2016. *Soil Liquefaction: A Critical State Approach,* 1st ed (Taylor and Francis: New York).

Been, K, 2016. Characterizing mine tailings for geotechnical design, *Australia Geomechanics Soceity,* p 41.

Been, K, Crooks, J H A, Becker, D E and Jefferies, M G, 1986. The cone penetration test in sands: part 1, state parameter interpretation, *Geotechnique,* 36(2):1751–17656.

Benz-Navarrete, M S, Bruel, P, Bacconet, C and Moustan, P, 2019. *The PANDA, Variable Energy Lightweight Dynamic Cone Penetrometer: Quick State of Art,* s.l.:s.n.

Cassan, M, 1988. Les essais in situ en méchanique des sols, *Réalisation et Interprétation,* 1:146–151.

Fell, R, MacGregor, P, Stapledon, D, Bell, G and Foster, M, 2018. *Geotechnical Engineering of Dams,* 2nd ed (CRC Press).

Fourie, A, Palma, J, Villavicencio, G and Espinace, R, 2013. Risk minimisation in construction of upstream tailings storage facilities based on in situ testing, in *Proceedings of the 18th International Conference on Soil Mechanics and Geotechnical Engineering, ICSMGE 2013,* pp 1471–1474.

French Standards Association (AFNOR), 2012. NF P 94–105: Grounds: Investigation and Testing, Measuring compaction quality, Method using variable energy dynamic penetrometer – Penetrometer calibration principle and method – processing results – interpretation, French Standards Association (AFNOR).

Global Tailings Review, 2020. Global Industry Standard on Tailings Management (GISTM) [online]. Available at: https://globaltailingsreview.org/global-industry-standard/ [Accessed February 2023].

Independent Expert Engineering Investigation and Review Panel (IEEIRP), 2015. Report on Mount Polly Tailings Storage Facility Breach, Province of British Columbia: Independent Expert Engineering Investigation and Review Panel.

Instituto Nacional de Narmalizacion (INN), 2012. Tailings Deposits-Control of compaction with light dynamic penetrometer, Nch 3261–12, Santigo, Chile: INN (Instituto Nacional de Narmalizacion).

Jefferies, M and Been, K, 2016. *Soil Liquefaction: A Critical State Approach,* 2nd ed (Taylor and Francis Group: Boca Raton).

Laurent, C, 1999. *Characterisation of Granular Mediaof Surface using a Penetrometer* (Blaise Pascal University: Clermont-Ferrand).

Morgenstern, N R, Vick, S G and Watts, B D, 2016. Fundao tailings dam review panel: Report on the immediate causes of the failure of the Fundao Dam.

Pineda, J, Kelly, R, Suwal, L L G T and Ouyang, K, 2018. Sampling Disturbance in Soft Ground: Implications in Geotechnical Design, Australia Geomechanics Soceity.

Robertson, P K, 2022. *Guide to Cone Penetration Testing for Geotechnical Engineering,* 7th ed (Gregg Drilling LLC: Signal Hill).

Robertson, P, 2010. Estimating in-situ state parameter and friction angle in Sand Soils from CPT, in CPT'10, the 2nd International Symposium on Cone Penetration Testing.

Scala, A, 1956. Simple Methods of Flexible Pavement Design Using Cone Penetroemeters, Proceedings of the Second Australia-New Zealand Soil Mechanics Conference.

Identification and analysis of a complex layer within the footprint of a tailings storage facility foundation utilising mini-block sampling

S H Lines[1], M Llano-Serna[2], R Rekowski[3] and W Ludlow[4]

1. Senior Geotechnical Engineer, Red Earth Engineering – a Geosyntec Company, Brisbane Qld 4000. Email: scott.lines@redearthengineering.com.au
2. Principal Geotechnical Engineer, Red Earth Engineering – a Geosyntec Company, Brisbane Qld 4000. Email: marcelo.llano@redearthengineering.com.au
3. Principal Geotechnical Engineer, Red Earth Engineering – a Geosyntec Company, Brisbane Qld 4000. Email: robert.rekowski@redearthengineering.com.au
4. Director, Red Earth Engineering – a Geosyntec Company, Brisbane Qld 4000. Email: wade.ludlow@redearthengineering.com.au

ABSTRACT

A thin, distinctly complex layer was identified within the foundations of a tailings storage facility located in northern Australia. The strength of the layer in comparison to surrounding layers meant that a potential preferential slope instability plane was present in the foundation, and thus the layer became a key target for additional investigation. A site investigation was conducted in 2022 whereby cone penetration testing, vane shear testing, seismic and dissipation testing were conducted targeting this layer within the foundation. The results of the *in situ* testing indicated a layer exhibiting a cone tip resistance of less than one megapascal. The layer was classified as clay-like contractive and clay-like sensitive using conventional CPTu techniques. CPT correlations indicated that the material behaved in a contractive manner. However, the pore water pressure response to CPTu testing was partially drained, often with negative pore pressure. Vane shear testing in the layer was challenging, often resulting in broken vanes due to a 'high strength' response. Four mini-block samples were obtained utilising the mini-block sampling apparatus based on the Sherbrooke sampler technique to obtain high-quality undisturbed samples. CT scanning was completed on these samples to assess the internal structure before testing. Further XRD and electronic microscope analysis were completed. Laboratory testing included index, triaxial, oedometer and simple shear testing. The case study presented herein follows the process of identification of a complex geotechnical unit in a tropical residual soil, obtaining high-quality mini-block samples at depth from the field, analysis of the sample disturbance, and the subsequent laboratory testing and analysis. Results are shown to enable a holistic evaluation. The aim is to demonstrate the role mini-block sampling can play in the site characterisation of complex layers while highlighting potential pitfalls and difficulties encountered throughout the process.

INTRODUCTION

Tailings Storage Facilities (TSF) are used to store the waste materials produced from mining operations, and they can pose a significant risk to human health and the environment. The failure of a TSF can result in catastrophic consequences, including loss of life, damage to property, and environmental contamination. According to a recent study by Islam and Murakami (2021) the frequency and severity of TSF failures have increased in recent years with the trajectory shifting from developed to developing countries. Overtopping, static liquefaction and dynamic liquefaction are the most common causes of TSF failure (Piciullo *et al*, 2022). In the wake of recent high-profile TSF failures, there has been increased attention on the need for better standards and regulations to govern their design and operation to reduce and eventually eliminate the risks of failure of existing and future TSFs (Oboni and Oboni, 2020).

Recent incidents, such as the Cadia collapse in Australia in 2018 and the Mount Polley tailings dam failure in Canada in 2014, have highlighted the dangers of TSFs and the need for improved management practices. The Cadia collapse resulted in the runout of approximately 1.3 million cubic metres of tailings, the various analysis indicated that the foundation yielded during the first stage of a buttress construction, with tailings liquefaction taking place as a result of a rapid increase of foundation deformations (Jefferies *et al*, 2019). The Mount Polley dam failure resulted in the release of 24 million cubic metres of tailings and water into nearby lakes and rivers, leading to widespread environmental damage. The failure of Mount Polley has been attributed to minor strain-weakening within a thin layer of glaciolacustrine unit, with simulations suggesting that strain localisation in the layer not exceeding 12–13 cm in thickness was necessary for the initiation of the failure (Zabolotnii, Morgenstern and Wilson, 2022). The failure of Mount Polley and Cadia highlights the importance of investigating and properly characterising potentially problematic and weak layers within the foundation of any TSF.

In any geotechnical investigation or problem obtaining detailed accurate knowledge of the potential soil behaviour is critical. While *in situ* methods such as cone penetration testing (CPT) are capable of obtaining a wealth of information, a more holistic approach includes laboratory testing. A key challenge when interpreting and describing material behaviour is that most of the soil mechanics literature and theories have been developed outside of the tropics, with a few exceptions, see Fookes (1997). Furthermore, challenges are encountered when approaching laboratory testing regarding sample quality; if the soil specimen obtained for testing loses its original structure (which is of particular relevance in tropical residual soils), the subsequent test can be unrepresentative and the results misleading. Where clays are considered, the most significant effects of sample disturbance are the reduction in the measured pre-consolidation stress, peak undrained shear strength and the increase in failure strain (Emdal *et al,* 2016). While sampling disturbance has been shown to be introduced at all stages of sampling, including retrieval, transportation, and laboratory handling, it has been shown that if steps are taken to avoid disturbance in the retrieval phase the influence of the other potential disturbing factors may be limited (Emdal *et al,* 2016; Lunne, Berre and Strandvik, 1997; Lunne *et al,* 2006; Ladd and DeGroot, 2004).

Obtaining high-quality soil samples with minimal disturbance is particularly difficult and an area of research that has been ongoing for decades (La Rochelle and Lefebvre, 1971; Hight *et al,* 1992; Amundsen, Emdal and Thakur, 2021). The mini-block sampling technique, derived from the traditional Sherbrooke-type of block sampling, has significant advantages, particularly in ensuring that the sample remains unaffected by shear distortion from the sampling tube which is more typical of traditional sampling methods such as thin-wall sampling (Lefebvre and Poulin, 1979). Pineda and co-workers have demonstrated the relevance of mini-block sampling techniques in the Australian industry (Lim *et al,* 2019; Pineda, Suwal and Kelly, 2014). One of the key benefits of this technique is that it provides high-quality samples that accurately represent the *in situ* conditions of the soil, allowing for more accurate measurements of soil properties that will lead to less conservative design parameters (Emdal *et al,* 2016; Pineda, Suwal and Kelly, 2014). However, despite the advantages the high costs and high time consumption remains a barrier to adoption. In this case study a thin, distinctly complex layer was identified within the foundations of a TSF in tropical residual soils located in Northern Australia. The strength of the layer in comparison to surrounding layers meant that a potential preferential slope instability plane was present in the foundation, and thus the layer became a key target for additional investigation. This case study documents the investigation and challenges, including the *in situ* testing and subsequent mini-block sampling and laboratory testing undertaken.

SITE BACKGROUND

A TSF in northern Australia was undergoing a routine site investigation for the purposes of collecting the requisite data for the construction of an upstream raise. The location of the TSF experiences an average annual rainfall of over 2 m and is located in a tropical monsoonal climate with a distinct wet and dry season. The TSF construction was completed in 2018 with the design intent to store tailings and decant water over the operating life of the facility. It comprises a 'paddock type' impoundment, of approximately 1000 Ha, with perimeter embankments and no upstream catchments. Tailings are discharged into the TSF from an arrangement of spigots at set intervals along the perimeter. The height of the perimeter walls ranges from 0 m to 12 m with downstream batter at 2.5H:1.0V and upstream batter of 2.0H:1.0V. Approximately 9–10 million dry tonnes per annum of tailings are deposited into the TSF. The first upstream raise occurred in 2022.

Regional geology

The regional geology consists of residual soils comprising the following:

- Ferruginous Duricrust: Aluminous laterite, including pisolitic zones, which forms parts of the Wyaaba beds. This is the dominant surface geological unit derived of heavily laterised sandstone of the Bulimba formation.

- Bulimba Formation: Clayey sandstones, conglomerate and sandy claystone commonly ferruginised. It outcrops along main creeks and includes the following lateritic profiles. Originally Sedimentary deposits via rivers comprising metamorphic and acid igneous rocks. Within its weathering profile it contains a Pallid and Mottled zone. The Pallid zone is dominated by kaolinite and sands, saturated for a longer period of time and bleached of oxide staining. The Mottled zone has considerable variability and mineralogical change (laterisation), partly due to significant groundwater table variance, including the dissolution of kaolinite to form bauxite, the dissolution of quartz, the dissolution and precipitation of iron and aluminium oxyhydroxides plus replacement of kaolinite with hematite.

- Rolling Downs Group: Shales and siltstones outcropping where the Bulimba Formation has been eroded. Saprolite: Lenses of highly weathered sandstones, siltstones and claystone. Stained with iron oxides with ferruginous gravel throughout.

- Gilbert River Formation: Sandstone and siltstone with some conglomerate.

The mottled zone within the Bulimba zone has groundwater table that rises up and down seasonally resulting in the potential for various reactions to take place. The voided Bulimba encountered is typically within the zone whereby seasonal variations occurs of groundwater table. The seasonal variation can be up to approximately 4 m in the vicinity of the TSF and can occur rapidly, responding to significant rainfall events.

INITIAL SITE INVESTIGATION (2021)

The complex layer was originally encountered during the 2021 site investigation using cone penetration testing with pore pressure (CPTu) methodology. The layer of interest was subsequently designated as voided Bulimba and appeared within the Bulimba formation which was later determined to contain both coarse- and fine-grained Bulimba. Figure 1 displays the voided Bulimba layer highlighted red and located between conventional Bulimba layers in green. The conventional Bulimba experienced a cone tip resistance of typically 5–15 MPa compared to the <1 MPa cone tip resistance for the voided Bulimba. Additionally, the sleeve friction (kPa) decreased from typically >100 kPa in the conventional Bulimba to <10 kPa for the voided Bulimba. The friction ratio within the voided Bulimba was generally greater than 4, indicative of clays.

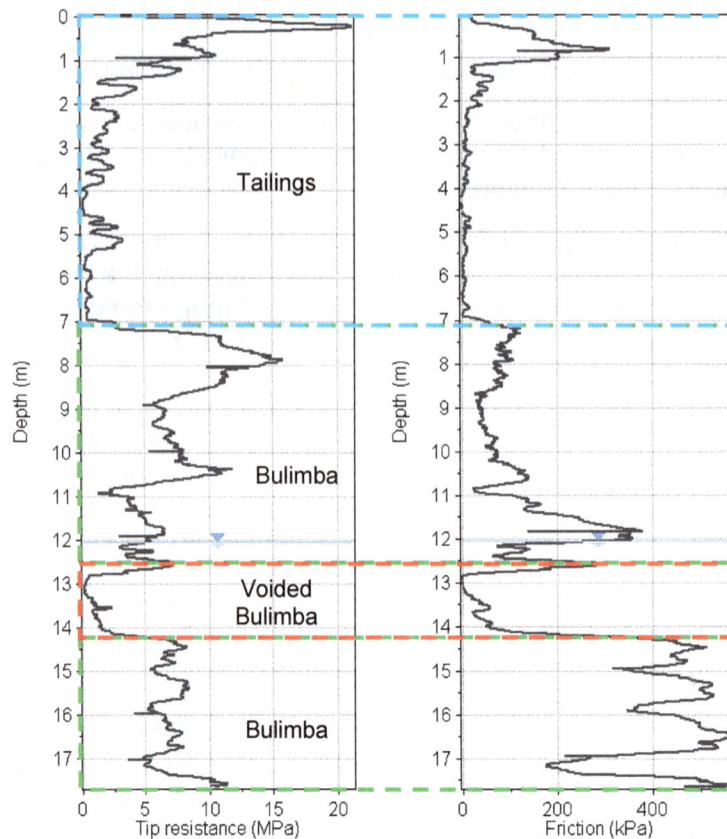

FIG 1 – CPTu profile of cone tip resistance (MPa) and sleeve friction (kPa) highlighting the weak layer.

A thin-wall tube sample of the voided Bulimba was recovered via a 75 mm diameter Geonor tube sampler at a depth of 12.80–13.25 m, pushed adjacent to the CPT location. A CT scan and post-extrusion images is shown in Figure 2 with the voided Bulimba material demarcated on the right-hand side, Figure 3 displays the extruded voided Bulimba being examined prior to testing. The material recovered within the thin-wall tube was deemed too disturbed for advanced laboratory testing.

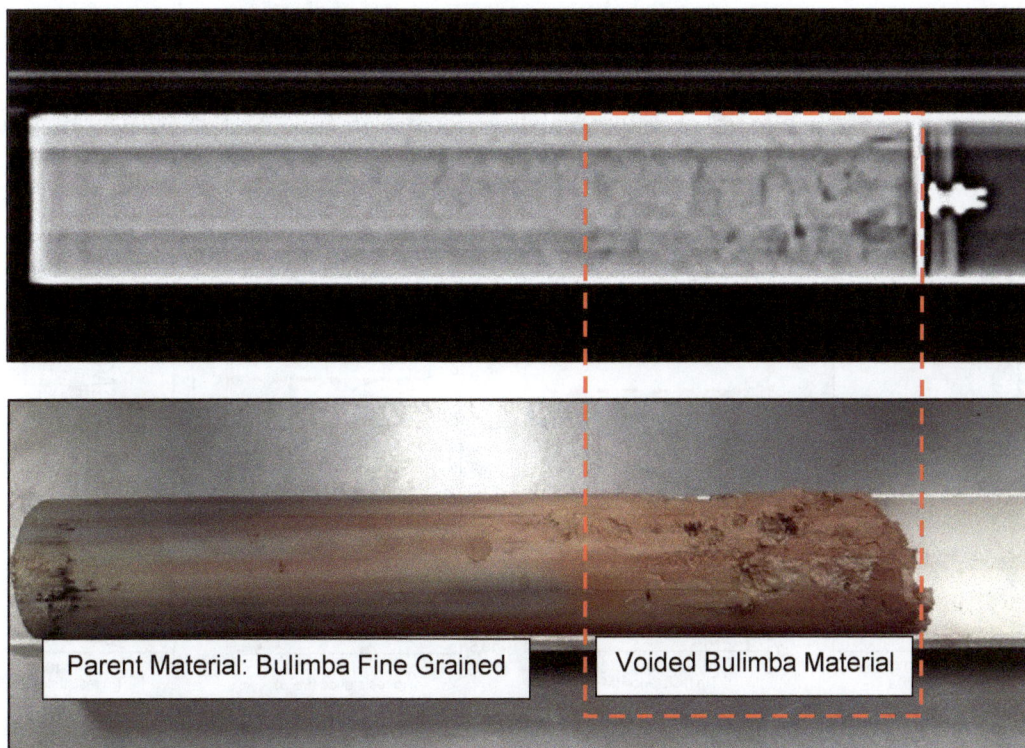

FIG 2 – Foundation: Voided Bulimba layer (CT scan, +40 per cent contrast, and post-extrusion).

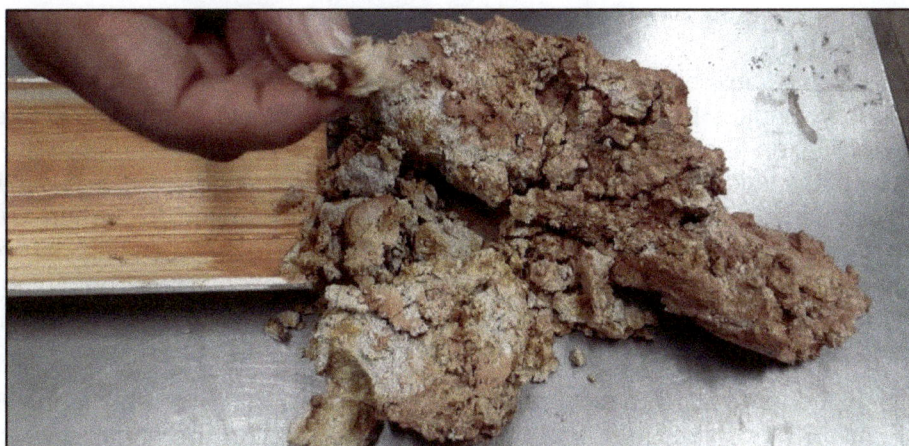

FIG 3 – Voided Bulimba layer broken down and examined.

TARGETED SITE INVESTIGATION (2022)

Following the discovery of the voided Bulimba in the 2021 CPTu investigation, a site investigation campaign was conducted in 2022 with the primary focus being the demarcation and further investigation of the voided Bulimba layer. The 2022 investigation involved both CPTu and seismic cone penetration testing with pore pressure (SCPTu), vane shear testing (VST), pore pressure dissipation testing (PPD) and obtaining mini-block samples of the voided Bulimba for laboratory testing. A 100 MPa compression cone was used to conduct the testing and a total of 32 CPTs, two SCPTs and 38 (VSTs) were completed as part of the investigation.

A total of 19 CPTu locations within the TSF footprint or at the foot of the TSF identified the presence of the voided Bulimba layer. These results determined that the voided Bulimba was not persistent across the TSF footprint and instead appeared in discrete 'pockets' within the foundation, including having several bands of voided Bulimba at multiple depths in the same CPTu hole. The relative thickness ranged from 0.01 m to 1.59 m. A selection of the CPTu profiles containing the voided Bulimba is shown in Figure 4.

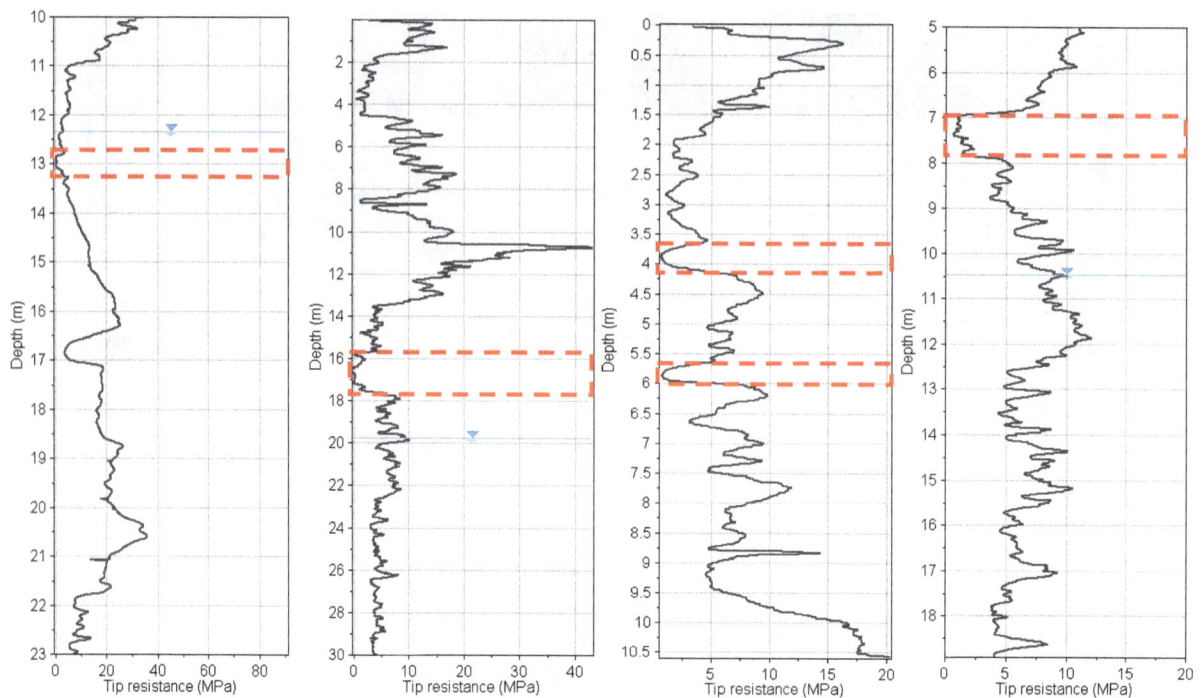

FIG 4 – Selection of CPTu cone tip resistance (MPa) locations displaying layers of voided Bulimba.

RESULTS

Site investigation

During the 2022 site investigation four mini-block samples were retrieved from adjacent to a SCPTu location to recover an undisturbed sample of voided Bulimba. The CPTu profile whereby the mini-block samples were taken is shown in Figure 5. This location was selected due to the shallow and continuous nature of the voided Bulimba. The experience at this site indicated that maximum depths at which mini-block samples could be recovered was practically limited to around 10 m. Previous experience with mini-block sampling has shown that the sampling technique is efficient for low plastic clays while high amounts of silts, sands and gravel present challenges for obtaining undisturbed testing through this methodology. Additionally, high-plasticity clays require additional time to retrieve the sample due to increased cutting (Emdal *et al*, 2016).

The SBTn chart (Robertson, 2016) typical results for the voided Bulimba results shown in Figure 5 are shown in Figure 6. It can be seen that the material plots within the contractive clay-like (CC) zone. Figure 7 displays the modified Schneider *et al* (2008) SBTn plot used to assist in classifying the soil based on data from the normalised cone tip resistance (Q_{tn}) and the penetration pore-water pressure (u_2). This figure highlights the negative pore pressure response plotting the in the transitional contractive (TC) region. A contractive soil (either transitional or clay-like) is expected to generate pore-water pressure during shearing, as such, the negative normalised pore-water pressure response was found to provide divergence of results. Robertson (2016) has previously indicated that it can be anticipated that negative pore pressure values tend to reflect large-strain dilative behaviour, whereby positive pore pressure values tend to reflect large-strain contractive behaviour. The differing classification results from Figures 6 and 7 imply that the voided Bulimba layer may have microstructure. This is further established through Figure 8 which displays the proposed Q_{tn}-I_g chart used to identify soils with microstructure (Robertson, 2016). This figure has been updated with the seismic data obtained over 100 mm intervals from a depth of 6.6–7.6 m during the investigation (see Figure 8). The choice of 100 mm intervals was to ensure a spread of data is available to minimise the impacts of having a small data sample, the shear wave velocity and shear strain modulus is shown in Table 1. Based on the normalised rigidity index results showing values of more than 300 it was concluded that significant microstructure exists in the voided Bulimba unit, hence, the traditional generalised CPT-based empirical correlation was anticipated to have less reliability.

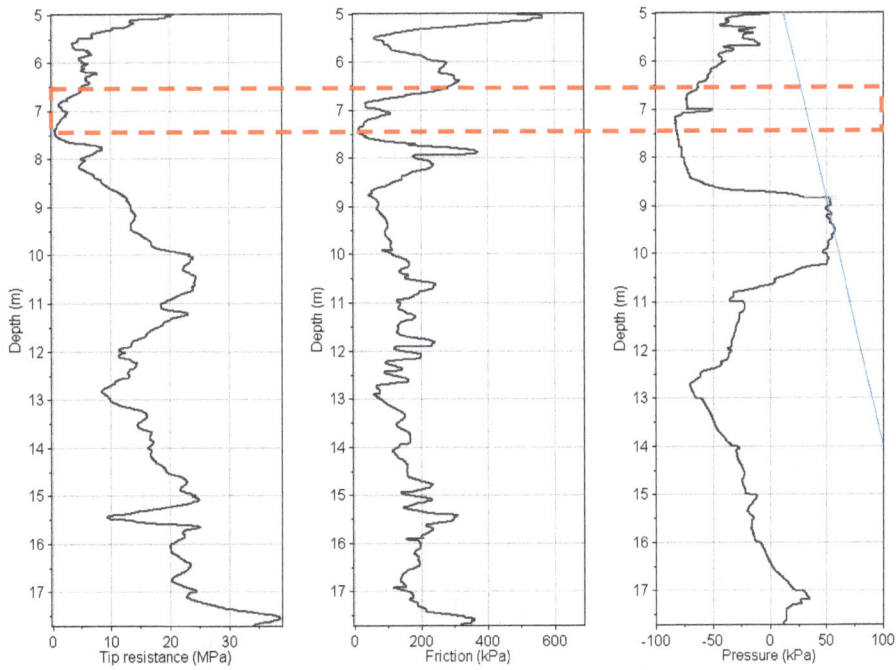

FIG 5 – CPTu profile for mini-block sampling location.

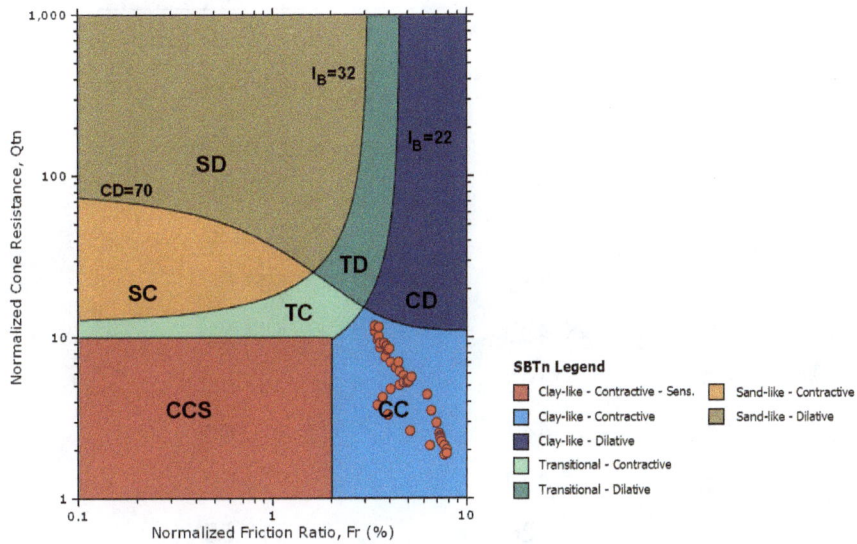

FIG 6 – Soil Behaviour Type normalised (SBTn) chart (Robertson, 2016).

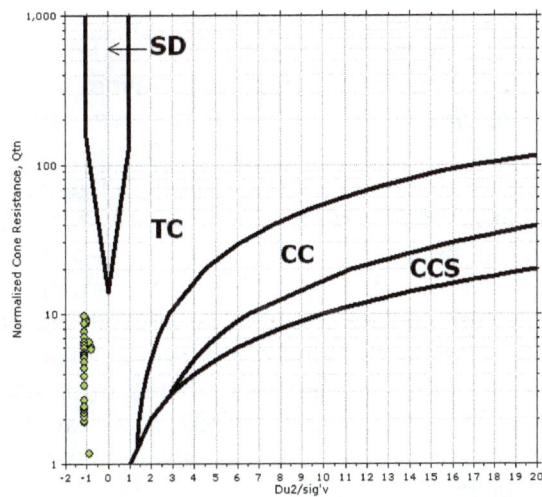

FIG 7 – Modified Schneider *et al* (2008) SBTn plot (Qtn – $\Delta u_2/\sigma'_v$), legend indicated in SBTn Legend.

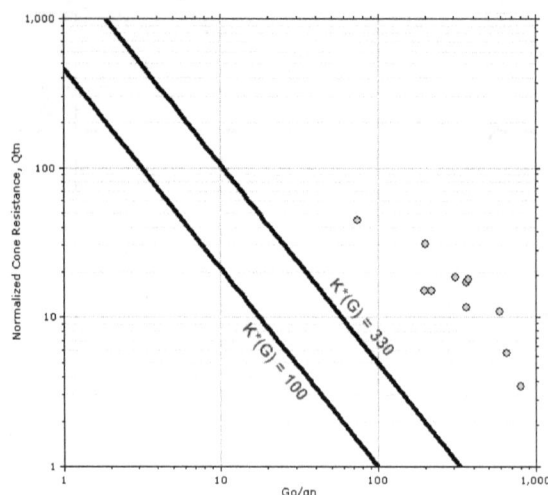

FIG 8 – Normalised Cone tip resistance and Small-strain rigidity index.

TABLE 1

Shear wave velocity and Shear strain modulus with depth of the voided Bulimba layer.

Depth (m)	Shear wave velocity, Vs (m/s)	Shear strain modulus, go (MPa)
6.6	402	328.40
6.7	560	615.40
6.8	584	630.70
6.9	614	658.60
7.0	614	705.30
7.1	562	597.90
7.2	499	444.80
7.3	483	397.80
7.4	441	310.90
7.5	422	291.60
7.6	420	318.50
7.7	443	392.00

During the site investigation, prior to any data interpretation and due to the low cone tip resistance and sleeve friction, it was decided to complete vane shear testing (VST) in the voided Bulimba layer. A 50 mm × 100 mm vane blade was initially selected, however, after several attempts yielded a broken vane blade it was decided to switch to 35 mm × 70 mm. The rotational speed of 1–2°/sec was selected. AS1289.6.2.1 recommends a rotation rate of 0.1°/sec, however, there was concern this rotation rate would result in a drained response. Figure 9 shows the results from *in situ* VST testing, VST-01 was measured at the same location as the mini-block testing (See CPTu results in Figure 5). The other VST results were from different locations, targeting the voided Bulimba layer. In total there were three broken vanes reached peak undrained shear strength values >125 kPa. VST-01 yielded the lowest peak undrained shear strength of approximately 15 kPa, whereby additional attempts ranged from 50 kPa to 120 kPa. Due to the limited thickness and unpredictability of the voided Bulimba only four pore pressure dissipation (PPD) tests were successfully undertaken within the layer, see Figure 10. All the material experienced a dilative response whereby the pore pressure continued to increase throughout the PPD test. Two of the tests experienced negative pore pressure, likely resulting from the voids within the material, and subsequently underwent repressurisation. The dilative nature of the results meant no permeabilities were calculated through this methodology.

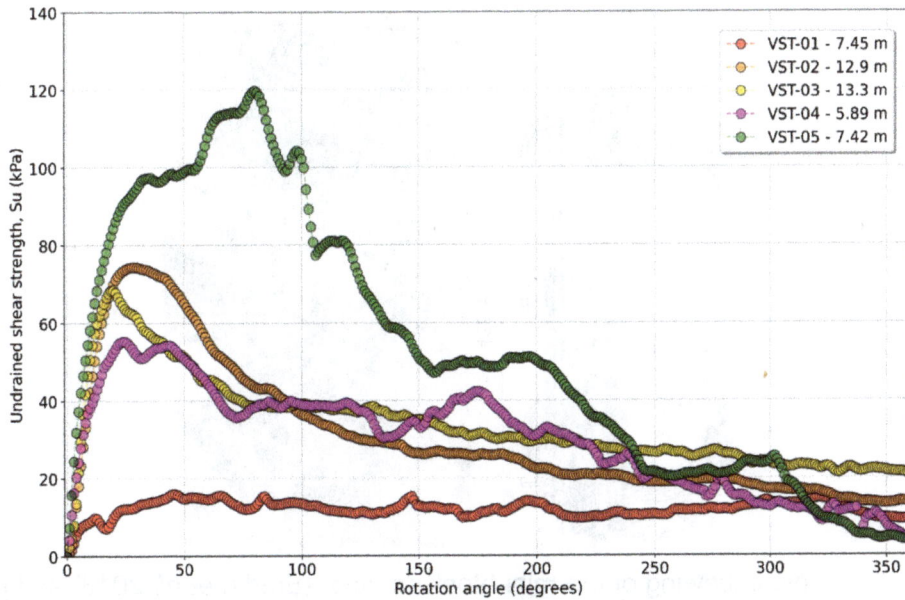

FIG 9 – Field vane shear results in the voided Bulimba layer.

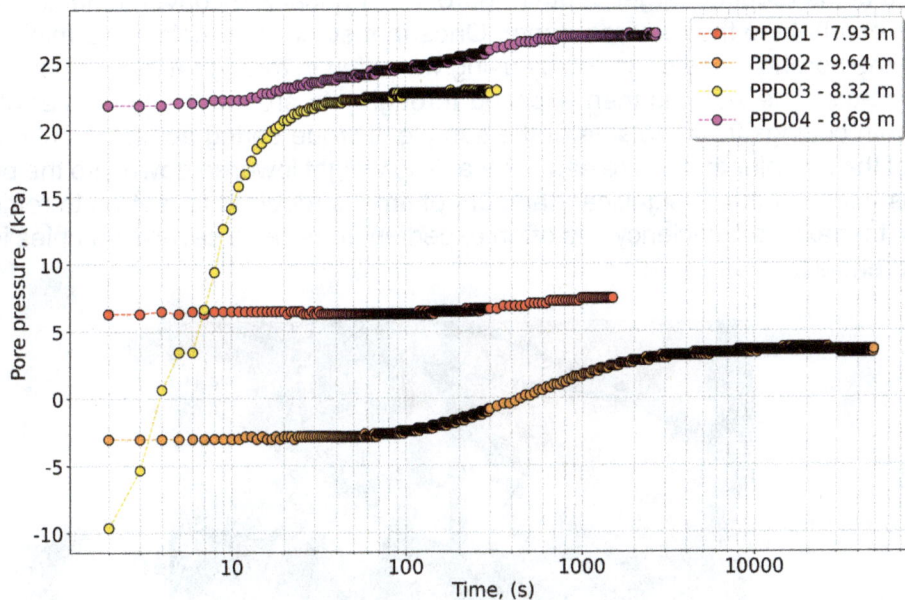

FIG 10 – Pore pressure dissipation results within the voided Bulimba.

Mini-block sampling

During the site investigation a total of four mini-block samples of the voided Bulimba formation were retrieved from adjacent to SCPTu locations. The mini-block sampler was developed by the Geotechnical Division of the Norwegian University of Science and Technology (NTNU) with the main difference being the reduced size of the mini-block sampler. The apparatus (Figure 11) was designed to retrieve samples up to 163 mm in diameter with a height of up to 300 mm. However, it should be noted that depending on the material being retrieved the actual sample size may vary. A material comprising more silts and sands would be expected to be more disturbed and hence less complete when being retrieved.

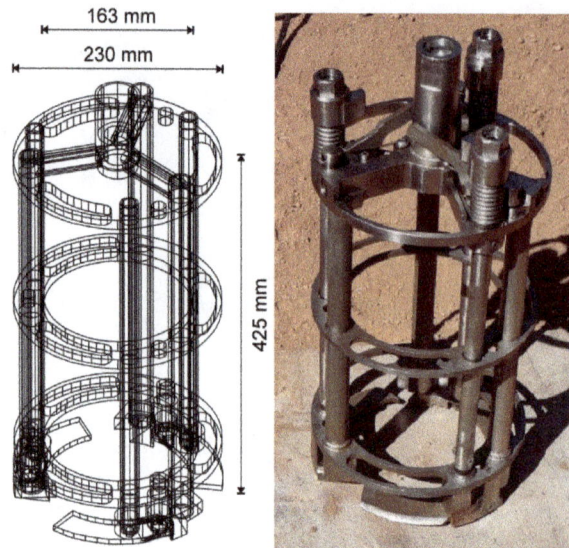

FIG 11 – Technical drawing of the mini-block sampler (Emdal *et al*, 2016) and photo.

To obtain a mini-block sample a large diameter borehole of approximately 300 mm is drilled and cased to the top of the required target depth (Figure 12). Typically, a conventional soil auger is used to advance the borehole to the required depth. Once that stage is complete the mini-block sampler is lowered into the borehole and operates by using water fed in through a mechanism located at the top of the apparatus. The water is then expelled through jets around the perimeter of the sampler and used to wash cuttings from the side of the sample. Torque spring activated cutters are located at the bottom of the sampler that are released by a drop weight lowered down into the borehole once the sample has been cut out. A significant amount of time is required for set-up to retrieve the mini-block samples, to maximise efficiency it is often expedient to retrieve several samples in succession at the same borehole location.

FIG 12 – Mini-block sampler being lowered into the case borehole.

Once the sample is retrieved, it is brought carefully to the surface and placed on a custom-made base to make the transfer of the sample easier with minimal sample disturbance. To retain the maximum amount of moisture the sample is initially wrapped in plastic, with several layers of aluminium foil and plastic wrap, followed by a coat of wax, see Figure 13 for semi-completed mini-block sample being prepared for transfer. The plastic wrap is added initially as aluminium has been shown to influence remoulded shear strength and the combination of aluminium and plastic wrap are used to minimise the influence of heat from the wax (Emdal *et al*, 2016). The sample is then transferred to a container and further wrapped with bubble wrap to minimise sample disturbance during transit. Due to the size of the sample often multiple laboratory tests can be conducted from a single sample.

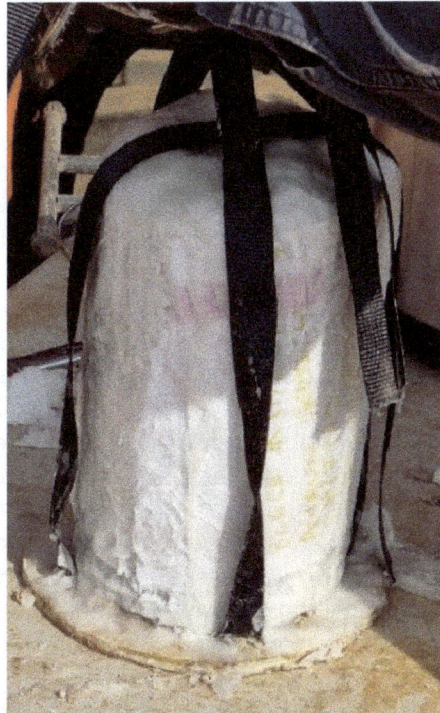

FIG 13 – Mini-block sample coated in wax and aluminium foil, located on custom-made base.

The four mini block samples were obtained (Figure 14), starting at a depth of 6.4 m and progressing to a depth of 7.5 m. The first sample was 200 m in height and the subsequent three samples were 300 mm in height. The samples visually appear more granular than the index test results as the finer particles within the cuttings wash out as the sampler progresses.

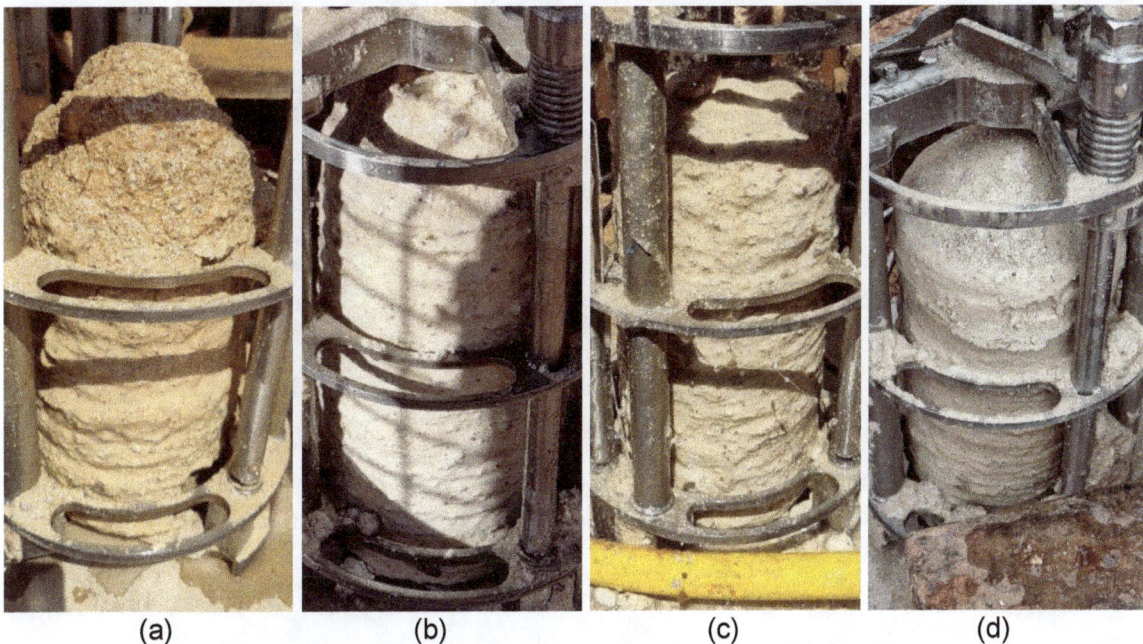

(a)   (b)   (c)   (d)

FIG 14 – Mini-block samples retrieved: (a) MBS01, 6.4–6.6 m, (b) MBS02, 6.6–6.9 m
(c) MBS03, 6.9–7.2 m, (d) MBS04, 7.2–7.5 m.

Laboratory testing

Prior to testing all four mini-block samples underwent X-ray computer tomography (CT) scans which is a non-destructive and non-invasive technique to investigate the spatial arrangement of the mineral components of the soil. The CT scanning was undertaken in 1.0 mm slices of the entire sample for all four mini-blocks from several angles, including top to bottom, 0 degrees, 120 degrees and 240 degrees. The tube current used was 310 mA with a tube voltage of 120 kV. The Dose length product (DLP) was 1564 mGy-cm and a total milliampere-second (mAs) of 4589. Images of the CT scan are shown in Figures 15 to 17. Displayed in Figure 15a is the outer structure of the mini-block sample, Figure 15b displays the inner structure with the gravel/denser components highlighted white. It can be seen that this material is dispersed throughout, this was evident when examining the material in the laboratory during sample trimming for advanced testing. Figure 16 displays a cross-section whereby the voided region can be clearly seen and Figure 17 displays a cross-section of the mini-block sample with 10 cm^2 cone (with an approximate diameter of 36 mm) overlaying the cross-section. The potential impact of the voided regions on the results of the CPTu can clearly be deduced from significant portion of voids evident. Should the cone be pushed through such a voided region this would result in a low cone tip resistance and sleeve friction, while if during the VST an area with gravel or denser material was hit, then a higher undrained shear strength would be the result. Semi-quantitative XRD analysis was undertaken with the concentration calculated using the normalised reference intensity ratio method whereby the intensity of the 100 per cent peak was divided by the I/Ic value for each mineral phase is summed and the relative percentages of each phase calculated based on the relative contribution to the sum. The results are shown in Table 2. Kaolinite and quartz are shown to be the dominate mineral within the layer. Classification testing was complete on the samples and included particle size distribution (PSD), Atterberg Limits (AL), Specific Gravity (SG), Moisture Content (MC). The summary results are shown in Table 3, with the PSD and AL shown in Figures 18 and 19, respectively. These results display that the material ranges from 26 per cent to 39 per cent clay with the majority of the remaining composition being silts. Additionally, the AL limits suggest the material is of medium plasticity.

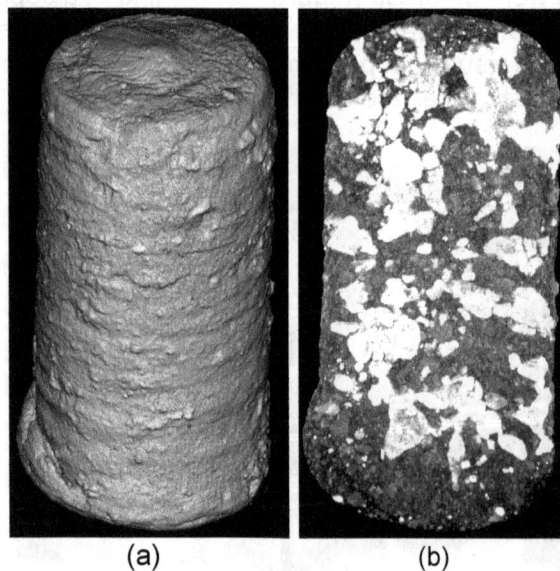

(a)   (b)

FIG 15 – Example CT scan of mini-block sample 2 (6.6–6.9 m) displaying the (a) outer structure, (b) the inner structure with the gravel/denser components highlighted white.

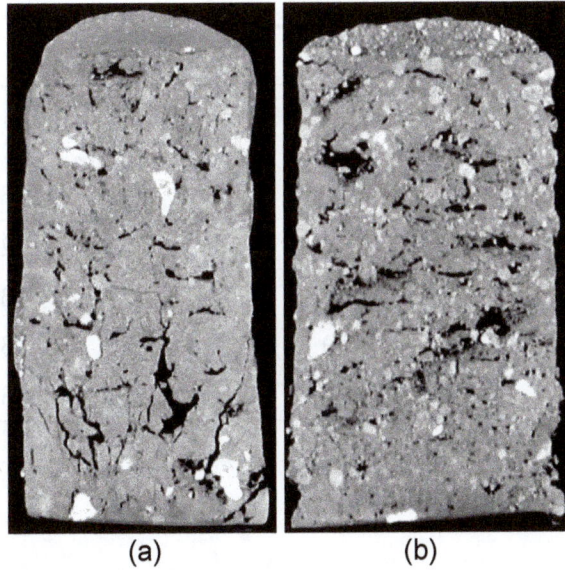

FIG 16 – Voided regions shown in the 2D space, (a) Sample 4 (7.2–7.5 m), (b) Sample 2 (6.6–6.9 m).

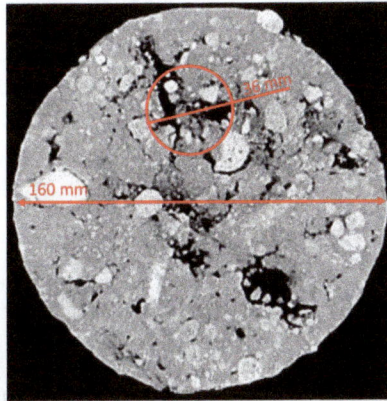

FIG 17 – Cross-section displaying sample thickness with an overlay of a 10 cm^2 cone (≈36 mm diameter).

TABLE 2

XRD results displaying mineral phase of voided Bulimba.

Crystalline mineral phase	Concentration (%)	ICDD match probability
Kaolinite-1A (Al$_2$Si$_2$O$_5$(OH)$_4$)	64	High
Quartz, syn (SiO$_2$)	34	High
Anatase, syn (TiO$_2$)	2	Low
Goethite (FeO (OH))	1	Low

TABLE 3

Classification test – summary results.

| Sample | Sample depth | | SG | MC (%) | PSD | | | AL | | |
	From (m)	To (m)			D_{50} (mm)	<75 µm (%)	<2 µm (%)	LL (%)	PL (%)	PI (-)
MBS01	6.4	6.6	2.64	21.2	0.062	51	26	39	21	18
MBS02	6.6	6.9	2.64	22.0	0.015	57	36	43	21	22
MBS03	6.9	7.2	2.64	20.9	0.016	57	39	42	18	24
MBS04	7.2	7.5	2.64	23.1	0.025	56	37	37	17	20

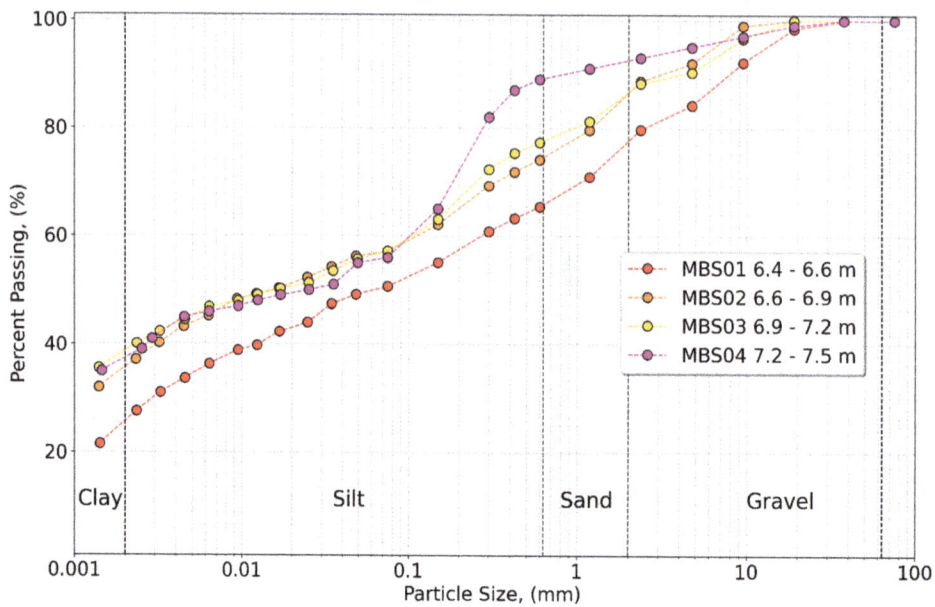

FIG 18 – Particle size distribution (PSD) for the mini-block samples.

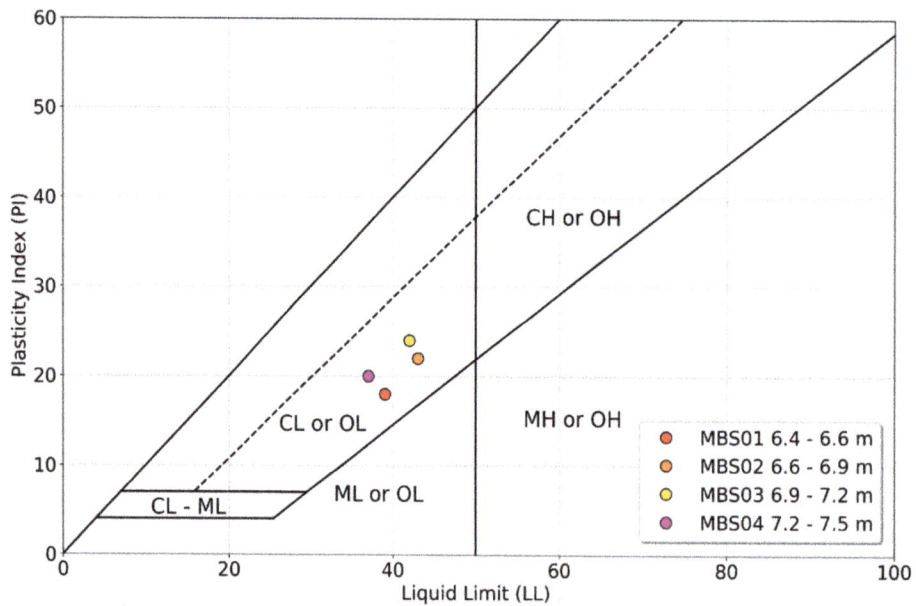

FIG 19 – Atterberg limits (AL) for the mini-block samples.

Oedometer testing was utilised for quantification of sample disturbance based on change in pore volume relative to the initial pore volume (Lunne *et al*, 2006). This procedure is widely adopted and has the advantage of being based solely on initial properties. This methodology has been updated by Berre, Lunne and L'Heureux (2022) to take into account false deformations, which have been highlighted to be more of a concern for oedometer testing when compared to triaxial testing due to the length to height ratio differences.

The procedure by which the false deformations for oedometer tests can be corrected, in the stress region up to the present *in situ* effective vertical stress, involves extending a tangent line to the stress-strain curve to the strain axis. The intersection point defines the amount of false deformation, which should be excluded in the strain when determining the sample disturbance (Berre, Lunne and L'Heureux, 2022). The only tests whereby the Berre, Lunne and L'Heureux (2022) methodology should be applied are those experiencing a concave upwards result on the stress-strain curve up to the present *in situ* effective vertical stress, when the axial stress is plotted on a linear scale with the strain-axis pointing downwards. Figure 20 displays the evaluation of the sample quality using both the Lunne *et al* (2006) and the Berre, Lunne and L'Heureux (2022) methods. During the application of the Lunne *et al* (2006) method it can be seen that three out of four samples are considered excellent quality. When adjusting for the false deformations through the Berre, Lunne and L'Heureux (2022) methodology then this corrects MBS02 and all samples are considered excellent quality.

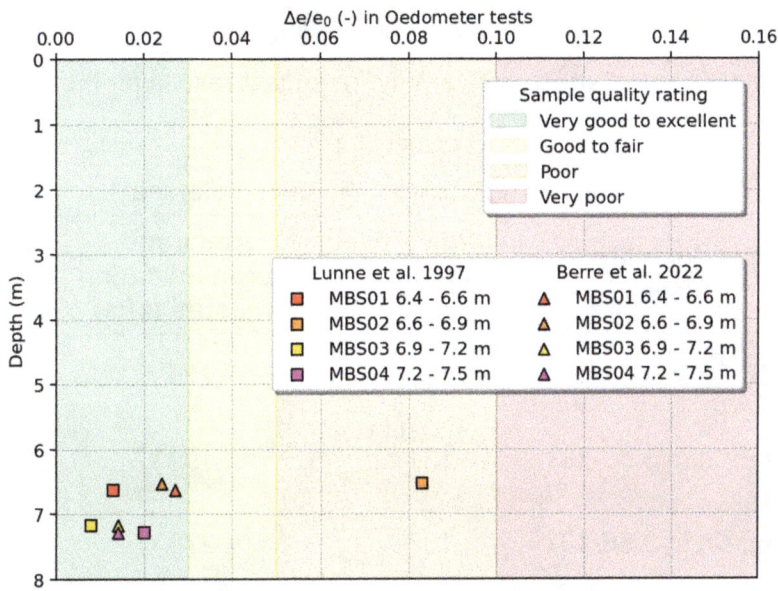

FIG 20 – Evaluation of sample quality based on $\Delta e/e_0$ using both Lunne, Berre and Strandvik (1997) and Berre, Lunne and L'Heureux (2022).

The results for the oedometer test are shown in Figure 21. The permeability estimates from the oedometer results range from 1.6E-08 m/sec to 4.6E-11 m/sec. Indicating a material of low permeability consistent with a clay-like material. A summary of the permeability results oedometer testing is shown in Table 4. An apparent OCR reported in Table 4 of between 1.1 and 2.0 further supports the existence of a lightly bounded microstructure. The apparent preconsolidation pressure varies between 160 kPa and 270 kPa.

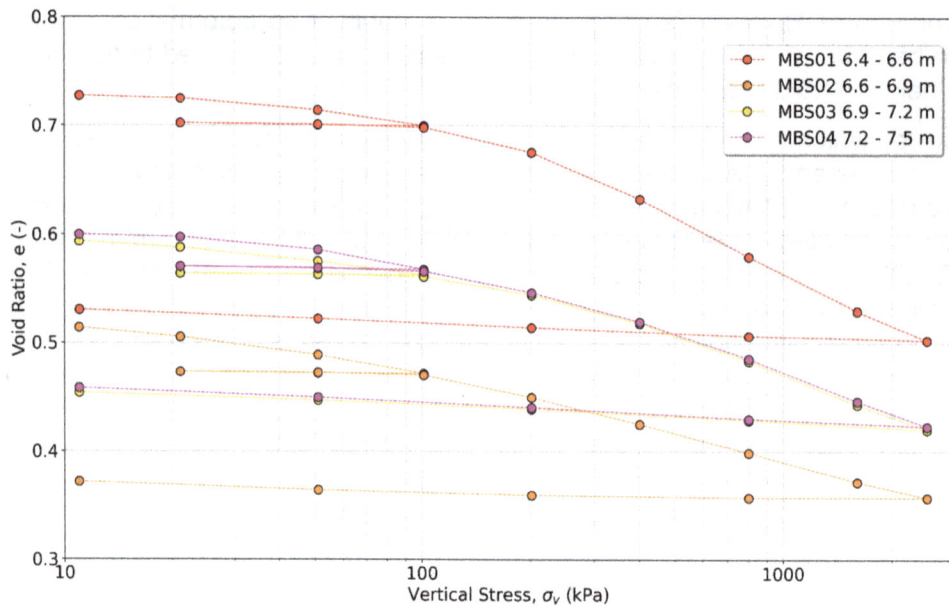

FIG 21 – Oedometer (Vertical stress versus Void ratio) test results for the mini-block samples.

TABLE 4

Permeability results from the oedometer tests.

Test	Permeability (m/sec)	*In situ* vertical effective stress (kPa)	Apparent preconsolidation pressure (kPa)	Apparent overconsolidation ratio (OCR)
MBS01 6.4–6.6 m	2.6E-09 to 4.2E-11	126	230	1.8
MBS02 6.6–6.9 m	3.8E-09 to 3.5E-10	131	190	1.5
MBS03 6.9–7.2 m	1.6E-08 to 3.3E-10	137	270	2.0
MBS04 7.2–7.5 m	1.0E-09 to 9.0E-10	145	160	1.1

Two triaxial (TX) and four simple shear (SS) were undertaken on the mini-block samples, shown in Figures 22 and 23 respectively. The TX results indicate a strain-hardening response at low stresses (below the apparent preconsolidation pressure) with an increasingly strain-softening response (though still very mild) at larger stresses. The SS testing occurring at lower normal stresses experienced strain-hardening responses. However, from approximately 150 kPa mean effective stress and 200 kPa normal stress (very close to the preconsolidation pressure), the strain-softening behaviour becomes apparent. Suggesting that the initially dilative material at stresses below the apparent preconsolidation pressure becomes contractive at larger stresses above the apparent preconsolidation pressure. SS results presented peak stress occurring at approximately 16 per cent strain when a normal stress of 400 kPa is applied. The shear mechanism expected to dominate a potential failure plane through a weak layer is that of SS, as such strength characterisation focused on this kind of testing. The SS results were used to compute the shear strength envelopes due to the stress orientation of the test closely resembling anticipated field conditions. The results are shown in Figure 24. The figure indicates lines of best fit and are provided herein as descriptive mechanical characterisation. The selection of adequate strength parameters for geotechnical design requires further consideration.

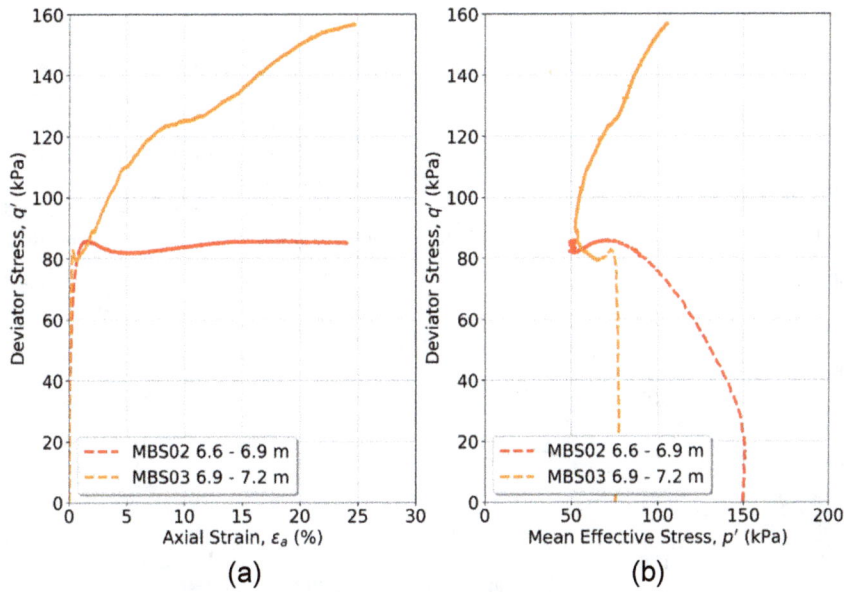

FIG 22 – Triaxial results; (a) axial strain versus deviator stress and (b) mean effective stress versus deviator stress.

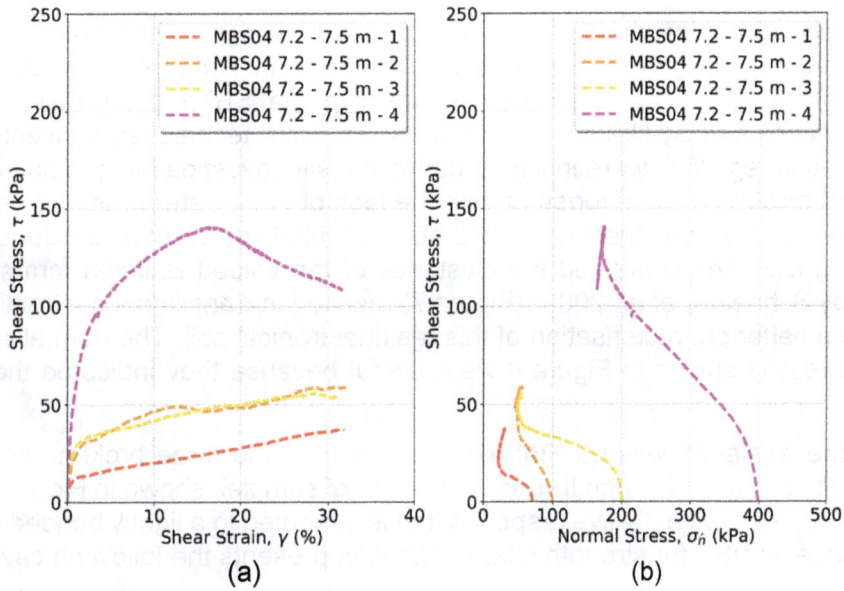

FIG 23 – Simple Shear results: (a) shear strain versus shear stress; and (b) normal stress versus shear stress.

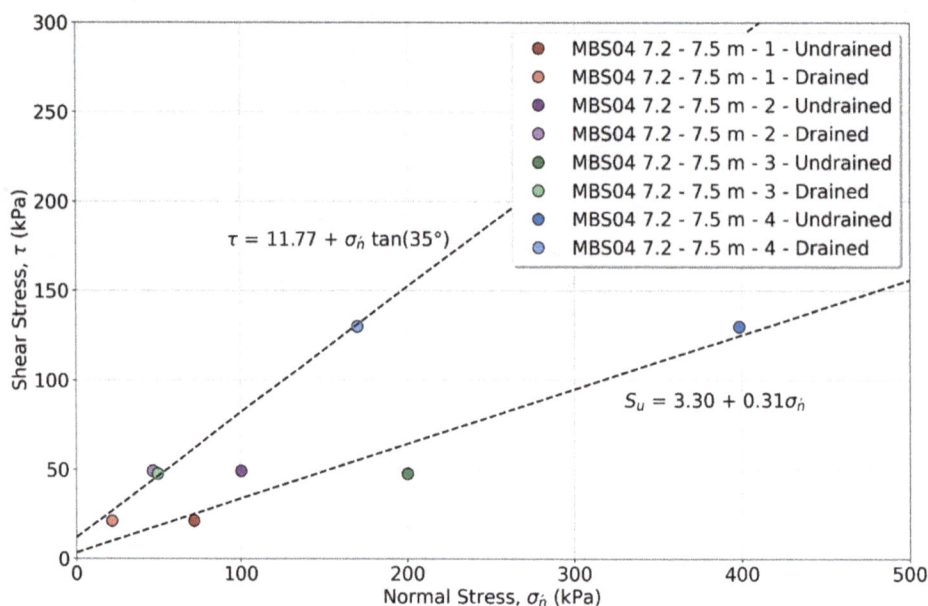

FIG 24 – Failure envelopes for both the undrained shear strength and the drained shear strength.

DISCUSSION

The characterisation of tropical residual soils remains a challenge for the practice of geotechnical engineering. During the interpretation of conventional CPTu, frictions ratios indicated the presence of clays, conventional CPT characterisation charts like the SBTn chart presented in Figure 6 indicated the presence of clay-like contractive soils, as such, techniques conventionally used for clays characterisation (eg VST) were included during the site investigation campaign. The low cone tip resistance and frictional sleeve together with the lack of pore-water pressure generation during shearing, seen in Figure 7, was first interpreted as a potential source of error. Subsequent findings at different testing locations confirmed the existence of the voided Bulimba formation. The SBTn chart proposed by Schneider *et al* (2008) (Figure 7) resulted in transitional contractive soil and was found to provide a better characterisation of this residual tropical soil. The normalised rigidity index concept and the results shown in Figure 8 were useful because they indicated the presence of a microstructure.

The microstructure existence was qualitatively confirmed by the three broken vanes encountered during the 2022 site investigation and through CT scans of samples shown in Figure 14. PPD testing results in Figure 10 indicated a dilative response that is attributed to a lightly bonded (microstructure) matrix. The reliance on VST for strength characterisation presents the following caveats:

1. Although the PSD shown in Figure 18 indicates a large presence of fine-grained soils (between 80 per cent to 90 per cent), the microstructure reduces confidence in the VST results; furthermore the presence of large particle sizes seen in the CT scan (Figure 14) undermines the quality of VST testing, it is believed that very low VST tests are the result of testing a heavily voided region, however the inverse is just as likely a reason for the higher VST results whereby the gravel or denser material was tested;

2. PPD resulted in dilative responses, creating difficulties for confirmation of a potentially undrained response during VST shearing.

However, PPD testing is useful to assess the rigidity index at the testing location.

The presence of large voids (see Figures 16 and 17) resulted in low to zero cone tip resistance and low to zero sleeve resistance (reported in Figures 1, 4 and 5). In addition to the aforementioned results, the lightly bonded (microstructured) matrix was quantified through SS testing. Strength envelopes indicated that approximately 12 kPa strength exists under drained conditions (at zero confinement, ie surface) and approximately 3 kPa strength at surface are present under undrained shear strength conditions. The oedometer compression results shown in Figure 21 and summarised in Table 4 show that the effects of the apparent preconsolidation pressure have an impact up to stresses of between 160 kPa and 270 kPa. Below these stresses the voided Bulimba formation is anticipated to behave strain hardening when sheared, see Figures 22a and 23a. A sample tested under triaxial conditions at the confinement effective stress of 150 kPa resulted in ductile response, this is interpreted as a transitional (between dilative to contractive) confinement state. Beyond the apparent preconsolidation stress the voided Bulimba formation is anticipated to behave strain softening.

The sample quality obtained using the 'Mini-block sampling technique was confirmed using the methods proposed by Lunne, Berre and Strandvik (1997) and Berre, Lunne and L'Heureux (2022), see Figure 20. The results and analysis presented in this paper was only possible due to the high quality of samples recovered. The bonding of between 3 kPa and 12 kPa at surface of a heavily voided sample posed challenges for sampling. The low strength nature of such bonds resulted in the collapse of the microstructure during thin-wall sampling (Figure 2). As a result of this information, thin-wall sampling as means of recovering undisturbed samples is not recommended for this kind of material.

CONCLUSIONS

Material characterisation of tropical residual soils remains a challenging endeavour. This work described a comprehensive characterisation journey that took several years of field and laboratory testing campaigns. The methods described herein have advantages and disadvantages. Some of the key learnings are summarised as follows:

- Thin-wall push tube sampling techniques are not adequate to recover undisturbed soil samples of lightly bonded tropical residual soils with the presence of voids.

- CPTu field testing was proven to reliably pick up the existence of the voided Bulimba formation.

- CPTu interpretative screening methods such as that one proposed by Schneider *et al* (2008) yielded acceptable results. The lightly bonded voided Bulimba formation was classified as transitional contractive using this methodology.

- PPD testing was useful for confirmation of the dilative response at existing stresses and also facilitated estimation of groundwater levels. However, resulting from this dilative response PPD testing was not useful in estimating *in situ* hydraulic conductivity for the voided Bulimba formation.

- VST resulted in highly variable results, several vanes were snapped during testing, indicating high stresses. Further VSTs resulted in undrained shear strength ratios of less than 20 kPa at 7.5 m depth. The voided Bulimba formation is highly heterogenous, and it is proposed herein that very low strength values were obtained when testing a voided location. Heterogeneity is a source of error for this kind of testing and does not allow development of the idealised cylinder required when performing VST interpretation. This kind of testing is problematic for a lightly bonded material containing large voids and denser clay (with a small amount of gravel) interspersed, such as that seen in the voided Bulimba.

- Mini-block sampling was instrumental to recover high quality samples. A successful laboratory campaign would not have been possible without using a sampling technique that enabled retention of the sample's microstructure during sample retrieval in the field and subsequent transportation to the laboratory.

- Oedometer testing was useful to quantify the range of apparent preconsolidation pressures for selected samples, furthermore, the hydraulic conductivity estimates derived indicated that the material had the potential to behave undrained (due to the low permeabilities estimated) despite the presence of large voids. This is particularly relevant for stresses beyond the apparent preconsolidation pressure when void connectivity is lost.

- SS testing was identified as the key shearing mechanism for the purpose of this work, and as such the results from this testing technique are recommended to characterise the mechanical behaviour of the unit. However, triaxial testing results are valuable to understand the overall mechanical behaviour of the voided Bulimba.

ACKNOWLEDGEMENTS

The authors are grateful for the assistance provided by Black Insitu Testing for their involvement in mini-block sampling, who ensured high quality samples were retrieved. Thanks also go to GTI Perth who with their professional skills ensured good quality data was obtained during laboratory testing.

REFERENCES

Amundsen, H A, Emdal, A and Thakur, V, 2021. A new method for storage of block samples: a pilot study, *Géotechnique Letters*, 11(3):179–186.

Berre, T, Lunne, T and L'Heureux, J S, 2022. Quantification of sample disturbance for soft, lightly overconsolidated, sensitive clay samples, *Canadian Geotechnical Journal*, 59(2):300–303.

Emdal, A, Gylland, A, Amundsen, H A, Kåsin, K and Long, M, 2016. Mini-block sampler, *Canadian Geotechnical Journal*, 53(8):1235–1245.

Fookes, P G, 1997. Tropical residual soils: A Geological Society Engineering Group working party revised report, Geological Society of London.

Hight, D W, Böese, R, Butcher, A P, Clayton, C R I and Smith, P R, 1992. Disturbance of the Bothkennar clay prior to laboratory testing, *Géotechnique*, 42(2):199–217.

Islam, K and Murakami, S, 2021. Global-scale impact analysis of mine tailings dam failures: 1915–2020, *Global Environmental Change*, 70:102361.

Jefferies, M, Morgenstern, N R, Van Zyl, D V and Wates, J, 2019. Report on NTSF embankment failure, Cadia Valley Operations, for Ashurst Australia.

La Rochelle, P and Lefebvre, G, 1971. *Sampling disturbance in Champlain clays*, ASTM International.

Ladd, C C and DeGroot, D J, 2004. Recommended practice for soft ground site characterization: Arthur Casagrande Lecture, Massachusetts Institute of Technology.

Lefebvre, G and Poulin, C, 1979. A new method of sampling in sensitive clay, *Canadian Geotechnical Journal*, 16(1):226–233.

Lim, G T, Pineda, J, Boukpeti, N, Carraro, J A H and Fourie, A, 2019. Effects of sampling disturbance in geotechnical design, *Canadian Geotechnical Journal*, 56(2):275–289.

Lunne, T, Berre, T and Strandvik, S, 1997. Sample disturbance effects in soft low plastic Norwegian clay, in Symposium on Recent Developments in Soil and Pavement.

Lunne, T, Berre, T, Andersen, K H, Strandvik, S and Sjursen, M, 2006. Effects of sample disturbance and consolidation procedures on measured shear strength of soft marine Norwegian clays, *Canadian Geotechnical Journal*, 43(7):726–750.

Oboni, F and Oboni, C, 2020. *Tailings dam management for the twenty-first century*, Springer International Publishing.

Piciullo, L, Storrøsten, E B, Liu, Z, Nadim, F and Lacasse, S, 2022. A new look at the statistics of tailings dam failures, *Engineering Geology*, 303:106657.

Pineda, J, Suwal, L P and Kelly, R, 2014. Sampling and laboratory testing of Ballina clay, *Australian Geomechanics*, 49(4):27–40.

Robertson, P K, 2016. Cone penetration test (CPT)-based soil behaviour type (SBT) classification system – an update, *Canadian Geotechnical Journal*, 53(12):1910–1927.

Schneider, J A, Randolph, M F, Mayne, P W and Ramsey, N R, 2008. Analysis of factors influencing soil classification using normalized piezocone tip resistance and pore pressure parameters, *Journal of Geotechnical and Geoenvironmental Engineering*, 134(11):1569–1586.

Zabolotnii, E, Morgenstern, N R and Wilson, G W, 2022. Mechanism of failure of the mount polley tailings storage facility, *Canadian Geotechnical Journal*, 59(8):1503–1518.

A paradigm shift in CPTu testing of extremely soft tailings

A J McConnell[1], E J C Wassenaar[2] and M K D Chapman[3]

1. MAusIMM, Founder, Insitu Geotech Services Pty Ltd (IGS), Brisbane Qld 4014.
   Email: allan@insitu.com.au
2. Regional Manager, Geomil Equipment B.V, Moordrecht, Netherlands.
   Email: wassenaar@geomil.com
3. Managing Director, Insitu Geotech Services Pty Ltd (IGS), Brisbane Qld 4014.
   Email: mark@insitu.com.au

ABSTRACT

CPTu (piezocone) testing is routinely used as part of modern site investigations in tailings storage facilities. The test is well-established and is celebrated as a convenient and rapid means to establish parameters used to analyse stability, settlement, potential for liquefaction, etc. The test was developed originally for natural soils. Many correlations exist for determining the various parameters, based on the three fundamental test outputs: cone resistance (q_c or q_t); sleeve friction (f_s); and pore pressure (u). An inconvenient truth however is that in extremely soft or very loose (ooze-like) materials, typical in many tailings situations, even the best quality conventional CPTu equipment cannot reliably or repeatably measure and report extremely small values of f_s. There are reasons for this, related to the physical design of CPTu sleeve systems, almost guaranteeing that this problem will exist. The in situ testing industry to some extent deals with this inconvenient truth as an 'elephant in the room'; a topic not talked about much, if at all, but rather agonisingly hanging around in the shadows in the background of CPT testing. Of course treating it this way means that parameters determined from the well-established correlations, and other parameters that might be directly measured, are compromised. In 2021 two of the authors, a testing practitioner and an equipment manufacturer, developed a new CPTu cone that has been proven to overcome this problem. This required a significant design paradigm shift that has evolved into a new innovative design that works. The new CPTu cone was described by the authors at the CPT'22 Conference in Bologna (McConnell and Wassenaar, 2022). Now it has been in use for more than a year and some meaningful comparative tailings testing data is available that demonstrates the efficacy of the new device, provided (100 per cent) that it is maintained and calibrated carefully. The new CPT cone is commercially available to all; it is not a secret in-house tool.

INTRODUCTION

In June 2022 the authors published and presented a paper at the conference CPT'22 in Bologna, titled '*An innovative new 3 MPa CPT – to detect and measure very small f_s values*' (McConnell and Wassenaar, 2022). That paper introduced the background and some design aspects of a 3 MPa special cone developed by the authors to detect and measure very small f_s values that could not (and still cannot) be detected and measured using conventional CPT compression cones – ie using the industry paradigm for high quality testing.

As that conference was a targeted CPT conference, and as a clear understanding of the new cone is essential to an appreciation of this current paper, the authors encourage the readers to refer to that paper for more information, and for an element of greater completeness. McConnell and Wassenaar's (2022) paper referred in places to the then-current international standard ISO 22476–1-2012, and where relevant those references are repeated here. It also referred to what was at that time a draft update of ISO 22476–1. This draft has now very recently evolved into the current international standard BS EN ISO 22476–1-2023. Where relevant, material from the new standard is referred to, and in some places this will be a little different to that from the draft previously referenced.

The CPT'22 introduction

Insitu Geotech Services (IGS) undertakes much CPT testing in soils and sediments that can be described in everyday terms as 'extremely soft', or even as 'ooze'; for clients who are seeking data that permits them to make confident designs.

As a consequence they are almost every day working in an arena where one would aspire to better than Application Class 1 testing quality, if being described according to (the past) ISO 22476–1-2012.

To achieve the highest quality that they can in testing these conditions and in satisfying clients they:

- Use known good quality CPT cones. In soft soils these have been previously usually of 25 MPa or 10 MPa tip capacity.

- Maintain/manage all CPTs meticulously.

- Undertake in-house calibration-checking-recalibration under a very stringent program, as described below.

The IGS calibration program is run in-house because, as you can see from the explanation below, it would be unworkable to contract it out:

- Every cone is calibrated before every job, then recalibrated after the job, each time using the cone's actual dimensions, not nominal values. If the job runs more than one week, then the cones are changed over with freshly calibrated cones on an approximate 7-day service cycle. All of this is carefully recorded and a calibration-drift-performance type of history is developed for each cone.

- Recalibration data is then always compared to the data from the previous calibration, to cross-check for any significant change.

- And, before-versus-after zero-load-drift comparisons (called 'reference values' in the new standard), are taken as part of each test's management.

- It's rigorous, it works, and it's business as usual, built into the company culture and cost structures.

Following the above approach IGS has been successful in regard to the ability to defensibly and repeatably measure very low q_c values, and to be as good as reasonably achievable at measuring f_s values.

LIMITATION RE MEASURING VERY SMALL SLEEVE FRICTIONS

In regard to f_s though, they have run up against the 'industry normal' limitation in measuring very small sleeve frictions; a limitation that is associated with the design of normal Compression Cones.

To help discussion of this, refer to Figure 1.

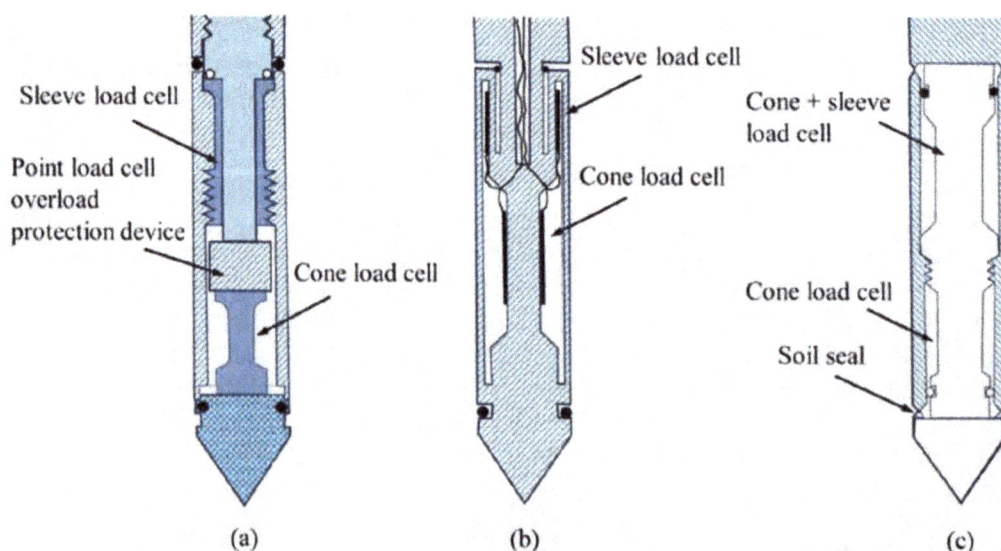

FIG 1 – Design of cone penetrometers (modified in Robertson and Cabal (2022) from Lunne, Robertson and Powell, (1997)).

In this paper type (a) is designated as a compression cone and type (c) as a subtraction cone.

For this discussion, the relevant difference between a compression cone and a subtraction cone is that:

- A compression cone has separate load cells for the tip and the sleeve. Hence these load cells can be sized for the purpose; typically a larger load cell for the tip and a smaller load cell for the sleeve.

- A subtraction cone has two load cells that must both be of the larger variety. Typically they would be identical or nearly identical. One of these, that just behind the tip, measures the tip load only and the other, above the screwed on connection to the sleeve, measures the combined tip-plus-sleeve load.

- Software subtracts one load cell value from the other to determine the sleeve friction. Hence a relatively big number is subtracted from another relatively big number to get a smaller number. If calibration of either or both load cells is not precise, or they drift differently during a test or after calibration, one can reasonably expect significant errors in the sleeve friction values determined this way.

Compression cones (these authors perceive) were developed in the first instance ostensibly to improve a CPT's ability to measure sleeve friction.

THE CURRENT PARADIGM

So this is the current industry-wide paradigm: *If you want to most accurately measure sleeve friction, you must use a compression cone.*

In support of this statement that this is the current paradigm, refer to Robertson and Cabal (2022), as follows:

> *For accurate sleeve resistance measurements in soft sediments, it is recommended that cones have separate (compression) load cells.* (p 22)

And in support of something that follows in this paper, quoting from the same document as above, the following paragraph, referring now to compression cones:

> f_s *measurements, in general, will be less accurate than tip resistance, q_c, in most soft fine-grained soils.* (p 23)

The problem is that, in extremely soft or ooze-like soils, this 'in general expected lower accuracy' in f_s typically ends up meaning no f_s measurement at all, or something too low to be credible.

EXPLANATION – WHY VERY SMALL SLEEVE FRICTIONS ARE ELUSIVE

A simple explanation

Referring to Figure 1:

- In a compression cone, the sleeve has to move slightly to permit it to apply load to the friction sleeve load cell.

- Dirt seals behind the friction sleeve resist this slight movement and use up some of the force applied by soil friction to the friction sleeve.

- Hence not all or any of the soil friction force 'gets to' the friction sleeve load cell. A significant error occurs if this applied soil friction force is very low (as it will be in extremely soft materials).

If one is testing stronger materials then it's not a significant issue at all, but if one is testing extremely soft material it becomes a problem.

Until recently this limitation was typically handled as an 'elephant in the room', not talked about much, but rather agonisingly hanging around in the shadows in the background of CPT testing.

But it's become a pretty big deal for organisations like IGS that nowadays do much testing in extremely soft or ooze-like soils; and for their clients.

References to this problem by others

This issue was discussed at CPT'14 in a paper by Santos, Barwise and Alexander (2014) where the authors correctly ascribed the problem to the friction in sleeve seals and presented an example sleeve calibration showing the difference in load cell readings compared to applied load (Figure 2a) without friction sleeve seals, and (Figure 2b) with seals.

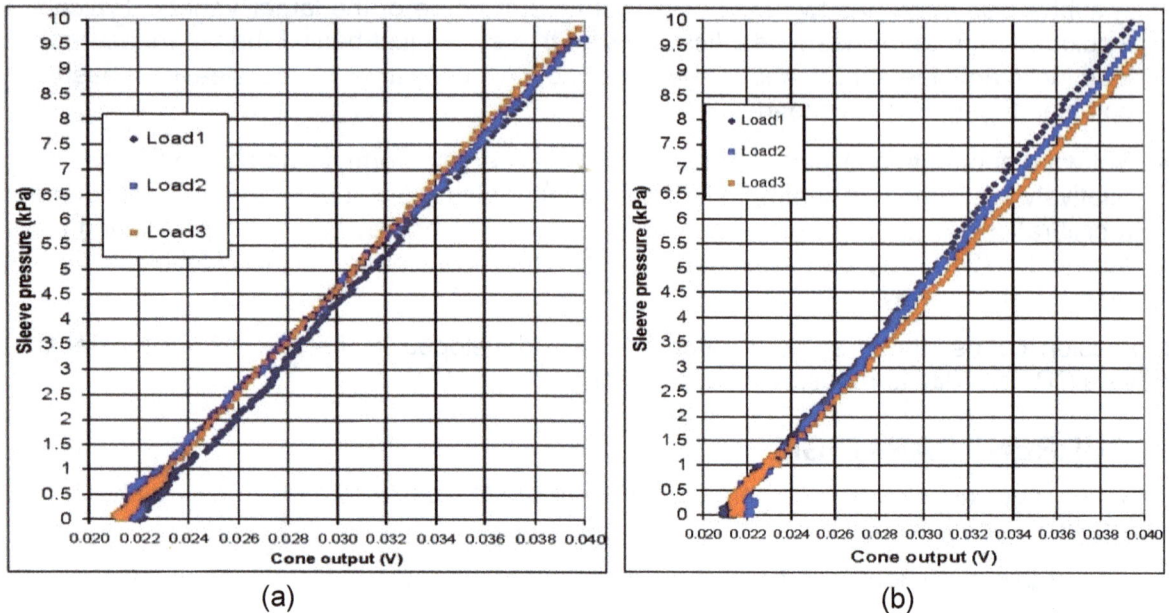

FIG 2 – (a) Sleeve calibration with no seals; (b) Sleeve calibration with seals.

Those authors discussed a novel spring-loaded (preloaded) sleeve seal design to attempt to solve the problem. This has/had been taken up by A P van den Berg for their cone design.

The issue was again mentioned in a paper at the 2021 Mine Waste and Tailings Conference, (Entezari *et al*, 2021). In that paper, the authors were discussing limitations in the use of CPT data to determine the fines content of extremely soft oil sands tailings. For the analysis they were undertaking, many data points were '*screened out in order to remove data where the soil-sleeve friction was less than internal o-ring (sic) friction*'. This meant discarding a lot of their data.

IGS WANTED A SOLUTION – NOT ANOTHER COMPROMISE

IGS wanted a solution, so talked at length to their supplier/partner Geomil as to what this might entail. They then jointly conceived and funded an atypical design as a trial.

The conversation and thinking went as follows:

1. As described in Sections 3 and 4, compression cones have friction sleeves that must move a small amount to be able to register an f_s reading.

2. The sleeves of subtraction cones do not have to move more than a miniscule amount to register friction load. But they have the historically-deemed problem of having to subtract one big number from another big number to get a small one, as discussed in Section 3.

3. As a solution, would it be possible to develop a subtraction cone with unusually high quality and sensitive load cells, and to calibrate these very rigorously to help overcome the problem in subtracting a big number from a big number?

4. And would it be possible to design for much lower cone capacity overall, hence making the two big numbers smaller, further reducing the problem?

5. And would it be possible to make the load cells more responsive than normal, by adopting different construction materials for the elastic bases on which the load cells' strain gauges would be fixed?

Of course this would all comprise a significant paradigm shift, compared to the existing one. Reiterating that paradigm here: *If you want to most accurately measure sleeve friction, you must use a compression cone.*

Some heart was taken in the knowledge that Fugro's very sensitive Fibre Optic Cone, announced at the prototype stage via Looijen *et al* (2018) at CPT'18 in Delft, also designed for testing soft materials, is a subtraction cone with unusually sensitive load cells; in principle the same idea.

So IGS and Geomil talked more, and eventually Geomil made a first small run of special cones to the agreed design. In accord with points 3 and 4 above, they opted for a cone size of 15 cm^2 and an extremely low tip capacity of only 3 MPa; this being adequate for the testing of extremely soft materials that were/are the target.

And, in accordance with point 5 above they opted for a special alloy base for the load cells, giving a physical strain gauge response approximately 300 per cent greater than it would be for a conventional steel base.

Of course, also in accordance with point 3 above, IGS has adopted the same calibration regime discussed in Section 1, with the enhancement of using dead weights for the application of load to the tip and sleeve.

HIGH HOPES AND EXPECTATIONS

Because of high hopes for these cones, IGS had set a policy in place that, to the extent possible, they will calibrate these cones to achieve the accuracy outcomes proposed by the (then) draft international standard ISO/DIS 22476–1 criteria for Class 1+ cones. Now that draft has become a standard, IGS has re-aimed to target, to the extent possible, the prescription described in the finalised standard, not the draft. The new standard BS EN ISO 22476–1-2023 has no Class 1+ but has changed the designation to Class 0.

To date, that objective is looking good. A full IGS calibration for one of these cones is shown in Appendix 1 (Figure A1).

In other words, these cones, under these rigorous calibration processes, are pretty remarkable by current measures.

Of course calibration is just one aspect of a CPT. We now want to ask, *how have these cones behaved in the field?*

FIELD TRIALS

IGS has had several 3 MPa cones in operation now since the beginning of 2022, in some places testing very soft natural soils but in more places testing ooze-like tailings sediments. The results are satisfying, demonstrating:

- Ability to repeatably detect, measure and record very low q_c values – say 10 kPa and even below.

- Ability to repeatably detect, measure and record very low f_s values that make sense when compared to the q_c values being measured – say 1 kPa and even below.

Figure 3 shows the improved response to sleeve friction compared to what was previously possible, see McConnell and Wassenaar (2022).

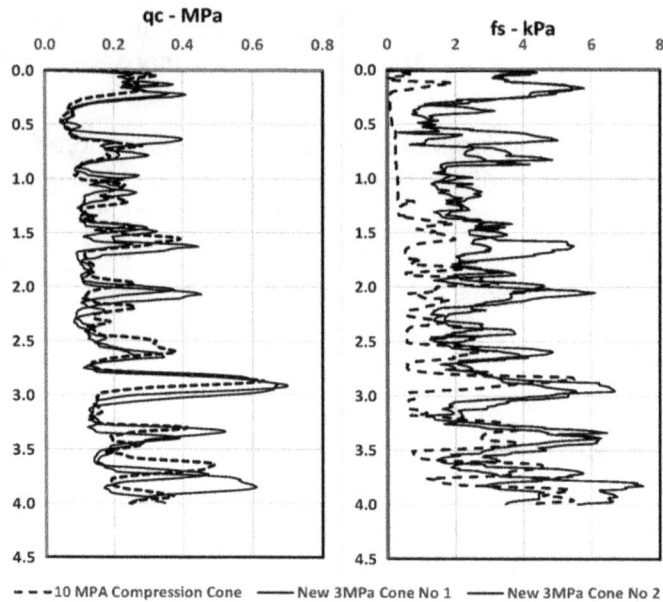

FIG 3 – q_c and f_s plots – new 3 MPa cone versus 10 MPa compression cone.

This is just a single illustrative example. All tests undertaken with these new cones have shown the ability to measure very low q_c and a much improved response in regard to extremely low sleeve friction f_s values.

REAL EXPERIENCE – WHAT'S THE BENEFIT FOR A TAILINGS INVESTIGATOR?

In pursuit of a method to answer this question in a clear and easily contrasted manner, data from a number of real project tests have been processed using the computer software CPeT-IT v.3.0 – CPT Interpretation Software (Ioannides, 2007) to plot CPT-based Soil Behaviour Type (SBT) using the (Robertson, 2016) Soil Behaviour Type Index I_B, that Robertson and Cabal (2022) states 'capture(s) the SBT boundaries better than the original circular Ic'.

Four examples are shown in Figure 4; three tests on tailings dams and one from a natural soil site. These clearly show that the Index I_B decreases (ie 'moves to the right') when testing was done using the new 3 MPa cone compared to side-by-side (or nearly so) tests using a conventional compression cone. It should be noted that all cones were in very good calibration during all of these tests. It should also be noted that, according to Robertson (2016), lower I_B values indicate more dilative soil behaviour than higher values; significant information for a tailings dam designer.

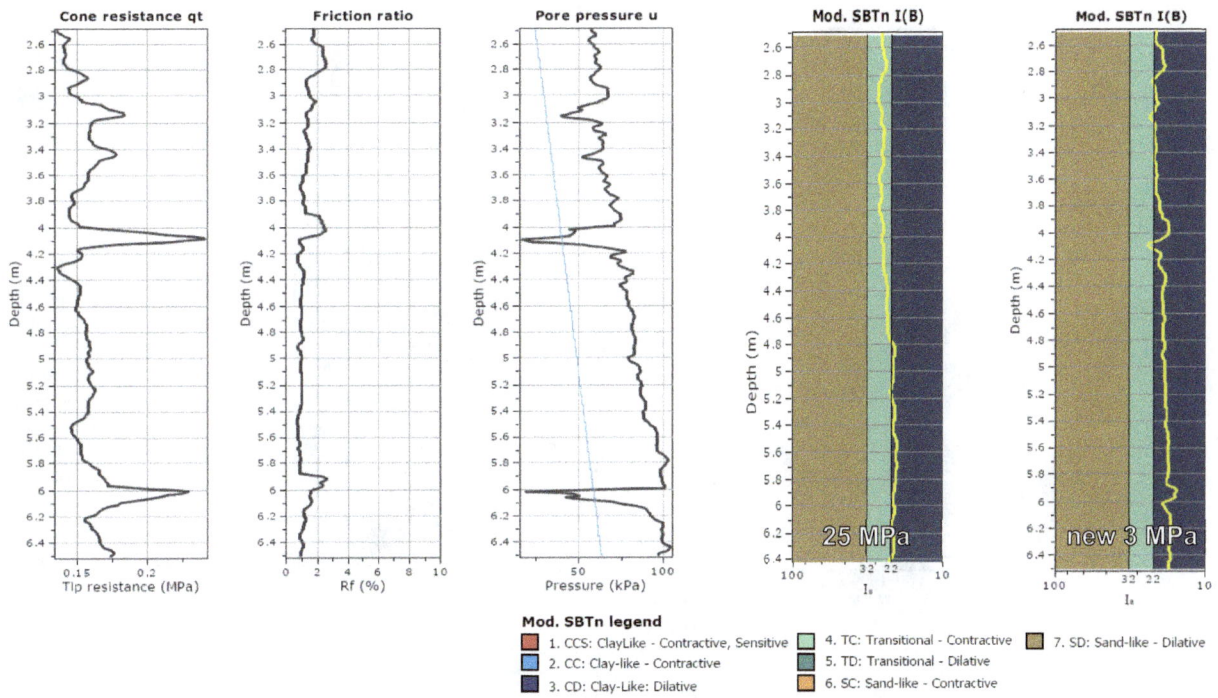

(a)

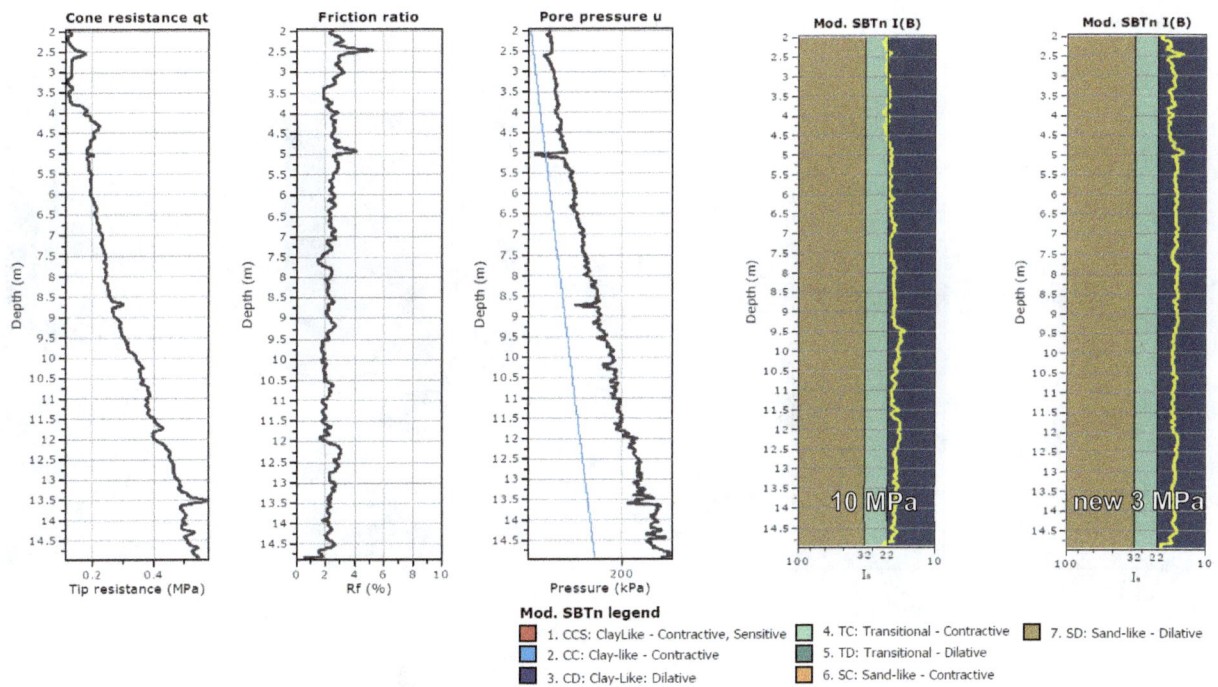

(b)

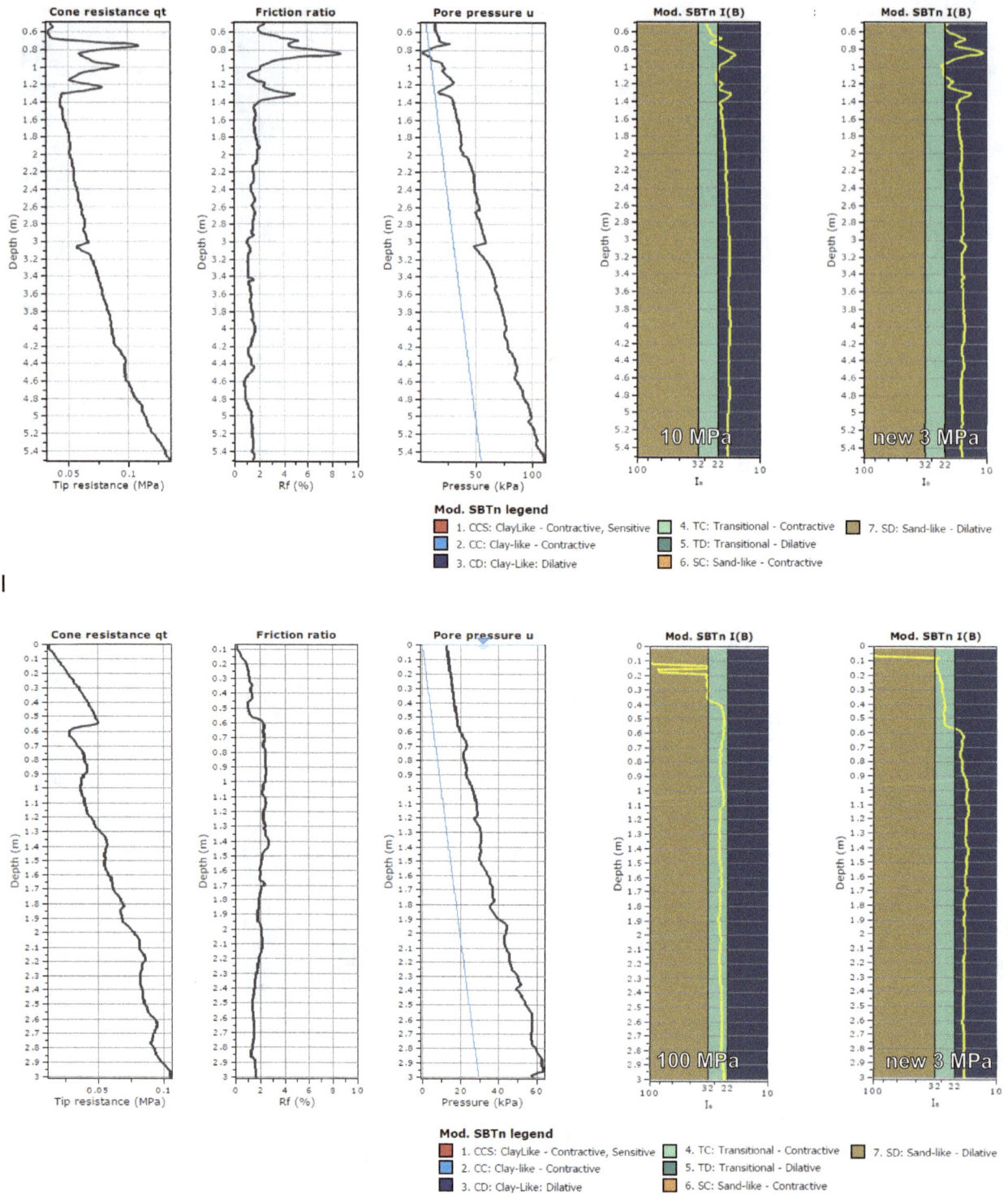

FIG 4 – (a) q_t, R_f, u and I_B plots for a coal tailings dam in The Bowen Basin (25 MPa compression cone versus new 3 MPa cone – q_t mainly 0–100 kPa);

(b) q_t, R_f, u and I_B plots for a coal tailings dam in The Hunter Valley (10 MPa compression cone versus new 3 MPa cone – q_t mainly 0–500 kPa);

(c) q_t, R_f, u and I_B plots for a metalliferous tailings dam in northern Australia (10 MPa compression cone versus new 3 MPa cone – q_t mainly 0–120 kPa);

(d) q_t, R_f, u and I_B plots for a natural very soft clay site in Tasmania (100 MPa compression cone versus new 3 MPa cone – q_t mainly 0–100 kPa).

FIELD TRIALS – REFERENCE VALUES

The CPT industry/profession is passionate on the issue of comparing 'before-versus-after' zero-load-offset values (ie reference values shift) as a method of determining/monitoring:

- the quality and actual condition of a cone itself
- the quality of a test that has been undertaken.

IGS's cone history recording system will one day allow a report on the reliability of this dependence on reference value shift; a quantified study to support the passion (or otherwise).

In the meantime however the authors acknowledge that this is an industry-accepted field indicator of the quality of a test undertaken, so monitoring and recording these zero-load-offsets is part of IGS's every-test practice.

The new standard BS EN ISO 22476–1-2023 relies heavily on this parameter in determining test quality, called 'Test Category', with different categories/qualities numbered A to D; A being the highest quality.

Under the new standard (Table 3), one decides the Test Category by:

- The Class of cone used.
- The drift/difference in the before-versus-after reference values.

In the standard, deployment of a Class 0 cone and a before-versus-after drift/difference of less than 15 kPa for the tip, combined with less than 5 kPa drift/difference for the sleeve, categorises the test as Category A, the highest test quality category.

This typically involves measuring the after-test-offsets once the cone has been cleaned and reassembled after testing. It cannot be a value extracted from the completed test's data file, as this immediately-after value is likely to be influenced by dirt in seals and gaps after cone extraction.

Figure 5 shows sets of before-versus-after drift/difference data from a randomly chosen 21 CPT tests using the new 3 MPa cone, The shaded parts represent the BS EN ISO 22476–1-2023 'Maximum allowable difference of reference values before and after test' for Test Category A (the standard's highest test category) Table 3, as described above.

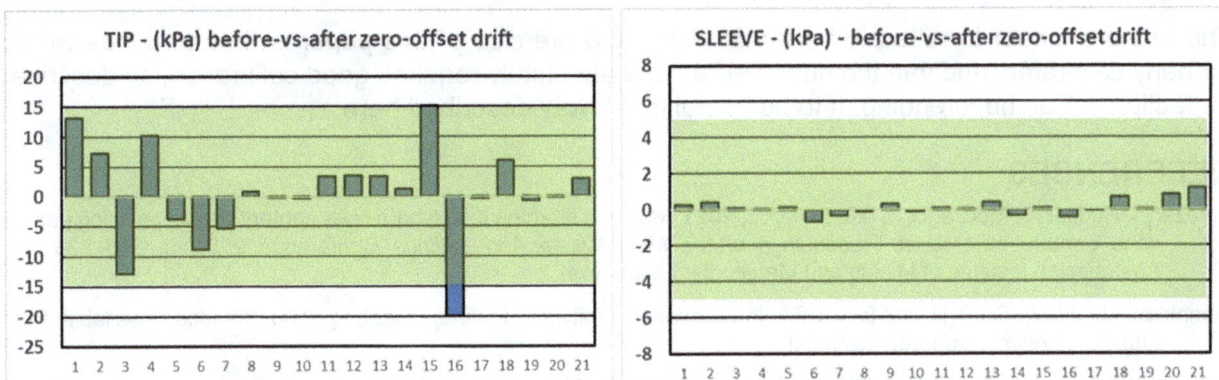

FIG 5 – (a) before-versus-after tip reference values; (b) before-versus-after sleeve reference values.

TRANSIENT TEMPERATURE EFFECTS

Transient temperature effect, arguably the least understood potential error in any CPT testing, is a matter that IGS takes seriously in all of their testing. Their default testing procedures do what is reasonably possible to minimise these effects during all stages of a test; including 'cool down or warm-up' pauses in testing in some profiles and circumstances.

The work that has been done on this aspect to-date in relation to the new 3 MPa cone has been essentially comparisons of the new cone against conventional cones that have been used for the same type of testing, and for which the management procedures have been proven with time. This is to judge whether the current management procedure to manage this issue can apply to the new cones.

Figure 6 demonstrates that the new 3 MPa cone is very similar to other cones as described above; essentially no different to a conventional cone in relation to management of this matter.

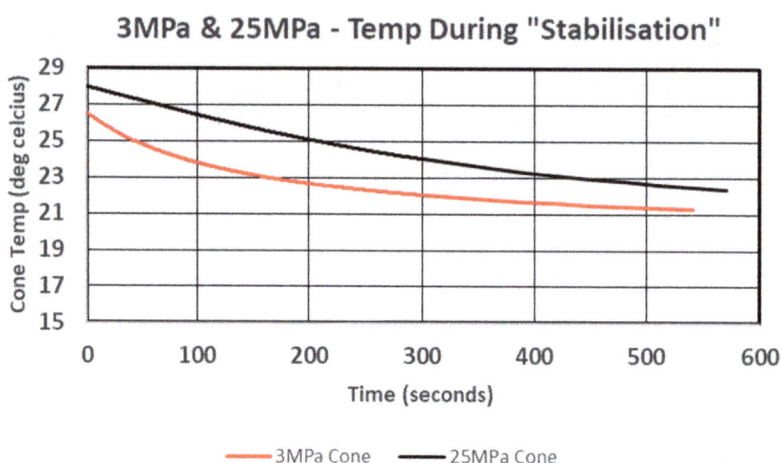

3MPa & 25MPa - Temp During "Stabilisation"

FIG 6 – Comparison of transient temperature stabilisation of the new 3 MPa cone versus a 25 MPa compression cone.

SUMMARY AND CONCLUSIONS

Accurate and repeatable detection and measurement of extremely low sleeve friction (f_s) values during CPT testing is an industry-wide problem, often treated as an 'elephant in the room'.

This paper describes the development of an innovative new CPT cone that the authors believe has solved this problem.

The paper goes on to demonstrate the difference in SBT determined by using this new cone in very soft tailings materials and natural soils. In essence, this difference is that lower I_B values (Robertson, 2016), which indicate more dilative soil behaviour than higher values, can be expected if using this cone.

The authors labour to reiterate here that the whole premise of this paper is that all cones will be properly calibrated, and that the new 3 MPa cone absolutely requires good calibration, as described in Section 1.1 of the foregoing. If to be useful in the way described here.

REFERENCES

Entezari, I, Boulter, T, McGregor, S and Sharp, J, 2021. Machine learning to estimate fines content of tailings using gamma cone penetration test, in *Proceedings of the Mine Waste And Tailings Conference 2021*, pp 218–226 (The Australasian Institute of Mining and Metallurgy: Melbourne)

Ioannides, J, 2007. CpeT-IT v.3.0 – CPT Interpretation Software [online]. GeoLogismiki, Greece. Available from: http://geologismiki.gr/products/cpet-it

Looijen, P, Parasie, N, Karabacak, D and Peuchen, J, 2018. Fibre optic cone penetrometer, in *Proceedings of the 4th International Symposium on Cone Penetration Testing (CPT'18)* (eds: M A Hicks, F Pisano and J Peuchen), pp 407–410 (Delft University of Technology: Netherlands).

Lunne, T, Robertson, P K and Powell, J J M, 1997. *Cone Penetration Testing In Geotechnical Practice* (Taylor and Francis: London).

McConnell, A J and Wassenaar, E J C, 2022. An innovative new 3MPa CPT – to detect and measure very small f_s values. In *Proceedings of the 5th International Symposium on Cone Penetration Testing (CPT'22)* (eds: G Gottardi and L Tonni eds), pp 198–203 (Taylor and Francis: London).

Robertson, P K and Cabal, K L, 2022. *Guide to Cone Penetration Testing*, 7th edition (Gregg Drilling LLC).

Robertson, P K, 2016. Cone penetration test (CPT)-based soil behaviour type (SBT) classification system – an update, *Canadian Geotechnical Journal*, 53(12), December 2016.

Santos, R S, Barwise, A and Alexander, M, 2014. Improved CPT sleeve friction sensitivity in soft soils, in *Proceedings of the Third International Symposium on Cone Penetration Testing (CPT'14)* (eds: P K Robertson and K L Cabal), pp 239–248.

APPENDIX 1

CONE IDENTIFICATION AND DIMENSIONS SHEET
3 MPa Special Purpose Piezocone

Note that this cone calibration has been undertaken taking these "actual" dimensions into account

cone area   15.22   cm2   sleeve area   228.91   cm2

NOMINAL TIP

Cone No	AS15CFIIPT.S21003
Type	SPECIAL PURPOSE
Tip Area (sq cm)	15	15.22	actual tip area
Tip Capacity (MPa)	3		deviation = 1.4%
Calibration Date	15 October 2021

MEASURED TIP DIMENSIONS			ISO 22476-1-2012 requirements			
CD	44.02	critical	43.20	to	44.10	mm
CH	36.50		29	to	38	mm
S	11.74		9	to	12	mm
A	60.00		55	to	65	degrees

NOMINAL SLEEVE

Cone No	AS15CFIIPT.S21003
Type	SPECIAL PURPOSE
Sleeve Area (sq cm)	225	228.91	actual sleeve area
Sleeve Capacity (kPa)	200		deviation = 1.7%
Calibration Date	15 October 2021

MEASURED SLEEVE DIMENSIONS			ISO 22476-1-2012 requirements			
SD-1	44.11	critical	44.02	to	44.37	max 44.2
SD-2	44.11	critical	44.02	to	44.37	max 44.2
SL	165.19		162.3	to	165.3	

CALIBRATED BY	AS
DATE	15/10/2021
NOTES	

CHECKED BY	
DATE	
NOTES	

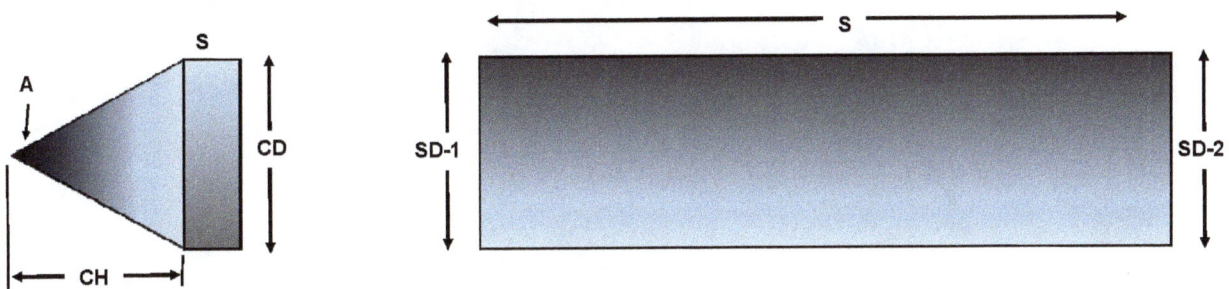

PORE PRESSURE

Cone No	AS15CFIIPT.S21003
Type	SPECIAL PURPOSE
Piezo Capacity (kPa)	2000
Calibration Date	15 October 2021

S

A

CD

CH

S

SD-1

SD-2

3MPa Special Purpose Piezocone Calibration

This cone has been re-calibrated. Use appropriately-dated calibration file. "Actual" cone dimensions used.

No: AS15CFIIPT.S21003

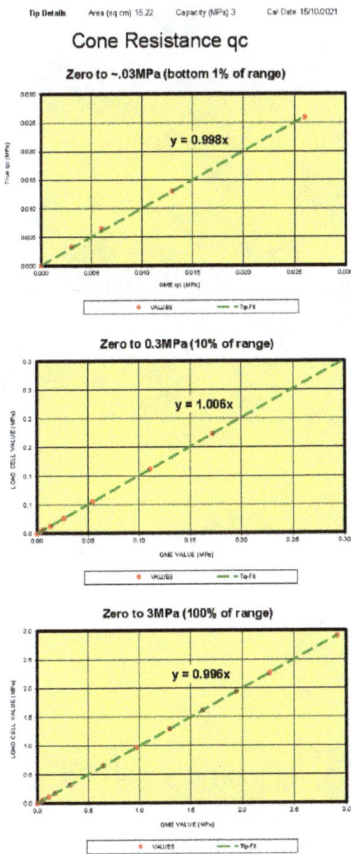

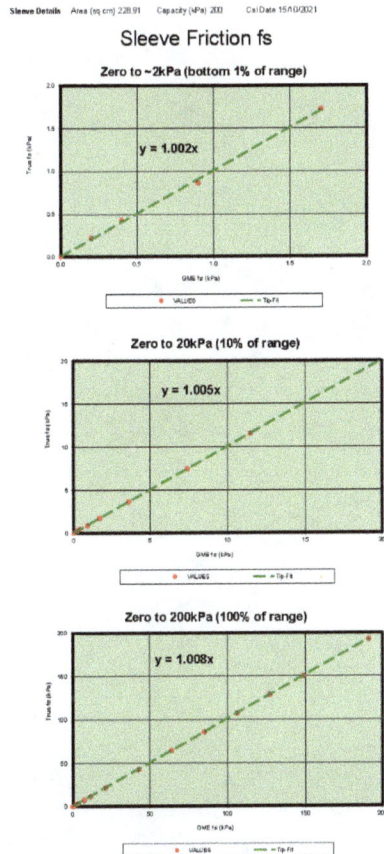

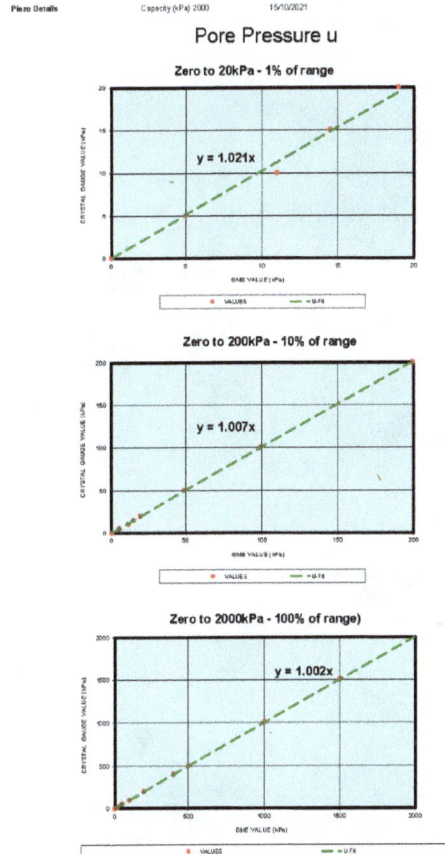

Tip Details   Area (sq cm) 15.22   Capacity (MPa) 3   Cal Date 15/10/2021

Sleeve Details   Area (sq cm) 228.91   Capacity (kPa) 200   Cal Date 15/10/2021

Piezo Details   Capacity (kPa) 2000   15/10/2021

Cone Resistance qc

Zero to ~.03MPa (bottom 1% of range)

y = 0.998x

Zero to 0.3MPa (10% of range)

y = 1.006x

Zero to 3MPa (100% of range)

y = 0.996x

Sleeve Friction fs

Zero to ~2kPa (bottom 1% of range)

y = 1.002x

Zero to 20kPa (10% of range)

y = 1.005x

Zero to 200kPa (100% of range)

y = 1.008x

Pore Pressure u

Zero to 20kPa - 1% of range

y = 1.021x

Zero to 200kPa - 10% of range

y = 1.007x

Zero to 2000kPa - 100% of range

y = 1.002x

Insitu Geotech Services Pty Ltd

Printed: 21/10/2021

FIG A1 – A full IGS calibration for one of the cones.

Analysis of the geotechnical characteristics of fibre reinforced tailings

S Quintero Olaya[1], D J Williams[2], C Zhang[3], M Llano-Serna[4] and G Gomes Silva[5,6]

1. Senior Research Technologist, School of Civil Engineering, Brisbane, Qld 4072.
   Email: s.quintero@uq.edu.au
2. FAusIMM, Professor of Geotechnical Engineering; Director, Geotechnical Engineering Centre, School of Civil Engineering, Brisbane, Qld 4072. Email: d.williams@uq.edu.au
3. Research Fellow, School of Civil Engineering, Brisbane, Qld 4072.
   Email: chenming.zhang@uq.edu.au
4. Principal Geotechnical Engineer, Red Earth Engineering – a Geosyntec Company, Brisbane Qld 4000. Email: marcelo.llano@redearthengineering.com.au
5. Professor of Geotechnical Engineering, Faculty of Engineering, Federal University of Catalão, Catalão, GO, Brazil. Email: gabriel_gomes@ufcat.edu.br
6. Visiting Research Student, School of Civil Engineering, Brisbane, Qld 4072.
   Email: g.gomessilva@uq.edu.au.

ABSTRACT

Understanding the behaviour of tailings deposited in layers is critical for estimating their long-term settlement and strength in tailings storage facilities. Well-consolidated tailings are stable and can help enhance the deposition efficiency and storage capacity of a tailings storage facility. Unfortunately, that does not take place on many storage facilities that struggle to meet their dry density targets as tailings are being deposited. Different measures have been developed to enhance the strength of tailings, such as adding chemical additives, dry stacking, and farming. As a physical additive, plastic fibres have been widely used to strengthen concrete and soil-cement mixes with very good and reliable results, while their effectiveness is yet to be tested. This is mostly due to the changing sources of tailings and ore extraction process which provide different sets of properties. This study aims to quantify the effect of fibre additives on the settling, consolidation and strength of tailings in order to have a more reliable understanding of how fibres can enhance tailings properties as they are poured and desiccate on storage facilities. Red mud tailings samples with varying percentages of fibre contents mixed with three different plastic fibre types were subjected to settling test in standard settling columns, consolidation, and direct shear tests on samples at different solid contents. Comparisons were made among the tests on tailings with fibre contents and a control test where no additive was added. This paper reports the results obtained from these tests.

INTRODUCTION

Tailings is the waste generated during the production and extraction processes of natural ore. This generally comes in the form of a liquid paste and it is poured into a tailings storage facility (TSF). One of the most common methods to place this material into the pond is called sub-aerial deposition. This is done by transporting the tailings through pipelines where the material is pumped and discharged at low solid concentrations (Reid, Fanni and Fourie, 2022) (25–28 per cent typically for red mud).

Australia is top worldwide bauxite producer and the second largest aluminium producer with an annual bauxite production of 101 million tonnes (Mt) by 2021 (Australia's Aluminium Council, 2022). The waste generated after the bauxite extraction and refinement is called red mud. These tailings are then store in large storage facilities that in the case of Australia, happen to be predominantly located in Western Australia and Northern Queensland. The typical landscape of these areas have a flat topography. This implies that no matter where the TSFs are located, it is likely these will not be very high, limiting the tailings total storage volume.

The moment the slurry is deposited into the TSF, it immediately starts to experience settlement, self-weight consolidation and evaporation (Zhang and Lei, 2017). The consolidation is one of the key factors to calculate in order to assess the long-term behaviour of tailings as further layers are discharged at the surface. The consolidation of tailings give an estimate of storage volumes (Shokouhi, 2015).

The evaporation is generally a slow process that relies on many factors such as average temperatures, annual rainfall and the rate at which the material settles and starts to desiccate. This slow evaporation combined with the mine sites high production of ore, may consequently trap soft partially desiccated tailings among layers preventing them to release water to any direction, by either gravity or capillarity. The layers of soft tailings that are stored or trapped before they dry out, occupy extra space within the TSF and have a very low strength to resist large upstream raises.

Many efforts to optimise the influence of water in geotechnical properties of tailings such as dewatering (Gálvez *et al*, 2014) farming and flocculants addition (Zhang *et al*, 2019), have been made. However, it is still a challenge to find reliable long-term solutions to increase strength of tailings with low drying rates (Islam *et al*, 2020).

Alternatively, fibres of various sources have been used on cemented soils and cemented tailings backfill with consistent results in the increase of mechanical strength at different dates of the mixes thought UCS testing (Consoli *et al*, 2004).

Slow settlement rates, settled dry density values and layers of tailings stored at high moisture content percentages are critical parameters that challenging to reach for many mine sites who constantly struggle to meet minimum design values (Aubertin and McKenna, 2016)This carries devastating consequences if corrective measures are not taken on time. Not only in terms of the structural stability of the tailings facility, but also the life span and total capacity of the storage pond.

This study aims to analyse the relationship of consolidation and the strength of red mud with the addition of plastic fibres.

EXPERIMENTAL METHODOLOGY

Material used

Red mud

The tailings used for this study is red mud sourced from south-east Queensland. This material is generally poured out to the facility at a 25 per cent solids percentage. The tested samples were collected directly out of the discharge point of the tailings facility. Figure 1 shows a representative red mud sample at 26 per cent solids.

FIG 1 – Dry sample of red mud.

Plastic fibres

Three types of plastic fibres manufactured by TEXO ReoCo are used in this study. Figure 2 shows each material used. This study refers to them as red, black and white fibres.

FIG 2 – Three types of plastic fibres.

The plastic fibres are made of different concentrations of polypropylene (PP), polystyrene (PE) and high density Polystyrene (HDPE). Table 1 shows a summary of the compositions and length of each fibre.

TABLE 1

Fibres composition.

Fibre type	Length (mm)	Specific Gravity (gr/cm^3)	Chemical composition
Red Fibre	22	0.88	PP (85%) + PE (10%) + hydrophilic additive (5%)
Black Fibre	45	0.91	PP (82%) + PE (10%) + antioxidant additive (3%)
White Fibre	22	0.94	High Density Polyethylene, HDPE (100%)

Test plan

Characterisation tests

Particle size distribution tests (PSD) were carried out following wet and dry method described on the Australian Standard AS 1289.3.6.1. Hydrometer test was also done according to AS 1289.3.6.3.

Atterberg tests were done on red mud to calculate the liquid limit, plastic limit and Index of Plasticity values, following the AS 1289.3.1.1, 1289.3.2.1 and 1289.3.3.1 respectively.

Scanning Electron Mycroscopy (SEM) images were taken on a cross-section of each fibre along with a SEM image for a representative red mud sample.

Settlement tests

The settlement tests selected for this study were carried out on red mud samples at the discharged solids percentage of 25 per cent solids using a 1000 cc bottle. The control sample has no fibre additions. Three fibre additions by dry mass were done on red mud samples at the same solids percentage. These fibre additions were 1 per cent, 2 per cent and 3 per cent for each fibre type. Each test was done twice to have more reliability in the results.

The red mud was mechanically stirred at a speed of 1500 rev/min for about 30 min prior to being introduced into the measuring cylinders. Fibres were added just prior to pouring into the measuring cylinders to replicate fibre addition as red mud is discharged to the TSF. Readings of the level of the sediment were taken at varying points in time at 0.5 min, 1 min, 2 min, 4 min, 8 min, 15 min, 30 min and 60 min, after 1 hr, 2 hr, 4 hr, 8 hr, 12 hr, 24 hr, 48 hr, and daily until the settling ceased.

Consolidation tests

The consolidation tests were carried out on 75 mm diameter rings following the Australian Standard 1289.6.6.1. The red mud solids concentration was the settled solids percentage of approximately 46 per cent solids. This is to replicate what happens on-site once the supernatant water evaporates. Before that settled solids percentage there is practically no weight the red mud needs to hold because the particles are still under suspension on the surface.

The load cycles used for this study are 2.5 kPa, 5 kPa, 10 kPa and 20 kPa. A 1 per cent addition of fibres by dry mass was introduced into each sample prepared for all three fibres. The red mud samples were left to dry by means of a fan blowing wind at a constant rate onto samples that had an initial solids percentage of 25 per cent. The drying stage was monitored for the samples to reach the settled solids percentage. The intention of this is to prevent the formation of bonded conglomerates that could change the tests results. The samples were also left on their process water to avoid that no red mud properties other than those found on the tailings pond were added.

Direct shear tests

The direct shear tests were carried out in accordance to the Australian Standard 1289.6.2.2. The sample dimensions used are 60 mm × 60 mm. A series of direct shear tests were done on red mud samples with and without fibres at 1 per cent addition. An initial moisture content of 85 per cent was used to prepare the samples. This is approximately 1.75 times the liquid limit of the red mud sample shown in the following section. Sub sequential tests were done on samples prepared at 1.5 times the liquid limit MC of the red mud. The normal loads used for the direct shear tests were 5 kPa, 10 kPa and 20 kPa. The shearing rate used for these tests is of 0.01 mm/min

TEST RESULTS

Red mud characterisation

The particle size distribution curve in Figure 3 shows that red mud is predominantly a silt like size material with coarser particles similar to fine sand fractions. The large difference between the wet sieving method and the dry sieving method is a good representation of the differences between red mud when it's discharged out to the pond and when the same material dries and sticks in larger blocks that gain cohesive strength. The wet method shows the actual particle size, whereas the dry method shows a conglomerated and very strong group of the same silt like particles bonded together.

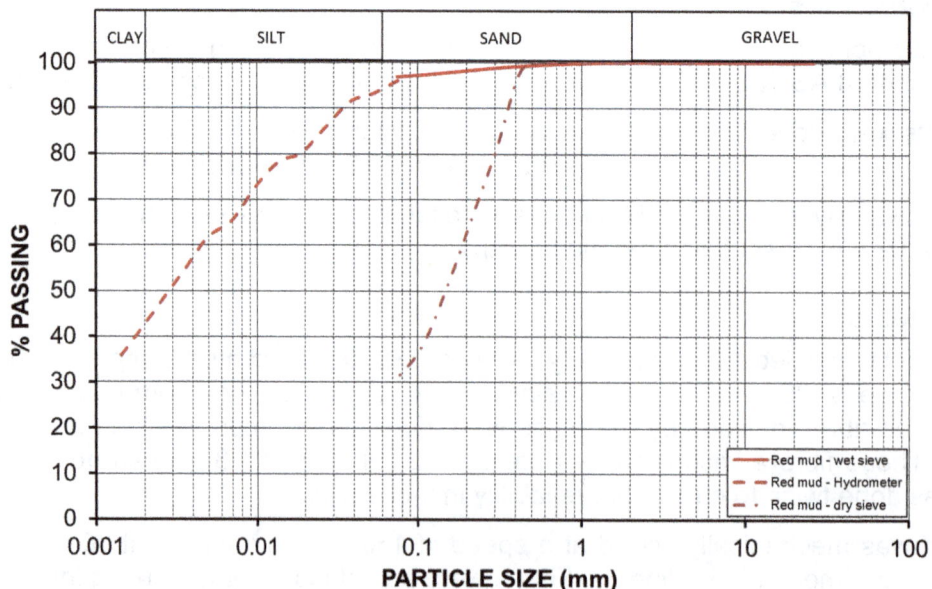

FIG 3 – Particle size distribution of Red mud.

The hydrometer test was carried out with deionised water on a representative sample of dried crushed red mud that passed the 0.075 mm sieve.

Table 2 shows the Atterberg limits and specific gravity of the material. It's worth noting that even though red mud is not a natural soil, the geotechnical characterisation tests also apply to this artificial 'soil' like waste. The specific gravity value of 2.9 gr/cm^3 suggests that red mud is intrinsically heavier than natural soil (Prakash *et al*, 2012) and much heavier than many other tailing types (Hu *et al*, 2017).

The liquid limit value can provide a quick insight into the MC expected for the red mud to start having air entry between particles.

TABLE 2

Red mud characterisation values.

Test type	Test result
Liquid limit – W_L	47.9%
Plastic limit – P_L	33.5%
Index of plasticity – IP	14.4%
Specific gravity (gr/cm^3)	2.9

SEM images

The shape of the cross-section for each plastic fibre can be found on Figure 4. These can be compared to the particle size found on the SEM done on red mud. The scale of 100 µm is shown at the bottom right of each image for comparison purposes. It is evident from the SEM on the red mud shown below and from the PSD shown on Figure 3, that the natural red mud particles are much finer than 100 µm, confirming that the red mud is predominantly fine in size despite its conglomerates found once it dries.

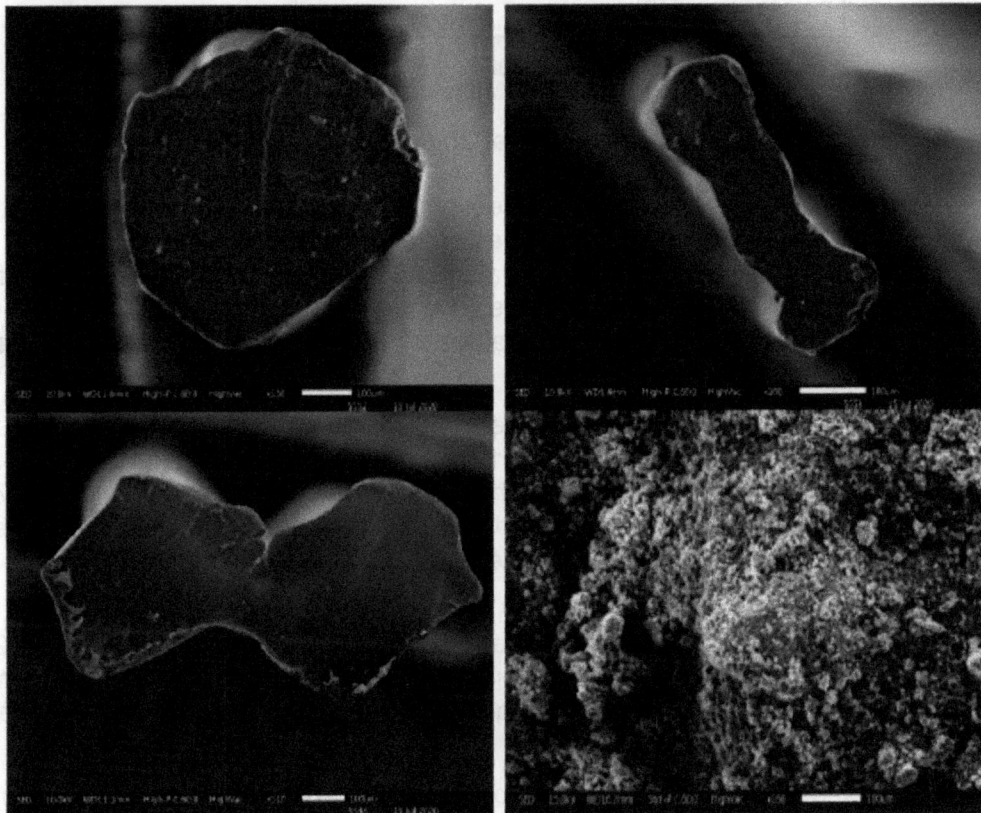

FIG 4 – SEM for red fibres, black fibres, white fibres and red mud.

Fibre differences

The cross-sections of the fibres above suggest that the different geometrical shapes of the fibres may play a role when they start interacting with the red mud as it settles. Both the particle size distribution and the SEM images of the red mud show that red mud is orders of magnitude smaller than the sizes of the plastic fibres.

How red mud sticks to the fibres during settling and consolidation will be crucial. Properties such as void ratio and dry density may determine if fibres enhance tailings geotechnical behaviour.

Settlement test results

The addition of fibres can be assumed to increase the solids of the red mud thus not enhancing the tailings settlement. However, the tests carried out show that the filament like shape of the texture of the fibres may allow for red mud to settle further. Figure 5 shows the settlement of all tests done on red mud with the addition of fibres in comparison to red mud alone.

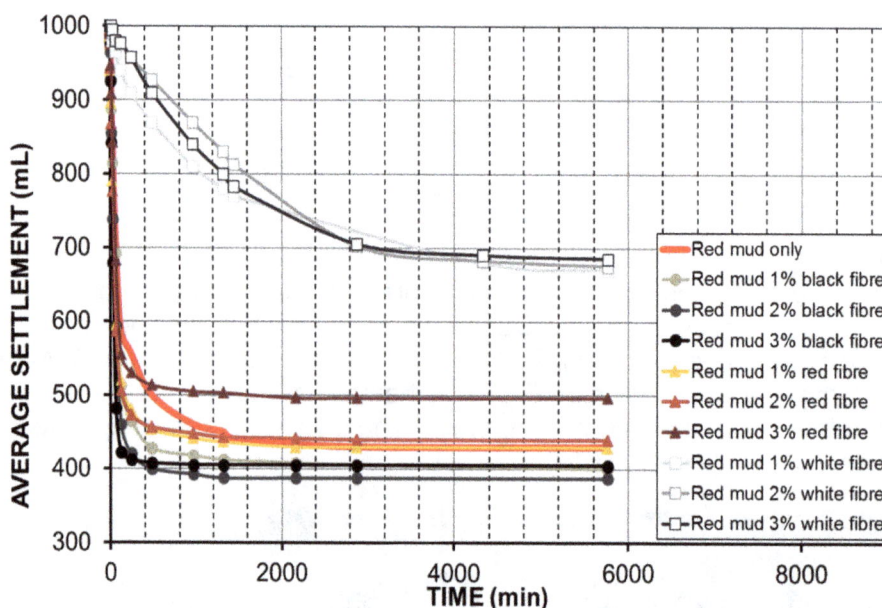

FIG 5 – Settlement of red mud at 25 per cent solids with fibres.

After testing, each sample was extracted in two sections for measuring the moisture contents. The upper sections for all samples tested were the wetter with an average moisture content of 155 per cent. The lower section had an average value of 95 per cent for red mud with no fibres and an average of 87 per cent with red and black fibres. The red mud samples with the addition of white fibres were overall much wetter with average values of 200 per cent on the top section and of 175 per cent in average for the bottom sections.

The settlement tests carried out for this study were left for 18 days. Almost all samples reached constant settlement except for the white fibres that showed very slow settlement over time. The addition of 2 per cent of black fibre on red mud shows the highest dry density values of 750 kg/m³ after settling. The red mud sample with the addition of 1 per cent in black fibre also shows an increase of dry density compared to red mud alone.

The red fibre addition shows no increase in the dry density of the red mud. Despite adding 1 per cent of the dry mass of red mud, the red fibres don't interfere with the red mud settlement. The 2 per cent addition of red fibres slightly decrease the dry density of the red mud from 663 kg/m³ which is the settled dry density of the material alone, to 649 kg/m³ settled dry density with 2 per cent addition of red fibre.

The white fibre altogether works to the detriment of the settlement of red mud adding volume to the material as it settles. Although Figure 6 shows that at some point in time, the dry density could reach a value similar to the settled red mud, it will be very impractical to attempt to keep adding white fibres in order to gain some settlement, or to wait longer for red mud to reach constant settlement.

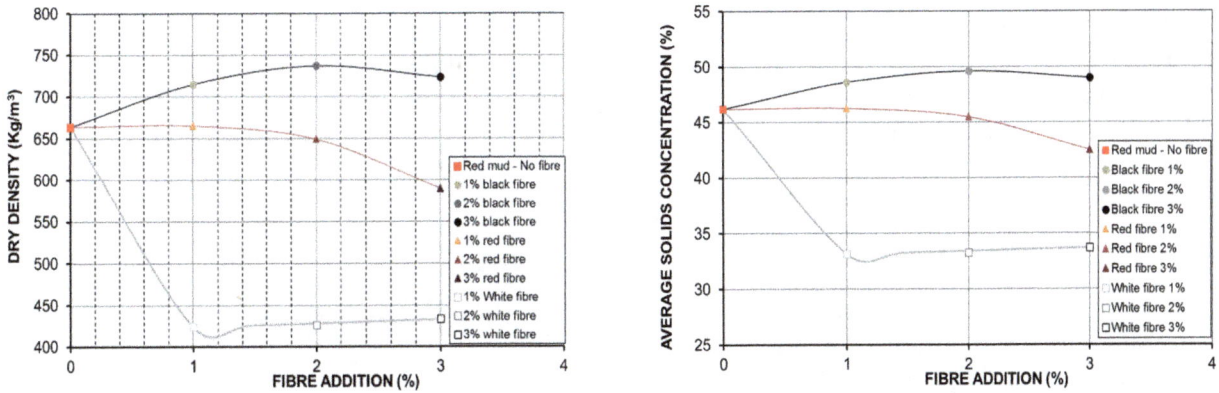

FIG 6 – Settled dry density and solids percentage of red mud.

Consolidation test results

The consolidation of the red mud with and without fibre addition can be analysed looking at different properties such as the increase of the change of settlement, increase of dry density or decrease of the void ratio. Figures 7 and 8 shows these properties as the red mud is subject to the proposed four cycles.

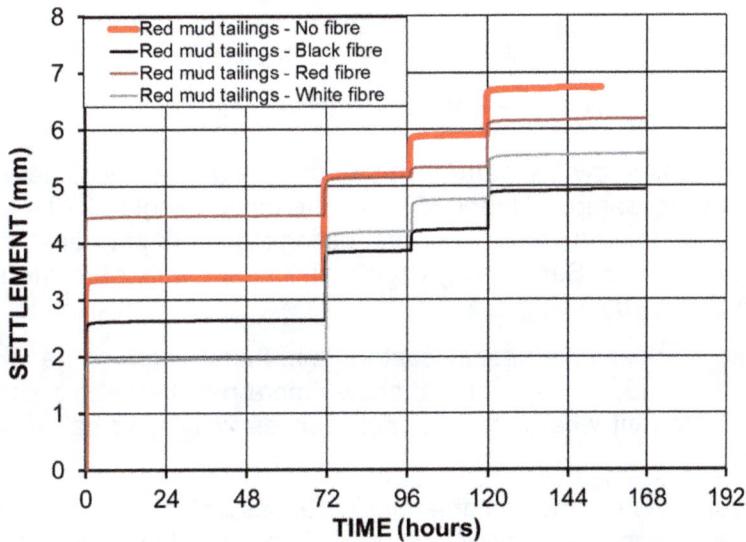

FIG 7 – Settlement results of consolidation test.

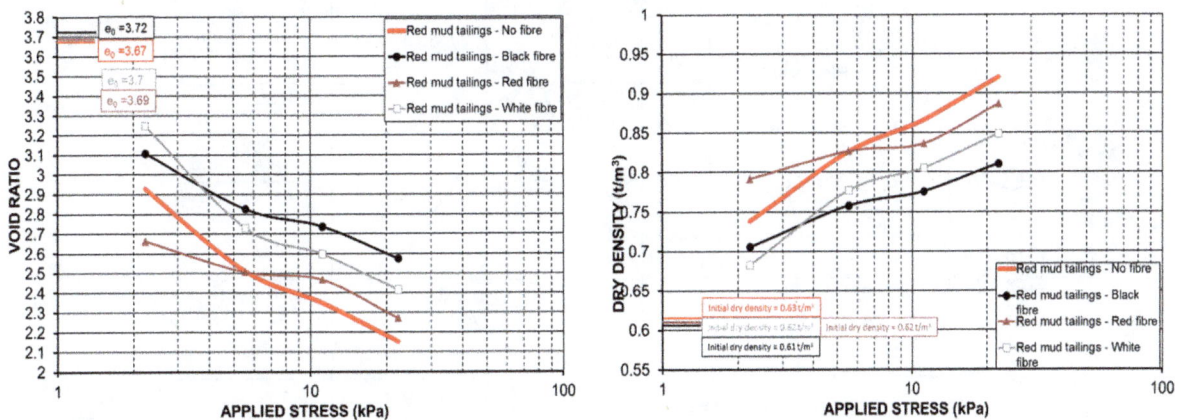

FIG 8 – Void ratio and dry density changes for consolidation test.

Overall, none of the fibre added red mud tests reach a higher dry density than the red mud alone. The slope may suggest that under higher normal stresses, the void ratios of the samples with fibres could eventually be greater than the red mud without fibre addition.

Direct shear test results

The peak shear test results on red mud samples are shown in Figure 9. Tests were varied by: (i) adding different types fibre or without fibre; and (ii) changing the initial gravimetric water content (either 75 per cent or 85 per cent).

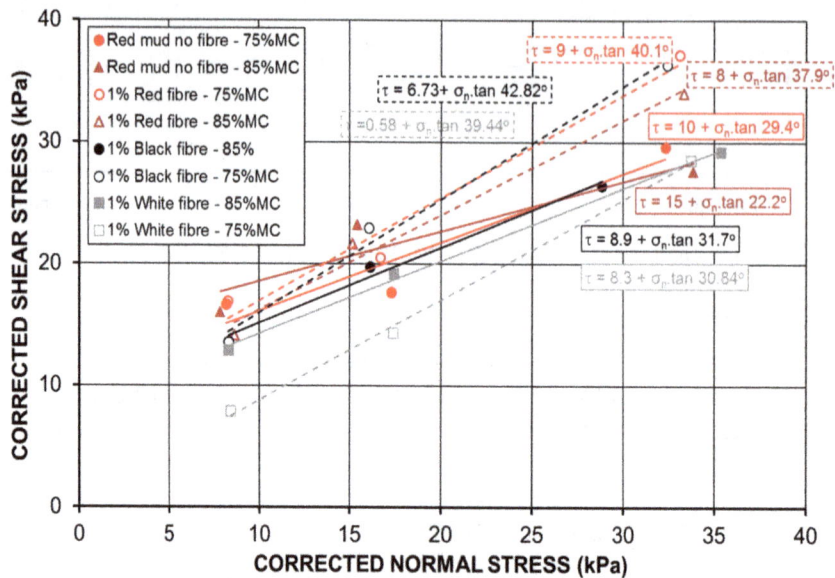

FIG 9 – Peak shear stress results.

The red coloured equations show the natural values of fibre as they desiccate without fibres. This are the results of the control sample. Under 85%MC, the friction angle of 29.4 shown on the control sample, is only higher than the red fibre added suggesting that at 85 per cent the red fibre is not able to bring strength to the sample. Same thing occurs at 75 per cent where the same red fibres are blow than the control sample by 2 degrees.

Both the white and the black sample directly contribute to the strength of the red mud from 85%MC with black fibres being the added material that shows more resistance to a shearing force. Similar situation occurs at 75 per cent where the material reaches values typical of sand material with a friction angle of 42.82.

Figure 10 shows the rate at which the friction angle increases as the moisture content of the red mud decreases. All fibres have a friction angle value larger than the value for red mud alone.

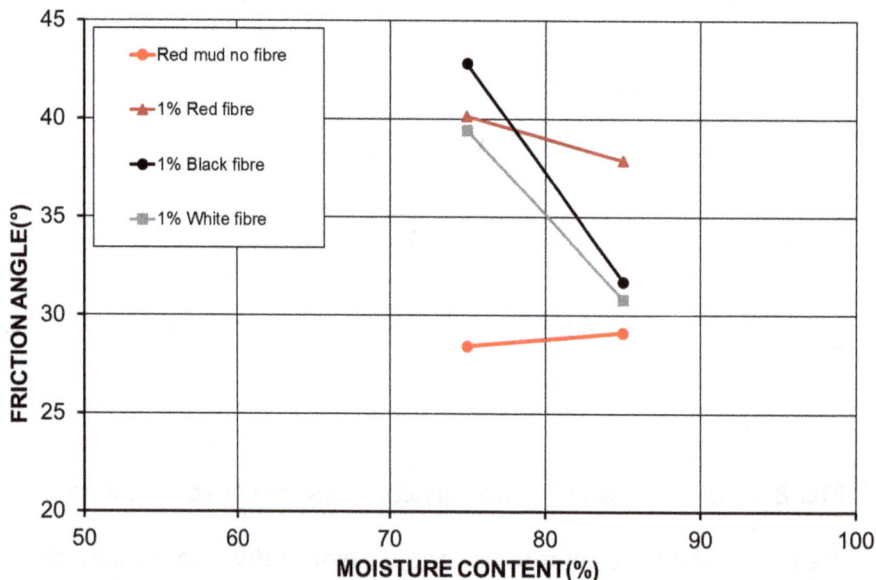

FIG 10 – Friction angle values with different moisture contents.

CONCLUSIONS

This study applies settling, consolidation and shear tests to investigate how the addition of plastic fibres can influence the geotechnical properties of red mud.

From the settlement test results, it is found that the increase of fibre content by mass does not lead to monotonic rise of dry density or reduction in void ratio. Rather, an optimum percentage of fibre addition were found to be 1 per cent for red and black fibres. The shape, texture and composition of the white fibres is not conducive to aid the settlement of the red mud.

Consolidation test results overall show that while the red mud is at a saturated state, fibres do not help reduce void ratio or enhance dry density.

The direct shear tests show that white and black fibres enhance the strength of the red mud. The friction angle increases as the fibre added red mud desiccates, because the fibre help enhance the bond of red mud.

The friction angles calculated from the peak shearing stresses show that fibres greatly enhance the strength of the red mud from a very wet loose state and the stress gain increases rapidly as the red mud desiccates.

The test shown in this study focus only on the fully saturated red mud. The results for consolidation and direct shear tests demonstrate that fibres increase the strength of the tailings as they desiccate. However, void ratio and dry density of the red mud with fibres may not carry significant changes during full saturation. Further studies should be carried out to analyse the contributions fibres can provide to red mud at a larger range of moisture contents, a process occurred in the tailings storage facility.

ACKNOWLEDGEMENTS

The authors of this paper thank Gillis Broinowski, Alex Lester, Stuart Neilson and all staff from TEXO ReoCo who provided the fibres for this study.

REFERENCES

Aubertin, M and McKenna, G, 2016. *Tailings Disposal Challenges and Prospects for Oil Sands Mining Operations* (American Society of Civil Engineers).

Consoli, N, Montardo, J, Donato, M and Prietto, P, 2004. Effect of material properties on the behaviour of sand-cement-fibre composites, *Ground Improvement,* 8:77–90.

Gálvez, E D, Cruz, R, Robles, P A and Cisternas, L A, 2014. Optimization of dewatering systems for mineral processing, *Minerals Engineering,* 63:110–117.

Hu, L, Wu, H, Zhang, L, Zhang, P and Wen, Q, 2017. Geotechnical Properties of Mine Tailings, *Journal of Materials in Civil Engineering,* 29:04016220.

Australia's Aluminium Council, 2022. Trade and competitiveness.

Islam, S, Williams, D J, Llano-Serna, M and Zhang, C, 2020. Settling, consolidation and shear strength behaviour of coal tailings slurry, *International Journal of Mining Science and Technology,* 30:849–857.

Prakash, K, Sridharan, A, Thejas, H and Swaroop, H, 2012. A simplified approach of determining the specific gravity of soil solids, *Geotechnical and Geological Engineering,* 30:1063–1067.

Reid, D, Fanni, R and Fourie, A B, 2022. Slurry deposition preparation method for tailings characterisation–history, debates, techniques and benefits. Tailings Conference 2022.

Shokouhi, A and Williams, D J, 2015. Settling and consolidation behaviour of coal tailings slurry under continuous loading, in *Proceedings of Tailings and Mine Waste 2015*, pp 263–274 (University of British Columbia).

Zhang C, Williams, D J, Lei, X, 2017. Design, manufacture and calibration of sensors for monitoring settling, consolidation and desiccation of tailings, in Proceedings of Tailings and Mine Waste 2017, pp (University of Alberta).

Zhang, C W, Zhang, C, Lei, X, Zhu, Y, Quintero, S, Williams, D J and O'Neill, M, 2019. Assessing salt migration from compacted bauxite residual into overlying cover using an instrumented column.

Assessing seepage during settling of iron tailings

S Quintero Olaya[1], D J Williams[2], C Zhang[3], A Tran[4], H Pan[5] and R Fan[6]

1. Senior Research Technologist, School of Civil Engineering, Brisbane Qld 4072.
   Email: s.quintero@uq.edu.au
2. FAusIMM, Professor of Geotechnical Engineering, Director, Geotechnical Engineering Centre,
   School of Civil Engineering, Brisbane Qld 4072. Email: d.williams@uq.edu.au
3. Research Fellow, School of Civil Engineering, Brisbane Qld 4072.
   Email: chenming.zhang@uq.edu.au
4. Bachelor of Civil and Environmental Engineering, School of Civil Engineering, Brisbane Qld
   4072. Email: amy.tran@uqconnect.edu.au
5. Master of Civil Engineering, School of Civil Engineering, Brisbane Qld 4072.
   Email: h.pan1@uq.net.au
6. CSIRO Mineral Resources, Urrbrae SA 5064. Email: rong.fan@csiro.au

ABSTRACT

Slurry tailings pumped to tailings storage facilities usually consist of 75 per cent of process water. The process water, usually hazardous if released to the ambient environment, will undergo one of the following processes: (1) evaporation; (2) forming overland flow and later merging in a decant pond; (3) mixing with settled tailings as porewater; and (4) infiltrating to the underlying tailings layer by gravity drainage and eventually forming seepage at the toe of the tailings dam. The conventional settling test may be able to quantify the ratio of overland flow to porewater storage, while incapable of estimating the gravity drainage as it has a sealed base. A purpose-made settling column allowing bottom drainage was constructed to investigate the outgoing process water during and after settling test. An outlet was designed at the bottom of the column to allow the drainage of process water while intercepting solids by placing 2-cm-thick settled tailings at the bottom with the particle size the same as the testing tailings samples. The test was started by depositing well-mixed tailings slurries into the column. During the experiment, the bottom drainage, and elevations of water surface and solid surface were monitored. The experiment was terminated when all the decant process water above was drained from the bottom outlet. After the initial settling, fresh water was poured above the settled tailings in the column to simulate rainfall-induced seepage. The outflow rate, EC, pH and total dissolved solids of the bottom drainage were monitored over time to quantify the leaching of chemicals/solids from the settled tailings. This paper reports the testing results of sandy and slime iron tailings.

INTRODUCTION

Acid and metalliferous drainage (AMD) is a chemical process that takes place when sulfide minerals start to oxidase. This natural oxidation then reacts to base minerals found on specific rocks which are exposed to transporting agents such as wind and water. The process itself found in the natural may not be catastrophic to the environment. However, when a nature altering event like mining production becomes part of this specific ecosystem, it increases the exposure of such sulfides to the same transporting agents of wind and water at a larger scale. This drained water that comes out of contact with sulfides starts to change its pH to more lower value thus decreasing the water quality of the ecosystem (INAP, 2017).

Measures to prevent acid drainage to form and contaminate water streams has been taken by mining authorities in Australia and around the globe (Department of Industry, Science, Energy and Resources, 2016). However, acid drainage continues to be a challenge to contain for current and abandoned mine sites (Wang *et al*, 2020; Moreno-González *et al*, 2022). It is imperative that academia continues to find ways to control AMD.

Tailings is one of the main sources of the occurrence of AMD and the material this study will focus on. Its chemical composition, very low drying rate and large volumes of water contained make it a major risk that needs to be controlled. Tailings properties such as chemical composition, pH, EC and permeability are critical to identify potential mine drainage and it is ideal to monitored this and other geotechnical properties constantly during operation.

Iron ore tailings are known to contain large amount of metal oxides such as Goethite and Hematite. Studies have shown that hematite can enhance the oxidation of pyrite, which is the most abundant sulfide mineral in nature and the main trigger of AMD (Tabelin *et al*, 2020). Iron ore is one of the fastest growing commodities worldwide with a forecasted increase of 2.7 per cent in production for the coming three years in opposition to the 1.3 per cent decrease occurred at the end of the 2010s decade. Iron tailings are also involved in the tailings dam failures of the Feijao dam in 2019 and the Fundao dam failure in 2015. Both dam failures took place in Brazil with devastating consequences to their civilisation and environment (Morgenstern, Viotti and Watts, 2016; Robertson, Williams and Wilson, 2019).

EXPERIMENTAL METHODOLOGY

Tailings used

Iron tailings were chosen for this study due to their large amount of oxides that can lead to AMD when in contact with other minerals such as pyrite. These tailings are also generally neutral in their pH in comparison to some that are neutralised with sea water, increasing their pH value.

The tailings are sourced from a mine site in Brazil with to sub-categories for tailings that are called sand tailings and slime tailings. Figures 1 and 2 show representative 'as received' samples. Both categories are produced from the same iron ore production. The waste that comes out of this specific ore extraction has been split into the two mentioned tailings. These tailings have different properties and physical aspect despite coming from the same ore.

FIG 1 – Sand tailings.

FIG 2 – Slime tailing.

Process water

The water used in this study was an acidic water with a pH of 1.8. This water was sourced from a tailings storage facility that predominantly deposits tailings of zinc and lead production. The very low pH was the main factor to use this water.

Filter aggregate

Several layers of commercially available sand and gravel of varying sizes was used as aggregate for the filter layers placed on the large columns. The material's representative diameter chosen was based on the PSD of the tailings with a nominal size of 0.02 mm.

Table 1 shows the D50 of each layer chosen. Layer 1 is the bottom and layer 4 is the top. Each layer thickness was 20 mm:

TABLE 1

Filter aggregate nominal sizes.

Layer number	Nominal size (mm)
1	10
2	5
3	2.36 > medium sand > 1.18
4	1.18 > fine sand > 0.6

Tests methodology

Characterisation tests of iron ore tailings

Both types of iron ore tailings used, called sand tailings and slime tailings were tested to find the particle size distribution and Atterberg limits. Particle size distribution tests (PSD) was carried out following the Australian Standard AS 1289.3.6.1. Hydrometer test was also done according to AS 1289.3.6.3.

Atterberg tests were done following the AS 1289.3.1.1, AS 1289.3.9.1, 1289.3.2.1 and 1289.3.3.1 to calculate the liquid limit using the casagrande and the cone penetration methods, plastic limits and index of plasticity respectively.

Settlement tests were carried out on samples at 25 per cent and at 40 per cent of solids percentage for both types of tailings.

X-ray diffraction tests

Four iron ore tailings samples were analysed using the quantitative mineralogy method of the bulk samples by X-ray powder diffraction (XRD). Two of each tailings type described in the above section. One sub sample per type of tailings was left as received with a low moisture content of approximately 12%MC for sand tailings and 21 per cent for slime tailings. One subsample per type was left soaking under the process water of a pH of 1.8. The pH of the paste for each type of tailings was recorded as 2.3

Approximately 1.5 g of the materials were micronised for 10 minutes with ~10 mL of ethanol and dried in an oven set to 60°C. Qualitative analysis was performed on the XRD data using in-house XPLOT and HighScore Plus (from PANalytical) search/match software. Quantitative analysis was performed on the XRD data using the TOPAS version 6 Rietveld analysis software from Bruker AXS. Amorphous content was determined using the internal standard method.

Consolidation tests

The consolidation tests were carried out on 75 mm diameter rings following the Australian Standard 1289.6.6.1. The tailings were prepared at the liquid limit MC. Two sets of samples were prepared per type of tailings. One set with process water on slime and sand tailings and a second set with deionised water on both tailings types. The load cycles used for this study are 2.5 kPa, 7.5 kPa, 15 kPa, 30 kPa, 60 kPa and 120 kPa. The test results will focus on void ratio, dry density and permeability changes.

Large Column leach tests

Two sets of large columns with a valve at the bottom were used to carry out the leach tests for sand and slime tailings. The two separate sets ran two columns each with a total of four large column tests. Both columns had equal layers of filter aggregate which were described in the heading above titled 'Filter aggregate'. A fine steel mesh wire with a nominal aperture size of 1 mm was placed at the bottom before the four layers of filter material were placed under manual compaction. Both rounds had two columns per round. Each column was filled with tailings mechanically mixed with process water, to reach a solids percentage of 25 per cent. This is to replicate a layer deposition at the storage facility near a drainage outlet. The first round was done on sand tailings (labelled as A-1 and A-2) and the second round was done on slime tailings (labelled as B-1 and B-2). Figure 3 shows the set-up of the two columns for one of the rounds.

FIG 3 – Leachate column set-up.

Each pair of columns had subsequent cycles of water flushed after the tailings settled and the water was drained through the drain valve at the bottom.

These cycles were done as follows:

- Sand tailings were deposited at 25 per cent solids. Settling time, pH and EC were monitored as the tailings settle. These cycles labelled A-1 and A-2.

- One flushing cycle containing process water was poured into column A-1. This is cycle B-1.

- Three flushing cycles with deionised water were poured into column A-2. These cycles are labelled as B-2, B-3 and B-4.

After assessing the cycles on sand tailings, the following cycles were determined:

- Slime tailings were deposited at 25 per cent solids on the two columns. Settling time, pH and EC were monitored as the tailings settle. These cycles labelled C-1 and C-2.

- One cycle containing deionised water was poured into each column (C-1 and C-2). These cycles are called D-1 and D-2 respectively.

The process water flushing cycles were done to replicate what a tailings layer will receive during operation. The deionised water was used to replicate a rainfall period that will fall over the tailings. pH and EC were monitored during these cycles at the same intervals as those used on the first leaching cycles per column. Further, the saturated hydraulic conductivity constant taken from the Darcy's law equation was calculated during flushing cycles with the falling head method.

TEST RESULTS

Characterisation test results

Figure 4 shows the PSD of both materials. The results show a clear difference in size of both types of tailings despite coming from the same iron ore production. The slime has predominantly a silt like size particle with some finer sizes and virtually no sand. The sand tailings are predominantly fine sand like in particle sizes with less than 50 per cent of its particle size made of silt size material.

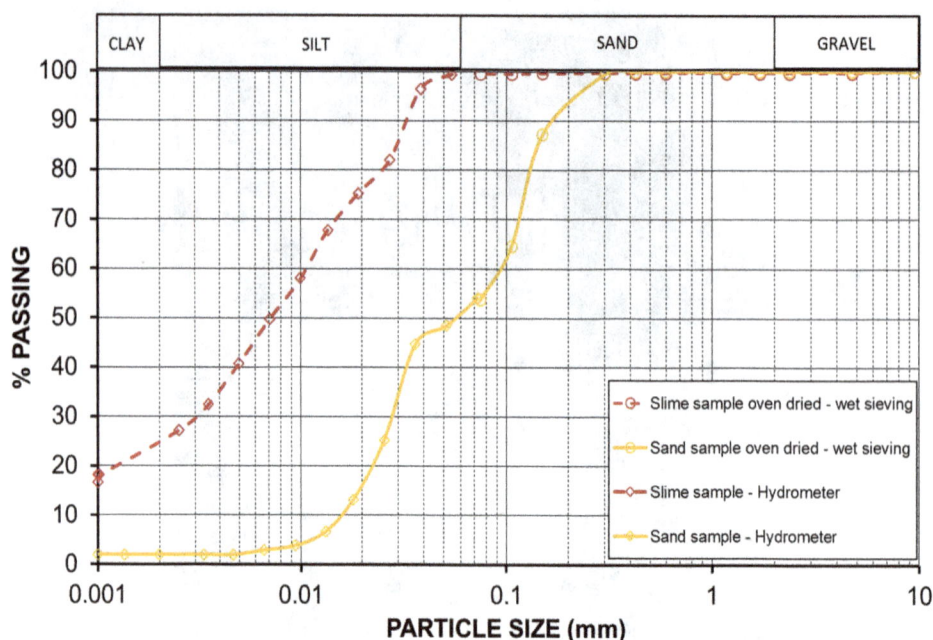

FIG 4 – Particle size distributions of tailings.

The Atterbergs limits can be shown in Table 2. Both samples have a rather similar liquid limit despite their differences in particle size. The sand tailings didn't present reliable results for the casagrande method. Plastic limits are as expected with slimes value higher than sands.

TABLE 2

Characterisation of sand and slime tailings.

Test type	Test result	
	Sand	Slime
Liquid limit Casagrande – W_L	N/A	28.73%
Liquid limit Cone – W_{cL}	25.91%	27.39
Plastic limit – P_L	NP	15.46%
Index of plasticity – IP	NP	13.27%
Specific gravity (gr/cm³)	2.85	3.89

The settlement tests results are found on Figure 5. The settlement on the sand is almost immediate which is expected due to its particle size. Within minutes already reaches its settled solids percentage of 75 per cent, equivalent to a settled dry density of 1.4 T/m³. The settlement time for slimes is over 24 hours to reach a settled solids percentage of approximately 61 per cent and a settled dry density of 1.1 T/m³. Both samples started at 25 per cent solids percentage. This is equivalent to 0.3 T/m³ for the initial dry density on a standard bottle of 1 L. This implies how rapidly both tailings gain dry density while settling, although the slimes are slower than sand and reach with only 72 per cent of the dry density reached by the sand tailings 24 hours later.

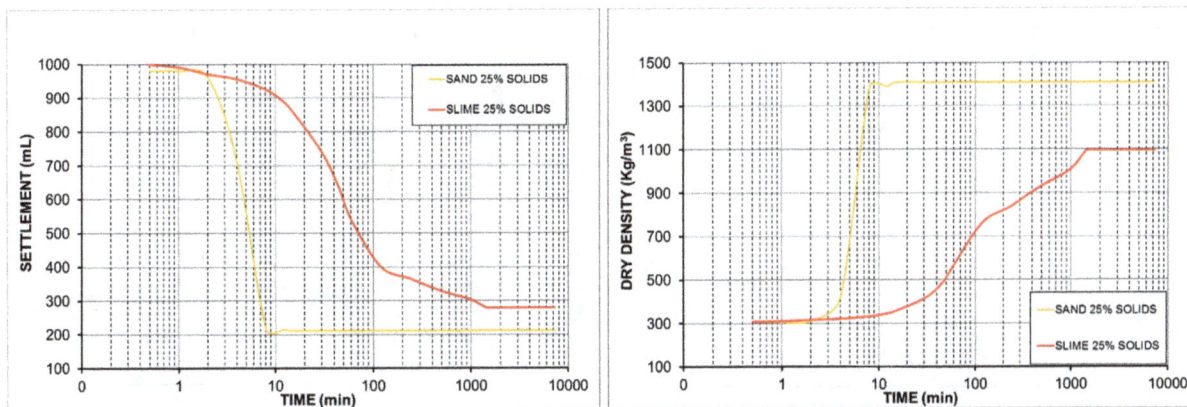

FIG 5 – Settlement plots of slime and sand tailings.

The final moisture content of the sand tailings is very close to the liquid limit moisture content with approximately 29%MC. The slime did retain some MC after extracting the supernatant water with a settled value of 55 per cent.

X-ray diffraction test results

The test results for the XRD tests is shown in Table 3. It can be seen from this table that the slimes are predominantly Goethite and Hematite, both oxides and predominant minerals from Iron ore which is obvious since that very same mineral is the source of these tailings. However, sand tailings are predominantly composed by Quartz which is still an oxide but from silicon. They have very little iron based minerals. Overall, the low pH of the process water doesn't alter the chemical composition of the tailings.

TABLE 3

XRD results on iron tailings.

Sample ID	Kaolinite	Quartz	Goethite	Hematite	Magnetite	Brushite	Amorphous
Slimes	4	13	33	35	1		14
Sands	1	77	3	10			9
Acid soaked Slimes	8	13	32	34	<1	<1	12
Acid soaked Sand	<1	80	2	7			11

Consolidation tests results

The consolidation test results show a subsequent phase of the settlement tests when analysing the dry density and void ratio as shown in Figure 6. The sand tailings settled at the liquid limit moisture content and the consolidation test starts at the same value, implying that the supernatant water doesn't need to evaporate for the material to start consolidating even without any load (further layers) on top. The difference of moisture content from the settled moisture content mentioned above in the settlement results and the prepared moisture content for the consolidation test at liquid limits, suggest that the slimes not only need to wait for the supernatant water to evaporate, but also it needs to dry a further 20 per cent to start consolidating. In practice this means that it is ideal to wait for the material to desiccate before depositing further layers of tailings on top, otherwise the wet material will remain soft and saturated with no option to release water.

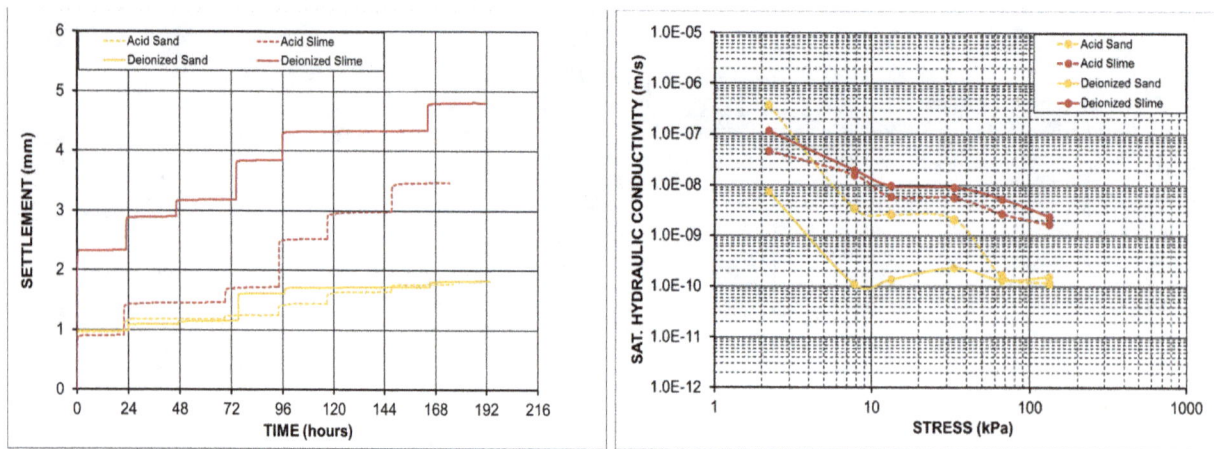

FIG 6 – Settlement results and hydraulic conductivity of consolidation tests.

Large column leachate test results

The settlement for the first cycle of the sand and slimes tailings is recorded in Figure 7. All four column tests had the same process water and the tailings mixed at 25 per cent solids.

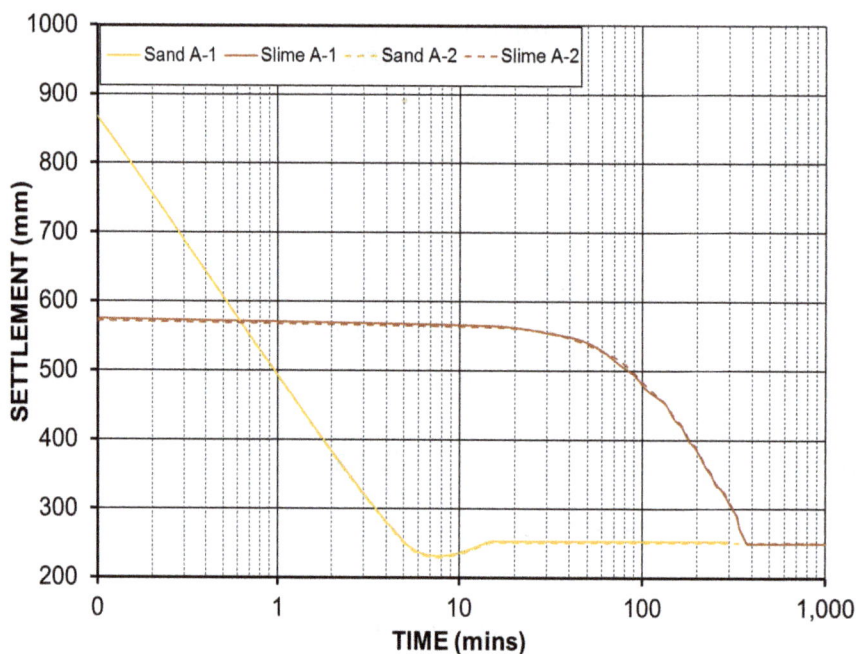

FIG 7 – Large column settlements.

The pH and EC values of all subsequent flushing cycles are found on Figure 8.

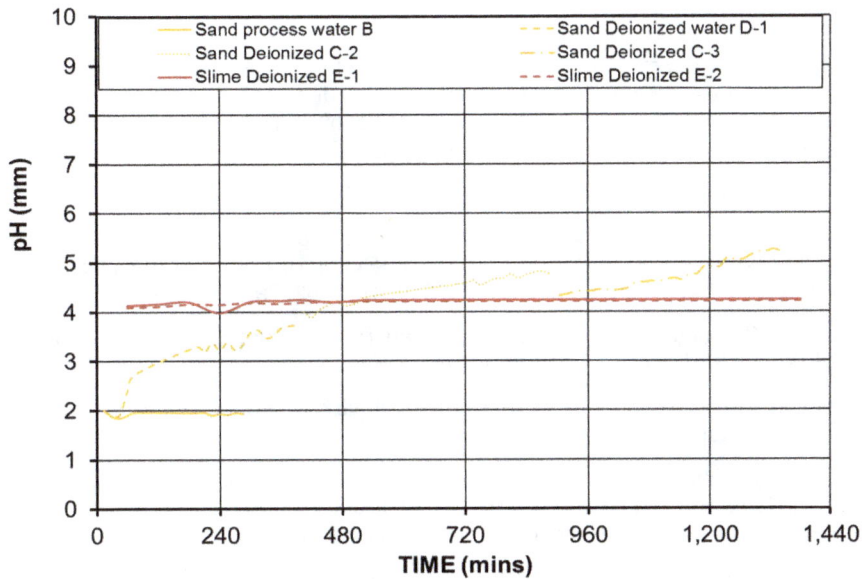

FIG 8 – pH values of all cycles.

Table 4 shows the saturated hydraulic conductivity values calculated from falling head on each flushing cycle.

TABLE 4

Hydraulic gradient of each flushing cycle.

Flush cycle	Sat. hydraulic conductivity (m/s)
Sand B	1.07 E-0.5
Sand C-1	1.05 E-0.5
Slimes E-1	7.5 E-0.8
Slimes E-2	8.3 E-0.8

CONCLUSIONS

This study was focused on identifying the effects of acid water in the chemical composition, permeability, pH and electrical conductivity of Iron ore tailings as these ones are subject to acid and deionised water cycles. The following can be concluded from the tests carried out.

The chemical composition of the sand and slime iron ore tailings is not affected by the acid water cycles. The X-ray diffraction tests show virtually no changes on the minerals quantity of each tailings.

The large column tests show that the pH still remains acid in both tailings after rainfall periods take place. Both the large column flushes and the consolidation tests show very little effect on the permeability of each material.

The filter material was shown to be effective when retaining the tailings and allowing the water to seep through to a drainage point. The more rainfall periods, the more neutral the pH becomes.

The permeability of the large column on the sand suggests these tailings are very porous and allow water to pass with little resistance. When this material densifies, the permeability reduces to three orders of magnitude making it difficult for the water to seep through the sand tailings.

ACKNOWLEDGEMENTS

The authors want to acknowledge and appreciate the manufacturing of the large columns done at the Advanced prototyping workshop from the UQ Innovate group.

REFERENCES

Department of Industry, Science, Energy and Resources, 2016. Preventing Acid and Metalliferous Drainage – Leading Practice Sustainable Development Program for the Mining Industry, Australian Government, Department of Industry, Science, Energy and Resources [Online]. Available: https://www.industry.gov.au/sites/default/files/2019–04/lpsdp-preventing-acid-and-metalliferous-drainage-handbook-english.pdf [Accessed September 2022].

International Network for Acid Prevention (INAP), 2017. *The Global Acid Rock Drainage Guide. The GARD Guide.* INAP.

Moreno-González, R, Macías, F, Olías, M and Cánovas, C R, 2022. Temporal evolution of acid mine drainage (AMD) leachates from the abandoned tharsis mine (Iberian Pyrite Belt, Spain), *Science Direct*. Available from: <https://www.sciencedirect.com/science/article/pii/S026974912102279X> [Accessed February 15].

Morgenstern, N R V, Viotti, C B and Watts, B D, 2016. Fundão Tailings Dam Review Panel. Report on the Immediate Causes of the Failure of the Fundão Dam, internal report.

Robertson, P K, Williams, D J and Wilson, G W, 2019. Report of the Expert Panel on the Technical Causes of the Failure of Feijão Dam I, internal report.

Tabelin, C B, Corpuz, R D, Igarashi, T, Villacorte-Tabelin, M, Alorro, R D, Yoo, K, Raval, S, Ito, M and Hiroyoshi, N, 2020. Acid mine drainage formation and arsenic mobility under strongly acidic conditions: Importance of soluble phases, iron oxyhydroxides/oxides and nature of oxidation layer on pyrite, *Journal of Hazardous Materials*, 399:122844.

Wang, Z, Xu, Y, Zhang, Z and Zhang, Y, 2020. Review: Acid Mine Drainage (AMD) in Abandoned Coal Mines of Shanxi, China, *Water*, 13:8.

On the accuracy of inferring residual undrained shear strengths from CPTu friction sleeve resistance

K Smith[1], R Fanni[2] and D Reid[3]

1. Senior Tailings Engineer, WSP, Perth WA 6000. Email: kyle.smith2@wsp.com
2. Principal Tailings Engineer, WSP, Perth WA 6000. Email: riccardo.fanni@wsp.com
3. Research Fellow, The University of Western Australia, Crawley WA 6009.
   Email: david.reid@uwa.edu.au

ABSTRACT

Accurate estimation of the residual (or liquefied) undrained shear strength of tailings is crucial to the stability assessment of tailings storage facilities (TSFs). Several approaches are available for the estimation of the residual undrained shear strength of tailings, with these approaches generally including cone penetration testing with pore pressure measurement (CPTu) and/or laboratory testing on tailings samples retrieved from within the TSF. One such approach, proposed in the literature and adopted in tailings engineering practice, assumes that the residual undrained shear strength is equal to the CPTu friction sleeve resistance, measured during standard rate CPTu with a standard CPTu cone, henceforth referred as the 'friction sleeve' method.

This paper presents a review of the literature relating to inferring the residual undrained shear strength of soils from CPTu friction sleeve resistance, including the theoretical considerations forming the basis for the method as well as the supporting empirical evidence. A case study from a TSF located in the Western Australian Goldfields-Esperance region is also presented by the authors. The literature review and case study indicate that the friction sleeve may be able to provide an estimate of the residual undrained shear strength of fine-grained soils and tailings, however, the accuracy of this estimate may not be sufficient for TSF design purposes.

The friction sleeve method is also compared to other CPTu based residual undrained shear strength estimation techniques, which primarily utilise CPTu cone resistance. This paper highlights that, for specific applications, the friction sleeve method provides similar estimates to accepted industry standard practice CPTu based approaches, eg Jefferies and Been (2016) state parameter-based semi-empirical method. Lastly, some considerations on the uniqueness of residual undrained shear strengths are presented, which should be acknowledged when undertaking stability assessments of TSFs.

INTRODUCTION

Recent catastrophic flow liquefaction failures of several large tailings storage facilities (TSFs) serve to highlight the importance of the accurate estimation of the residual (or liquefied) undrained shear strength of tailings. Several approaches are available for the estimation of the residual undrained shear strength of tailings ($s_{u,r}$), with these approaches generally including cone penetration testing with pore pressure measurement (CPTu) and/or laboratory testing on tailings samples retrieved from within the TSF (Jefferies *et al*, 2019; Robertson *et al*, 2019). One such approach, proposed in the literature and adopted in tailings engineering practice, assumes that $s_{u,r}$ is equal to the CPTu friction sleeve resistance (f_s), measured during standard rate CPTu with a standard CPTu cone, henceforth referred as the 'friction sleeve' method.

Modern standard (35.7 mm diameter, 10 cm^2 end area) CPTu cones include an equal end area friction sleeve of 134 mm length (150 cm^2 surface area) located immediately behind the CPTu tip and filter element (for u_2 pore pressure measurement) (ASTM D5778-20, 2020). It is widely accepted that the peak undrained shear strength ($s_{u,p}$) of saturated, clayey or fine-grained soils, for which CPTu penetration is essentially undrained, can be estimated by dividing the net cone resistance ($q_t - \sigma_v$, where q_t is the corrected cone resistance and σ_v is the *in situ* vertical total stress prior to CPTu) by an empirical cone factor (N_{kt}) which varies with soil sensitivity amongst other factors (Shuttle and Jefferies, 2016; Monforte *et al*, 2021). There is less consensus in the literature as to what f_s is a measure of, however it is often stated that f_s represents the approximate $s_{u,r}$ of a soil as both occur at large strains (Robertson, 2021; Jefferies and Been, 2016) and as in fine grained soils CPTu penetration is essentially undrained (Robertson, 2021).

The concept that f_s represents the approximate $s_{u,r}$ of a soil is examined in this paper, through a review of the literature including case studies and theoretical considerations. A case study from a TSF within the Western Australian Goldfields-Esperance region is presented to assess application of the friction sleeve method to tailings. Lastly, some considerations on the uniqueness of residual undrained shear strengths are presented, which should be acknowledged when undertaking stability assessments of TSFs.

LITERATURE REVIEW

Case studies

A friction sleeve was first implemented on a cone penetration test (CPT) cone by Begemann (1953), who introduced the 'friction jacket cone' which included a mechanical friction sleeve that was advanced incrementally following advancement of the CPT tip, which was also advanced incrementally. The test provided both cone resistance and the sum of cone and sleeve resistance at discrete intervals. Begemann (1964) found that the friction sleeve resistance was similar to the $s_{u,p}$ of clayey soils as measured through *in situ* vane shear testing (VST). It should be noted that the friction sleeve used on the friction jacket cone is located approximately 160 mm behind the top of the CPT tip and is influenced by return flow of soil behind the cone jacket which leads to greater sleeve resistances for Begemann friction jacket cones compared to electric cone CPTu (Meisina *et al*, 2017).

Schmertmann (1978) proposed that f_s approximated the average of the $s_{u,p}$ and $s_{u,r}$ of a soil, and that soil sensitivity ($S_t = s_{u,p}/s_{u,r}$) could be related to the ratio of f_s to cone resistance (q_c) through Equation 1 (where R_f is the CPT friction ratio $f_s/q_c \times 100$ per cent). Equation 1 was reported to be based on data from testing undertaken using Begemann friction jacket cones, although supporting evidence was not provided for the equation or the assertion that f_s approximated the average of the $s_{u,p}$ and $s_{u,r}$ of a soil:

$$S_t = 15/R_f \tag{1}$$

Following the widespread adoption of 'Fugro style' (ie standard) electric CPTu cones, Robertson and Campanella (1983) proposed that f_s from an electric cone CPTu is close to the $s_{u,r}$ of clays, and that S_t could be related to the ratio of friction sleeve resistance to cone resistance through Equation 2 (where FR is the CPT friction ratio $f_s/q_c \times 100$ per cent), although no supporting evidence was provided. Equation 2, was generally consistent with Equation 1 (Schmertmann, 1978), considering that Begemann friction jacket cones measure greater f_s compared to electric cone CPTu (Meisina *et al*, 2017). Robertson and Campanella (1983) state that Equation 2 'provides only a rough estimate' and that the numerator should be 'determined through local experience':

$$S_t = 10/FR \tag{2}$$

Quiros and Young (1986) presented data that demonstrated that f_s measurements in a marine clay showed good agreement with $s_{u,r}$ values obtained from remoulded unconsolidated undrained (UU) triaxial compression tests (see Figure 1). Inspection of the data indicates that f_s measurements were approximately 83 per cent of $s_{u,r}$ from the remoulded UU triaxial compression tests, with a correlation coefficient (r^2) of approximately 0.99. This indicates a strong positive correlation between f_s and $s_{u,r}$, although they were not indicated to be equal. It should be noted that the accuracy of the correlation is likely influenced by the use of remoulded UU triaxial compression tests, which are not commonly utilised in current practice due inherent limitations of the test procedure (Ladd and DeGroot, 2003).

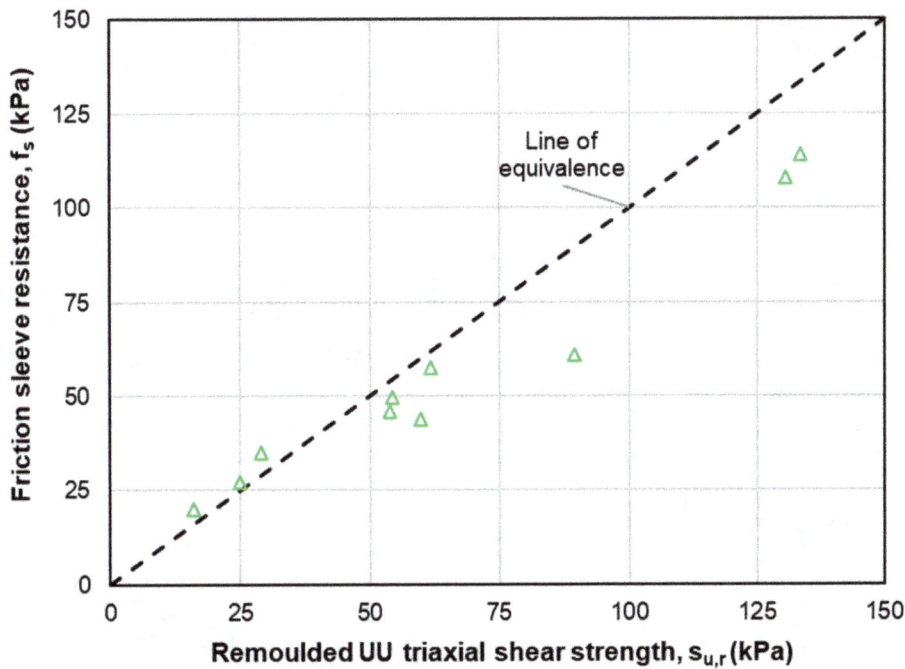

FIG 1 – Comparison of remoulded UU triaxial compression tests and CPTu f_s from Quiros and Young (1986).

Robertson *et al* (1986) presented data that demonstrated that f_s measurements for soils in the Vancouver area were 'generally close' to shear strengths obtained from remoulded VST (Figure 2) and proposed that S_t could be related to the ratio of f_s to q_t through Equation 3 (where FR% is the CPT friction ratio $f_s/q_t \times 100$ per cent). Interpretation of the VST data and corresponding CPTu data, presented in Figure 2, indicates f_s was equal to approximately 112 per cent of $s_{u,r}$ from *in situ* remoulded VST, with a r^2 of approximately 0.84. Robertson *et al* (1986) attributed discrepancies between the two sets of results to inherent inaccuracy when measuring small friction sleeve values, such as those included in the study.

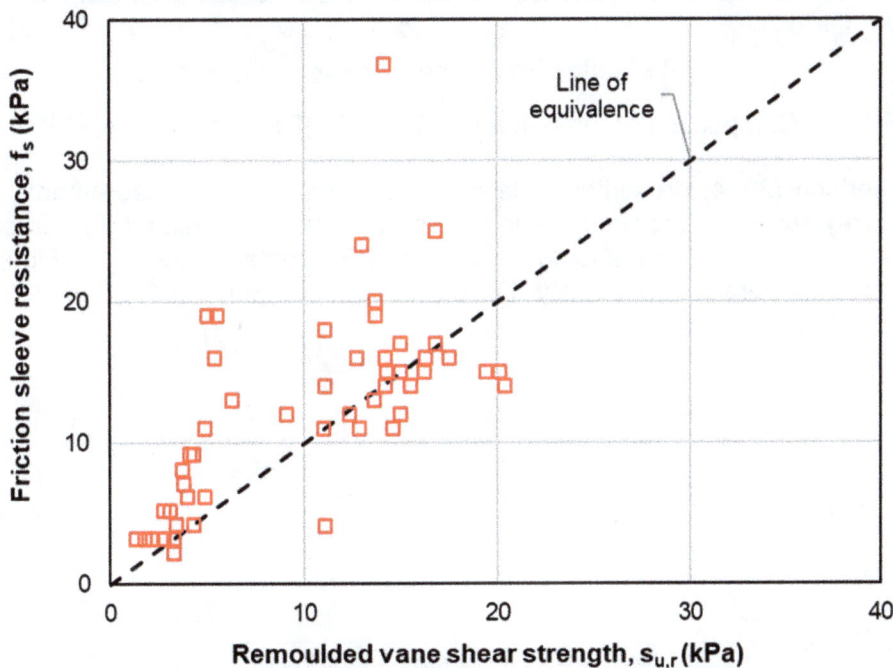

FIG 2 – Comparison of remoulded VST and CPTu f_s from Robertson *et al* (1986).

$$S_t = 6/FR\% \qquad (3)$$

Lunne, Robertson and Powell (1997) proposed that f_s is approximately equal to $s_{u,r}$, and that S_t could be related the ratio of f_s to q_t through Equation 4, with N_s varying between six and nine due to mineralogy, over-consolidation ratio (OCR), and other factors, with a value of 7.5 suggested for use.

$$S_t = N_s/R_f \tag{4}$$

Farrar (2010) presented CPTu and VST data from testing of a stiff clay foundation soil, which purportedly demonstrates that CPT f_s values are equivalent to $s_{u,r}$. In total, 82 pairs of VST were presented in the paper, with VST undertaken following both 360° and 1800° of remoulding. Interpretation of the VST data and corresponding CPTu data, presented in Figure 3, indicates that f_s measurements were approximately 105 per cent of $s_{u,r}$ from VST following 360° of rotation, and 133 per cent of $s_{u,r}$ from VST following 1800° of rotation, with an $r^2 = 0.93$ and 0.91, respectively. It should be noted that ASTM standard D2573-08 states that $s_{u,r}$ should be measured after five to ten vane rotations. Furthermore, drainage during the VST may have influenced the test results, owing to adoption of standard rate VST in an over-consolidated foundation soil.

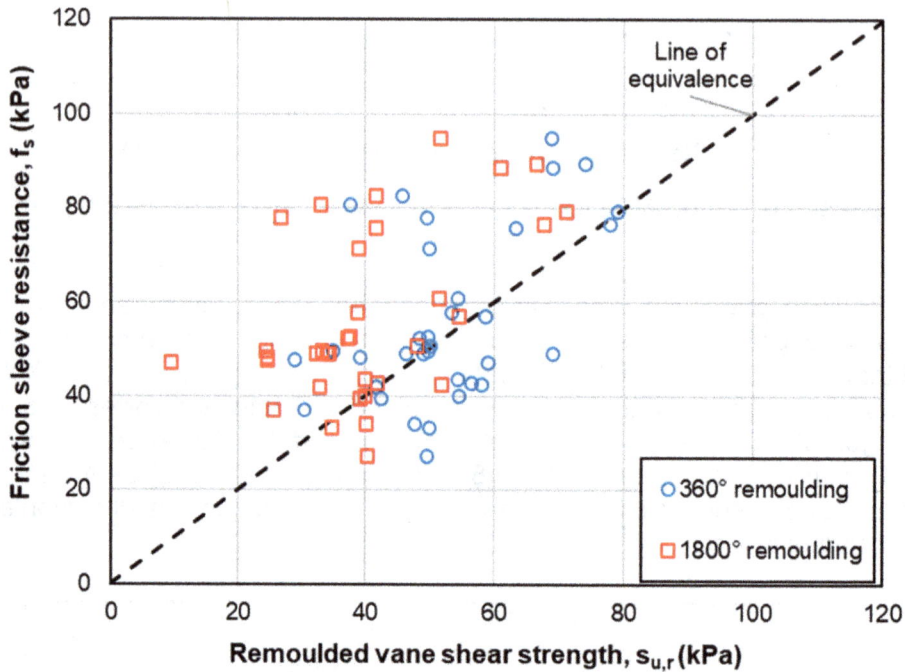

FIG 3 – Comparison of remoulded VST and CPTu f_s from Farrar (2010).

Cabal and Robertson (2014) presented data demonstrating that f_s measurements at a normally consolidated clayey soil site were 'very similar' to shear strengths obtained from *in situ* remoulded VST. Interpretation of the VST data and corresponding CPTu data, presented in Figure 4, indicates that f_s measurements were approximately 120 per cent of $s_{u,r}$ from remoulded VST, with a r^2 of approximately 0.89.

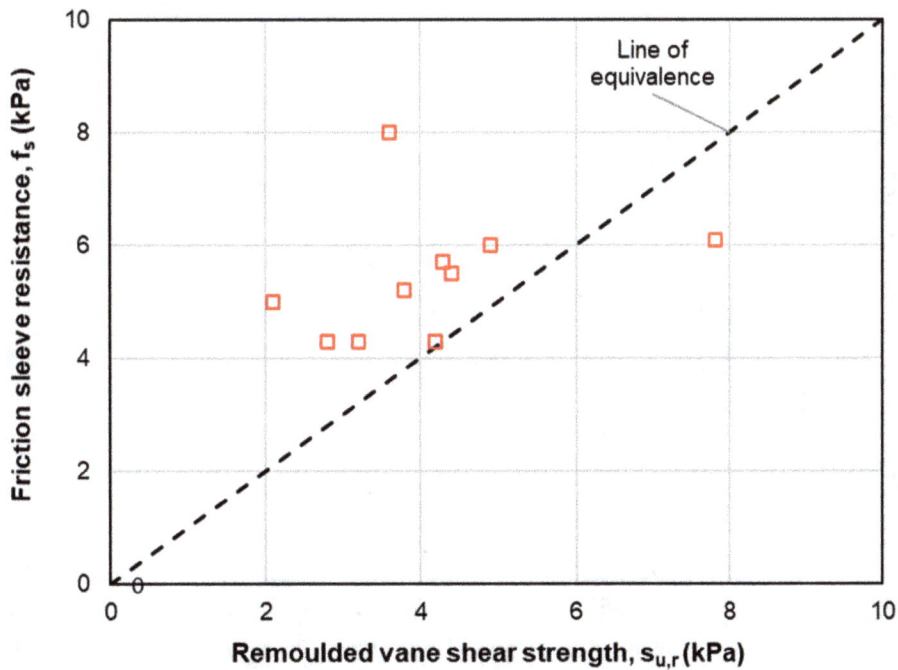

FIG 4 – Comparison of remoulded VST and CPTu f_s from Cabal and Robertson (2014).

Robertson (2021) proposes that for clay like soils (those with soil behaviour-type index ($I_{c,RW}$) greater than 3.0) f_s is approximately equal to $s_{u,r}$. Estimated residual strength ratios from failure case histories (Robertson, 2010) are provided to support of the proposal. Furthermore, Robertson (2021) stated that $s_{u,r}$ estimated in such a manner is generally consistent with the semi-empirical, state parameter-based $s_{u,r}$ estimation technique proposed by Jefferies and Been (2016), which the authors of this paper believe is currently an accepted industry standard practice approach.

The data from all of the aforementioned studies that used remoulded VST along with electric CPTu cones is presented in Figure 5, with tests following 1800° of remoulding selected from Farrar (2010) as the number of remoulding vane rotations complies with ASTM D2573-08. Interpretation of the data indicates that f_s measurements are approximately 131 per cent of $s_{u,r}$ from remoulded VST, with a r^2 of approximately 0.90. This indicates that f_s and $s_{u,r}$ from remoulded VST are strongly positively correlated and that in general f_s may overestimate $s_{u,r}$ in fine grained soils. Interestingly, the differences between f_s and $s_{u,r}$ increase with increasing f_s, which may indicate that friction sleeve inaccuracy at low values of f_s may not be the primary cause of the differences observed between f_s and $s_{u,r}$ estimated from VST.

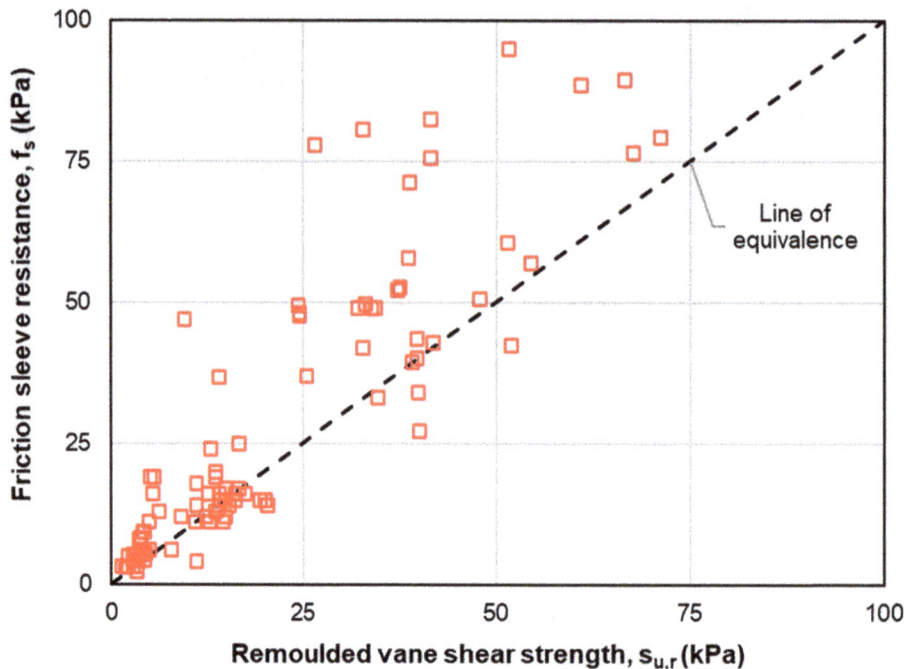

FIG 5 – Comparison of remoulded VST and CPTu f_s.

Theoretical considerations

The friction sleeve measurement is analogous to an interface shear test, albeit in an axisymmetric orientation (DeJong, Frost and Cargill, 2001). Therefore, all of the factors which have been found to influence the shear strength mobilised in interface shear tests should influence f_s. The most significant factors are the normal effective stress on the soil/interface boundary, the relative roughness and hardness of the interface (both relative to the soil), the angularity of the soil, and the density of the soil (Lee, 1998). The later of these two factors are not discussed further in this paper, as their effect on interface shear strength should be intuitive to geotechnical engineers.

Surface roughness

With regard to the relative interface (surface) roughness, researchers have demonstrated that for smooth interfaces the interface shear strength is less than that of the soil, with soil particles observed to slide at the interface (Yoshimi and Kishida, 1981; Uesegi, Kishida and Tsubakihara, 1988). The difference is most significant for coarse grained soils, as the relative (normalised) roughness ($R_n = R_a/D_{50}$, where R_a is the average roughness, and D_{50} is the median soil particle size) is lower than that for fine grained soils for the same interface material. Researchers have also shown that as the relative roughness of the interface is increased, the interface shear strength increases, eventually approaching a limiting value equivalent to the shear strength of the soil (termed the critical roughness, see Figure 6), at which point soil sliding no longer occurs at the interface, but rather shearing occurs within the soil itself (Brumund and Leonards, 1973; Uesegi and Kishida, 1986). Increases in interface shear strength with relative roughness has also been demonstrated for CPTu friction sleeves of varying roughness (DeJong and Frost, 2002) and through numerical modelling of CPTu (Mozaffari and Ghafghazi, 2019; Monforte *et al*, 2021). ISO 22476-1:2022 states that friction sleeves shall be manufactured to a surface roughness of 0.15–0.65 µm; the range of relative roughness for different soil types are shown on Figure 6, based on this standard. It is apparent that for almost all soils the normalised roughness may not be sufficient to mobilise the full frictional strength of the soil on the friction sleeve interface, although it should be recognised that both friction sleeve wear and manufacturing not meeting specified tolerances can result in higher values of normalised roughness, approaching the frictional resistance of the soil (DeJong, Frost and Cargill, 2001).

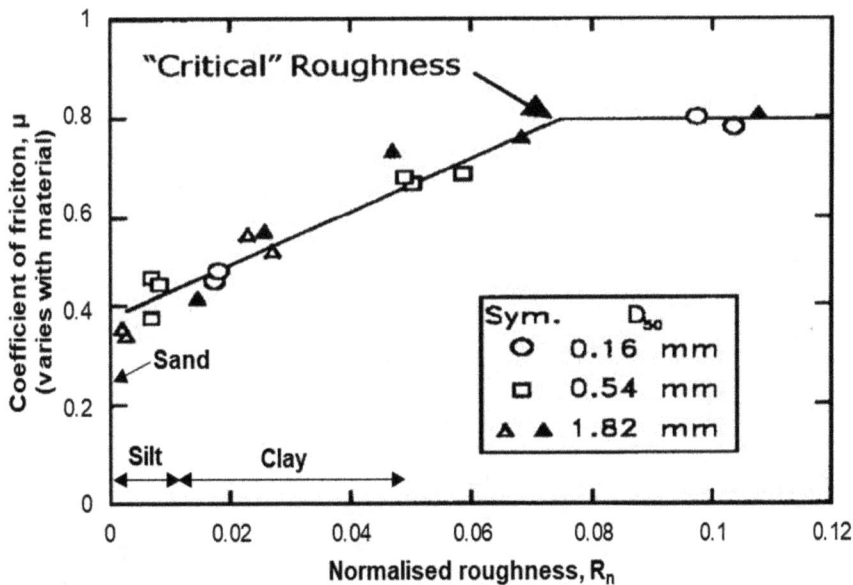

FIG 6 – Coefficient of friction versus normalised roughness (original figure from Uesegi and Kishida (1986)).

Effective stress acting on friction sleeve

The effective stress acting on a CPT friction sleeve is complex, as it results from both the initial 'baseline' horizontal effective stress acting in the soil body prior to CPT insertion (which is generally not known with any certainty) as well as the change in horizontal stress resulting from insertion of the CPT tip (Huntsman *et al*, 1986; Jefferies, Jonsson and Been, 1987).

Many CPTu interpretation techniques rely upon the assumption of undrained (or near undrained) conditions at the CPTu tip during soundings in fine grained soils (at the standard penetration rate of 20 mm/s). However, such assumptions are not valid at the friction sleeve as even in clays, significant dissipation of excess pore pressures (ie consolidation) occurs at the cone shoulder and along the friction sleeve (Campanella, Robertson and Gillespie, 1982; Konrad, Bozozuk and Law, 1985; Robertson *et al*, 1986) (see Figure 7), which generally result in the effective stress acting on the friction sleeve increasing along its length away from the CPT cone shoulder. Such behaviour has also been demonstrated through numerical modelling of CPTu penetration eg Monforte *et al* (2021) who demonstrated that consolidation along the friction sleeve is a result of both the consolidation parameters (eg coefficient of consolidation (c_v, c_h)) of the soil as well as its sensitivity.

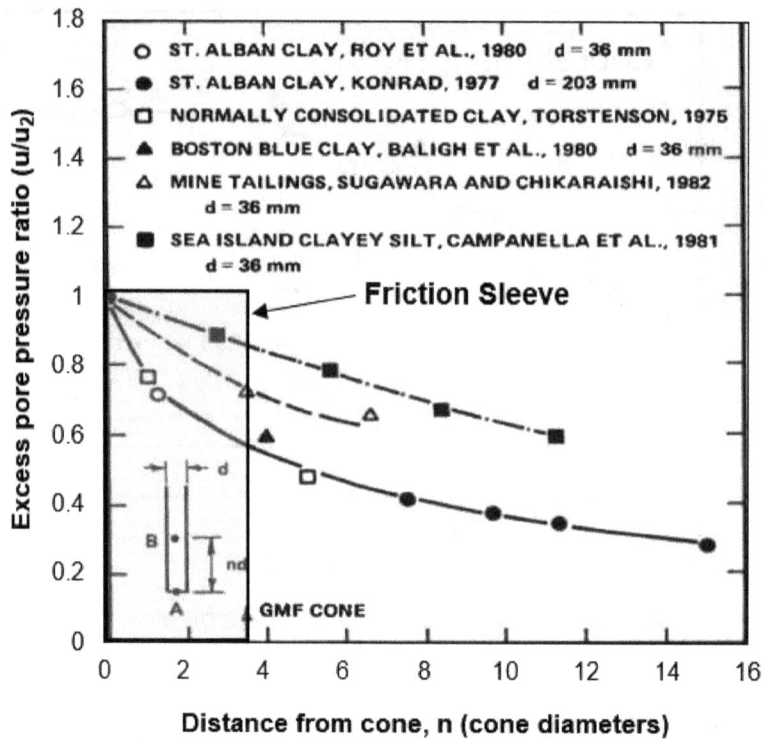

FIG 7 – Distance from cone versus excess pore pressure ratio (original figure from Konrad, Bozozuk and Law (1985)).

The result of consolidation occurring behind the CPTu tip has been demonstrated experimentally by several researchers who have shown that in fine-grained soils f_s increases with distance away from the CPTu tip, including along the length of the friction sleeve (Campanella, Robertson and Gillespie, 1982; Konrad, Bozozuk and Law, 1985; Konrad, 1987).

Furthermore, as the pore pressure acting at the top of the friction sleeve (u_3 location) is seldom measured during commercial CPTu soundings, f_s cannot be corrected for the different pore pressures that act on each end of the friction sleeve. This can result in overestimation of friction sleeve resistances by greater than 20 per cent in soft, fine-grained soils when using equal end-area friction sleeves (Lunne, Robertson and Powell, 1997).

Equipment and procedure limitations

In addition to the lack of pore pressure measurements at the top of the friction sleeve, industry standard CPTu cones are subject to limitations which can materially impact f_s measurements. These factors are presented at length by Peuchen and Terwindt (2014); only those factors that are stated to have the most significant impact on f_s measurements are discussed in further detail.

ASTM D5778-20 states the friction sleeve diameter must be equal to the base of the cone with a tolerance of +0.35 mm and -0.0 mm, whilst ISO 22476-1:2022 limits the tolerance to +0.25 mm and -0.0 mm. This tolerance is interpreted by many as an allowance for friction sleeve wear prior to replacement. However, researchers have demonstrated that use of a friction sleeve with a larger diameter than the CPT cone, though within the ASTM D5778-20 tolerance and with equal sleeve end areas, can result in an overprediction of friction sleeve exceeding 50 per cent in sandy soils (Holtrigter, Thorp and Hoskin, 2014), and exceeding 100 per cent in very soft clayey soils (Cabal and Robertson, 2014; Lunne *et al*, 2018), with this attributed to end bearing (of soil) on the lower end of the friction sleeve as well as an increase in the radial stress acting on the friction sleeve. Furthermore, wear on the friction sleeve during use can result in a reduction in the diameter of the friction sleeve, with greater reductions at the top of the sleeve (Jekel, 1988; ASTM D5778-20, 2020). Reduction in diameter of the friction sleeve can impact the measured friction sleeve resistance, through both unequal bearing on the sleeve ends that may obtain different areas, as well as reduction of the radial stress due to reduced diameter, with the combined effect resulting in a reduction of sleeve friction by up to 10 per cent (Peuchen and Terwindt, 2014) or greater (Cabal and Robertson, 2014).

Temperature can influence the accuracy of friction sleeve measurements, with both temperature differences between the ambient temperature during calibration and the *in situ* ambient temperature, as well as transient temperature developed in the cone during penetration limiting accuracy. The strain gauges on modern CPTu cones are generally temperature compensated, somewhat mitigating the influence of ambient temperature differences. Some CPT cones also include temperature sensors that allow for corrections to be made based on laboratory calibrations (Kim *et al*, 2010; Soage Santos *et al*, 2022). Furthermore, operational practices including calibration at the *in situ* temperature (Konrad, 1987), as well as penetration interruptions following transition from soils with high cone resistance to those with low cone resistance (Post and Nebbeling, 1995), can mitigate the influence of ambient temperature differences, and transient temperature differences, respectively.

For soft, sensitive soils, which may exhibit very low friction sleeve values, the accuracy of friction sleeve measurements can be materially influenced by a combination of temperature differences and force transfer to the dirt seals at the top of the friction sleeve (Barwise, Santos and Alexander, 2014). This is often observed as negative or zero f_s measurements in soft soils, as the inaccuracy of the measurements is greater than that of soil shear strength (Barwise, Santos and Alexander, 2014). Issues with force transfer to the dirt seals can be overcome by modifications to the CPTu cone design including preloading the friction sleeve (Barwise, Santos and Alexander, 2014; McConnel and Wassenaar, 2022), however such cones are not always used in practice.

Lastly, the design of the cone (ie subtraction versus compression) can influence the accuracy of friction sleeve measurements, with subtraction type cones often considered to be less accurate (Lunne, Robertson and Powell, 1997; Peuchen and Terwindt, 2014; ASTM D5778-20, 2020) and capable of introducing an error in sleeve friction of ±3 per cent (Peuchen and Terwindt, 2014). However, the development of improved subtraction cones eg McConnel and Wassenaar (2022) may lead to comparable accuracy between the two cone designs.

COMPARISON TO INDUSTRY PRACTICE TECHNIQUE

Considering the theoretical and practical (equipment and procedure) limitations of the CPTu friction sleeve in accurately measuring $s_{u,r}$ the strong positive correlation between and similar values of f_s and $s_{u,r}$ presented in some of the case histories in the literature may appear to be somewhat puzzling. Such study outcomes may be the result combination of inaccuracies that sometimes fortuitously balance out one another.

To further assess the friction sleeve method, a comparison was undertaken with an accepted industry standard practice CPTu based approach for estimation of $s_{u,r}$, that proposed by Jefferies and Been (2016). The method proposed by Jefferies and Been (2016) uses a combination of strengths inferred at critical state (which are applied to very loose soils with state parameter (Ψ) greater than the slope of the critical state line of the soil (λ_{10})) as well as conservatively fit trends to CPTu data and $s_{u,r}$ values inferred from the back analysis of liquefaction failure case histories. The later was done to account for the influence of the quasi-steady state (QSS) as well as behaviours that occur during field scale failures of lightly contractive to dilative soils, but that may not be observed in laboratory element testing (eg strain localisation, void ratio redistribution). It is for this reason, in addition to its ease of use, that the method is widely adopted in the tailings industry and is often preferred to the estimation of residual strengths directly from the results of laboratory element testing.

To enable comparison of the two methods, the Plewes, Davies and Jefferies (1992) Ψ estimation method (with updates proposed by Shuttle and Cunning (2007)) has been adopted, with input values of geostatic stress ratio (K_0) of 0.7, and excess pore pressure ratio (B_q) = $0.08(F_r)$), consistent with those utilised by Jefferies and Been (2016). The comparison has been undertaken for two values of the critical friction ratio (M_{tc}), namely 1.4 and 1.0, to represent CPTu in tailings (Reid, 2015; Smith *et al*, 2019; Marcedo and Vergaray, 2022), and clayey soils (Schofield and Wroth, 1968), respectively.

Approximate contours of $s_{u,r}/\sigma'_v$ derived from the two methods are presented on soil behaviour-type charts (Figures 8 and 9) for comparison purposes. It is apparent from Figures 8 and 9 that for a large proportion of fine grained soils (ie those with $2.76 < I_{c,BJ} < 3.6$) $s_{u,r}/\sigma'_v$ estimated using the two methods are very similar for soils with low residual undrained shear strength ratios (ie $s_{u,r}/\sigma'_v \leq 0.1$), consistent with the findings of Robertson (2021). However, for soils with higher $s_{u,r}/\sigma'_v$ the friction sleeve method predicts higher strength ratios than the Jefferies and Been (2016) method, with this effect more pronounced in the tailings than clayey soils. For coarse grained soils (ie those with $I_{c,BJ}$ < 2.76) the friction sleeve method predicts higher $s_{u,r}/\sigma'_v$ than the Jefferies and Been (2016) method. For soils with very high $I_{c,BJ}$ (ie > 3.6) the Jefferies and Been (2016) method appears to produce very strange estimates of $s_{u,r}/\sigma'_v$, due to the inversion parameter equations adopted in the Plewes, Davies and Jefferies (1992) Ψ estimation method. For such materials the friction sleeve method may provide more accurate estimation of $s_{u,r}$ or alternate methods to obtain CPTu inversion parameters (eg Shuttle and Jefferies, 2016) should be utilised.

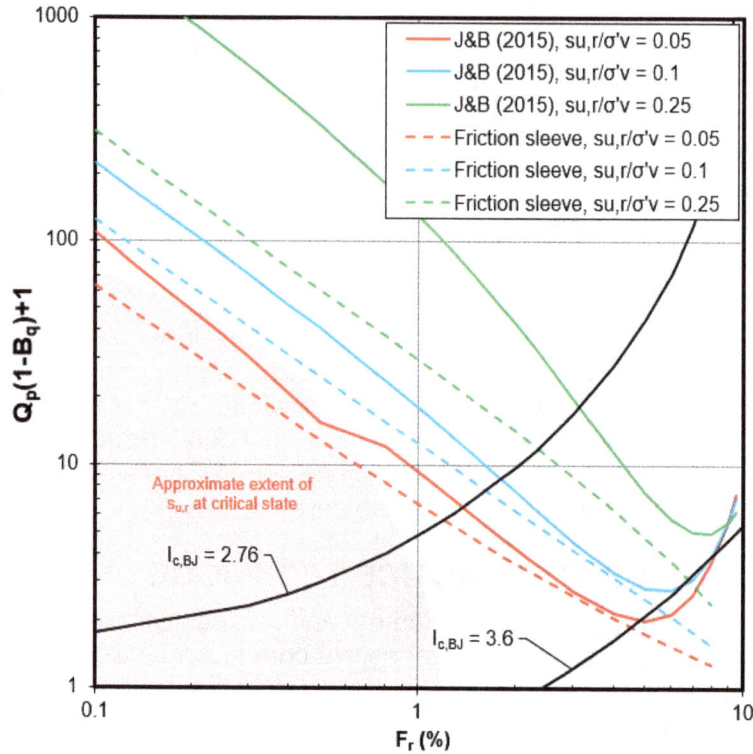

FIG 8 – Comparison of friction sleeve and Jefferies and Been (2016) $s_{u,r}$ estimation techniques – tailings.

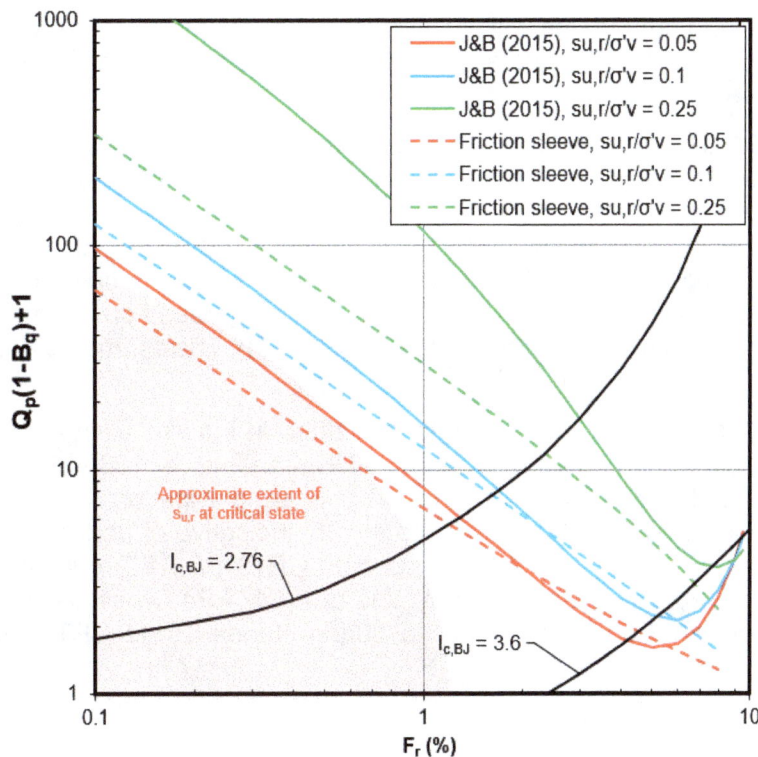

FIG 9 – Comparison of friction sleeve and Jefferies and Been (2016) $s_{u,r}$ estimation techniques – clayey soil.

It should be noted that inaccuracy in measurement of f_s will not only affect results of the friction sleeve method, but also the estimation of Ψ, as f_s is an input to the estimation of Ψ in the Plewes, Davies and Jefferies (1992) method. Therefore, inaccuracy in f_s can significantly influence the $s_{u,r}$ estimated from the Jefferies and Been (2016) method.

TAILINGS CASE STUDY

Site and tailings description

The site is an upstream raised TSF at a gold mine located in the Goldfields-Esperance region of Western Australia. The tailings considered in this case study is located beneath the current crest of the TSF, at 30.5–33.5 m depth (near the base of the TSF), and is located beneath the phreatic surface. The tailings is believed to have been deposited subaqueously at a high rate of rise, during the initial operations of the TSF. The tailings is classified as a low plasticity CLAY, having a liquid limit of approximately 40 per cent, plasticity index of approximately 15 per cent, fines content of 94 per cent, and clay sized fines content of 21 per cent. The tailings is primarily comprised of kaolinite (50 per cent), as well as quartz and muscovite (approximately 25 per cent each). Triaxial compression testing (using lubricated specimen ends, and end of test freezing) on specimens reconstituted using the slurry deposition technique indicated the following critical state line (CSL) parameters: $\lambda_{10} = 0.206$, altitude at 1 kPa mean effective stress (Γ) = 1.334, M_{tc} = 1.23.

Field investigation

A field investigation of the TSF was undertaken, which included CPTu at standard penetration rate as well as non-standard 'slow' (2 mm/s) and 'fast' (50 mm/s) penetration rates, CPTu dissipation tests, downhole seismic tests, VST, and sampling using a thin-walled stationary piston sampler (82 mm diameter, 500 mm length).

The CPTu soundings were undertaken with a standard compression cone (10 cm^2 end area) with 100 MPa q_c capacity, 2 MPa f_s capacity, and 3.5 MPa u_2 capacity, achieving Application Class 1 (ISO 22476-1:2012). Dissipation tests were carried out in duplicate CPTu soundings (located in close proximity) to avoid their influence on CPTu response and interpreted parameters. VST were undertaken following CPTu soundings, within a radius of 2 m, using a rectangular vane (50 mm diameter, 100 mm height) conforming with ASTM standard D2573-08, driven by an electric motor at rotation rates of 300°/min, 1200°/min, and 1440°/min in order to achieve undrained conditions (Chandler, 1988). Five vane rotations were undertaken to remould the tailings prior to testing.

Following CPTu and VST, multiple tube samples were retrieved from 31.0–31.5 m depth, within a radius of 4 m, to enable comparison between *in situ* and laboratory test results. The tube samples masses (tube and sample) were measured immediately following sampling to preclude leaks influencing results. The void ratio of each tube sample was estimated from the whole of tube moisture content and the average particle density of each tube. The Ψ of each tube sample was estimated by comparing the tube sample void ratio to the CSL of the tailings, based on the estimated *in situ* stress (assuming $K_0 = 0.7$, consistent with Jefferies and Been (2016)). The estimated Ψ of the tube samples were 0.19, 0.19, and 0.17.

Data from the standard rate CPTu sounding is presented in Figure 10; over the depth considered (30.5–33.5 m), median $q_t = 1.68$ MPa, $f_s = 26.1$ kPa, and $u_2 = 1033$ kPa. Dissipation testing undertaken at these depths indicated a time for 50 per cent dissipation (t_{50}) of approximately 40 s, and a normalised cone velocity (V) of approximately 35, indicative of undrained to partially drained penetration conditions (DeJong *et al*, 2012). Fast-rate CPTu (V ~ 85) over the depth considered produced median $q_t = 1.82$ MPa, $f_s = 30.3$ kPa, and $u_2 = 647$ kPa, which reinforces that undrained conditions were likely achieved at the cone tip during standard rate CPTu, however, significant differences in f_s and u_2 were noted.

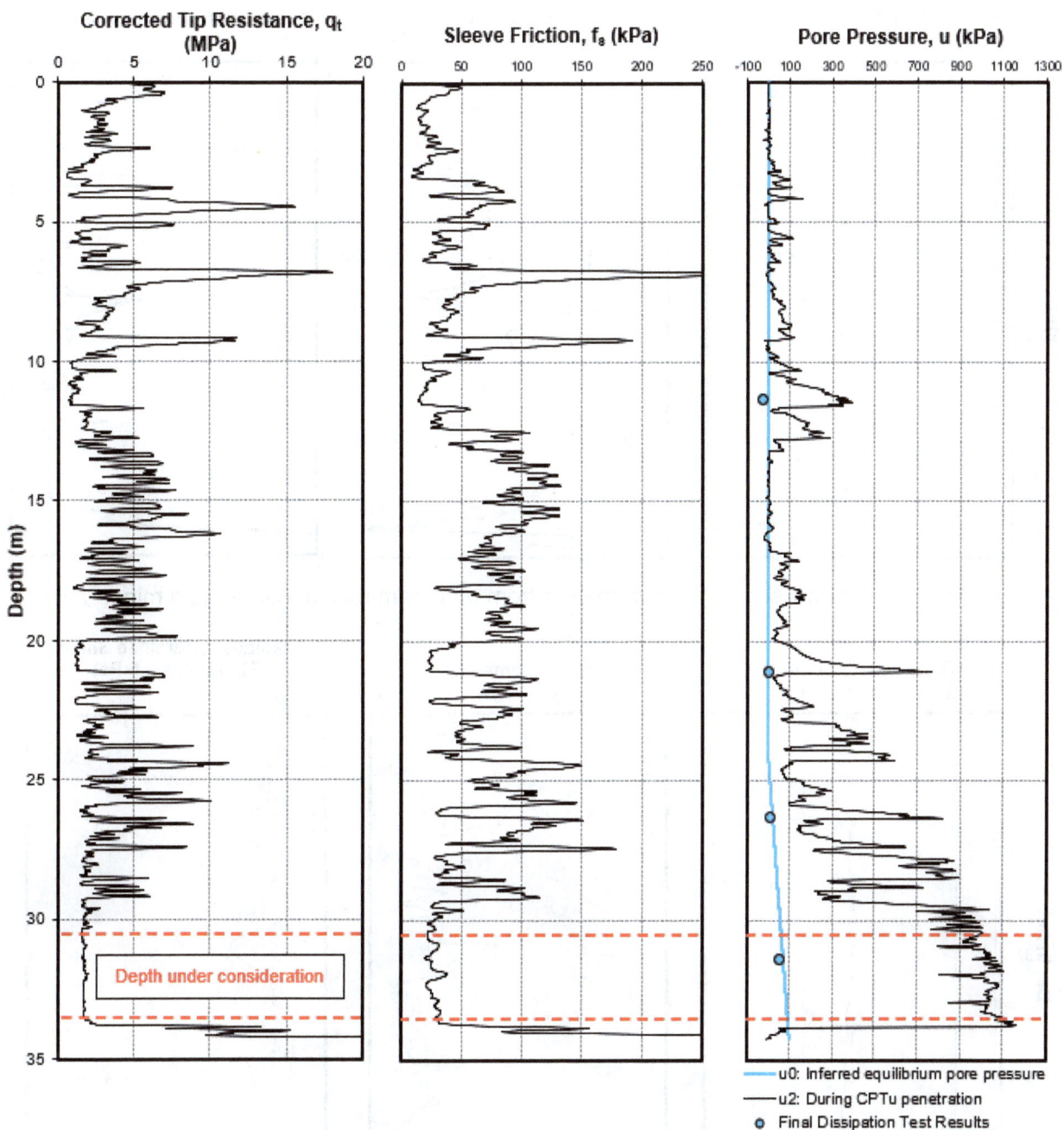

Corrected Tip Resistance, q_t (MPa)

Sleeve Friction, f_s (kPa)

Pore Pressure, u (kPa)

Depth under consideration

- u0: Inferred equilibrium pore pressure
- u2: During CPTu penetration
- ○ Final Dissipation Test Results

FIG 10 – CPTu data from upstream raised TSF at gold mine.

The normalised CPTu parameters, normalised cone resistance (Q_p), friction ratio (F_r), and pore pressure ratio (B_q) are presented in Figure 11. Over the depth considered, median Q_p = 2.98, F_r = 2.51, and B_q = 0.79. The interpreted parameters, soil behaviour-type index ($I_{c,BJ}$) and Ψ are presented in Figure 12. Consistent with Figures 8 and 9, Ψ was estimated using the method proposed by Plewes, Davies and Jefferies (1992), however, the λ_{10} from the laboratory testing was adopted as the λ_{10} estimated from F_r was found to be inconsistent with the λ_{10} from laboratory testing. Over the depth considered, median $I_{c,BJ}$ = 3.53 and Ψ = 0.21, whilst the 80th percentile Ψ = 0.22. This indicates a reasonable agreement with the Ψ estimated from the tube samples.

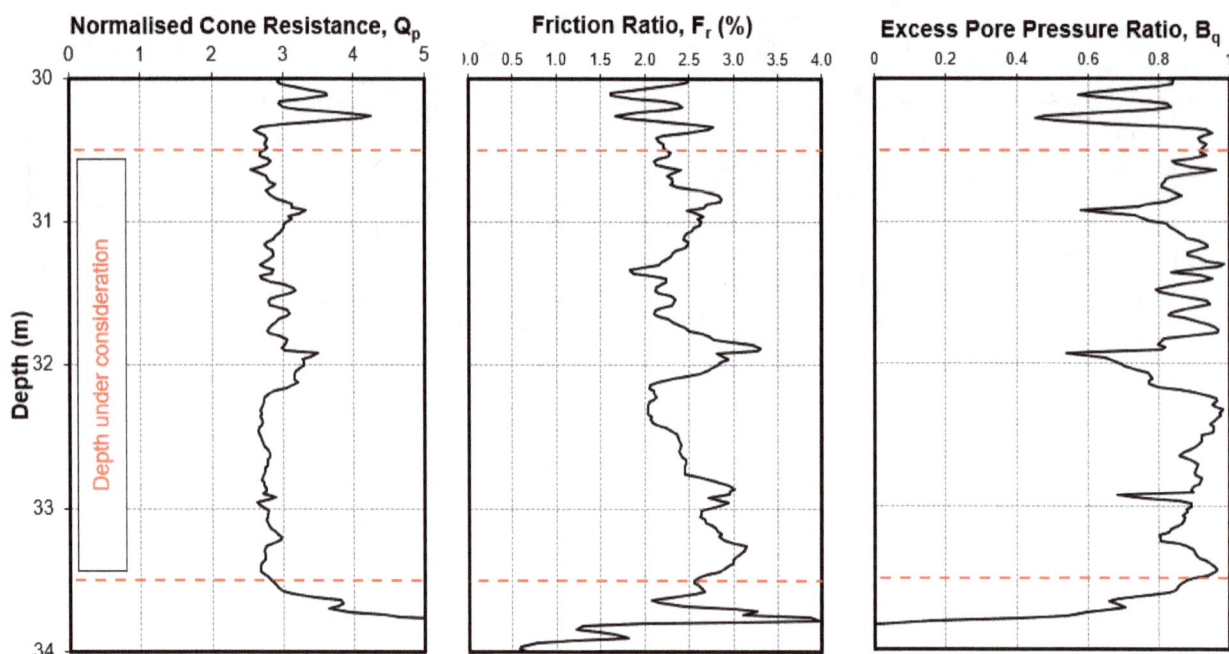

FIG 11 – Normalised CPTu parameters from upstream raised TSF at gold mine.

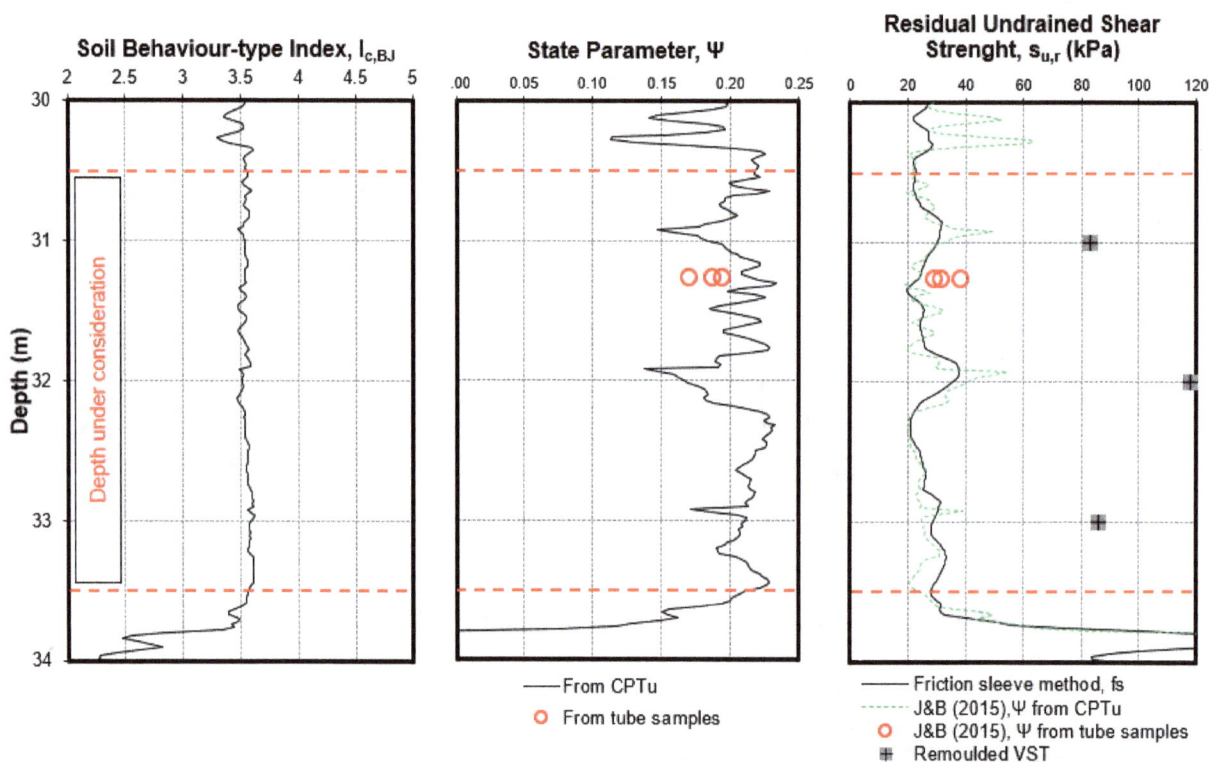

FIG 12 – Interpreted CPTu parameters from upstream raised TSF at gold mine.

A comparison of f_s with $s_{u,r}$ estimated from VST and Ψ (using the Jefferies and Been (2016) $s_{u,r}$ estimation method) is presented in Figure 12 and Table 1. The results show good agreement between f_s and $s_{u,r}$ estimated using the Jefferies and Been (2016) $s_{u,r}$ estimation method for both the tube samples and CPTu, however, $s_{u,r}$ estimated from remoulded VST greatly exceed f_s. This may have been due to the failure to achieve undrained conditions during the VST owing to a time delay between vane insertion and rotation (Morris and Williams, 2000).

TABLE 1

Summary of estimated $s_{u,r}$ over depth under consideration.

Method	$s_{u,r}$ (kPa)	$s_{u,r}/f_s$ (%)
Friction sleeve method, median	26.1	100
Jefferies and Been (2016), median Ψ from CPTu	25.1	96
Jefferies and Been (2016), mean Ψ from tube samples	32.7	125
Remoulded VST, mean	95.8	367

CONSIDERATIONS ON THE UNIQUENESS OF SU,R

Whilst this paper has assessed the accuracy of the friction sleeve method in estimating $s_{u,r}$ against complementary *in situ* and laboratory testing, it should be noted that $s_{u,r}$ estimated from complementary testing may be non-unique. Laboratory testing using hollow cylinder torsional shear (HCTS) equipment has demonstrated the QSS, which is speculated to control $s_{u,r}$ during field scale slope failures (eg Sivathayalan and Vaid, 2002; Jefferies and Been, 2016), is not only a function of soil state, but also of the mode of shearing (eg Sivathayalan and Vaid, 2002). Furthermore, $s_{u,r}$ may be a function of the rate of shearing. For example, Yamamuro and Lade (1998) present triaxial compression testing results on Nevada Sand (with 25 per cent fines) indicating that the QSS varies with the rate of shearing, whilst Schule, Moerz and Kreiter (2010) present VST results in a harbour mud that indicate $s_{u,r}$ is strongly rate dependent, over a range of vane rotation rates assessed to be undrained. Lastly, as $s_{u,r}$ is a function of the friction angle of a soil (eg Sivathayalan and Vaid, 2002; Jefferies and Been, 2016), the impact of friction softening on the $s_{u,r}$ of friction softening materials should be considered. Engineers should be aware of these considerations when selecting a unique $s_{u,r}$ to characterise a soil in design analyses, in particular when comparing the results of different testing methods.

CONCLUSIONS

Several researchers have proposed that f_s represents the approximate $s_{u,r}$ of a soil. A review of the literature has revealed f_s and $s_{u,r}$ estimated from complementary *in situ* and laboratory testing, are strongly positively correlated. However, f_s often provides a biased estimate of $s_{u,r}$, which should be recognised if the data are to be utilised in TSFs design analyses.

Considering the theoretical and practical limitations of the CPTu friction sleeve, it is unsurprising that f_s does not provide a highly accurate estimate of $s_{u,r}$ in many cases, in particular in sensitive soils that mobilise low values of f_s. Although it may be possible to improve the accuracy of CPTu f_s measurements whilst retaining the current standard format of the cone and efforts are being made toward this end, inherent limitations will remain unless significant changes are made to the CPTu cone design.

Acknowledging the limitations of the CPTu friction sleeve, the friction sleeve method was compared to the $s_{u,r}$ estimation method proposed by Jefferies and Been (2016) (using Plewes, Davies and Jefferies (1992) to estimate Ψ), which is widely adopted in the tailings industry. Interestingly the two methods were found to provide very similar estimates of $s_{u,r}$ in soils with $2.76 < I_{c,BJ} < 3.6$ and $s_{u,r}/\sigma'_v \leq 0.1$. Furthermore, for soils with $I_{c,BJ} > 3.6$ the friction sleeve method may provide more accurate estimation of $s_{u,r}$ than the Jefferies and Been (2016) method when used in conjunction with the Plewes, Davies and Jefferies (1992) method, due to limitations of that method at high values of $I_{c,BJ}$.

An example from a TSF was provided, which demonstrated good agreement between f_s and $s_{u,r}$ estimated using the Jefferies and Been (2016) $s_{u,r}$ estimation method, however, $s_{u,r}$ estimated from remoulded VST greatly exceed f_s, potentially due to drainage influences on the VST. It should be cautioned that this example of agreement between f_s and $s_{u,r}$ inferred from complementary testing may be a fortuitous coincidence, selected for inclusion in this paper due to confirmation bias.

Lastly, some considerations on the uniqueness of residual undrained shear strengths were presented, which should be acknowledged when undertaking stability assessments of TSFs.

REFERENCES

ASTM, 2008. D2573-08 Standard Test Method for Field Vane Shear Test in Cohesive Soils, October 2008.

ASTM, 2020. D5778-20 Standard Test Method for Electronic Friction Cone and Piezocone Penetration Testing of Soils, June 2020.

Barwise, A, Santos, R and Alexander, M, 2014. Improved CPT sleeve friction sensitivity in soft soils, in *Proceedings of the Fifth International Workshop on CPTU and DMT in soft clays and organic soils* (Adam Mickiewicz University, Poland).

Begemann, H K S, 1953. Improved method of determining resistance to adhesion by sounding through a loose sleeve placed behind the cone, in *Proceedings of the 3rd International Conference on Soil Mechanics and Foundation Engineering* (Switzerland).

Begemann, H K S, 1964. The friction jacket cone as an aid in determining the soil profile, in *Proceedings of the 6th International Conference on Soil Mechanics and Foundation Engineering* (Montreal).

Brumund, W F and Leonards, G A, 1973. Experimental study of static and dynamic friction between sand and typical construction materials, *Journal of Testing and Evaluation*, 1(2):162–165.

Cabal, K and Robertson, P K, 2014. Accuracy and repeatability of CPT sleeve friction measurements, in *Proceedings of the 3rd International Symposium on Cone Penetration Testing (CPT'14)*, (Las Vegas).

Campanella, R G, Robertson, P K and Gillespie, 1982. Cone penetration testing in deltaic soils, *Canadian Geotechnical Journal*, 20:23–35.

Chandler, R, 1988. The in-situ measurement of the undrained shear strength of clays using the field vane, in *Vane Shear Strength Testing in Soils: Field and Laboratory Studies* (ASTM, Baltimore).

DeJong, J T and Frost, D J, 2002. A multisleeve friction attachment for the cone penetrometer, *Geotechnical Testing Journal*, 25(2):111–127.

DeJong, J T, Frost, D J and Cargill, E, 2001. Effect of surface texture on CPT friction sleeve measurements, *Journal of Geotechnical and Geoenvironmental Engineering*, 127(2):158–168.

DeJong, J T, Jaeger, R A, Boulanger, R W, Randolph, M F and Wahl, D A J, 2012. Variable penetration rate cone testing for characterization of intermediate soils, in *Proceedings of The Fourth International Conference on Geotechnical and Geophysical Site Characterization* (The International Society for Soil Mechanics and Geotechnical Engineering: Porto de Galinhas).

Farrar, J A, 2010. Post earthquake shear strengths of clay by CPT and vane testing methods at Scoggins Dam, in *Proceedings of the 2nd International Symposium on Cone Penetration Testing (CPT'10)*, (Huntington Beach).

Holtrigter, M, Thorp, A and Hoskin, P W O, 2014. The effect of sleeve diameter on f_s measurements, in *Proceedings of the 3rd International Symposium on Cone Penetration Testing (CPT'14)*, (Las Vegas).

Huntsman, S R, Mitchell, J K, Klejbuk, L W and Shinde, S B, 1986. 'Lateral stress measurement during cone penetration, in *Proceedings of ASCE Specialty Conference In Situ'86: Use of In Situ Tests in Geotechnical Engineering* (New York, NY).

International Organization for Standardization, 2012. ISO 22476-1:2012 Geotechnical investigations and testing – Field testing – Part 1: Electrical cone and piezocone penetration test, withdrawn.

International Organization for Standardization, 2022. ISO 22476-1:2022 Geotechnical investigations and testing – Field testing – Part 1: Electrical cone and piezocone penetration test, December 2022.

Jefferies, M G and Been, K, 2016. *Soil Liquefaction – A Critical State Approach, 2nd edn* (CRC Press, Boca Raton).

Jefferies, M G, Jonsson, L and Been, K, 1987. Experience with measurement of horizontal geostatic stress in sand during cone penetration test profiling, *Géotechnique*, 37(4):483–498.

Jefferies, M G, Morgenstern N R, Van Zyl, D and Wates, J, 2019. *Report on NTSF embankment failure, Cadia Valley Operations for Ashurst Australia*. Available from: <https://www.newcrest.com/sites/default/files/2019-10/190417_Report on NTSF Embankment Failure at Cadia for Ashurst.pdf>, [Accessed: 30 April 2019].

Jekel, J W A, 1988. Wear of the friction sleeve and its effect on the measured local friction, in *Proceedings of the First International Symposium on Penetration Testing (ISOPT-1), (Orlando)*.

Kim, R H, Lee, J S, An, S W, Bae, K D and Lee, W J, 2010. Real-time temperature compensation technique on CPT using FBG sensor, in *Proceedings of the 2nd International Symposium on Cone Penetration Testing (CPT'10)*, (Huntington Beach).

Konrad, J D, 1987. Piezo-friction-cone penetrometer testing in soft clays, *Canadian Geotechnical Journal*, 24:645–652.

Konrad, J M, Bozozuk, M and Law, K T, 1985. Study of in-situ test methods in deltaic silt, in *Proceedings of the 11th International Conference on Soil Mechanics and Foundation Engineering* (San Francisco).

Ladd, C C and DeGroot, D J, 2003. Recommended practice for soft ground site characterization, *in Proceedings of the 12th Panamerican Conference on Soil Mechanics in Geotechnical Engineering (Soil and Rock America 2003), (La Serena).*

Lee, S W, 1998. Influence of surface topography on interface strength and counterface soil structure, PhD thesis, School of Civil and Environmental Engineering, Georgia Institute of Technology, Atlanta, GA.

Lunne, T, Robertson, P K and Powell, J J M, 1997. *Cone Penetration Testing in Geotechnical Practice* (SPON Press: Taylor and Francis).

Lunne, T, Strandvik, S, Kasin, K, L'Heureux, J S, Haugen, E, Uruci, E, Veldhuijzen, A, Carlson, M and Kassner, M, 2018. Effect of cone penetrometer type on CPTU results of at a soft clay test site in Norway, in *Proceedings of the 4th International Symposium on Cone Penetration Testing (CPT'18), (Delft).*

Marcedo, J and Vergaray, L, 2022. Properties of mine tailings for static liquefaction assessment, *Canadian Geotechnical Journal*, 59(5):667–687.

McConnel, A J and Wassenaar, E J C, 2022. An innovative new 3MPa CPT – to detect and measure very small f_s values, in *Proceedings of the 5th International Symposium on Cone Penetration Testing (CPT'22)*, (Bologna).

Meisina, C, Persichillo, M G, Francesconi, M, Creatini, M and Lo Presti, D C, 2017. Differences between mechanical and electrical cone penetration test in the liquefaction hazard assessment and soil profile reconstruction, in *Proceeding of the 2017 International Conference of Civil Engineering (ICCE 2017)* (Tirana).

Monforte, L, Gens, A, Arroyo, M, Manica, M and Carbonell, J D, 2021. Analysis of cone penetration in brittle liquefiable soils, *Computers and Geotechnics*, 134.

Morris, P H and Williams, D J, 2000. A revision of Blight's model of field vane testing, *Canadian Geotechnical Journal*, 37:1089–1098.

Mozaffari, M and Ghafghazi, M, 2019. A numerical study of the effects of contact friction angle on CPT results, in *Proceedings of the 72nd Canadian Geotechnical Conference (GeoStJohn's 2019)*, (St. John's, Newfoundland and Labrador, Canada).

Peuchen, J and Terwindt, J, 2014. Introduction to CPT accuracy, in *Proceedings of the 3rd International Symposium on Cone Penetration Testing (CPT'14)*, (Las Vegas, NV).

Plewes, H D, Davies, M P and Jefferies, M G, 1992. CPT based screening procedure for evaluating liquefaction susceptibility, in *Proceedings of the 45th Canadian Geotechnical Conference* (The Canadian Geotechnical Society: Toronto).

Post, M L and Nebbeling, H, 1995. Uncertainties in cone penetration testing, in *International Symposium on Cone Penetration Testing (CPT'95), (Linkoping, Sweden).*

Quiros, G W and Young, A G, 1986. Comparison of field vane, CPT, and laboratory strength data at Santa Barbara Channel Site, in *Vane Shear Strength Testing in Soils: Field and Laboratory Studies* (ASTM, Baltimore).

Reid, D, 2015. Estimating slope of critical state line from cone penetration test – an update, *Canadian Geotechnical Journal*, 52:46–57.

Robertson, P K and Campanella, R G, 1983. Interpretation of cone penetration tests. Part II: Clay. *Canadian Geotechnical Journal*, 20:734–745.

Robertson, P K, 2010. Evaluation of flow liquefaction and liquefied strength using the cone penetration test, *Journal of Geotechnical and Geoenvironmental Engineering*, 136(1):842–853.

Robertson, P K, 2021, Evaluation of flow liquefaction and liquefied strength using the cone penetration test: an update. *Canadian Geotechnical Journal*, 59(4): 620–624.

Robertson, P K, Campanella, R G, Gillespie, D and Greig, J, 1986. Use of piezometer cone data, in *Proceedings of Use of In-situ Testing in Geotechnical Engineering (IN-SITU'86) (ASCE).*

Robertson, P K, de Melo, L, Williams, D J and Wilson, G W, 2019. *Report of the Expert Panel on the Technical Causes of the Failure of the Feijão Dam I.* Available from: <http://www.b1technicalinvestigation.com/>, [Accessed: 13 December 2019].

Schmertmann, J H, 1978. Guidelines for cone penetration test – performance and design, US Department of Transport, Federal Highway Administration, FHWA-TS-78–209.

Schofield, A and Wroth, C P, 1968. *Critical State Soil Mechanics* (McGraw-Hill, London, UK).

Schule, B D, Moerz, T and Kreiter, S, 2010. Influence of shear rate on undrained vane shear strength of organic harbour mud, *Journal of Geotechnical and Geoenvironmental Engineering*, 36(10):1437–1447.

Shuttle, D A and Cunning J C, 2007. Liquefaction potential of silts from CPTu, *Canadian Geotechnical Journal*, 44:1–19.

Shuttle, D A and Jefferies, M G, 2016. Determining silt state from CPTu, *Geotechnical Research*, 3(3):90–118.

Sivathayalan, S and Vaid, Y P, 2002. Influence of generalized initial state and principal stress rotation on the undrained response of sands, *Canadian Geotechnical Journal*, 39:63–76.

Smith, K, Fanni, R, Chapman, P and Reid, D, 2019. Critical State Testing of Tailings: Comparison between Various Tailings and Implications for Design, in *Proceedings of Tailings and Mine Waste 2019* (The University of British Columbia: Vancouver).

Soage Santos, R, Gomez Myer, E, Peuchen, J, Yetginer, G, Lunne, T and Carrington, T, 2022. Calibration of cone penetrometers according to International Organization for Standardization requirements, in *Proceedings of the 5th International Symposium on Cone Penetration Testing (CPT'22)*, (Bologna).

Uesegi, M and Kishida, H, 1986. Frictional resistance at yield between dry sand and mild steel, *Soils and Foundations*, 26(4):139–149.

Uesegi, M, Kishida, H and Tsubakihara, Y, 1988. Behavior of sand particles in sand-steel friction, *Soils and Foundations*, 28(1):107–118.

Yamamuro, J A, and Lade, P V, 1998. Steady-state concepts and static liquefaction of silty sands, *Journal of Geotechnical and Geoenvironmental Engineering*, 124(9):868–877.

Yoshimi, Y and Kishida, T, 1981. A ring torsion apparatus for evaluating friction between soil and metal surfaces, *Geotechnical Testing Journal*, 4(4):145–152.

Assessing various *in situ* techniques to infer remoulded strength in coal tailings

B Tiwari[1], D Reid[2], F Urbina[3], A B Fourie[4], M K D Chapman[5] and A J McConnell[6]

1. PhD Candidate, The University of Western Australia, Crawley WA 6009.
   Email: bandana.tiwari@research.uwa.edu.au
2. Research Fellow, The University of Western Australia, Crawley WA 6009.
   Email: david.reid@uwa.edu.au
3. PhD Candidate, The University of Western Australia, Crawley WA 6009.
   Email: felipe.urbinaespinosa@research.uwa.edu.au
4. MAusIMM, Professor, The University of Western Australia, Crawley WA 6009.
   Email: andy.fourie@uwa.edu.au
5. Managing Director, Insitu Geotech Services Pty Ltd, Banyo Qld 4014.
   Email: mark@insitu.com.au
6. MAusIMM, Founder, Insitu Geotech Services Pty Ltd, Banyo Qld 4014.
   Email: allan@insitu.com.au

ABSTRACT

With the increasing number of tragic failures of tailings storage facilities (TSFs), the reliable determination of remoulded strength for stability analyses has become essential. However, the determination of remoulded strength for contractive strain softening materials includes many uncertainties. Considering the difficulty of recovering undisturbed samples of contractive fine grained soils, and limitations associated with laboratory testing, different correlations have been developed for the determination of remoulded strengths. One widely used correlation for fine grained soils is between the sleeve friction from cone penetration tests and the remoulded strength of fine grained soils. In an attempt to re-evaluate that correlation, a wide range of *in situ* tests and sampling was carried out in a coal tailings deposit to enable comparisons of inferred remoulded strength from a variety of techniques. In particular, both conventional friction sleeves and an innovative new 3 MPa CPT were used to assess the performance of the new sleeve system in the context of remoulded strength estimation. Vane shear tests (VST) at a range of rotation rates were then used to compare to the friction sleeve measurements and critically assess their performance in the measurement of remoulded strength. Finally, as reliable VST measurements of remoulded strength require undrained conditions to be maintained, an assessment of the potential for drainage to have occurred in these tests is made and only remoulded values deemed to be undrained were used in comparisons to the sleeve measurements.

INTRODUCTION

The proper estimation of the remoulded strength of tailings is a key challenge given the recent failures of TSFs resulting in devasting effects on human life, economic losses and irreversible damage to the downstream environment. Mount Polley's TSF failure in 2014 caused around 25 Mm^3 release of wastewater and tailings into the environment. Failure of the Feijão TSF in 2019 caused over 260 fatalities and a huge impact on the downstream environment (Fourie *et al*, 2022). With all the recent failures, the proper characterisation of contractive tailings and sensitivity behaviour is unarguably important.

Different laboratory tests and correlations are used to determine the remoulded strength. However, laboratory testing may fail to replicate the *in situ* liquefiable soil condition due to issues such as strain limitations of many experimental devices and sample disturbance/densification (Verdugo and Ishihara, 1996; Yamamuro and Covert, 2001). For the estimation of remoulded shear strength, *in situ* tests are arguably more commonly applied owing to the previously-noted difficulties regarding laboratory measurements. Many empirical approaches and relationships have been developed to calculate the residual shear strength based on the *in situ* results (Seed, 1987; Seed and Harder, 1990; Olson and Stark, 2002; Idriss and Boulanger, 2015; Kramer and Wang, 2015). Two approaches are common for case history-based residual strength estimation, the direct approach and the normalised strength approach (Kramer and Wang, 2015). The direct approach considers the residual strength as a direct function of penetration resistance. Similarly, in the normalised strength approach, the ratio of residual strength to initial vertical effective stress is considered to correlate better to the penetration resistance based on Castro (1987). Olson and Stark (2002) presented a relationship between the liquefied shear strength ratio (LSR) by normalising liquefied shear strength with pre-failure vertical strength from CPT and standard penetration test (SPT) from the back calculation of past failures. Similarly, Idriss and Boulanger (2015) recommended SPT and CPT based relationships between the residual shear strength ratio of liquefied non-plastic soils for significant void redistribution and without void redistribution.

Lunne, Robertson and Powell (1997) stated that the sleeve friction values are very small and similar to the remoulded undrained shear strength for sensitive soft clays. Based on the statement from Lunne, Robertson and Powell (1997) and the results of *in situ* VST, Robertson (2009) suggested that the remoulded shear strength could be measured from the sleeve friction for normally consolidated fine grained soils. Farrar (2010) compared the results of friction sleeves with field vane tests from the investigation at Scoggins Dam in Northwest Oregon and concluded that the CPT sleeve friction and the remoulded strengths of VST are similar. Alternatively, according to Lunne (2007), remoulded shear strength and sensitivity cannot be evaluated from sleeve friction and friction ratio as other factors can influence the value of sleeve friction, such as pore pressure effects, surface roughness of the sleeve, rate of penetration, instrumentation effects, and the amount of remoulding.

The major issue during the measurement of remoulded strength in VST is the drainage effect, which is a function of the rate of rotation of the vane and the soil permeability and coefficient of consolidation (Sharifounnasab and Ullrich, 1985). If partial drainage occurs during testing, the undrained shear strength measured is overestimated and the validity of the results becomes questionable (Fourie *et al*, 2022). The current engineering practice and standard is mostly focused on maintaining undrained conditions at peak strength but not considering the effect of drainage on measured remoulded strengths. Reid (2016) assessed the effects of drainage considering the time of failure using multiple rates for the estimation of undrained shear strengths. For silty tailings with high permeability, low plasticity and high brittleness index, it is particularly important to select the rotation rate to achieve an undrained condition, not only for the peak strength but for the remoulded strength at high rotations.

The purpose of this paper is to better understand the remoulded strength obtained with the vane shear testing under multiple rotation rates for the coal tailings. Friction sleeve measurements using an innovative special design 3 MPa cone and the conventional compression cone are compared with the remoulded strengths to review the correlation between the remoulded strength with sleeve friction. The effect of multiple rotation rate VSTs and consolidation parameters on the drainage condition were assessed considering the difficulty in achieving an undrained condition in the field.

GEOTECHNICAL INVESTIGATION

Site and material description

The *in situ* tests presented for this study are from the same location mentioned in McConnell and Wassenaar (2022), which is a coal tailings dam located in Australia. These tailings have low plasticity, with plasticity indices (PI) ranging around 6–9 per cent. The index properties obtained for this material are presented in Table 1.

TABLE 1

Index test characteristics of coal tailings.

Specific gravity	1.75
Plastic limit (%)	38
Liquid limit (%)	32
Plasticity index (%)	6–9
% <75 μm (wet sieve)	24%

CPTu testing

The CPTu results from a conventional 10 MPa capacity compression cone and an innovative special design cone with a 3 MPa capacity cone, designed by Insitu Geotech Services (IGS) and Geomil, was used in this study (McConnell and Wassenaar, 2022). The purpose of the special 3 MPa cone is to overcome the fundamental problem that normal industry-standard compression cones cannot properly detect and measure sleeve friction in extremely soft materials (Santos, Barwise and Alexander, 2014; Entezari *et al*, 2021). This new 3 MPa cone is a subtraction cone, with tip capacity of 3 MPa and sleeve capacity of 200 kPa. A special alloy base is used in the load cell for enhanced physical strain gauge response for higher sensitivity of the cone. Careful cone calibration was done before and after testing using dead weights for the application of load to the tip and sleeve.

In situ dissipation tests were part of the investigation to assess the phreatic level and coefficient of consolidation at regular depth intervals. This was carried out by measuring the time needed for the excess pore pressure to return to equilibrium pressure.

Vane shear test

A field vane shear test is used extensively to measure peak and remoulded strengths. A series of VSTs were performed using an AP Van den Berg electronic down the hole vane with 75 mm diameter and 150 mm height at multiple rates of rotation for different depths to assess the impact of rotation rate on the drainage condition and remoulded strength. For the first approach, the vane was continuously rotated at the same initial rate of rotation to measure the remoulded strength. Continuous rotation was performed at two different rates (a slower rate of 60°/min and a faster rate of 360°/min). For the second approach, the vane was initially rotated at 60°/min to measure the peak strength, followed by a faster rate of 360°/min for five rotations to remould the soil and finally rotation at the initial rotation rate of 60°/min to measure the remoulded strength.

There are different approaches for the assessment of drainage in VSTs (Morris and Williams, 2000; Blight, 1968; Chandler, 1988). An approach based on Blight (1968) and Chandler (1988) was used to measure the dimensionless time factor, T. This approach assumes that the excess pore water pressure in the VST is generated during the vane rotation with subsequent reduction in undrained shear strength and directly links the degree of drainage to the time factor. This method suggests T should be smaller than 0.05 to ensure an undrained condition.

The dimensionless time factor, T to assess the drainage condition is calculated as:

$$T = c_v\, t_f /\, D^2 \qquad (1)$$

where c_v is the coefficient of consolidation, t_f is the time to failure, and D is the vane diameter.

RESULTS

In situ phreatic conditions and rate of consolidation

The horizontal coefficient of consolidation (c_h) was determined using the equation proposed by Teh and Houlsby (1991) from the results of dissipation tests with time for 50 per cent of dissipation. The value of the coefficient of consolidation (c_h) ranged from 4 m^2/a to 16 m^2/a from the results of dissipation tests, as presented in Figure 1. The c_h values of 5 m^2/a and 15 m^2/a were adopted, given it likely represents the most relevant value to ensure undrained conditions for the VST calculations. A plot of equilibrium pore pressure profile with depth from the dissipation test results is presented in Figure 2.

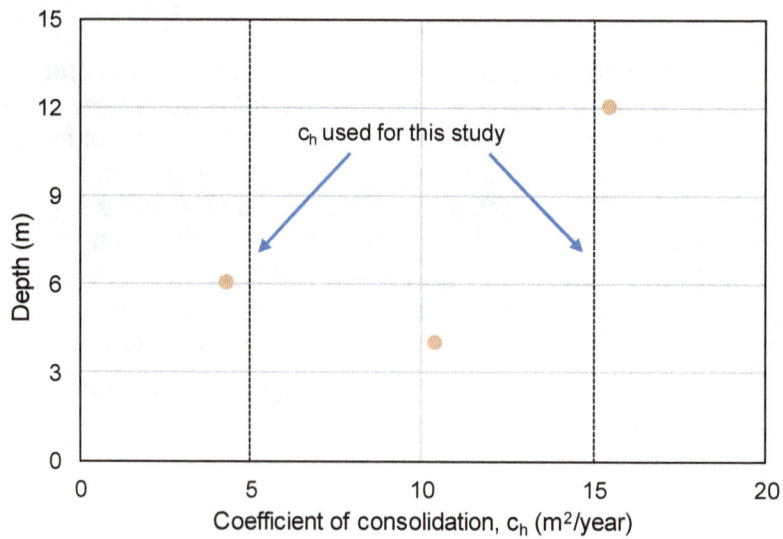

FIG 1 – Summary of coefficient of consolidation.

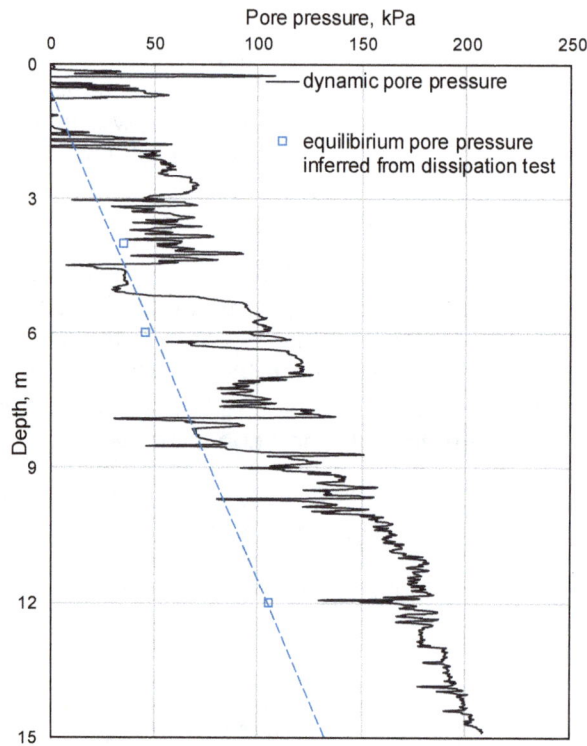

FIG 2 – Pore water pressure with depth.

Comparison of different sleeve friction measurements

The results of the CPT tests undertaken using two different cone types is presented in Figure 3. The result clearly shows variation of the friction sleeve measurements. The sleeve friction values measured by the special 3 MPa cone (which is specifically designed for accurate measurement of low sleeve frictions) are slightly higher in comparison to values measured by the conventional 10 MPa compression cone at most depths. The 3 MPa cone measuring higher sleeve friction values with the line of equivalence is clearer in Figure 4 for the comparison friction sleeve values measured by different types of cone.

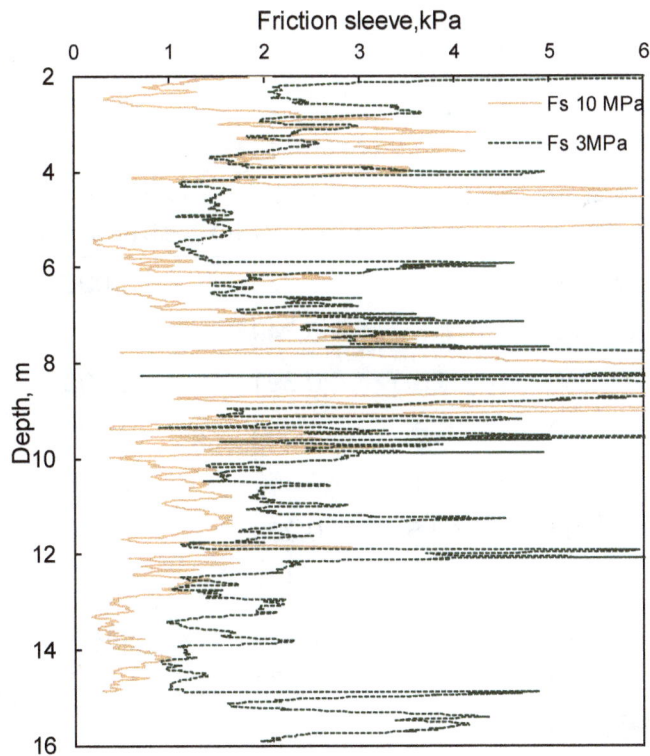

FIG 3 – Depth versus friction sleeve measurements for 3 MPa and 10 MPa capacity cones.

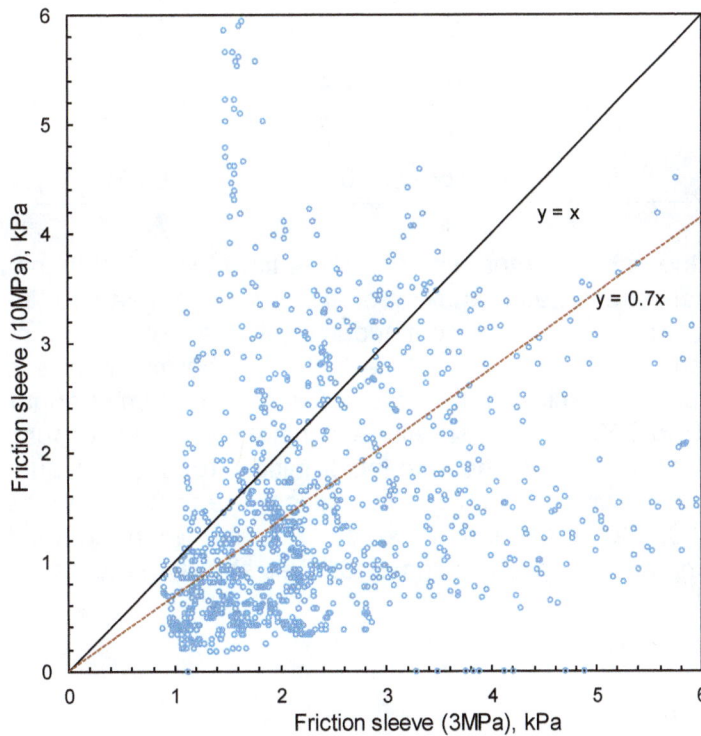

FIG 4 – Comparison of friction sleeve using different cone capacities.

Vane shear test interpretation

The summary of peak and remoulded shear strength of the VST, normalised by effective overburden stress at different depths for multiple rotation rates is shown in Table 2. Here, the remoulded strength refers to the lowest strength obtained during the vane rotation. The peak strengths at the faster rotation rate (360°/min) are higher than the peak obtained for the lower shear rate (60°/min) whereas the remoulded strengths are higher at lower rotation rates in most of the vane tests performed at the same depths. However, the discrepancy in the peak stress ratio is higher than that for the remoulded strength ratio while comparing the results from different rotation rates for the same depth value.

TABLE 2

Summary of vane shear test.

Rate (°/min)	Depth (m)	Peak stress ratio	Remoulded stress ratio
360	3.5	0.320	0.093
360	4.5	0.349	0.099
360	5.5	0.391	0.103
360	6.5	0.353	0.105
360	11.5	0.350	0.078
360	12.5	0.269	0.050
360	14.5	0.167	0.041
360	15.5	0.193	0.049
60/360/60	4.5	0.374	0.112
60/360/60	6.5	0.313	0.117
60/360/60	12.5	0.196	0.062
60/360/60	15.5	0.170	0.048
60	3.5	0.405	0.084
60	5.5	0.302	0.117
60	11.5	0.309	0.096
60	14.5	0.165	0.039

A typical illustration of the change in the dimensionless time factor with vane rotation rate in terms of the measured shear stress ratio and c_h of 15 m^2/a is presented in Figure 5, using the Blight/Chandler method. For the VST with continuous rotation at 360°/min, the strength is decreases continuously until the end of rotation, with a time factor for the minimum strength of around 0.03. When the vane is rotated at 60°/min, the time factor for the minimum strength is around 0.04–0.05, which is at around 540 to 750° of rotation with a constant strength for further rotation of vanes. Similarly, for rotation at multiple rates, the time factor initially reaches a value around 0.0037 within the first 40° peak. However, with the increase in rotation rate to 360°/min the time factor changes from 6.3×10^{-5} to around 0.0025 at the end of five rotations and when the rotation rate is adjusted to the initial rate of 60°/min, the time factor becomes 0.18 as shown in Figure 5c.

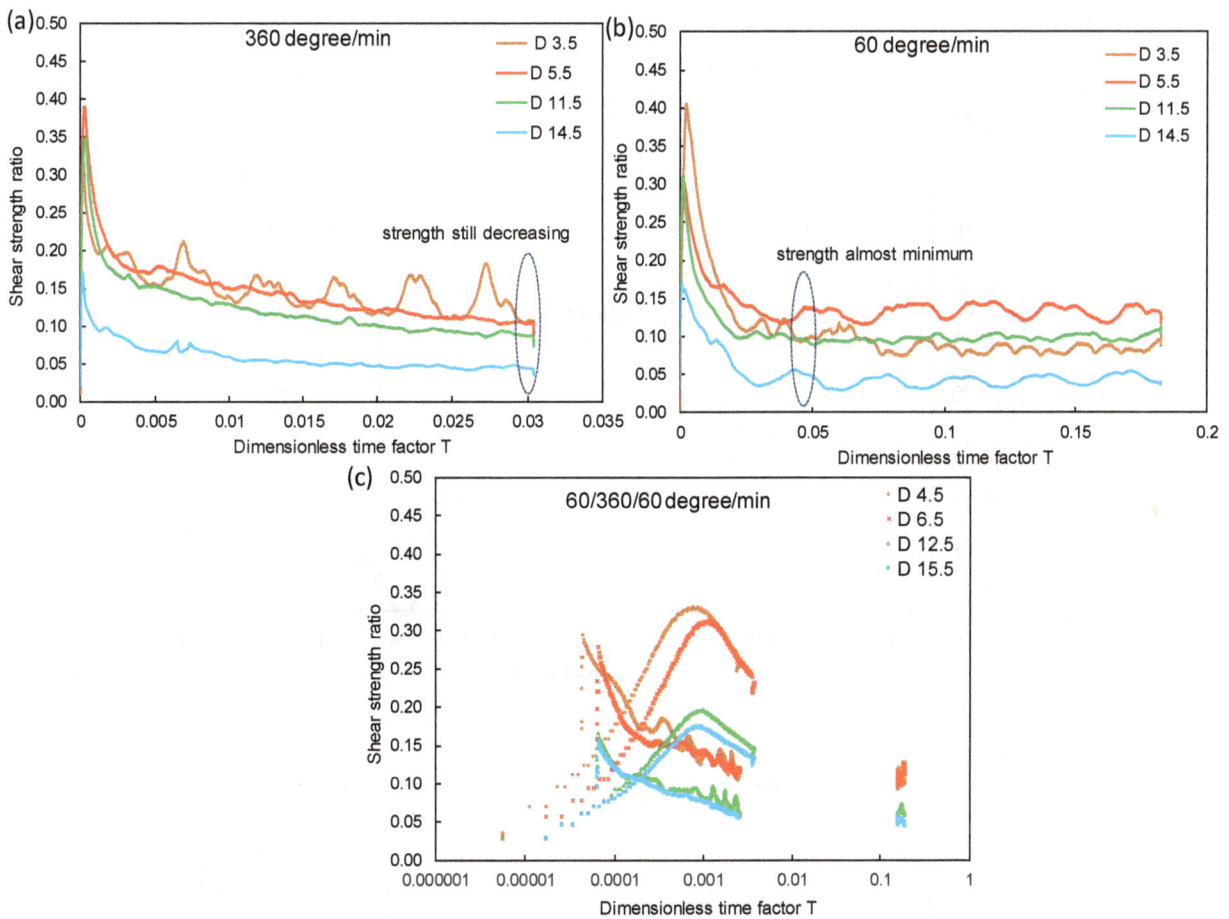

FIG 5 – Typical illustration of variation of undrained strength ratio with time factor for different rates of rotation for c_v of 15 m²/a (based on Blight/Chandler): (a) 360°/min; (b) 60°/min; (c) multiple rates.

For a better understanding of the effect of the c_v on the drainage condition for multiple rotation rates of VST, the dimensionless time factor based on Blight/Chandler was calculated adopting values of c_v of 5 m²/a and 15 m²/a, respectively, as presented in Figure 6. In both conditions, the peak strength appears to be in undrained condition for all rotation rates. For a higher c_v of 15 m²/a, the possibility of partial drainage is amplified for the lower rate of rotation in VST in comparison to c_v of 5 m²/a. The remoulded strength determined from a partially drained condition may not represent the behaviour of tailings, particularly with regards to the influence on the brittleness and sensitivity behaviour. This highlights the importance of a better understanding of the consolidation parameters and rotation rate for obtaining fully undrained remoulded strengths.

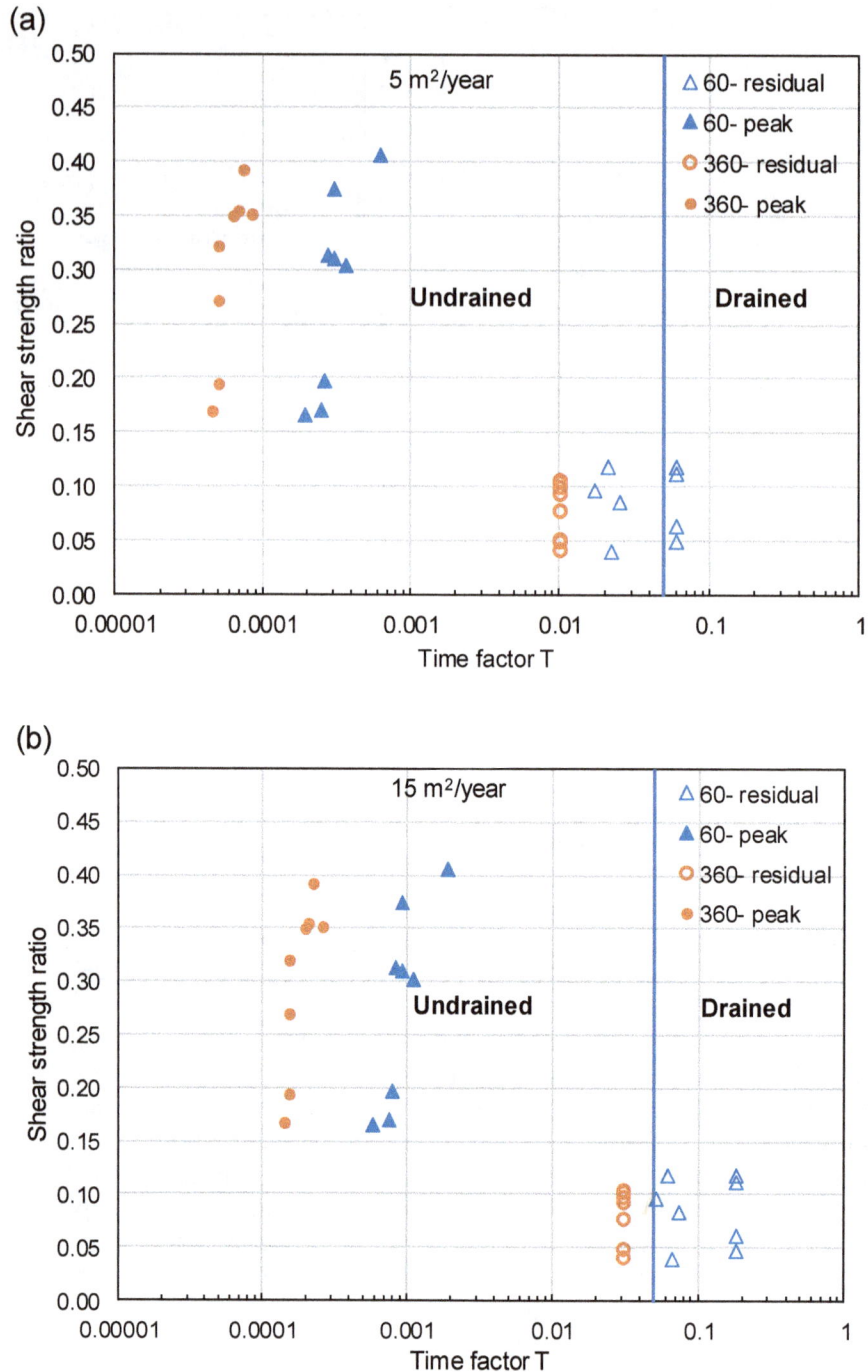

FIG 6 – Change in time factor and drainage condition (based on Blight/Chandler): (a) c_v of 5 m²/a; (b) c_v of 15 m²/a from vane results.

Synthesis of *in situ* data

The results of the CPT undertaken using a conventional compression cone, a special 3 MPa cone and VST in the coal tailings are presented in Figure 7. The results indicate that the remoulded strength measured using the VST is slightly higher in comparison to the sleeve friction value measured by both types of cones. However, the remoulded strength values measured by the VST are closer to the sleeve friction measured by the special 3 MPa cone rather than the conventional 10 MPa cone.

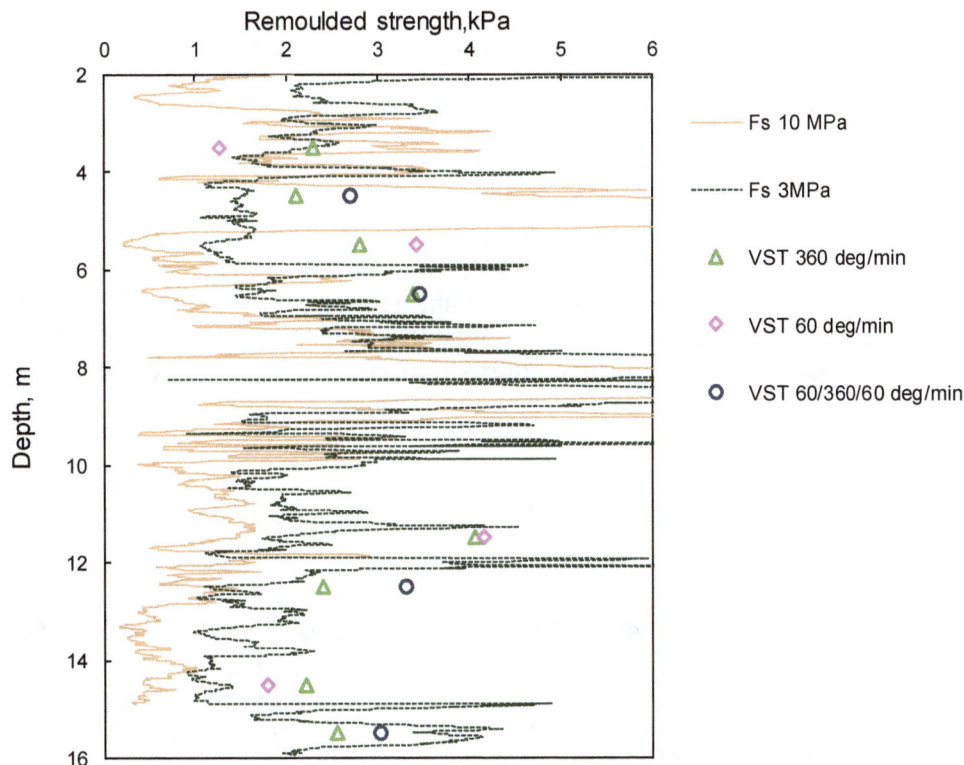

FIG 7 – Summary of remoulded strength inferred from VSTs and friction sleeve.

A comparison between the remoulded strength from the VST with the friction sleeve from 3 MPa and 10 MPa cones is presented in Figure 8. The values obtained using the 3 MPa cone correlate better with the remoulded shear strength from the VST than with the results of the 10 MPa cone, as mentioned above.

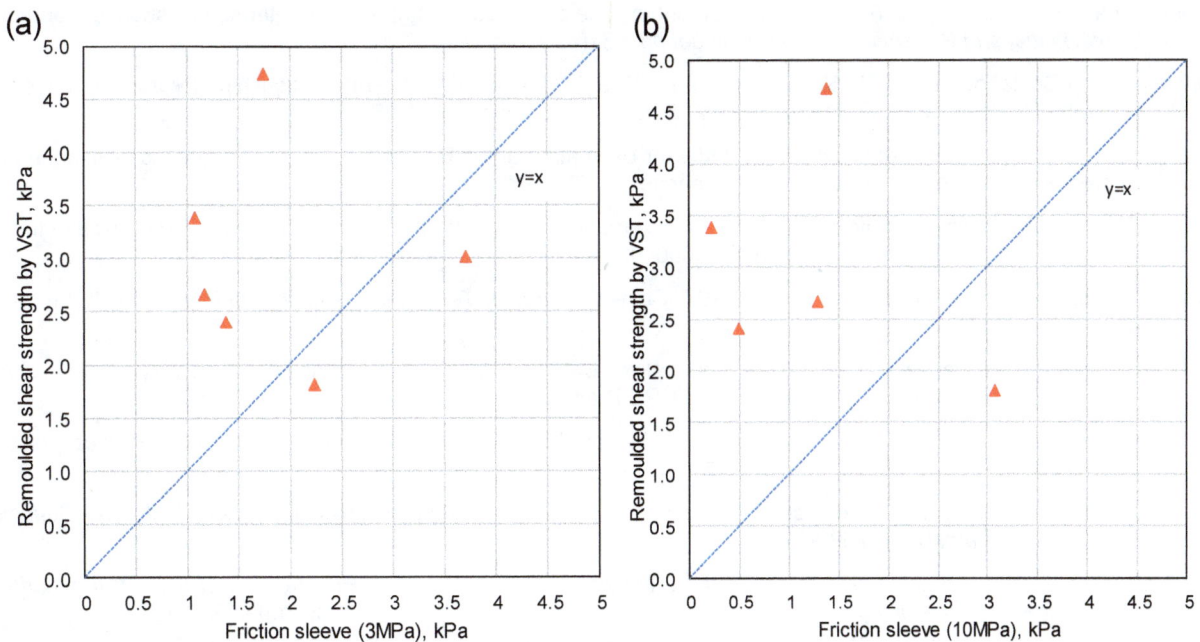

FIG 8 – Comparison of remoulded strength from different *in situ* tests with: (a) 3 MPa cone friction sleeve; (b) 10 MPa cone.

Sleeve friction results from the CPT results can be influenced by various factors such as geometry, interface roughness of the cone, the extent of remoulding of soil, unequal end areas of the cone, or drainage conditions due to the rate of penetration and viscous effects (when a fully undrained condition is maintained) all of which may contribute to the measured friction sleeve, as explained by

Lunne (2007). Also, Frost and DeJong (2005) have verified that the interface roughness between the sleeve and the soil surface potentially influences the sleeve friction measured.

CONCLUSIONS

In situ techniques including VST and CPTu were used to infer the remoulded strength in coal tailings. The correlation between the remoulded strength and friction sleeve was assessed by using two cones of different mechanical designs. The values of sleeve friction measured using a special 3 MPa cone were higher than the values obtained with a conventional 10 MPa compression cone and were found to have better agreement with the remoulded strength values from the VST. A discussion was presented for the propensity of a drained condition in the VST for a material with higher coefficient of consolidation and its impact on the remoulded shear strength, when lower rates of rotation are used in the VST.

REFERENCES

Blight, G E, 1968. A note on field vane testing of silty soils, *Canadian Geotechnical Journal*, 5(3):142–149.

Castro, G, 1987. On the behavior of soils during earthquakes–liquefaction, *Developments in Geotechnical Engineering*, 42:169–204.

Chandler, R J, 1988. The in-situ measurement of the undrained shear strength of clays using the field vane, ASTM International West Conshohocken, PA, USA.

Entezari, I, Boulter, T, McGregor, S and Sharp, J, 2021. Machine learning to estimate fines content of tailings using gamma cone penetration test, in *Proceedings of the Mine Waste And Tailings Conference 2021*, pp 218–226 (The Australasian Institute of Mining and Metallurgy: Melbourne).

Farrar, J A, 2010. Post earthquake shear strengths of clay by CPT and vane testing methods at Scoggins Dam, in *Proceedings of CPT'10, 2nd International Symposium on Cone Penetration Testing*, 7 p. Available from: <https://www.scribd.com/document/542132385/Post-earthquake-shear-strengths-of-clay-by-CPT-and-vane#>

Fourie, A B, Verdugo, R, Bjelkevik, A, Torres-Cruz, L A and Znidarcic, D, 2022. Geotechnics of mine tailings: a 2022 State of the Art, in *Proceedings of the 20th International Conference on Soil Mechanics and Geotechnical Engineering (ICSMGE) – State of the Art and Invited Lectures* (eds: M Rahman and M Jaksa), pp 121–183 (Australian Geomechanics Society: Sydney).

Frost, J D and DeJong, J T, 2005. In Situ Assessment of Role of Surface Roughness on Interface Response, *Journal of Geotechnical and Geoenvironmental Engineering*, 131(4):498–511.

Idriss, I M and Boulanger, R W, 2015. 2nd Ishihara Lecture: SPT – and CPT-based relationships for the residual shear strength of liquefied soils, *Soil Dynamics and Earthquake Engineering*, 68:57–68.

Kramer, S L and Wang, C-H, 2015. Empirical Model for Estimation of the Residual Strength of Liquefied Soil, *Journal of Geotechnical and Geoenvironmental Engineering*, 141(9):1–15.

Lunne, T, 2007. NGI Technical note: Can we measure remoulded shear strength or sensitivity with CPTU sleeve friction and friction ratio. Annex A, Report number 20061023-1.

Lunne, T, Robertson, P K and Powell, J J M, 1997. *Cone Penetration Testing in Geotechnical Particle*, 312 p (Blackie Academic & Professional).

McConnell, A J and Wassenaar, E J C, 2022. An innovative new 3 MPa CPT—to detect and measure very small fs values, *Cone Penetration Testing 2022*, pp 197–202, CRC Press.

Morris, P H and Williams, D J, 2000. A revision of Blight's model of field vane testing, *Canadian Geotechnical Journal*, 37(5):1089–1098.

Olson, S M and Stark, T D, 2002. Liquefied strength ratio from liquefaction flow failure case histories, *Canadian Geotechnical Journal*, 39(3):629–647.

Reid, D, 2016. Effect of rotation rate on shear vane results in a silty tailings, in *Proceedings of the 5th International Conference on Geotechnical and Geophysical Site Characterisation, ISC 2016*, pp 369–374.

Robertson, P K, 2009. Interpretation of cone penetration tests – A unified approach, *Canadian Geotechnical Journal*, 46(11):1337–1355.

Santos, R S, Barwise, A and Alexander, M, 2014. Improved CPT sleeve friction sensitivity in soft soils, Las Vegas, USA, in *Proceedings of the 3rd International Symposium on Cone Penetration Testing. Geotechnical Journal*, 53(12).

Seed, H B, 1987. Design problems in soil liquefaction, *Journal of Geotechnical Engineering*, 113(8):827–845.

Seed, R B and Harder, L F, 1990. SPT-based analysis of cyclic pore pressure generation and undrained residual strength, in *Bolton Seed Memorial Symposium Proceedings* (BiTech Publishers Ltd: Vancouver).

Sharifounnasab, M and Ullrich, C R, 1985. Rate of shear effects on vane shear strength, *Journal of Geotechnical Engineering*, 111(1):135–139.

Teh, C I and Houlsby, G T, 1991. An analytical study of the cone penetration test in clay, *Geotechnique*, 41(1):17–34.

Verdugo, R and Ishihara, K, 1996. The steady state of sandy soils, *Soils and Foundations*, 36(2):81–91.

Yamamuro, J A and Covert, K M, 2001. Monotonic and cyclic liquefaction of very loose sands with high silt content, *Journal of Geotechnical and Geoenvironmental Engineering*, 127(4):314–324.

Assessing the enhancement of settling, consolidation, and desiccation of bauxite residue after a combination of physical and chemical treatments

C Zhang[1], S Quintero Olaya[2], D Williams[3] and A Ranabhat[4]

1. Research Fellow, School of Civil Engineering, Brisbane Qld 4072.
   Email: chenming.zhang@uq.edu.au
2. Senior Research Technologist, School of Civil Engineering, Brisbane Qld 4072.
   Email: s.quintero@uq.edu.au
3. FAusIMM, Professor of Geotechnical Engineering. Director, Geotechnical Engineering Centre, School of Civil Engineering, Brisbane Qld 4072. Email: d.williams@uq.edu.au
4. PhD candidate at the School of Civil Engineering, Brisbane Qld 4072.
   Email: a.ranabhat@uqconnect.edu.au

ABSTRACT

To improve the settling, consolidation, and desiccation efficiency of bauxite residue in tailings storage facilities, a series of physical and chemical treatments were proposed before their disposal, including:

- whether or not to add fly ash and flocculent to thicken the residue

- whether or not to use caustic to reduce the pH

- pH neutralisation by either acid or sea water

- whether or not to dewater the sample through filtration and repulping.

Six types of residue samples subjected to various processing methods were identified, including one as Control. To identify the best combinations of these treatments, a series of laboratory and field tests were carried out on bauxite residue samples that underwent different sequences of treatments. The proposed tests include:

- Settling tests, particle size distribution tests, consolidation tests, and Atterberg limit tests to identify the impact of these treatments on the geotechnical parameters of tailings.

- Basin desiccation tests to identify how these treatments influence the desiccation behaviour of the tailings.

This paper discusses the outcome of the range of tests to date.

INTRODUCTION

Slurried tailings deposited to the sub-aerial storage facilities are subjected to settling, consolidation and desiccation. These physical processes can help dewater the tailings, and increase their dry densities and strengths, hence enlarging the storage capacity and enhancing the safety of the facilities. The optimisation measures to enhance these processes could be undertaken either before or after the tailings are deposited at the facility. The post-deposition optimisation methods aim to maximise the dewatering of wet tailings by extending their exposure time to the air, by periodically alternating the discharge cells, amphiroling, bulldozing and rolling. These methods are usually effective if the facility has a large footprint, and is situated in a dry climate where potential evaporation is high. On the other hand, pre-deposition optimisation methods aim to reduce the water content of the tailings before their discharge, or alternate their composition and properties so the physical processes can be accelerated after the tailings are deposited into the facility. These pre-deposition methods include thickening, centrifuge, hydrocyclones, and filtration.

Alternative ways of processing bauxite residue before its deposition in the proposed Bauxite Residue Management Area 2 (RMA 2) were proposed to achieve improved and safer residue dewatering outcomes. Currently, residue processing at Residue Manage Area 1 (RMA 1), as illustrated in Figure 1, involves:

- The last washer underflow from A260 neutralisation reactor is neutralised by sea water in the A270 neutralisation reactor.
- The underflow from A270 is then thickened by adding flocculant, and fly ash is added in the A580 thickener.
- The underflow from A580 is pumped to Residue Storage Area.

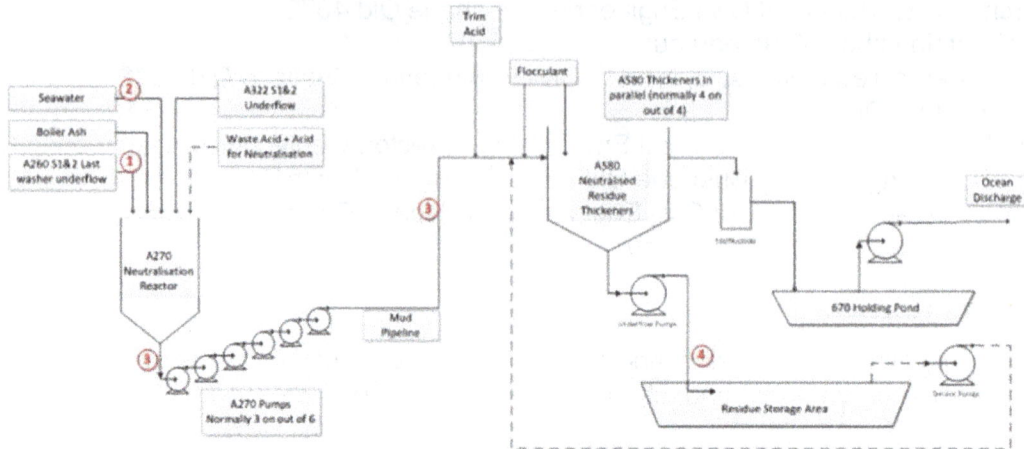

FIG 1 – Schematic of residue processing at RMA 1.

The proposed dewatering method, beginning with the underflow from A260 will involve selection among the following densification, neutralisation and thickening processes:

- whether or not to add fly ash and flocculant to thicken the residue
- whether or not to use caustic to reduce the pH
- either sea water or acid is used to neutralise the residue
- whether to thicken the residue by filtration and repulping.

To quantify the impact of the different processes on the geotechnical and hydrological parameters of the residue, and to identify the optimal processes, a series of laboratory-based tests were carried out at The University of Queensland Geotechnical Engineering Centre Laboratories (UQ-GEC) on residue processed by the conventional and alternative methods, including:

- initial gravimetric moisture content and solids content, electrical conductivity (EC), and pH
- Atterberg limits
- particle size distribution (PSD) and specific gravity (SG)
- settling time and settled dry density
- desiccation behaviour under constant heat and wind conditions, including settlement, cracking, densification, moisture content and suction, and changes in evaporation rate
- consolidation of the slurry residue using an instrumented slurry consolidometer
- instrumented column testing (not yet commenced).

Six different types of residue samples were nominated for testing (Table 1). This paper summarises the testing results and provides recommendations on the proposed processing methods.

TABLE 1

Description of the tested residue samples. The cells shadowed by light blue suggest the treatments were also adopted in the Control sample.

Sample ID	Detailed processing	Simplified processing
Control	Underflow from 580 neutralised residue thickener	580 Thickener UF
	Add 2% solids by mass fly ash, with flocculant (Flopam AN 913 VHM) at 140 g/t	Ash and flocculant
	No caustic dilution	No caustic dilution
	Neutralised with sea water	Sea water neutralisation
	Do not subject to Filtration and Repulping	No Filtration and repulping
SP01	Underflow from 580 neutralised residue thickener	580 Thickener UF
	Add 2% Solids by mass fly ash, with flocculant (Flopam AN 913 VHM) at 140 g/t[1]	Ash and flocculant
	No caustic dilution	No caustic dilution
	Neutralised with sea water	Sea water neutralisation
	Filtered and repulped with sample filtrate	Filtration and repulping
SP04	Last washer underflow from 260 neutralisation reactor	260 LWUF
	No fly ash or flocculant added	Neither ash nor flocculant
	Diluted to 75% of current caustic and re-thickened to 450 g/L solids	Caustic dilution and rethickening
	Neutralised with sea water	Sea water neutralisation
	Filtered and repulped with sample filtrate	Filtration and repulping
SP06	Last washer underflow from 260 neutralisation reactor	260 LWUF
	No fly ash or flocculant added	Neither ash nor flocculant
	No caustic dilution	No caustic dilution
	Neutralised with H_2SO_4	Acid neutralisation
	Filtered to 68% Solids, Repulp with sea water	Filtration and repulping
SP09	Last washer underflow from 260 neutralisation reactor	260 LWUF
	Add 2% Solids fly ash, with flocculant (Flomin AL P 40 VHM) at 25 g/t	Ash and flocculant
	Diluted to 75% of current caustic, and re-thickened to 450 g/L solids	Caustic dilution
	Neutralised with sea water	Sea water neutralisation
	Filtered to 68% Solids, and Repulp	Filtration and repulping
SP11	Last washer underflow from 260 neutralisation reactor	260 LWUF
	Add 2% Solids fly ash, with flocculant (Flomin AL P 40 VHM) at 25 g/t	Ash and flocculant
	No caustic dilution	No caustic dilution
	Neutralised with H_2SO_4	Acid neutralisation
	Filtered to 68% Solids, Repulp with Sea water	Filtration and repulping

TESTING METHODOLOGY AND RESULTS

Initial sample state and test procedures

Six buckets of each of the six residues were delivered for testing. The EC and pH of the supernatant water in each bucket were measured. Before the characterisation tests, samples in the bucket were well mixed using a hand-held paddle powered by a rotary drill. The methods used for the characterisation tests are detailed in Table 2.

TABLE 2

Characteristic testing of residue samples.

Testing		Methodology	Standard
Initial state	Gravimetric moisture and % Solids	Weighing sample before and after oven drying at 60°C for 24 hours	-
	pH and EC	pH and EC metre Hannah™ IC-HI2020–02	-
Atterberg limits	Liquid limit	Casagrande method	AS 1289.3.1.1 (2009)
		Falling cone method	AS 1289.3.9.1 (2015)
	Plastic limit	Thread rolling method	AS 1289.3.2.1 (2009)
	Plasticity Index and Unified Soil Classification (USC)	-	AS 1289.3.3.1 (2009) AS 1726 (2007)
PSD	Particles passing 0.075 mm	Wet sieving	AS 1289.3.6.2 (1995)
	Particles finer than 0.075 mm	Hydrometer	AS 1289.3.6.3 (2003)
	SG	Pycnometer using Helium	AS 1141.6.2 (1996)
Settling from 25% or 35% Solids		Settling column test	Islam *et al.* (2020)
Desiccation from the initial state		Basin desiccation method	Islam *et al.* (2021)
Consolidation from the initial state		Slurry consolidometer	Williams and Zhang, (2017)

Basin desiccation test

The instrumented basin desiccation tests (Figure 2) carried out on each residue type were intended to provide a preliminary assessment of the drying behaviour of the surface residue from its initial state, under constant laboratory drying conditions. Well-mixed residue samples at their initial solids content and of known initial masses (with initial slurry masses of 11.54 kg, 16.8 kg, 16.59 kg, 15.22 kg, 13.9 kg and 14.6 kg, solids mass of 322 g, 753 g, 703 g, 677 g, 569 g and 672 g, and solids contents of 27.9 per cent, 44.8 per cent, 42.4 per cent, 44.5 per cent, 41.0 per cent, and 46.0 per cent respectively, for Control, SP01, SP04, SP06, SP09 and SP011) were tested in six separate basins instrumented with UQ-GEC moisture and suction sensors (see Figure 3) at their mid-depth.

An electric fan was installed above each basin to enhance the desiccation process (Figure 2). The loss of water to evaporation from each basin was recorded by a digital balance at 30-second time intervals. The settlement, crust and crack development and salt precipitation in each basin were captured by high-resolution images at 4-hour time intervals using a camera installed above each basin. The moisture content and suction of the residue over time were obtained from the calibrated moisture and suction sensors via a UQ-GEC data logger. The test was terminated when evaporation had practically ceased.

FIG 2 – Instrumented basin desiccation tests, showing: (a) test set-up, and (b) basins filled with residue samples.

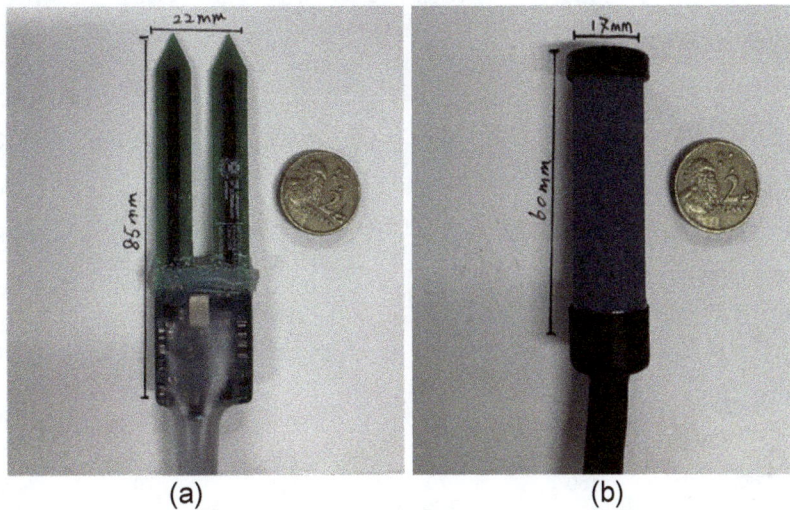

FIG 3 – (a) Dielectric moisture, and (b) Thermal suction sensors.

Test results

A summary of the test results to date of all the residue samples is given in Table 3, in which treatments different from the Control sample are not shaded under the simplified descriptor, and the highest and lowest values of each parameter are highlighted in red and blue, respectively.

TABLE 3

Characterisation testing results. The cells shadowed by light blue suggest the process was also adopted in the control sample.

Sample ID		Control	SP01	SP04	SP06	SP09	SP11
Simplified descriptor		UF 580	UF 580	LWUF 60	LWUF 260	LWUF 260	LWUF 260
		Ash and flocculant	Ash and flocculant	No ash or flocculant	No ash or flocculant	Ash and flocculant	Ash and flocculant
		No caustic dilution	No caustic dilution	Caustic dilution	No caustic dilution	Caustic dilution	No caustic dilution
		Sea water	Sea water	Sea water	Acid	Sea water	Acid
		No Filtration and Repulping	Filtration and Repulping	Filtration and Repulping	Filtration and Repulping	Filtration and Repulping	Filtration and Repulping
Initial state	GMC (%)	258.7	123.9	136.1	124.8	143.6	117.5
	% Solids	27.9	44.8	42.4	44.5	41.0	46.0
	pH	8.0	10.0	11.0	6.5	11.0	6.0
	EC (mS/cm)	43.4	68.8	32.4	55.5	40.6	57.7
Atterberg limits	Liquid Limit (%, Casagrande)	80.7	90.7	65.5	65.8	64.3	66.9
	Liquid Limit (%, Falling cone)	85.6	81.6	69.3	67.1	69.6	67.4
	Plastic Limit (%)	55.3	46.7	45.6	44.8	45.1	32.8
	Plasticity Index (%)	25.5	44.0	19.9	21.0	19.3	34.0
	USC	MH	MH	MH	MH	MH	MH
PSD	% Clay-size	52	65	55	63	79	52
	% Silt-size	45	27	40	30	18	41
	% Sand-size	3	7	5	7	3	7
	SG	2.73	2.68	2.90	2.83	2.90	2.82
Settling from 25% Solids	Settling time (days)	16	14	17	7	13	13
	Settled % Solids	29.2	31.0	32.0	36.8	30.1	29.7
	Settled dry density (t/m^3)	0.361	0.385	0.405	0.483	0.375	0.367
Settling from 35% Solids	Settling time (days)	16	10	17	10	12	10
	Settled % Solids	29.2	36.2	35.5	35.8	39.5	36.2
	Settled dry density (t/m^3)	0.361	0.451	0.463	0.467	0.534	0.473
Desiccation	GMC at crack initiation (%)	130	90	85	85	105	85
	GMC at the rise in suction (%)	150	95	80	90	90	80
	GMC at desaturation (%)	150	70	80	70	90	90
	GMC at decreased evaporation (%)	100	55	45	55	65	50
	Size of crack blocks	Smallest	Large	Small	Small	Small	Large

Initial gravimetric moisture and solids concentration

The control sample has the lowest solids content of 27.9 per cent while the highest initial gravimetric moisture (GMC) Subjected to filtration and repulping, the other five residues have relatively similar GMC and solids concentration, with average values of 129 per cent and 43.7 per cent, respectively. The sample with the highest initial solids content is SP11 (46 per cent Solids and GMC of 117.5 per cent). This suggests filtration and repulping increased the initial solids concentrations of the processed residue.

pH and EC

The Control sample has a pH of 8.0, indicating the effectiveness of using sea water in neutralising bauxite residue from a pH of ~11. Among the non-control samples that were subjected to two neutralisation methods, acid neutralisation is much more effective in reducing the pH of the residue than sea water. Acid neutralisation reduces the pH of the residue to *circum* neutral between 6 and 6.5 (SP06 and SP11), while sea water reduces the pH only marginally to remain alkaline between 10 and 11 (SP01, SP04 and SP09). It is not evident how caustic dilution can help reduce the pH of the samples as the samples subjected to this treatment (SP04 and SP09) has the highest pH of 11.

The EC values of all residue samples are close to that of sea water (50 mS/cm), suggesting that sea water is not a significant contributor to their EC. The two residue samples neutralised by acid have very similar EC values (55.5 mS/cm for SP06 and 57.7 mS/cm for SP06), while the EC values of the residue samples neutralised by sea water range between 32.4 mS/cm and 68.8 mS/cm, implying that their EC values before neutralisation varied more widely. Due to the dilution of ions, samples subjected to caustic dilution have less EC values (32.4 and 40.6 for SP04 and SP09, respectively) than that without caustic dilutions.

Atterberg Limits

Atterberg limits are an indicator of the plasticity of a material, and the higher the plasticity the more difficult the material is to dewater and handle. The SP01 sample has the highest Liquid Limit using the Casagrande method (90.7) while the control sample has the highest Liquid limit when using the Falling cone method (85.6). In contrast, the Liquid Limits of all other samples (SP04, SP06, SP09 and SP11) are similar, averaging 67 per cent with only ±3 per cent variation. The Control sample has the highest plastic limit of 55.3 per cent. Regarding the samples after filtration and repulping, the Plastic Limits of most are around 45 per cent, apart from SP11 with a lower value of 32.8 per cent. Sample SP01 has the highest Plasticity Index of 44 per cent, followed by SP11 of 34 per cent, then Control, SP04, SP06 and SP09 with values of about 20 per cent forming a cluster in the. All residue samples are classified as MH, high plasticity silty clay-sized materials, as shown in the plasticity chart Figure 4. However, samples SP04, SP06 and SP09 form a cluster with slightly lower liquid limits and plasticity index, suggesting they are comparably easier to dewater and handle in a facility than the other materials (Control, SP01 and SP11).

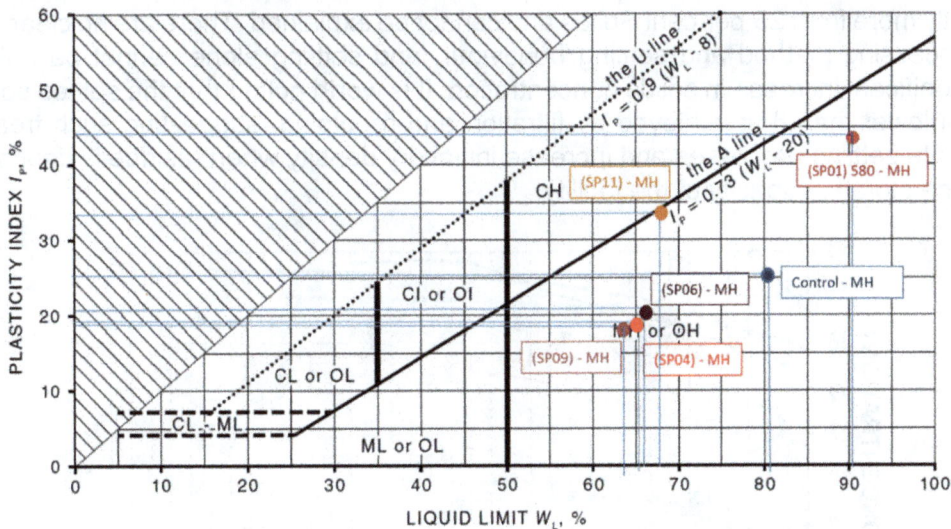

FIG 4 – Plasticity chart and Unified Soil Classification of residue samples.

Particle Size Distribution and Specific Gravity

The particle size distribution curves of the residue samples are shown in Figure 5, and the % clay-, silt- and sand-size are given in Table 3, along with the specific gravity values. All the residue samples classify as silty clay-sized, with 52 to 79 per cent clay-size and less than 7 per cent sand-sized. There does not appear to be a difference in PSD with residue processing, suggesting that the difference in PSD may relate more to the variability of the bauxite ore feed. The small addition of fly ash is unlikely to impact the PSD of the residue. The addition of flocculant would agglomerate fine particles, however, flocculants are likely to 'unwrap' on wet sieving, leading to the breakdown of agglomerates.

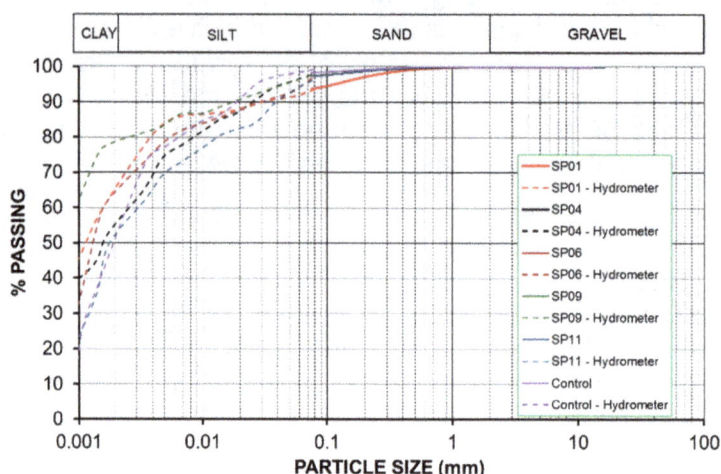

FIG 5 – Particle size distribution curves of residue samples.

The SGs of the residue solids ranged from 2.68 (SP01) to 2.90 (SP04 and SP09). Filtration and repulping seem to impact negatively the SGs of the samples. The small addition of fly ash is unlikely to impact the PSD of the residue.

Settling

Settling tests were conducted from nominal initial 25 per cent Solids and 35 per cent Solids for each residue sample. The settling results versus time are plotted in Figures 6 and 7 in terms of solids concentration and dry density, and the final settled % Solids and dry density values are given in Table 2. The duration of the settling tests varied from 7 to 17 days, being generally shorter from the higher initial solids concentration, which is postulated to be the result of coarser particles forcing finer particles to settle with them. The final solids concentration and dry density were generally higher than the higher initial solids concentration, apart from the result for residue sample SP06, which settled slightly more from 25 per cent Solids and cannot be explained. There are no clear correlations between processing method and settling behaviour, and settling alone cannot be relied upon to achieve a significant increase in solids concentration. It is worth noting that the settled solids content is in general lower than that achieved by filtration and repulping, suggesting such treatment may help accelerate settling processes and increase initial dry density when residue is deposited, hence enhancing the storage capacity of the facility.

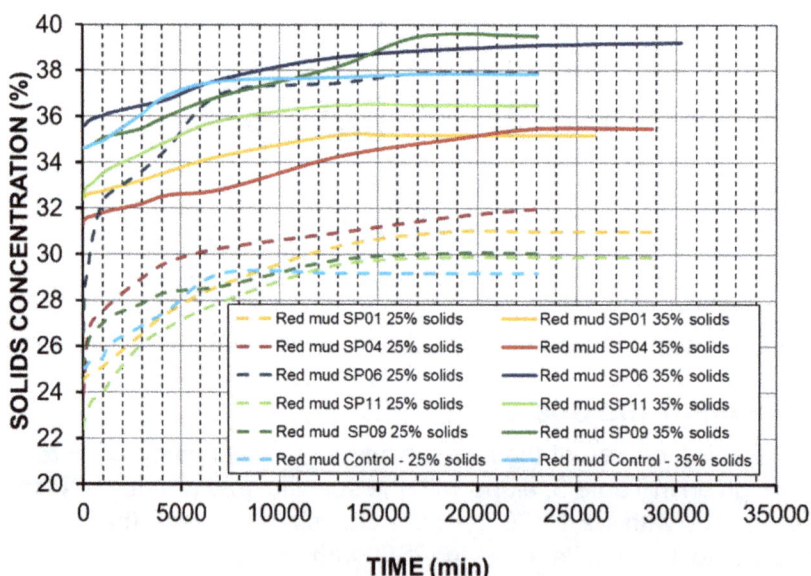

FIG 6 – Solids concentration versus time during settling tests.

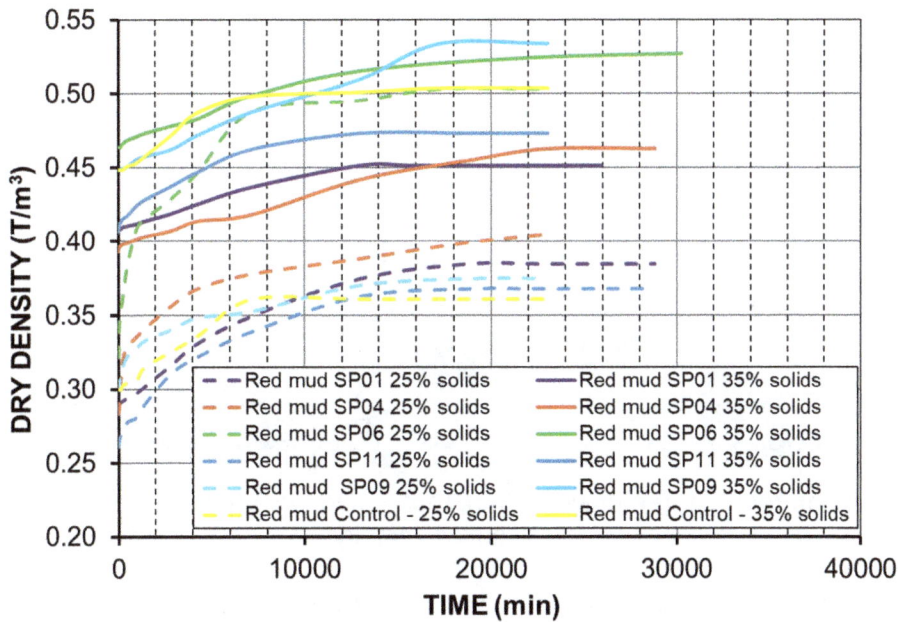

FIG 7 – Dry density versus time during settling tests.

Desiccation

Figure 8 shows images of the residue samples during desiccation testing, and Table 3 and Figure 9 summarise the results of the desiccation testing. In their initial slurry state, evaporation was accompanied by the settlement of the residue samples, as seen in Rows 1 and 2 in Figure 8 and Row 4 in Figure 9. The evaporation rates at high GMC remained relatively high and constant at 3 to 4 mm/day, with daily fluctuations induced by changes in the laboratory temperature and humidity and positioning of the fans, as seen in Row 1 in Figure 9. The residue initially remained fully water saturated with no suction development, as seen in Rows 2 and 3 of Figure 9.

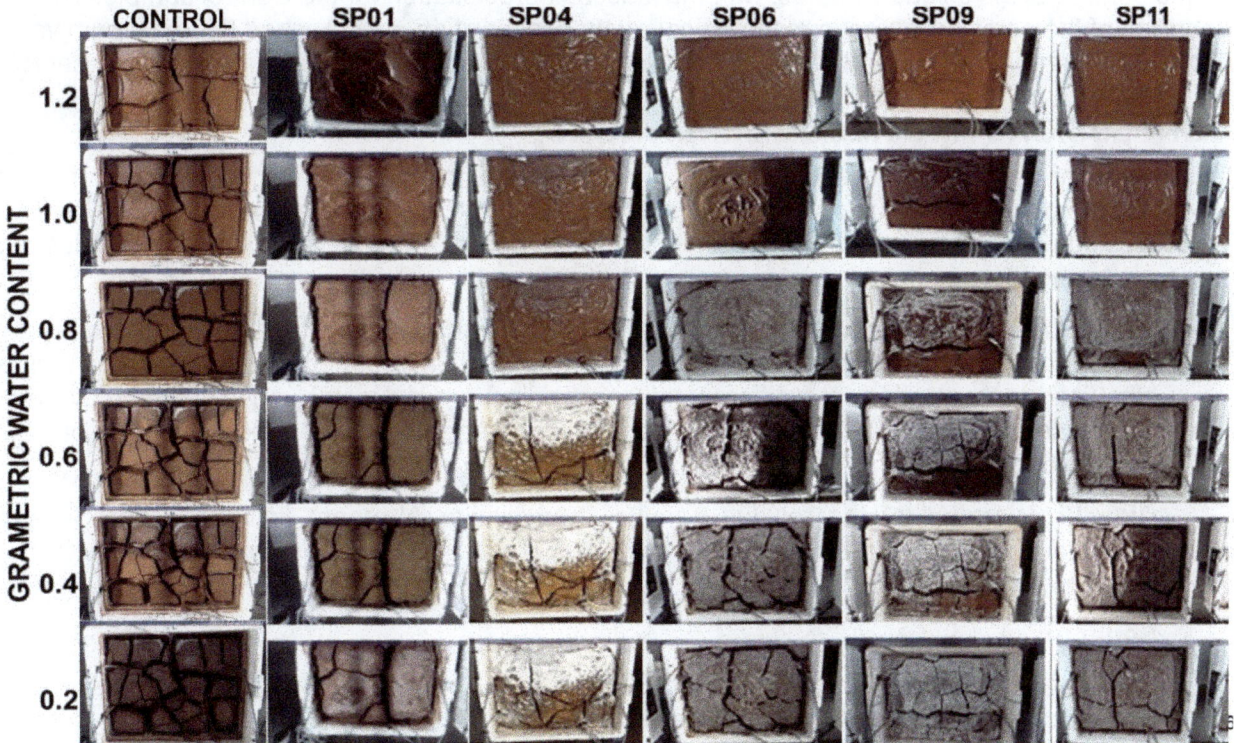

FIG 8 – Images of the residue samples during desiccation testing.

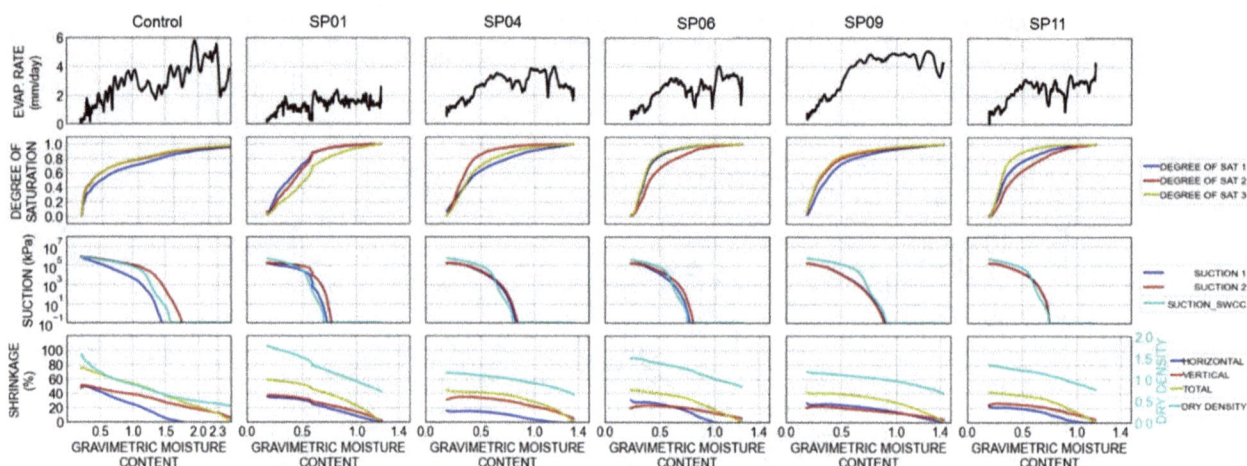

FIG 9 – Evaporation rate, degree of saturation, matric suction, shrinkage and over GMC during desiccation testing.

For the non-control samples, cracks formed in the residue surface as desiccation continued, starting at 105 to 85 per cent GMC, followed by suction development from 95 to 80 per cent GMC, desaturation from 90 to 70 per cent GMC, and a drop off in evaporation rate from 65 to 45 per cent GMC. On the contrary, the control sample experiences these events at higher GMC, specifically cracking at 130 per cent GMC, desaturation and developing suction at 150 per cent, and evaporation decline at 100 per cent GMC. Such drastic difference in desiccation behaviours between the Control and Non-control samples is likely due to the influence of filtration and repulping, which generate slurries with much higher initial GMC (~44 per cent for samples with filtration and repulping and 27.9 per cent without it), and facilitate agglomerating fine particles. Such particle binding effect can also be identified by the size of blocks after cracking, as samples with lower tensile strength tend to crack with smaller block sizes. The size of the crack block was smallest for the Control sample and largest for residue samples SP01 and SP11. Evaporation caused salts to accumulate and precipitate on the surface of all residue samples, with the least salt precipitation for the Control sample. The dry densities of the samples increase monotonically with desaturation, and reached the maximum value between 1.2 and 1.6 at the end of the desiccation test, when the GMW is lower than 20 per cent.

CONCLUSIONS AND RECOMMENDATIONS

To quantify the impact of different processes on the geotechnical and hydrological parameters, and to identify the optimal processes, a series of laboratory-based tests are being carried out at UQ-GEC on six residue samples processed by the conventional and alternative methods. The conclusions arising from the test results are:

- **Initial state:** Samples subject to filtration and repulping had higher initial gravimetric moisture contents (or solids concentrations) than sample without this treatment. All samples had similar electrical conductivities, but different pH depending predominantly on whether they were neutralised using sea water or acid. Caustic dilution seems to impact negatively on pH, but helps reduce EC.

- **Classification:** All of the residue samples had similar Atterberg limits and particle size distributions, and all were classified as MH, high plasticity silty clay-sized materials with less than 7 per cent sand-size.

- **Settling:**

  o None of the settling tests from nominal 25 per cent Solids and 35 per cent Solids achieved the as-supplied 41 to 46 per cent Solids.

  o On settling from a nominal 25 per cent Solids, all but one of the residue samples achieved a modest increase in the % Solids to between 29 per cent and 32 per cent Solids, while residue sample SP06 was an outlier, achieving over 36 per cent Solids. Samples after filtration and repulping achieved a slightly higher % solids than samples without this treatment.

- On settling from a nominal 35 per cent Solids, all the residue samples showed little increase in solids concentration.

- On settling from nominal 25 per cent Solids and 35 per cent Solids, samples after filtration and repulping achieved a slightly higher % solids than samples without this treatment.

- There are no clear correlations between processing method and settling behaviour and settling alone cannot be relied upon to achieve a significant increase in solids concentration.

- **Desiccation:** Leads to a two- to three-fold decrease in gravimetric moisture content from the initial as-supplied state, although this occurs to only limited depth and ceases on desaturation, with significant crack development limiting the loss of volume. Samples that experience filtration and repulping tend to desaturate, develop suction, form cracks and reduce evaporation at lower gravimetric moisture content than the sample without this treatment. This is likely because filtration and repulping help agglomerate solids, hence forming stronger bindings between particles.

Overall, the laboratory-tested hydrological and geotechnical parameters of all the residue samples are similar, likely because the proposed physical and chemical treatments do not significantly alter the compositions of the residues. However, filtration and repulping effectively reduced the solids contents of the residue, which may lead to different consolidation and desiccation behaviours when the residue is deposited in the facility. In particular, the control sample with higher water content is more prone to the development of excess pore water pressure, leading to poorer consolidation. Additionally, as desiccation by natural weather can only start from the surface and gradually progress to the deeper residues, the control sample with higher water content may be subjected to inefficient desiccation if the sun exposure is too short or each deposition is too thick. Therefore, it is recommended to carry out field tests to identify how each sample is subjected to settling, consolidation and desiccation in natural weather, which could unravel their behaviour in the proposed new residue management area.

REFERENCES

Islam, S, Williams, D J, Llano-Serna, M and Zhang, C, 2020. Settling, consolidation and shear strength behaviour of coal tailings slurry, *International Journal of Mining Science and Technology*, 30(6):849–857. https://doi.org/10.1016/j.ijmst.2020.03.013

Islam, S, Williams, D J, Zhang, C and Llano-Serna, M, 2021. Geotechnical characterisation of coal tailings down the beach and constant rate of loading consolidation in a slurry consolidometer, *Mining Technology: Transactions of the Institute of Mining and Metallurgy*, 130(2):67–80. https://doi.org/10.1080/25726668.2021.1880351

Williams, D J and Zhang, C, 2017. A suite of tests for designing slurried tailings deposition, in *Proceedings of the Fourth International Seminar on Tailings Management*, 8 p.

Governance and standards

Vale's roadmap for the future – integrating multiple tailings compliance frameworks

F da Silva[1], G Puro[2], S Wood[3], O Iwakun[4] and S Fernandez[5]

1. Senior Geotechnical Governance Specialist, Vale, Canada. Email: fabricio.dasilva@vale.com
2. Base Metals Manager for Dams, Vale, Canada. Email: greg.puro@vale.com
3. CEO, Forwood Safety, Brisbane Qld 4101. Email: steve@forwoodsafety.com
4. Principal Geotechnical Engineer, Vale, Canada. Email: olumide.iwakun@vale.com
5. Senior Geotechnical Governance Specialist, Vale, Canada.
   Email: sebastian.fernandez@vale.com

ABSTRACT

The last decade has been characterised by some of the worst environmental and loss-of-life disasters in the history of tailings dam failures. As a result, the Global Industry Standard on Tailings Management (GISTM) has been introduced as a multi-stakeholder effort to prevent future tailings storage facilities (TSF) related disasters from happening again. Many mining companies, ICMM (International Council on Mining and Metals) members and non-members have committed to comply with the Standard.

Operationalising this governance framework will require regular self-assessments to identify and manage gaps toward standard compliance. In that respect, different requirements are provided for implementing and auditing the standard. This undertaking, combined with the additional effort to comply with jurisdictional legal requirements, other standards and guidelines as well as internal policies can be extensive and resource laden.

Even though there are many requirements derived from all these standards-related demands, there are also many resemblances. In that direction, Vale has taken this challenge as an opportunity to streamline the efforts and, in collaboration with Forwood Safety, built a platform to simplify the compliance journey against multiple standards.

This paper presents an innovative approach to merge requirements from multiple standards into a single integrated verification process, thereby reducing the number of assessments and introducing a new attempt to differentiate the best practice from standard practice. The equivalency study completed by ICMM and MAC (Mining Association of Canada) as well as Vale's internal normative documents and local regulatory compliance were used to develop this solution.

Vale also designed an active action management approach to link the resulting action plans from the gap analyses with the compliance score, gaining real-time compliance visibility, allowing the system to forecast when a TSF will be 100 per cent compliant against the chosen standard.

INTRODUCTION

History shows that, several catastrophic tailings dam collapses have highlighted the urgent need for mine operators to address the safety of these structures. In recent years, significant efforts have been made to enhance the standards for tailings storage facilities (TSFs) and water storage dams. Vale acknowledges the importance of these standards and improvements, which stem from insights gained from international industry bodies.

Tailings management frameworks vary from country to country and even within jurisdictions, which make it challenging for organisations to navigate and comply with the different requirements. For example, in Canada, the Mining Association of Canada (MAC) Towards Sustainable Mining (TSM) sets out specific requirements for tailings management, while Global Industry Standard on Tailings Management (GISTM) sets out several requirements across a range of topics for tailings management.

Some standards emphasise on technical aspects, while others have a broader focus aimed at improving transparency for internal and external stakeholders. Vale acknowledges the strengths from different standards and commits to proactively working towards conformance of multiple industry tailings standards including GISTM, MAC TSM and the organisation's own internal tailings management systems.

Operationalising multiple tailings management frameworks is a complex and resource-intensive process that requires careful planning and coordination to ensure compliance, consistency, and effective stakeholder engagement. This paper discusses the innovative approaches that Vale has implemented to overcome the challenges and effectively operationalise their commitment to comply with multiple tailings management frameworks.

Vale's roadmap goes beyond the compliance to meet the true intent of the requirements of the standards.

OPERATIONAL CONFORMANCE CHALLENGES

Conformance with tailings management frameworks and local, national, and international regulations requires significant investment in technology, infrastructure, and resources. It also requires a commitment to continuous improvement and transparency, which may require changes in organisational culture and management practices. These are the key challenges with operationalising conformance with multiple tailings management frameworks:

- **Coordination and Consistency:** Consistent implementation of multiple tailings management frameworks can be a complex and challenging process that requires significant coordinated efforts between different teams and departments within an organisation.

- **Public Disclosure:** Manual processes to ensure that public disclosed information is accurate, substantiated, accessible, understandable, current, and approved for release presents its own set of difficulties.

- **Adherence Tracking:** Managing verification findings in multiple systems without a direct relation to compliance goals can lead to inefficiencies and inconsistencies in the reporting process. It can also result in an increased risk of non-compliance if verification findings are not properly addressed or prioritised.

- **Verification Fatigue:** Regular self-assessments, external verifications on active and legacy sites, maintaining the assessment data, managing identified gaps and areas of non-conformance, proves to be a formidable task for all TSF and dam owners.

- **Communication and Engagement:** The implementation of multiple frameworks can hinder the effective communication and engagement with all stakeholders including employees, regulators, and local communities, involved in tailings management.

- **Organisational Standards and Procedures:** Complying with various tailings management guidance and conformance protocols, as well as adhering to internal standards and procedures, can be an arduous task for TSF and dam owners.

- **Resource Allocation:** Implementing multiple tailings management frameworks requires significant resources, including financial and human resources. This can be especially challenging for smaller companies with limited resources.

- **Technical Complexity:** Different tailings management frameworks may have different technical requirements, which can make it difficult to ensure that all sites are being managed effectively and safely. This can be especially challenging for companies with multiple sites located in different regions or countries, as local regulations and standards may also differ.

- **Systems Integration:** Conformance to the standards requires stronger interaction and integration between the range of disciplines and various systems dealing with engineering, social and environmental monitoring.

- **Monitoring and Review:** A challenge that organisations may encounter is effectively assigning responsible individuals for identified gaps and monitoring performance against internal and external targets. This impacts the organisation's ability to efficiently address gaps and meet established targets.

- **Document Control:** Along with performing regular verifications, the verification findings need to be consolidated and visually reported to the relevant authorities which is a major challenge. The maintenance and upkeep of verification records is also a challenge.

A ROADMAP FOR INTEGRATED TAILINGS COMPLIANCE

Embarking on the journey of operationalising a tailings management framework entails establishing consistent processes and continually advancing towards achieving full compliance.

As shown in Figure 1, the journey can be broadly classified in four phases:

1. **Verification Phase:** Independent verifications are a crucial part of the process to ensure compliance with multiple global and internal standards. After an independent verification is conducted, it is essential to identify any gaps or non-conformances and develop action plans to address them. Tracking compliance and collecting supporting evidence is essential to demonstrate compliance.

2. **Reporting Phase:** The next phase after the independent verification is to report the verification findings. This phase involves several key goals, including executive compliance reporting, action management, overall adherence tracking and plans for ironing out site-level differences to achieve deadlines.

3. **Integration Phase:** Establishing Standard Operating Procedures (SOPs) with multiple standards combined into an integrated verification is a key deliverable of this phase. This involves analysing and developing integrated verifications that cover multiple standards, rather than conducting separate verifications for each standard. Regular reviews are undertaken on the existing governance structures and programs to identify any areas for improvement to meet compliance requirements. This involves evaluating the effectiveness of current processes and procedures, identifying any gaps or weaknesses in transparent reporting to the internal and external stakeholders.

4. **Beyond Compliance Phase:** Going beyond compliance means implementing practices that exceed the minimum requirements set by the tailing's standards. This can help organisations to achieve higher levels of sustainability, social responsibility, and ethical behaviour by setting ambitious goals, investing in innovation, collaboration with other organisations. Building trust with stakeholders and investors through transparent reporting and produce evidence that demonstrates the organisation is going beyond compliance.

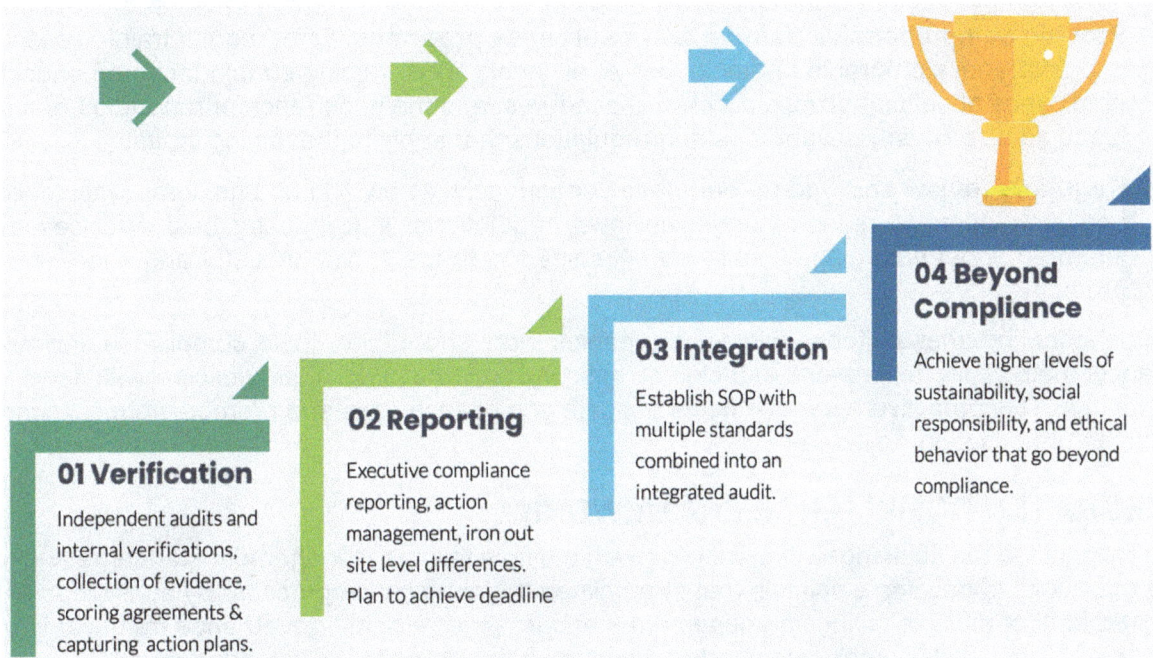

01 Verification

Independent audits and internal verifications, collection of evidence, scoring agreements & capturing action plans.

02 Reporting

Executive compliance reporting, action management, iron out site level differences. Plan to achieve deadline

03 Integration

Establish SOP with multiple standards combined into an integrated audit.

04 Beyond Compliance

Achieve higher levels of sustainability, social responsibility, and ethical behavior that go beyond compliance.

FIG 1 – Roadmap for integrated tailings compliance.

Compliance audit and tracking

In a Harvard Business Review, the author examines 'Why Compliance Programs Fail – and How to Fix Them' (Chen and Soltes, 2018). The paper summarises that the firms should start by linking compliance initiatives more closely to specific objectives: preventing misconduct, detecting it, or aligning policies with laws and regulations. Companies routinely had policies and procedures but did not track breaches, or even a formal conformance score.

Mining companies can take several steps to link compliance initiatives more closely to specific objectives and improve their tracking of breaches. The following steps describe in general framework applied at vale to rollout compliance audit and tracking.

- **Conduct a comprehensive compliance assessment:** Begin by conducting a thorough assessment of existing compliance policies, procedures, and controls. Identify any gaps or areas for improvement that need to be addressed. This assessment will serve as a baseline for further actions.

- **Establish clear compliance objectives:** Define specific compliance objectives related to preventing misconduct, detecting it, and aligning policies with laws and regulations. These objectives should be measurable, realistic, and aligned with the company's overall goals and values.

- **Develop a robust compliance framework:** Create a well-defined compliance framework that outlines the key elements of compliance, including policies, procedures, risk assessments, training programs, and internal controls. Ensure that this framework aligns with the specific compliance objectives established in the previous step.

- **Evidence based reporting and tracking mechanisms:** Implement robust systems and processes to capture and track compliance **breaches** effectively. This may involve leveraging technology solutions, such as compliance management software that supports evidence-based tracking. Encourage employees to attach substantiating evidences to demonstrate the adherence of compliance

- **Regular monitoring and auditing:** Conduct regular monitoring and auditing activities to assess compliance with policies and regulations. This includes reviewing data, conducting internal audits, and performing risk assessments. Regular review of breaches to ensure they are completely resolved and supporting evidence are provided. This allows the internal assessments frequency to be kept more reasonable without leading to audit fatigue.

- **Provide comprehensive training and awareness programs:** Offer regular training sessions and awareness programs to employees at all levels. This training should focus on educating employees about compliance policies, procedures, and the importance of reporting breaches. It should also cover relevant laws and regulations that apply to the mining industry.

- **Regularly review and update policies:** Continuously review and update compliance policies and procedures to reflect changes in laws, regulations, and industry best practices. Stay informed about emerging compliance risks specific to the mining industry and address them promptly.

By implementing these steps, mining companies can strengthen their compliance initiatives, enhance their ability to prevent and detect misconduct, and align their policies with laws and regulations. This proactive approach helps mitigate compliance risks and promotes ethical conduct within the organisation.

INTEGRATED COMPLIANCE FRAMEWORK

Vale recognised the challenges of complying with multiple tailings management standards and took the opportunity to develop a solution that streamlined the compliance process. By consolidating the requirements of multiple tailings management standards into a single governance framework, Vale was able to simplify its compliance efforts and improve the efficiency of its operations.

In developing this solution, Vale relied upon equivalency assessments conducted by different industry bodies to compare various tailings management standards and identify areas of alignment and divergence. The equivalency assessments are aimed to identify the key elements of a credible tailings management systems and evaluate the equivalency of different tailings management standards, such as the GISTM and the MAC TSM.

The study's findings provided insights and recommendations to help TSF owners to improve their tailings management practices and ensure compliance with relevant standards. Vale conceptualised the integration of the GISTM, MAC TSM and their internal Tailings and Dams Management System (TDMS) to create a single verification, covering these three standards. Bringing the conceptual idea of an integrated verification to fruition was a challenging and multi-step process that involved:

- Analysis of the resemblances of the requirements derived from 77 GISTM requirements, 96 TSM requirements, 36 TDMS requirements to identify commonalities, and five legal requirements applicable to a specific jurisdiction (not covered in any applicable standard); and streamline the compliance process. More than 200 requirements were analysed.

- Creation of mappings against the requirements of GISTM versus TSM, GISTM versus TDMS, TSM versus TDMS to derive the commonalities, and mapping all the local regulations that would still require an additional effort to comply with.

- Development of 107 integrated verification questions based on the common requirements identified through the exercise of mapping individual standards. The count of questions developed for the integrated verification was reduced by nearly 50 per cent.

- Linkage of the 107 integrated verification questions to the relevant GISTM, TSM and TDMS requirements.

This was a time and resource-intensive process for Vale as it required a thorough understanding of each standard and its specific requirements.

Conceptual design

Figure 2 illustrates the conceptual design in how different standards have common requirements and an integrated verification can cover multiple requirements in a single verification.

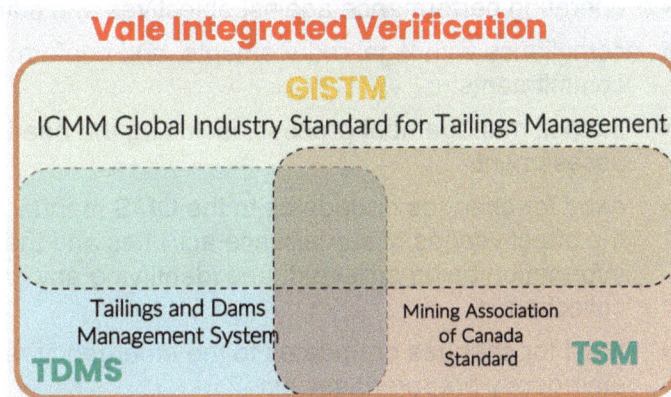

FIG 2 – Integrated verification – conceptual design.

Table 1 maps the commonalities between the requirements of GISTM 7.2, 7.4 and 7.5 and TSM 2.27 and the merging of these requirements into a single verification question.

TABLE 1

Mapping of GISTM and TSM requirements example.

GISTM requirements	**Topic III: Design, Construction, Operation and Monitoring of the Tailings Facility**
	PRINCIPLE 7: Design Implement and Operate Monitoring Systems
	7.2 Design, implement and operate a comprehensive and integrated engineering monitoring system that is appropriate for verifying design assumptions and for monitoring potential failure modes. Full implementation of the Observational Method shall be adopted for non-brittle failure modes. Brittle failure modes are addressed by conservative design criteria.
	7.4 Analyse technical monitoring data at the frequency recommended by the EOR, and assess the performance of the tailing's facility, clearly identifying and presenting evidence on any deviations from the expected performance and any deterioration of the performance over time. Promptly submit evidence to the EOR for review and update the risk assessment and design, if required. Performance outside the expected ranges shall be addressed promptly through Trigger Action Response Plans (TARPs) or critical controls.
	7.5 Report the results of each of the monitoring programmes at the frequency required to meet company and regulatory requirements and, at a minimum, on an annual basis. The RTFE and the EOR shall review and approve the technical monitoring reports.
TSM requirements	**Indicator 2**
	Performance Evaluation
	2.27 Do performance evaluations include results of surveillance and reviews (both internal and independent) and address:
	• operating performance against objectives and critical controls.
	• compliance with legal requirements, and conformance with plans and commitments.
	• the risk management process, including the need to update the risk assessment.
	• need for changes or updates to the OMS manual, including evaluating the effectiveness of surveillance activities and the utility of the information being collected, and identifying any gaps in information collection.
	• need for changes or updates to the emergency response plan and emergency preparedness plan?
TDMS requirements	**People**
	1 – TDMS Training
	1. Geotechnical employees trained in the TDMS
	Performance
	25 – Engineer of Records
	25 The EOR or DOR (if the structure is in the design or construction phases and the EOR is from Vale) contracted, appointed, and managed according to the guidelines in PNR – 000106.
Integrated Verification requirements	1. EoR states that the facility is performing according to its design intent

CHALLENGES

The journey in these last 22 months among Vale Base Metals business units using the new developed platform, aiming to deliver the compliance against all applicable standards has requested a uniformed and aligned efforts from different areas within the company. Some of the challenges and lessons learned from the process will be listed below.

Scarcity of expertise

The main challenge faced was related to the shortage of dams and tailings experienced professionals with the required knowledge and time to perform the comparison exercise and hence compile the verification assessment checklist that would aim to encompass all the applicable standards. This situation can be even more emphasised, when that professional needs to be knowledgeable in multiple jurisdictions where Vale Base Metals business unit operates.

Reporting and scoring variations

The inconsistent or non-existent reporting scores for each standard. While one can be binary, other is a letter ranking categorisation, and a third one would be in a decimal format or percentage. Work in the system background to find a common conformance score denominator that would support he evidence-based approach to reflect the conformance reality, and still tracking progress along the journey was an exciting overcome problem.

Conformance correlation issues

The conformance correlation from one standard to another is another major challenge in applying a unified approach. In the integrated verification approach, there many occasions where we would have encountered one to many correlations, this means that one single requirement in one standard, is equivalent to many similar requirements in other. As can be imagined, this brings a problem of error allowance. For exemplification simplification, percentage will be used as a scoring report metric. From the integrated verification approach, if in a standard A you can achieve 95 per cent conformance, the same Standard A, but in the stand-alone verification, that score can very base on company error tolerance (4 per cent), meaning the score can be between 91–99 per cent. It is up to each company to defined which tolerance to use, recognise this, and continue improve the process to the error can be minimise at each review cycle.

INTEGRATED FRAMEWORK BUSINESS BENEFITS AND LEARNINGS

Vale has saved a significant amount of time and effort by combining various tailings management standards into a single verification. This integration has resulted in a reduction of regular self-assessments, and reporting, culminating on the company concentrating its efforts on managing the newly created and existing action plans to address the identified gaps.

As shown in Figure 3, Vale experienced an approximate 50 per cent reduction in verification time and effort through this innovative approach and optimised its tailings management practices across its entire operations. In conjunction with the integrated verification, the solution was developed to address another time-consuming routine task, which is the reporting capabilities, in which another approximately 50 per cent-time reduction is being achieved.

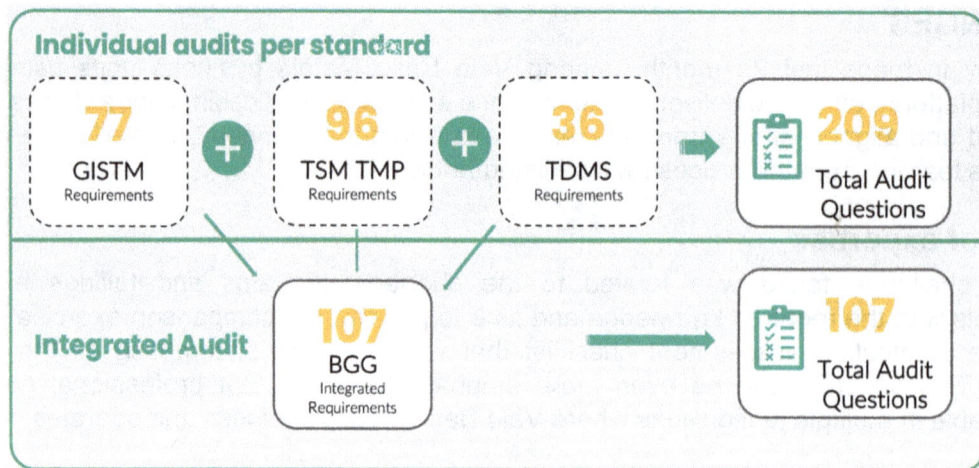

Individual audits per standard

77 GISTM Requirements **+** **96** TSM TMP Requirements **+** **36** TDMS Requirements → **209** Total Audit Questions

Integrated Audit **107** BGG Integrated Requirements → **107** Total Audit Questions

Approx. **50%** reduction in audit time and effort

FIG 3 – Optimisation through integrated verification.

The incorporated dashboards, provide users the ability to assess their progress visually and keep track of their findings and associated actions. Apart from simplifying the compliance process, the system's focus is on identifying gaps, creating action plans for all requirements, displaying incremental adherence progress, and linking existing actions to complex requirements.

The integrated verification solution has led to substantial cost reductions and increased efficiency since fewer resources are needed to oversee the verification procedure.

Through effective integration of industry best practices and requirements from various standards, Vale has established an integrated approach to tailings management. This has resulted in a comprehensive understanding of tailings management, including a holistic perspective of the associated risks and opportunities.

Vale can efficiently recognise and rectify gaps in its tailings management procedures, leading to continuous compliance improvement.

Reducing the verification cycles

The integrated verification allows Vale to streamline its compliance and avoid duplication of efforts. Instead of conducting separate verifications for each standard, which leads to verification fatigue, Vale can now perform a single verification to assess compliance with all relevant requirements. This has resulted in significant cost savings and efficiency gains, as fewer resources are required to manage the verification process.

Vale has come up with an innovative approach to assign compliance weightage for each gap identified during the self-assessment process. This approach aids in managing compliance adherence and reducing verification fatigue by assigning a compliance weightage for each gap identified during the self-assessment process. This allows Vale to prioritise the most critical compliance gaps, rather than trying to address all gaps simultaneously.

By assigning a weightage score to each gap, it enhances compliance visibility and enables efficient monitoring and tracking of adherence.

Adopting a data-driven approach to compliance management enables Vale to make well-informed decisions regarding resource allocation and efforts, while continuously enhancing its compliance adherence.

MOVE BEYOND USING SPREADSHEETS

Capturing the entire compliance process in multiple spreadsheets can present various challenges when it comes to maintaining, managing multiple verifications, action plans, and compliance tracking. While excel spreadsheets can be a useful tool for tracking and managing compliance data, they have several limitations that can hinder effective compliance management.

Spreadsheets are not flexible in capturing changes in data, which is one of the primary limitations of using them for compliance management. Since spreadsheets rely on manual data entry and manipulation, they are prone to errors, omissions, and inconsistencies, which can undermine the integrity and accuracy of compliance data.

In a joint effort to enhance the efficiency of regular self-assessments, ensure seamless compliance with multiple standards, and effectively manage gaps, Vale and Forwood Safety collaborated to develop WeComplAI Tailings, an innovative software solution.

This comprehensive system helps track, manage, and achieve compliance across multiple tailings management standards with a single verification and provide compliance measurement against each standard.

The system is capable of presenting each verification question with all the relevant mapped requirements from different standards along with additional guidance to the user to fully understand the context of the question and how to achieve compliance.

As previously mentioned, the path to achieving continuous compliance begins with standardised verifications and gap analysis before progressing towards integrated operational frameworks. If an organisation is commencing its compliance journey, the software system can carry out independent verifications against a single standard. As the organisation progresses the compliance journey, the software system can enable a more comprehensive integrated compliance framework.

The software system includes dashboard visualisations that allow users to gauge the status of the various verifications and determine the progress achieved quickly and easily. Findings for each requirement can be captured and reviewed in the verification summary. Associated gaps are captured in an action plan and assigned to a responsible individual to address the gap

Key features and benefits of the software solution

- Centralised Compliance Management provides a single platform to manage and monitor compliance across multiple tailings management frameworks, enabling organisations to streamline their compliance processes.

- Real-Time Adherence Reports against internal, GISTM, and MAC TSM standards without conducting separate verifications against each standard. This helps organisations to identify gaps and take corrective actions quickly.

- Accelerated Verifications with a unique user-centred verification design that provides maximum context for faster and consistent verification results, reducing verification time and efforts.

- Reduced Verification Fatigue as verification questions are linked to multiple tailings management standards, eliminating the need for multiple verifications and reducing verification fatigue.

- Site-Based Performance enables organisations to analyse multiple sites and establish performance benchmarking, helping them to identify focus areas for improvement.

- Comprehensive Verification Capability provides a comprehensive verification capability for both internal and external verifications, ensuring that verifications are conducted efficiently and effectively.

- Risk Assessment and Remediation establishes processes to assess risks on a regular basis and work through the remediation process, enabling organisations to identify potential risks and take proactive measures to mitigate them.

- Improved Visibility across all the tailings assets and the nearby environment and social profile, enabling organisations to manage their operations responsibly and sustainably.

- Reporting capabilities for both internal and external stakeholders, providing organisations with the necessary information to make informed decisions and demonstrate compliance.

CONCLUSION

In summary, multiple tailings management standards have been introduced as a multi-stakeholder effort to prevent TSF and dam disasters from happening. The operationalisation of tailings and dam governance frameworks is a complex and resource-intensive process that requires regular self-assessments to identify and manage gaps towards standard compliance. In collaboration with Forwood Safety, Vale has taken this challenge as an opportunity to streamline efforts and built a platform to simplify the compliance journey against multiple standards. The innovative approach presented in this paper merges requirements from multiple standards into a single integrated verification, reducing the number of self-assessments and introducing a new way to differentiate best practice from standard practice.

Vale's active action management approach links the resulting action plans from gap analyses with compliance scores, gaining real-time compliance visibility, allowing the system to forecast when a tailings storage facility will be 100 per cent compliant against the chosen standard. These efforts serve as an excellent example for other organisations seeking to enhance their tailings management practices and improve their overall compliance performance. By adopting this innovative approach and streamlining efforts, organisations can not only ensure compliance with multiple standards and regulations across multiple jurisdictions.

Effective implementation of integrated tailings management roadmap goes beyond compliance, achieving higher levels of transparency across the system and setting a solid foundation towards broader organisational sustainability and (Environmental, Social, and Governance) ESG commitments.

REFERENCES

Chen, H and Soltes, E, 2018. Why Compliance Programs Fail – and How to Fix Them, *Harvard Business Review*, pp 116–125.

TSM – driving continual improvement in tailings management and governance

S Fitzgerald[1] and C Dumaresq[2]

1. MAusIMM, Senior Manager – Towards Sustainable Mining, Minerals Council of Australia, Forrest ACT 2603. Email: simon.fitzgerald@minerals.org.au
2. VP – Science and Environmental Management, Mining Association of Canada, Ottawa ON, Canada. Email: cdumaresq@mining.ca

ABSTRACT

Minerals Council of Australia (MCA) has adopted the Towards Sustainable Mining® (TSM®) system in recognition of evolving community expectations of the Australian minerals industry's environmental, social and governance (ESG) performance. TSM was established by the Mining Association of Canada (MAC) and is recognised globally with adoption in 11 countries.

TSM builds on existing industry commitments in Enduring Value – the Australian minerals sustainable development framework – by providing a consistent approach to assess, demonstrate and communicate site level performance in a transparent and accountable way, building community confidence and trust. TSM protocols have been adapted to suit the Australian context through the inclusion of local guidance, reference to regulatory requirements and input from key industry stakeholders.

Tailings management has been central to the TSM program since its inception in 2004. The MAC Tailings Guide and the Operation, Maintenance and Surveillance Manual (OMS) Guide pre-date TSM, having been introduced in 1998 and 2003, respectively. The TSM requirements and guidance for tailings management provide a systematic approach to tailings management enabling companies to benchmark against good practice, supporting improved performance. Tailings requirements have continued to improve. For example, significant revisions were completed after the 2014 failure at the Mount Polley Mine in Canada, which was informed by an independent review.

Following the release of the Global Industry Standard for Tailings Management (GISTM), MAC undertook a gap analysis that informed further improvements while noting there is a high degree of alignment between TSM and the GISTM. TSM takes an integrated approach, helping to ensure a systematic, holistic approach to tailings management. TSM also has well-established tools for detailed performance criteria, reporting, and third-party verification, providing a high level of rigour.

INTRODUCTION

Towards Sustainable Mining (TSM) is an international standard for responsible mining. Developed by the Mining Association of Canada (MAC) and launched in 2004, the objective of TSM is to enable mining companies to meet society's needs for minerals, metals, and energy products in the most socially and environmentally responsible way. TSM is a globally recognised standard that has been adopted in 11 countries.

Through the TSM Guiding Principles, mining companies exhibit leadership by engaging with communities, driving world-leading environmental practices and committing to the safety and health of employees and surrounding communities. TSM also provides companies with a set of tools, referred to as protocols, to measure, report, and improve their performance in eight key areas:

1. Tailings management
2. Water stewardship
3. Biodiversity conservation management
4. Climate change
5. Indigenous and community relationships
6. Safety and health
7. Crisis management

8. Prevention of child and forced labour.

These areas were chosen by industry and stakeholders as material issues with gaps not adequately addressed through regulation or where the industry had identified an opportunity to drive significant improvement through sharing of leading practice (eg Safety and health).

The protocols and indicators are designed to measure the quality and breadth of facility-level management. Sites are graded from level C to Level AAA, with Level A broadly defined as good practice and Level AAA defined as excellence and leadership. Mining facilities undertake annual self-assessments to measure performance against each indicator of the TSM protocols, the results of which are independently verified every three years. All results are made publicly available and as the protocols measure performance where the mining activity actually takes place – at the facility (site) level and the results can provide local communities with a meaningful view of how a nearby mine is faring.

TSM program governance and decision-making involves internal and external groups and committees. The role of the national association board is to set the overall objectives and approve any major decisions related to the TSM program (eg revisions to protocols or new policies) following recommendation from the relevant internal committee. Each company implementing TSM selects an Initiative Leader who is responsible for leading implementation, coordination of reporting and key contact for the program to ensure that the latest requirements are communicated. The TSM Initiative Leaders meet regularly to discuss their own experiences, best practices and challenges.

Externally, the TSM program is guided by an independent 'Community of Interest' Advisory Panel. The panel serves as a platform for communities and members to discuss and collaborate on issues of mutual concern. The independent, multi-stakeholder group includes individuals from Indigenous groups, communities where the industry is active, environmental and social NGOs, and labour and financial organisations. In Canada, the Panel has played a key role in the program's design from the very beginning and continues to be integral to its evolution (eg Climate Change protocol) and ongoing implementation.

TSM has also played a leading role in collaboration with other standards as they have emerged and has formal equivalency agreements in place with four other global standards, International Council on Mining and Metals' Mining Principles, World Gold Council's Responsible Gold Mining Principles, International Copper Alliance's The CopperMark, and Responsible Minerals Initiative's Risk Readiness Assessment. The Responsible Sourcing Alignment Supplement provides the additional criteria that allow member companies to use their TSM performance to meet these standards without duplication.

AUSTRALIAN ADOPTION AND ADAPTATION

The Australian minerals industry recognised that societal expectations around the industry's safety and ESG performance were continuing to evolve. Investors, rating agencies and insurers were also increasingly focused on ESG performance in assessing risk. The industry identified the need to enhance its licence to operate shore-up community support, mitigate regulatory risk and ensure ongoing access to land, capital and markets. As a result, in 2018, MCA investigated options to enhance its Enduring Value sustainability framework to build stakeholder confidence in the industry's safety and ESG performance.

The MAC TSM system was identified as a credible and mature mining-specific system to demonstrate accountability on safety and ESG commitments. TSM would enable companies to operationalise and demonstrate performance against key Enduring Value commitments.

In 2019 the MCA Safety and Sustainability Committee assessed and piloted TSM which informed the value proposition for Australian adoption. TSM builds on existing industry commitments in Enduring Value – the Australian minerals sustainable development framework – by providing a consistent approach to assess, demonstrate and communicate site level performance in a transparent and accountable way, building community confidence and trust. TSM would provide:

- A credible platform that demonstrates industry leadership on safety and ESG performance, focused on the most material issues for Australia and applicable to all MCA members.

- Brand recognition among investors and downstream initiatives and the ability to integrate requirements with other initiatives (eg World Gold Council, Responsible Steel).

- A mechanism to integrate broader MCA commitments (eg the MCA climate change action plan) and that can be adapted to meet evolving expectations and circumstances (eg from the Taskforce for Nature-related Financial Disclosures).

MCA adopted TSM in 2021 and members are in the process of implementation. The implementation consists of annual cycles that allow for continual improvement through additional training, support and shared learning.

Adaption of TSM

The first stage of implementation was the adaption of TSM to the Australian context. The existing framework addressed the material risks and opportunities identified by Canadian industry and stakeholders. In adapting and then endorsing the Australian protocols, the aim was to strengthen TSM while ensuring consistency with the intent of the Canadian protocols. The adaption of TSM was guided by the following principles:

- The overarching purpose of the protocols should remain consistent.

- The overall protocol structure should be retained, including the type and number of indicators.

- Australian regulations should be referenced where they satisfy indicator requirements.

- Additional Australian requirements should be included in supporting policy and guidance.

- Language in the protocols should be updated to suit Australian nomenclature, aligned with the understanding of industry, representative groups and communities.

These principles would maintain already recognised equivalency and minimise any additional effort above regulation to avoid unnecessary duplication. The endorsed Australian protocols include local guidance, reference to regulatory requirements and input from key industry stakeholders.

The review of the Tailings Management Protocol took advantage of both member companies' subject matter expertise and an independent review. GHD (2021) noted that Australian guidelines and practices are generally of a high standard and comply well with the technical aspects of the protocol. The protocol provided improved guidance on the governance of tailings management and would require a more rigorous approach in applying and documenting the criteria.

TSM AND TAILINGS MANAGEMENT

Responsible tailings management is integral to TSM. The objective of the tailings management component of TSM is to continually work towards minimising harm from both the physical and chemical risks associated with tailings facilities, including having no catastrophic failures, and no significant impacts on the environment or human health.

The tailings management component of TSM consists of the Tailings Management Protocol, two guidance documents, and a Table of Conformance, a detailed tool for performance measurement. The protocol measures performance in five key areas related to tailings management:

1. Developing and implementing a corporate policy on tailings management.

2. Assigning corporate level and site-level accountability and responsibility for tailings management.

3. Developing and implementing site-specific:

   o Tailings management system

   o Plans for emergency preparedness

4. Developing and implementing an Operations, Maintenance and Surveillance (OMS) manual.

5. Conducting an annual review of tailings management to drive continual improvement.

The Tailings Management Protocol refers to and is supported by:

- A Guide to the Management of Tailings Facilities (the Tailings Guide) provides guidance on responsible tailings management, helps companies develop and implement site-specific tailings management systems, and improves the consistency of application of engineering and management principles to tailings management.

- Developing an Operation, Maintenance and Surveillance Manual for Tailings and Water Management Facilities (the OMS Guide) provides guidance on the development and implementation of operation, maintenance, and surveillance (OMS) activities, described in a site-specific OMS manual.

- Table of Conformance which identifies elements of the Tailings Guide and the OMS Guide that must be implemented to meet the performance criteria for each of the indicators in the Tailings Management Protocol.

TSM also considers the potential impact of tailings activities on Indigenous and community relationships, crisis management, climate change and water stewardship through the requirements within the related protocols:

- Indigenous and community relationships are an important source of information to support responsible tailings management, including risk assessment, planning and design, and closure planning.

- Crisis management and communications planning facilitates the physical response to an emergency and help prevent or lessen impacts on communities and the operation.

- Climate Change and potential impacts are an important consideration for tailings management and closure planning.

- The Water Stewardship Protocol addresses site-wide water management and tailings facilities should not be planned or implemented in isolation from the broader site.

Continual improvement in tailings management and industry performance

Regular reviews and updates of Tailings Management guidance by MAC's Tailings Working Group have been supplemented by targeted reviews following incidents or significant changes in industry or stakeholder guidance. The continued engagement of experts and industry stakeholders provided members training, access to communities of practice and reporting of facility-based performance to demonstrate collective improvement in performance.

Tailings management

The Tailings Management Protocol, Table of Conformance, Tailings Guide, and OMS Guide are collectively referred to as the tailings management component of TSM. The development and evolution of the tailings management component of TSM began in the 1990s. The tailings management component has a demonstrated history of continual improvement as shown in Table 1.

The first edition of MAC's Guide to the Management of Tailings Facilities, released in 1998, was developed in response to a series of international tailings-related incidents that occurred in the 1990s. The purpose of the first edition was to:

- provide information on safe and environmentally responsible management

- help companies develop tailings management systems

- improve the consistency of application of sound engineering and management principles.

The first edition reflected sound management practices already in place at that time. It adopted principles and approaches from sources that included mining company manuals, proceedings of two MAC workshops, the MAC Environmental Policy and Environmental Management Framework, the ISO 14000 standards related to environmental management, the Canadian Dam Association's draft Dam Safety Guidelines (1997), and international guidelines and standards.

TABLE 1

Continual improvement in Tailings Management Guidance.

Year	Summary of changes in guidance
1998	The first edition of the Tailings Guide was released.
2003	The first edition of the OMS Guide was released, the companion tool to provide guidance to support day-to-day implementation of a tailings management system.
2004	The first version of the Tailings Management Protocol was released. First year of public reporting of performance against the Tailings Management Protocol was 2006.
2011	The second edition of the Tailings Guide was released.
2017	The third edition of the Tailings Guide was released, together with an updated version of the Tailings Management Protocol. The Table of Conformance was introduced. Drawing from internal and an independent review of the tailings management component of TSM following the 2014 failure of a tailings facility at the Mount Polley Mine in British Columbia.
2019	The second edition of the OMS Guide was released, together with Version 3.1 of the Tailings Guide, an updated version of the Tailings Management Protocol. The Table of Conformance was revised and expanded.
2021	Version 3.2 of the Tailings Guide and Version 2.1 of the OMS Guide were released with the goal of continual improvement and increasing alignment with the Global Industry Standard on Tailings Management, released in 2020.
2022	Updated Tailings Management Protocol and Table of Conformance were released. Aligned with latest versions of the Tailings and OMS Guide. Expanded mandatory application of the Protocol to include inactive tailings facilities.

Building on the implementation of the Tailings Guide and lessons learned, MAC introduced a companion document in 2003 the OMS Guide. This guide focuses on the need for a site-specific OMS manual as an integral component of an overall tailings management system. The document was designed to help companies comply with legal requirements and corporate policy, demonstrate voluntary self-regulation and due diligence, practice continual improvement, and protect employees, the environment and the public.

In 2011, the second edition of the Tailings Guide was released. This edition reflected information and experience gained through the course of developing and implementing the tailings management component of TSM and working with tailings management systems around the world.

In August 2014, a tailings dam foundation failure occurred at the Mount Polley Mine in British Columbia. Soon after this incident, the MAC Board of Directors initiated a review of the tailings management component of TSM. The review, formally launched in March 2015, consisted of an independent review and an internal review by the MAC Tailings Working Group. The final report included 29 recommendations addressed in Version 3 of the Tailings Guide:

- The requirement for an independent review of site investigation and selection, design, construction, operation, closure, and post-closure of tailings facilities.

- Evaluate how best to include in the Tailings Guide assessment and selection of both Best Available Technology (BAT) and Best Applicable Practices (BAP).

- Develop and include guidance related to managing a change of Engineer-of-Record and a change of ownership.

- Include a risk-based ranking classification system for non-conformances and have corresponding consequences and guidance.

- Include more specific technical guidance related to site selection and design, including how to select objectives and set design criteria.

In 2019 the second edition of the OMS Guide was released in conjunction with Version 3.1 of the Tailings Guide and included:

- Detailed guidance for the preparation of emergency response plans and emergency preparedness plans.

- Affirmation that the performance evaluation and management review included the site-specific tailings management system, emergency response plan, emergency preparedness plan, and the OMS manual.

- Guidance on post-incident analyses.

The latest updates to the Tailings Guide completed in March 2021, include refinements that improve TSM identified from a gap analysis undertaken following the release of the Global Industry Standard on Tailings Management (GISTM) and additional improvements identified by the Tailings working group.

The most significant changes in Version 3.2 were:

- Expanding on aspects to be considered in developing a corporate policy and/or commitment.

- Increasing the level of detail of descriptions of the roles and responsibilities of the Accountable Executive Officer and Responsible Person.

- Expanding on aspects to be addressed in a management review for continual improvement.

Following the update MAC (2021) documented their assessment of the equivalency between TSM and GISTM. The assessment shows TSM largely meets or exceeds 63 of 77 requirements, partially meets nine and does not address five, some of which are addressed through jurisdictional guidance from the Canadian Dam Association. The requirements in the Table of Conformance and the Tailings Management Protocol (Mining Association of Canada (MAC), 2022) are aligned with the latest guidance and apply to inactive tailings facilities.

Industry performance

Continual improvement in tailings management has also been demonstrated on an industry level. The governance process for TSM has an important role to play in ensuring the credibility, transparency, accountability and continual improvement of the program. The MAC Tailings Working Group has conducted most of the technical work associated with the Tailings Management Protocol and guidance and continues to advise the Initiative Leaders and company subject matter experts on implementation and updates.

Companies implementing TSM publicly report their own progress annually and commit to demonstrating continuous improvement over time. MAC's objective in tailings management is for members to achieve level A (good practice) or higher within three years of adopting TSM. MAC tracks industry-wide performance trends and Figure 1 (MAC, 2023) demonstrates the improved performance in Tailings Management over time and across facilities. This improved performance has been in addition to the evolving expectations and changed tailings requirements.

Percentage of Facilities at a Level A or Higher − 2006, 2021 and 2022

Level A ●   Level AA ●   Level AAA ●

Last Updated: Mar 17, 2023 (86% Facilities Reporting)

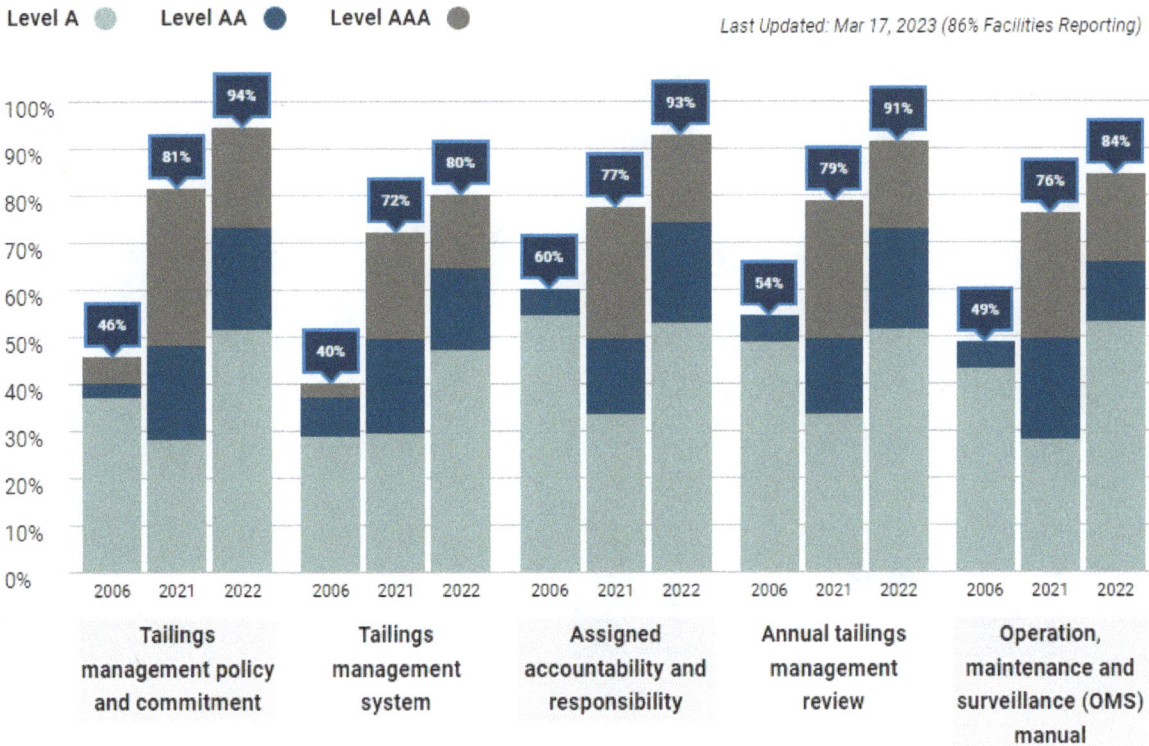

FIG 1 – Improvement in tailings management – industry performance.

CONCLUSIONS

MCA adopted TSM in 2021, to provide a consistent approach to assess, demonstrate and communicate site level performance in a transparent and accountable way, building community confidence and trust. The protocols have been adapted to suit the Australian context and member companies are currently in various stages of implementation with public reporting expected by 2025.

The requirements and guidance of TSM as a whole provide a comprehensive, integrated approach to tailings management, reinforced by a detailed, rigorous approach to performance measurement. TSM provides an effective vehicle to address the complex, ever changing nature of tailings management and manage perhaps the largest single risk in tailings management – the human factors.

The TSM requirements and guidance for tailings management have continued to evolve and improve in the almost 25 years since the First Edition of the Tailings Guide was released in 1998. Continual improvement is one of the TSM Guiding Principles, and in keeping with this, the requirements and guidance for tailings management will continue to improve. Industry-wide performance trends also demonstrate continual improvement and progress against the performance indicators over time.

TSM provides a strong, comprehensive approach to tailings management that is based on decades of implementation experience and continual improvement. TSM provides an effective vehicle to achieve the goal shared by the mining industry, communities, regulators, investors, and the public – preventing catastrophic failures and minimising harm.

REFERENCES

GHD, 2021. Review of TSM – Tailings Management Protocol, internal MCA report, unpublished.

Mining Association of Canada (MAC), 2021. Equivalency between requirements of Towards Sustainable Mining and the Global Industry Standard on Tailings Management. Available at: https://mining.ca/download/1038/

Mining Association of Canada (MAC), 2022. Tailings Management Protocol. Available at: https://mining.ca/wp-content/uploads/dlm_uploads/2022/08/TSM-Tailings-Management-Protocol-June-2022.pdf

Mining Association of Canada (MAC), 2023. Environmental Stewardship: Aggregate Performance. Available at: https://mining.ca/towards-sustainable-mining/tsm-progress-report/environmental-stewardship-aggregate-performance/

A critical review of the tailings dam design process

K T Mandisodza[1], D J Williams[2] and C Zhang[3]

1. PhD Candidate, Geotechnical Engineering Centre, School of Civil Engineering, The University of Queensland, St Lucia Qld 4072. Email: k.mandisodza@uq.edu.au
2. FAusIMM, Professor Geotechnical Engineering, Geotechnical Engineering Centre, School of Civil Engineering, The University of Queensland, St Lucia Qld 4072. Email: d.Williams@uq.edu.au
3. Research Fellow, Geotechnical Engineering Centre, School of Civil Engineering, The University of Queensland, St Lucia Qld 4072. Email: chenming.zhang@uq.edu.au

ABSTRACT

Tailings waste material from processed ore stored in tailings dams in a slurry consistency within a tailings storage facility (TSF) system is generally ubiquitous globally. A significant number of key engineering and management considerations are required to determine the successful selection of waste management strategies, often deliberated in the planning stages of a tailings dam storage facility and through subsequent raises during the life of the storage facility, closure planning or post-closure in perpetuity. Waste management considerations are important to prevent unwanted events of extreme consequences, such as the recently experienced tailings dam failures in the last decade. Management of waste disposal strategies is often guided through design criteria that define the design process. This paper critically reviews the design process and inherent uncertainties in design parameters in the wake of devastating global tailings dam failures and the subsequent introduction of governance regulations.

INTRODUCTION

Mine waste in the form of slurry consistency must be stored safely within storage facilities systems, which mainly exists in the form of dam impoundments (ANCOLD, 2012). Mine waste production has significantly increased globally due to improved mineral recovery processes and technological advancements (Blight, 2009; Kossoff et al, 2014). With increased tailings production, estimated to be 50 billion cubic metres (Bm^3) between 2019 and 2023 (Franks et al, 2021), mine waste management requires strategic planning, risk management, and accountability to prevent unwanted events with extreme consequences. The design and construction of tailings dams are based on conditional structural stability and foundational performance determined through analyses, judgments, experiences, regulations, and guidelines to select stability factors that define the safety criteria. However, these criteria should integrate the inherent uncertainties, often associated with the input design parameters and material variability during construction and operational processes, the contribution of the human element to design uncertainty and model uncertainty to determine design acceptability.

The tailings dams design process, an appendage to water dam design, is often an integrated multi-criteria analysis (MCA) combining the conditional performance of the dams through the life cycle stages from design, closure, and post-closure into perpetuity. Other associated contributions to the performance of the tailings dams are inherent to the site's geographical location, the topographical setting, the climatic conditions, and the seismic environment of the area of siting of the tailings dams. In most stability assessments for tailings dams, the material is assumed to be homogenous, intending to simplify the imputed inhomogeneity associated with material variability from construction stages, material blending for mechanical improvement and depositional inconsistencies. Applying the homogeneity hypothesis to evaluate the mechanical stability of tailings can lead to unrealistic safety evaluations (Villavicencio et al, 2011). This paper aims to appraise and critique the tailings dam design process and highlight some areas that require improvement.

TAILINGS DAM DESIGN PROCESS

As currently known and applied, the tailings design process has its legacy from the design of water dams or more an appendage to the design of water dams (Morgenstern, 2018). Tailings dam construction dates back to the pre-nineteenth century, with early construction recorded in Japan and

France (Warbuton *et al*, 2019). Structural analyses characterise dam design criteria to assess the dam's structural stability and the foundational interface under select loading conditions. However, technology has significantly improved water dam engineering and upholds higher safety levels than tailings dams. This improved safety performance for water dams has been developed incrementally from experiences related to recorded accidents and incidents (ICOLD, 1988). Conversely, the high frequency of embankment failures for dams has often resulted in low public confidence for many centuries (Jansen, 1983). Tailings dam failures recorded in the last decade, shown in Figure 1, and sourced from the CSP2 database (CSP2, 2023), have cumulatively agitated the public and the entire mining industry, which invited focused scrutiny into the design process of tailings dams, culminating in more governance strategies being introduced and implemented to regulate the design of tailings dams to minimise associated risk to the public, particularly those located downstream of facilities.

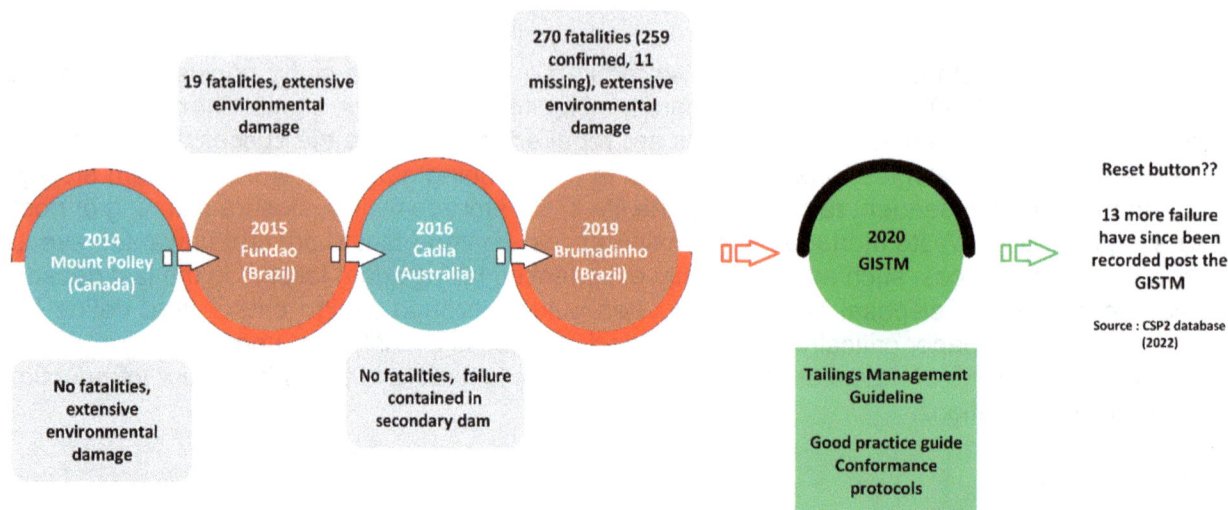

FIG 1 – Tailings dam failures timeline record for the last decade led to the introduction of the Global Industry Standard on Tailings Management (GISTM) in 2020. The failures recorded after the introduction of the GISTM were sourced from the CSP2 database (CSP2, 2023).

There are several design considerations undertaken in the design process for tailings dams. Some tailings properties that impact design are in-place and relative density, an important factor in determining the impoundment volume and size, permeability, plasticity, compressibility, consolidation, shear strength and stress parameters (Vick, 1990). However, lacking holistic design criteria, most designs blend different guidelines and consider the application of best available practices (BAPs) and best available technologies (BATs) to achieve a sound design with no failure as opposed to a tolerable failure rate (Morgenstern, 2018). The structural design of tailings dams should conform to the BAPs and BATs and must satisfy governing regulatory and jurisdictional requirements. The definition of hydrology and hydraulic conditions, structural design, instrumentation, and monitoring are critical in the design criteria. Other considerations include the social licence to operate (SLO), risk tolerance, and the acceptance of design criteria, which depends on the proper selection of input parameters, behaviour and response for structure and foundation materials, and reliable deformation prediction and displacement.

The tailings design process considers a hybrid approach often initiated through the set-up of storage capacity, a function of the production requirements and processing capacity and varies with the construction method (Franks *et al*, 2021, op. cit). The depositional method in the tailings facility is primarily a determinant of the tailings dam siting, which is a function of geographical location, topographical and climatic conditions, the seismic setting and the tailings material type and inherent properties. These properties are critical in the overall structural and foundational stability during operational stages through closure and post-closure. Once the depositional method has been ascertained as the economical option in the conventional slurry deposition method, a starter embankment is designed and preferably constructed with local or borrowed and blended materials. Having locally borrowed material for starter embankment construction and subsequent raises provides an economical option for most operations. However, this approach is attributable to causing increased structural failures, especially the upstream method, alongside the high maintenance cost (Gomes, De Toni and Assis, 2016).

The design of tailings dams should be done to withstand the most severe and unfavourable combination of static and dynamic loads. These may include loads that may occur only during the construction stage and loads that develop due to variation of loading conditions through structural aging for the embankment or due to asynchronous system evolution, where one part of a system changes or deteriorates without the related necessary changes in other parts of the system (Dekker and Pruchnicki, 2014). Spatial variability due to the dispersion of properties of the same material or structure in space, ie depositional variability, which affects the consolidation of tailings material leading to non-single consolidation line, and temporal variability due to time-dependent aging are the main drivers for uncertainty in geotechnical analyses. The index properties mainly drive the variability and construction processes used to establish the tailings facilities. The consequence category, used to determine various design and operational requirements, is a function of the parametric inputs with inherent uncertainties that are not exhaustively analysed and incorporated into the resulting social, environmental and financial outcomes. The non-exhaustive analysis and integration of uncertainties in the resulting outcome invoke weaknesses in the current design process and determination of operational requirements for tailings dams.

Hence, the often applied and preferred assumption that tailings material has a prevalent homogeneity consistency in the evaluation of tailings dam design during stability and deformations analyses is a misconception that has implications on the overall design outcome and performance. This is mainly attributable to the challenges and inability to test all the element tests required for material components that are part of the tailings dam system and the assignment of material parameters.

Risk analysis of tailings dams

Risk analysis for dam safety is fundamentally a characterisation of the uncertainties in the performance capability of dams under loading conditions of interest (Hartford and Baecher, 2004). Though this assertion is mainly for water dams, it is applicable and relevant in tailings dams design. The traditional approach of conducting a dam safety analysis is often called deterministic or standards-based for tailings design, starting with the potential hazard or consequence classification to ensure that the system components of a dam structure conform to a deterministic set of principles, rules and requirements, provided in Guidelines.

Risk is defined as the effect of uncertainty on objectives (AS ISO, 310000:2018 Risk Management – Guidelines). Regarding tailings, the aim is to ensure that they meet the design intent, which is to contain mine waste safely without failure that threatens public and environmental risks. Tailings dam failures are a good example of the effect of uncertainty on design, considering that most of the applied design practices and guidelines dictate the use of deterministic procedures, where any significant loss of life is a possible failure consequence (Williams, 2021). Any over-reliance on the prescribed 'guideline' design values is inadequate to eliminate the failure (Whitman, 1984). The number and risk factors of tailings dam failures have increased in the last two decades. A representation of global tailings dam failures as a function of the construction method indicates that most failures have occurred on dams constructed using the upstream construction method (Halabi et al, 2022). Considering that most tailings dam facilities worldwide have been constructed using the upstream method, it is somewhat inevitable that more tailings dam failures will occur due to the aging of most legacy facilities. This assertion invokes a revamp to the tailings dam design process to a process that acknowledges inherent deficiencies in the existing design criteria to address the associated risks and uncertainties. Table 1 shows global tailings dam failures as a function of the construction method.

TABLE 1

A summary of global tailings dam failures, adopted from (Halabi *et al*, 2022).

Construction method	Total dams	Total failures	Total failures (%)	PoF(Annual Estimate)
Upstream	7926	113	43	6.11×10^{-4}
Downstream	5031	27	27	2.30×10^{-4}
Centrelline	1583	15	9	4.06×10^{-4}
Other	3860	34	21	3.77×10^{-4}
Total	18 400	168	100	3.91×10^{-4}

The reasons probabilistic analyses should be critical for tailings dams include, as stated by Vick, Atkinson and Wilmort (1985 op. cit):

- The limited life of tailings dams makes them less susceptible to extreme design events.

- There is a corporate variable risk that is not included in the current design criteria.

- The economic implications of seismic design criteria are not included in the current dam design criteria.

The principles interpretation for probability in analyses is its function as a frequency and a degree of belief. Probabilities describing rates of occurrence are interpreted as frequencies, while probabilities describing states of nature of truth or the truth of hypotheses are interpreted as degrees of belief (Hartford and Baecher, 2004). Tailings dam risks are complex within each contributing system's components, and this complexity should be reflected in the criteria used for tailings dam design. One of the most applied methods for stability assessments of tailings dams is the factor of safety (Morgenstern, 2018 op. cit). The selection of factors of safety must include influencing factors such as:

- Consequence of failures.

- Uncertainty of material properties and subsurface conditions.

- Variable constructing and operating conditions.

- Comprehensive site investigation and geotechnical monitoring.

- Soil response (contractive/dilative and its variation with confining stress and shear stress laws), including the potential for brittle behaviour.

- Time-dependent, deformation-dependent and stress-path-dependent processes that may affect the critical material properties.

- Seismic loading as appropriate.

- Implementation of the effective risk management system, ie the observational method.

The incorporation of expert opinions considering risk, regardless of a specific definition, is influenced by the character of the probability information used for determining the risk (Borovcnik, 2015). Meaning risk components are prone to subjective interpretations. In the design and operations of tailings dams, the involvement of corporate risk concerning the public, particularly those downstream of the facilities, should be incorporated in the design criteria. This can be driven by effective governance strategies, a property of social systems, providing coherence between diverse parts of the compound, and complex systems, enabling coordination and cooperation. Governance means coping with uncertainty (Willke, 2007).

Risk and uncertainty in tailings design

Risk is the statistical expectation value of an unwanted event that may occur (Hansson, 2007). It can be preferably used in technical environments dealing with the reliability of systems as it is more objective and widely communicable or compulsive. However, its relevance depends on the actual probabilities involved (Borovcnik, 2015 op. cit). Risk can also be represented as an equation below (Morales-Torres et al, 2019).

$$Risk = \int P(loads) \bullet P(response|loads) \bullet C(loads, response)$$

Where **P**(loads) is the probability of the different load events, P(response|loads) is the conditional probability of the structural response for each loading event, and C(loads,response) is the consequence of the system response to each loading event.

The distinction between aleatory and epistemic uncertainties (also called risk and uncertainty, respectively) has a long history grounded in the definition of probability and the development of expected utility decision theory (Paté-Cornell, 1996). Risk and uncertainty differ by the quality of probability. For tailings dams, the information availability frequency and respective underlying probabilities of failures, the more objective decisions can be made about the tailings dams and the determination of inherent risks. Uncertainty analysis in the design and risk analysis of tailings dams can provide estimates of uncertainty distributions for risk analysis, provide estimates of confidence in meeting tolerable risk guidelines, and provide an enhanced way to represent risk assessment outcomes to decision-makers. Conversely, the best estimates input parameters analysis provides the best judgement and a single point of the spectrum from industry experts. The sensitivity analysis provides a range around the best estimates of input parameters but has limitations in estimating their confidence levels. These all fall short of the strength and capabilities offered by uncertainty analysis.

The tailings dam design process is an expert system that relies on expert judgements to quantify design input parameters. This process has inherent uncertainties whose measurement can be done through interpretation, imprecision, calculus, consistency, assessment and computation (Walley, 1996). Walley continues to state that de Finnetti's Bayesian theory and the coherent lower previsions (expectations) theory (foresight and foreknowledge) emphasise interpretations and consistency uncertainty measures, whilst the Dempster-Shafer theory of belief functions emphasises a simple rule of combination. Fuzzy logic emphasises judgment in natural language (ie qualitative assessment). Considering that the Bayesian theory is the most highly developed and best understood theory of uncertainty, it cannot adequately model ignorance, partial information, assessment of uncertainty in natural language or conflicts between expert opinions, implying that the Bayesian sensitivity analysis interpretation is both misleading and unnecessary. It is misleading because, in most problems, such as tailings dams, no useful meaning is given to underlying linear prevision. It is unnecessary because upper and lower previsions can be given a direct behavioural interpretation.

For a quantitative safety criterion to be acceptable to those concerned, four requirements must be satisfied (Whitman, 1984 op. cit), these four requirements:

1. The criterion must be defined on a logical and understandable basis.

2. There must be reasonable, acceptable methods for demonstrating the criteria.

3. The criteria must reduce risk over the inherent current practice.

4. The fixes required by imposing the criteria must not lead to severe economic penalties.

The overall design process for tailings dams and their respective quantitative safety criterion highlights some flaws and gaps that can be explored to make it more effective and robust in minimising failures and a growing loss of confidence and mistrust in the mining industry. A statistical analysis of water dam embankment failures indicates that the leading organisational causes of failures are always designing inefficiencies, with most having a failure context dominated by flood conditions characterised by unusual events, operational conditions, earthquake conditions, other extreme loading conditions and unknown conditions (ICOLD, 2019). In his 17th Rankine Lecture, De Mello (De Mello, 1977) stated:

> *Design decisions, implying action despite uncertainty, may face exponential risks in dam engineering and must be sorted into two categories, those controlled by statistics of extreme value conditions and those amenable to treatment under statistics and probabilities of averages. Absolute priority belongs to the first group, and it is*

emphasised that any computations connected with them are delusions. The designer must run away from the problem by choice of change of the statistical universe.

The use of extreme input values in the design process is prevalent in the consequence-based method widely applied in the tailings dam design. De Mello highlights a striking coincidence in the observed historical dam failures, all associated with phenomena controlled by extreme value statistics. Though the focus was primarily on water dams, the fact that the design processes for tailings dams are appended to that of water dams makes the statement true for tailings dams.

Tolerability and acceptability of risk in the tailings design process

The tolerability of risk context has many difficulties related to how uncertainties are considered under risk analysis, compared to traditional implicit treatment in state-of-the-art dam safety practice (Morales-Torres *et al*, 2019 op. cit). Considering that risk can never be eliminated, adopting a design criterion that reduces the likelihood of failure to such a low level for the risk to be acceptable is fundamental. From this perspective, an alternative tailings dams design criterion should be pivoted, focusing more on the input design parameters, whose associated uncertainties should be exhaustively accounted for in the associated safety factors and inherent consequences.

Tolerability of risk (TOR) is defined as the framework that capitalises on the strength of the decision-making criteria, equity, utility, and technology as the main building blocks, according to the UK Health and Safety at Work Act 1974 (Section 2.5). The framework was formulated based on the method used to control the risk at nuclear power stations and has generally been accepted as universally applicable.

Figure 2a shows the TOR framework, indicating the level of risk. The region between the negligible and intolerable regions represents the region of tolerability, which should be as low as reasonably practicable (ALARP). This region dictates the tolerability criteria when risks lower than the limit of tolerability are tolerable only if risk reduction is impracticable or the cost is grossly disproportionate (ANCOLD, 2022). In the ALARP region, there is a general risk tolerance for the inherent benefits against associated risk, with an expectation that:

- The nature of risk is adequately assessed, and results are used correctly to determine the control measures based on BATs or BAPs.

- Residual risks are kept ALARP.

- Continuous process reviews are done to validate those risks that are ALARP through BATs and BAPs.

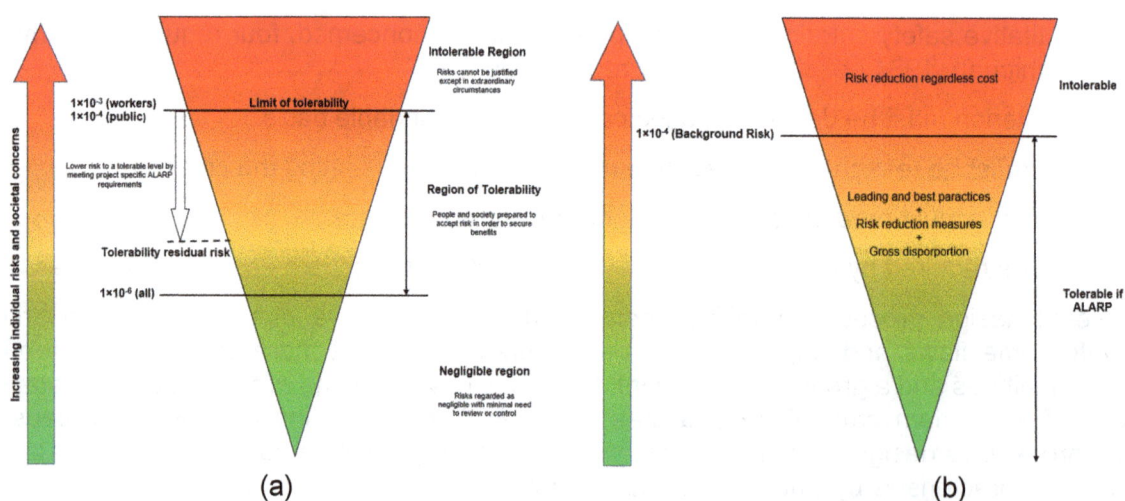

FIG 2 – (a) Tolerability of risk (TOR) after the UK Health and Safety Executive (HSE), 1974; (b) TOR with the application of ALARP below the background risk after ANCOLD Guidelines (ANCOLD, 2022 op. cit).

In defining the tolerability and acceptability of risk, voluntary and involuntary should be differentiated. Voluntary risk is what we assume about ourselves, and involuntary risk is imposed (Rowe, 1975).

Risk acceptance criteria generally accept that risk levels are based on involuntary risk (Riddolls and Grocott, 1999; Hopkins and Kemp, 2021; Williams, 2021). Recent updates in the guidelines on risk assessment by ANCOLD recommend that ALARP is applied indefinitely below the limit of tolerability, which abandons the broadly acceptable level or objective level. The objective level emanates from the need to apply the ALARP principle to decide how far the background level risk is tolerable, as shown in Figure 2b.

Evaluation of ALARP, which is well adapted to the common law legal system (Bowles *et al*, 2003), is done through various methodologies for individual and societal risks, ie the gross disproportion method. Disproportionality is evaluated and tested using risk estimates through loss of life analysis (*individual risk, societal risk, annualised potential loss of life, PLL, cost to save a life, cost per statistical life, CSSL*) or financial analysis (*cost-benefit analysis, financial capability, annualise cost or frequency of accidents*) (Riddolls and Grocott, 1999). The quantification of risk through risk assessments (QRA) is generally accepted through its objectification of risk that facilitates comparison between risk reduction and control measures (Stallen, Geerts and Vrijling, 1996). However, it is generally known that QRA's are generally uncertain due to subjectivity, human judgement and perception of risk. Regardless of the inherent shortfalls of the QRA risk evaluation process, it retains its potency for risk quantification.

Considering that risk is associated with loss of life, the level of risk is represented by f-N or F-N diagrams, which have proven to help represent the meaning of probabilities that the public appreciates as framed around other discernable risks. Tailings design guidelines use the f-N and F-N diagrams to represent individual and societal risks (ANCOLD, 2012; CDA, 2013; ANCOLD, 2022). Other methods that have been used include the semi-empirical approach proposed through the Silva, Lambe and Marr (SLM) method (Silva, Lambe and Marr, 2008), which estimates slope failure probabilities (annualised) against the factor of safety. The method is based on expert judgment anchored in three ways of estimating event probabilities:

1. Based on the frequency of observations (historical data).

2. Derived from probability theory and mathematical modelling.

3. Quantification of expert judgement subjective probabilities.

The SLM method proves that expert judgement and subjective probabilities can be used in risk evaluation and decision-making processes, though its effectiveness should be refined through uncertainty evaluation. However, there are limitations to the SLM method where annual exceedance probabilities (AEPS) are being evaluated for hazards to avoid combining two annual probabilities as it is not mathematically correct (ANCOLD, 2022).

The consequence-based risk approach for the tailings dam design process

Conventional water and tailings dam design practice dictates the use of deterministic procedures where any significant loss of life is a possible failure consequence (Vick, Atkinson and Wilmort, 1985). However, newer guidelines for risk assessment, ie ANCOLD (2022), advocate a risk-based approach focusing more on probabilistic analysis. Most dam failure consequence assessments are conducted in first-order probabilistic risk analysis that results in a single value, consequential loss. It is proposed that the probability distribution of consequences should be obtained by analysis of epistemic uncertainties (Morales-Torres *et al*, 2019). The use of consequence-based as a risk-based approach should depend on the probability of failure, not on the potential loss of life or economic losses induced by dam failures. This lack of uncertainty analysis in the approach potentially results in underestimation and overestimation of inherent risk, directly impacting risk tolerability. Integration of uncertainties should be propagated through the complete risk analysis evaluation to effectively deal with their effect on input parameters and subsequent outcomes.

The methods used for consequence estimation for water dams and tailings dams, ie the United States Bureau of Reclamation (USBR) and the Reclamation Consequence Estimation Method (RCEM), are not suitable for estimating individual risk because they use average-based input parameters in the estimation of the fatality rate, which discounts evacuation and vulnerability of the individual at risk. The individual stepwise process to estimate the model assumptions is affected by knowledge uncertainty from breach parameter prediction, which does not consider a quantitative assessment. The relative uncertainties to the parameter are not incorporated into most risk assessment studies (Wahl, 2004). Other affected assumptions are the failure modes and effects analysis (FMEA) process and its drawbacks (Santos, Caldeira and Serra, 2012), flood wave travel times, floating debris, warning diffusion and evacuation process. Considering that the consequence-based risk estimates are hinged on the effectiveness of hazard warning dissemination and evacuation process, not having uncertainty analysis in the very pivotal contributions to the results of the estimates induces limitations in the effectiveness of the consequence estimates. Another shortfall of the process is its lack of portability to other regions. The case histories are mostly region specific and based on land flooding caused by rainfall or water dam failures, which may not be broadly applicable in other regions. However, despite the various shortfalls of the existing design process, it remains in use due to its easiness and assimilation in the design guideline to generate consequence estimates (Beadenkopf, 2013).

CONCLUSION

The tailings design process is a multi-criteria analysis involving risk analysis and decision-making at different process intervals. Similarly to the design of water dams, the tailing dams design process is well documented and researched mainly as an appendage. However, the different construction methods between water dams and tailings dams, their life expectancy, operational and maintenance approaches, material types, and depositional cycle warrant a different design approach to water dams. Though the water dam design process formed the foundational criteria, criteria incorporating the differences will be preferred for tailings dams.

The concept of uncertainty analysis is fundamental in engineering design processes. The tailings design process is dominantly based on judgement and subjectivity to determine the design factor of safety magnitudes. Its inherent consequence-based risk estimation processes do not fully integrate uncertainty analysis. However, the factor of safety appropriateness in classifying dam stability conditions is determined by the uncertainties in the conditions being analysed (USACE, 2003), which can impose misleading design factors if not integrated as in the tailings dam design process. Sensitivity analysis, a preferred method to check the effect of variability in input material parameters or loading conditions on design, also has limitations due to the lack of estimation of the distribution of the output uncertainties from the joint distribution of input uncertainties. Similarly, using extreme loading conditions such as probable maximum precipitation or flood and maximum credible earthquake to evaluate worst-case design scenarios and simplify the solution does not include the uncertainties of those events, which can lead to congenital overestimation or underestimation of inherent risk. Hence, geotechnical investigations, stability and deformation analyses, risk analyses, and decision-making processes that feed into the design process of tailings dams should be premised on reducing uncertainty and increasing the confidence level in the design and engineering of tailings dams.

ACKNOWLEDGEMENTS

The author would like to thank the Geotechnical Engineering Centre (GEC) at the University of Queensland for the opportunity to research an exciting topic and for the support offered to supervise and direct my research programme.

REFERENCES

ANCOLD, 2012. Guidelines on Tailings Dams–Planning Design, Construction, Operation and Closure, ANCOLD.

ANCOLD, 2022. Guidelines on Risk Assessment, ANCOLD, pp 1–212.

Beadenkopf, E, 2013. Federal Guidelines for Inundation Mapping of Flood Risks Associated With Dam Incidents and Failures, United States Government.

Blight, G E, 2009. *Geotechnical engineering for mine waste storage facilities* (CRC Press).

Borovcnik, M, 2015. Risk and decision making: The 'logic' of probability, *The Mathematics Enthusiast,* 12(1):113–139.

Bowles, D S, Anderson, L R, Glover, T F and Chauhan, S S 2003. Dam safety decision-making: Combining engineering assessments with risk information, Proceedings of the 2003 US Society on Dams Annual Lecture.

Canadian Dam Association (CDA), 2013. Dam safety guidelines 2007, CDA.

CSP2, 2023. TSF failures from 1915. Available from: http://www.csp2.org/tsf-failures-from-1915 [Retrieved 05/05/2023].

De Mello, V F, 1977. Reflections on design decisions of practical significance to embankment dams, *Géotechnique,* 27(3):281–355.

Dekker, S and Pruchnicki, S, 2014. Drifting into failure: theorising the dynamics of disaster incubation, *Theoretical Issues in Ergonomics Science,* 15(6):534–544.

Franks, D M, Stringer, M, Torres-Cruz, L A, Baker, E, Valenta, R, Thygesen, K, Matthews, A, Howchin, J and Barrie, S, 2021. Tailings facility disclosures reveal stability risks, *Scientific reports,* 11(1):1–7.

Gomes, R B, De Toni, G and Assis, P S, 2016. Iron ore tailings dry stacking in Pau Branco mine, Brazil, *Journal of Materials Research and Technology,* 5(4):339–344.

Halabi, A L M, Siacara, A T, Sakano, V K, Pileggi, R G and Futai, M M, 2022. Tailings Dam Failures: A Historical Analysis of the Risk, *Journal of Failure Analysis and Prevention,* 22(2):464–477.

Hansson, S O, 2007. *Risk and ethics: Three approaches* (Routledge).

Hartford, D N and Baecher, G B, 2004. *Risk and uncertainty in dam safety* (Thomas Telford).

Hopkins, A and Kemp, D, 2021. *Credibility Crisis: Brumadinho and the Politics of Mining Industry Reform* (CCH Australia).

International Commission on Large Dams (ICOLD), 2019. ICOLD Incident Database Bulletin 99 Update: Statistical Analysis of Dam Failures, International Commission on Large Dams.

International Commission on Large Dams (ICOLD), 1988. *Dam design criteria the philosophy of their selection Bulletin 061.*

Jansen, R B, 1983. *Dams and public safety,* US Department of the Interior, Bureau of Reclamation.

Kossoff, D, Dubbin, W E, Alfredsson, M, Edwards, S J, Macklin, M G and Hudson-Edwards, K A, 2014. Mine tailings dams: characteristics, failure, environmental impacts, and remediation, *Applied Geochemistry,* 51:229–245.

Morales-Torres, A, Escuder-Bueno, I, Serrano-Lombillo, A and Rodriguez, J T C, 2019. Dealing with epistemic uncertainty in risk-informed decision making for dam safety management, *Reliability Engineering & System Safety,* 191:106562.

Morgenstern, N R, 2018. Geotechnical risk, regulation, and public policy.

Paté-Cornell, M E, 1996. Uncertainties in risk analysis: Six levels of treatment, *Reliability Engineering & System Safety,* 54(2–3):95–111.

Riddolls, B and Grocott, G, 1999. Quantitative risk assessment methods for determining slope stability risk in the building industry, Building Research Association of New Zealand.

Rowe, W D, 1975. *An Anatomy of Risk* (Environmental Protection Agency).

Santos, R N C D, Caldeira, L M M S and Serra, J P B, 2012. FMEA of a tailings dam, *Georisk: Assessment and Management of Risk for Engineered Systems and Geohazards,* 6(2):89–104.

Silva, F, Lambe, T W and Marr, W A, 2008. Probability and risk of slope failure, *Journal of Geotechnical and Geoenvironmental Engineering,* 134(12):1691–1699.

Stallen, P J M, Geerts, R and Vrijling, H K, 1996. Three conceptions of quantified societal risk, *Risk Analysis,* 16(5):635–644.

US Army Corps of Engineers (USACE), 2003. Slope Stability, Engineer Manual, EM 1110-2-1902 (USACE Washington, DC).

Vick, S G, 1990. Planning, design, and analysis of tailings dams.

Vick, S G, Atkinson, G M and Wilmort, C I, 1985. Risk analysis for seismic design of tailings dams, *Journal of Geotechnical Engineering,* 111(7):916–933.

Villavicencio, A G, Breul, P, Bacconnet, C, Boissier, D and Espinace, A R, 2011. Estimation of the variability of tailings dams properties in order to perform probabilistic assessment, *Geotechnical and Geological Engineering,* 29(6):1073.

Wahl, T L, 2004. Uncertainty of predictions of embankment dam breach parameters, *Journal of Hydraulic Engineering,* 130(5):389–397.

Walley, P, 1996. Measures of uncertainty in expert systems, *Artificial Intelligence,* 83(1):1–58.

Warbuton, M, Hart, S, Ledur, J, Scheyder, E and Levine, A J, 2019. The Looming Risk of Tailings Dams. Available from: <https://graphics.reuters.com/MINING-TAILINGS1/0100B4S72K1/index.html> [Accessed 04 November 2021].

Whitman, R V, 1984. Evaluating calculated risk in geotechnical engineering, *Journal of Geotechnical Engineering,* 110(2):143–188.

Williams, D J, 2021. Lessons from Tailings Dam Failures—Where to Go from Here?, *Minerals,* 11(8):853.

Willke, H, 2007. *Smart governance: governing the global knowledge society* (Campus Verlag).

The Engineer of Record role in a capital project framework – can we do better?

B Tiver[1]

1. Principal Tailings Engineer, WSP Australia Pty Ltd, Adelaide SA 5000.
   Email: brad.tiver@wsp.com

ABSTRACT

This paper presents a retrospective commentary of two case studies of the Engineer of Record (EOR) role in the context of projects undertaken within a 'capital project' framework and presents areas of the process that could be improved in the context of dam safety. The capital project framework is largely focused on proving the feasibility of a project and gaining funding approval, with motives such as accurate cost estimation and value optimisation, where the design consultant is primarily encouraged to demonstrate solutions that reduce capital expenditure. A feasible, compliant design is required but risk management design improvements are typically only considered if there is an accompanying cost reduction. Under this focus, the role of EOR (as defined by the Global Industry Standard on Tailings Management (GISTM) by the International Council on Mining and Metals (ICMM, 2020)) is often to confirm alignment with standards, with the integrity of the facility during design, construction, operations and closure having a diminished focus.

The first case study was for detailed design of an embankment raise to a TSF where the author was the study lead and the EOR was from the same consultancy as the author. Two design features are discussed. One feature reduced risk but added cost. The second feature reduced costs for the embankment raise, increased tailings and flood storage capacity, but resulted in a perceived increase in the risk profile to the owner. The case study demonstrates that the value optimisation framework needs improvement to consider risks and the EOR's perspective.

The second case study was for a design of a buttress around an upstream raised TSF where the author was the study lead and the EOR was from another consulting firm. An example is discussed where the author's team proposed solutions they considered would significantly reduce capital expenditure, whilst maintaining the integrity of the facility, that were ultimately not endorsed by the EOR at the time of publication. The case study demonstrates that, in the author's view, the design of a TSF, and any improvements, should be undertaken by the EOR. However, where this is not feasible, due to resource constraints or other reasons, the capital project framework needs to improve how the design consultant is able to interact with the EOR and the owner to achieve a 'best for project' outcome.

INTRODUCTION

This paper presents a retrospective commentary of two case studies of the Engineer of Record (EOR) role in the context of tailings storage facility (TSF) projects undertaken within a 'capital project' framework and presents areas of the process that could be improved in the context of dam safety. The capital project framework is largely focused on proving feasibility of a project and gaining funding approval, with motives such as accurate cost estimation and value optimisation, where the design consultant is primarily encouraged to demonstrate solutions that reduce capital expenditure. A compliant design is required but risk management design improvements are typically only considered if there is an accompanying cost reduction.

The definition of the EOR role, in accordance with the Global Industry Standard on Tailings Management (GISTM; ICMM, 2020), is as follows:

> The qualified engineering firm responsible for confirming that the tailings facility is designed, constructed, and decommissioned with appropriate concern for integrity of the facility, and that it aligns with and meets applicable regulations, statutes, guidelines, codes, and standards. The Engineer of Record may delegate responsibility but not accountability.

This implies that the EOR should be involved in the entire life cycle of the TSF – this adds value because they understand the current operations, future life of the TSF, its interaction with other processing considerations across the site and its relationship to closure. The scope of focus is not just the isolated design feature being studied in the project.

The EOR is accountable for verifying that the owner's responsibilities have adequately been addressed or provides a path forward to reduce risk and achieve a safe and dependable TSF. Time-bounded risk mitigation measures must be achieved; the EOR is responsible for developing measures and timelines in conjunction with the Responsible Tailings Facility Engineer (RTFE). It is the responsibility of the owner to affect any recommendations made, and within the required time period, by the EOR. The effective relationship between the owner and EOR and a clear acceptance of their roles and responsibilities is important and at the core of the safe and proper management of mine waste.

THE CAPITAL PROJECT FRAMEWORK

In the case studies presented herein, the engineering was undertaken within a capital project framework which drives investment decisions. Note that the author works for a consulting company, and hence the descriptions herein are presented in that context. In the framework, projects are initiated with a set of requirements that are based on a business case.

In a capital project framework, in the case of there being no existing TSF, the owner's project team would identify the need for a new TSF, which then initiates a capital approval process. When an expansion, mitigation measure or improvement project is required during operations, it is usually the RTFE who initiates the business case, typically as a result of a recommendation from the EOR or Independent Tailings Review Board (ITRB) or due to expansion requirements to meet changing tailings storage requirements. In the author's experience, the RTFE is capacity constrained to deliver a capital study, and so a capital projects team is assembled to deliver the study for the business. Usually, at a conceptual level, a preferred option is developed and selected, which is then carried through to pre-feasibility study (PFS) and feasibility study (FS) level design. In the case of risk mitigation measures, for example a buttress to an upstream-raised TSF, the study usually commences at feasibility study level as the preferred risk mitigation technique is usually well understood.

These three steps are typically at levels of Class 4, 3 and 2, respectively, in accordance with the American Association of Civil Engineering (AACE, 2020) cost estimate classification system. Class 4 is pre-feasibility study, Class 3 is feasibility study and Class 2 is detailed design, although different terminology is used across mine owners. At the end of each stage of engineering, the gated delivery process includes technical and commercial review of proposed investment decision to assess its viability. In the two case studies presented herein, the concept development stage was not undertaken – for case study one, it was a TSF due for another embankment raise, which therefore doesn't require a PFS as there was only one preferred solution. For case study two, it was expanding a buttress around an already buttressed TSF, so the preferred option for improving the stability of the TSF perimeter embankments had already been established.

In the PFS and FS stages, there is a task for the engineering consultant to demonstrate cost reduction measures of the proposed solution, typically called 'value optimisation' or 'value engineering'. Value optimisation registers using simple spreadsheet style tools are typically developed in the study, and in the experience of the author, focus on key drivers such as 'CAPEX', 'OPEX' and 'NPV' and have a specific column about the estimated 'trend', ie cost saving. In the author's view, the current tools provide little value in demonstrating that an idea could cost more money but significantly reduce risk.

CASE STUDY ONE

Background

The first case study was for detailed design of an embankment raise to a TSF where the author was the study lead and the EOR was from the same consultancy as the author. During the study, an early trade-off assessment indicated a preferred investment for an upstream embankment raise, on the basis of available fill materials and the cost of pipeline removal and reinstatement. This design then contributed to design of the TSF, and its adjoining TSF (ie there are two TSFs with a common dividing embankment) for submission to the regulatory authority for approval. The 'approvals design' included a spillway within the internal dividing embankment from the higher to lower TSF, and an external spillway on the lower TSF. The spillway was designed to increase tailings storage and flood retention capacity in the TSF basin. The EOR proposed that a spillway was included for good engineering practice, not because it was needed from a risk management perspective in the view of the EOR. The spillway was proposed to not be lined with erosion resistant materials, principally due to the very low of risk of spillway activation and that the risk of catastrophic release of tailings was assessed as having an extremely low probability, noting that there is no industry accepted definition for assigning 'non-credible' failure modes. Furthermore, in the quantitative risk assessment (QRA) undertaken during the study, it was identified that overtopping of the embankments was not a key risk-driving failure mode.

The consequence category of the TSF (both TSFs as the combined structure) is High C under the ANCOLD Guidelines (Australian National Committee on Large Dams (ANCOLD), 2019) and High in accordance with the GISTM (ICMM, 2020).

Example one

Summary of value optimisation process

The embankment raise design included design of a spillway that was focused primarily on increased storage capacity in the basin – both for tailings and flood storage. For compliance with the adopted consequence categories, the freeboard requirements for the TSF would be the 1:100 AEP, 72-hour event under ANCOLD and the 1:2475 AEP 72-hour event under the GISTM. The TSF spillway was designed such that the invert level resulted in the TSF storing an event greater than the 1:100 000 AEP event, and less than the 24-hour Probable Maximum Precipitation (PMP). On top of this, the spillway was then sized to pass the Probable Maximum Flood (PMF), assuming that the TSF was full to the invert level of the spillway at the time for PMP occurs.

This design philosophy led to two important decisions in the VO process – the first was that the spillway geometry was long and wide, ie the measure that increases the flood storage capacity in the basin. This design measure was judged as a capital cost saving to the project of A$0.5M based on not constructing the same type of spillway as was existing, which was a typical ~30 m wide concrete mattress, with a flow depth of about 1 m. The proposed spillway was 0.15 m deep and 600 m long.

The second design feature considered in the VO process was to not line the spillway (both the crest of the embankment and the downstream chute) on the basis that the embankment construction materials were relatively robust (gravelly materials), and some erosion would be accepted in the case of activation of the spillway. The long and wide spillway was estimated to have relatively low discharge velocities, which would reduce the risk of erosion. This was considered reasonable in the context of the available flood capacity and the design storm event for the spillway.

The owner's capital projects team had approved the design in the 'approvals design' report, which was also subject to peer review by a Senior Independent Technical Reviewer (SITR) (For TSFs with consequence categories of High or below, a SITR can be appointed rather than an ITRB). Notwithstanding acceptance at this stage from all parties, following the capital project framework, further assessment occurred over the following months, and the final value optimisation assessments included in a report that was approved by the capital projects team.

During the project, there was a change of staff at the owner's team for dam governance, which triggered re-consideration of the spillway erosion protection. The owner's team was concerned with the consequence of erosion, not the likelihood, and therefore further assessments were undertaken to support not lining the spillway chute. If the spillway chute was to be lined, installation of riprap on the 600 m long downstream batter was judged to cost A$1M to A$2M, and due to project timing and the tailings storage benefits of maintaining the higher spillway invert, it was unfavourable to return to the concrete mattress style design.

Areas for improvement

In this example, the author's view is that there is room for improvement on the consistent governance and agreement on design criteria for TSF spillway design, particularly acceptance of a risk-based approach for all aspects of the design (not just components). The EOR proposed a solution on the basis of the QRA and risk being a combination of likelihood and consequence. The consequence of spillway erosion was proposed to be accepted on the basis of the likelihood being very low and the delegated authority of the Accountable Executive accepted the risk. However, the author is aware of proposed changes that would require placement erosion protection measures on the spillway. In the context of the QRA, the EOR's view is that the money spent on potentially lining the spillway would be better spent on risks that have a higher likelihood of occurrence.

Example two

Summary of value optimisation process

The second example within this study relates to decant infrastructure. The TSF had been fallow for a few years and the owner had assumed that the existing decant access and associated infrastructure could be reused. This assumption turned out to be incorrect – the access to the decant pumping infrastructure was too steep and did not comply with site requirements, and the pumping infrastructure was assumed to be able to be reused, which would have required significant, costly refurbishment.

The author's team then provided two options to provide access to the decant and designed new pumping infrastructure. The two decant access causeway options that were conceptually designed varied by approximately A\$1M, and the cheaper and more traditional design was accepted by the owner. The cost of the new pumping infrastructure and causeway was approximately estimated at A\$0.7M during the study, and was accepted on the basis of reduced risk – access to the decant causeway is directly related to control of pump operations and consequently the size of the decant pond, which is a key safety parameter.

Areas for improvement

The author's view is that the scope of work for the study was developed by a separate team, without enough interaction between the RTFE and the EOR. An improved amount of planning at the project initiation stage, with input from site-based personnel, could have prevented adding cost to the project baseline.

CASE STUDY TWO

Background

The second case study was for PFS and FS level design for an expanded buttress around a TSF. The author was the study lead and the EOR was from a different consultancy to the author. This case study is focused on issues around project development and basis of design, as well as value optimisation.

At the commencement of the project and based on information available at the time, the EOR judged that the residual shear strength of the tailings was lower than the previous designs had adopted, and that therefore an expanded buttress was required for compliance with the owner's internal guidance, and industry guidance (ANCOLD, 2019). Two key changes were made during the study – a reduced ultimate design height for the TSF, and a slight increase in the residual undrained shear strength of the tailings. Due to these changes, the project was re-framed twice during the study by the owner.

Whilst the author acknowledges that the risk would be reduced with increased buttressing, in the context described above, the additional buttressing being designed in the studies described herein was for compliance, and not necessarily about significantly reducing risk. This framework and context led to the value optimisation process being mis-used during the study.

The consequence category of the TSF (both TSFs as the combined structure) is High A under the ANCOLD Guidelines (ANCOLD, 2019) and Very High in accordance with GISTM (ICMM, 2020).

Example

Summary of value optimisation process

A value optimisation register was developed during the PFS in which the author suggested undertaking a deformation analysis to indicate the performance of the embankments with the existing buttress. An initial step was undertaken by the author's team to undertake a one-dimensional seismic ground response analysis (SGRA), which indicated that the tailings would unlikely be susceptible to cyclic liquefaction based on the design earthquake demands at site. However, the principal focus of the study was a compliant design with ANCOLD factor of safety recommendations, so the SGRA was not carried further at that time. The unfortunate outcome was that the SGRA was not used as a tool to help judge the performance of the TSF, and hence a potential saving to reduce the amount of buttress fill based on risk was missed, with the focus remaining on factor of safety compliance.

A year later, after the first project change for a reduced height of the TSF, the SGRA was revised. The assessment reconfirmed that the tailings were unlikely to be susceptible to liquefaction. Again, it is questionable whether this truly added value in the project context. Notwithstanding, compliance was first and foremost, and so the traditional limit-equilibrium slope stability analyses were updated, and indicated a reduced buttress size, commensurate with the reduced height of the TSF. The value optimisation register/process was used by the owner to document the trend of savings to the project as is encouraged by the framework.

Thereafter, the EOR judged that the residual undrained shear strength had improved, and so, once again, the slope stability analysis and corresponding size of the buttress was updated, again reducing the fill volume and saving the project capital expenditure. Only at this time, over a year into the study, did the author's design team get to interact with the EOR and discuss the outputs of the slope stability assessment. Then a further year transpired with iterative discussions about the assumptions in the slope stability models regarding features such as characterisation of the foundation and the phreatic surface.

Alongside this final year of slope stability modelling, the EOR undertook a numerical deformation analysis, considering both static and cyclic liquefaction triggering assessments. As the design consultant, the author's team was asked to review the analyses. The analyses indicated that liquefaction triggering events did not result in unacceptable deformations, as expected as the buttress had been sized to meet a minimum factor of safety of 1.1 under residual (post-peak) strength conditions for the tailings. It also indicated that the residual undrained shear strength of the tailings that the design team had adopted in the design of the buttress could not be realised under reasonable circumstances, and that an event not able to be conceptualised would be required to trigger residual shear strength. Notwithstanding, the basis of design for the project was for the residual shear strength and so deformations were indicated to be acceptable under the design case.

Areas for improvement

Being the design consultant and being challenged to present ideas for reduced costs, principally in this example by reduced buttress size, led the author's team to suggest performance-based design of the TSF. The idea was never adopted by the EOR or the owner at the time of publication, as it was irrelevant in the context of the compliance with factors of safety. During the project, the numerical deformation analysis was adopted as a 'check' due to the consequence category of the TSF, not as a design tool. There needs to be improved clarity in value optimisation tools about management of risk instead of only just cost reduction.

In this project example, the value optimisation process was undertaken by the design consultant, with the RTFE included, but no interaction with the EOR. The example indicates that having the design consultant separate from the EOR resulted in misalignment of the purpose and outcomes of the stability and deformation analyses, initially proposed as a value optimisation, and then later adopted as a check for compliance with ANCOLD (2019).

During a design project that lasted nearly three years, there were only a handful of meetings between the design consultant and the EOR, and mainly interaction via a 'comments register' on the slope stability assessment report. In the project cases where the EOR is not the designer, there needs to be much improved interaction between the parties to improve efficiency and maintain alignment on relevant value optimisation studies.

CONCLUSIONS

Both studies described in this paper were ultimately successful in that they met the study objectives and ultimately reduced risk. The capital project framework is a useful one for owners to make well-informed decisions about significant capital investments. However, there is room for improvement in tailoring the framework for tailings studies, which are onerous based meeting the requirements of ANCOLD and GISTM. In these industry frameworks, there is a large focus on risk management, but this is not incorporated into the capital project framework. The capital project framework should therefore be adjusted to match. Based on the author's experience, the following recommendations are made:

1. Development of project scopes should include significant input from the EOR and RTFE before they are initiated by capital projects. The EOR and RTFE are those closest to the issues and operation of the TSF, and project definitions and requirements would be well-served to include their input and review. This recommendation is in accordance with Requirement 8.5 of the GISTM (ICMM, 2020):

   Appoint a site-specific Responsible Tailings Facility Engineer (RTFE) who is accountable for the integrity of the tailings facility, who liaises with the EOR and internal teams such as operations, planning, regulatory affairs...'

2. Consistent governance of risk-based versus compliance requirements would reduce confusion and improve project efficiency.

3. Use of the QRA framework to judge risks and where value optimisation should be focused is supported. Each element of the failure pathway in a QRA has a certain level of confidence and the EOR can quickly identify where the biggest uncertainties are and where tailings management and dam safety can truly benefit from capital investment. Current value optimisation processes are not equipped to indicate reduced risk against capital investment.

4. Linked to the QRA, value optimisation tools for tailings projects undertaken in the capital projects framework should be adjusted from those otherwise adopted to allow focus on risk. There is a need for a framework so that risk and financial decisions about TSFs can be made together.

Recommendations 2 to 4 above are in accordance with the general theme in the GISTM related to risk-informed design, and specifically the following requirements:

Requirement 4.4 of the GISTM (ICMM, 2020):

   Select, explicitly identify and document all design criteria that are appropriate to minimise risk for all credible failure modes for all phases of the tailings facility life cycle.

Requirement 5.4 of the GISTM (ICMM, 2020):

   Address all potential failure modes of the structure, its foundation, abutments, reservoir (tailings deposit and pond), reservoir rim and appurtenant structures to minimise risk to ALARP (As Low As Reasonably Practicable). Risk assessments must be used to inform the design.

Requirement 9.1 of the GISTM (ICMM, 2020):

   Engage an engineering firm with expertise and experience in the design and construction of tailings facilities of comparable complexity to provide EOR services for operating the tailings facility...

5. The EOR should be the design consultant. In some cases, this is not possible, but in those instances, the EOR itself should be closely involved with the project decisions of the design consultant to significantly improve efficiency. This recommendation is not specific to a guideline but is a general observation based on the author's experience – given the intimate

knowledge the EOR has with the facility and its connection to the RTFE, being an external party to that relationship is challenging.

REFERENCES

American Association of Civil Engineering (AACE), 2020. Cost Estimate Classification System – As Applied in Engineering, Procurement, and Construction for the Process Industries, TCM Framework: 7.3 – Cost Estimating and Budgeting, AACE International Recommended Practice No. 18R-97, 7 dated August 2020.

Australian National Committee on Large Dams (ANCOLD), 2019. Guidelines on Tailings Dams, Planning, Design, Construction, Operation and Closure, Rev1.

International Council on Mining and Metals (ICMM), 2020. The Global Industry Standard on Tailings Management (GISTM), March.

Importance of site climatic, topographic, seismic and social settings

Considerations in tailings storage expansions in tropical environs

C S Hogg[1] and S V Tolu[2]

1. Senior Principal – Tailings, CMW Geosciences Pty Ltd (CMW), Perth WA 6018.
   Email: chrish@cmwgeo.com
2. TSF Project Senior Engineer, PT Meares Soputan Mining (PT MSM), Indonesia.
   Email: stenly.tolu@archimining.com

ABSTRACT

This paper explores the considerations and constraints in the implementation of a tailings storage expansion project in Indonesia. An existing tailings storage is being operated at the site. The existing TSF is a valley fill type storage with upslope catchments. Currently the TSF has a main embankment with a height of around 80 m. The site has a tropical climate with high rainfall exceeding evaporation, and medium to high seismicity.

An options study was conducted to assess the future storage needs for the life-of-mine. This study considered a variety of site options including the existing site. At the time of the study, the site had a mine life nine years with 45 Mt of projected ore production. The existing TSF at the time of the study had 13 Mt of storage capacity. The existing TSF has a consequence rating of 'High B' in accordance with ANCOLD (2019), due to potential loss of life at the plant and 'major' consequence of an embankment failure.

Five site options were considered for future tailings storage including raising the existing facility and in-pit storage. A ranking of the site options was performed and considered project location/cost, engineering risk, environmental risk and social risk. Each option was scored based on outcomes for the criteria. From these scores the options were ranked under each category. Each category was assumed to have equal weighting. From the ranking of the options for each category, an overall ranking was assessed. Based on the assessment the best ranking, was raising the existing facility however an alternative site which had the next ranking was also investigated. Following investigations it was decided that the option of raising of the existing facility was preferred.

The study concluded that the additional 32 Mt can be stored in the existing TSF by raising the embankment between 20 and 30 m. The paper presents the design adopted and presents a discussion of the design constraints, and design and operational controls adopted.

The project is in the process of being approved by government regulators.

INTRODUCTION

This paper seeks to outline details of the process of implementation of the expansion of tailings storage facilities at the Toka Tindung Mine in Indonesia. Toka Tindung is located in northern Sulawesi approximately 40 km from the city of Manado.

In 2019, the mine embarked on an options study to examine the preferred tailings storage site for future tailings storage. At that time, Toka Tindung had a mine life to 2029 at nominally 4.5 Mt/a or approximately 45 Mt of projected ore production. Based on an estimated storage capacity of the existing Tailings Storage Facility (TSF) to RL290 m of 13 Mt, future tailings storage for 32 Mt (minimum) of tailings was still required.

BACKGROUND

Topography

The topography in the project area comprises low hills and valleys with steep slopes encountered. Figure 1 shows a drone photo of the existing tailings storage facility set in a broad valley.

FIG 1 – View of existing TSF with plant site in centre of photo.

Climate

The site has a tropical climate with high rainfall exceeding evaporation. Based on available records the average annual rainfall is 3087 mm (as measured at site) and the pan evaporation as measured at Manado is 1458 mm. The site is subject to intense rainfall, the 1:100 annual exceedance probability (AEP) 24 hr storm event, has a rainfall depth of 505 mm (ref Coffey, 1997).

Seismicity

Indonesia lies in an area of complex interaction between tectonic plates. North Sulawesi, is influenced by seismic activities by active subduction zones and transform faults. Movement along both these fault zones will result in earthquakes impacting the project site. Peak ground accelerations at the site are 1:1000 AEP 0.29 g operating basis earthquake (OBE) and 1:10 000 AEP 0.4–0.6 g safety evaluation earthquake (SEE) as determined from the Indonesian Earthquake Hazard Map for the Toka Tindung TSF location (PusGen (Center of Indonesian Earthquake Study), 2017).

Geology and ground conditions

The bedrock geology generally comprises andesites, however basalt was encountered in part of the original TSF site. The ground conditions are summarised in Table 1. The embankment foundation comprised stiff brown clay up approximately 5 m thick overlying very stiff silty clay.

TABLE 1

Typical subsurface ground profile.

Unit	Description	Depths
Tephra	Sand – black, grey, coarse, rootlets in upper 300 mm	to 1.5 m
Silty Clay	Firm to stiff, light brown, high plasticity, with cobbles/boulders (two boreholes, some test pits)	to 3 m
Sandy Clay	Stiff to very stiff, mottled white, pink, yellow, medium to high plasticity	to 8.0 m
Andesite	Low strength, Light brown, grey, completely to moderately weathered	>18 m

Tailings properties

The tailings are derived from the processing of a mix of weathered ore types (ie mainly oxides and transition) from multiple pits. The tailings properties were last confirmed in 2022 by conducting a suite of tests on samples of tailings from the beaches. The results indicated the following:

- Moisture content between 35.3 per cent and 51.9 per cent.

- The tailings samples are low plasticity and plotted below the 'A' line.

- The tailings were sandy clayey silt to clayey silt with a fines content (material passing 75 micron) between 75.0 per cent and 98.2 per cent.

- The strength and consolidation tests on tailings samples indicated:

  o Very low strength <25 kPa. Note the strength of the tailings does not impact directly on the embankment stability.

  o Coefficient of Consolidation of between 7.6 m^2/a. and 9.2 m^2/a. Indicating relatively poor consolidation characteristics.

  o Permeability generally between 1×10^{-5} m/s and 5×10^{-7} m/s (unconsolidated surficial tailings).

OPTIONS STUDY

Options considered

The mine began a process of looking for alternative tailings storage sites in 2019. A list of sites was proposed with the list critically examined by both mine management and the design consultant CMW. The options considered in the documented options study (CMW, 2019) were:

- Raising of the existing TSF to RL300 m – Option 1.

- Construction of a new above ground TSF in the Pangisan valley – Option 2.

- Construction of a new above ground TSF in the Maen valley – Option 3.

- Construction of a new above ground TSF in the Maesa valley – Option 4.

Each of the options that were considered were valley storage type options in keeping with the terrain. Each option would involve construction using downstream construction methods. The operations for each option would be similar with tailings deposition using multi-point spigots from the perimeter embankment with a decant system in the upstream section of the valley, similar to the existing TSF. The existing decant facilities comprise a floating pontoon pump behind a filter wall.

Other alternatives tailings storage types were also considered. In-pit tailings storage in the Toka Pit is also a possible future tailings storage option and has been the subject of previous studies. A separate study was also commissioned to examine dry stacking, which would have challenges in the tropical environment.

The locations of the options are shown in Figure 2.

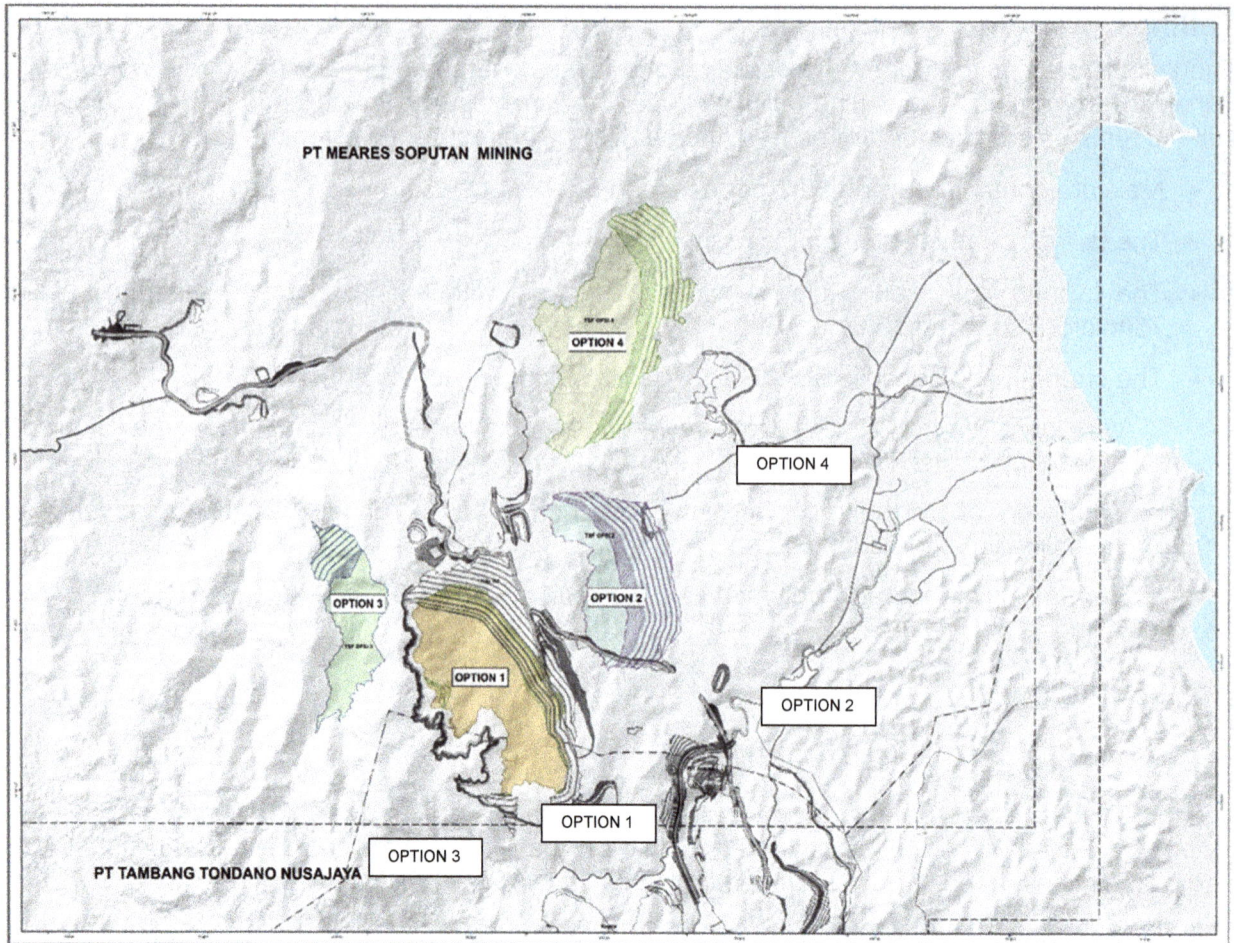

FIG 2 – Option locations.

Ranking of options

A desk study of the options was carried out. Estimates were made of TSF characteristics including tailings area, storage capacity and earthworks volumes and maximum embankment height. Each of the options were assessed in terms of advantages and disadvantages. Table 2 summarises the results of that assessment.

TABLE 2

Options, advantages and disadvantages.

Advantages	Disadvantages
Option 1 – Raising of the existing TSF	
• Limited land disturbance by raising an existing facility. No land acquisition required. Amendment of existing permitting required. • The option will utilise existing infrastructure for construction and partly for existing operations.	• It is understood pumps will need to be upgraded due to the increase in pumping head between the plant and RL320 m. • The new footprint will likely impact on forestry land.
Option 2 – new TSF, Pangisan valley	
• Location is downstream of the plant site. • Land is largely owned by PT MSM. • Proximity to plant site and pit areas.	• A creek diversion required. • TSF area relatively small, will need to be operated with another TSF for the proposed production rate of 4.5 Mt/a. • New permitting required. • TSF interferes with other mine site requirements, eg hot water management and waste dump requirements.
Option 3 – new TSF, Maen Valley	
• Efficient storage of tailings (ie higher storage/earthworks ratio compared to other options). • Site close to the existing TSF, sharing of existing infrastructure for construction and partly for operations possible.	• Major diversion of large upstream catchment required. • Longer pump distance than existing TSF, pump upgrade likely required. • TSF area relatively small will need to be operated with another TSF for the proposed production rate of 4.5 Mt/a. • New permitting required. • Land acquisition required. Required land rezoning from forest. • Downstream water collection and use by surrounding communities.
Option 4 – new TSF, Maesa Valley	
• Location is downstream of the plant site. • Site lower than plant site. • The site will have a large tailings area and can contain the life-of-mine tailings.	• Large amount of waste required on a long lead, potentially making IWL construction expensive. • Site close to existing mine camp and resettlement area. • Land acquisition required. • New permitting required.

Following assessment of the advantages and disadvantages of each option, further assessment of the options was performed by ranking each option to the following categories, refer to Table 3.

TABLE 3

Ranking categories and criteria.

Category	Criteria
Project location/cost	Storage/earthworks ratio, distance from borrow/construction materials and distance from plant
Engineering risk	Stability (height), water diversion (area to be diverted), operational difficulty
Environmental risk	Area of disturbance, land type, seepage, closure
Social risk	Distance to nearest village, land acquisition

A preliminary assessment of the cost of the options was performed using the criteria outlined above (storage/earthworks ratio, distance from borrow/construction materials and distance from plant). A low cost option will have a high storage/earthworks ratio, have a short lead distance from an earthworks source to the TSF and a short distance to the plant (ie for piping and pumping) and a low pumping head.

A preliminary assessment of the engineering risk of the options was performed using the criteria presented in Table 3 (stability (height), water diversion (area to be diverted), operational difficulty). In general, a rule that the higher the facility (embankment) the greater the risk of failure. The area of run-off diversion provides an indicator of the risk of failure from embankment overtopping (ie a large diversion area would lead to a larger run-off flow/volume to be diverted and hence greater risk). A qualitative assessment of the operational risk has also been made.

A preliminary assessment of the environmental risk of the options can be performed using the criteria presented in Table 3 (area of disturbance, land type, seepage). In general a facility with a larger footprint could potentially have a greater impact on the environment downstream and will be more costly to rehabilitate. Other issues that were evaluated included land type (Forestry) and whether the TSF is located upstream of an existing wetland. A qualitative assessment of these risks was made.

A preliminary assessment of the social risk of the options can be performed using the criteria above (distance to nearest village, land acquisition). Distance to nearest village downstream provides an indication of the risk to local population due to a dam break. Land Acquisition has also been identified as a major social factor (ie as advised by PT MSM).

Each option was scored based on outcomes for each criteria, with ratings of 1 for best outcome up to 4 for worst outcome. From these scores the options were ranked under each category (ie project location/cost, engineering risk, environmental risk and social risk). Each category was assumed to have equal weighting. From the ranking of the options for each category, an overall ranking was assessed as presented in Table 4.

TABLE 4

Overall ranking.

Various ranking	Option 1: Raising TSF to RL320 m	Option 2: New TSF Pangisan	Option 3: New TSF Maen west	Option 4: New TSF Maesa
Cost – ranking	1	2	2	1
Engineering risk – ranking	3	1	4	2
Environmental risk – ranking	1	2	2	4
Social risk – ranking	1	3	2	4
Score	6	8	10	11
Overall ranking	1	2	3	4

In terms of project location/cost, the various options, had similar rankings. Option 1 had the preferred (lowest score) ranking for environmental and social risk, and likely the lowest cost. Option 2 had the preferred (lowest score) ranking for engineering risk.

Option 3 was less preferred than Option 2 due to the large catchment area to be diverted which would substantially increase the facility risk. The least preferred option was Option 4 'New TSF Maesa Valley', mainly due to its proximity to the camp and a re-settlement area. The Option 4 site location is away from the plant and Toka pit area.

Options 1 and 2 were taken forward and geotechnical investigations were performed at these sites (only). Drilling and sampling was carried out on the main embankment of the existing TSF (Option 1) to allow further assessment the embankment stability. A preliminary investigation involving test pits and some drilling was performed at the Option 2 site.

The results of the investigation at the Option 1 site were used in stability assessments (ie SPT testing and sampling, and laboratory testing including triaxial testing). The investigation of the Option 2 site revealed varied depths of sediments across the site up to 30 m in-depth. The ground conditions at the Option 2 site were considered less favourable than the existing site and hence design of a TSF at the Option 2 site was not taken further.

Operator perspective

The update of the Mine of Mine plan (LOMP) identified the need for additional tailings storage facilities. The previous concept was that it was probable that an in-pit tailing storage facility would be permitted in the Toka pit. As part of the life-of-mine process this was no longer an option due to timing of the completion of an extended open pit from the updated LOMP and the available tailings storage capacity.

The requirements for permitting of activities can take time to communicate plans and obtain approvals from multiple agencies. This time consideration is an important element of the decision-making process.

In addition the area of the mine operates has high demands on land use by the surrounding communities and minimising the mine's footprint is a significant factor for consideration.

TSF DESIGN

The mine commenced operations in April 2011. The existing TSF was designed in 2009/2010. The TSF comprises a valley type storage with a containment embankment on the northern side. The TSF is enclosed on all other sides by sloping valley sides and is therefore 'isolated' from other drainage catchments.

The TSF, Cell 1 (original cell) to RL278 m provided approximately 11 Mm^3 of storage volume. The TSF Expansion (Cell 1 and 2, to RL290 m) provided an additional 11.5 Mm^3 of storage volume. The total storage capacity based on 22.5 Mm^3 at 1.25 t/m^3 (dry) is approximately 28 Mt.

The existing TSF embankment was constructed in seven stages by downstream construction methods. The Stage 1 (starter) main embankment crest RL236 m and Stage 2 embankment crest RL247 m were constructed in 2010/2011. The embankments were raised progressively. The construction of Stage 7 embankment raise to RL278 m was completed in January 2018.

The TSF Embankments were raised and extended to RL290 m, between the August 2018 and October 2022 under a separate approval for raising and expansion of the TSF. The raising and extension of the TSF to RL290 m involved construction of a new cell to the south of Cell 1. This cell was raised and extended along with Cell 1. The TSF cells have now been integrated into one cell, at crest RL290 m.

Design considerations

The following considerations were adopted in the raising and expansion design (CMW, 2023):

- Additional geotechnical investigation for the raising and extension of the existing TSF, including boreholes and test pits, sampling and laboratory testing.

- Confirm embankment design concept by undertaking additional geotechnical investigations to confirm embankment stability and embankment geometry. The stability relies on a functioning embankment drainage system, seepage analyses were conducted to confirm the system meets capacity requirements. The existing internal drainage system was performing adequately.

- Water management: in order to manage excess water on the facility, diversion drainage will be required and temporary spillways designed for each stage. Water balance analyses were performed to check the pumping requirements to remove excess water which is pumped to the site wetland system.

- Design the expanded TSF with the flexibility to store up to 45 Mt, by adopting three stage levels (RL300 m, 310 m and 320 m) is based on the following parameters:

  o Annual tailings production 4.5 Mt/a (2019–2029).

  o Tailings deposited at 45 per cent solids. Future tailings grind 80 per cent passing 53 μm (current 80 per cent passing 75 μm).

- The cyanide within the tailings will be removed in a 'detox' treatment plant before the tailings are deposited in the TSF.

- Design in general accordance with ANCOLD Guidelines (2019) and the design also must comply with Indonesian Dam Safety requirements.

A plan of the TSF design is shown in the Figure 3.

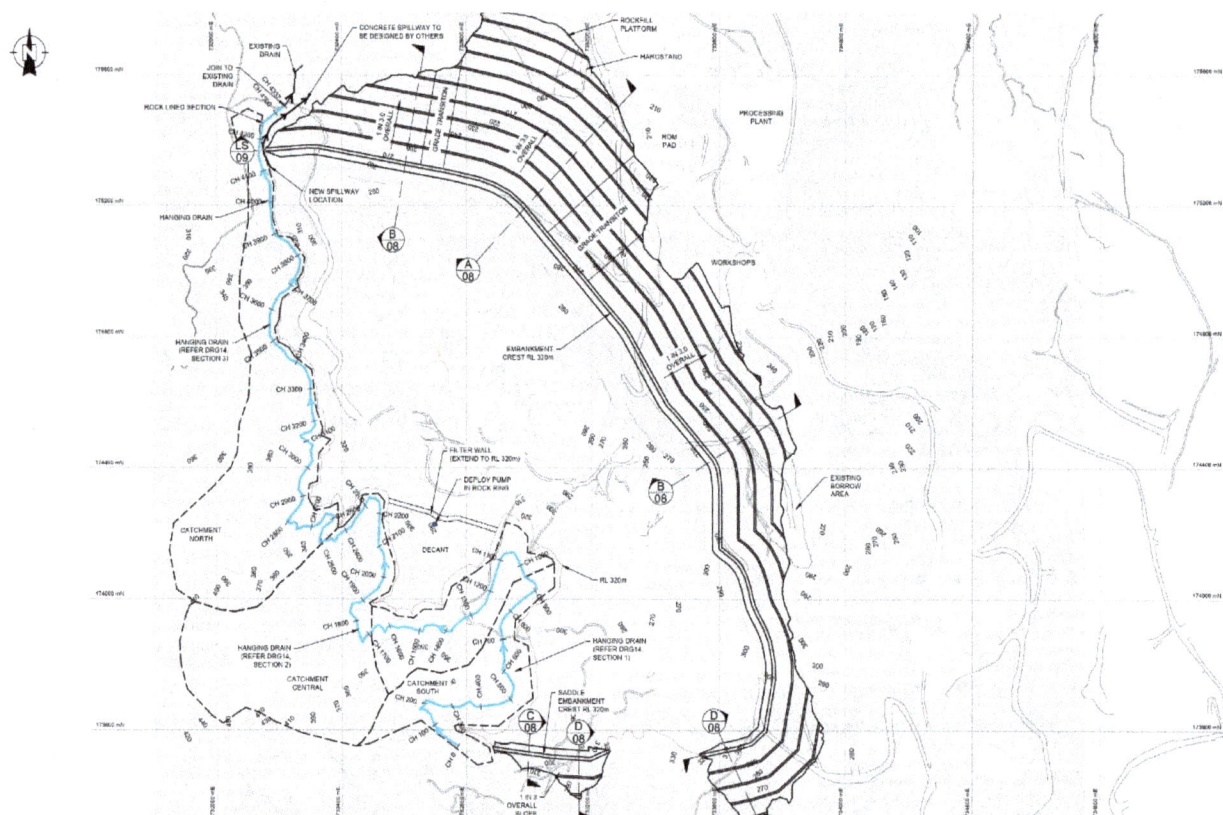

FIG 3 – Plan of TSF raised and Extended to RL320 m.

Hazard rating

The raising of the TSF to RL320 m has been assigned a consequence (hazard) rating of 'High A', based on the potential consequences of an embankment failure as outlined in Tables 1 and 2 of the ANCOLD Guidelines (2019). In this assessment of the hazard rating, it was considered that loss of life may be expected (population at risk (PAR) >100 to 1000 ie mine personnel) from an embankment failure with 'major' consequence. For a major consequence, there may be appreciable economic loss (ie severe business disruption). There is unlikely to be any major environmental impact or social dislocation due to a TSF embankment failure as tailings are likely to be contained within the mine site location (ie Toka Pit).

As the TSF is considered as a 'High-A' consequence category storage, the following criteria were adopted in the design.

- Earthquake design:
  - ○ 1:1000 AEP (OBE).
  - ○ 1:10 000 AEP (SEE).
- Water management:
  - ○ Storage of 1 in 10 000 AEP 24 hr rainfall event (935 mm) (ref Coffey, 1997) on top of the TSF.
  - ○ Spillway design probable maximum precipitation (PMP) event.
  - ○ Freeboard to ANCOLD (2019) requirements, 1 m to 1.1 m.

Embankment design

The raised TSF is a valley fill type storage with upslope catchments. A main embankment (maximum 112 m height) forms the storage boundary on the north-eastern and eastern side. The raised TSF will be enclosed on the western side by natural slopes forming the valley and on the southern side by a ridge line. The raised TSF is 'isolated' from other drainage catchments.

The main embankment of the raised TSF will be a zoned embankment which can be raised in several stages from RL290 m to RL320 m as shown in Figure 4. The embankment zoning comprises:

- An upstream (or inner) clayey mine waste zone, which will be a low permeability zone.

- A downstream (or outer) zone of mine waste which provides bulk/strength and buttresses the low permeability upstream zone.

- The existing internal drainage (or finger drains) between the upstream and downstream zones of the main embankment will be raised a further 30 m, above RL290 m. The internal drains will be wrapped in geotextile to prevent fines migrating into the drainage system.

- The embankment drainage system comprises three types of drains. The finger drains join internal gravel drains, which feed into perforated pipe drains. The perforated pipe drains will divert seepage water to drains downstream of the main embankment towards to the Toka Pit. All existing drains were extended downstream by installation of new internal drainage.

- These internal drains have been incorporated into the design in order to control the phreatic surface in the main embankment to increase the factor of safety and reduce the risk of main embankment instability.

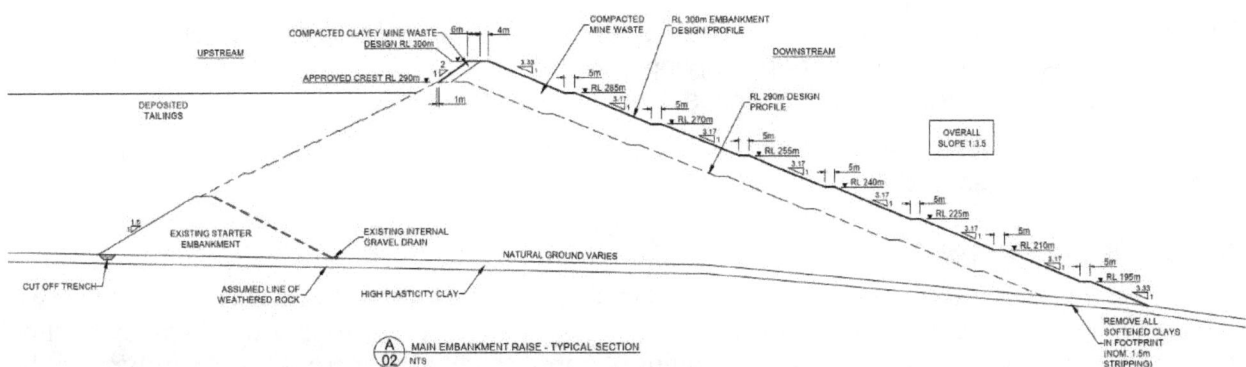

FIG 4 – Typical section of raised TSF embankment.

The embankment internal drainage is shown in Figure 5.

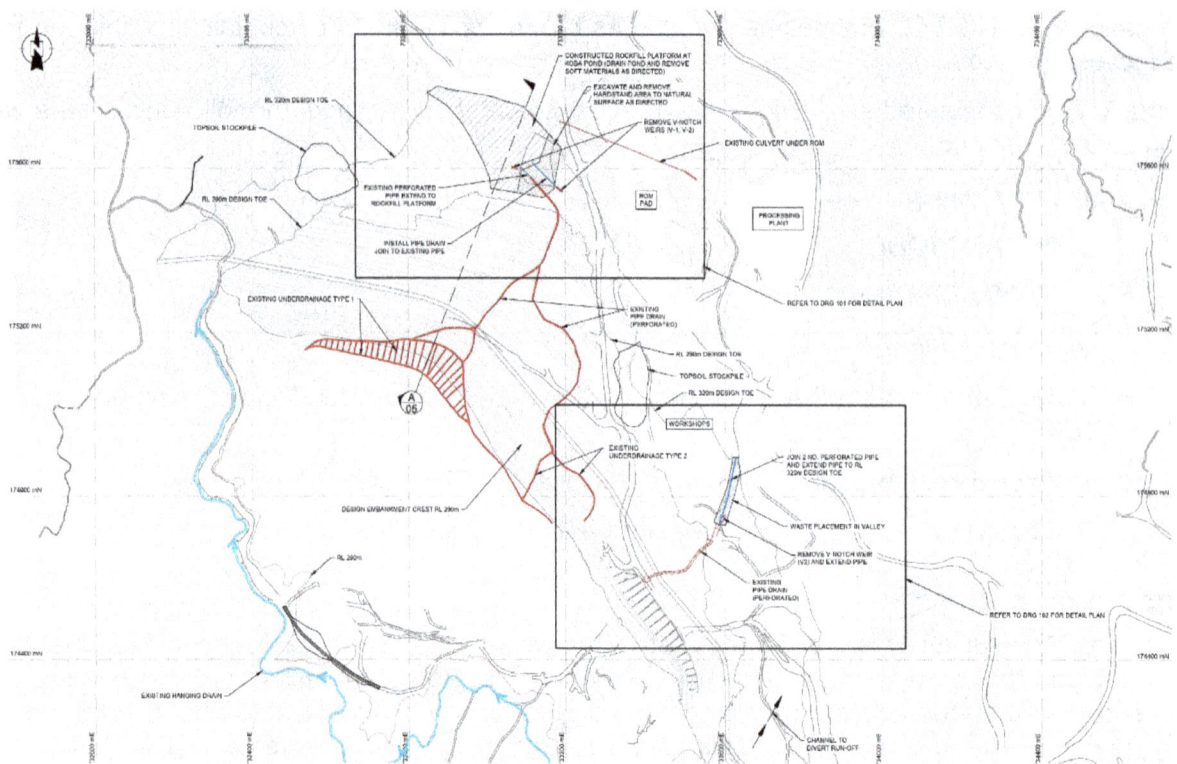

FIG 5 – Plan of TSF with internal drainage.

It is envisaged that the construction work for raising of the TSF will occur as one embankment raising construction (ie the TSF will not be split into cells). The inner upstream zone of the main embankment raise (above crest RL290 m) will utilise compacted clayey borrow material from within or adjacent to the TSF. The outer downstream zone of the main embankment raise (above crest RL290 m) will be constructed using waste dump construction techniques utilising waste from the Toka Pit.

The embankment will be extended to the north (up the steep abutment) and to the south along a ridge line. A new cut-off trench underneath the embankment will be extended up the western abutment and along the ridge line forming part of the eastern side of the TSF.

A saddle embankment will be required on the southern boundary of the site. The maximum saddle embankment height will be approximately 25 m at RL320 m.

Additional standpipe piezometers, vibrating wire piezometers and survey prisms have been installed on the TSF embankments. Existing instruments include: three standpipes, nine VWP piezometers. Instrumentation will also include inclinometers at the final embankment crest level.

Embankment geometry

The design adopted for the embankment raising has an upstream or internal slope of 1:2 (V:H) and an overall downstream or external slope varying between 1:3 (V:H) to 1:3.5 (V:H). The geometry of the downstream slope will have 5 m benches at 15 m vertical intervals.

The design adopted for the saddle embankments has an upstream or internal slope of 1:2 (V:H) and a downstream or external slope of 1:3 (V:H).

Water management

The raised TSF will have a decant at a central location on the side of the valley. The decant will comprise a floating decant pump behind a rock filter wall constructed across a small embayment on the side of the valley. This decant arrangement has the potential to increase water return as the pond can be located within the ring with the aim that slimes (fine tailings) are kept away from the pumps.

The expanded TSF is a valley type storage facility with upslope catchments. At present, an existing diversion drain diverts run-off from upstream catchments to the north. A new diversion (hanging) drain will be constructed in the future above the RL320 m level. The upstream catchment will be diverted to the north around the main TSF embankment (refer to Figure 3).

The design capacity adopted for the new diversion drain is 1:100 year AEP within the drain 'cut' and a 'full' channel capacity approaching 1:1000 year AEP. Watershed into the TSF will thus be reduced and run-off will mostly comprise the actual tailings storage area.

The TSF has been designed in accordance with ANCOLD (2019) and sized to temporarily contain stormwater from a 1 in 10 000 AEP 24 hour rainfall event which is estimated to be 935 mm.

Spillways have been designed for the expanded TSF at RL300 m, RL310 m and RL320 m to reduce the risk of water overtopping the embankment. The spillways have been sized for a PMP 24 hour rainfall event in line with the 'High' hazard rating for the TSF.

The spillways at RL300 m will be constructed adjacent to the south-western embankment abutment, and the spillway at RL310 m, will be constructed adjacent to and to the east of the south saddle embankment. The base of the inlet is to be lined with 0.5 m thick rock protection with underlying geotextile. The base of the outlet chute is to be lined with Reno Mattresses of the same thickness. Energy dissipator rock will be deployed downstream of the spillway chute to reduce erosion. The spillway at RL320 m will be constructed at the north-east corner of the TSF and will also become the closure spillway. The closure spillway will likely be a concrete spillway to meet regulatory requirements.

Water balance analyses for the TSF were undertaken in order to examine the water balance during the operation. Inflows and outflows for the tailings storage were estimated on a monthly basis. The results of the analyses indicate a larger positive water gain can be expected from the raised TSF above RL300 m and particularly above RL310 m to RL320 m, with excess water being required to be disposed of safely 'downstream' (ie via a wet land). The results of the analyses indicate that the positive water gain is likely to increase as the TSF is raised and the tailings area is expanded. The capacity of current pumping/treatment system will need to be increased in order to ensure the total excess water can be removed over the life of the project.

Closure

Closure and rehabilitation of the TSF will comprise the following:

- Establishment of a partial permanent pond on top of the TSF. In order to achieve this, a spillway(s) will be constructed such that any excess water will be removed from the facility at closure. The spillway(s) will take the form of channel cut (ie cut earthworks) with a water pond located well away from the main embankment.

- Capping of the tailings near the main embankment with non-acid forming (NAF) mine waste. This should be achievable as the tailings near the embankment should comprise silty sands of higher strength than the finer tailings located near the decant location.

- Decommissioning of the decants (ie removal of pumps and pipework).

- The rehabilitation of the waste dump outer slopes in accordance with PT MSM/PT TTN rehabilitation proposals for waste dumps at Toka Tindung.

- The embankment drainage system will be left open at decommissioning and closure in order that the phreatic surface within the embankment is managed and the factors of safety against embankment failure are adequate. The embankment drainage system will need to be re-assessed for closure.

- The diversion drain upslope of the TSF will be required upon decommissioning in order to reduce the amount of excess water to be removed from the TSF for treatment. However ultimately the TSF spillway will be designed for flows from the entire upstream catchment and a probable maximum precipitation (PMP) event. It is assumed that the diversion drains will not be serviceable on a 'permanent' basis.

- Embankment monitoring (ie piezometers and movement monitoring) together with environmental monitoring will be required post closure. The instrumentation system will need to be re-assessed for closure.

The above closure recommendations were based on consideration of the tailings properties (ie tailings strength at closure) and climatic considerations. The government Regulators are concerned about closure risk and legacy issues relating to a large dam. More studies will be required to further assess closure design and risk. Including seeking alternatives such as a 'dry' cover to reduce closure risk.

FACILITY MANAGEMENT

Company management

Full time supervision of construction on day shifts is conducted by a Civil Engineer from PT MSM/PT TTN. The Civil Engineer is assisted by a Superintendent and two supervisors. The Civil Engineer also provides input into construction and operational planning (ie deposition and decant management). Monthly reports are produced by the Civil Engineer.

Soil Technicians from PT MSM/PT TTN carried out the QA/QC testing of the works on the constructed embankment. PT MSM/PT TTN has a well-equipped laboratory on-site with an oven, balances, Atterberg limit apparatus, sieves and a sand replacement kit.

The TSF team maintains active communication to all relevant personnel with regard to TSF activities. This enables the design specification to be achieved and improvements to be made to on-site systems and processes.

Reviews and oversight

The design engineer (CMW) also conducts annual site visits to conduct an audit of construction and operations. The annual visits are supplemented with quarterly reviews by the designer or 3rd parties.

Regulatory requirements

New laws in Indonesia governing tailings storage facilities require:

- Construction Services are to be performed by a business entity based on Indonesian Law via collaboration by PT MSM with a national construction service business entity.

- For TSFs with heights >75 m, the mine owner and manager are required to appoint an independent expert panel to oversee the planning and execution of TSF construction. The panel should include an earthquake expert and a geology/geotechnical expert and will be involved during construction and operation as well.

This independent design panel provides additional feedback to the government dam safety commission.

CONCLUSIONS

Despite the options study considering site location, cost, engineering risk, environmental risk and social risk, a flexible approach is required and this may mean carrying forward more than one option.

Based on technical reviews of the TSF raising and expansion design, it has been concluded that the option adopted from the 2019 study was appropriate. Raising and expansion of tailings storage facilities need careful consideration both from a designer and company perspective. One should not underestimate the level of complexity, implementation of the raising and expansion of a large TSF involves. The project needs to be adequately resourced and planned both from a design and construction perspective.

The requirements for permitting of activities can take time to communicate plans and obtain approvals from multiple agencies. This time consideration is an important element of the decision-making process.

In addition the area of the mine operates has high demands on land use by the surrounding communities and minimising the mine's footprint is a significant factor for consideration.

With a larger TSF dam particularly in a tropical setting with high rainfall and seismic risk, comes additional management requirements that need to be implemented.

ACKNOWLEDGEMENTS

The assistance and contributions from colleagues in PT MSM working on various aspects of the TSF Raising Project and contribution to the preparation of this paper are gratefully acknowledged. The authors would like to thank the company PT MSM for approval to publish this paper.

REFERENCES

ANCOLD, 2019. Guidelines on Tailings Dams – Planning, Design, Construction, Operation and Closure.

CMW, 2019. Options Study, Future Tailings Storage, Toka Tindung Mine, North Sulawesi, Indonesia, Geotechnical Assessment and Preliminary Design Report, ref PER2019–0078AE Rev 0, prepared for PT Meares Soputan Mining (PT MSM) and PT Tambang Tondano Nusajaya (PT TTN).

CMW, 2023. Raising and Expansion of TSF to RL320 m, Toka Tindung Gold Project, Indonesia, Design Report, ref PER2020–0077AE Rev 2, prepared for PT Meares Soputan Mining (PT MSM) and PT Tambang Tondano Nusajaya (PT TTN).

Coffey, 1997. Geotechnical and Water Management Feasibility Studies, Report G6021/1-RA.

PusGen (Center of Indonesian Earthquake Study), 2017. Indonesian Earthquake Hazard Map for the Toka Tindung TSF location.

In-pit waste rock dumping as a mitigating strategy for waste rock management

J Isarua[1] and S Elit[2]

1. Senior Short-term Planning Engineer, Ok Tedi Mining Limited, Tabubil, Western Province 332. Email: joe.isarua@oktedi.com
2. Short-term Planning Engineer, Ok Tedi Mining Limited, Tabubil, Western Province 332. Email: susanne.elit@oktedi.com

ABSTRACT

Ok Tedi Mining Limited (OTML) has an annual production rate of about 110 Mt total material movement (TMM) and 95 Kt Cu production with an ultimate pit design depth of 1 km with an approved mine life end in 2033. As one of the wettest mines in the world with 10 m per annum rainfall, the site's climatic conditions are both humid and hot, with a rugged topography covered by tropical rainforest. Towards the start of Quarter 4 of 2022, the company embarked on in-pit dumping in one of its depleted pits as part of a waste rock management strategy and environmental commitment. The 2023 Budget has 40 Mt waste rock from its West wall pits planned for dumping into the southern and northern dumps.

Due to high rainfall, steep topography, and landslides, an erodible dumping method was approved, for dumping, by the mine's regulators since operation commencement in 1984. In 2020, the Engineered Waste Rock Dump (EWRD) Project came onto the scene as part of the mine waste rock management. EWRD plans are already in place, with construction work underway. The strategic intent of the EWRD is to extend mine life to 2033 by storing waste in a geotechnically stable waste rock dump and minimising erosion rates into the river systems. Given the ongoing construction phase status, EWRD is yet to receive different rock types due to capital constraints. While waiting to purchase additional trucks, most of the planned West wall wastes, for the northern dumps, are sent to Taranaki 1 in-pit dump as an alternate solution. In-pit dumping will support space requirements for stockpiles, crib-hut, ready line, and other business improvement projects.

The in-pit dumping project has progressed, with Phase 1, construction of efficient West wall ore haulage and backfill of pit entry access completed. Phase 2 covers the dumping of large volumes of wastes from West wall pits until EWRD is ready to receive waste rocks. The dump will be constructed by lifts, with an overall slope angle under 25°, drainage, well maintained, and while simultaneously maintaining West wall ore haulage, achieving budget, forecast, and short-term schedules that sustain overall operation profitability.

INTRODUCTION

In-pit waste rock dumping, is the process of dumping uneconomical rock materials within a mined out pit. The cash flow value is usually zero or below with a lower than break-even cut-off grade. Current site pit shell verification shows dumping activity as being outside of the pit shell, with future cutbacks in this area having to pay for the rehandling of additional waste to be economically viable. Figure 1 shows a cross-sectional view of the in-pit dumping concept.

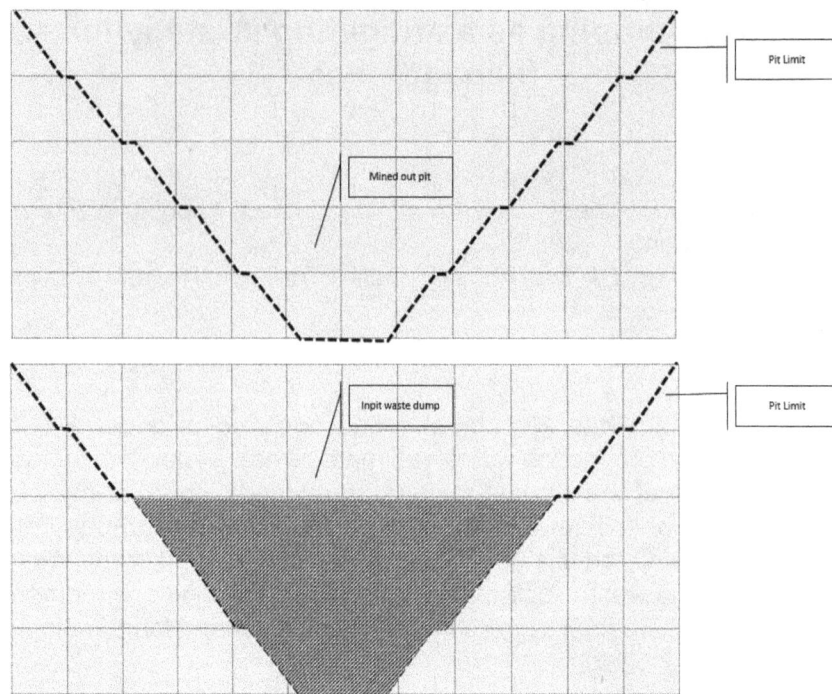

FIG 1 – A cross-section showing in-pit dumping concept.

Erodible dumping is the current practice on-site where waste rocks are dumped at the pit edge into steep gullies, discharged into tributaries to the main Ok Tedi River, and flows downstream. The dumping process is controlled by strict geotechnical controls, inspection, and rainfall triggers with demarcated dozer work areas where remote dozing is allowed in fill areas, and drainage is maintained to minimise dump progression. The concept works well for a high rainfall region like Ok Tedi with steep dump faces.

OTML has always envisioned a stable waste dumping option apart from the current practice with reduced environmental impacts. In 2020, EWRD Project was introduced, with a technical team put together, and designs progressed. Trucking constraints have limited waste dumping at EWRD, allowing waste placement at the northern and southern erodible dumps. Taranaki 1 in-pit dumps are also set-up and receiving wastes.

In principle, the best way to do in-pit dumping is to build lifts with specifications such as a 37° repose angle, berms, and bench height that generates an inter-ramp angle under 25° slope. Construction methodology should account for lift access, dumping width, and drainage by bench in such a manner as overall drainage is diverted away from the pit bottom. The competent materials are limestone and blocky siltstone from East wall and West wall pits. Strict design compliance is required to ensure the safety of the dump build.

Mine Plans for 2022 and 2023 focused on sending wastes to Taranaki 1 dumps with less attention on limits to sulfide creek. The faster waste dump advancement towards the sulfide creek has prompted an independent geotechnical review, whereby restrictions issued from further dumping north, triggering the need for an alternate dumping solution for West wall wastes. The Taranaki 1 waste backfill, as a waste rock management strategy, was considered upon considering haulage, bench advancement, ore profile, truck fleet, drainage, EWRD, and business improvement projects. The short-term planners completed a sign-off of the remnant reserve with marginal profit, where the scrap metal burial process completed enabling waste rock dumping.

PHASE 1 – WEST WALL ORE HAULAGE AND PIT ACCESS BACKFILL

Phase 1 of the in-pit dumping project included maintaining efficient West wall ore haulage to the crusher and building a fill ramp connecting the mine pit office to the pit regions. The access construction would impact the pit–office traffic, hence was carried out in segments. Regrade from Phase 1A was carried out primarily, with the pit access maintained from Phase 1B. Upon completion of Phase 1A, backfill on Phase 1B commenced. Progress on Phase 1 was slow due to low availability of road base rocks, equipment downtimes, varied construction methodology, and crew change. A Minesight snapshot is shown in Figure 2 showing the final outlook of the phase 1 construction. The key lessons learnt include; ensuring a selected team owns the project, one set of instructions, timely survey checks, and stockpiling quality rocks where required.

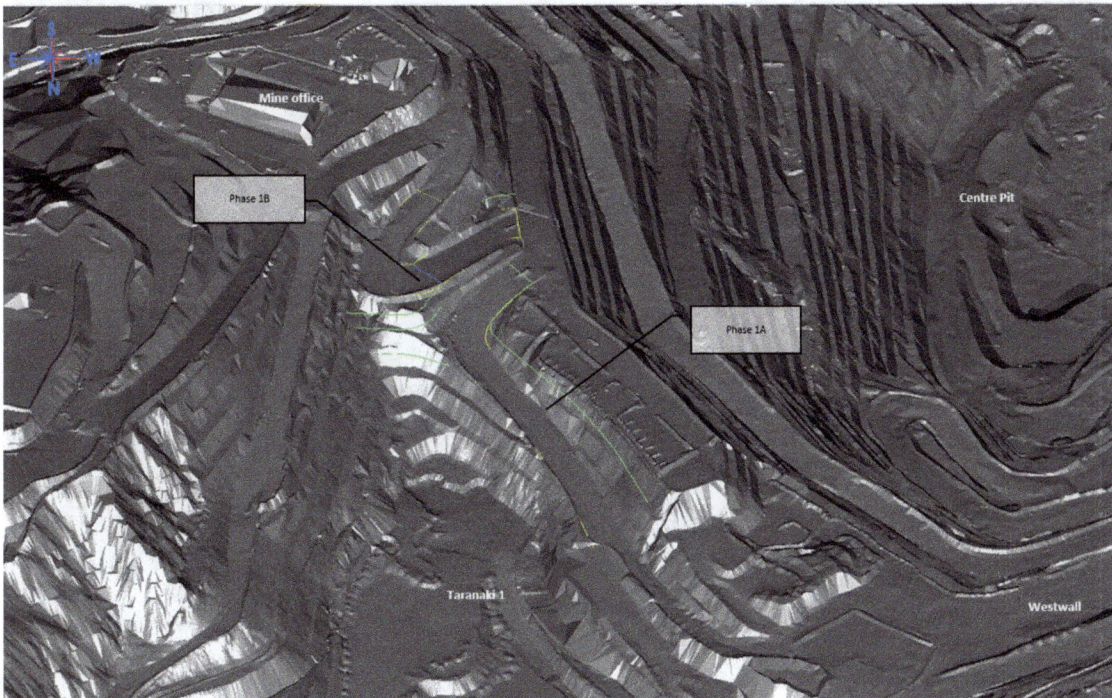

FIG 2 – Minesight snapshot showing the final outlook of the phase 1 construction.

PHASE 2 – IN-PIT DUMPING AT TARANAKI 1 PIT

Phase 2 covers in-pit dumping of a large volume of waste into the Taranaki 1 area. The complete backfill of the current build progression at 1633RL will enable the next lift to 1663RL. The set-up work on the erodible portion, will also be progressed to provide dumping options during dump remediation. Traffic management is critical to ensure haulage and EWRD access are maintained. Figure 3 is a snapshot showing completion of phase 1, and expected outlook of phase 2 of the Taranaki 1 in-pit dumping project.

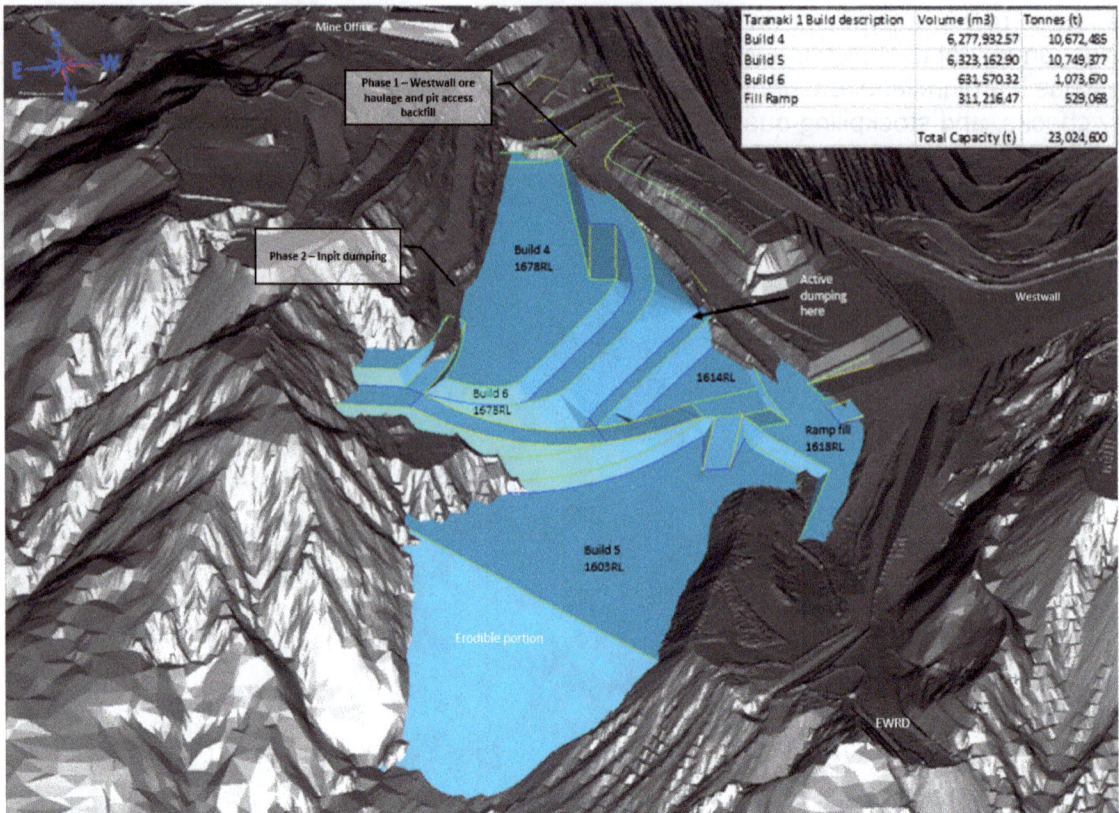

Taranaki 1 Build description	Volume (m3)	Tonnes (t)
Build 4	6,277,932.57	10,672,485
Build 5	6,323,162.90	10,749,377
Build 6	631,570.32	1,073,670
Fill Ramp	311,216.47	529,068
Total Capacity (t)		23,024,600

FIG 3 – Showing completion of phase 1 and expected outlook of phase 2 of the in-pit dumping project.

The dump build sequence of Phase 2 will be such that both dump build 4A 1633RL and the erodible portion fill 5 1603RL are progressed concurrently. After completion of 4A 1633RL, the next lift 4B 1663RL gets filled. With dumping progress at 1663RL, the West wall design ramp fill, must be completed to provide adequate drainage control. Final fill 6 1678RL accounts for haulage connection to the overall West wall design and drainage control to the centre pit. Figure 4 demonstrates different stages of the phase 2 in-pit dump build sequence.

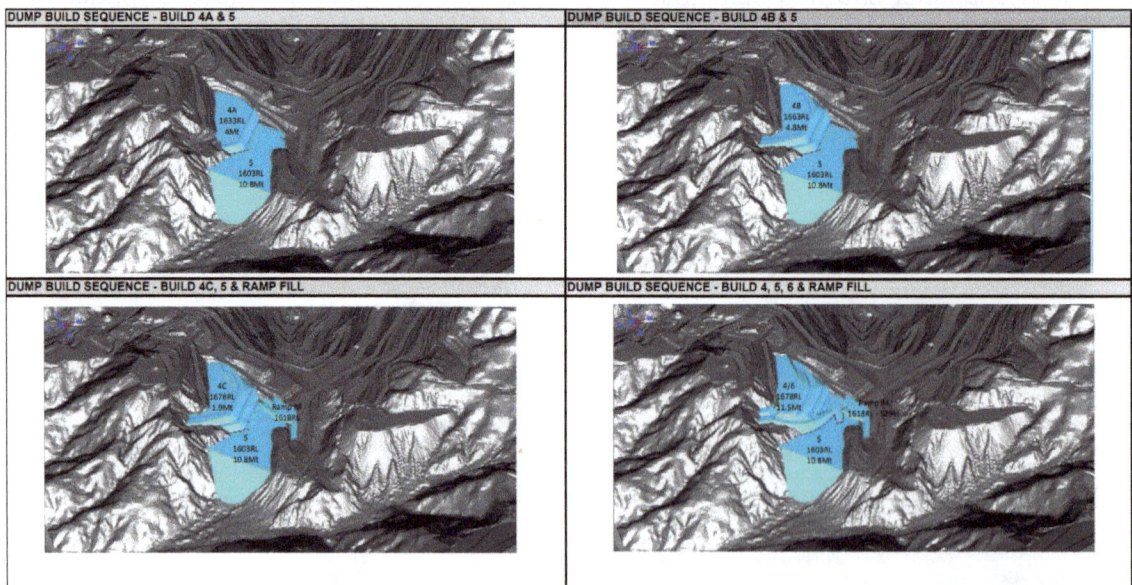

FIG 4 – Showing different stages of the phase 2 in-pit dump build sequence.

CHALLENGES AND LESSONS

The scope of this paper discusses the challenges and lessons learnt by OTML from in-pit waste rock dumping, which provides a good transition into EWRD and overall waste management and impacts on the environment. Although faced with numerous environmental issues such as dieback effects, over the years the company has managed the waste rocks discharge well with a Bige Dredging set-up. The idea of dumping waste on the land has been a long time coming, and the EWRD commencement in 2020 has started this process, with the expectation of a controlled waste material discharge into the river system when fully operational. Active in-pit waste dumping is progressing well. The challenges and lessons learnt from the current in-pit waste deposition are captured below.

High rainfall events and drainage control

OTML receives an average of 10 m of rainfall annually, making it one of the wettest mines in the world. Washouts, rockfalls, wall failures, slope movements, dump erosion, and closures are all driven by rainfall events. Recent rainfall events have washed out alternate access tip head, significantly slowing good progress. The operation, however, has continued to utilise the high rainfall it receives by practicing an erodible dumping strategy with a drain set to strategically erode waste dumps. This strategy has worked well over the years, given the high dumping face and good drainage.

Drainage control, is one of the main challenges faced when turning to Taranaki 1 pit as a dumping solution for West wall waste rocks. The need for ongoing waste rock dumping, washout clean-up, and haulage to and from Taranaki 1 stockpile puts pressure on delivering planned drainage construction works at Taranaki 1, especially with the installation of culverts, drainage maintenance and drain diversions. Alternate routes have been constructed to divert traffic and enable drainage work commencement. Weekly discussions are carried out between the mine geotech hydrogeology team, EWRD, mine planning, and mine operations refining existing drainage plans based on field conditions and priorities.

With EWRD in perspective, the challenges posed by high rainfall events are witnessed with the slow progress to the sulfide creek floor, due to washouts at the tip head, roads, and intersections. Drainage diversion at higher elevations is the key to allowing access construction to the toe. However, EWRD requires control of surface water away from its active dumping face making it challenging to manage drainage and access. Furthermore, improper drainage plans and implementation have always affected road conditions resulting in reworks, increased equipment hours, and decelerated progress of strategic project works.

Limited dumping at EWRD and alternate waste dump solution

EWRD is the planned dump destination for most of the West wall waste rock. Given the toe drain construction challenge, less limestone presence at West wall, equipment procurement delays, and the current site truck constraint situation, it has pushed dumping into the future. In the interim, stage 11 of the EWRD at 1652RL has been set-up and is receiving limestone rocks when truck availability presents. About 200 kt of limestone waste rocks, has been dumped so far, with alternate access in progress. Upon completion of construction, the new track will provide a route to the EWRD toe for drain construction, set-up of pore pressure measurement equipment, and a pathway to exploration sites in the future.

Mine plans for the previous two years have scheduled more waste to Taranaki 1 dumps, with dump sections S14, S15, S18 and S19 open. Given the shallow dumping face angle, high volume of waste coming through from West wall pit, and lack of adequate drainage water pathways for Taranaki 1, erodible dumps resulted in significant advancement of the dump face towards the sulfide creek. An independent geotechnical review has restricted dumping past the sulfide creek. The 2023 Budget has up to 15 Mt waste rock for the northern dumps, and given the restrictions, Taranaki 1 in-pit dumping provides the alternate waste dumping solution. The in-pit dumping exercise, at present, has generated dump sections S20 and S21 with a new lift in progress at 4A 1633RL.

EWRD dumping is constrained by a small truck fleet to supply limestone to EWRD, timing to expose quality rocks, and unavailability of limestone rocks within proximity. Truck procurement completion is due in 2025, with a geotechnical model update in progress with drilling and face mapping of the rock mass and sub-rock mass. Alternate waste dumping solutions, should always be captured in the mine plans. It is always good to understand the trends in an erodible dump, dump rates, and constraints and get factored into the schedules for a realistic business plan.

Ore sterilisation, scrap metal burial and material selection

Taranaki 1 was an ore pit supporting the business from 2017 to 2020. After depletion of pyrite material, the mine pit became obsolete, and remnant ore material was left below on its AD Bench. Despite efforts to recover the ore, it was not successful. A cash flow analysis was performed and reflected as marginal with a high strip ratio hence was subsequently written off and sterilised. Scrap metal disposal was on the radar as well. After discussions and engagements with the mine's environment department and Conservation Environment Protection Authority (CEPA), the approval, was given for burial in the *in situ* rock on AD Bench with a slower erosion rate.

Although in-pit waste dumping at Taranaki 1 has no material restrictions, blending, is carried out for good haulage and drainage pathways. Different waste materials, including the weak thrust materials from the chasm crown, are sent to the two sections of in-pit dumps with blocky siltstone as the blend material. The northern portion exposed to the pit edge is supplied with limestone for net acid producing potential (NAPP) balance and tracked monthly.

Haulage

The operation is truck-constrained, given availabilities generally below budget annually, although higher performances are a product of productivity and utilisation. West wall haulage, for both ore and waste to the crusher and waste dumps, has been a real challenge, given the need for concurrent activities. These activities include drill and blast, access shifting, final ramp development, stockpile mining, stockpiling, pit final wall trims, and traffic interactions. The truck cycle times fluctuate with changes in dump destinations, with trucks being distributed to other pit regions when digging equipment experiences downtimes.

The dynamic site haulage network is a product of stage mining, changing dump destinations, rockfall hazards, crushing circuit constraints, blending requirements, and haulage detours. For production continuity, all these challenges are well managed as presented. Shown in Figure 5 is a snapshot of the haulage network within and around the Taranaki 1 in-pit dumping project.

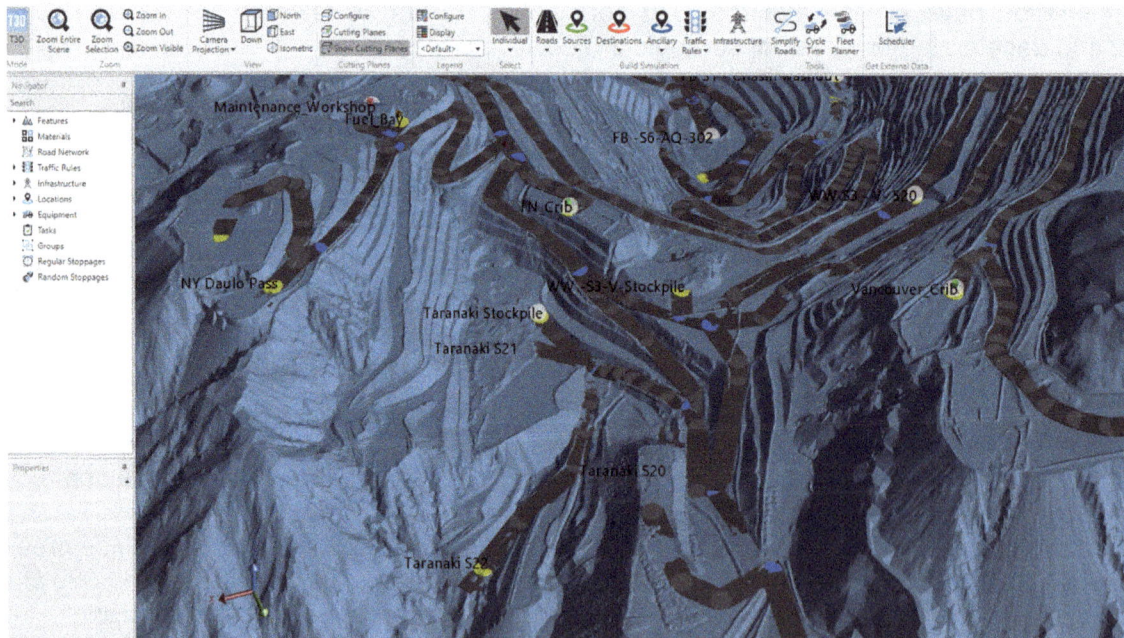

FIG 5 – Showing a snapshot of the haulage network within and around the Taranaki 1 in-pit dumping project.

In-pit waste dumping and environmental impacts

Large-scale in-pit waste rock dumping in an obsolete pit has not been done in OTML since operation commencement in 1984, although backfill of gullies such as land bridge, waste dumps, deflection bund fill, and temp ramp fill are evident. From the existing plans, the only major in-pit dumping expected at the later stage of life-of-mine (LOM) is at the Fubilan pit. The Taranaki in-pit waste dumping in the *in situ* has come about due to constraints with waste deposition past sulfide creek and limited dumping status of EWRD for the planned waste rocks.

OTML has come a long way in managing waste rocks, with the establishment of the Bige Dredging Project, which pumps waste rock in the river back onto the land as reclamation and allows for rehabilitation programs. These efforts continue to serve the business well. The introduction of EWRD recently, further aims to control waste rock deposition into the river system minimising environmental effects, and plays a significant role in mine life extension to 2033. In-pit waste rock dumping at Taranaki 1 provides a good transition for EWRD.

Other sites

Other open cut mines in Papua New Guinea, such as Lihir Gold Mine and Hidden Valley Gold Mine, have different waste rock management strategies. At Lihir, waste rocks get deposited in the sea with only low-grades stockpiled in-pit and are reclaimed later for pit development. At Hidden Valley Mine, according to one of Harmony's mining engineers then, Jackery Aiwa (December 2018, personal communication), wastes are managed through a stable waste rock dump outside the pit shell, and are rehandled with cutbacks. The Hamata pit conversion into a TSF is an option for the future.

CONCLUSIONS

In-pit waste rock dumping is a waste-rock management strategy that acts as an alternate solution for planned waste destinations for the next few years until EWRD is ready. After backfill, business improvement projects such as readyline, crib-hut location, ore stockpile estate, and fuel bay get placed on the platform. Erodible waste rock dumping will continue throughout the life-of-mine, balancing NAPP, even after completion of in-pit dumping and EWRD. Site challenges of high rainfall, drainage control, truck-constraint haulages, dumping capacities, and material selections are well managed as presented. With expected truck procurement completion in 2025, the in-pit waste dumping solution, with other dumping options, will keep digging equipment advancing planned face positions, meeting ore targets for 2023 and 2024. The company has an in-pit waste dumping plan at a later stage of LOM after depleting Fubilan and East wall ore pits with a continued vision to minimise environmental impacts.

ACKNOWLEDGEMENTS

The author takes this opportunity to thank Mining Business Unit headed by General Manager Derrick Kelly and the mine management team for the guidance in completing this paper.

The author also acknowledges mine planning engineers in designing and tracking plan compliance of the in-pit dumping project.

West wall team members, are acknowledged in commencing the project, particularly Desmond Ohsan in taking the lead.

Lastly, Alban Reupana of mine geotech, is acknowledged for his work on sub-rock mass mapping, improving current rock mass data understanding.

REFERENCES

Bong, A, 2021. Engineered Waste Rock Dump Designs/Plans, Weekly EWRD Project Management Dashboard Review (internal company document).

Isarua, J, 2022. Taranaki 1 Scrap Metal Back-Fill Plan and Dumping Capacity, Taranaki 1 dumping strategy Meeting (internal company document).

Meapa, Y, 2023. Mine Budget, Budget 2023 Run9e Physicals (internal company document).

McKinnon, E, 2002. The environmental effects of mining waste disposal at Lihir Gold Mine, Papua New Guinea, *Journal of Rural and Remote Environmental Health*, 1(2):40–50.

Reupana, A, 2023. Inpit waste rock classification and utilization – Ok Tedi Mining Limited, Oxidized Zone – Paris Geotechnical Review, pp 2-6 (internal company document).

Use of multi-criteria alternatives assessment for TSF option selection – lessons learned from recent applications

J Penman[1] and J Casey[2]

1. Senior Geotechnical Engineer, Klohn Crippen Berger, Perth WA 6005.
   Email: jpenman@klohn.com
2. Senior Geotechnical Engineer, Klohn Crippen Berger, Vancouver, British Columbia, Canada V5M 4X6. Email: jcasey@klohn.com

ABSTRACT

The Global Industry Standard on Tailings Management (GISTM; Global Tailings Review (GTR), 2020) requires the use of a multi-criteria alternatives assessment (MAA) for a new TSF with the goal to select an alternative that: (1) minimises risks to people and the environment throughout the tailings facility life cycle; and (2) minimises the volume of tailings and water placed in external tailings facilities. The use of TSF alternatives trade-off studies in one form or another is well-established; however, in the context of GISTM roll-out, the widespread application of the often more rigorous MAA process is a significant emerging trend in tailings management.

Application of the MAA process can be a complex undertaking with requirement for engagement with stakeholders from technical and non-technical backgrounds, and consideration for a broad range of alternatives including tailings dewatering and alternative disposal methods such as commingling waste. This paper provides an overview of applying the MAA process and lessons learned from six recent applications of the MAA process covering a diverse range of ranging of commodities (coal, copper, gold, iron ore, and poly-metallic), project statuses (proposed, operational, and closed), and locations (Australia, Africa, and the Americas).

The typical MAA process applied in the reference projects included confirmation of the 'must haves' (or design basis), the 'can't haves' (or exclusions/'fatal flaws'), and 'nice-to-haves' (or objectives) for identification and assessment of TSF alternatives. The nice-to-haves/objectives are the foundation for the MAA framework and, in line with guidance provided from Environment Canada (Environment Canada, 2013), these were typically grouped under a set of 'accounts' covering technical, environmental, and social (or socio-economic) considerations alongside project economics. The comparison of MAA frameworks from the reference projects provides useful insights into commonalities between projects as well as the site-specific objectives driven by nuances in the site's physical (climate, topography, geology etc) and social settings.

INTRODUCTION

The identification and selection of an optimal tailings management solution requires consideration for a wide range of technical, environmental, and social factors as well as compatibility with overall project economics. In the authors' assessment, project economics has traditionally been treated as the dominant factor when choosing a tailings management solution and this has likely led to suboptimal outcomes for technical, environmental, and social considerations in some cases. The due consideration of 'alternative' tailings management solutions – such as filtered tailings which are typically of relatively high capital and operating cost but, with relatively favourable environmental and social benefits, in some cases – has suffered where the focus has primarily been on project economics. A method to appropriately consider the positive and negative externalities from a range of alternatives is required to ensure an optimal solution is arrived at. To this end, a multi-criteria alternatives assessment (MAA) can be used to reduce the bias and improve transparency in the decision-making process.

The Global Industry Standard on Tailings Management (GISTM) prepared by the Global Tailings Review (GTR, 2020) outlines that the goal for a tailings management solution should be one that: (1) minimises risks to people and the environment throughout the tailings facility life cycle, and (2) minimise the volume of tailings and water placed in external tailings facilities. With these goals in mind, the GISTM has recognised the benefits of using of a multi-criteria alternatives assessment for selection of tailings management solutions.

An outline of the general approach taken in tailings management alternatives assessments using an MAA process and lessons learned from recent applications are presented herein.

APPROACH TO ALTERNATIVES SELECTION USING MAA

The desired outcome of the alternatives selection process is to generate an 'optimal' site-specific tailings solution; however, in the authors' experience, the site-specific outcome can be achieved through following a common approach. The common approach includes nine key steps as outlined in further detail in this paper. The steps generally are taken in sequence; however, it is possible to complete steps in parallel or out of sequence and achieve the desired outcome.

Step 1 – Establish design basis

The design basis outlines what conditions a solution must meet to satisfy the overall project requirements. Collectively the design basis presents the 'must haves' for a tailings management solution and any proposed solutions that do not meet the 'must haves' are not considered as viable alternatives.

For a greenfield project with limited/no prior assessments, the design basis can be as simple as a statement on the intended purpose of the tailings management solution, the anticipated life-of-mine and tailings production rate/total amount. A more advanced project could include minimum design criteria as defined in relevant design standards and guidelines.

The design basis should also outline the battery limits for the alternatives assessment including a maximum distance from the site to be considered in TSF siting, eg locations up to 20 km radius form the orebody to be considered. Other battery limits outlined in the design basis can include technical considerations, eg minimum water recovery from raw tailings, etc.

The design basis should also outline the exclusions (also known as 'fatal flaws' or 'can't haves'). Where the exclusions should reflect known constraints that would render a potential alternative as non-viable, for example:

- Physical constraints, eg no facilities located over orebodies, no facilities located over known active faults etc.

- Regulatory/approvals constraints, eg no facilities within national/regional parks, no facilities on culturally significant sites etc.

- Corporate policy/values constraints, eg no use of upstream raised facilities, no population-at-risk (PAR = 0) downstream etc.

Step 2 – Establish knowledge base

GISTM emphasises the development and maintenance of a facility Knowledge Base that is used to inform decision-making over the full facility life cycle beginning with the alternatives assessment for new facilities.

In the authors' experience, the Knowledge Base for the MAA should – as far as practical – cover the following areas/topics:

- Project Overview (project status/history, locality, target commodities, mine development plans etc).

- Tailings characteristics (proposed ore treatment process(es), classification properties, assumed *in situ* dry density, geochemistry, potential water recovery etc).

- Spatial/Topographical (survey, project datum etc).

- Land Tenure/Ownership (mining leases, public lands etc).

- Land Use (pastoral/agricultural, residential/commercial, industrial etc).

- Infrastructure (roads, water/power/gas supply lines, public services (hospitals, schools etc)).

- Social (population density, ethnographic groups, heritage/culturally-significant sites etc).

- Climate (rainfall, evaporation, wind etc).

- Environmental (flora/fauna habitats, national/regional/local parks, and reserves etc).

- Hydrology (catchment sizes/boundaries, watercourses, sensitive receptors etc).

- Hydrogeology (regional groundwater levels/flows, existing monitoring bores/wells, sensitive receptors (eg stygofauna, water supply) etc).

- Geology/Geotechnical (bedrock geology, surficial geology, seismicity etc).

- Closure (expectations, post-mining land use etc).

The level of detail included in the Knowledge Base will be a function of the project status, eg greenfield versus brownfield, and potential facility locations, eg on-tenure versus off-tenure. Early-stage and greenfield projects will likely rely on public-domain data and information, eg SRTM for topography (versus site-specific LiDAR survey), national/regional heritage site registers, national/regional park boundaries etc. Similarly, public domain data and information may be relied upon for operational projects where candidate locations are remote from existing mine development.

The use of public-domain data and information introduces some risks to the identification of an optimal tailings solution; however, it is often unavoidable until preferred potential locations can be identified for subsequent site-specific data and information gathering. Geographical information systems (GIS) platforms are a useful repository for much of the Knowledge Base and often include readily-accessible public-domain data and information.

An understanding of tailings characteristics is important to identify viable alternatives and key design considerations. Where very little or no information is available on the potential tailings characteristics, guidance presented in Bulletin #181 (ICOLD, 2021) can be used as a reasonable approximation until project-specific data and information is available. Original equipment manufacturers (OEMs) can provide simplified assessments of tailings dewatering system requirements, (eg size and number of filtration units, thickener/cyclone requirements) when provided with basic tailings classification properties and a tailings production profile. Test work for tailings characteristics should include bench-scale testing to assess the feasibility of tailings thickening and filtration which will guide viable alternatives identification.

Step 3 – Facility location assessment

With the design basis and knowledge base established, a process of identifying potential facility locations is then carried out. The spatial domain for the potential location search area should be established using the best available topographical survey. Obvious constraints, such as significant population centres, and agreed constraints, such as maximum distance from process plant, should be used to limit the search area. The additional exclusions identified during the design basis development can also be used to tighten the search area, eg no facilities within national/regional/local park boundaries, major roads. An example of the spatial search domain is illustrated in Figure 1.

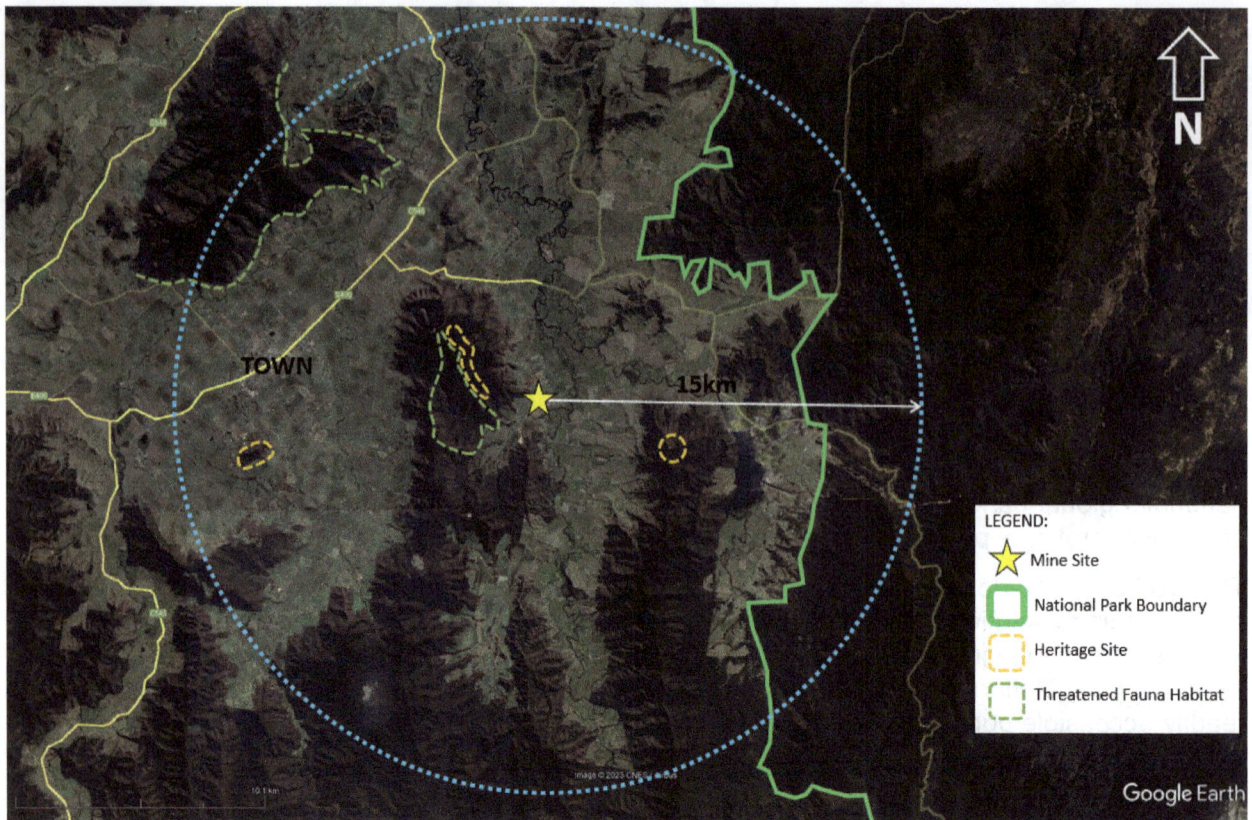

FIG 1 – Illustration of facility location spatial domain.

With the spatial search domain established, the facility siting process typically involves the use of geometrical modelling software such as Minebridge's *Muk3D* which allow for a rapid, iterative assessment of facility footprints required to meet the design basis of total tailings capacity. Typically, the authors initial approach is to identify locations suitable for conventionally thickened slurry tailings with the assumption that a solution employing alternatives such as paste or filtered tailings would require a smaller footprint and therefore the 'slurry tailings' location could also be suitable.

Following the initial location identification (and using the judgement of the design engineer), consideration should also be given to locations that could accommodate smaller but potentially more favourable facilities, eg only 60 per cent of total tailings accommodated but in a location with minimal upstream catchment and reusing already disturbed ground, and for sites that are amenable to dry-stacked tailings but not slurry tailings, eg valleys that would have a poor storage-to-fill ratio for a dam. A combination of smaller facilities may end up presenting the optimal solution to tailings management and therefore should not be discounted.

In addition to 'new' facilities within the search area, consideration should also be given to expanding existing on-site facilities (where applicable) and off-site disposal such as backfilling of nearby abandoned mine voids (open pit and underground), use of existing waste management facilities, and/or transport of material to a facility outside of the search area.

An example output showing the outcomes of a facility siting assessment is shown in Figure 2.

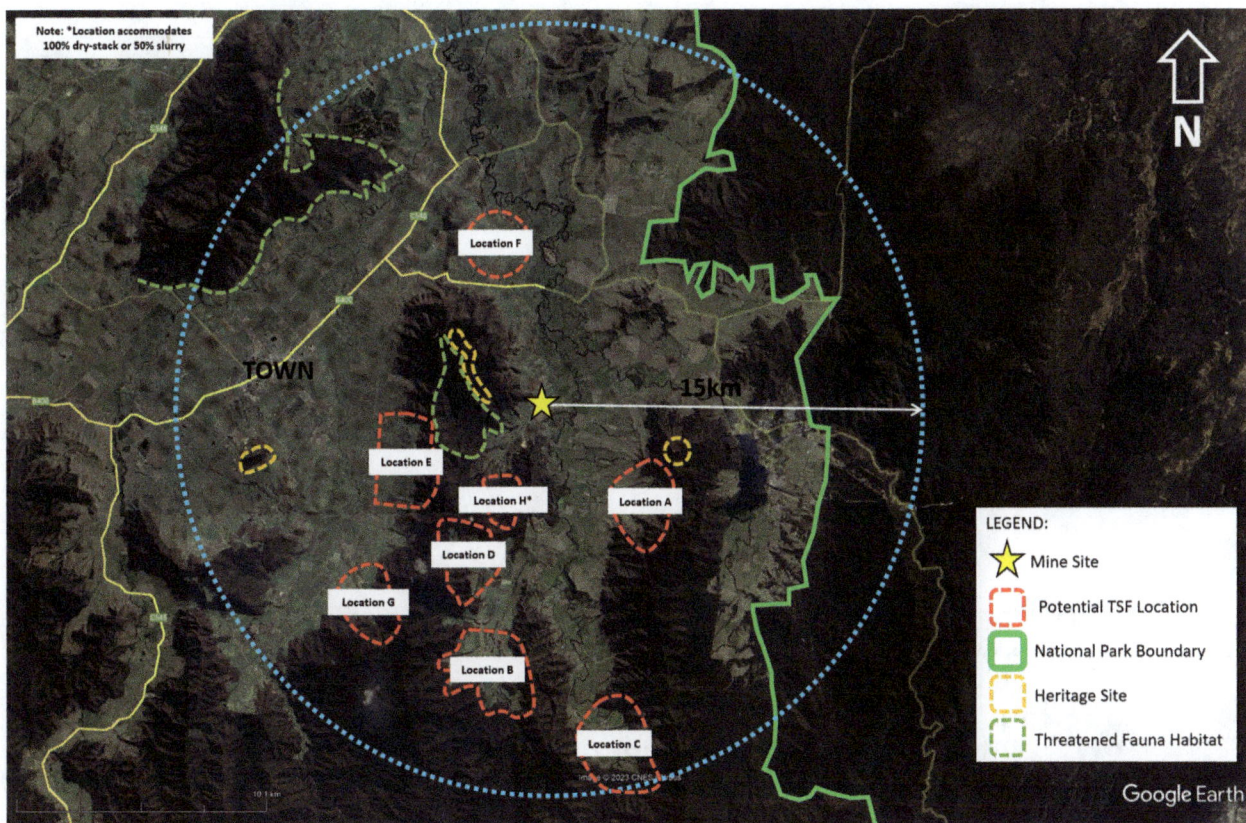

FIG 2 – Example output from facility location assessment.

Step 4 – Alternatives identification

The GISTM emphasises the consideration for alternatives in tailings management outside of the typical slurry tailings stored behind a tailings dam. Potentially feasible alternatives include all combinations of location (from Step 3) and tailings technology that could achieve the requirements of the decision statement and design basis.

Potential alternative tailings processing, transportation, placement, and management strategies should be identified and reviewed at this stage. The assessment of alternative strategies must align with the mine plan and the characteristics (geotechnical and geochemical) of the tailings and waste rock. Examples of alternatives that – alongside conventional facility alternatives – should be considered, include central discharge facility; downslope discharge facility; integrated waste landform; comingled waste landform; pit or underground void backfill; and dry-stack landform. Consideration should be given during this step to requirements for specific design features to manage known risks, eg requirement for a low-permeability liner to address geochemical risks.

Tailings dewatering OEMs can provide an assessment on the feasibility for application of their range of tailings thickening and filtration equipment to the project tailings. However, given the potentially significant risks to project economics and viability, mining companies are often reluctant to be the 'first-mover' in applying new dewatering technologies and techniques. To address concerns, the authors often rely on 'industry precedent' for application of tailings dewatering technologies. The MEND (2017) study of tailings management technologies is a useful tool to illustrate if the project is within industry precedent or not. The specific output from MEND (2017) that is often used is illustrated in Figure 3.

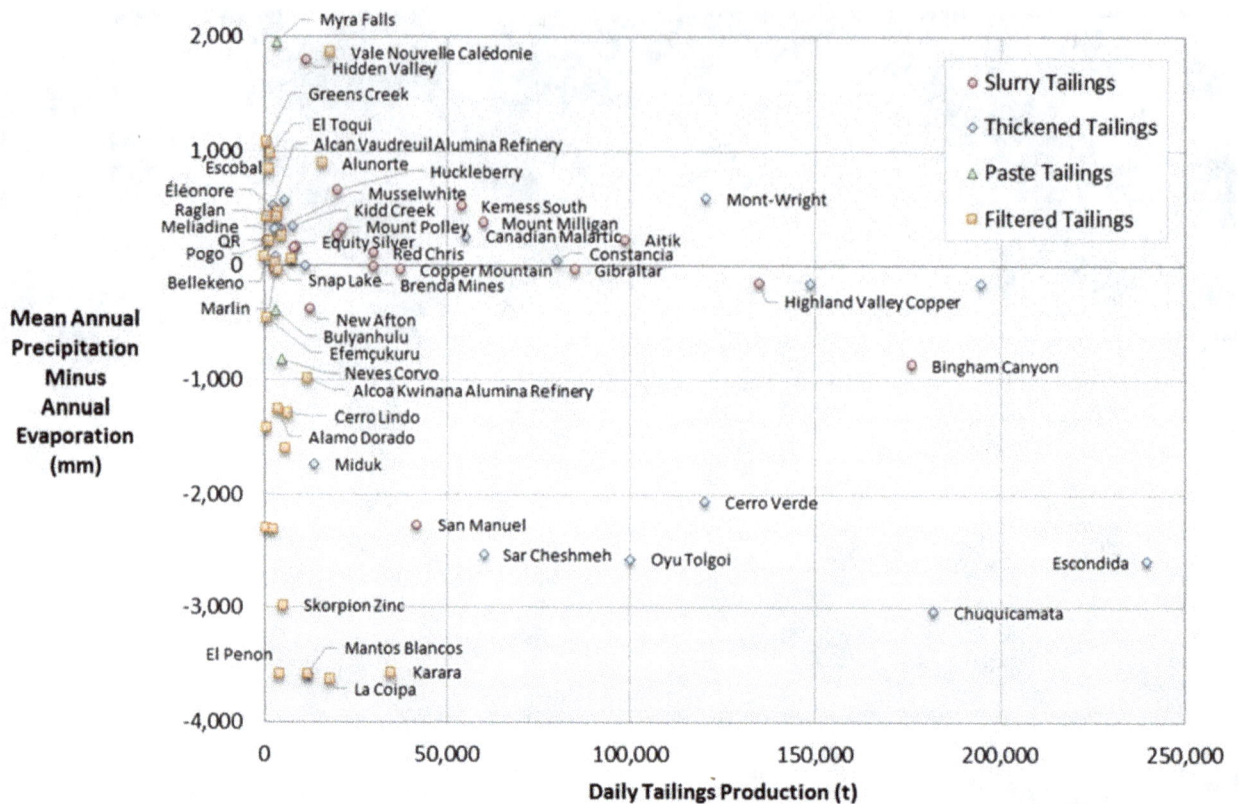

FIG 3 – Industry precedent for tailings dewatering (after MEND (2017; Figure 4.2)).

Step 5 – Preliminary option screening

Once the potentially feasible alternatives are identified, a preliminary screening should be completed to eliminate those that have critical flaws. A typical definition for a critical flaw is an attribute an alternative possesses that is so unfavourable that it eliminates that alternative from further consideration. Examples of critical flaws include:

- Unable to meet design criteria or requirements of the facility purpose/intent.
- Conflict with agreed alternative exclusions.
- Site conditions that are not suitable for a tailings technology or construction method (eg terrain not suitable for tailings/waste rock co-disposal or construction of conveyors for filtered tailings transportation).

Alternatives that do not possess a critical flaw can be brought forward to the next step, the MAA.

If the list of alternatives following preliminary screening is too long, a high-level qualitative ranking could be performed to eliminate alternatives that, while not possessing a critical flaw, are clearly less favourable than several other feasible alternatives, or are very similar to other alternative(s) but less favourable. Although these alternatives are eliminated from further consideration, they can be reintroduced at a later stage if additional alternatives are desired.

Step 6 – MAA framework development

Identifying an optimal preferred solution is best achieved using a project-specific MAA framework ledger that is developed using a list of objectives (or 'nice-to-haves') against which potential alternatives can be evaluated.

The objectives should reflect considerations that are 'nice-to-have' or desirable attributes of an alternative, ie an alternative that meets all attributes is expected to be the most favourable but failure to achieve a given objective does not preclude the alternative from being potentially viable.

Following the Guidelines for the Assessment of Alternatives for Mine Waste Disposal (Environment Canada, 2013), the development of objectives is typically grouped into three key 'accounts' (plus project economics):

1. Technical Considerations: Focus on characterising and comparing the engineered elements of each alternative.

2. Environmental Considerations: Focus on characterising positive and negative impacts to the environment surrounding each alternative.

3. Socio (or Socio-Economic) Considerations: Focus on positive and negative impacts to local/regional/national land users or other impacted stakeholders, from each alternative.

Project economics can be included as a separate account with its own objectives during this step or simply presented as total estimated cost for each alternative. In the authors' experience, performing the MAA first, based on technical, environmental, and social considerations, and then introducing cost estimate afterwards, is a good approach and avoids biasing the discussions towards the lowest-cost alternative.

Each objective is required to have a clear rationale for inclusion in the framework that outlines why it is important to the project. If no clear rationale can be provided, the objective should be critically reviewed and potentially removed from the framework if no rationale can be provided. Developing objectives and accompanying rationale requires input from stakeholders covering a diverse skillset including (but not limited to): geology, mine planning, process engineering, community engagement, environmental, approvals/permitting, finance, logistics/construction, legal etc.

The phrasing of each objective is an important consideration when developing the framework. Objectives should be worded using 'minimise or maximise' statements which reflect the favourable attributes of an alternative. Some examples of objectives following this approach include:

- Technical: Minimise reliance on strength of tailings for dam raising.
- Technical: Minimise elevation increase between process plant and facility.
- Environmental: Maximise reuse of existing areas of mining development disturbance.
- Environmental: Minimise disturbance of existing ecosystems.
- Social: Minimise number of dwellings requiring relocation.
- Social: Minimise negative visual impact on the area.

To adjust for bias and ensure that the key factors are given appropriate importance, stakeholders are asked to weight the relative importance of each objective. Objective weightings are typically based on a scale of 1 to 10. An objective weighting of 10 is considered the highest level of importance for the project, while an objective weighting of 1 is assigned to objectives that are still important, but the least important relative to the full range of the objectives.

For each objective, one or more measurement criteria must be defined that can be used to gauge the attainment of a given objective. The measurement criteria should be unique attributes for each alternative that can be assessed from conceptual design work and/or engineering/ESG-practitioner judgement. Criteria should, as far as practical, be quantitative as opposed to qualitative to allow for a more transparent assessment of objective attainment. For example, under the objective to 'minimise disturbance of existing ecosystems', the actual area (hectares, Ha) of native vegetation impacted can be used to quantitatively 'measure' the impact as opposed to using qualitative measurement such as 'no/minor/major impact'. A focus on quantitative measurement criteria is useful to mitigate against bias in the decision-making process; however, sometimes the use of qualitive criteria is unavoidable.

For each measurement criteria, a scoring range must be defined to allow for assessment of how well an alternative attains the defined objective. Typically, a six-point best-to-worst (6-to-1) scale is used. In the authors' opinion, a zero score should not be used as it can have an outsized impact on the calculation of the MAA score using the arithmetic presented in Step 8. Examples of scoring scales are provided in Table 1.

TABLE 1

Examples of scoring range.

Objective:	Minimise disturbance of existing ecosystems	Minimise number of dwellings requiring relocation
Scoring Range:		
1 (Worst)	≥100 Ha	≥20 relocations required
2	<100–80 Ha	<20–15 relocations required
3	<80–60 Ha	<15–10 relocations required
4	<60–40 Ha	<10–5 relocations required
5	<40–20 Ha	<5–0 relocations required
6 (Best)	<20 Ha	No relocations required

The framework is intended to be a comparative assessment tool, and in the authors' experience it is therefore often easiest to generate the scoring range after the measurement criteria values for each alternative are generated under Step 7 such that the maximum, minimum, and range of values are known. Defining a scoring range that does not reflect the (often minor) variances between alternatives risks resulting in all alternatives being given the same score, despite being clearly different, thereby reducing the ability to differentiate between alternatives.

Step 7 – Alternatives characterisation

To define scoring criteria and assign scores, each alternative to be assessed using the MAA framework must be characterised. The alternative characterisation typically includes developing and evaluating conceptual layouts; assessing impacts of site characteristics (eg geology; terrain; catchments) on dam stability, water management and seepage control; water balance; quantifying possible environmental and social impacts; public safety; and cost among other considerations. Mechanical assessments may be required to support cost estimates and to provide a better understanding of the requirements for tailings dewatering and transport (eg conveyor versus truck; pumping requirements; location of dewatering plant).

Costings should include consideration for capital expenditure (CAPEX), operating expenditure (OPEX) and closure/long-term care and maintenance, to inform the assessment of project economics. If possible, CAPEX and OPEX cost estimates should include the net benefits for a given technology, eg reduced make-up water demand using filtered tailings which returns a higher fraction of process water. Normalising of these costs to a unit cost ($/t) is a useful method to compare alternatives and check for reasonableness in estimates against other operations.

Step 8 – Apply and review MAA framework

Once the alternatives to be assessed and characteristics of those alternatives are established, the MAA framework ledger can be populated. The measured values are used to assign a score to each alternative for each objective. The scores are then weighted based on the average weighting for each objective as determined through stakeholder engagement. The weighted scores are then used to rank the alternatives from highest (most likely to attain objectives) to lowest (least likely to attain objectives). A conceptual ledger is shown in Figure 4. A typical ledger is likely to have several additional objectives and alternatives. Subtotals and rankings for each of the 'accounts' can also be tabulated and presented to support discussion on alternatives that are more favourable for specific subset of objectives.

Objective		Scoring Criteria						Alternative Scores			MAA Scores		
Objective Description	Objective Weighting	1	2	3	4	5	6	Alt. 1	Alt. 2	Alt. 3	Alt. 1	Alt. 2	Alt. 3
Account 1 – Technical													
Minimize/Maximize...	5							3	4	5	15	20	25
Minimize/Maximize...	7							4	2	4	28	14	28
Minimize/Maximize...	8							1	3	5	8	24	40
Account 2 – Environmental				Qualitative or quantitative description developed based on alternative characterization									
Minimize/Maximize...	3							5	4	3	15	12	9
Minimize/Maximize...	6							1	2	4	6	12	24
Minimize/Maximize...	2							3	2	1	6	4	2
Account 3 – Social													
Minimize/Maximize...	9							4	3	3	36	27	27
Minimize/Maximize...	10							2	3	4	20	30	40
								Total MAA Score			134	143	195
								Rank			3	2	1

$$Total\ MAA\ Score = \sum (Objective\ Weighting\ X\ Alternative\ Score)$$

FIG 4 – Illustration of populated MAA framework ledger.

The outcome of the populated ledger should be used to guide discussion and decision-making for the preferred tailings management solution. Ideally, the highest ranked alternative would be the preferred solution; however, consideration for costs of the alternative must be given and it may be that the highest ranked alternative is economically unfeasible. At least the top 2–3 ranked alternatives should be advanced for further study to account for the potential that the top ranked alternative becomes non-viable.

Step 9 – Sensitivity, threats, and opportunities assessment

Sensitivity assessments can be used to see what changes could result in a reordering of the alternative ranking within the MAA ledger. These assessments can include changes to scoring range, objective weighting, and to the designs themselves (eg increased facility height for a reduced facility footprint).

A risk assessment should be carried out for the alternatives to be advanced with the intent to identify threats to the viability of the top ranked alternatives that should be addressed early on in future study phases. This process can also be used to identify opportunities for refinement of the top ranked alternatives to improve the attainment of objectives and/or to reduce cost.

IMPORTANCE OF STAKEHOLDER ENGAGEMENT

The process outlined above can theoretically be carried out with limited interaction between the MAA facilitator and the client; however, the importance of stakeholder engagement cannot be understated in developing a site-specific, credible, and useful MAA. The stakeholder group should be as diverse as practical to ensure relevant viewpoints covering all design criteria, accounts and objectives are well represented.

Stakeholder engagement typically includes a minimum of three facilitated workshops with the primary objective of each workshop outlined below:

- Workshop #1: Introduce the stakeholders to the project and the MAA approach; confirm that representatives of all relevant stakeholder groups are engaged; establish and agree on the design basis/'must haves' and exclusions/'can't haves' (Step 1); and discuss extent of available data and information to inform the knowledge base (Step 2).

- Workshop #2: Present the inputs and outputs of the facility location assessment (Step 3); present the inputs and outputs of the alternatives identification (Step 4); present the outcomes of the preliminary screening (Step 5) and agree on any additional alternatives to be discounted (if required); and establish and agree on the objectives/'nice-to-haves' including the measurement criteria to allow for development of the MAA framework (Step 6).

- Workshop #3: Present the outcomes of the alternatives characterisation (Step 7); and present the populated MAA framework ledger and agree on the preferred alternatives to advance for further study/consideration.

In the authors' experience, Workshop #2 is often split into two separate workshops with one covering the MAA framework development (Step 6) and the other covering Steps 3–5. Additional workshops may be required where the list of alternatives to be considered is long or complex, or to present the outcomes of sensitivity and risk assessments (Step 9). As many stakeholders as practical should be included throughout the process including attendance/input to the workshops. Maintaining presence of a diverse stakeholder group reduces the potential for the process to be misguided by bias from individuals where a specialist counterargument cannot be provided, eg vocalising preference for a location with technical merits but without considering the significant impacts to local land users.

Prior to Workshop #3, a survey should be sent out to the stakeholders requesting them to weight each objective in isolation for their importance to the overall success of the project. The request for stakeholder weighting is not necessarily limited to the stakeholder group, eg inclusion of senior management/board can provide viewpoints outside of those at the project level.

A copy of the populated MAA ledger should also be circulated as soon as practical to allow for stakeholder review in support of the scoring. During Workshop #3, the range of survey results can be reviewed with the stakeholders to agree on weightings and decide on sensitivity cases to assess. It is unlikely that complete consensus will be reached among a diverse group of stakeholders, but the key objective is to hear from all voices and document all input. Sensitivity analyses can help assess the impact of contentious weightings and avoid impasse. Discussions that don't end in consensus could be indicators of uncertainties or gaps that need to be addressed.

The discussions occurring during the workshops should be well documented to provide for a credible record of the decision-making process.

Outside of the workshops, regular engagement with the stakeholder group will be required to review and comment on the outcomes of the workshop; provide weighting of objective importance; and provide input to the alternatives characterisation.

RECENT APPLICATIONS

The general MAA approach outlined above was recently applied to five projects of varying design basis (Step 1) as summarised in Table 2.

TABLE 2
Design basis of sample projects.

ID	Commodity	Location	Project status	Facility purpose/purpose of MAA	Rate of tails prod	Tailings total qty
01	Gold	Tanzania	Active	Providing additional storage capacity for an extended life-of-mine.	3.5 Mtpa	40 Mt
02	Iron Ore	Australia	Proposed	Provide LOM storage capacity for proposed new mine expansion.	10.8 Mtpa	200 Mt
03	Copper	Australia	Proposed	Provide LOM storage capacity for proposed new greenfield site mine.	2.0 Mtpa	20 Mt
04	Coal	Australia	Active	Providing additional storage capacity for an extended life-of-mine	0.6 Mtpa	12 Mt
05	Copper	USA	Active	Providing additional storage capacity for an extended life-of-mine.	14.0 Mtpa	450 Mt
06	Copper	USA	Closed	Evaluate alternatives for closure and facility risk reduction	n/a	n/a

The number of locations (Step 3) and dewatering technologies/disposal techniques (Step 4) to be considered within the MAA for the sample projects was significantly reduced following the preliminary screening (Step 5):

- Project 01: 36 alternatives initially identified → 18 alternatives assessed using MAA ledger.
- Project 02: 32 alternatives initially identified → five alternatives assessed using MAA ledger.
- Project 03: 24 alternatives identified → nine alternatives assessed using MAA ledger.
- Project 04: 21 alternatives identified → 11 alternatives assessed using MAA ledger.
- Project 05: 100 alternatives identified → 17 alternatives assessed using MAA ledger.
- Project 06: 13 alternatives identified → 13 alternatives assessed using MAA ledger.

The varying complexity of the site-specific MAA ledgers developed for each project (Step 6) is illustrated in Table 3. The variation in the number of objectives within each account demonstrates the importance of not using an 'off-the-shelf' ledger.

TABLE 3
Overview of MAA ledger complexity.

ID	Number of objectives in ledger			
	Technical	Environmental	Social	Total
01	12	11	8	31
02	8	5	4	17
03	10	4	4	18
04	4	2	2	8
05	11	12	5	25
06	10	10	8	28

A summary of the highest ranked and preferred alternative for each project is summarised in Table 4. As shown, there is notable diversity in the top ranked alternatives which demonstrates the utility of the MAA process in identifying and assessing alternatives. In some cases, the top ranked alternative was not the chosen preferred alternative; however, the justification for the decision was established and documented which supports transparency in the decision-making process.

TABLE 4
Summary of MAA outcomes.

Id	Description of top ranked alternative	Description of chosen preferred alternative	Justification for chosen preferred alternative
01	Thickened slurry tailings deposited into existing open cut mine void.	Thickened slurry tailings deposited into new-build conventional facility.	Preferred alternative presented lowest complexity for approval and implementation over relatively short LOM extension.
02	Thickened slurry tailings deposited into existing open cut mine void and then into new open cut mine void.	Thickened slurry tailings deposited into existing open cut mine void and then into new open cut mine void.	Preferred alternative was highest ranked.
03	Filtered tailings deposited into a dry-stack landform	Thickened slurry tailings deposited into an integrated waste landform	Capital costs of the top ranked alternative were considered non-viable for project.
04	Filtered tailings deposited into expansion of existing facility footprint.	Filtered tailings deposited into expansion of existing facility footprint.	Preferred alternative was highest ranked.
05	Conventional slurry tailings with embankment construction from cycloned sand.	Five top ranking alternatives were carried forward to the next phase of study, including the top-ranked alternative.	The stakeholders agreed that further work was required to evaluate top-ranking alternatives.
06	Relocate tailings to a mined-out open pit	Relocate tailings to a mined-out open pit	Preferred alternative was the highest ranked and provided an opportunity to eliminate the risk economically.

LESSONS LEARNED

Key lessons and recommendations have been learned from applying the common MAA approach to the reference projects including:

- Stakeholder diversity is very important to ensure an accurate representation of environmental and social considerations are provided alongside technical and project economics. Ensuring the stakeholder group is established, complete, and engaged early in the process will support a credible process and outcome. Stakeholders should be given the opportunity to provide feedback and endorsement of the MAA framework along the way including providing the populated ledger before the final workshop if possible.

- Developing a Knowledge Base with as much detail as possible as early as possible is important to ensure that effort is not wasted on options that are clearly unfavourable and to reduce the uncertainty in the outcome. Pausing a MAA process until better data and information is available is something that should be considered if the inputs are severely limited.

- Outputs from Report 2.50.01: Study of Tailings Management Technologies (MEND, 2017) are inherently useful in demonstrating industry precedent; however, the data set used in the study is likely to be outdated given the significant shifts in focus since the study was published for projects to utilise tailings dewatering at increasing scales. Industry precedent supported by published studies only is likely an inappropriate approach to screening out potentially viable alternatives. An industry push to develop a reference library of tailings management approaches for a much wider and more current data set would be useful for future MAAs.

- The GISTM goal to minimise the volume of tailings and water placed in external tailings facilities requires creativity in solutions that consider the full development cycle of a mining project. The flexibility of the MAA approach to consider aggregated solutions combining multiple alternatives, eg small above-ground facility before transitioning to in-pit, should be used.

- Objectives can be strong indicators of what is important to a project proponent and do not necessarily need to be differentiators in assessing alternatives. For example, with an objective to 'minimise number of dwellings requiring relocation' all alternatives may 'score' the same if zero dwellings require relocation. This outcome would demonstrate to external stakeholders that this objective was very important to the proponent. Objectives can also be very useful in guiding subsequent steps in the process and defining these as early as possible can streamline the process.

- A robustly developed MAA is a flexible tool that can be used to assess (or reassess) a wide range of alternatives in response to changing design basis for the project and/or an expanded and updated Knowledge Base. Consistent with the GISTM, the MAA framework should remain in use for continued assessment of tailings management solutions in response to (almost) inevitable changes in a project's life-of-mine.

- Workshops are a core part of the process, and an experienced facilitator is necessary to ensure that all stakeholders understand the process and feel comfortable providing input to the framework development. The workshops should focus on discussion on the alternatives in the context of the scoring and ranking rather than spend time discussing individual measurements and weightings. Typically, a minimum of three formal stakeholder workshops are required; however, should allow for four or five to ensure adequate time. Documenting the rationale behind decisions arising from the workshops is critical for transparency in the decision-making process. The authors' strongly encourage in-person workshops to fully engage the stakeholders and maximise the workshops' value.

REFERENCES

Environment Canada, 2013. Guidelines for the Assessment of Alternatives for Mine Waste Disposal, Mining and Processing Division. Available from: <https://www.canada.ca/en/environment-climate-change/services/managing-pollution/sources-industry/mining-effluent/metal-diamond-mining-effluent/tailings-impoundment-areas/guidelines-alternatives-mine-waste-disposal.html>

Global Tailings Review (GTR), 2020. Global Industry Standard on Tailings Management (GISTM). Available from: <https://globaltailingsreview.org/global-industry-standard/>

International Committee on Large Dams (ICOLD), 2021. Bulletin #181: Tailings Dam Design Technology Update. Available from: <https://www.routledge.com/Tailings-Dam-Design--Conception-des-Barrages-de-Steriles-Miniers-Technology/Icold/p/book/9780367770464>

Mine Environment Neutral Drainage Project (MEND), 2017. Report 2.50.01: Study of Tailings Management Technologies. Available from: <https://mend-nedem.org/mend-report/study-of-tailings-management-technologies-2017/>

Innovative mine waste management

Performance of a flow-through tailings dam

R Longey[1], D Tonks[2] and E Salas[3]

1. MAusIMM, Technical Director Tailings, GHD, Hobart Tas 7000. Email: rob.longey@ghd.com
2. Engineering Superintendent, Grange Resources, Burnie Tas 7320.
   Email: dan.tonks@grangeresouces.com.au
3. Technical Director Geotechnical Engineering, GHD, Burnie Tas 7320.
   Email: edgar.salas@ghd.com

ABSTRACT

The water quality and ecology in the Savage River located on the north-west coast of Tasmania has previously suffered due to historical mining operations from past operators as a result of legacy mine waste rock and tailings storage facilities contributing significant acid and metalliferous drainage to the receiving environment. A series of projects jointly run by the current owners Grange Resources and the government have resulted in synergies and net positive benefits using the current mine operations to manage and remediate legacy mining issues. One such positive example of this synergy is the South Deposit Tailings Storage Facility (SDTSF). The 140 m high facility will provide storage for approximately 37 Mm^3 of tailings, equating to approximately 20 years' tailings storage. The SDTSF embankment is constructed entirely of waste rock won from mining operations and provides Grange with an economical storage solution for waste management (both tailings and waste rock) minimising the sites environmental footprint. The embankment features a permeable filter face which is sufficiently fine to retain tailings and allows passing of normal catchment flows, while larger flood events are stored within the facility and slowly released. Outflows from the filter face are passed to a 'flow-through' drain constructed from alkaline, waste rock. The outflows discharged through the 'flow-through' will provide a long-term source of alkalinity to Main Creek, subsequently feeding into the Savage River. The facility is showing the first signs of environmental benefits to the downstream ecology that has long been degraded due to the effects of legacy acid and metalliferous drainage (AMD), caused by historical mining operations. Based on initial monitoring following commissioning of the facility in 2017, the filter performance compared with design, along with water quality from the 'flow-through' drain.

INTRODUCTION

The Savage River open cut magnetite mine is situated on Tasmania's west coast, approximately 100 km south-west of the coastal port of Burnie. The area is characterised by high rainfall and steep topography, making water management critical to enable safe, stable storage of tailings and waste rock.

The site comprises open cut pits on both the north and south sides of the Savage River. The mine facilities, workshops and primary crusher are located on the north side of the river, with ore fed to the concentrator plant on the south side of the river via a 1.3 km long conveyor. The concentrator plant produces approximately 2.9 Mtpa of magnetite concentrate. The concentrate is transported in a slurry pipeline, 83 km north to be further refined at Granges' Port Latta pellet plant on the north coast of Tasmania, prior to the pellets being transported by ship for use in steel production. The residual material (tailings) from the ore passed through the concentrator plant consists of sand and silt sized particles in a slurry, which require safe disposal in a tailings storage facility (TSF).

The site has acid forming historic tailings as well as some potentially acid forming (PAF) waste rock production, which make risk reduction for generating acid and metalliferous drainage (AMD) from new mining activities a priority (Grange Resources, 2013). The mine's previous storage, Main Creek Tailings Dam (MCTD) has reached its final filling capacity, thus a new facility, the South Deposit Tailings Storage Facility (SDTSF), has been designed and constructed.

The mine operator, together with the state government, jointly manage legacy AMD issues through the Savage River Rehabilitation Project (SRRP). The AMD is caused by historic waste rock dumps (WRDs) constructed by previous owners, which have previously caused extensive damage to the downstream ecology of Savage River. The Mine operator through the SRRP has been able to significantly reduce this impact and largely remediate the ecology of Savage River since resuming operations. These legacy issues are now largely contained to the Main Creek catchment on which the SDTSF is located, prior to Main Creek entering the Savage River.

The SDTSF embankment is a significant structure totalling approximately 140 m in height, constructed in a three year period and commissioned in 2017. The facility will provide approximately 20 years of tailings storage (~37 Mm^3) and is constructed entirely of waste rock from mining operations, providing storage for approximately 15 Mm^3 of waste rock within the embankment construction. The use of waste rock generated from the adjacent mining occurring at the South Deposit Pit as an embankment construction material has made for an economical integrated waste management solution for the mine.

This paper describes the key features of the design, namely the alkaline 'flow-through' rock drain designed to pass regular inflows entering the TSF and provide a long-term source of alkalinity to Main Creek, assisting in improving the water quality in the historically degraded environment.

The SDTSF design is unique in adopting a permeable rock fill 'filter face' on the upstream face of the embankment, designed to prevent ingress of tailings into the 'flow-through' and downstream environment.

The SDTSF provides environmental benefits particularly for Main Creek, whilst offering a safe and economical storage of tailings during its operational life, and offers long-term closure benefits in enabling capture and transfer of AMD seeps from legacy WRD's and TSF's for transfer to the South Deposit Pit for treatment.

Future construction work on SDTSF is required to complete a remaining portion of the embankment to RL300 m, construct the clay core and filters on the uppermost section and cut the closure spillway in natural ground on the right abutment with capacity to safely pass a Probable Maximum Flood in accordance with ANCOLD Guidelines (ANCOLD, 2019).

KEY FEATURES

Location

The SDTSF is sited downstream of the existing Main Creek Tailings Dam (MCTD), also within the Main Creek catchment. It is situated approximately 1 km from the South Deposit pit, the source of waste rock for the embankment construction.

A general arrangement of the site is provided in Figure 1. This also shows the context of the overall mine site, including the location of the Broderick Creek 'flow – through' WRD that is a successful prototype for the proposed operation of the SDTSF (Brett and Hutchison, 2003).

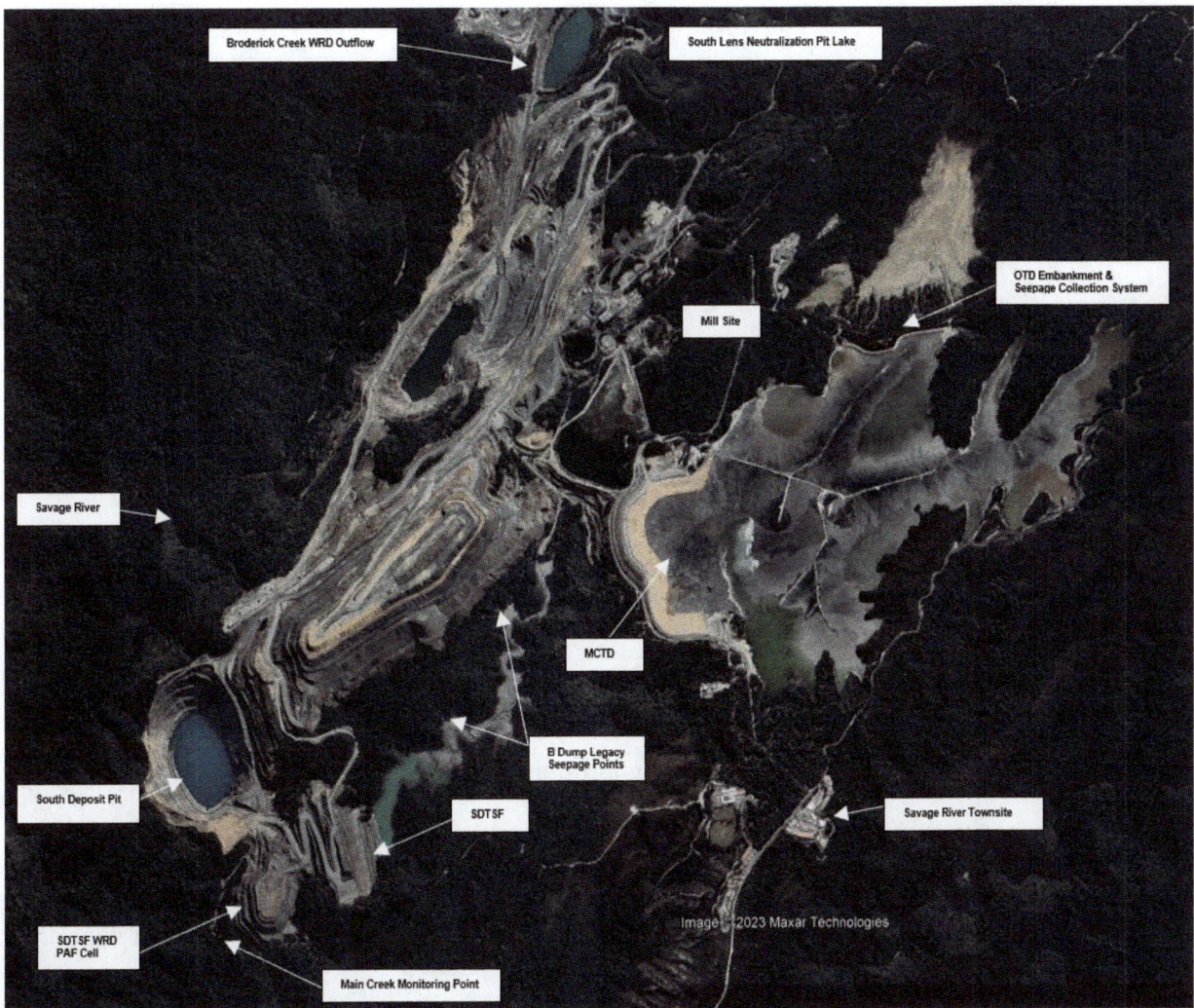

FIG 1 – General arrangement of site (Google Earth 2023).

General

The SDTSF has been constructed entirely from waste rock materials produced from the mining of South Deposit Pit. Access to the dam site was gained by constructing ramps using waste rock from the pit. The ramps developed tip heads for dumping, working towards the upstream face and filling the valley from the right abutment side (right side of valley when looking downstream).

The design features two significant innovative engineering elements:

1. A 'flow-through' rock drain to be constructed of coarse (D50 ~200 mm), alkaline A Type waste rock, designed to pass regular inflows entering the TSF storage, utilising designs and performance data from the Broderick Creek flow-through project. The facility has been designed to store flood inflows temporarily, before releasing water through the 'flow-through' drain.

2. A 'Filter Face' on the upstream face of the embankment. This has been designed to prevent ingress of tailings into the 'flow-through' and downstream environment, whilst filtering water flows into the flow-through drain.

The two zones were constructed utilising tip-head placement techniques, whereby large mining dump trucks place waste from a minimum height of 20 m at the angle of repose for the rock. This effectively causes segregation in the mine waste, which results in the coarsest rock being placed in the creek bed forming the 'flow-through' drain. Examples of the segregated particle size distributions are provided in Figure 2.

Filter Face / Flow-Through Gradings

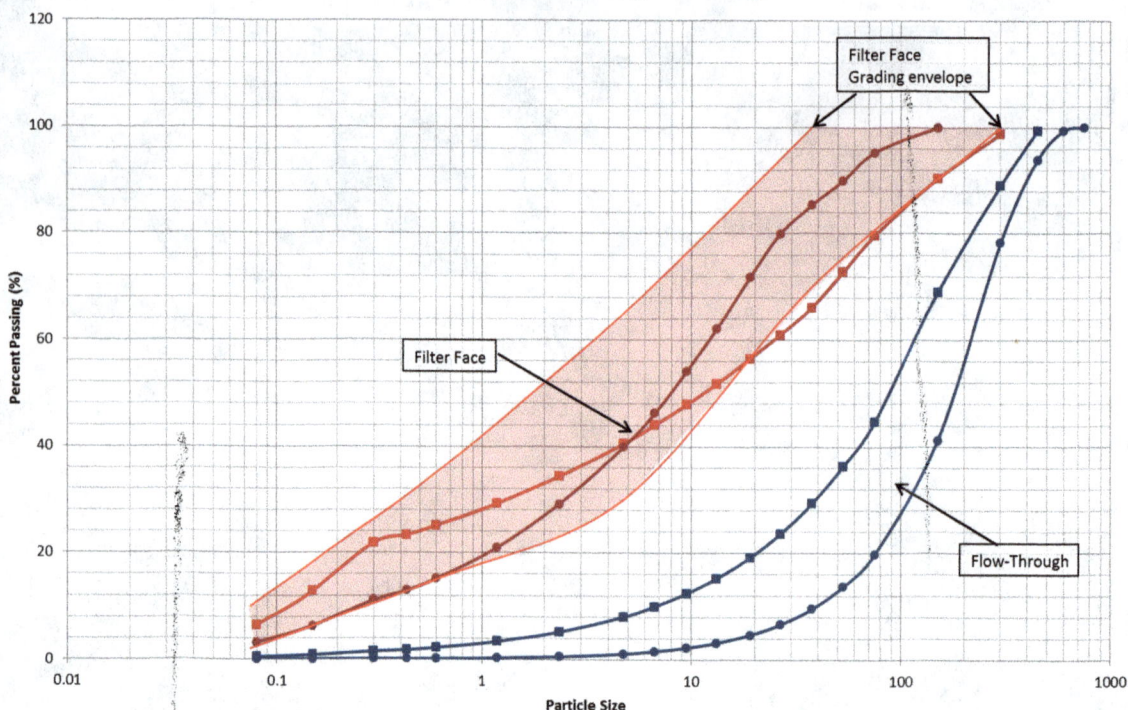

FIG 2 – Filter face/flow-through grading examples.

The remainder of the embankment has been be constructed from waste rock (Type A and B), through a combination of paddock dumping in maximum 2 m lifts (Consolidation Section in Figure 3), and tip-head placement in 10–20 m lifts (D/S Shell in Figure 4). The paddock dumping and compaction in lifts within the consolidation zone is to limit settlement of the section supporting the filter face and the closure section.

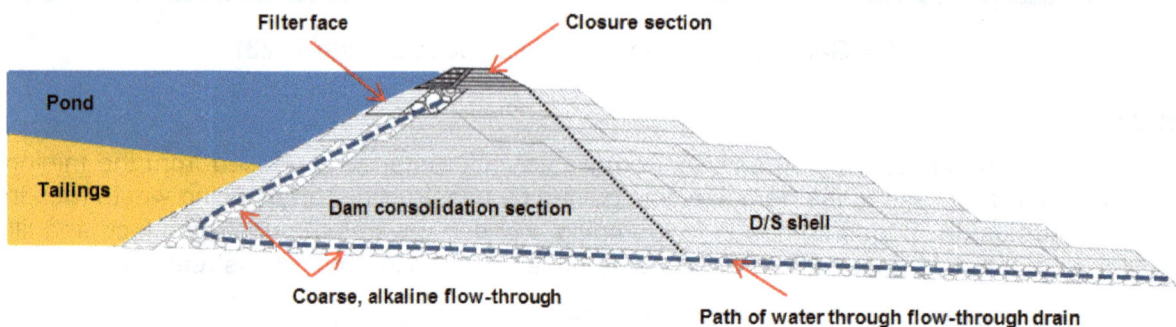

FIG 3 – Embankment cross-section.

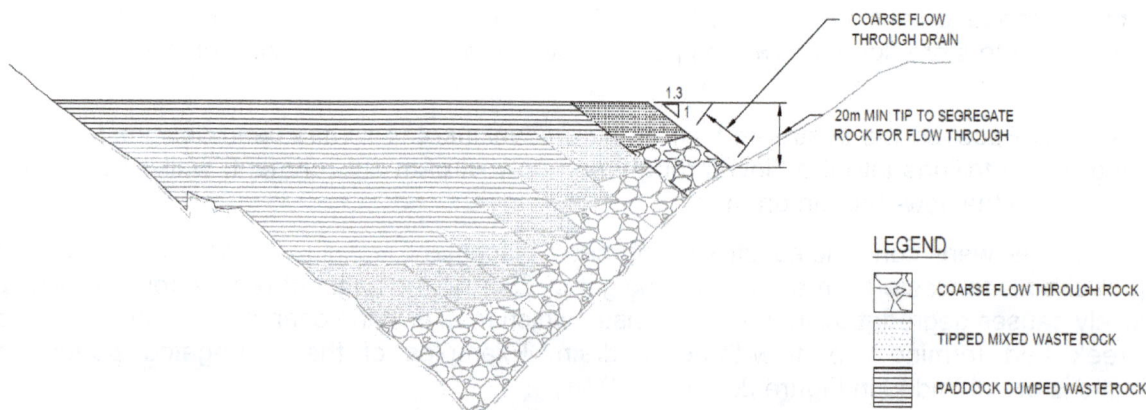

FIG 4 – Section across valley.

The embankment was constructed using a conventional 'downstream construction' methodology; ie stages of the embankment will be constructed in the downstream side of the embankment crest on natural ground.

The SDTSF is unique for a large, water retaining embankment as it does not feature a conventional clay core. It is intended that the pond level will be controlled by the tailings stored – as the tailings level builds up against the filter face, they block off the lower levels of the filter face and create a low permeability barrier along the face of the embankment and foundations. The level of the pond will rise as the tailings level rises.

This type of permeable embankment is similar to that used in management of coal wastes, whereby coarse coal rejects are used to construct permeable embankments that retain the fines, while allowing water to pass. However, the authors are not aware of an application for tailings retention with a filter face and flow-through drain of the scale of the SDTSF in a wet climate such as the Savage River mine.

The final stage of the embankment which is yet to be constructed has been designed as a conventional water retaining embankment, with a clay core and engineered filter zones, in order to maintain a full water cover over the tailings in perpetuity.

FLOW-THROUGH ROCK DRAIN

A key design element of the SDTSF is the flow-through rock drain. This was constructed from coarse alkaline A-Type waste rock. Constructing a flow-through drain has a number of benefits, including:

- Water flowing through the spillway will pick up alkalinity, improving the water quality as per the Broderick's Creek flow-through drain. As the TSF is constructed on Main Creek which has previously been contaminated with historic AMD issues), this is beneficial to the existing water quality.

- The TSF was able to be continuously constructed, without the need for cutting staged spillways into natural ground. Due to the naturally steep topography of the area, cutting spillways would result in significant cost.

The flow-through rock drain has been designed to cater for regular flows. The embankment has been designed with enough dry freeboard (approximately 30 m) to allow for storage of the 72 hour Probable Maximum Precipitation (PMP) flood event, estimated to be approximately 5.8 Mm^3. The flow capacity during flood events of the flow-through drain is expected to be between 2–3 cumecs; however, the flow will be controlled by the capacity of the Filter Face, which is not expected to exceed 2 cumecs (GHD, 2013). Having the inflow limited by the Filter Face limits the risk of developing a high phreatic surface within and on the downstream face of the TSF. The store and release methodology for dealing with large flood events, results in elevated ponding above normal operating levels. The worst case scenario allows passing of a PMF event within 30 days.

The 'tip-head' construction method causes segregation of the A-Type waste rock and is an essential methodology for construction of the flow-through drain. Grange have many years' experience in carrying out this type of construction. The drain was developed along the left abutment of the TSF from valley floor to the crest, and along the front face behind the 'Filter Face' zone to create the necessary flow paths.

The 'tip-head' construction methodology requires a minimum height of 20 m to achieve segregation of the rock fill, based on historical and current site experience at Savage River in the construction of waste rock dumps, and the flow-through drain at Broderick Creek Waste Rock Dump (Brett and Hutchison, 2003). The height of the flow-through will necessarily be 20 m below the crest height on the left abutment until the final stage of construction when the water retaining closure embankment section is constructed (see Figure 4).

Design precedent – Broderick's Creek waste rock dump

The Broderick Creek WRD is a successful example of 'flow-through' rock drain construction at Savage River, and serves as basis for performance monitored design parameters used in the SDTSF flow-through design.

The Broderick Creek WRD consists of an initial flow-through spillway structure comprised of a permeable zone constructed within a 'dam' of dumped rock. The permeable zone consists of selected hard rock with open grading produced by segregation during placement. Outflow is limited by the cross-section area of the drain. A long section and cross-section of Broderick Creek is shown in Figure 5. The flow-through was designed to pass normal creek flows, with a pond building up under flood conditions. The storage area upstream of the waste rock dump has the capacity to store in excess of the Probable Maximum Flood (PMF) event. The pond also acts as a silt trap with sediment being deposited upstream of the waste rock dump, thus mitigating the risk of blocking the permeable zone.

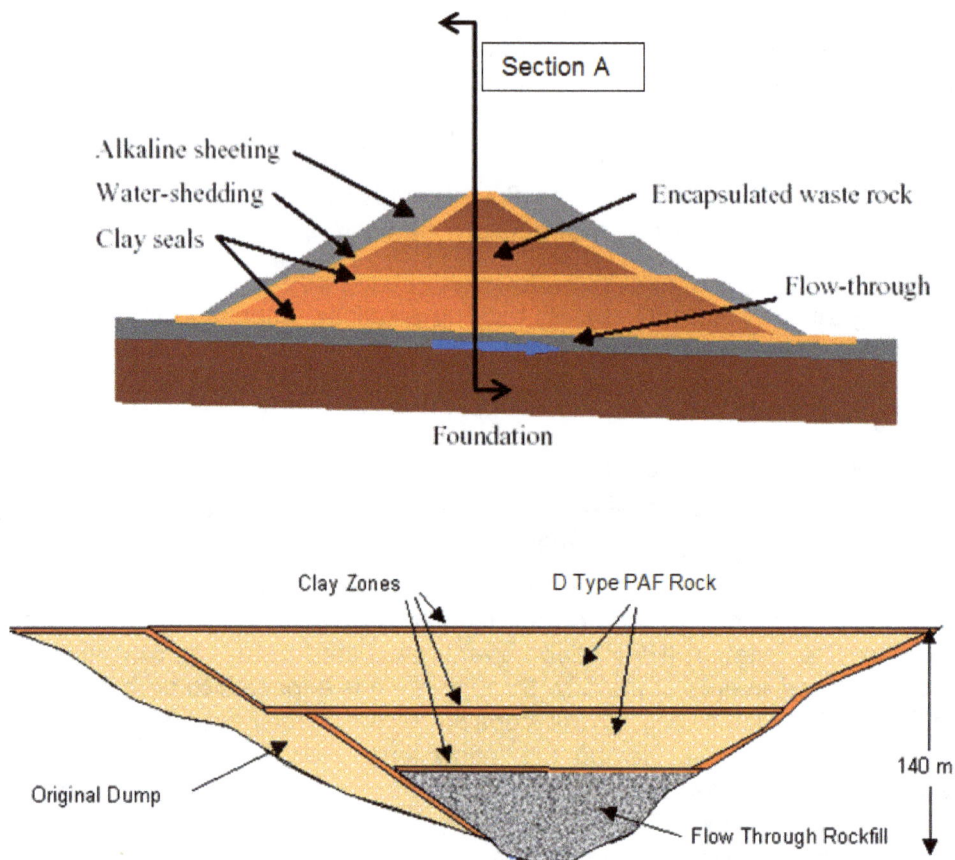

FIG 5 – Broderick Creek long section and cross-section.

Based on monitoring data maintained by Grange, the characteristic permeability of the flow-through has been calculated to be 0.26 m/s, which is within the predicted range in the original design (0.2–0.4 m/s).

The success of the original Broderick Creek 'flow-through' concept has allowed the development of an extended flow-through zone under expanded dumps, which now spans the complete length of Broderick Creek in excess of 4 km. As the Broderick Creek flow-through has catered for the majority of planned waste rock disposal for the life-of-mine since the mid-1990s, it has effectively made the mine viable (Brett and Hutchison, 2003).

Filter face

The 'filter face' located on the front face of the embankment has been designed for the following purposes:

- To prevent migration of tailings into the flow-through rock drain and subsequent possible release to the downstream environment.

- To be permeable enough so that water in the storage does not build-up during normal operations and overtop the flow-through drain.

The SDTSF Filter Face does not perform some of the functions that a traditional filter might in other embankment dams, that is:

- There is no soil zone within the embankment that it is designed to protect from internal erosion (piping failure).

- It is not required to act as a drain for the TSF/WRD, as the zone immediately downstream of the filter is the flow-through rock.

As such, the filter face has been classified as a non-critical filter (Fell *et al*, 2005).

Prevention of tailings migration

The following design criteria have been adopted to prevent migration of tailings into the flow-through:

- $D15F$ (ie particle size of the filter material for which 15 per cent by weight is finer) ≤ 0.7 mm. This is the $D15F$ design criterion for a Type 2 soil (Sherard and Dunnigan, 1989). The $D15F$ criterion is used to achieve a partial size distribution in the filter so that the voids are sufficiently small to prevent migration of the base soil, which in this case is the tailings slimes. Testing on MCTD slimes have found the fines component ($<75\ \mu m$) to be approximately 77 per cent, and are therefore classified as a 'Type 2' soil.

- $D15F/D85B < 5$ (where 'F' is the filter and 'B' is the tailings slimes). This is a criterion used as a measure of whether fines will be trapped within the filter; ie the soil is considered self-filtering if this criterion is met (Schuler and Brauns, 1993).

- Filter to be of sufficient width to negate the effects of segregation during construction (segregation in this case would be defined as coarse particles aligning in the filter resulting in a conduit within the filter). The filter face has been designed to be of 10 m minimum thickness.

- Filter to be constructed of well graded rock fill with a maximum size of 150 mm (as per requirements for a non-critical filter).

- QA/QC monitoring during construction was undertaken to confirm the filter face placement specification was achieved.

Permeability

The permeability of the filter face is a critical element of the design, as the SDTSF does not feature a conventional decant structure. Therefore, the filter face has been designed to be permeable enough to allow the flow of water into the flow-through drain, so as to prevent water from backing up significantly within the storage causing potential for overtopping of the flow-through down the left abutment. The SDTSF upstream face, pond and beached tailings can be seen in Figure 6.

FIG 6 – SDTSF storage and upstream face.

A further consideration is that if a significantly large pond was maintained due to the filter face not being permeable enough, then it may affect the storage capacity as beach development would be hindered due to lower tailings density.

The design permeability of the Filter Face has been estimated using three methods:

- Empirical estimation of the permeability (k), based on the soil type. Based on the geotechnical investigations, the soil has been classified as generally well-graded gravel with some fines. As such, it is expected that the permeability should be in the range of 1×10^{-2} and 1×10^{-4} m/s.

- Using Hazen's Formula: $k = C(D_{10})^2$, where the D_{10} value has been estimated from the grading curves, and the factor C has been assigned a value of 0.004 (considered conservative). This gives a permeability of 1.6×10^{-4} m/s.

- Field permeability testing: four permeability tests were conducted on A Type rock fill materials within the MCTD embankment (during geotechnical investigations), constructed out of materials that are considered to be similar to those that will be used for the Filter Face. The tests were conducted using a 100 mm diameter hand auger and as such are not considered to be representative of the overall permeability of the material; however, they do give an indication. The results of the permeability tests ranged from 2.0×10^{-6} to 4.4×10^{-4} m/s.

- Larger scale permeability testing on a trial embankment. A mini storage was constructed with a plastic lining, and tailings sluiced into it. Permeability was measured using the Darcy's formula $Q = kiA$, where Q (m³/s) was determined through the rate the water level in the storage dropped, A was the cross-sectional area of the upstream face, and i was estimated based on seepage location. The permeability obtained ranged from approximately 2×10^{-5} to 1×10^{-4} m/s.

From these observations and calculations, it is determined that the permeability would be approximately between 1×10^{-5} to 1×10^{-4} m/s.

Based on measuring the flow-through drain outflows and back calculating the wetted area of the filter face an indicative permeability of the filter face has been calculated in the order of 5×10^{-4} m/s which is within the design range.

Despite the abovementioned theoretical and field test work showing reasonable correlation of expected performance, there are contingency plans in place should adjustment to the filter face be deemed necessary, to ensure optimal permeability. Contingency scenarios which can be adopted during operations include:

- If the filter face is lower permeability than design, the storage will fill with water more rapidly, as excess water above the design permeability will accumulate within the storage. This will cause less beaching of tailings, resulting in reduced storage capacity due to a lower tailings density. If construction monitoring showed this to be the case, the compaction of the filter face was proposed be reduced to increase its permeability.

- If the filter face is higher permeability than design, it can be more heavily compacted or can be partially sealed with clay post construction.

Hence, the filter face design, construction and operations have inbuilt flexibility and contingency to allow for the observational design approach.

ENVIRONMENTAL BENEFITS

The flow-through drain consists entirely of A-Type alkaline waste rock. All water entering the catchment of the SDTSF reports to the flow-through, introducing a long-term source of alkalinity to Main Creek downstream of the embankment, which feeds into the historically degraded Savage River which has been impacted by acid and metalliferous drainage. SDTSF is providing a source for improving the overall water quality of Savage River, in addition to other improvement programs provided by the mine operator and state government.

The Savage River Rehabilitation Project has been monitoring water quality for over two decades at the site. Based on monitoring directly below the SDTSF outflow (MCbSD) the SDTSF is considered effective at neutralising and retaining metals contained in the mine tailings, legacy B Dump seepage and in the discharge from the MCTD. Several metals/metalloids and other parameters are consistently at or near the minimum level of laboratory reporting (arsenic, cadmium, chromium, lead, molybdenum and selenium) downstream of SDTSF. Key metals of concern are shown in Figure 7 to be trending down and have reduced since 2014 by approximately two orders of magnitude. Key changes in the graph represent when SDTSF construction commenced in 2014 which introduced a mechanism to capture and neutralise legacy AMD. Secondly further reduction was realised once SDTSF was commissioned to receive tailings in 2018 which introduced further alkalinity to SDTSF from the tailings process water.

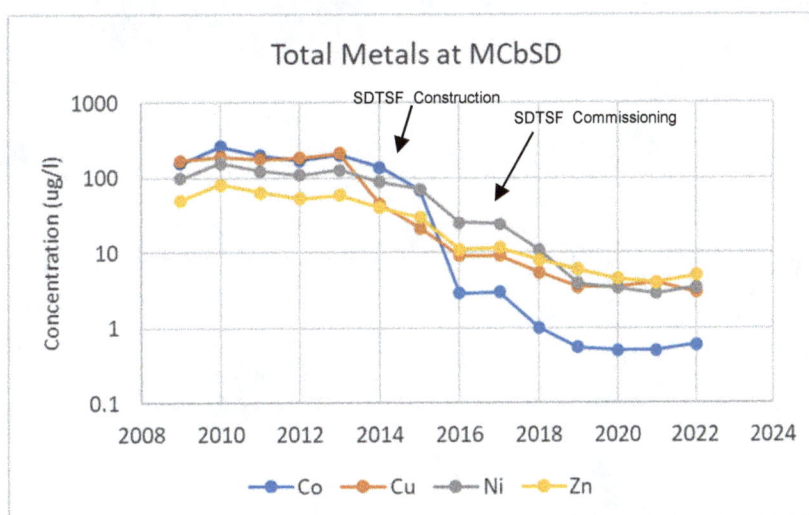

FIG 7 – Total metals monitoring below SDTSF.

On the balance of evidence, the SRRP is presently achieving its goal of establishing a modified but healthy ecosystem in the Savage River. Main Creek has also shown a marked improvement, with both rivers shifting towards reference condition as compared to previous monitoring. The improved environmental condition parallels the decrease in metals in the river (Koehnken, 2022).

CONCLUSIONS

The design of the SDTSF aimed to integrate waste management at the Savage River mine site, by utilising waste rock in the construction of the dam embankment, thereby creating a storage for both tailings and waste rock. This integration is effective in both limiting the site's environmental footprint, as well as reducing costs associated with waste management.

The innovative flow-through drain and filter face allow for cost savings associated with water management, as the facility does not require infrastructure such as staged spillways and decant structures. In addition, the alkalinity introduced to the downstream environment through the flow-through drain has been shown to provide environmental benefits to the downstream ecology in the long-term since construction.

ACKNOWLEDGEMENTS

The authors express their gratitude to the management of Grange Resources Ltd for permission to present this paper and to staff, consultants and contractors who have contributed to the success of the project.

REFERENCES

ANCOLD, 2019. Guidelines on Tailings Dams – Planning, Design, Construction, Operation and Closure – REVISION 1, ANCOLD.

Brett, D and Hutchison, B, 2003. Design and Performance of a 'Flow – Through' Spillway at Broderick Creek Waste Rock Dump - Savage River mine, *Australian Journal of Water Resources*, 6(2):2003.

Fell, R, MacGregor, P, Stapledon, D and Bell, G, 2005. *Geotechnical Engineering of Dams*. 912 p.

GHD, 2013. South Deposit Tailings Storage Facility Design Report, GHD Report 57083.

Grange Resources, 2013. Development Proposal and Environmental Management Plan: South Deposit Tailings Storage Facility, Grange Resources (Tasmania) Pty Ltd report, submitted to Tasmanian Environmental Protection Agency.

Koehnken, L, 2022. Savage River Rehabilitation Project Water Quality Review 2017–2022.

Schuler, U and Brauns, J, 1993. Behaviour of coarse and well graded filters, *Filters in geotechnical and hydraulic engineering* (eds: Brauns, Helbaum and Schuler), pp 3–17 (Balkema: Rotterdam).

Sherard, J L and Dunnigan, L P, 1989. Critical filters for impervious soils, *J Geotech Eng ASCE*, 115(7):927–947.

Mining waste in Chilean mines – a big challenge to solve

J Rayo[1]

1. Principal Consultant, JRI, Santiago, Chile. Email: juan.rayo@jri.cl

ABSTRACT

JRI estimated that all Chilean mines have generated more than 50 000 million tons (Mt) of waste (tailings, slags, rocks, among others) which have been stored on the mainland, and the actual mining production of waste is around 5 Mt per day.

Most of the Chilean ore deposits have an expected life of over 100 years, some of them over 200 years, and it estimates that the mining industry will produce more than 150 000 Mt of waste by the end of this century.

More than one-third of the future residues will be stored in the Central Zone of Chile. However, the environmental regulation and communities' sensibility (majorities against the mining industry) will not allow new residues deposits. Consequently, new technologies should be applied to solve this issue: HD thickening, filtering, over-storage in existing ponds, or backfill.

Existing residues have potential economic value using a polymetallic view (not only copper). A big portion of the existing residues deposited and future residues should be possible to retreat profitably by Secondary mining.

Abandoned pits and underground mines could store a big portion of retreated residues if the government regulations change to permit that easily.

Recently, experimental metallurgical studies permit achievement recoveries of over 70 per cent in all the valuable elements inside residues if they have good commercial grades. It is remarkable the values obtained with rare earths.

This paper shows the statistics of Chilean mining residues, the forecast for the next decades, and alternatives to retreat and restore.

In general, Chile needs strong mining investment to process mining residues profitably.

INTRODUCTION

Chilean mines have generated over 50 000 Mt of waste materials to date. Most of these are tailings from concentrators, large quantities of waste from open pits, and lixiviated ores and slags from smelters.

Waste is directly related to the ore treatment in the largest mining companies in Chile because 1 ton of ore produces 2–3 tons of waste. The orebodies for the main Chilean mines normally have an expected life of 100–250 years, and the throughput is between 80.000 to 250.000 ton/day, as shown in Figure 1.

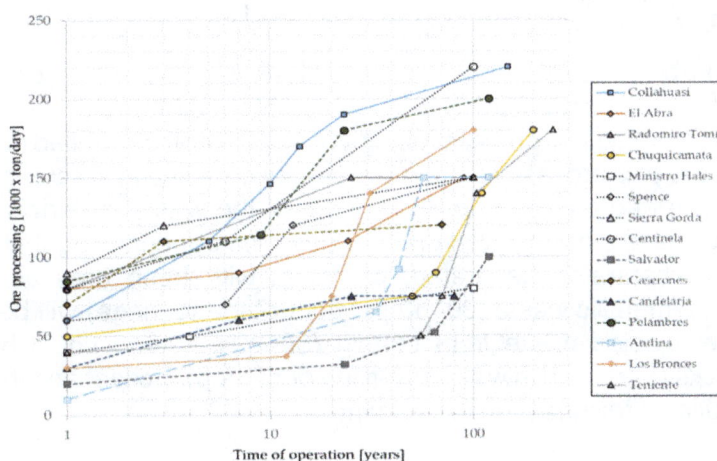

FIG 1 – Chilean mines operation rates and expectancy life.

Most tailings were deposited in the north zone of Chile (60 per cent) in isolated and desertic areas. However, a huge amount is still produced in the central zone of Chile (40 per cent), affecting communities and agricultural areas. Minor quantities had (less than 1 per cent) flowed to the sea, creating coastal beaches of tailings.

The environmental impact is high because the current mining waste production is around 5 Mt/day. Considering the increasing rate from 3 to 5 per cent of waste per annum due to new mines and expansion caused by low-grade ore metal.

The Chilean Agency for Mining (SERNAGEOMIN) reported 757 tailings dams from copper mines in which 26 300 Mt of tailings are authorised to be deposited. JRI Engineering estimated that more than 20 per cent of additional tailings are deposited (very old) or from other kinds of ores.

At the end of this decade, Chile will have to manage the following wastes productions:

- Mine waste (open pits): 2.5–3.5 Mt/day.

- Tailings from concentrators: 2.0–2.5 Mt/day.

- Waste of leached ores lixiviated: 1.0–1.2 Mt/day.

- Slages for smelter: 0.01–0.02 Mt/day.

Current operations and future mining projects have permits to dispose of waste in traditional methodologies until the next decade or, at most, the 2040s. As a result, much of the existing or future waste will need to be disposed of using different and innovative schemes.

Chile's options for managing mining waste in the near future are outlined in this paper.

MINING RESIDUES MANAGEMENT

Chile is moving towards secondary polymetallic mining to recover valuable metals in existing and future mining residues. Government policy is being changed to facilitate removing and retracting all mining residues with fewer restrictions. As a result, mining companies expect the current number of permits for new projects (500–600) to be reduced to less than 200 for secondary mining.

In general, the characteristics and properties of mine waste are variable:

- Slags deposits (old and new) from smelters have high-grades of Cu (0.5 to 1.2 per cent Cut) and other valuable metals. A big amount is removed with loader and trucks for grinding and flotation. It is feasible that 90 per cent or more of the slags will be retraited, and its residues (90–95 per cent) will be fine tailings that normally, as blended, will be primary tailings ore deposited separately (underground mines/small deposits).

- There are two types of hydrometallurgical waste: *old rubble* produced from leaching after grinding and *new rubble* deposited in layers after crushing and leaching. Both types of tailings contain minerals not recovered by leaching and could be recovered by grinding and flotation after pH adjustments. It is estimated that more than 50 per cent of the waste from leaching could be reprocessed to produce additional tailings. After reprocessing the residues, this secondary mining will produce 96 to 98 per cent of tailings.

- Mining waste can be divided into two groups of similar tonnage in common: earth without metals and waste with metals, usually between 0.05 per cent to 0.25 per cent Cut, plus other valuable metals. A selection process using X-ray technology, called *sorting,* is used after removing the stocks and processing them by primary crushing and screening. This technology allows the selection and retrait 10 to 30 per cent of the waste materials with a higher metal content, which will have a commercial grade (above 0.3 per cent Cut) that could be processed in the existing concentrator, separated, or mixed with fresh ores as shown in Figure 2. Consequently, non-metallic waste (30–50 per cent of the tonnage) will be retained as a ROM deposit, and the rejected waste from sorting (30–40 per cent) could be managed by belt conveyors or trucks to be used in wall earth-filled or other purposes, and the low-grade process will produce additional tailings (20–40 per cent).

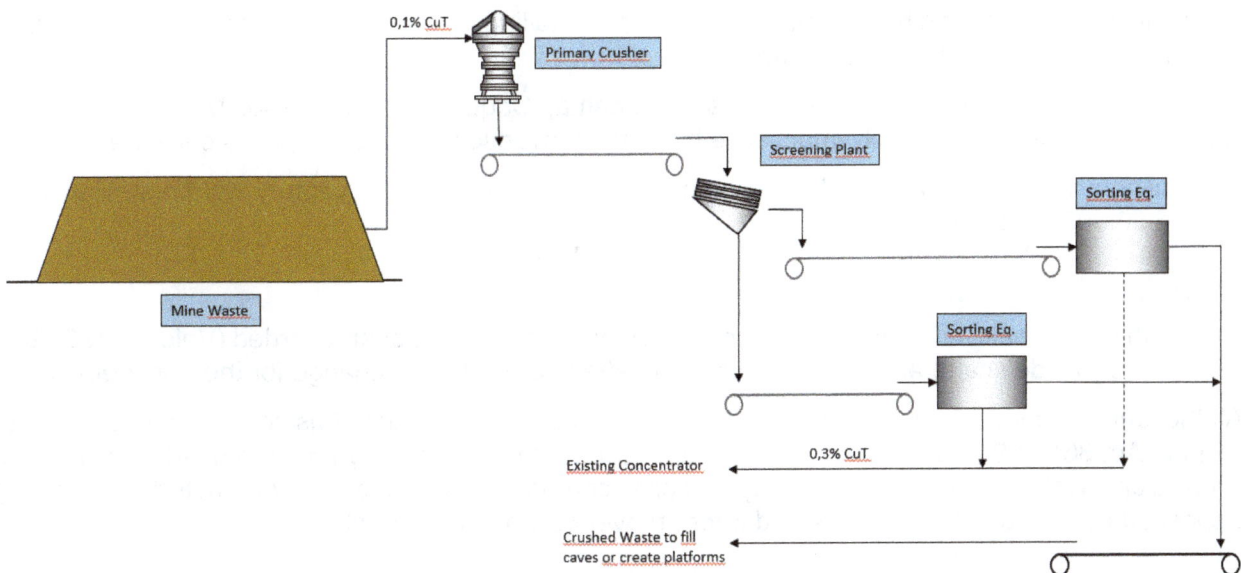

FIG 2 – Waste retreatment by sorting.

The main problem is how to deposit tailings in Chile because around 50 per cent of the existing and future residue non-tailings will be transformed into tailings by Secondary Mining. Figure 3 shows the increasing amount of deposited tailings over the decades.

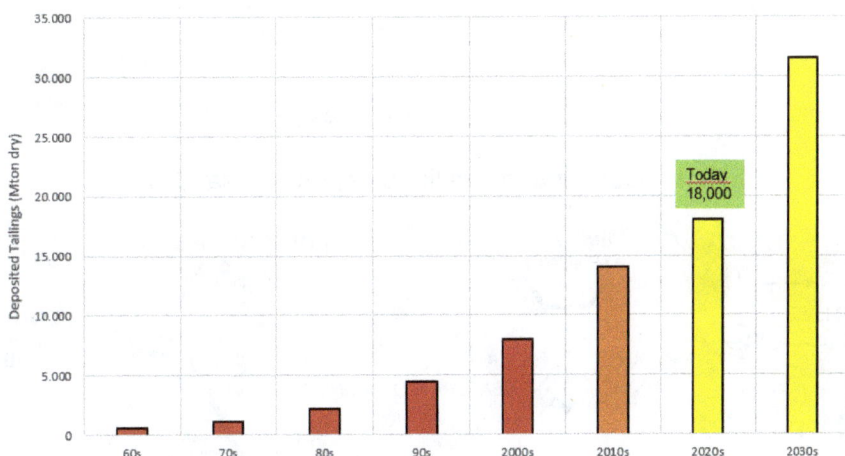

FIG 3 – Cumulated tailings deposited in Chile.

As shown in Table 1, Chile's challenge is depositing the following quantities of materials.

TABLE 1

Total of tailings to be deposited.

Description	Primary tailings	Secondary tailings	Total
Existing tailings (Mt)	18 000	7000	25 000
Future tailings (to the end of the Century)	57 000	20 000	77 000
		Total to be deposited	100 000

Consequently, Chile needs to define how to store 100 000 Mt until the end of the century.

LOCALISATION

Chilean tailings storage systems in the north zone differ from those in the central-north zone. The first zones correspond to the states of Arica-Parinacota, Tarapaca, and Antofagasta, where the conventional deposits can be maintained. Additionally, tailings dams are accepted by the authorities and the population for at least a few decades.

The other mining states in the country are Atacama, Coquimbo, Valparaiso, Metropolitan, and O'Higgins, where it is practically impossible that new conventional tailings deposits could be erected.

The tailings expected to accumulate in both zones by the end of the century will be similar, with around 50 000 Mt each.

CONVENTIONAL DAMS

It is worth noting that Chile registered the most powerful earthquake ever recorded (Valdivia 1960/9.5 Richter magnitude scale) and has been severely affected by climate change for the past decade.

All the Chilean conventional dams have been designed and constructed using safe methodologies by law (Act 86/1969 and updates). Since then, tailings dam technology has improved to build safe down-slope walls of sand recovered by cyclons, compacted by heaps or compacted earth-filled, deposited by trucks. See Figures 4 and 5 for an overview of this technology.

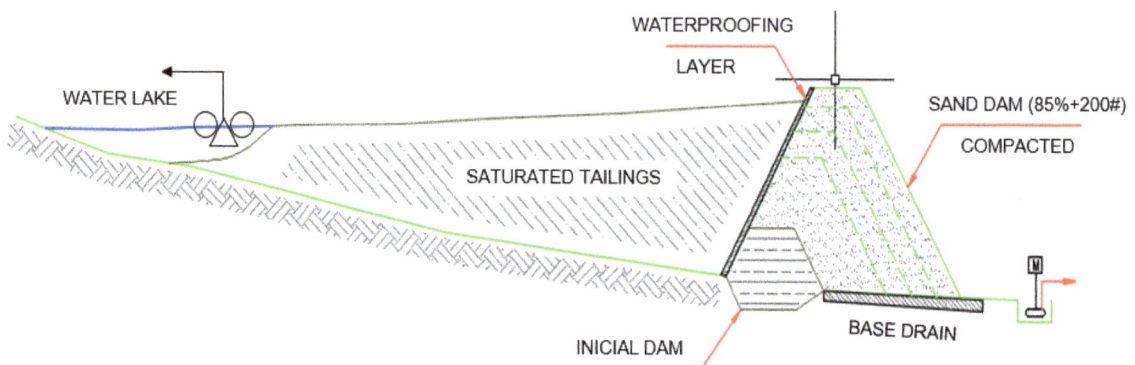

FIG 4 – Conventional tailings deposition sand wall.

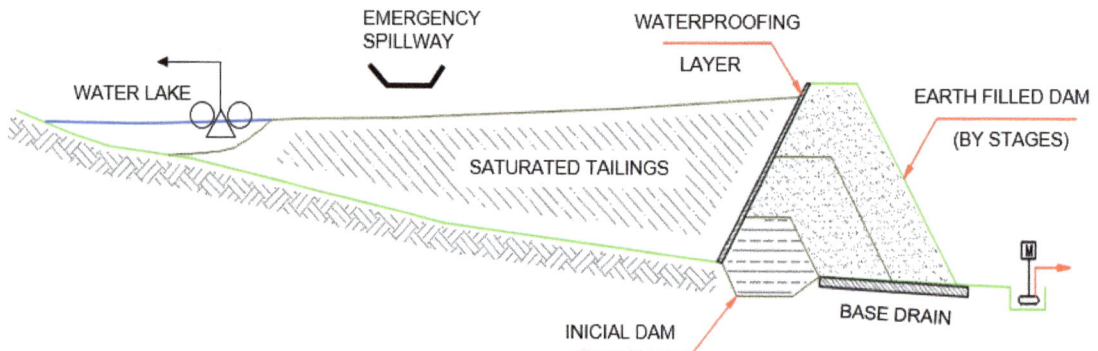

FIG 5 – Conventional tailings deposition earth-filled wall.

In general, the Chilean experience with conventional dams suggests that they are designed and operated according to the best engineering practices in the world. The risks of collapse could never be considered solved or negligible for the following reasons:

- Earthquakes could be stronger than the design. Geotechnical information could have weaknesses, or construction could have undetected failures.

- Severe storms could be higher than expected by the uncertainties of Climate Change, and the emergency spillways may not be safe for such flood events.

- The trend is to increase the height of the walls to store more tailings in the same place to over 300 m. The geotechnical performance of this huge volume of materials could be unknown.

- People living near tailings dams don't like them, mainly because they produce dust, contaminate groundwaters, and the risks of collapse is notable after recent dam failures in Brazil, Canada, and China.

Therefore, the expositor considers that conventional dam technology should be abandoned before the 1950s.

Statistics of failures of conventional tailing in the world show that the causes are multiple, as shown in Figure 6.

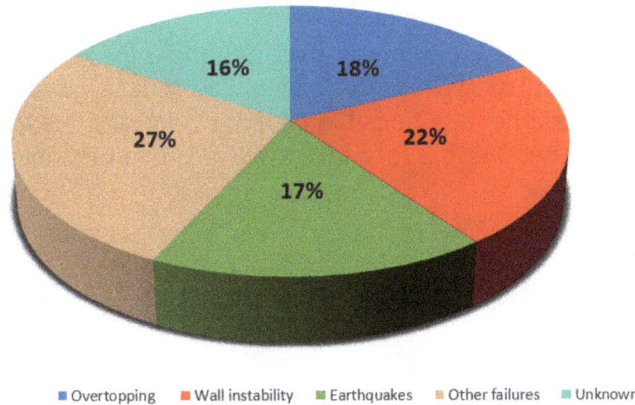

■ Overtopping ■ Wall instability ■ Earthquakes ■ Other failures ■ Unknown

FIG 6 – Failure mode of conventional tailings deposits.

ALTERNATIVES TO CONVENTIONAL DEPOSITS

Chile started the evaluation of six alternatives for the management of existing tailings to be removed and future tailings (primary or secondary):

1. Deep-sea tailings disposal (DSTD)

2. High-thickened tailings disposal: over 65 per cent of solids.

3. Filtered tailings: less than 18 per cent humidity.

4. Consolidated tailing: cemented or geopolymer.

5. Backfilled mining cavities: pits, caves, or dumps.

6. Backfill simultaneous with an underground operation: WLM technology.

Each alternative is analysed as follows.

Deep sea tailings disposal (DSTD)

The Chilean coast is well suited to use DSTD because of an oceanic trench called *Fosa de Atacama*, which is over 8000 m deep at 150 km off the coast.

Fosa de Atacama could store the entire tailings production without affecting the environment and commercial fishing if the minimum depth to the deposit is over 300 m. A scheme of this technology is shown in Figure 7.

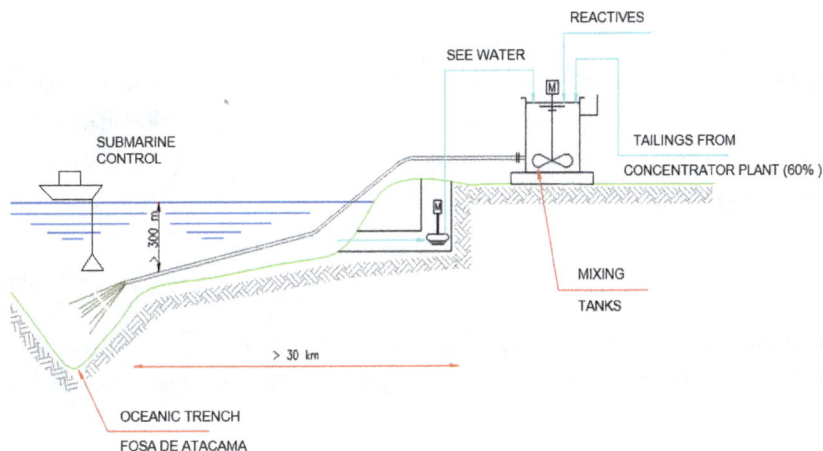

FIG 7 – Deep-sea tailings disposal (DSTD).

The author believes this is the best long-term environmental solution for Chilean mining. Still, all joint efforts by the mining industry to develop DSTD technology have been stopped by political pressure. Most Chilean believe that the tailings would contaminate the sea, and consequently, it is not a feasible option.

High thickened tailing disposal

High concentrations slurries have a yield stress over 30 or 50 Pa, and the tailings are deposited relatively flat using spigotling technology. The deposit could produce beaches with an average slope of 2 to 5 per cent. In such situations, it is possible to reduce water consumption and use flat areas with low to highwalls. See Figures 8 and 9 for a description of this technology.

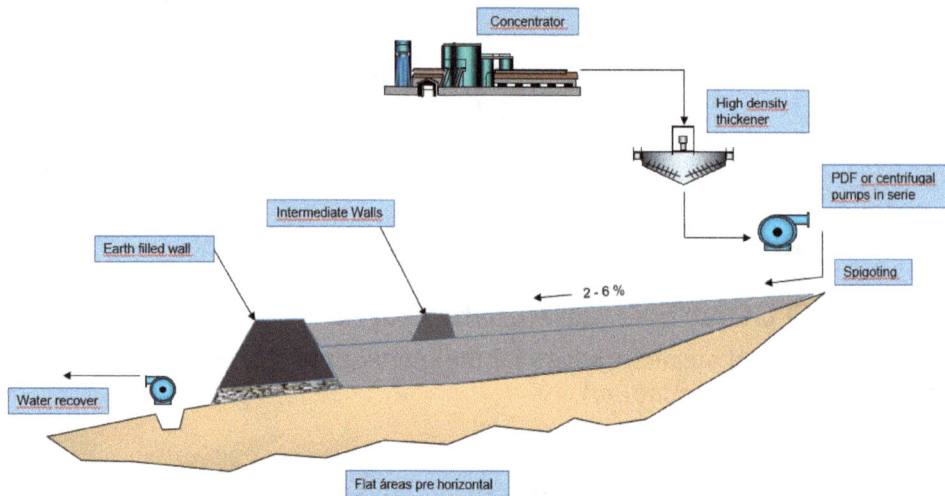

FIG 8 – High thickener tailing.

FIG 9 – Laboratory test at CIMS-JRI: Tf ~180 Pa: paste with a vertical limit of penetration on the rock fill wall.

Chile has a few cases of high thickened tailing with low throughput (below 5 Kt/d): one successful case in the great mining (Esperanza mine: 105 Kt/d), and other cases in progress.

Unfortunately, this option is only suitable in the northern zone of Chile, in the middle of the Atacama desert. It is practically impossible to use this technology in the Central Zone because it requires large semi-flat areas.

Filtered tailings

Filtered slurries is an old technology, mostly associated with relatively low throughput of tailings, with exceptions where the water recovered from the slurry contains dissolved metals that can be recover profitably, in most cases in Africa.

The market currently offers a large filter capacity that handles more than 8000 t/d per equipment.

The problem of how to deposit filtered tailings is more complex. One solution is to deposit filtered tailings on top of an existing conventional tailings dam, as shown in Figure 10. The other alternative is to create a new deposit by compacting at least one-third of the material, as shown in Figure 11.

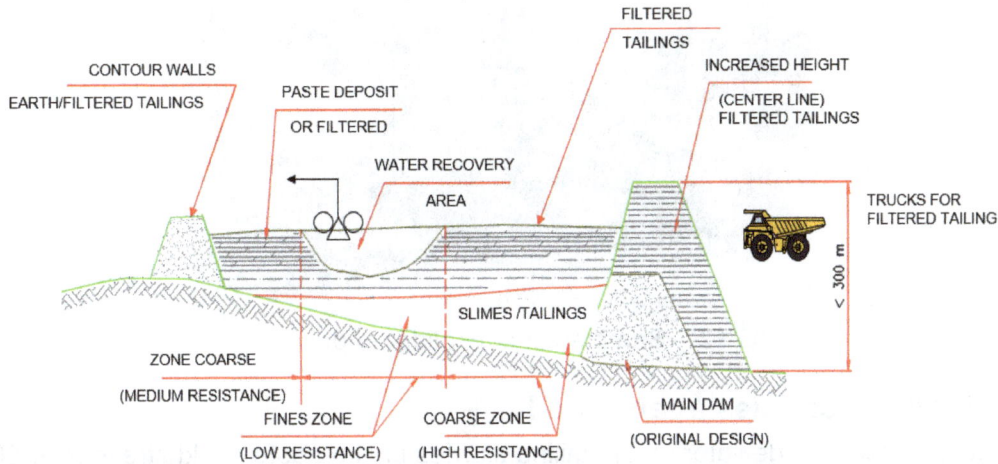

FIG 10 – Rising (super-elevation) of existing tailings dam walls – filtered tailings.

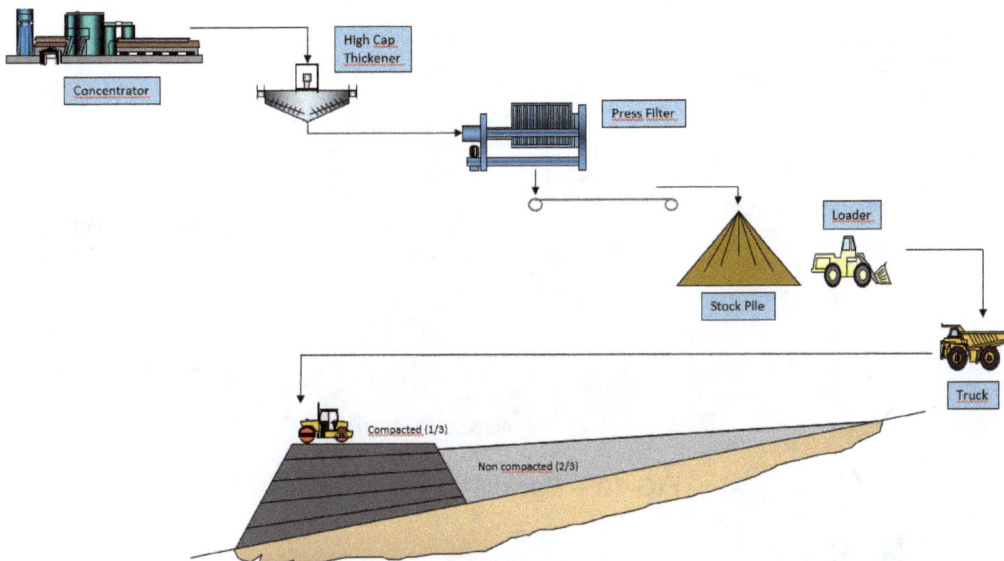

FIG 11 – Filtered tailings deposition.

Filtering large volumes of tailings is expensive, in terms of Capex and Opex, but reduces the risk of catastrophic collapse. Still, in the long-term, this type of deposit has the risk of dust and acid drainage contamination.

Consolidated tailings

Tailings can be consolidated until a mechanical resistance of 1 MPa. Then, they could be used for various purposes: backfill underground mining, a platform for roads or industrial facilities, and anchors for river and coastal defenses.

Tailings could be consolidated in two ways:

1. High concentration of slurries mixed with cement (6 to 10 per cent). This technology has been used in small mines for a long time but is expensive in Capex and Opex. However, this technique eliminates the problem of tailings.

2. Geopolymers, consisting of tailings mixed with dry materials and temperature. This is a technology R&D with a level of development TRL 5–6 uses dry material plus temperature and cure time. This methodology makes it possible to achieve resistances between 1 to 10 MPa of the tailings, almost as rocks. In the Research Center CIMS-JRI, with a grant from the Chilean

Government, it was possible to create 'rocks' from tailings, some than nature do with 'sedimentary rocks.' See the photography in Figure 12.

FIG 12 – Samples of consolidated polymers.

Backfill of mining cavities abandoned

This alternative consists of dewatering the tailing slurries until a paste (yield stress over 50 Pa) and filling mine cavities as pits, caves of old/existing underground mines, or the space between rocks in old damp. See Figures 13 and 14 for a description of this technology.

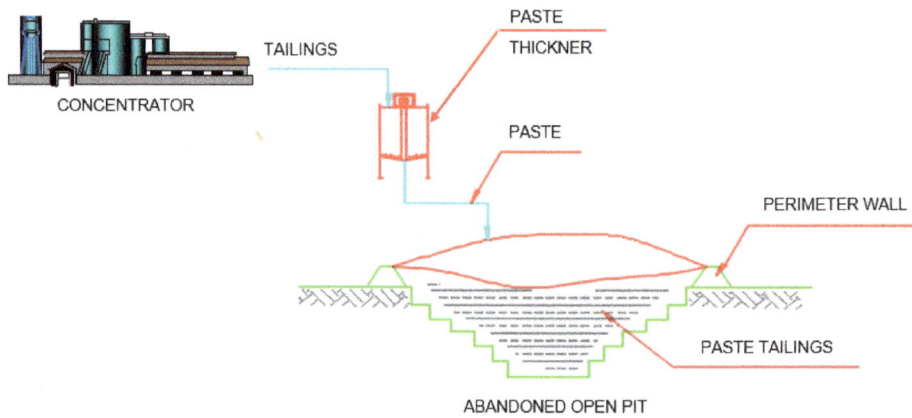

FIG 13 – Backfill open pits.

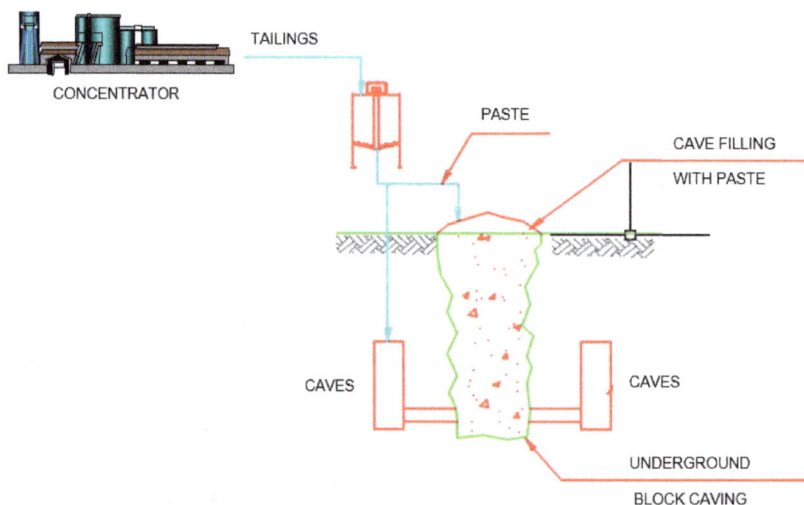

FIG 14 – Backfill underground mine.

This option has been tested in Spain at 2 Kt/d in an open pit mine and the Middle East.

This technology had been successfully used in small mines, but no large pit has been filled, or a big underground mine has been partially filled.

The main problem with this technology is moving the material long distances and establishing the distribution around the pit perimeter or inside the underground mine.

There is a big chance to use abandoned pits and underground mines in Chile, but the available volume is low.

Backfill simultaneous with underground operations

JRI was commissioned by a mining company to develop mathematical models and laboratory/pilot tests to demonstrate that it is possible to operate a large mine using the block or panel caving method while simultaneously depositing pastes over the cave created by the mining operation. This method has been patented in several countries (license number 62.272 Chilean Government).

The process is shown in Figures 15 to 20.

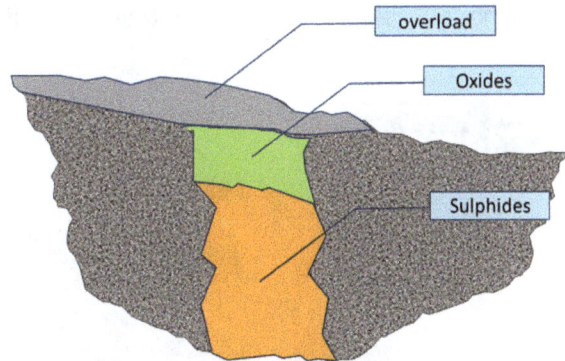

FIG 15 – WLM system – typical orebody.

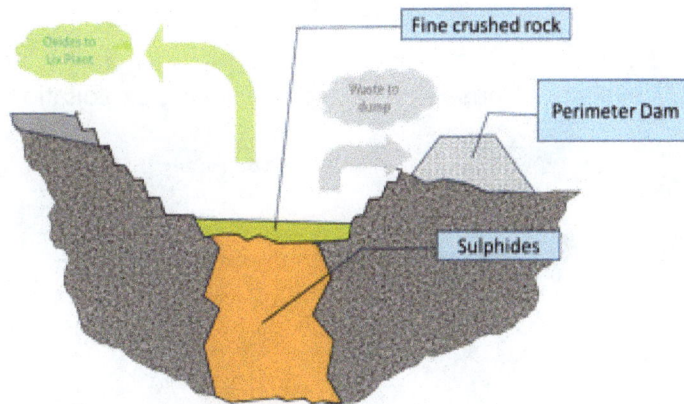

FIG 16 – WLM system – Phase 1: pre-stripping and/or oxides open pit.

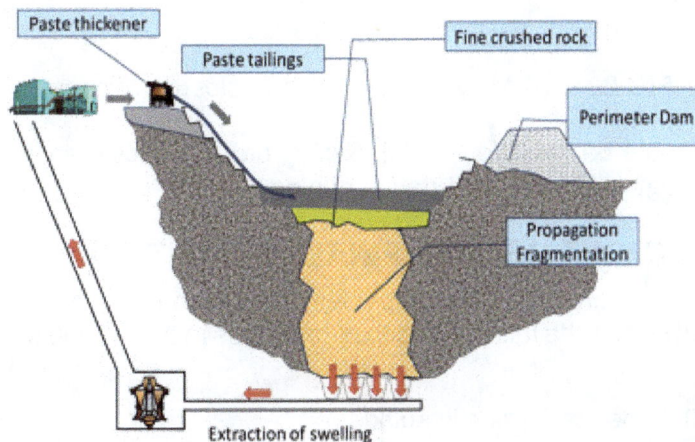

FIG 17 – WLM system – Phase 2: block caving initial stage (propagation of cave back).

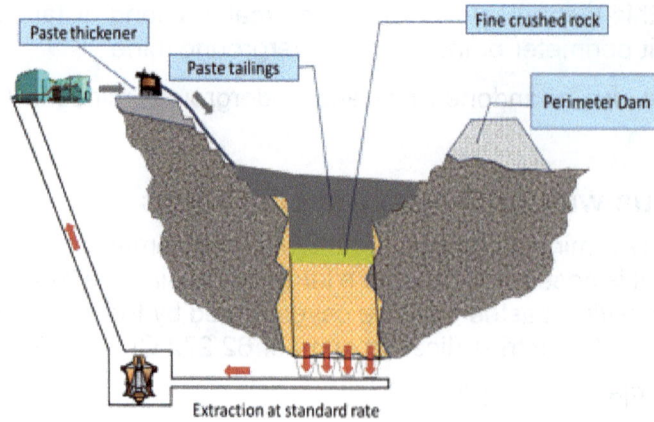

FIG 18 – WLM system – Phase 3: block caving exploitation of sulfides (standard rate).

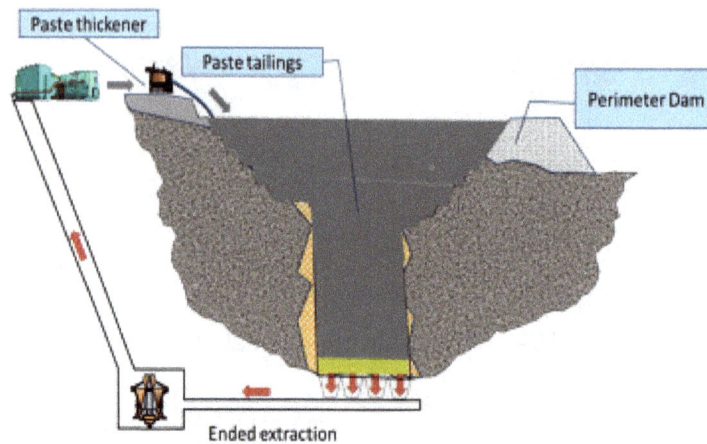

FIG 19 – WLM system – Phase 4: end of block caving exploitation of sulfides.

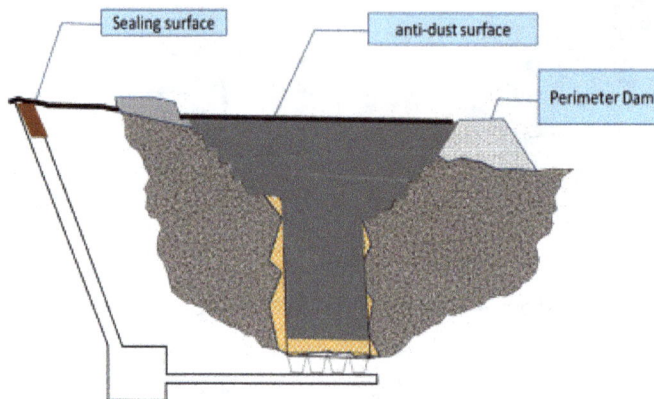

FIG 20 – WLM system – Phase 5: closure of mine.

Today this technology could be considered TRL 6 or 7, but it is necessary to develop an industrial demo case at a low throughput of 3 to 5 Kt/day of mining.

TAILINGS RETREATMENT BEFORE FINAL DEPOSITION

Most of the Chilean tailings contain valuable species in addition to copper. A summary of the data collected and processed shows the following ranges of grades in the 18 000 Mt of currently deposited tailing.

- Copper: 0.15 to 0.35 per cent (mostly sulfides).
- Molybdenum: 50 to 250 ppm.
- Iron Magnetic: 0.5 to 8.0 per cent.

- Rare earth: 50 to 200 ppm.

Proven industrial technologies can recover over 70 per cent of Cu, Mo, and Fe.

Rare earth recovery technology has been developed in Chile by a joint venture between JRI (private) and Ecometales (public), which has been analysing tailings for more than six years to recover rare earth. The process was successful, and it is possible to recover more than 60 per cent of the rare earth with several techniques in series. See the flow diagram in Figure 21.

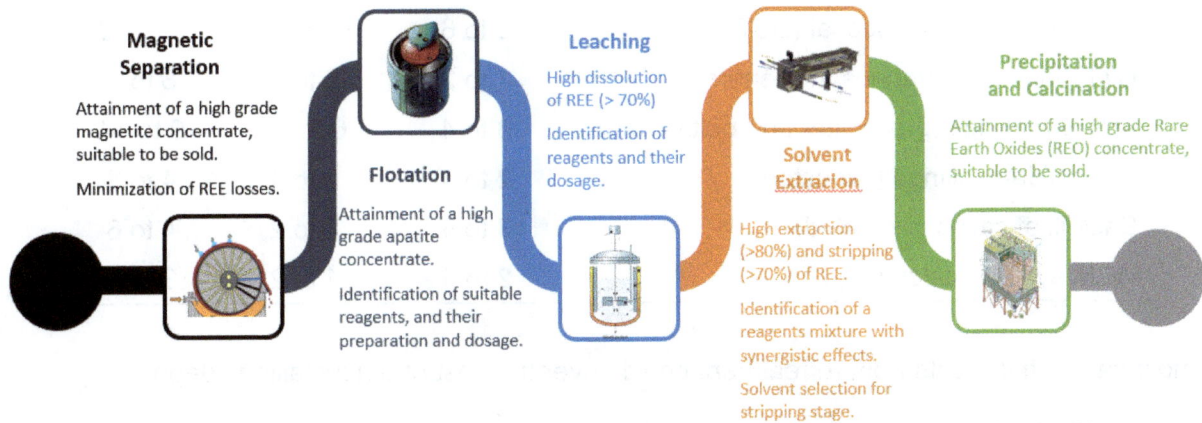

FIG 21 – Tailings reprocessing to recover Cu/Fe/Mo/Reo.

Today rare earth process is in TRL 4–5, and JRI is seeking funding to build a 100 t/d demonstration plant to validate the technology and to find investors.

ECONOMICAL ANALYSIS FOR RETREATED TAILINGS

The expected economic values for an old tailings deposit to be removed and reprocessed are clear:

- Value from copper content: 5 to 10 US$/ton

  (5 p/t × 70% rec × 50% margin, 4US$/p)

- Molybdenum and iron: 1 to 3 US$/ton

  (normally 20–30% of the copper)

- Rare earth: 7 to 25 US$/ton

  (200 ppm × 70% rec × 200 US$/kg × 50% margin (±50%)

It should be noted that the rare earth process is at the R&D level TRL 5.

ECONOMICAL ANALYSES TO DEPOSIT TAILINGS

Deposit costs depend on the throughput and location of the mines and the tailings. As shown in Table 2, the cost is expressed in unit Capex, which is the total nominal Capex divided by the maximum storage capacity. In general, Opex includes energy, labor, controls, pipes replacement, equipment, oil, etc. Both parameters include transportation, dewatering, distribution, civil works, water reclamation, and controls.

TABLE 2

Estimation of Capex, Opex and Total Costs.

Technology	Unit Capex (US$/Ton)	Opex (US$/Ton)	Total (US$/Ton)
Conventional deposits (cyclone sands)	1 to 3	3 to 5	4 to 8
Conventional deposits (Earth compacted)	2 to 8	2 to 3	4 to 11
Deep sea tailings disposal (DSTD)	2 to 6	2 to 4	4 to 10
High-thickened (various thickeners)	1 to 2	2 to 4	3 to 6
Filtered tailings (press filter and belt/trucks)	3 to 4	6 to 10	9 to 14
Consolidate tailings (expected)	2 to 4	2 to 4	4 to 8
Backfill of caves (expected)	3 to 4	1 to 2	4 to 6
WLM system (calculate)	2 to 4	1 to 2	3 to 6

In most cases, the profits from retreatment could cover the cost of a new tailings deposit.

CONCLUSIONS

The following can be concluded:

- Chile faces the challenge of depositing around 100 billion tons of tailings by the end of the century.

- Many old tailings could be mined at a reasonable profit to recover copper, molybdenum, iron, and rare earths.

- Conventional tailings using compacted earth or cyclone sand are limited in Chile (20 to 30 per cent), and they are located in the country's north zone.

- Most tailings should be deposited with new techniques (high-density thickening or filtering), probably 20 to 30 per cent.

- Tailings should be deposited in abandoned pits or underground caves, probably 20 to 30 per cent.

- An additional percentage, 20 to 30 per cent, must be transported from the mines as consolidated materials (cemented or geopolymers) to use in civil works.

- It is necessary to continue doing R&D in rare earth recovery technologies and consolidated tailings. Both have a good chance of success.

Dam safety and efficient tailings dewatering through enhanced and adapted drainage solutions

P Saunier[1], S Fourmont[2] and A Warwick[3]

1. Business Development Manager NA and Pacific, Afitex-Texel Geosynthetics Inc, Sainte-Marie QC, Canada. Email: psaunier@afitextexel.com
2. Business Development Manager Canada, Afitex-Texel Geosynthetics Inc, Sainte-Marie QC, Canada. Email: sfourmont@afitextexel.com
3. Product Manager, Global Synthetics Ltd, Virginia Qld 4014. Email: andy@globalsynthetics.com.au

ABSTRACT

Over the past years, dam's breach disasters, fortunately rare, have been investigated revealing that, in many cases, the lack of efficient drainage was one of the main causes of these failures. The accumulation of water into the tailings dam or at the base of them is critical situation engineers and operators want to stay away from to ensure the safety of the work. As with an adequately thick homogeneous layer covering the entire base or at interlifts of the dam, or when being constructed in French drains as finger drains directly placed into the structure during construction, it is not always feasible to set-up a granular drainage system. The granular material can be unavailable or expensive on the site and therefore might make its use impossible or financially irrelevant. This is where geosynthetic solutions offer a valuable alternative. However, two problems arise:

- The extreme loads to which these materials are exposed to, which for most of them will lose their drainage properties because of the creep effect.

- The large quantity of very fine particles in the material itself, which can lead to rapid clogging of the geotextile fibres and thus lead to a complete loss of its permeability, and by extension their efficiency.

The particularity of Multi-Linear Drainage Geocomposites (MDLG) is that they resist particularly well to the load, even extreme, once they are confined, and their manufacturing process allows the selection of geotextiles previously tested to be compatible with the materials placed in contact with them. Therefore, creep and clogging are no longer an issue and those manufactured products offer a stable cost year after year, making easier investment management.

This paper presents, through case studies, the different applications where these materials are particularly efficient, the strategies implemented for their characterisation, and the techniques implemented for their efficient installation and use.

INCREASED REGULATION OVER THE YEARS

History

Since the 18th century, when we can consider the mining industry started in Canada, the regulation never stopped getting more and more drastic, especially to ensure the security of works and people as high as possible. This tendency is even more important since the beginning of the 21st century were major concerns such as global warming and the protection of the environment is at the centre of the political and social concerns. Reducing at a minimum the potential impact of mining operations on the environment led the entire industry to adjust its way of operating and closing their operational sites. Limiting the potential infiltration of processed water into the soil, through a better drainage below a lined pond or a dam, reducing the amount of ore oxidation or leachate generation by building a look proofed cap during a final closure, or reducing the water pressure into the dam for an increased stability and therefore safety during construction and operation.

Importance of drainage

One key factor for safe and environmentally respectful designs is having efficient drainage. This is true for an increased safety in the works where water can be the most challenging element that needs to be managed, otherwise disasters may occur, but also in ponds or final remediation simply because a liner or low permeability layer is not efficient without a good drainage. For decades, gravel was used to offer high permeability layers, acting as drainage elements, and located at the Dam's base, below pond, or for the final cover. The two biggest issues granular materials present are:

- They tend to be rare, and therefore more and more expensive to use.

- They generate a heavy quantity of GEG, and especially when the pit is located far away from the construction site.

This is why, since several decades, geosynthetics are more and more used in replacement of those granular materials.

MULTILINEAR DRAINAGE GEOCOMPOSITE TECHNOLOGY

Description of the technology

The use of geomembranes in mining applications has been widely documented. However, geocomposite compatibility studies with mined material are scarce and very limited information is available.

The Multi-Linear Drainage Geocomposites (MLDG) used in this study is developed by Afitex-Texel and called DRAINTUBE®. It is composed of (Figure 1):

- A nonwoven polypropylene or polyester geotextile acting as a filter.

- A series of corrugated polypropylene mini-pipes spaced at regular intervals (from 0.25 m to 2 m on centres). These perforated mini-pipes provide most of the drainage capability of the product.

- A nonwoven thick polypropylene or polyester geotextile acting as the drainage medium and as a cushion to protect the underlying geomembrane.

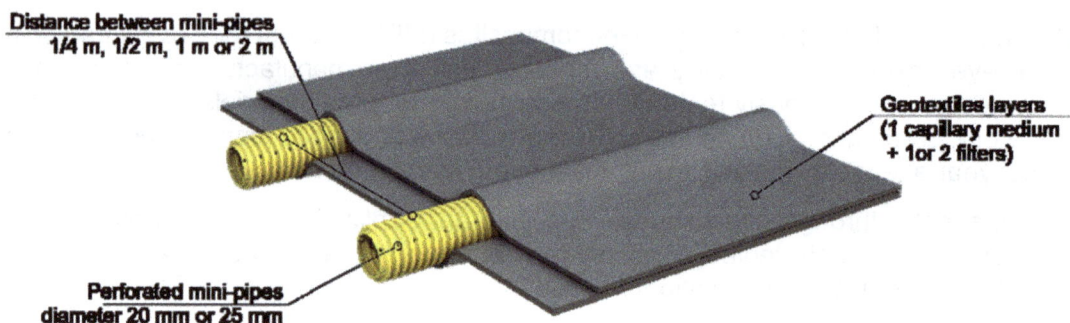

FIG 1 – DRAINTUBE® Multi-Linear Drainage Geocomposite.

Filtration applications with mine residues may be among the most challenging filtration applications. First, the high seepage forces and suspended particles that must be filtered can lead to clogging. Second, leachate is typically a highly loaded solution and mineralisation can lead to chemical clogging (Faure *et al*, 2006; Fourie *et al*, 2009; Legge *et al*, 2009). Although it is likely that a clogging problem would also occur with mineral drainage systems (such as gravels, see Giroud, 1996). In order to check if MLDG are able to fulfil the function of a drainage and dewatering layers, long-term hydraulic properties, soil retention, and chemical resistance must be evaluated. Results of experimental studies aiming at checking these points are presented in the following sections.

Behaviour under high compressive load

With an ore density between 1.5 and 1.8, the compressive load on the drainage layer can reach 2 MPa (Thiel and Smith, 2004; Castillo *et al*, 2005). For traditional planar geocomposites involving a planar drainage core (such as biplanar or triplanar geonet), it has been shown by several authors that the hydraulic properties of these geosynthetics are adversely affected by such high compression stresses.

However, Saunier, Ragen and Blond (2010) have shown that the particular structure of Draintube MLDG is favourable to the development of an arching effect around the pipe. As a consequence, the transmissivity (the volumetric flow rate per unit width of specimen per unit gradient in a direction parallel to the plane of the specimen; see ASTM D4716 and GRI GC15 standards) is not affected by the compression stress, nor by time, as no creep can develop in the pipe. Their results are reported in Figure 2.

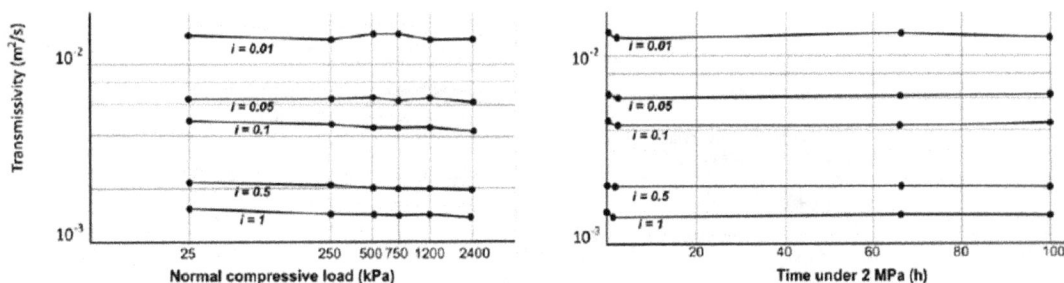

FIG 2 – Transmissivity under different loads up to 2400 kPa and 100 h (i = hydraulic gradient) (after Saunier, Ragen and Blond, 2010).

The lack of sensitivity of the product to compression loads up to 2400 kPa suggests that these observations are likely to be applicable to the high normal loads which are typically experienced in tailings ponds and dams (and can exceed 100 m in height).

The applied combined reduction factors (intrusion of the geotextile into the drainage core RF_{IN}, creep of the drainage core RF_{CR}, chemical clogging of the drainage core RF_{CC} and biological clogging of the drainage core RF_{BC}) for Draintube MLDG are at least half of those applied to standard geonet geocomposites (Maier and Fourmont, 2013). In other words, for the same index transmissivity, Draintube MLDG offers at least two times higher long-term flow capacity than a geonet geocomposite. The ASTM D7931-18 standard provides recommendations to determine the allowable flow rate of drainage geocomposites including MLDG.

COMMON APPLICATIONS IN THE MINING INDUSTRY

Ponds

Drainage under single and double lined ponds (Figure 3) is crucial to ensure their containment integrity. Indeed, the pressure developed by the high-water table or the accumulation of gas under the geomembrane creates swelling and irreversible degradations of the liner. The use of MLDG allows to protect the geomembrane against puncture (Blond *et al*, 2003), dissipate water or gas pressures and maintain the performance of the structure over time.

FIG 3 – Double lined produced water pond drained with a MLDG below the secondary liner (groundwater) and the primary liner (leak detection layer).

In double lined ponds, Draintube MDLG is also used between the two liners as leak detection layer. In order to control the integrity of the primary liner with geoelectrical methods (such as ARC test,

Water Puddle or Dipole), conductive Draintube MLDG, including an electrically conductive grid, have proven to be effective and are now approved by most inspection agencies (Figure 4).

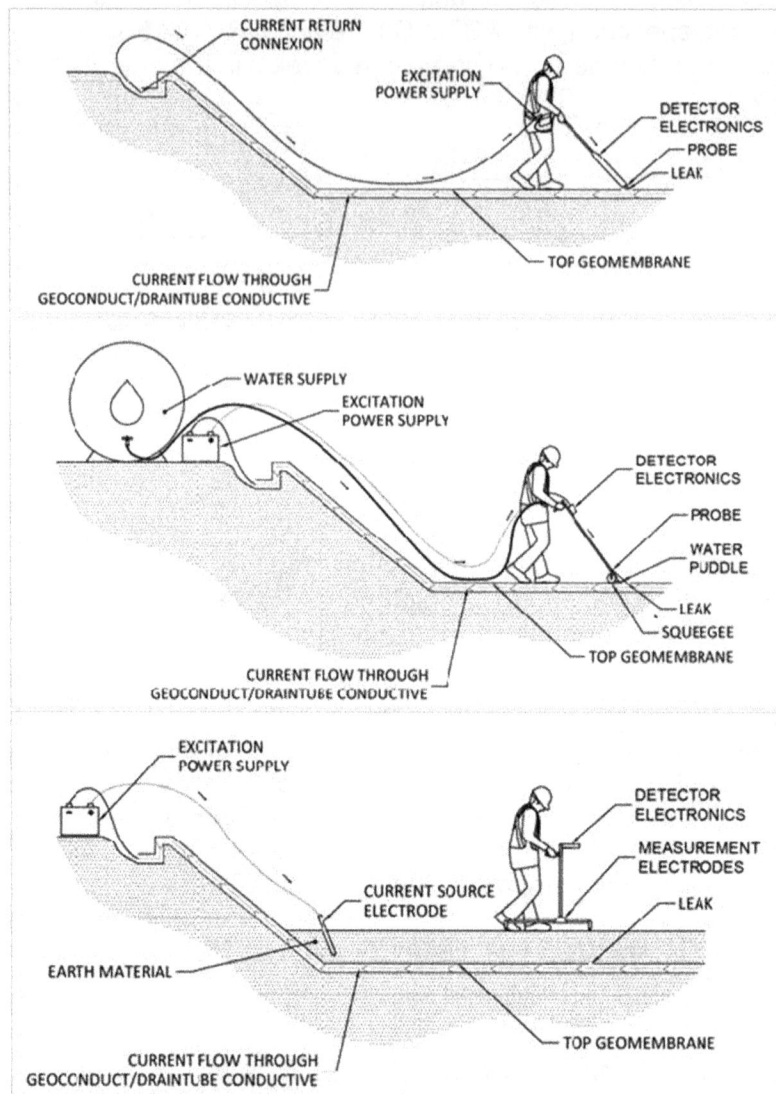

FIG 4 – Use of Conductive Draintube MLDG to run Leak Location Surveys (ARC test, Water Puddle and Dipole).

Final cover

Covers on mine waste disposal sites are required to minimise the infiltration of precipitation. Limiting water infiltration reduces the creation of leachates that can be problematic to manage afterwards, and in some cases, limiting air intrusion into the waste limits the oxidation of the tailings. Finally, the vegetative coverage of these structures allows the mine site to have a more neutral appearance that blends naturally into its environment (Figure 5).

FIG 5 – Mine Waste disposal site during and after construction of its final cover.

As per Del Greco *et al* (2012), Draintube MLDG evacuates the infiltrated water faster and with a higher flow rate than a 500 mm thick gravel drainage layer (coarse clean gravel ⅓ to 1½ inches). Figure 6 shows the typical cross-section of the two test pads, 10 m long and 4 m wide.

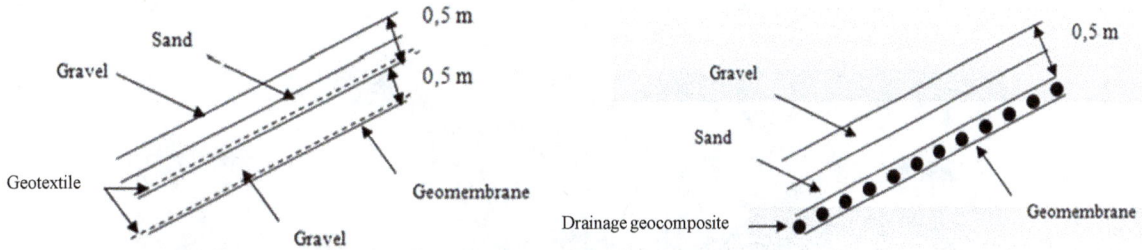

FIG 6 – Test pad description.

Flow rate was measured for two initial hydraulic conditions, dry (d) and partially saturated (ps). Regardless of the initial conditions, the Draintube MLDG had a faster response time than the gravel drainage layer, about 30 per cent faster.

Also, the Figure 7 shows the flow rate measured for the four configurations:

1. Draintube MLDG in dry conditions, MLDGd.

2. Draintube MLDG in partially saturated conditions, MLDGps.

3. Gravel layer in dry conditions, Gd.

4. Gravel layer in partially saturated conditions, Gps.

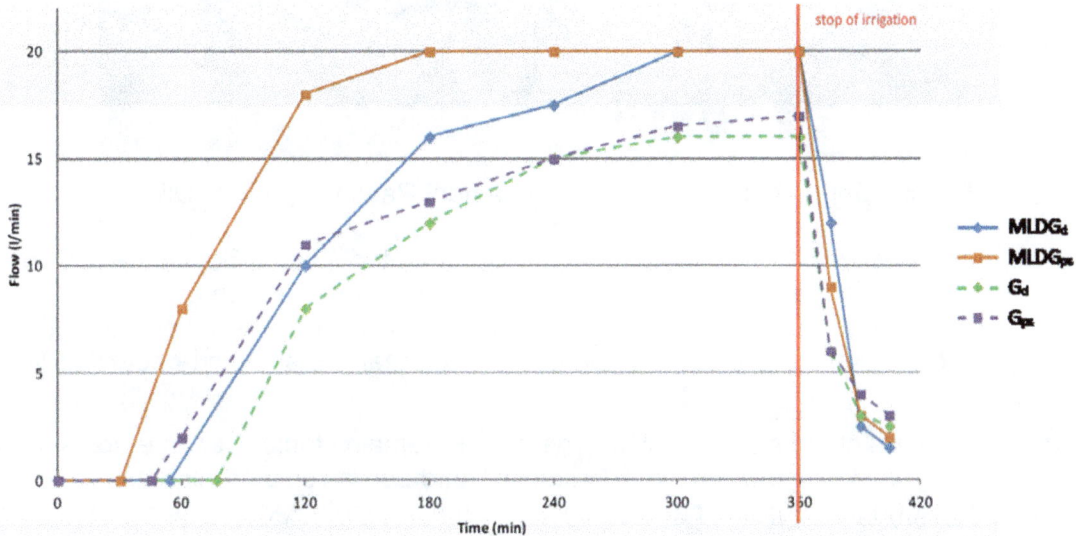

FIG 7 – Flow rate comparison over time between MLDG and gravel layer.

As shown in the Figure 7, the total amount of drained water was greater with Draintube MLDG. Indeed, the gravel drainage layer retained between 20 per cent and 25 per cent more water than the MLDG. The water remaining into the gravel layer may increase the infiltration rate through any defect in the geomembrane or though the low permeability layer.

The mini-pipes of the MLDG collect and evacuate the fluid in one given direction (the direction of the mini-pipes) even if the slope equals to zero. This directional aspect of the MLDG helps to reduce the impact of differential settlements that will occur on a cover and causes reverse slopes. A homogeneous drainage layer will drain in the direction of the reverse slope into the land subsidence whereas the MLDG will drain the water in the direction of its tubes to the collector trenches or ditches.

Tailings dewatering

The tailings deposition in a TSF leads to two major problems when saturated: they are unstable and they occupy a gigantic space, which increases their environmental footprint. It is therefore necessary to reduce their water content as much as possible. Granular drainage layers, if they are efficient, present the problem of increasing cost and reducing the storage capacity of the TSF or the pond. The geosynthetic solution is therefore a promising avenue, both financially and in terms of optimising the available volume, but only if the management of fine particles is properly addressed. The filtration compatibility of the geotextile filters and the tailings can be confirmed by running a Gradient Ratio test (Figure 8).

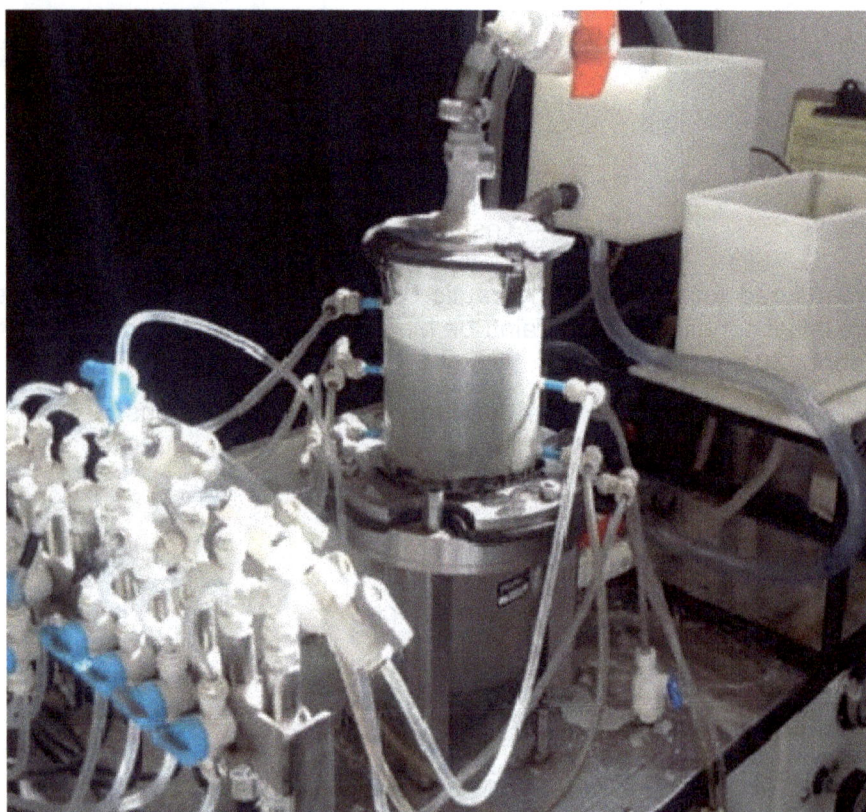

FIG 8 – Set-up of the filtration test (Gradient Ratio test, ASTM D5101).

A complete methodology has been developed and documented (Blond, Saunier and Dolez, 2018) for MLDG using a specific range of filters offering a Filtration Opening Size (FOS) ranged between 60 and 120 microns (as measured per CAN/CGSB 148.1 No. 10). The objective of the applied methodology is to determine, from the gradient ratio test, that the selected geotextile filter is not blocked/clogged.

The Permittivity values of the system (tailing + geotextile) remain stable over time and similar to the tailing permeability itself (Figure 9a). The Gradient Ratio curve remains also stable through the duration of the test (Figure 9b) and can be concluded that the selected filter is compatible with the filtered tailings and will behave in a proper manner against fast clogging or pipping.

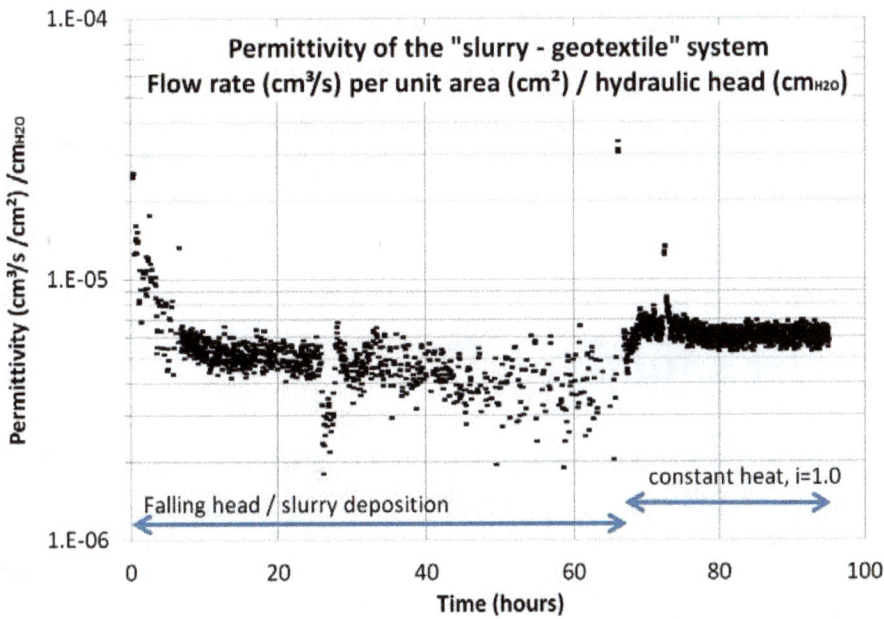

(a)

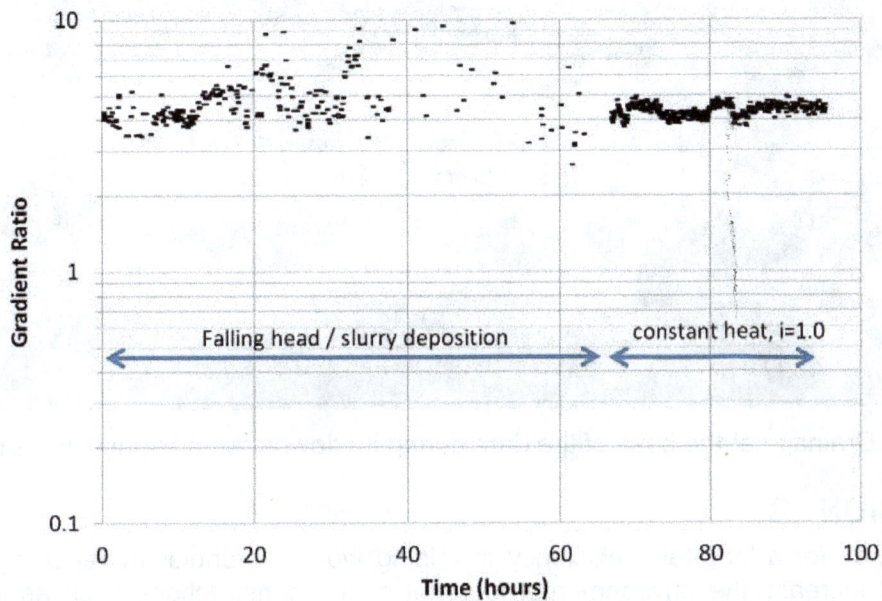

(b)

FIG 9 – (a) Permitivity versus time; (b) Gradient Ration versus time.

Base dam drainage

Recently, several dam failures have occurred in Canada and Brazil. After analysis of these disasters, it was found that the lack of drainage was one of the main factors. Downstream expansion is a very popular technique to increase the storage capacity of the dam. However, the risk of water seepage into the structure is high and it is therefore necessary to ensure that effective drainage is carried out at the base of the dam to release the interstitial pressures created.

Draintube MLDG are an advantageous alternative to granular materials because, like them, they are not sensitive to creep when confined and therefore offer a constant flow capacity in time, even under high loads (Figure 10).

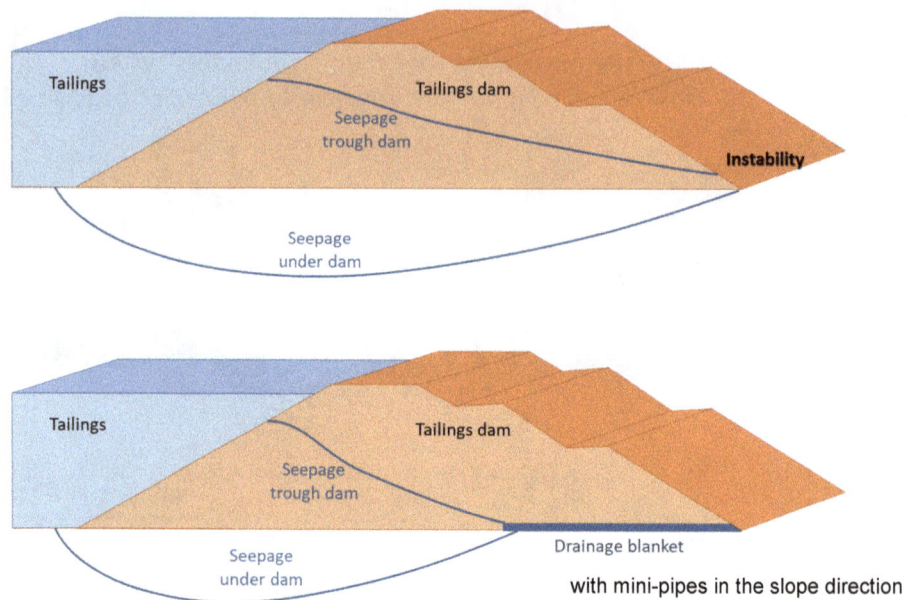

FIG 10 – Drainage at the base of the dam during its downstream expansion using a MLDG.

CONCLUSION

Drainage is key for a long-term efficiency in mining works. Granular material is no longer cost-effective and increase the environmental footprint of the constructions such as dam expansion, ponds, tailings dewatering and final closures. That is why geosynthetics are use more often to act as thin drainage layers.

Draintube Multi-Linear Drainage Geocomposites are one of the most efficient solutions, especially under high load, because they are not suffering from creep nor intrusion when confined into a soil matrix, because of their core structure. They also offer a great flow capacity, and because some of them are made of components with very well designed filters, they can stay away from fast clogging and therefore keep the works in good condition of use and performance. Finally, thank to their capability to easily incorporate conductive elements in their structure, they are offering an extremely performant solution for Leak Location Surveys on HDPE, LLDPE as well as BGM geomembranes.

REFERENCES

ASTM International, 2013. ASTM D4716 / D4716M-14, Standard Test Method for Determining the (In-plane) Flow Rate per Unit Width and Hydraulic Transmissivity of a Geosynthetic Using a Constant Head, ASTM International, West Conshohocken.

ASTM International, 2017. ASTM D5101, Standard Test Method for Measuring the Filtration Compatibility of Soil – Geotextile Systems, ASTM International, West Conshohocken.

ASTM International, 2018. ASTM D7931-18, Standard Guide for Specifying Drainage Geocomposites, ASTM International, West Conshohocken.

Blond, E, Bouthot, M, Vermeersch, O G and Mlynarek, J, 2003. Selection of protective cushions for geomembrane puncture protection, 56th Annual Canadian Geotechnical Conference, Canada.

Blond, E, Saunier, P and Dolez, P, 2018. Filtration of oil sands tailing slurries, *Geosynthetics Magazine*, October/November 2018, pp 13–20.

Canadian General Standards Board, 1994. CAN/CGSB 148.1 No 10-94, Methods of Testing Geosynthetics – Geotextiles – Filtration Opening Size. Standard by Canadian General Standards Board, Canada.

Castillo, J, Hallman, D, Byrne, P and Parra, D, 2005. Dynamic Analysis of Heap Leach Pad under High Phreatic Levels, Proceedings of 27th Mining Convention, Peru.

Del Greco, O, Fey, A, Oggeri, C and Fourmont, S, 2012. Behaviour of drainage geosynthetics for landfill capping systems, in *Proceedings of the 5th European Geosynthetics Congress*, Spain, 2:114–118.

Faure, Y H, Baudoin, A, Pierson, P and Plé, O, 2006. A Contribution for Predicting Geotextile Clogging during Filtration of Suspended Solids, *Geotextiles and Geomembranes*, 24:11–20.

Fourie, A B, Bouazza, A, Lupo, J and Abrao, P, 2009. Improving the Performance of Mining Infrastructure through the Judicious Use of Geosynthetics, in *Proceedings of 9th International Conference on Geosynthetics*, pp 193–219.

Geosynthetic Institute, 2017. GRI GC15, Determining the Flow Rate per Unit Width of Drainage Geocomposites with Discrete High Flow Component, Folsom, 9 p.

Giroud, J P, 1996. Granular Filters and Geotextile Filters, *Geofilters'96,* Canada, 1:565–680.

Legge, K R, Baxter, P, Meyer, P J, Sander, M G and Viljoen, D T V, 2009. Keynote Lecture: Geosynthetics for Africa, *Proceedings 1st African Regional Geosynthetics Conference*, pp 1–40.

Maier, T B and Fourmont, S, 2013. How Tubular Drainage Geocomposite Was Used in Landfill Final Cover, *Geosynthetics,* 31(3):48–51.

Saunier, P, Ragen, W and Blond, E, 2010. Assessment of the resistance of drain tube planar drainage geocomposites to high compressive loads, 9th International Conference on Geosynthetics, Brazil.

Thiel, R and Smith, M E, 2004. State of the Practice Review of Heap Leach Pad Design Issues, *Geotextiles and Geomembranes,* 22:555–568.

Development of coal tailings-based aggregates for road construction

A Khazaie[1], D Osborne[2], S Jahandari[3], M Rahme[4], A Harriman[5], Z Chen[6] and Z Tao[7]

1. PhD student, Centre for Infrastructure Engineering, Western Sydney University, Sydney NSW 2747. Email: a.khazaie@westernsydney.edu.au
2. FAusIMM(CP), Industry Advisor, Somerset International Australia Pty Ltd, Brisbane Qld 4000. Email: dosborne@somersetpty.com
3. PhD student, Centre for Infrastructure Engineering, Western Sydney University, Sydney NSW 2747. Email: S.Jahandari@westernsydney.edu.au
4. Managing Director, Nu-Rock Technology Pty Ltd, Sydney NSW 2137. Email: maroun.rahme@nu-rock.com
5. Project Manager, Altonx, Brisbane Qld 4000. Email: harri0437@gmail.com
6. Professor, School of Science, Western Sydney University, Sydney NSW 2753. Email: z.chen@westernsydney.edu.au
7. Professor, Centre for Infrastructure Engineering, Western Sydney University, Sydney NSW 2753. Email: z.tao@westernsydney.edu.au

ABSTRACT

The quantity of coal tailings (CT) currently discharged annually in Qld and NSW is estimated to be between 14 and 16 million dry tonnes, which is normally disposed in tailings storage facilities (TSF). Due to the continuous mining activities over the last 50 years, the existing tailings storage in Australia has reached an order of 500 million tonnes (Mt), which is emerging as a long-term serious threat to the surrounding environment (Katwal *et al*, 2022). Despite efforts have been made so far regarding the potential application of mine tailings as construction materials, very little progress has been achieved in the large-scale commercial application of tailings-based products. The reason can be attributed to the limitations of available technology in making cost-effective commercial products.

In this research, the manufacture of road-based aggregates from coal tailings has been investigated as a practical solution, which could not only reduce the volume of tailings that needs to be disposed of but also create a potential new revenue stream for coal producers toward effective dry disposal. It also fits well in the current plans for eliminating TSFs and returning process water for reuse. Dry solids disposal is potentially easier to justify but the quantities involved require other solutions for using the material where this can be shown to be commercially viable. In this regard, the bespoke pelletisation technology developed via a joint research program was used to make aggregates from high concentration solid cake by adding a small amount of cementitious or supplementary cementitious binders/chemicals. By optimising the pelletisation parameters such as the moisture content, rotation speed and tilting angle of the disc, coarse aggregates were successfully produced from coal tailings for the specific use as road-subbase materials. The surface treatment was also conducted to reduce water absorption and improve the durability of aggregates. The proposed solution could provide an economical solution for coal mining companies to address problems associated with coal tailings.

INTRODUCTION

Coal production is one of the main industries in Australia, contributing to the Australian economy and national energy supplies. Coal tailings are the mineral waste remaining after the processing. Many of the current coal tailings discharged contain clay minerals that have degraded to sub-micron sized particles affecting both dewatering and deposition. Since about 3–15 per cent of the coal must be managed as tailings, tailings management is very costly particularly when regulations of coal waste are to be met to minimise the environmental impacts. Recently, researchers have tried to develop new tailings management solution to reduce cost and improve environmental outcomes. Some cement manufacturers in China have included dried coal tailings in cement production (Qiu *et al*, 2011). However, only a small amount of tailings can be consumed in this process. Fly ash has been mixed with coal tailings and fired at above 1000°C to produce bricks and blocks in India, Turkey and China. Although this production can consume a large volume of coal tailings, the combustion of coal particles inside the bricks or blocks will lead to high porosity and a high water absorption rate. This has a negative influence on the long-term durability of the fired bricks or blocks (Lafhaj *et al*, 2008).

Meanwhile, the manufacture of fired bricks/blocks has high energy consumption and cost. Rose (1982) and Mansouri, Tahmooresi and Ebrahimi (2019) produced lightweight aggregates by using coal tailings alone or mixing with 30 per cent of clay or above. They found that using coal tailings as a raw material requires less than one-third of the amount of added fuel as is required when native clays or shales are sintered conventionally. But a high coal content in raw tailings would lead to strong bloating characteristics and low crushing strength of aggregates. In general, these studies proved the feasibility of manufacturing coal tailings-based lightweight aggregates and the use of artificial aggregates in structural concrete, pavement and masonry blocks. However, further research is required to study the influence of raw materials and manufacturing process, and to further optimise the product performance. Therefore, it was found that previous solutions have either only consumed a small amount of the tailings generated (eg coal-water slurry fuel and cement clinker), or have high energy consumption and cost for their reuse/repurposing (eg bricks). In this research, production of low-cost tailings-based aggregates from coal tailings was investigated as a feasible solution to overcome the environmental issues associated to tailings disposal and create a new revenue stream from coal tailings.

EXPERIMENTAL PROGRAM

Materials

For this research project, coal tailings were collected from a coal preparation plant near a mining site in Queensland, Australia. Cementitious materials including ordinary Portland cement (OPC), ground granulated blast furnace slag (GGBFS), lime, and low calcium fly ash (FA) were provided by local suppliers in Australia. All these materials were commercial products in powder form. Powdered sodium metasilicate nonahydrate used as the alkaline material was purchased from Sigma Aldrich, Australia. EPO100G® Glaze Epoxy was purchased from *All Purpose Coatings Pty Ltd* for the surface treatment of aggregates.

Mix design, sample preparation, and curing

To determine the optimum mix design for the production of coal tailings-based aggregates (CTBAs), various mix designs were developed in which coal tailings were replaced by different cementitious materials and sodium silicate as an alkali activator up to 5 per cent by total dry weight. Raw coal tailings were oven dried for 24 hrs and used for specimen preparation in powder form. Samples were prepared using a California Bearing Ratio (CBR) test machine with a uniform load considering the volume of mould (the cylindrical mould 50 × 100 mm) and maximum dry density obtained from the standard proctor test (AS 1289.5.4.1, 2007) as 18 per cent and 1462 kg/m^3. Prepared samples were placed in plastic bags for curing at ambient temperature. The summary of mix designs is presented in Table 1.

TABLE 1

Mix design of CTBAs for road-base application.

Mix #	ID	Tailings	Binder	Activator
1	Ref	Coal tailings (100%)	-	-
2	S3	Coal tailings (97%)	Slag (3%)	-
3	L3	Coal tailings (97%)	Lime (3%)	-
4	C3	Coal tailings (97%)	Cement (3%)	-
5	FA3	Coal tailings (97%)	Fly ash (3%)	-
6	S5	Coal tailings (95%)	Slag (5%)	-
7	L5	Coal tailings (95%)	Lime (5%)	-
8	C5	Coal tailings (95%)	Cement (5%)	-
9	FA5	Coal tailings (95%)	Fly ash (5%)	-
10	S2-SS1	Coal tailings (97%)	Slag (2%)	Sodium silicate (1%)
11	L2-SS1	Coal tailings (97%)	Lime (2%)	Sodium silicate (1%)

Mix #	ID	Tailings	Binder	Activator
12	C2-SS1	Coal tailings (97%)	Cement (2%)	Sodium silicate (1%)
13	FA2-SS1	Coal tailings (97%)	Fly ash (2%)	Sodium silicate (1%)
14	S3-SS2	Coal tailings (95%)	Slag (3%)	Sodium silicate (2%)
15	L3-SS2	Coal tailings (95%)	Lime (3%)	Sodium silicate (2%)
16	C3-SS2	Coal tailings (95%)	Cement (3%)	Sodium silicate (2%)
17	FA3-SS2	Coal tailings (95%)	Fly ash (3%)	Sodium silicate (2%)
18	S1-FA2-SS2	Coal tailings (95%)	Slag (1%) + fly ash (2%)	Sodium silicate (2%)
19	S2-FA1-SS2	Coal tailings (95%)	Slag (2%) + fly ash (1%)	Sodium silicate (2%)
20	L1-FA2-SS2	Coal tailings (95%)	Lime (1%) + fly ash (2%)	Sodium silicate (2%)
21	L2-FA1-SS2	Coal tailings (95%)	Lime (2%) + fly ash (1%)	Sodium silicate (2%)
22	S1-L2-SS2	Coal tailings (95%)	Slag (1%) + lime (2%)	Sodium silicate (2%)
23	S2-L1-SS2	Coal tailings (95%)	Slag (2%) + lime (1%)	Sodium silicate (2%)

After the optimisation of the mix design, the selected mix design was used for the production of CTBAs. Aggregate production was carried out through the pelletisation process using our fabricated disc pelletiser. Based on several trial tests, the angle of the disc, the rotation speed, moisture content, and the duration of the test were set as 45°, 28 rev/min, 21 per cent, and 15 min respectively.

Testing procedures

To determine the particle size distribution (PSD), sieve analysis and hydrometer test were conducted on coarse fraction and fine fraction of coal tailings in accordance with AS 1289.3.6.1 and AS 1289.3.6.3 respectively. The liquid limit and plastic limit were also measured following AS 1289.3.1.1 and AS 1289.3.2.1. Ash content of coal tailings was determined according to ASTM D 3174–02. SEM-EDS analysis using SEM JEOL 6510 LV (Low Vacuum with EDS microanalysis) was conducted on coal tailings to identify the chemical compositions. To identify the mineralogical composition of coal tailings, XRD analysis using Bruker D8 Advance Powder Diffractometer was also performed on the tailings samples. The compressive strength of cylindrical samples was measured after 7 days and 28 days of curing using the Instron universal testing machine in accordance with AS 4456.4. Water absorption rate of CTBAs was obtained after soaking in water for 24 hrs following AS 1141.6.1. To prove the feasibility of using CTBAs as road construction materials, wet and dry strength tests were performed for different particle fractions in accordance with AS 1141.22.

RESULTS AND DISCUSSION

Characterisation of coal tailings

Based on the PSD and the Plasticity Index (PI = 13), coal tailings are categorised as medium plastic fine-grained soils. SEM-EDS analysis results shows that coal tailings are mostly composed of silica and alumina which are the main elements in geopolymerisation process. The XRD analysis also revealed that the phase compositions of the tailings include quartz, kaolinite, muscovite, and other clay minerals. The summary of the SEM-EDS analysis and ash content test are shown in Table 2.

TABLE 2

Chemical composition of coal tailings using SEM-EDS analysis.

Coal tailings	SiO_2 (%)	Al_2O_3 (%)	Fe_2O_3 (%)	CaO (%)	K_2O (%)	MgO (%)	Na_2O (%)	SO_3 (%)	TiO_2 (%)	Ash content (%)
	61.5	22.3	5.2	1.6	3.5	1.3	0.5	2.7	1.4	69

Mix design optimisation results

In the first stage of optimisation, coal tailings were replaced by different cementitious materials with an amount of between 3 per cent to 5 per cent without adding alkali activators (Mixes 2–9). Reference samples (Mix 1) without adding any binders or activators were also prepared to compare the results. The compressive strengths after 7 and 28 days of curing are shown in Figure 1. As can be seen, adding cement and lime to coal tailings had a substantial impact on improving the compressive strength compared to adding slag and lime after 7 days of curing. Moreover, increasing the amount of binders from 3 per cent to 5 per cent did not have a noticeable effect on 7-day compressive strength. However, compressive strength improved significantly after 28 days of curing in all mix designs, which can be attributed to the completion of chemical reactions over time. For the 28-day samples, adding cement showed the maximum strength improvement, followed by adding lime.

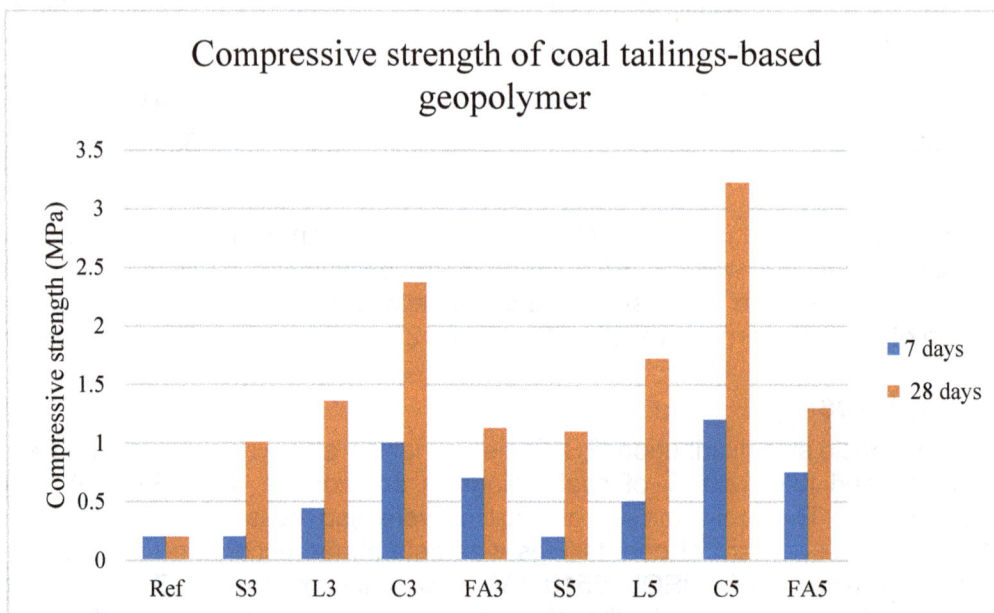

FIG 1 – Compressive strength of coal tailings-based geopolymer containing different binders.

In the second stage, cementitious binders were partially replaced by sodium silicate as the alkali activator between 1 per cent to 2 per cent in a way that the total replacement of coal tailings was fixed at 5 per cent (Mixes 10–17). The compressive strengths are compared in Figure 2 for mixes with a total of 3 per cent binder and activator and Figure 3 for mixes with a total of 5 per cent binder and activator.

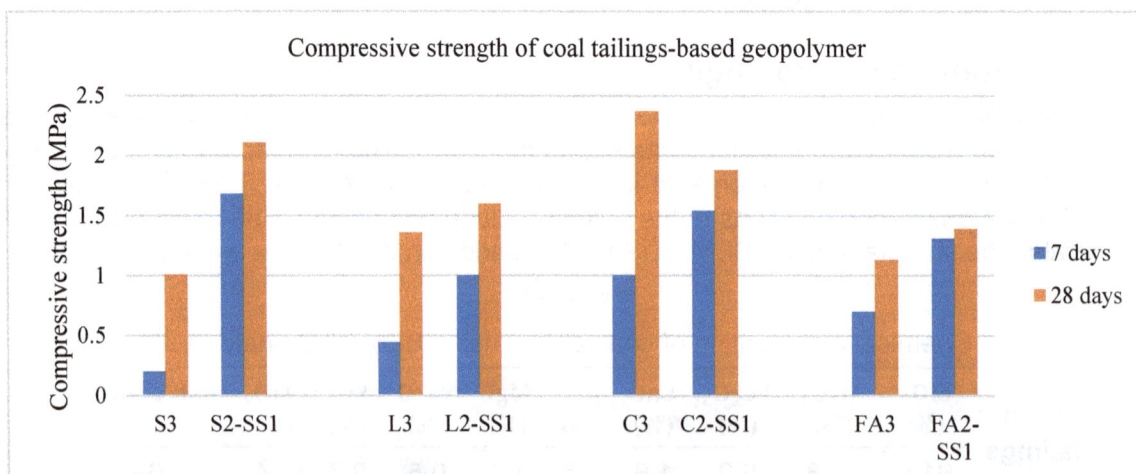

FIG 2 – Compressive strength results of coal tailings-based geopolymer containing 3 per cent binder and activator.

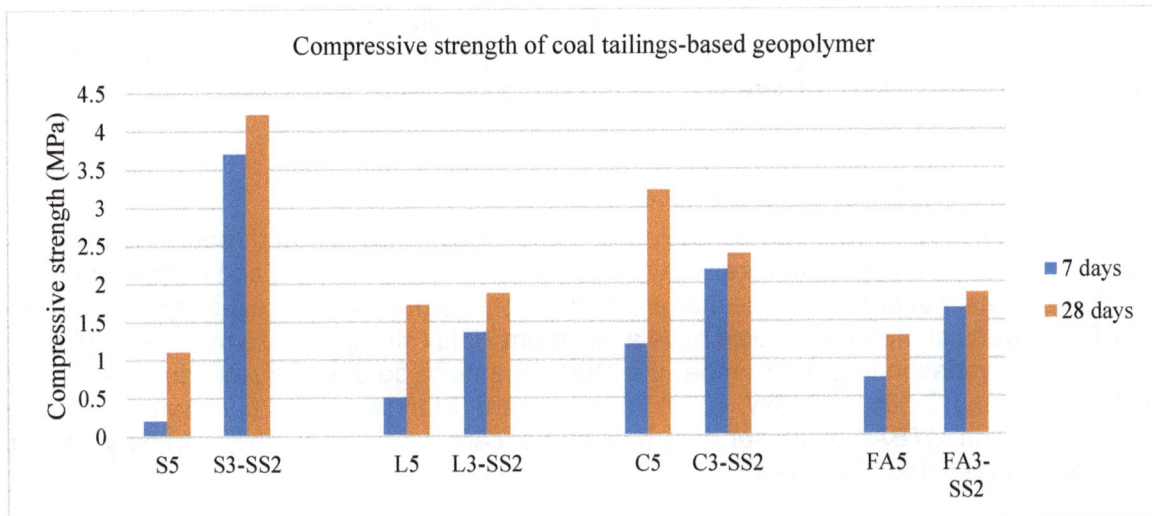

FIG 3 – Compressive strength results of coal tailings-based geopolymer containing 5 per cent binder and activator.

According to the compressive strengths presented in Figures 2 and 3, it can be found that the replacement of cementitious binders such as slag, lime, and fly ash with sodium silicate results in a considerable increase in compressive strength. In addition, the combined use of 3 per cent slag and 2 per cent sodium silicate provides the highest compressive strength of 3.7 MPa and 4.2 MPa after 7 and 28 days of curing, respectively. The compressive strength results show that the replacement of cement with sodium silicate improves the early strength of coal tailings. However, it negatively affects the development of strength after 28 days of curing. While increasing the curing time from 7 days to 28 days has a considerable impact on improving the compressive strength for the mixes containing cementitious binder only, the curing time has a less significant influence on the strength development of mixes containing both binder and sodium silicate.

Considering that the current market price of fly ash and lime is cheaper than slag in Australia, synergic effects of different binders have been investigated by developing new mixes with combined use of different binders. This effort has been made with an aim to reduce the cost of making CTBAs. In Mixes 18–23, the total replacement of coal tailings and the sodium silicate content were fixed at 5 per cent and 2 per cent, respectively. The 7 and 28-day compressive strengths are shown in Figure 4. The results revealed that the samples with 2 per cent slag and 1 per cent fly ash achieve a good compressive strength (3.58 MPa at 28 days), while no obvious synergic effect is found for the combined use of lime and fly ash.

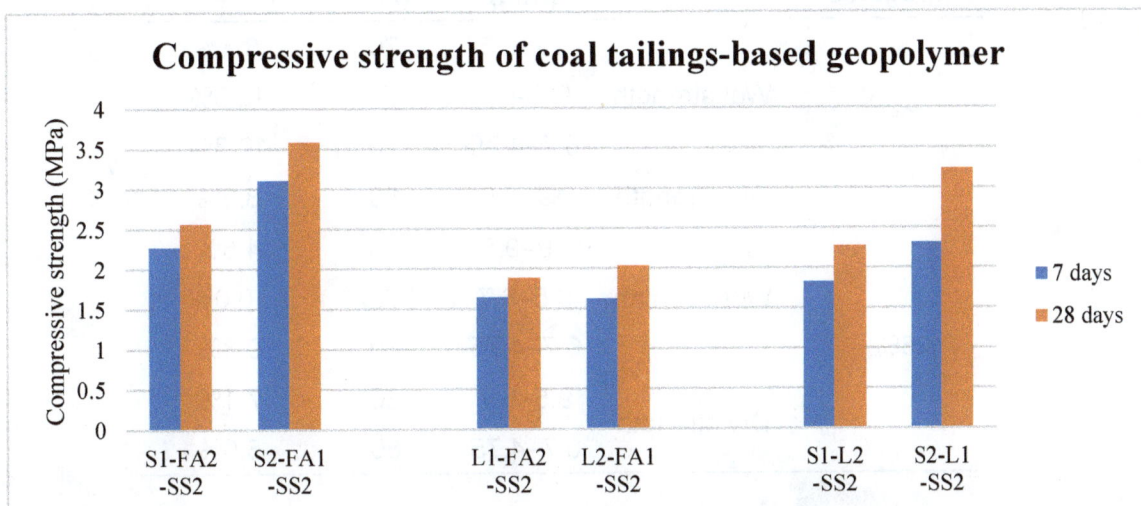

FIG 4 – Synergic effects of different binders on compressive strength of coal tailings-based geopolymer.

The mix design optimisation results showed that the combination of 3 per cent slag and 2 per cent sodium silicate provided the highest compressive strength of 3.7 MPa and 4.2 MPa after 7 and 28 days of curing, respectively. Therefore, this mix design was selected for the production of CTBAs in the next stage.

Wet and dry strength of CTBAs

Bulk production of CTBAs was carried out using the optimum mix design through the pelletisation process. The produced aggregates are shown in Figure 5. The dry strength and the wet strength are the crushing forces that produce fines that are 10 per cent by mass of the test portion. According to TfNSW T215, granular materials should have a minimum 70 kN strength and a maximum 35 per cent wet/dry strength variation in order to meet the requirements to be used as road subbase materials (Specification D and C 3051, TfNSW, 2020). In this regard, Standard DGS20 as natural aggregates were replaced with CTBAs at 25 per cent and 50 per cent by volume. The summary of wet/dry strength test is presented in Table 3.

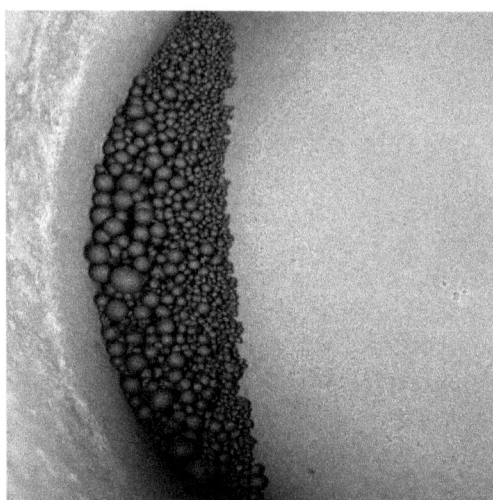

FIG 5 – Produced CTBAs using disc pelletiser.

TABLE 3

Wet/dry strength analysis results.

Replacement level of natural aggregates	Test	Particle fraction (mm)	Applied load (kN)	Fine percentage
50%	Wet strength	19–9.5	70	8.5%
		9.5–6.7	70	12.5%
		6.7–4.75	70	12.3%
	Dry strength	19–9.5	80	8.5%
25%	Wet strength	19–9.5	70	4.5%
		9.5–6.7	70	10.0%
		6.7–4.75	70	9.3%
	Dry strength	9.5–6.7	80	7.1%
		6.7–4.75	80	5.0%

According to the wet/dry strength test results, natural aggregates with particle sizes between 19–9.5 and 9.5–4.75 can successfully be replaced by 50 per cent and 25 per cent CTBAs respectively to meet the requirements for road subbase construction application.

Surface treatment

To increase the replacement level, surface treatment was employed using epoxy glaze to reduce the water absorption and improve the wet strength of CTBAs. For this purpose, CTBAs were soaked in the epoxy glaze for 2 mins and placed for drying at ambient temperature for 24 hrs. The test results show that coating the CTBAs with epoxy glaze results in decreasing the water absorption from 24 per cent to 2 per cent. The wet strength test was conducted on a full batch of CTBAs with particle size between 19 to 9.5 by applying 70 kN force and the fine percentage was calculated as 6.8 per cent which meets the Australian standard for road subbase application. It can be concluded that by surface treatment, natural aggregates can be fully replaced by CTBAs for road subbase application.

CONCLUSIONS

This research project has investigated the feasibility of using coal tailings for the production of synthetic aggregates for road construction applications. Characterisation tests are conducted on raw coal tailings to identify the physical, chemical, and microstructural composition of raw coal tailings. Coal tailings were replaced by cementitious materials up to 5 per cent and the compressive strength of samples was measured to obtain the optimum mix design for the development of CTBAs. Finally, synthetic aggregates were produced through the pelletisation process using the optimum mix design. Surface treatment has also been employed to improve the mechanical properties of aggregates. The main conclusions derived from this research are listed below:

1. Coal tailings are categorised as medium plastic fine-grained soils mostly composed of silica and alumina. The phase compositions of the tailings include quartz, kaolinite, muscovite, and other clay minerals.

2. The mix design containing 95 per cent coal tailings, 3 per cent slag, and 2 per cent sodium silicate provides the highest compressive strength of 3.7 MPa and 4.2 MPa after 7 and 28 days of curing, respectively.

3. 50 per cent and 25 per cent of natural aggregates with particle sizes between 19–9.5 mm and 9.5–4.75 mm can be successfully replaced by CTBAs to meet the requirements for road subbase construction.

4. Surface treatment of CTBAs using the epoxy glaze can decrease the water absorption and improve the wet strength of CTBAs which results in increasing the natural aggregate replacement level.

ACKNOWLEDGEMENTS

The authors of this report would like to acknowledge the financial support from the Australian Coal Association Research Program (ACARP) to conduct this project. Also, the support from Nu-Rock Pty Ltd., Somerset Pty Ltd, and Western Sydney University is greatly acknowledged.

REFERENCES

Katwal, U, Khazaie, A, Yong, M T, Jahandari, S, Babla, M, Osborne, D, Rahme, M, Harriman, A, Chen, Z and Tao, Z, 2022. Value-Added Products from Coal Tailings.

Lafhaj, Z, Samara, M, Agostini, F, Boucard, L, Skoczylas, F and Depelsenaire, G, 2008. Polluted river sediments from the North region of France: Treatment with Novosol® process and valorization in clay bricks, *Construction and Building Materials,* 22(5):755–762.

Mansouri, A I, Tahmooresi, M and Ebrahimi, A, 2019. Production of Lightweight Aggregate Using Zarand Coal Processing Plant Tailings, *Journal of Mineral Resources Engineering,* 3(4):26–29.

Qiu, G, Luo, Z, Shi, Z and Ni, M, 2011. Utilization of coal gangue and copper tailings as clay for cement clinker calcinations, *Journal of Wuhan University of Technology-Mater. Sci. Ed,* 26(6):1205–1210.

Rose, J G, 1982. Processing and material properties of energy-efficient sintered coal refuse lightweight aggregate, *Resources and Conservation,* 9:119–129.

Transport for NSW (TfNSW), 2020. Specification D&C 3051 – Granular Pavement Base and Subbase Materials, NSW government.

Metal recovery for reduced environmental liability from VMS tailings

M Thorpe[1] and B Hill[2]

1. CEO, EnviroGold Global, Toronto, Canada. Email: mark.thorpe@envirogoldglobal.com
2. CTO, EnviroGold Global, Brisbane Qld 4000. Email: brock.hill@envirogoldglobal.com

ABSTRACT

Processing Volcanogenic Metal Sulfides (VMS) ore is complicated. In many cases, large quantities of strategic, critical, and precious metal remain in the tailings – often trapped within the pyrite matrices. These potentially acid generating tailings may require ongoing, in perpetuity, management. Acidic drainage from such tailings often mobilises metals that, if released to the downstream receiving environment, may be of concern to broader stakeholders. Such mines are complicated to permit, and external stakeholders raise questions about the post-closure management of the potential environmental liability.

Following years of research, the EnviroGold Global team has developed a process to profitably recover precious, strategic, and critical metals from live tailings, tailings being reprocessed, or as a stand-alone project to reprocess tailings from a tailings storage facility. This procedure is part of the circular economy using waste for the betterment of society.

Volcanogenic Metal Sulfides Tailings have been studied for many years with various processes being tried to liberate the copper, zinc, gold, and silver from the tailings. More recently, EnviroGold Global completed an evaluation of VMS tailings – in which it was found that the gold was submicroscopic in size. EnviroGold Global then developed a process that uses a catalysing acid to start the reaction to break apart the various pyrite species within the tailings. The resulting reaction, which takes place at about 85°C and at atmospheric pressure, is further catalysed by the sulfuric acid and powerful ferric ion (Fe^{3+}) oxidant produced. Following the initial reaction, the residue containing the gold and silver is separated from the acid. The residue is washed, and the gold and silver recovered by conventional leaching. The zinc and copper can then be recovered from the acid via a combination of solvent extraction and ion exchange resins, widely used in the industry.

The history of the study along with the experimental results and pre-pilot plant testing data will be presented. The initial assessment of the process indicates that it is widely applicable to VMS ores and can be used as ad-on technology for processing plants in a similar way that the INCO SO_2 process is used for CN destruction. The process recovers metals and reduces the overall environmental liability of VMS ores making permitting easier and closure less costly.

ENVIROGOLD GLOBAL

EnviroGold Global (NVRO) partners with mining companies to efficiently recover metals from complex ores while alleviating the environmental and financial burdens of mine rehabilitation. Our hydrometallurgical chemistry technology solves the key technical problems preventing the recovery of billions of dollars of trapped resources. Our proprietary process reduces acid-generating potential of waste by oxidising sulfides, and can be applied to primary 'live tailings' (tailings that are being produced directly from an operating plant) in operating mines as well as legacy tailings, transforming waste into vast new resource stockpiles and reducing environmental liabilities. The NVRO process is an update to existing metallurgical processes (similar to that formerly used at the Bakyrchik mine) with an emphasis on recycling of the catalysts.

MINERAL PROCESSING AND METALLURGICAL TESTING

NVRO has developed a patent pending hybrid, two-stage leach process for reprocessing VMS tailings using a powerful mixed acid and ferric ion (oxidant) coupled with low to moderate reaction temperatures, and normal atmospheric pressure. These key parameters support the potential for low capital costs for construction materials and ameliorate concerns of material corrosion observed previously in oxidative leach pilot plants.

NVRO has undertaken, both at its own laboratory and through Core Resources (Core) laboratory (in Brisbane), a series of alternative and combinations of mixed acid leaches with a view to optimising the recovery of both precious (gold and silver) and base metals by-products (zinc, lead and copper).

Most of the metallurgical program was devoted to the acid pre-leaching stage of the process, more particularly acid pre-leach optimisation. The initial scouting work explored blended acid concentrations and residence times on extraction of lead (Pb), zinc (Zn), copper (Cu), silver (Ag), and gold (Au) on re-floated zinc scavenger tailings (ZST). The purpose was to determine the extraction kinetics of Pb, Zn, Cu, and Ag, while leaving the gold in the residue for further downstream processing.

The second stage of testing observed extraction of precious metals in the blended acid leach residues. Process pathways on the acid-leach residues explored lixiviants such as cyanide, glycine, thiosulfate and hypochlorite on Ag and Au recovery. A cyanide leach of acid leach residues proved to be the preferred option for maximum precious metal extraction.

The initial cyanide leach tests were conducted as a two-hour stirred beaker test. Gold recovery was approximately 59 per cent and silver recovery was approximately 90 per cent.

In December 2021, EnviroGold Global, along with Core Resources undertook additional test work on the residue material from the acid pre-leach test work toward investigating the possibility of additional precious metal recovery through longer cyanidation residence time. Gold recovery was increased to indicatively 83.5 per cent and silver recovery was indicatively 94.6 per cent. The results of the test program were used to determine the preferred leach configuration, together with expected cyanide leach recoveries for gold and silver. Deductions to the test work extractions were applied to expected zinc and copper reporting to the acid pre-leach Pregnant Leach Solution (PLS) and recoveries to simulate scale-up to a commercial production facility were modelled using METSIM modelling software.

At the time of writing, pre-pilot plant testing is ongoing at the ALS Laboratory in Perth Australia. The current work aims to validate the previous test work and provide input to the pilot plant design that will start pilot plant construction in May 2023. Sedgman has been contracted by EnviroGold Global as the design engineering company. Following the pilot plant work, there will be an update to the Technical Report followed by detailed engineering for a definitive feasibility study. It is anticipated that the commercial plant will be installed in 2024.

RECOVERY METHODS

EnviroGold Global plans to operate a hydrometallurgical process to reprocess tailings from former and current Pb/Zn flotation operations. The initial stages of reprocessing will be an add on to an existing processing plant.

The processing facility will consist of the following main sections:

- Feed dewatering/thickening from floating, cutter-suction dredge.
- Mixed Acid regeneration and gypsum precipitation.
- Acid pre-leach.
- Acid pre-leach solid-liquid separation.
- Acid pre-leach solid residue cyanidation for gold and silver recovery.
- Goethite (iron) and scorodite (arsenic) precipitation from acid pre-leach filtrate.
- Copper recovery (97 per cent efficient) from acid pre-leach filtrate by solvent extraction.
- Zinc recovery (98 per cent efficient) from acid pre-leach filtrate by precipitation.
- Tailings dewatering and filtration for deposition in the TSF.

Essential to developing NVRO's process for oxidation of test tailings ore was understanding the nature of the gold-sulfur bond, since this chemical bond competes with the conventional use of carbon in the recovery of gold and silver from pregnant cyanide liquors.

The NVRO process can integrate the oxidation pre-treatment process with existing infrastructure, and not interfere with mine owner's ongoing operations.

To accomplish these goals, NVRO developed a hybrid two-stage oxidative leach process for the tailings using low mixed acid concentrations, low to moderate reaction temperatures, and normal atmospheric pressure. These key parameters support potential for low capital costs for materials of construction and ameliorate concerns of material corrosion observed previously in oxidative acid leach pilots (perhaps in a processing plant that avoids the use of metallic materials in favour of reinforced advanced plastics).

A block flow diagram for the proposed flow sheet to re-treat Zinc Scavenger Tailings (ZST) is shown in Figure 1.

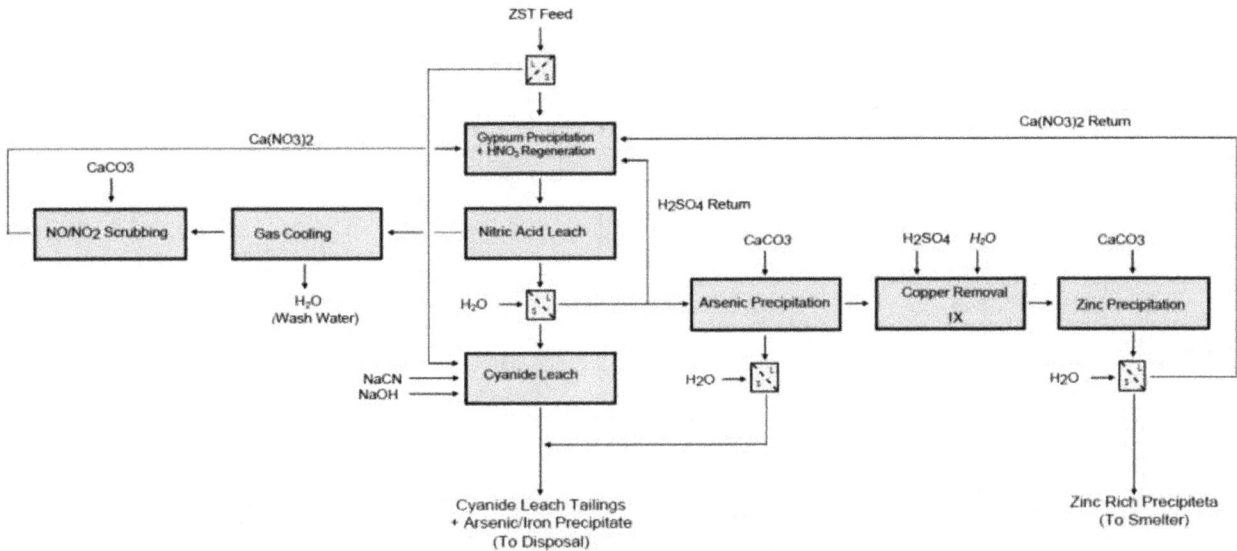

FIG 1 – Simplified block diagram to re-treat zinc scavenger tailings.

Updates to the process flow design are ongoing with more detailed updates currently in review by the engineering team.

ENVIRONMENTAL BENEFITS

The production of metals from wastes has inherent benefits of closing the circular economy. In the case of the EnviroGold Global work, the production of metals is completed with a 96 per cent reduction in the production of CO_2 when compared to conventional metal production. By establishing the facility on a brownfields site, minimal environmental effects are anticipated and the placement of the tailings into an existing facility reduces the time to operational production.

The liberation of the metals from within the pyrite matrix through oxidation has environmental benefits. There are several environmental benefits as follows:

- The removal of the metals results a lower level of heavy metal presence within the tailings, and hence the reduced risk of the mobilisation of these metals into the downstream environment.

- The environmental risk of acidic drainage is greatly reduced through the oxidation of the pyrite.

- Arsenic can be precipitated as ferric arsenate, (scorodite), insoluble in water and environmentally benign in the closure scenario.

- Option for a dry closure of the tailings storage facility. The current management plan for the tailings requires a wet closure (2 m of water) to prevent acidic drainage. However, with the removal of most of the pyrite in the EnviroGold Global process, there is the option for a dry closure and the subsequent land use for the area would be more compatible with the existing surrounding current land use.

SUMMARY

The EnviroGold Global technology can be applied to complex pyritic tailings for the recovery of strategic, critical, and precious metals. The use of a waste product to produce the metals required for the transition of the economy to a low carbon model has the additional benefit of the reduced energy requirement to produce the metal and a reduced overall environmental liability for the site owner.

Mine closure

Mine closure – repurposing mine sites and tailings for solar power

A Bone[1] and D Hng[2]

1. Managing Director APAC, Alion Energy, Sydney NSW 2000.
   Email: aidanbone@alionenergy.com
2. Former Director of Engineering, Alion Energy, Sydney NSW 2000.

ABSTRACT

Legacy mine sites are becoming attractive locations to install Variable Renewable Energy (VRE), as developers are losing social licence to build solar PV power stations on agricultural and farming land.

Mining operations typically include high voltage electrical infrastructure for powering the site which with some engineering and augmentation to the network can be repurposed as a generator and/or energy hub.

There are several mining sites across Australia that have been repurposed after mine closure for the purpose of solar PV generation, most notably the Kidston gold mine tailings dam (Figure 1), which now supports a 50 megawatt (MW) solar farm. There are several other solar PV projects that have been announced on mining sites and a few more which are adjacent to power stations on either the ash dams or former mining sites.

FIG 1 – 50 MW Kidston Solar built on the Kidston tailings dam.

The challenge with most mine tailings and mine closure sites for the repurposing of their site for other infrastructure is the cost of engineering the fill to provide a stable hard stand that will support the new infrastructure, which is especially difficult on mine tailings, mine spoil or just land that has been mined or may be contaminated.

The purpose of this paper is to explore the technical capability of a novel concrete ballasted solar PV tracking structure to be installed totally above ground on mine sites, tailings dams, ash dams and other contaminated sites.

The technology uses a concrete rail similar to what is installed for residential street kerb and guttering. The structure is built on four legged tables which are connected by a coupling which allows for some construction and operational movement.

The technology is patented by Alion Energy Pty Ltd which has fundamentally redesigned the way solar power stations are built, how bifacial energy gains are generated, and how these solar facilities are maintained.

INTRODUCTION

Australia's Climate Change Act came into effect on 14 September 2022 and acts as an umbrella legislation to implement Australia's two main commitments under the Paris Agreement to reduce net Green House Gas (GHG) emissions to 43 per cent below 2005 levels by 2030; and to reduce net GHG emissions to zero by 2050.

Approximately 35 per cent of Australia's GHG emissions are a direct result of burning fossil fuels to generate electricity, a figure which does not include transport, fugitive emissions and stationary energy. Therefore, this is a key sector targeted to achieve our legislated emissions reduction targets of 82 per cent of all electricity to be supplied by VRE.

To most likely roadmap under the Integrated System Plan (ISP) shown in Figure 2 is the 'step change' scenario which requires the National Electricity Market (NEM) to treble its existing 16 GW of Variable Renewable Energy (VRE) by 2030 and add another 32 GW. It is then required to double that capacity by 2040 and then double it again by 2050.

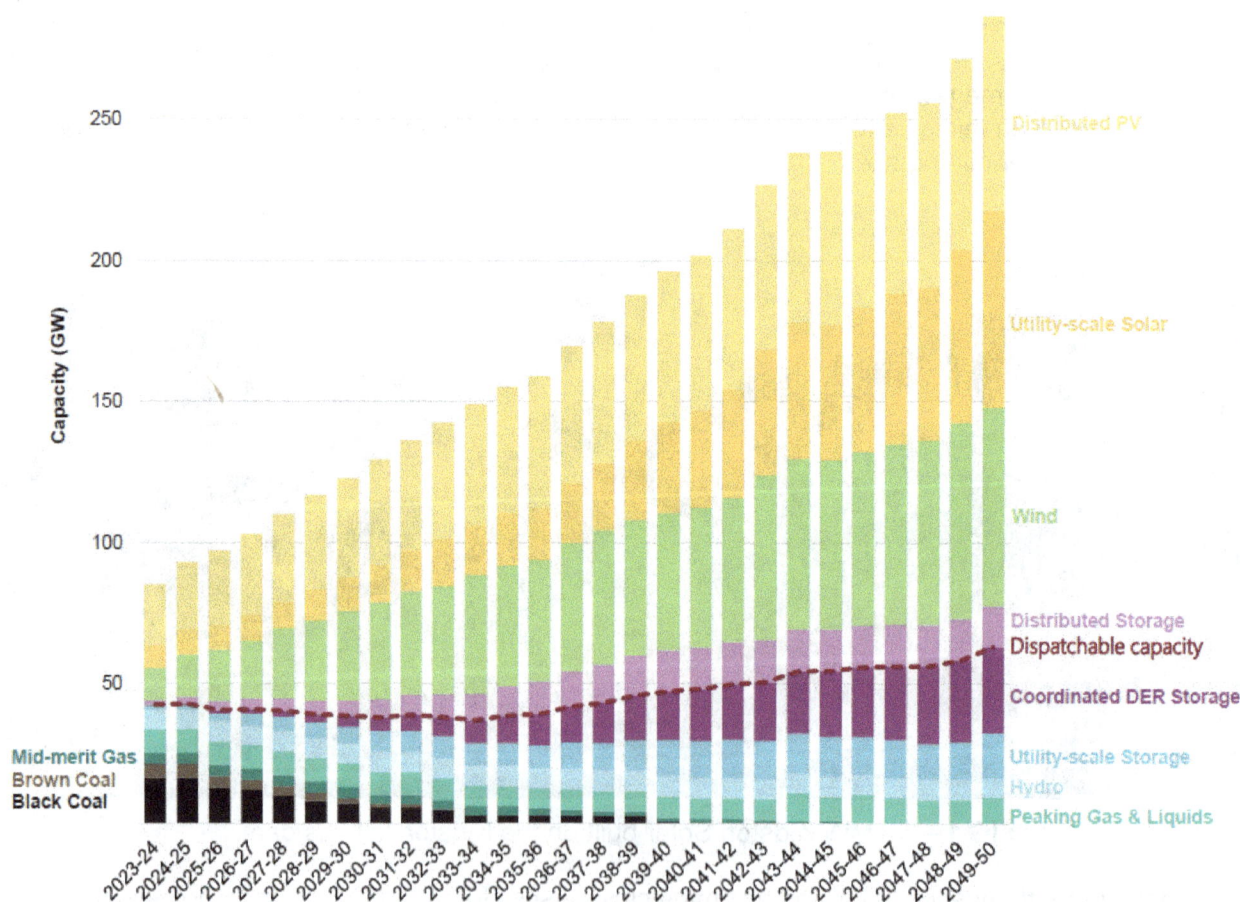

FIG 2 – 2022 Integrated System Plan forecast NEM capacity to 2050, 'step change' scenario.

One of the most pressing issues in the energy transition is securing a social license to the build the electrical infrastructure with the support of the community. There are growing concerns that delays in development approvals, coupled with coal fired generators retiring early means that construction must be begin immediately at a rate of 300 MW a month to meet our target.

This has created a great opportunity for retiring power stations and mine waste and mine sites to be seen as attractive locations to install new energy generators or energy hubs, already seen at Eraring, Liddell Munmorah and Swanbank power stations.

There are a number of papers and articles on the use of mine tailings for the purpose of constructing a solar farm. The author of this paper has personal experience on the development and installation of a 50 MW solar farm on the Kidston gold mine tailings dam, near Kidston, Queensland as well as providing input for the development approval and the design and construction of a 70 MW solar farm on the coal ash dam at the Vales Point power station in Mannering Park, New South Wales. They were also engaged to undertake a site investigation for the feasibility of solar power station at the Northern Power Station, Port Augusta, South Australia.

Constructing solar farms on mine sites, mine waste and tailings is a win for all stakeholders including miners, governments and communities. If we can educate the mining industry on technical solutions which enable mine waste and tailings to become solar power stations it will assist to deliver the following benefits:

- Meet our climate goals.
- Create a new use for a mining legacy asset, which generates income and delays some remediation costs. These costs can be significant for both private owners and governments who are often left footing part of the remediation costs.
- Desperately needed new VRE in the NEM.
- Quick approvals due to fewer social licence issues than greenfield sites.
- Growth in regional mining areas with ongoing employment.

This paper will focus on common geotechnical problems in design and construction of a solar PV power station utilising tracker structures as their main mounting structure. Problems that will be discussed are corrosion, reactive soils, rocky ground, soft ground, mine waste, as dams and undulating terrain will be addressed. This paper will then go on further to discuss how these problems are typically resolved like land reshaping, deep foundations, specialised foundations and corrosion protection.

This paper will consider an alternative foundation system, namely a concrete ballasted solar mounting solution. This paper will show how a concrete ballasted design will alleviate most of the common problems encountered in solar PV foundation design and installation. With a clear understanding of these issues, building a solar power station can be done easily and with minimal risk.

SITING CONSIDERATIONS FOR SOLAR POWER STATIONS

Utility and commercial size ground mount solar power stations are being designed and built around the world in record numbers. They are being constructed in less time and at more competitive prices than ever before. With this increasing growth, the risks in getting the engineering and construction wrong are enormous, and erroneous calculations can be detrimental to the success of any given the project.

Solar power station developers, designers, builders, asset owners and operators are often faced with numerous challenges ranging from site selection, site constraints, technology selection, geotechnical conditions, logistics, construction, grid connection, and maintenance.

Of the various challenges, the three most unpredictable and elusive challenges are social license to construct, grid connection, and geotechnical conditions. The former risk has become more and more challenging to navigate, whilst the engineering and construction risk associated with the latter two challenges are often poorly understood and underestimated.

Social license

Identifying and addressing environmental and social impacts of solar development are not only crucial to building relationships with local communities but are becoming increasingly important to secure both investment and regulatory approvals.

There is a growing movement against the use of agricultural and productive land for the use of solar PV as demonstrated by the NSW Farmers Association putting together an energy transition working group for farmers and country communities to voice their concern that they are being sacrificed in the energy transition.

There are several examples of mine sites being repurposed as solar power stations, such as the Maxwell solar farm, developed by Malabar Resources as part of the rehabilitation of the Drayton open cut coalmine, and the conversion of the Kidston gold mine tailings dam to a solar power station which both have strong community support.

Repurposing mine waste, mine tailing and mine sites for the purpose of solar PV is a very attractive alternative to delay expensive remediation costs and at the same time develop an electrical generation asset with continuing employment for mine workers.

Grid connection

There is a huge bottleneck in the grid connection process, and as reported by Parkinson (2022) that in some parts of the grid, AEMO reverted to 'sequencing' for connections, processing one or two projects at a time in a bid to ensure it can maintain grid stability. Many projects have been delayed for up to two years. Clean Energy Council Chief Executive, Kane Thornton (Aubrey, 2023) says that it has proposed a set of rule changes to the Australian Energy Market Commission (AEMC) that would give clarity to a process that historically has created serious delays for renewable investors and developers. As 'currently, some generators have faced open-ended delays to connection and unmanageable risks, creating a culture of uncertainty for renewable investors.'

Mine sites usually involve large mining equipment to extract or process materials and these are usually energy intensive and require high voltage (HV) power lines and substations at the mine site. Having electrical infrastructure already at the proposed power station site will significantly reduce the risk and cost of connecting to the grid, and if the mine is still operational, the electricity will be able to be consumed 'behind the meter'.

The is a huge advantage over greenfield projects seeking approval to connect to the grid.

Geotechnical/ground risk

The following geotechnical characteristics are considered in the design and construction of a solar PV power station utilising tracker structures as their main mounting structure. Every site has its own unique geological and geotechnical features which if insufficient studies are undertaken can cost a contractor/or client a significant financial penalty on top of the initial project costs to re-engineer the structural solution. This has seen ASX listed engineering companies such as RCR Tomlinson forced into administration.

When deciding on a location to build, developers undertake a pre-feasibility geotechnical investigation to provide the developer with some certainty over the risks they will be assuming for the construction of solar power structures. The usual geotechnical issues that solar developers are taking a risk on include, but are not limited to the following:

- reactive or expansive soils
- rocky ground
- soft soils
- agricultural/horticultural land
- undulating/uneven ground
- corrosive soil
- landfill
- mine waste or tailings are just another variation or combination of the above ground risks.

With the mining industry being heavily regulated and the risks undertaken during mining requiring a very high level of safety both during operation and then under the mine closure plans, it would be reasonable to suggest that the physical properties of the mine waste and tailings would be better understood than the solar power stations that are currently being developed.

Solar power stations have been built and are operational on reclaimed landfill sites at Summerhill near Newcastle and Albury, NSW and have been approved on other sites such as Springvale South near Melbourne, Victoria. Landfill sites share many physical characteristics with mine waste and tailings, and it is only through engineering the site capping to support the solar structures, or more likely, engineering the structures to suit the geotechnical characteristic of the landfill site.

With every mine waste and tailings site being different, but very well understood by the asset owner, it is reasonable to assume that a structure can be engineered to suit the site just as easily as any greenfield site.

Depending on the type of mine waste or tailings it may be a combination of these risks which are briefly described below.

Reactive/expansive soil

The nature of expansive soil is characterised by the expansion of soil as a result of an increase of moisture content in a dry soil. Conversely when the moisture content of wet soil decreases, shrinkage of soil occurs.

In Australia the severity of expansive soil sites are commonly classified by design standards into reactivity classes ranging from class S (ys <= 20 mm) to class E (ys > 75 mm) where ys 4 is the surface movement anticipated for the site due to soil reactivity. Table 1 describes in more detail the different site classes with respect to their surface movements.

TABLE 1

General definition of site classes.

Site Class	Foundation	Characteristic Surface Movement
A	Most sand and rock sites with little or no ground movement from moisture changes	
S	Slightly reactive clay sites, which may experience only slight ground movement from moisture changes	0 - 20mm
M	Moderately reactive clay or silt sites, which may experience moderate ground movement from moisture changes	20 - 40mm
H1	Highly reactive clay sites, which may experience high ground movement from moisture changes	40 - 60mm
H2	Highly reactive clay sites, which may experience very high ground movement from moisture changes	60 - 75mm
E	Extremely reactive sites, which may experience extreme ground movement from moisture changes	> 75mm
P	Sites which include filled sites (refer to AS 2870 2.4.6), soft soils, such as soft clay or silt or loose sands; landslip; mine subsidence; collapsing soils; soils subject to erosion; reactive sites subject to abnormal moisture conditions or sites which cannot be classified otherwise.	

This classification system has helped structural engineers and designers quickly identify the best footing system to use in a building design according to the severity of soil reactivity.

Rocky ground

In the geological world, Figure 3 describes the classification of rocks as one of the following three categories:

1. Sedimentary, where the rock is formed as a result of the layered accumulation of sediments or small particles over a long period of time, compacted and the cementation of these particles.

2. Metamorphic, where the original rock undergoes a transformation process into a new rock with the presence of intense heat and pressure.

3. Igneous, where rocks formed from the cooling and solidification of lava or magma.

| Sedimentary | Metamorphic | Igneous |

FIG 3 – Types of rocks.

Very often in a solar farm, the presence of rock can vary from a small isolated quantifiable location right through to a large area with more than one rock type of varying strength and depth.

Igneous rock like basalt has the tendency to have varying depth mimicking lava flows overlain by reactive soils. Sedimentary rock like limestone develops Karst regions or pinnacles, again with varying depths.

Having an accurate understanding of what lies beneath the surface and being able to carefully quantify the works and risk involved to build on a site with rock can often make or break a solar farm project.

Soft soils

Solar farms with soft soils have a similar level of ground risk but are more easily identifiable and manageable. Soft soils are defined as soils possessing low undrained shear strength, containing relatively large amounts of moisture with an allowable bearing pressure of less than 50 kPa.

It is common to see on a solar farm with very low strength a requirement for soil engineering, having the need for either earthworks carried out to strengthen the soil or have larger or more piles installed to spread or lessen the loads transferred to the soil. In addition to load minimisation on the soil, additional *in situ* testing will need to be carried out to gauge the performance of the ground.

Agricultural/horticultural land

It is becoming common to see landowners repurposing/selling their land for the construction of solar power stations. This is particularly true if the crop is coming to its end of life and is ready to be replanted. Recent examples have seen palm oil plantations, mango plantation and almond plantations being repurposed.

In the case of plantation land that is to be repurposed as a solar farm, the traditional foundation solution has always been a pile-driven foundation solution, which require the ground to be free of any vegetation above ground or organic matter (mass subterranean root ball) in the soil that would decay and weaken the soil. When soil weakens from decaying of organic material, soil supporting the foundation loosens, resulting in excessive displacement or instability of the underlying foundation.

To overcome the above-mentioned problems, clearing and removal of plant and vegetation matter will need to be carried out. Deeply penetrated root balls may also require removal and soil re-compacted and tested to a density matching the native ground.

Undulating/uneven ground

Undulating or uneven ground is often a prominent landscape feature of any solar site. The challenge of these features during design and construction is to leave them in their original form with no change to natural drainage or topography and yet still have a functioning solar farm.

Solar trackers often have strict limitations on allowable slopes, change in grades and distance to ground. To conform to the limitations set by the solar tracker manufacturers, the designer would carry out upfront engineering design works in order to minimise or eliminate earthwork volumes or engineer for longer steel piles to smooth out ground undulations.

Corrosive soil

A long-term problem faced in the design of solar farm support foundations is, the ability of the foundation to endure being in corrosive soils for its design life (20–30 years) without affecting performance of the structure. The aggressivity of soil directly relates to how corrosive soil affects the durability of concrete and steel structures. The few main exposure conditions causing corrosion in soils are its, pH, resistivity, Chloride and sulfates in the soil. These exposure conditions occur when the structure is embedded into the ground, as in the case of steel or reinforced concrete footings. Conversely, these exposure conditions will have minimal effects on a concrete ballast that is placed alongside the surface of the soil.

Landfill

Building on landfill requires careful consideration due to its unstable founding material, hazardous gases, and near surface lining material just to name a few.

Voids on mine sites are often filled with organic material, or material other than the mine spoil produced on-site, this may lead to the mine waste sites having similar properties to a landfill site such as the settlement demonstration in Figure 4.

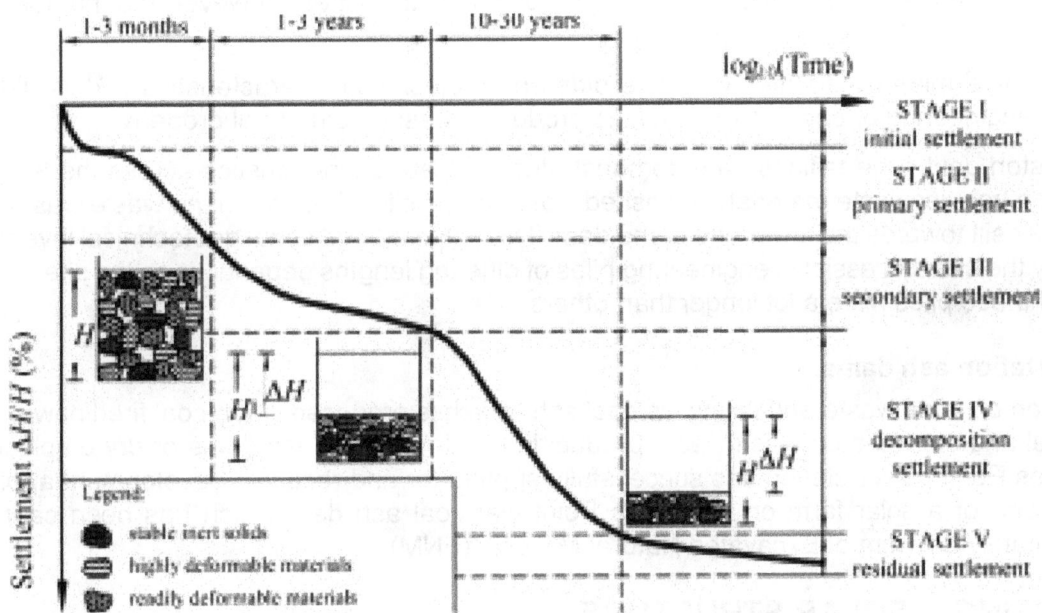

FIG 4 – General settlement curve of municipal solid waste with time.

A landfill typically consists of a large hole that is filled with waste material from households, commercial and/or industrial site and demolition works. Once this hole is full, it is capped over with a membrane, covered with topsoil and revegetated.

Over time, organic matter in landfills breaks down causing ground settlement. This settlement occurs over a period of 20 years or more and occurs five stages:

1. Initial/immediate
2. Intermediate/primary
3. Secondary/long-term settlement
4. Decomposition settlement
5. Residual settlement.

Initial settlement occurs quickly and non-uniformly across the site, surface settlement of up to 40 per cent has been observed after 20 years. Intermediate stage of settlement is noticeably slower and shorter. Finally, long-term settlement is a slow and steady settlement.

This very problem on settlement renders the land unsuitable for building, as a result the landfill is converted into parks or sporting fields. However, in recent times, building solar farms on landfills is gaining popularity where councils are finding methods to convert unusable land into an asset for the community and generate excess energy for sale into the grid.

Building a solar farm on landfill means careful consideration of foundations need to be made as the protective layer over the landfill needs to be preserve, which means piled foundation is off limits due to the surface penetrative method of installation for a piled foundation. However, a ballasted foundation is ideal for landfills as it is situated on the surface and does not involve any surface penetrative measures.

Mine waste and mine tailings sites

Tailings are mine waste and can be described as the waste materials left after the target mineral has been extracted from ore. They can consist of crushed rock, trace metals, water, and additives from processing such as petroleum by-products, sulfuric acid and cyanide.

Tailings, most often in the form of a wet slurry, are conventionally stored above-ground behind earthen dams.

There is no comprehensive global registry of tailings dams. But the Earthworks website (2023) estimates there are around 18 000 dams, 3500 of which are active. However, that number may be much higher.

Each of these mine waste sites will have different geotechnical characteristics and would require different engineering or capping materials to produce a desired structural property.

The Kidston gold mine tailings dam demonstrated different characteristics across the legacy dam site as the heavier waste materials deposited closer to the site where the mine waste was deposited and a finer silt towards the lower end of the dam. This required additional geotechnical investigations to qualify the site and assist in engineering piles of differing lengths dependent on the site conditions. Some of these piles were a lot longer than others.

Power station ash dams

A variation on mine waste and tailings is coal ash, which is produced when coal fired power stations burn coal and store the coal ash waste product in either wet coal ash dams or dry emplacements. The Vales Point power station has successfully applied for and received development approval for construction of a solar farm on the Vales Point wet coal ash dam which has been capped with approximately 200 mm of Excavated Natural Material (ENM).

SELECTING A SOLAR STRUCTURE

For the purpose of this paper, it is assumed that the solar power station will utilise a sun-tracking solution as the majority of new solar power stations utilise this technology due to the 30 per cent additional energy yield over fixed systems.

Complexities in assembly

Trackers come in all types of configurations and complex designs, often resulting in the need for special tools and equipment and upskilling of labour in order to competently assemble a tracker. Others require working at heights or having parts that are excessively heavy that cranes are needed to assist in the assembly of a tracker.

In constructing a solar structure, ease and speed of tracker assembly plays utmost importance during the mechanical installation phase of construction. A tracker that is lightweight, easy to assemble with basic tools, and without the need for mechanical assistance is ideal and would go a long way in minimising labour on-site.

Complicated pile designs

The design of foundation piles for a solar structure often needs to consider and rely significantly on geotechnical, site information and parameters before a pile design package can be issued.

Many of the problems with pile design are ground associated, the most common issues include corrosion, soil reactivity, hard/soft soils and pile length. Other issues include the need to optimise pile design for the entire site based on geotechnical information.

Trackers that are inflexible, without the capability to follow site undulations will result in extra different pile lengths (above and below ground).

A fully optimised pile design means significant savings in materials and shipping of piles; however, and if not careful in design, this saving can be quickly eroded away by the additional earthworks, additional logistics, and a complex set of design drawings with a large number of pile sizes and gauges. Hence a balance between optimising steel tonnages and the limiting the number of earthworks to be carried out will need to be made in order to have an optimised steel and earthworks design.

Pile testing

Pile performance testing is an important step in a solar PV farm design and construction. Pile performance tests are a must for not only ensuring pile performance, but also for design optimisation, mitigating risk and gaining confidence during construction and design.

When a pile test is carried out early in the development phase of a solar farm, the results of the test can be used to assist in the design of piles by providing actual insight into the pile and soil interaction. With this insight, the designer is less likely to over engineer the pile and provide an optimise design.

Although pile testing provides lots of benefits, it does come with problems of its own. Some of the more common problems include, formulation of a pile test procedure to comply with standard as well as the intended purpose of the test, as it is not uncommon to see a fix tilt mounting structure pile test procedure be used in a tracker style pile test. It is also common to see the adoption of pile test for buildings be used in solar farm style piles, resulting in an unnecessary conservative test procedure.

With the above in mind, time and cost is the next thing to balance against the risk in of carrying out the works incorrectly. Deciding the frequency of testing while balancing time to complete pile tests and the cost to carry out the works before and during construction are important decisions that impact time, quality and cost. Hence, it is important that an experienced and competent contractor/designer be engaged in carrying out the testing works.

STRUCTURAL FAILURES

Galloping

Tracker galloping mainly occurs in flexible lightly dampened structures that are self-induced by low velocity crosswind across the module as demonstrated in Figure 5. This cross wind flowing across the module/structure generates aerodynamic forces that are in-phase with the motion of the structure causing it to oscillate. Galloping is characterised by this progressive increase in amplitude of oscillation with wind until failure of the structure occurs.

FIG 5 – Aerodynamic forces on a module.

While galloping has been a long-established dynamic wind phenomenon; it is very often overlooked during wind studies of a tracker structure, the omission of this phenomenon often results in the addition of unnecessary risk for the project owner to bare and rectify. Well respected and established tracker companies over-come galloping by carrying out the necessary wind studies, with results to assist in either increasing the stiffness of a tracker by the addition of dampers, stowing trackers at specific angles to minimise the cross-wind impact or to stiffen the structure adequately enough to avoid galloping.

Differential settlement/deflection

Differential settlement or deflection can occur when pile foundations have plastically deformed the soil. This deformation of soil occurs when either there is an over stress on the soil or in the case of reactive soils, the deformation is due to the moisture variation in the soil. This moisture variation results in the ground to undergo volumetric changes, either expanding or shrinking, this volumetric change results in the embedded pile out of alignment with an adjacent pile.

The misalignment if left unchecked can lead to voiding of tracker warranties, reduce the life expectancy of motors and bearings, and in extreme cases, the functionality of tracker may be interrupted leading to tracker failure.

TRADITIONAL SOLUTIONS AND DRAWBACKS

Typical foundation types

Typical foundations used to secure solar structures are (Figure 6):

- Screw Piles – these are typically a hollow steel tube with a helix head at the end of the pile. These piles acts as a screw and are fixed into the ground via a screwing action.

- Driven Piles/Rod – these are the most commonly adopted form of solar tracker foundation, they consists mainly of a steel 'H' shaped section that are directly driven into the ground by way of a hammering action.

- Ballast Foundation – these type of foundation purely rely on the mass of its own to function as a footing and is placed into position either as a precast footing or extruded into place.

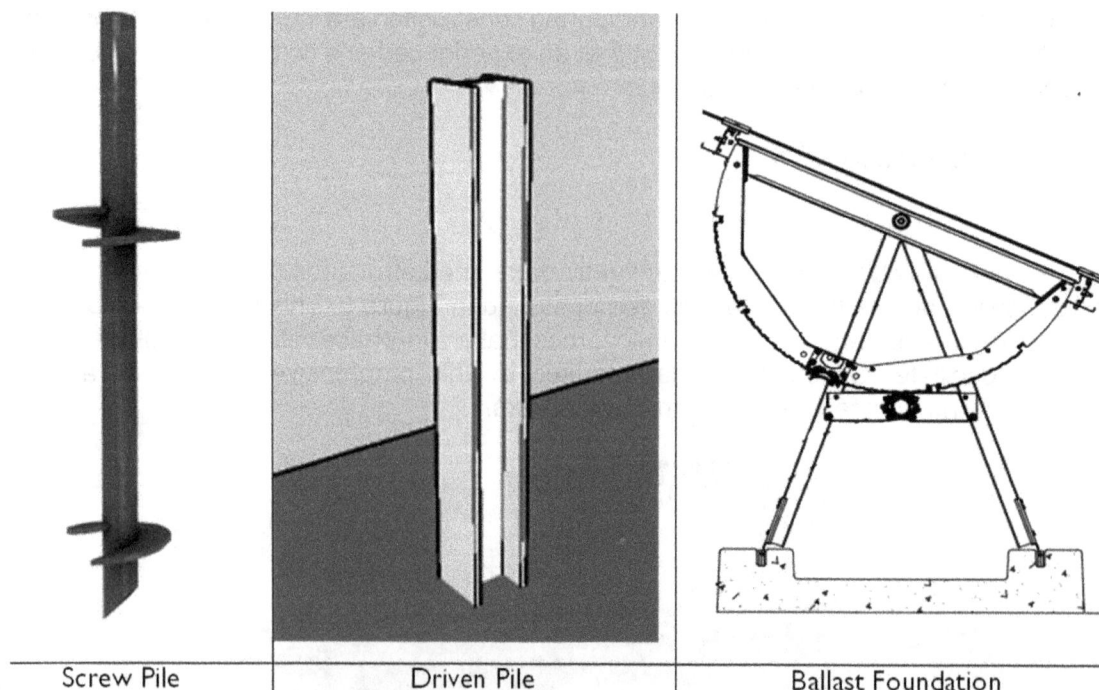

| Screw Pile | Driven Pile | Ballast Foundation |

FIG 6 – Types of foundations.

Foundation

Foundations to support a tracker come in several different forms and each have their own advantages and disadvantages depending on the solar farms' topology, geotechnical conditions and wind region. The three types of foundations considered in Figure 6 are screw piles, driven piles and ballasted foundation.

Piled foundations have traditionally been the most commonly adopted solution used in overcoming all types of ground conditions. There are many reasons why a piled solution is adopted as the preferred solution and justly so due to its ability to be so customisable to suite most if not all geotechnical and site topographic issues. However this flexibility comes at a price, there is a price to pay to have a pile made durable through the use of galvanising to overcome corrosion, there is a price to pay to have a pile made longer and heavier due to soft or expansive soils, there is a price paid if piles cannot be driven into the ground due to rock, there is a price to be paid when the piles are too varied and require additional logistics to over-come installation complexities.

Screw piles have also been used as foundations for solar farms and similarly to piled foundations is adaptable to suite most geotechnical and site topographic issues. Like a piled foundation the piles ability to be so adaptable also come at a price.

Ballast foundations for solar structures are somewhat new and are mainly found in fixed tilt module arrangements and recently in tracker style foundations. Ballast style foundations offer the convenience of using locally sourced material, with practically no lead on the material and the avoidance of having to consider many of the geotechnical problems faced by a steel piled solution.

Trackers

A good tracker supplier would employ the services of specialist wind consultants. These consultants carry out wind studies on structures and report on any deficiencies in a tracker. One common problem in less experienced or reputable trackers is the effect of galloping on the tracker structures, this occurs when the torsional stiffness of torque-tube based trackers have been overly optimised to the point where the torque tube becomes too flexible. This flexibility in the torque tube may be adequate in overcoming static wind load but not necessarily address a dynamic wind event. If the dynamic response is ignored or not properly engineered for, catastrophic failure of the structure is possible even at wind speeds during normal service.

SOLUTION FOR MINE WASTE AND TAILINGS SITES

Alternative

With all the problems that come with the design and construction of a solar structure with traditional tracking systems, the above highlighted geotechnical challenges and wind effects cannot be ignored. By not considering these challenges during design and construction, the plant owner is often left owning these risks in the form of additional costs, added construction time, operations and maintenance costs, and lost energy production.

An alternative solution has now been engineered which eliminates most risks associated with traditional tracker solutions and hence goes a long way in reducing the overall cost and time associated with a solar PV farm construction. This alternative solution involves the use of a ballasted foundation with light gauge, A-frame, steel trackers on top (Figure 7).

FIG 7 – Vales Point ash dam trial foundations.

The advantages and how this alternative tracker overcome traditional tracker issues are discussed further.

Above ground ballast foundation

A tracker sitting on ballast foundations will eliminate virtually all the ground related problems mentioned. Being a ballasted solution, there will be no ground penetration needed, thus eliminating soil related corrosion, hard/soft ground and soil cracks during dry weather. By not having to penetrate the ground, a ballasted tracker is ideally suited for used in landfill sites where preserving the integrity of the capping membrane if of upmost importance.

With an above ground solution, what you see on the ground is what you get, meaning that from an operation and maintenance point of view, there are no hidden surprises below ground that could potentially go unnoticed as with a screwed or piled solution.

The ballasted concrete foundation being an inert and stable material has a design life expectancy of up to 50 years if a specific mix design is adopted, installed correctly and designed in accordance with local concrete standards.

Construction begins almost immediately with a ballasted foundation when concrete arrives on-site, as it avoids the time and cost of comprehensive Geotech reports, pile design and long shipping times of piles, furthermore ballast is made using commonly available kerbing machines operated by local contractors. Depending on the site specific pile design, specialised contractors may need to be brought in.

Wind lock/stow

During high winds, the Alion Storm Tracker stows the tracker into a neutral position, in this position the motor is protected by way of disengagement of the drive line to prevent any load transfer into the motor and drive line. With this feature, the tracker can withstand wind speeds of up to 58 m/sec. In the stow position, wind forces on the tracker are at the minimum, hence enabling an optimised design for both the structure and footing.

A frame structural stability and construction

The Alion Storm Tracker main support is made up of an 'A' frame construction that directly transfers design loads via tension and compression forces into the footing and foundation. The 'A' frame support enables the structure to be structurally stiffer than traditional cantilever style post supports which increases stability in the tracker.

The structure is made from light gauge steel sections. Structural sections are made from pressed cold formed steel sections with G90 coating for durability. There are no parts that are more than 11 kg hence it can be easily assembled by two people and does not require power tools for assembly.

The structure is broken up into 6 m sections, having supports at 4 m centres, thus possessing a natural frequency of the structures is approximately 7 Hz. Most torque tube-based trackers have a natural frequency of 0.8 Hz due to their low torsional stiffness in the torque tube.

Adaptable to terrain

One main feature of the drive tube is that it uses a flexible coupling between tracker sections which, enables changes in grade of 6 per cent at every 6 m by allowing the tracker to closely follow natural ground profile, in doing so, the need for any earthworks can be eliminated or significantly reduced, even on severely undulating land lots.

No pile testing

Alion trackers on ballasted foundations provide the benefit of having a tracker while not requiring the need for piles, hence avoiding many of the problems associated with piles in the ground. Designing to construct without piles means there is:

- no need for pile testing needed to optimise design

- no testing require during pile refusal or pile embedment does not meet the design depth

- no testing needed to validate design or demonstrate quality of construction.

Ballast testing

While driven piled solutions require elaborate testing in order to demonstrate a pile ability in meeting its performance in soil, foundation testing for a ballasted foundation involves checking whether if there is adequate ground bearing pressure available to resist the ballast applied pressure via the use of a simple dynamic cone penetrometer (DCP) test.

Mounting systems overview

Fix tilt or tracker, that is the question faced by the solar farm designer. Depending on land shape, availability and energy yield requirements, the designer will have to assess the best mounting system to meet and optimise the given constraints.

Tracker prices have reduced to the point where it is now becoming the mounting structure of choice considering the additional yield it can achieve over a fixed tilt structure. Given that most solar farms mounting systems are tracker based, choosing the 'right' type of foundation that would minimise complexities in construction would go a long way in delivering an optimise solar farm (Table 2).

TABLE 2

Ballasted foundation versus pile-driven foundation buyer's guide.

Challenge	Steel pile	Ballast foundation
Corrosion	Thicker steel or thicker galvanising, $$	No steel to corrode
Rock refusal	Predrilling and concreting, $$$	Not affected by rock
Reactive soils	Longer piles to anchor pile, $	Tracker movement
Ground undulations	Longer pile and/or earthworks, $$$	Follows undulations
Low ground bearing pressure	Longer and wider piles, $	Good in low bearing pressure
Testing	Elaborate expensive pile load test, $$	Simple DCP test to prove ground bearing pressure
Shipping and transport	Long lead item with inland transport, $$	Local concrete with minimal lead time

Ballast design

The design of a ballast foundation is standard and does not significantly vary between one site to another. The sizing of a ballast foundation mainly involved adequately sizing the concrete to overcome uplift forces, downward allowable bearing pressure on soil and sliding resistance of the ballast foundation. This one solution fits all allows the one design to be used in all similar wind region sites, thereby standardising and eliminating the need to re-engineer foundations.

CONCLUSIONS

Mine waste and tailings sites have the potential to be the ideal location for the installation of a solar power station. There are financial, social, regulatory, legislative and technical benefits to explore each and every site for its potential.

With a strong social licence to repurpose the mine, the reuse of existing HV electrical infrastructure and its close proximity to the grid, and the well understood mine waste tailings and a structurally sound ballasted solar structure make this a highly desirable site location to explore.

Australia has a real challenge to meet our legislated emissions reduction targets and mine waste and tailings sites could become a major solution to this problem, at the same time as addressing the growing costs of mine closure.

ACKNOWLEDGEMENTS

The author acknowledges the previous report written for Alion Energy Incorporated by former Director of Engineering, Mr David Hng.

REFERENCES

Parkinson, G, 2022. AEMO suspends new wind and solar projects as it battles to deal with market crisis, *Renew Economy*, https://reneweconomy.com.au/aemo-suspends-new-wind-and-solar-projects-as-it-battles-to-deal-with-market-crisis/

Aubrey, J, 2023. Australia's energy transition needs new rules for grid connections, *Clean Energy Council*, https://www.cleanenergycouncil.org.au/news/australias-energy-transition-needs-new-rules-for-grid-connections

Earthworks, 2023. Tailings are Mine Waste, *Earthworks*, https://earthworks.org/issues/tailings/

Design and construction of a combination soil and water cover on a tailings storage facility in Tasmania

C Cahill[1], R Longey[2] and D Tonks[3]

1. Senior Civil Engineer, GHD, Burnie Tas 7320. Email: clem.cahill@ghd.com
2. Technical Director Tailings, GHD, Hobart Tas 7000. Email: rob.longey@ghd.com
3. Engineering Superintendent, Grange Resources, Savage River Tas 7320.
   Email: dan.tonks@grangeresouces.com.au

ABSTRACT

The Main Creek Tailings Dam (MCTD), located at Grange Resources Pty Ltd's (Grange) Savage River Mine in north-west Tasmania has been in operation since 1985 and is transitioning from an operational TSF to closure. The MCTD is an upstream constructed facility with a maximum height of approximately 83 m. Tailings stored in the MCTD are potentially acid forming (PAF) and require careful management through operation and closure to minimise the risk of Acid and Metalliferous Drainage (AMD) forming in the TSF.

The site is situated on the west coast of Tasmania with rainfall significantly exceeding evaporation. Therefore, a water cover would typically be most suitable, however, due to the upstream constructed embankments a soil cover is required adjacent to embankments to meet long-term stability requirements.

During operations, three trial covers were constructed and instrumented to monitor the performance over several years. The information obtained was used to evaluate cover performance and calibrate numerical transient seepage models. The preferred cover based on the trial cover performance was a combination of clay and rock fill cover which maintained a high degree of saturation in the clay, minimising oxygen ingress to the underlying tailings, and reducing the likelihood of AMD formation.

The preferred clay and rock combination cover was assessed and optimised during the detailed design phase by undertaking 2-dimensional transient unsaturated seepage modelling in SVFlux, considering a conservative climate scenario.

Construction of the clay component of the combination cover has recently been completed. The construction process, challenges, and QA/QC is discussed in this paper.

INTRODUCTION

The Savage River open cut magnetite mine is situated on Tasmania's west coast, approximately 100 km south-west of the city of Burnie. The mine has been in operation since 1967 and is currently operated by Grange Resources Pty Ltd (Grange). The Main Creek Tailings Dam (MCTD) was the primary active tailings storage facility on the site until recently when the South Deposit Tailings Storage Facility (SDTSF) was commissioned downstream of the MCTD.

The MCTD comprises primarily of valley type storage with several embankments to impound the tailings. The main embankment utilises upstream construction methods and has a maximum height of 83 m. The tailings stored within the MCTD are potentially acid forming (PAF), therefore prevention of acid and metalliferous drainage (AMD) is a key consideration for the closure design.

The closure planning for the MCTD has evolved through the life of the facility involving concept closure planning, construction and monitoring of trial covers, detailed design, and construction which is currently underway.

The closure cover comprises a combination of soil and water cover to minimise oxygen ingress to tailings. Rainfall at the site significantly exceeds evaporation which maintains a water cover in perpetuity. A soil cover is provided adjacent to embankments to meet long-term stability requirements.

This paper presents the various components of planning undertaken to transition the MCTD from operations into closure.

MCTD HISTORY

MCTD is a zoned earth and rock fill embankment constructed (using conventional downstream construction methods) by previous mine operators between 1982 and 1985 using waste clay and rock fill materials from mine development operations. The main embankment was constructed to RL310 m, giving in a maximum height of approximately 60 m.

The main embankment crest was raised 5 m to RL315 m in 1994, as part of the closure works for mine abandonment by the previous operators.

The storage was then reopened, with the embankment raised by upstream construction methods to RL319 m in 2002, RL324.5 m in 2004, RL328.5 m in 2006, RL 333 m in 2012 and RL 336 in 2014/15. The closure raise to RL338 m was completed in 2019, which resulted in a total of 23 m height using upstream construction.

Throughout the life of the facility a number of embankments have been constructed to maintain impoundment, these embankments include the NW Pond Embankment, Spillway Embankment and the Saddle Dam. The Old Tailings Dam (OTD) is situated to the north of the MCTD, the Emergency Tailings Dam (ETD) is situated to the west of the MCTD. The general arrangement of the MCTD is presented in Figure 1.

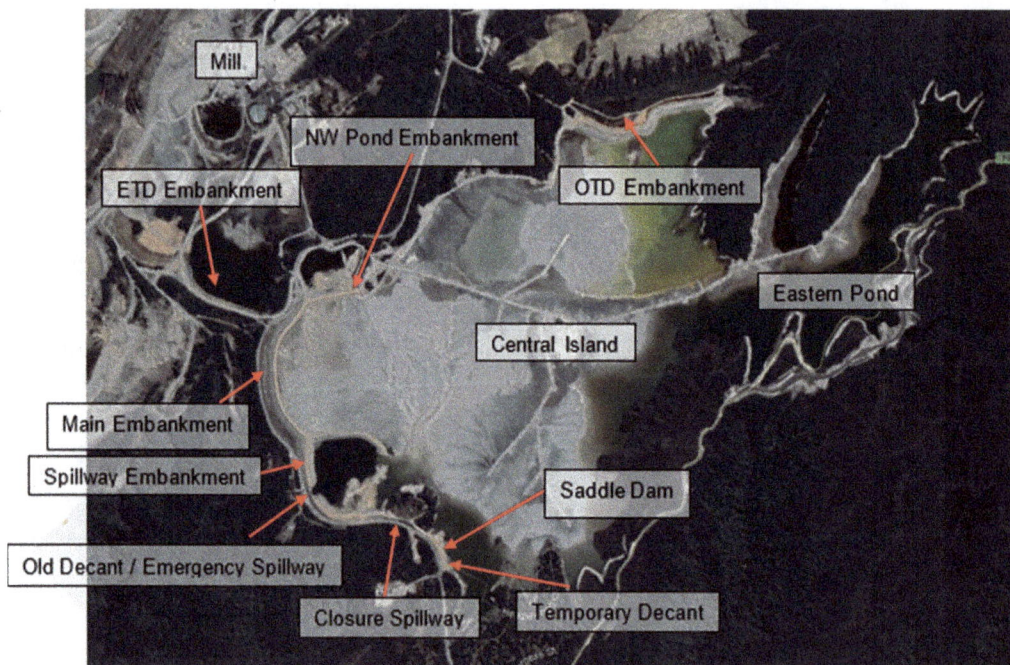

FIG 1 – MCTD general arrangement.

CLOSURE CONCEPT

The climate on the west coast of Tasmania is well known for high rainfall and low evaporation with an approximate 2:1 ratio, therefore the intent of the closure cover design is to use the wet climatic conditions to advantage to limit oxygen ingress to the stored tailings at closure to reduce the likelihood of the formation of AMD. The net positive rainfall environment enables a full water cover, however, a full water cover would result in water directly against upstream constructed embankments which are not designed to be water retaining.

Both soil and water cover types which limit oxygen ingress suit the climatic conditions when plotted on the GARD cover guide as shown in Figure 2. The closure concept involved a combination cover system comprising a 'dry' or 'soil type' cover 150 m wide adjacent to the upstream constructed embankments with a water cover of 1 m minimum depth elsewhere.

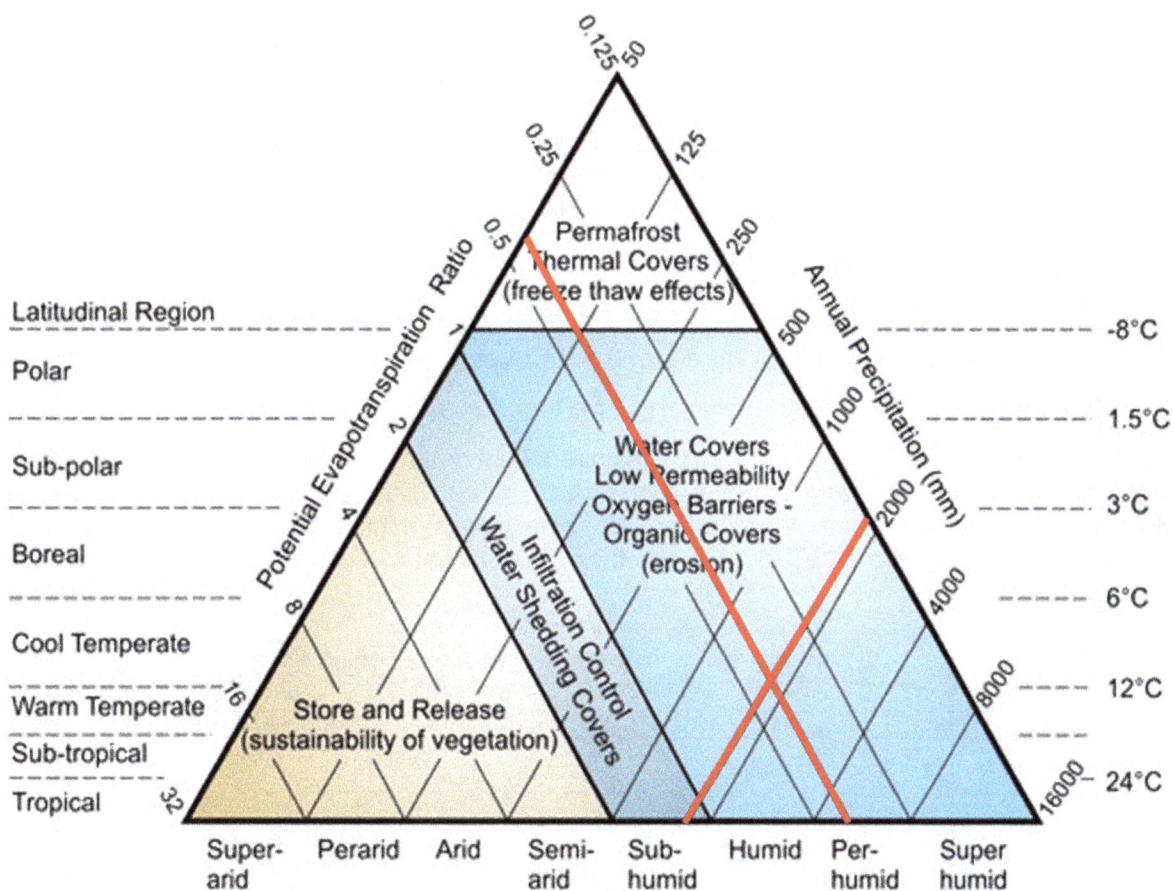

FIG 2 – Climate and cover types (INAP, 2014).

During the various stages of raising MCTD the tailings management plan and conceptual closure plan were updated based on the expected life-of-mine. Premature closure plans were also developed at multiple raise heights such that MCTD could be closed successfully should an unplanned stoppage or closure occur.

TRIAL COVERS

The long-term performance of the closure cover is critical, therefore trials with field performance monitoring were undertaken comprising a section of water covered tailings and three trial soil covers approximately 40 m wide by 60 m long. The following trials were constructed on the NW Pond area:

1. Homogenous Clay Cover 2 m thick.

2. Geosynthetic Clay Liner (GCL) underlying a rock fill protection layer 1 m thick.

3. Clay 1 m thick underlying a 1 m thick rock fill protection layer.

The intent of all three covers was to maintain a high level (>80 per cent) saturation within the cover and therefore saturation was the key parameter being monitored. Instrumentation was installed in the underlying tailings and covers to monitor oxygen, soil matric suction, and volumetric water content. Standpipes with vibrating wire piezometers and a weather station were also installed.

Test pits were excavated into the trial covers and material samples were collected for laboratory testing including Soil Water Characteristic Curve and Permeability.

The collected data from all three covers was used to calibrate finite element models of the cover in the software package SVFlux. The performance from all three covers was acceptable, with >80 per cent saturation being met at all times, however, the close proximity to a pond (~60 m away) may have impacted saturation levels.

Other cover performance elements such as erosion due to rainfall run-off and wave erosion at the pond edge were assessed visually and found to be minimal on all trial covers. Erosion performance was also assessed based on the performance of the upstream batters of the MCTD embankments.

COVER DESIGN

Based on the results of the trial cover performance and model calibration, the larger MCTD main embankment was modelled at closure with SVFlux software to determine the optimal cover design for the MCTD. The climate model considered average conditions and also the driest 365-day period on record at Savage River to assess the extreme case where cover recharge is low resulting in worse case cover saturation. Four covers were modelled in 2D sections as follows:

1. 300 m wide – 1 m thick Clay Cover.
2. 150 m wide – 2 m thick Clay Cover.
3. 150 m wide – 1 m Clay underlying a 1 m Rock Cover.
4. 150 m wide – 1.7 m Clay underlying a 0.3 m Rock Cover.

The target saturation for the 1 m and 2 m thick covers was determined based on the chart shown in Figure 3 (Miller, 1998). A 1 m thick cover required a minimum saturation of 80 per cent and a 2 m thick cover required a minimum saturation of 70 per cent, which provides a reduction in Acid Sulphate Generation Rate (ASGR) of more than 95 per cent by limiting oxygen ingress to the underlying tailings.

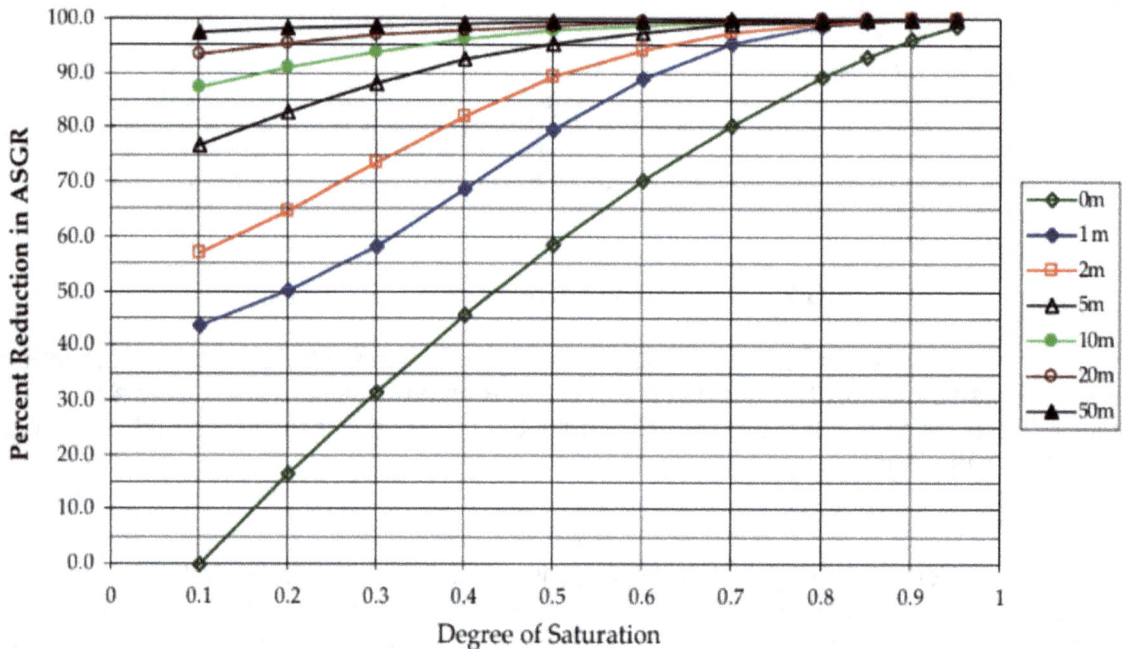

FIG 3 – Cover saturation versus reduction in ASGR (Miller, 1998).

A typical cross-section from the 2D model is presented in Figure 4 which shows that the modelling takes into account seepage vertically and horizontally throughout the tailings and MCTD Embankment.

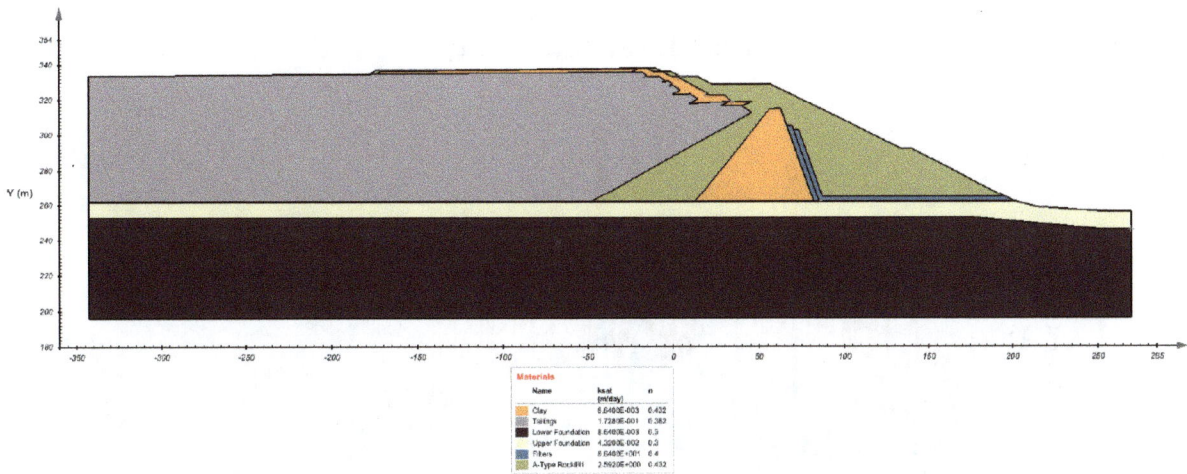

FIG 4 – Cover model typical cross-section.

Climate data set

The climate data is a key input to the SVFlux model as rainfall infiltration and evaporation have a significant impact on the cover saturation.

Daily rainfall data has been recorded by the Bureau of Meteorology (BOM) from February 1966 to 2018 at the Savage River Mine. The daily data was reviewed to determine the driest 365-day period on record. The driest period recorded was 25/08/1966 to 25/08/1967, with a total rainfall of 1242.5 mm. The average annual rainfall is 1952.7 mm.

Rainfall data from the driest 365-day period is presented in Figure 5. The time period starts from August 25, 1966.

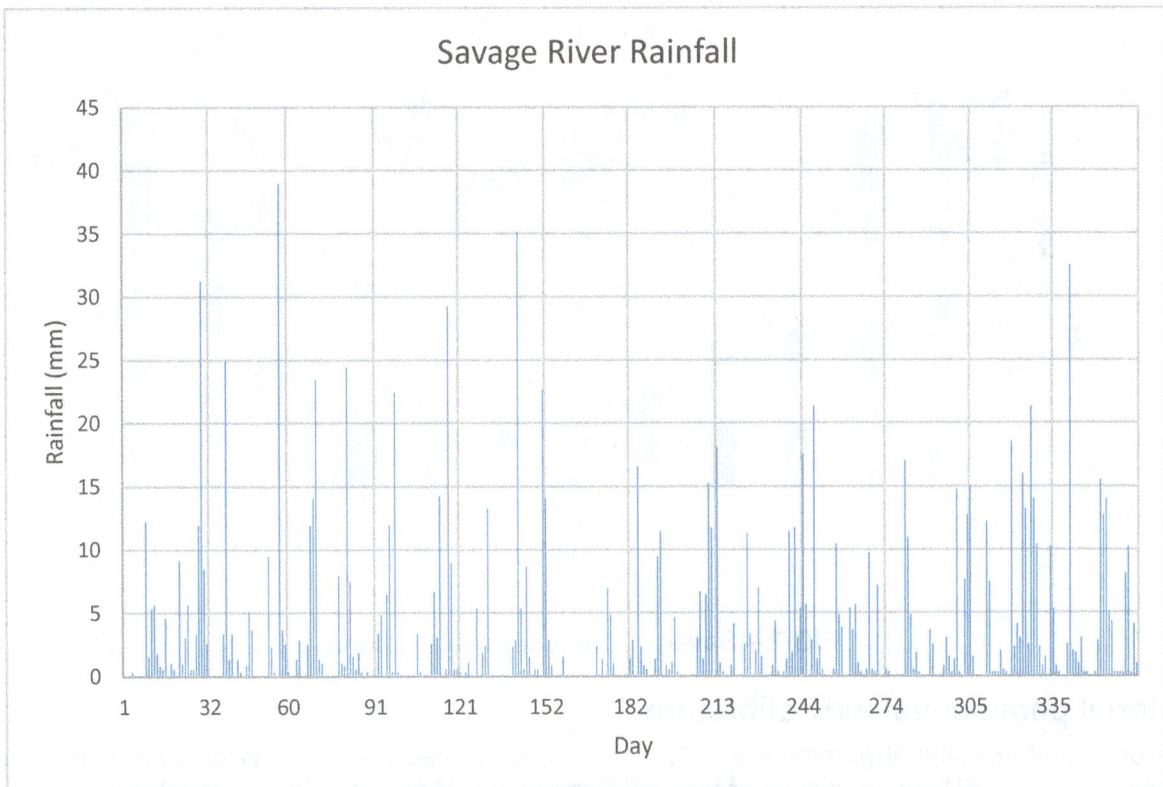

FIG 5 – Savage River rainfall.

Temperature data is also recorded at the Savage River Mine by BOM, data for the driest 365-day period has been collated and is presented in Figure 6. Temperature data is required to calculate Actual Evaporation in SVFlux using the Modified Wilson Empirical Equation in SVFlux.

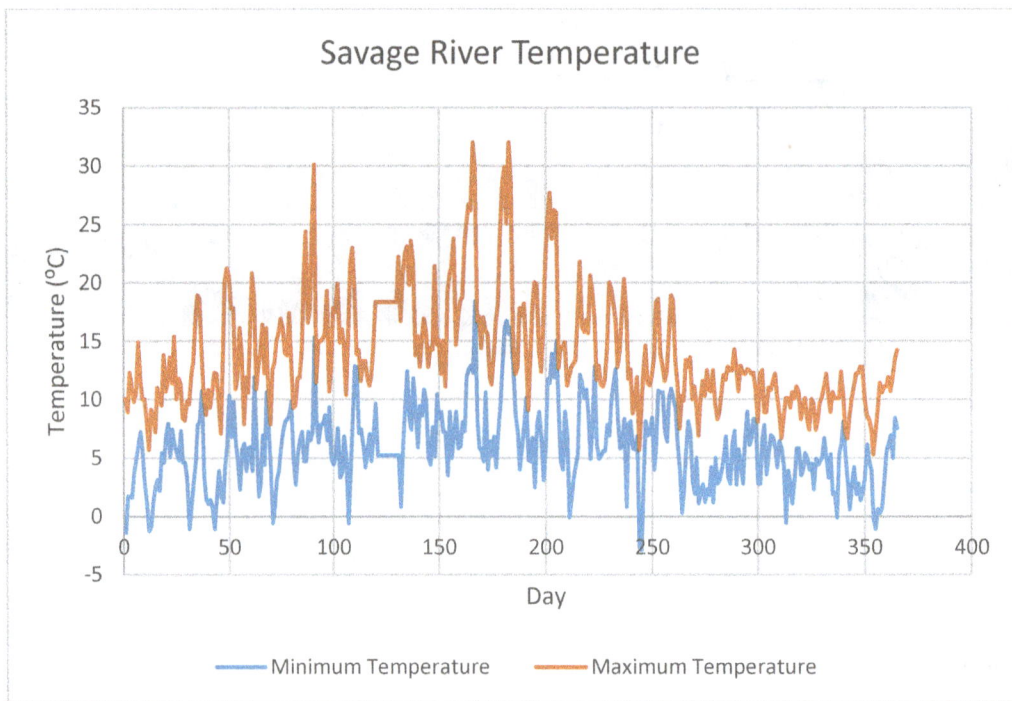

FIG 6 – Savage River temperature.

Potential Evaporation is used to calculate actual evaporation in SVFlux using the Modified Wilson Empirical Equation. Average Daily Evaporation for each month is recorded by the BOM. The data is presented in Figure 7. The data is used to calculate the actual evaporation in SVFlux.

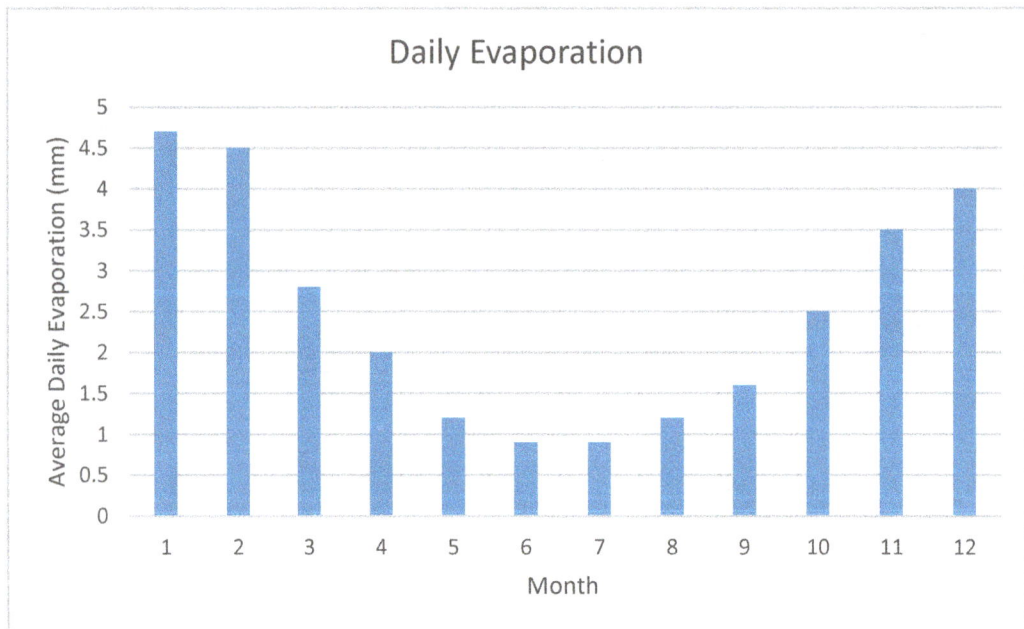

FIG 7 – Savage River daily evaporation.

Material parameters and calibration

The unsaturated material parameters (SWCC and unsaturated permeability curves) were initially adopted from the SVFlux modelling of the NW Pond Cover Modelling. The material parameters for the foundation and filter materials have been adopted from previous design phases, saturated parameters have been assumed only. The permeabilities are presented in Table 1.

TABLE 1
Material permeability.

Material	Permeability, K (m/s)	Anisotropic Ratio
Lower Foundation	1×10^{-7} m/s	1
Upper Foundation	5×10^{-7} m/s	1
Filters	1×10^{-3} m/s	1
Tailings	2×10^{-6} m/s	10
Embankment rock fill	3×10^{-5} m/s	1
Embankment clay	1×10^{-7} m/s	1
Cover rock fill	3×10^{-5} m/s	1
Cover clay	1×10^{-7} m/s	1

An initial model set-up with the material parameters indicated a low phreatic surface through the embankment. The phreatic surface was lower than the phreatic surface indicated by the installed vibrating wire piezometers. To model a phreatic surface that represented actual conditions the permeability of both the rock fill and tailings were adjusted. The adjusted permeabilities are presented in Table 2.

TABLE 2
Calibrated Parameters.

Material	Permeability (m/s)	Anisotropic Ratio
Tailings	2×10^{-6} m/s	10
Rock fill	3×10^{-5} m/s	1

The current pond is approximately 300 m from the embankment, with a level of RL331.2 m. At closure, the pond will be 150 m from the main embankment, with a level of RL336 m. Therefore the final phreatic surface is expected to be higher than currently indicated by the piezometers. The calibration considered a phreatic level slightly above the current phreatic surface.

The tailings are expected to have a higher permeability horizontally and as such an anisotropic ratio Kh:Kz of 10 has been applied. The permeability of the tailings is significantly higher than that adopted for the NW Pond assessment. The tailings at the main embankment are typically coarse due to being spigotted and have a higher permeability due to the increased sand content. Previous falling head tests undertaken in the tailings beach near the embankment crest showed permeability ranging from 2×10^{-6} to 1×10^{-5} m/s.

Results

The 300 m wide cover enabled the pond to remain further from the main embankment, lowering the phreatic surface within the tailings and embankment and improving stability outcomes, however, the minimum target saturation levels of 80 per cent were not achieved for this case.

A series of 150 m wide covers with a total thickness of 2 m were also modelled as the thicker cover requires a slightly lower saturation level of 70 per cent to achieve a similar reduction in oxygen transfer rate (Miller, 1998). The combination clay and rock cover's both met the target saturation levels throughout the 365 day model period. The final design was selected as the 1 m Clay and 1 m Rock cover due to limited clay availability on-site. The rock cover allows for collecting recharge and limiting evaporation to keep the underlying clay cover at a satisfactory saturation level and provides erosion protection and a growth medium. A summary of the results is presented in Table 3.

TABLE 3

Cover modelling results.

Cover	Clay saturation			
	Target	Minimum	Maximum	Average
1 m Clay Cover	80%	65.0%	74.3%	69.4%
2 m Clay Cover	70%	69.8%	75.8%	72.9%
1 m Clay + 1 m Rock	70%	71.3%	74.0%	73.0%
1.7 m Clay + 0.3 m Rock Cover	70%	70.6%	76.3%	73.3%

The stability of the MCTD embankments and the selected closure cover was assessed and achieved acceptable factors of safety under ANCOLD guidelines (ANCOLD, 2019).

CONSTRUCTION

Construction of the closure cover commenced in December 2020 with the placement of the clay layer. The construction methodology was unproven to this point. Before construction, it was not certain how placing clay over the beached tailings would perform due to the variable nature of the tailings beach and variation of bearing capacity. However, prior experience on-site with constructing access roads over the tailings had shown it was feasible. A safe work method was developed and to limit the risk of bearing failure and the potential to bog machines in the tailings. The method comprised a 1 m thick clay layer which was placed and compacted in a single lift. The design and specification called for multiple clay layers of 300 mm thickness. Placing a single 1 m thick layer is unconventional, therefore field and laboratory testing was undertaken at various depths to ensure the clay cover particle size distribution, density, and permeability specifications were achieved.

In summary, the lab testing showed that minor segregation occurred with the material being marginally coarser at the bottom of the layer, however, this achieved the required permeability. The placement methodology was effective throughout the whole clay placement with only minor bearing failures observed in extremely weak or wet areas of the tailings beach. Grange plan to complete the overlying rock layer once materials become available from the mine. Photos of the construction are presented in Figures 8 and 9. Instrumentation will be installed similar to that used in the trial cover systems to monitor closure performance, particularly the soil cover and tailings saturation levels in addition to reviewing for any increase in pore pressures in the embankment when the pond is allowed to rise to meet the soil cover.

FIG 8 – Clay cover placement.

FIG 9 – Clay cover placement.

TAILINGS MANAGEMENT

In the years preceding closure, tailings management required careful consideration to ensure that settled density was maximised, while storage capacity was fully utilised and the closure objectives could be met. The key closure objectives to be considered through tailings management were ensuring the large tailings beach maintained during operations was maintained at least 1 m below the future closure pond level to ensure that a minimum 1 m water cover would be established at closure. Tailings deposition modelling was utilised to plan deposition through to closure, allowing Grange to forecast available storage volumes and inform appropriate discharge locations to maximise the storage potential of MCTD whilst transitioning to a new facility.

ASSOCIATED INFRASTRUCTURE

Before closure, several key pieces of infrastructure associated with the MCTD were required to be designed and constructed before transitioning to closure. Two of the key infrastructure items are detailed in the following subsections.

Closure spillway

The MCTD is designed to meet the Australian National Committee on Large Dams (ANCOLD) Guidelines on Tailings Dams. The guidelines recommend that for closure all TSFs can safely pass a Probable Maximum Flood (PMF) event. A new spillway was designed comprising a low flow channel in the centre of a larger 60 m wide open channel spillway. The mine access road crosses the spillway, therefore the low flow channel was included to pass regular flows up to a 1:140 Annual Recurrence Interval event through a single concrete culvert below the mine access road. The larger 60 m open channel was designed to pass a peak flow rate of 100.32 m^3/s in the PMF event. The spillway features a concrete nib wall at the inlet to maintain a consistent pond level. The spillway is lined with rip-rap with a minimum D50 of 100 mm, with shotcrete applied to the spillway excavation prior to rip-rap placement to prevent erosion of the *in situ* material.

OTD seepage collection bund and pipeline

The Old Tailings Dam (OTD) is a legacy TSF located upstream of the MCTD. The OTD is the major source of AMD due to its large surface area of exposed PAF tailings. Seepage from the toe of OTD embankment flowed into the MCTD, which was neutralised by the alkalinity in Grange's tailings slurry during MCTD operations. However, for closure the continual seepage of AMD into the MCTD pond would have turned the pond acidic, leading to AMD formation in MCTD also.

To capture the AMD seepage, the OTD seepage collection bund and transfer pipeline was designed and constructed. The scheme comprised the following:

- A 6 m high earth fill embankment collection bund at the base of the OTD to collect seeps from the OTD at the interface with the OTD embankment and MCTD Tailings.

- Surface water diversion drains on the OTD embankment to reduce clean water inflows into the collection bund.

- An intake structure at the OTD seepage collection pond to allow precipitation to settle before entering the transfer pipeline.

- A gravity polyethylene DN280 mm PN10 transfer pipeline from the OTD intake to the SDTSF Storage with allowance for pressure pigging of the pipeline to enable precipitation cleanout if required.

The items above were constructed in stages from 2014 to 2018 and the scheme is now transferring AMD seepage from the OTD to the SDTSF where it is neutralised through the higher pH active tailings deposition.

CONCLUSION

Planning for the closure of MCTD began in approximately 2010, approximately ten years before commencing construction of the closure cover. The planning process undertaken allowed key infrastructure needs to be designed and constructed in a staged manner in the lead-up to the cessation of tailings deposition and transition to closure.

As the knowledge base for closure has continually improved over time, the closure design was also progressively refined. Maintaining premature concept closure plans at various stages ensured the MCTD was operated in a state which could successfully enter a care and maintenance phase or early closure if required. The staged closure planning development approach and closure cover construction have thus far proved to be successful in that the cover has been able to be constructed in accordance with the design specification. At the time of writing, the post closure performance monitoring was being developed to validate the closure design.

REFERENCE

ANCOLD, 2019. Guidelines on Tailings Dams – Planning, Design, Construction, Operation and Closure – REVISION 1, ANCOLD, Australia.

INAP, 2014. Global Acid Rock Drainage Guide, INAP [online], viewed 24 May 2022. Available from: <http://www.gardguide.com/index.php?title=Main_Page>

Miller, S, 1998. Theory, Design and Operation of Covers for Controlling Sulphide Oxidation in Waste Rock Dumps, ACMER, Proceedings of the Third Australian Acid Mine Drainage Workshop.

Decommissioning and reclamation of the central tailings dam

F A R Do Carmo[1] and Y M Azevedo[2]

1. MAusIMM, Technical Services Manager, Mineração Usiminas SA, Itatiaiuçu, Minas Gerais 35685-000, Brazil. Email: frederico.carmo@usiminas.com
2. Environment Manager, Mineração Usiminas SA, Itatiaiuçu, Minas Gerais 35685-000, Brazil. Email: yuri.azevedo@usiminas.com

ABSTRACT

The central dam was decommissioned in February 2022. The decommissioning process of this upstream dam was total removal of the 7.5 million cubic metres (Mm³) of the tailings in three years, which were processed in a processing plant built with the objective of using all the tailings already deposited in old dams and the actual tailings generated by the plants in operation. The operation was an integrated effort between the geotechnical, mine planning, hydrology, and environmental teams. The activities took place in parallel, where the areas were mined, reshaping slopes and revegetation was carried out with 12 000 species native to the region. The vegetation cover in tailings dams increases the amount of organic matter and biological activity, in addition to improving aesthetics, controlling erosion processes, and enabling more post-closure uses.

Another fundamental issue for revegetation is the choice of species, which must be made according to climate, soil, drainage conditions and biotic conditions. In this case, pioneer species and legumes were sown, which fix nitrogen in the soil, an advantageous characteristic in areas that were initially occupied by tailings, which are normally devoid of this nutrient.

Mining was done with trucks and excavators and the mining planning activities was carried out to ensure trucks trafficability and geotechnical premises. Reinforcement was also made downstream of the dam during this process to increase safety factors.

INTRODUCTION

The Central tailings dam is in Itatiaiuçu city, in the state of Minas Gerais in Brazil. This dam was intended to contain the iron ore waste generated at the Central processing plant and the recovery and return of water for the process.

This structure began with the construction of a starting dyke in a compacted landfill and had its upstream raising by upstream using the own tailings released into the reservoir as construction material.

The operation of this structure began in 1992 and ended in 2014. The dam was 59 m high and the volume of tailings inside the reservoir was 7.56 Mm³ in an area of approximately 30 ha.

FIG 1 – Downstream slope of the Central tailings dam 2017.

The decommissioning of the Central Dam, which started in 2014, was interrupted in the same year and then restarted in 2019 and completed in 2022. The tailings that were removed were reprocessed in a plant built for this purpose, taking advantage of all the tailings generated in the past and also those generated in current operations.

The waste from this plant is piled up after going through a dewatering process. Specifically in this dam, it was the total removal of the tailings that were in this dam and part of the starting dyke. In summary, the sequence defined below presents the activities that were carried out for the decommissioning of the structure:

- Implementation of a downstream reinforcement berm.

- Construction of a peripheral channel to divert surface water away from the drainage area at elevations higher than those of the reservoir, with the aim of minimising water discharge in the region.

- Installation and operation of wells to lower the water level in the tailings deposited in the reservoir and the water table in the dam, to safely enable tailings mining operations.

- Mining operation of the tailings deposited in the dam's reservoir and removal of its dam.

- Automated monitoring of pore pressure control instruments, water table elevation and displacements – piezometers, water level indicators and displacement milestones.

During decommissioning, all monthly mining plans were evaluated by the geotechnical team, to verify the safety of the dam and workers both downstream and upstream.

The hydrological studies developed for the Central Dam's contribution basin indicated the construction of three sumps that are interconnected by a central channel that has the objective of slowing down the rainfall events and sedimentary particles to control the turbidity of the waters.

After the completion of the removal of the tailings, the revegetation was carried out. The vegetation cover in tailings dams increases the amount of organic matter and biological activity, in addition to improving aesthetics, controlling the erosion process, and enabling more post-closure uses.

METHOD

The operation of the Central Dam began in 1992 with the construction of the starting dyke in compacted soil, cresting the El. 928.00 m. The elevations were carried out upstream, using the coarser proportion tailings generated in the Ore Treatment Facilities, with bench of 5.0 m in height and slope inclination of 1.0V:2.5H, with berms of variable width. This structure operated until 2014 and the tailings were once deposited in its reservoir.

In its final configuration (before the mining of tailings), the crowning quota was at El. 973.00 m, making a dam height of 58.0 m and a total volume of tailings deposited in the reservoir of around 7.5 Mm^3.

To guarantee the safety of the structure during the mining stages of the dam during its decommissioning, a downstream reinforcement berm was built, consisting of rock fill and contact transitions on a treated foundation, with the removal of soft soil. The project was carried out in four stages, up to an elevation of 949.00 m, to ensure the structure has a Factor of Safety (FoS) greater than 1.5; both stages considering undrained condition. The work started in August 2019 and was completed in November of the same year. Both the reinforcement and the removal of the tailings were carried out with 40-ton trucks and excavators with 2 m^3 bucket.

Material Name	Color	Unit Weight (kN/m3)	Sat. Unit Weight (kN/m3)	Cohesion (kPa)	Phi (deg)
Dique de partida		16.5	18	50	20
Solo residual maduro		19.1	19.85	5.5	30
Solo residual jovem		19.1	19.85	11	32
Berma		20	22	0	35
Alteamentos - rejeitos		20	22	8	30
Rejeito - seco		20.75	25.31	0	35

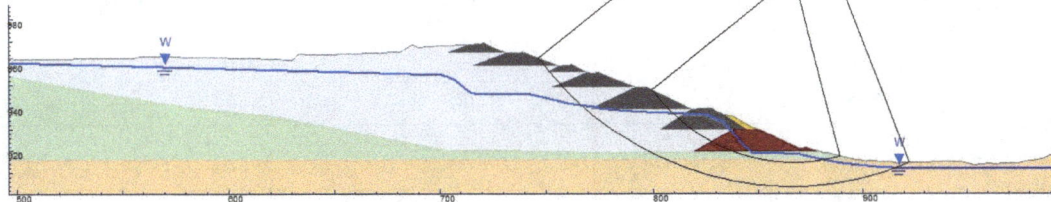

FIG 2 – FoS before reinforcement.

Material Name	Color	Unit Weight (kN/m3)	Sat. Unit Weight (kN/m3)	Cohesion (kPa)	Phi (deg)	Vertical Strength Ratio
Dique de partida		16.5	18	50	20	
Solo residual maduro		19.1	19.85	5.5	30	
Solo residual jovem		19.1	19.85	11	32	
Berma		20	22	0	35	
Alteamentos - rejeitos		20	22	8	30	
Rejeito - Drenado		20.75	25.31	0	35	
Rejeito - Não drenado		20.75	25.31			0.26
Berma reforço_Enrocamento		22		0	38	

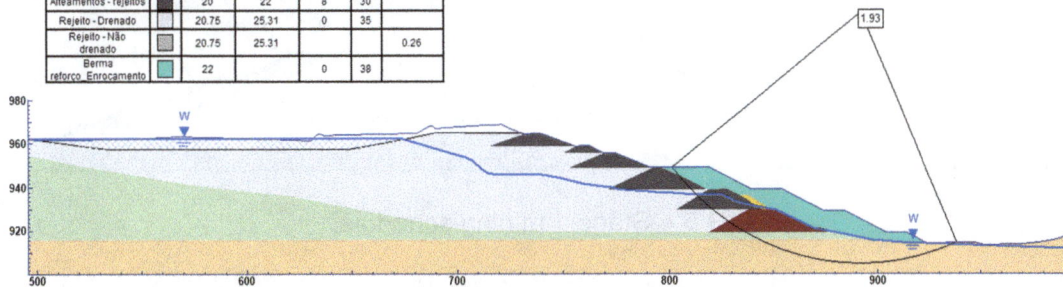

FIG 3 – FoS after reinforcement.

Figure 4 shows the reinforcement work downstream of the Central Dam.

FIG 4 – Downstream reinforcement, November 2019.

The decommissioning project included the total removal of the tailings deposited in its reservoir and the entire tailings upstream dykes, as well as part of the original started dyke. The mining schedule was divided into four stages, starting from the August 2019 mining plan. These stages were prepared by a consulting firm and are referenced in the documents MJ00B05000BRTP0001 and MJ00B05000DRGP0029.

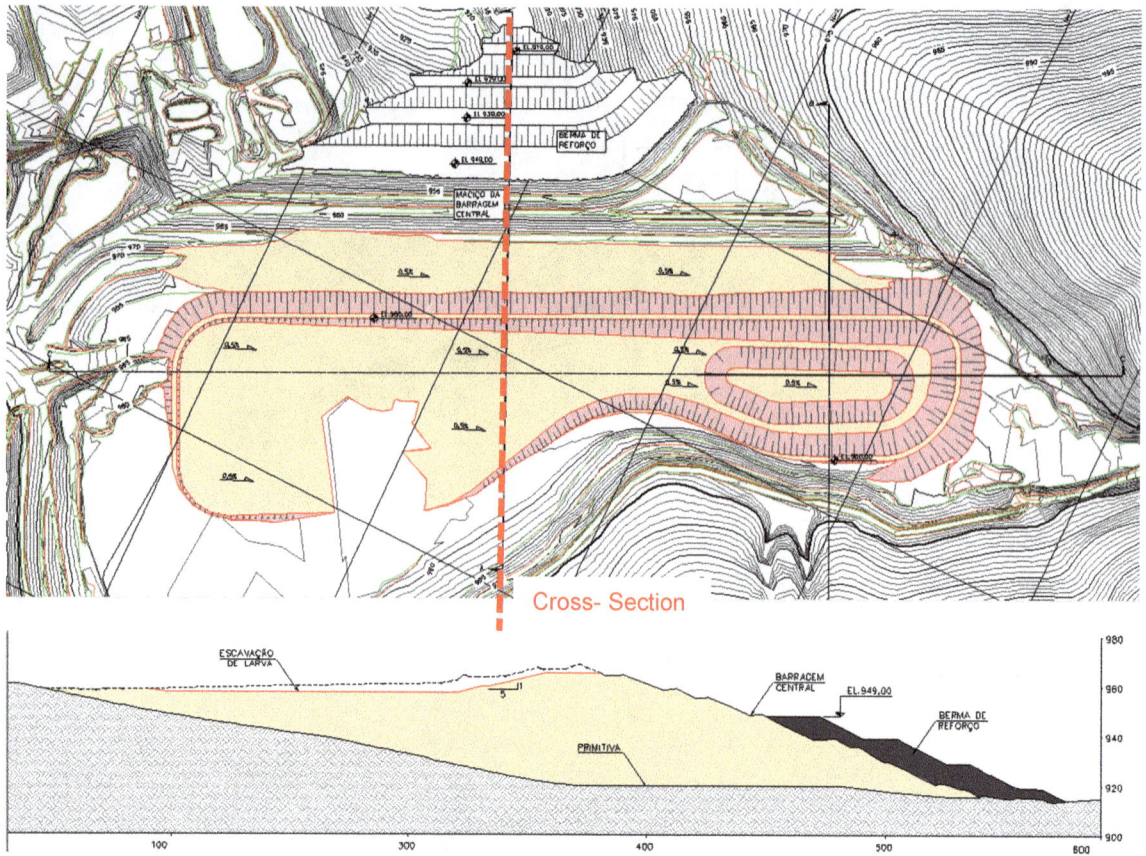

FIG 5 – Stage 1 mining schedule.

Cross- Section

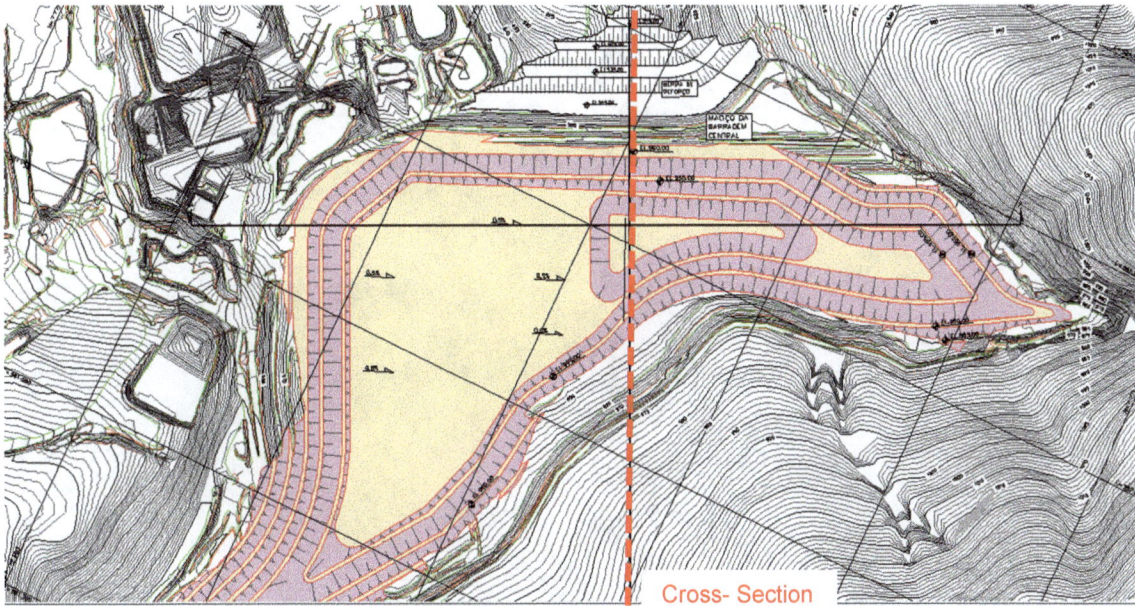

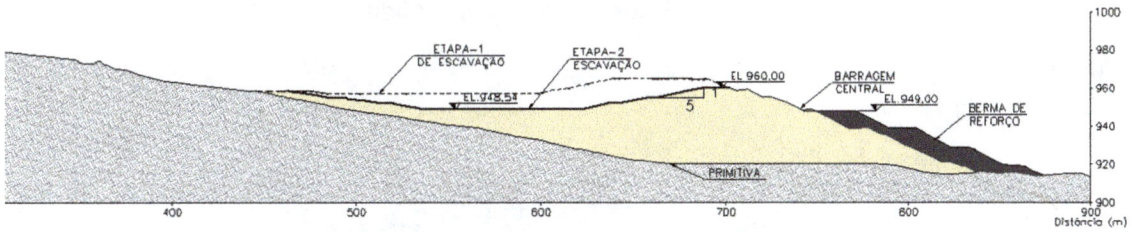

Material Name	Color	Unit Weight (kN/m3)	Sat. Unit Weight (kN/m3)	Cohesion (kPa)	Phi (deg)
Dique de partida		16.5	18	50	20
Solo residual maduro		19.1	19.85	5.5	30
Solo residual jovem		19.1	19.85	11	32
Berma		20	22	0	35
Alteamentos - rejeitos		20	22	8	30
Rejeito - seco		20.75	25.31	0	35
Berma reforço_Enrocamento		22		5	38

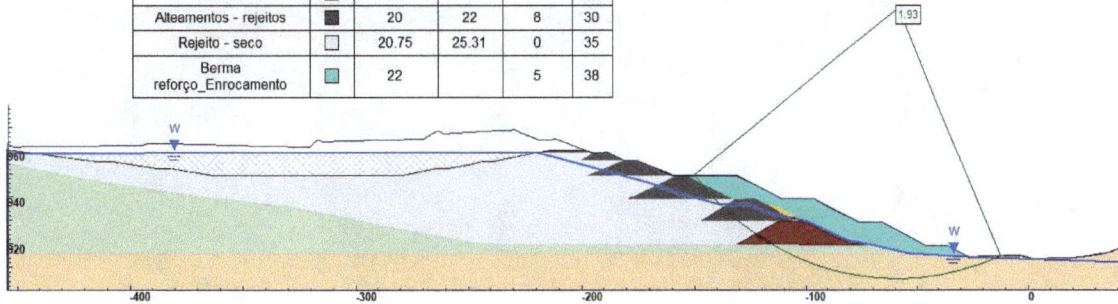

FIG 6 – Stage 2 mining schedule.

FIG 7 – Stage 2 mining operation.

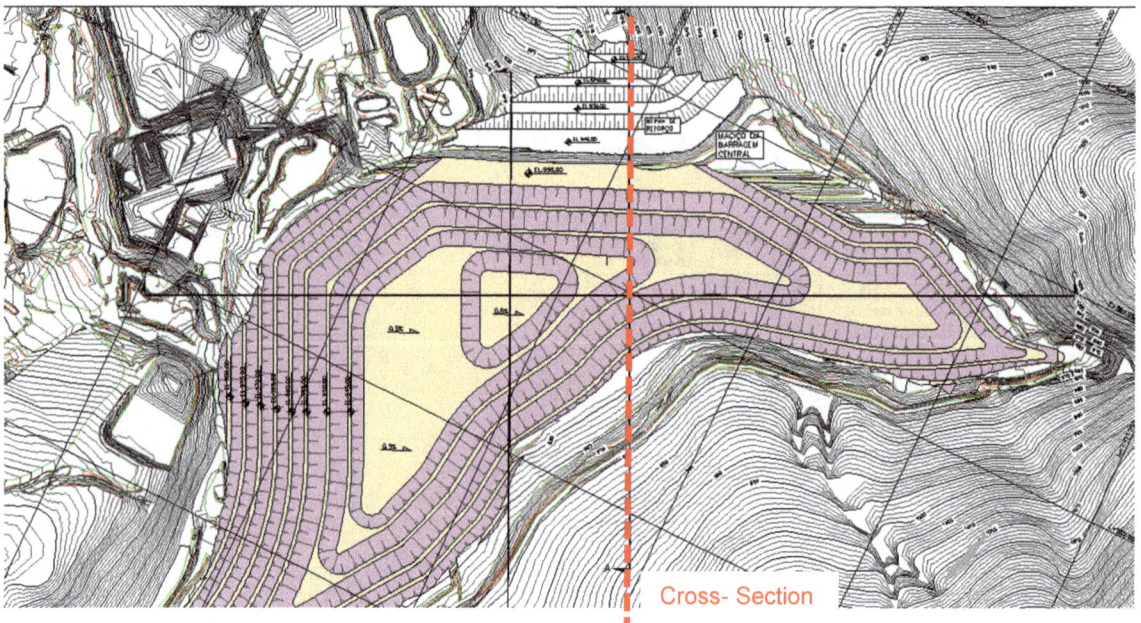

Cross- Section

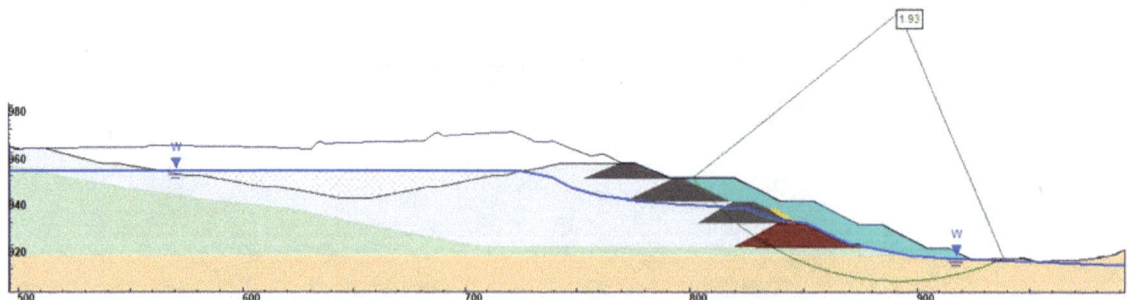

FIG 8 – Stage 3 mining schedule.

FIG 9 – Stage 3 mining operation.

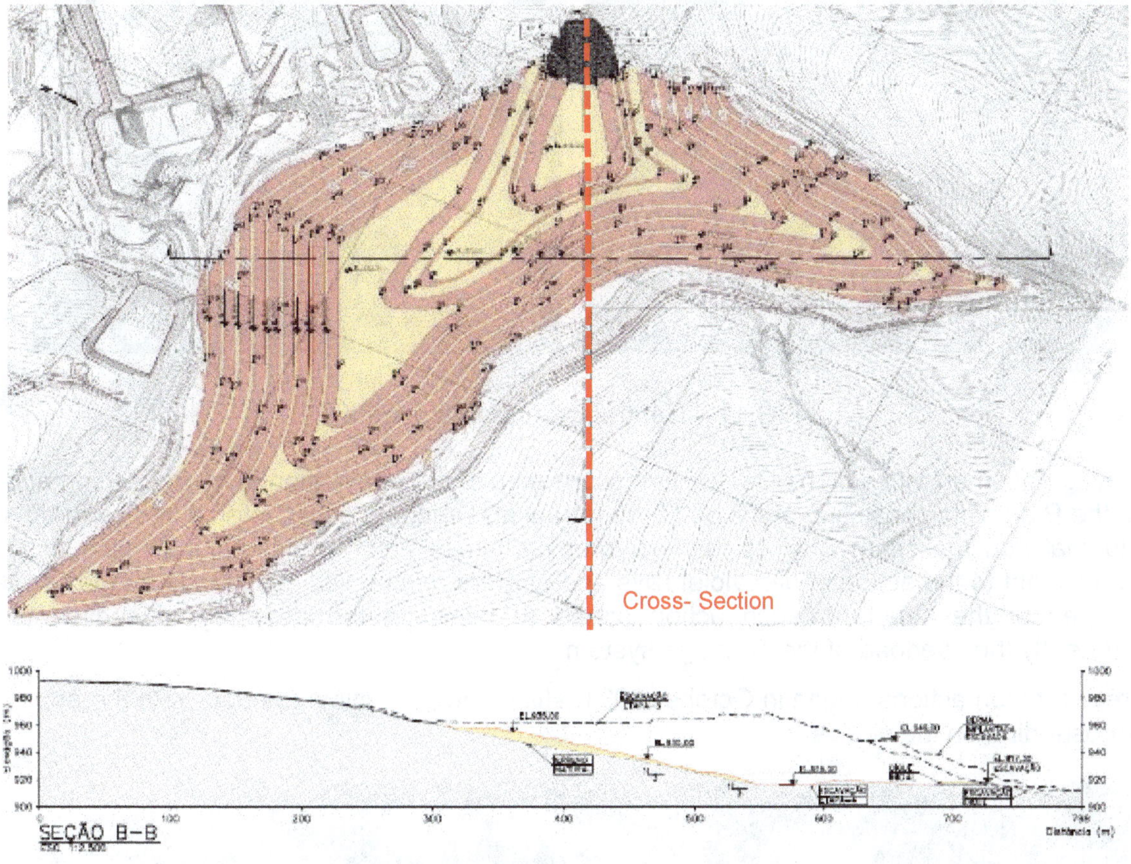

FIG 10 – Final stage mining schedule.

During the execution of the project, it was necessary to carry out a revision. The problem in the original design was the water velocity in the drains which was dragging material into the downstream stream. Thus, some sedimentation sumps were created to improve turbidity control.

The hydrological studies were developed for the contribution basin of the Central Dam, according to document MJ20B05020DRTP0034, consisting of the evaluation of the transit of floods through the sumps, through the structures of the sumps and the channels that interconnect these structures; and the slope capacity of the spillway system, in order to identify the characteristics of rainfall events with the potential to generate floods capable of promoting the overtopping of the structure and also the sedimentological studies evaluated the annual contribution of sediments carried to the hydraulic structures, such as channels and sumps.

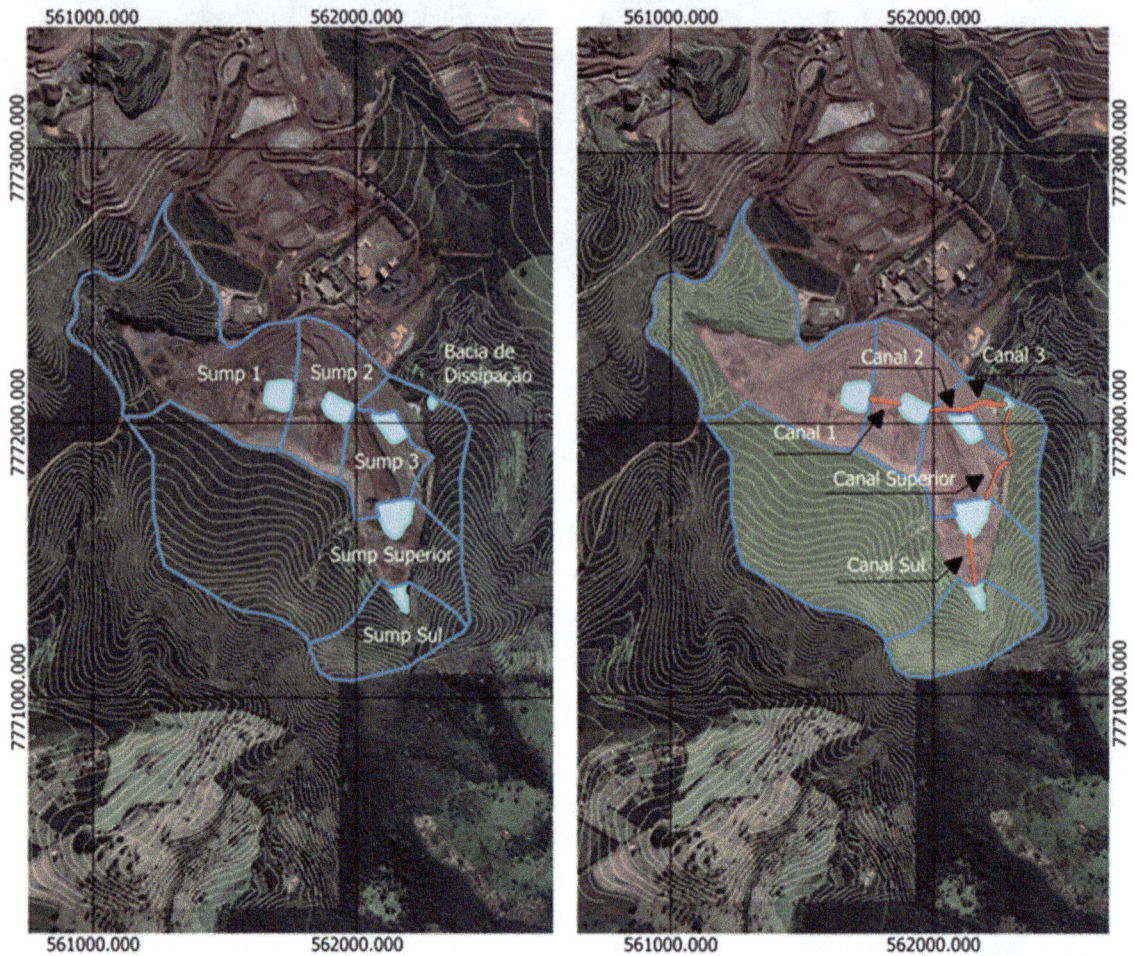

FIG 11 – Central Dam drainage area.

The revegetation activities of the Central Dam consisted of the application of hydroseeding and Bio-mats, the Bio-textile consisting of mixed fibres (straw and coconut fibre), intertwined by means of a longitudinal industrial seam, with resistant polypropylene nets, and two-dimensional mesh, this mesh being resistant to the action of ultraviolet rays and of great mechanical resistance and durability on the slopes of the structure, such action aimed to minimise the dragging of sediments and consequently the overload of the drainage system.

The revegetation actions began in October 2021, still in the dam mining period, with the application of hydroseeding on the slopes.

FIG 12 – Hydroseeding process.

To increase the conditions of revegetation, it was used organic soil to cover the shortage of nutrients that area. A rich soil it is favourable to seed of species that attract pollinators (insects). The remaining soil from the dam had an alkaline pH and a low concentration of nitrogen, phosphorus, potassium, sodium, and sulfates.

A fundamental point for revegetation is the choice of species, in this case they were made according to climate, soil, drainage conditions and biotic conditions.

Pioneer species and legumes were used, which increased nitrogen in the soil, an advantageous characteristic for waste, which normally lacks this nutrient.

The use of grasses at the Central Dam was tested, species were selected due to their rapid development, and it was found that the native species adapted better to low levels of nutrients, even in direct planting in the most nutrient-poor soil.

In the revegetation process, natural products were applied to assist in the drainage system and accelerate the revegetation process, these were as follows.

Sediment retainers (Figure 12) – flexible and extremely resistant cylinders made of dehydrated vegetable fibres that have the characteristic of retaining sediments and enormous capacity to absorb and retain moisture. In the Central Dam, retainers measuring 40 cm in diameter and 1.60 m in length were used. More than 6000 m of retainers were installed in the structure.

FIG 13 – Sediment retainers.

Anti-erosion bio-blanket (Figure 14) – it has the function of mitigating environmental disturbances, protecting the applied vegetation, creating a microclimate susceptible to vegetation growth, and reducing rainfall in the first years of planting while the vegetation does not reach the necessary maturity. More than 220 000 m² of anti-erosion bio-mats were installed in the structure.

FIG 14 – Anti-erosion bio-blanket.

In addition to all the work to enrich the soil with grasses and legumes, more than 12 000 native seedlings and tree species were planted in the Central Dam with the aim of creating a large forest and reintroducing the local fauna.

Due to the rehabilitation of the Central Dam and surrounding area, it was decided to rename the structure to Central Park. Continuous monitoring must be carried out to ensure long-term success and sustainability of the process. Likewise, post-revegetation monitoring has been carried out in Central Park, with a view to guaranteeing the objectives of mine closure (Figures 15–18).

FIG 15 – Vegetation growth process, March 2022.

FIG 16 – Vegetation growth process, April 2022.

FIG 17 – Vegetation growth process, May 2022.

FIG 18 – Vegetation growth process, March 2023.

CONCLUSIONS

Planning and design for closing tailings facilities reduces costs, reduces risks, and allows mines to meet the agreed upon goals and objectives. Construction and reclamation of Central Park were completed successfully, underpinned by good governance and collaboration between the company, the regulator, and local communities.

Landscape performance monitoring continues through regular observation (Figures 19–21) as the vegetated cover matures, and a maintenance plan will result in a positive legacy for generations to come.

FIG 19 – Aerial view of the area called Central Dam, 2014.

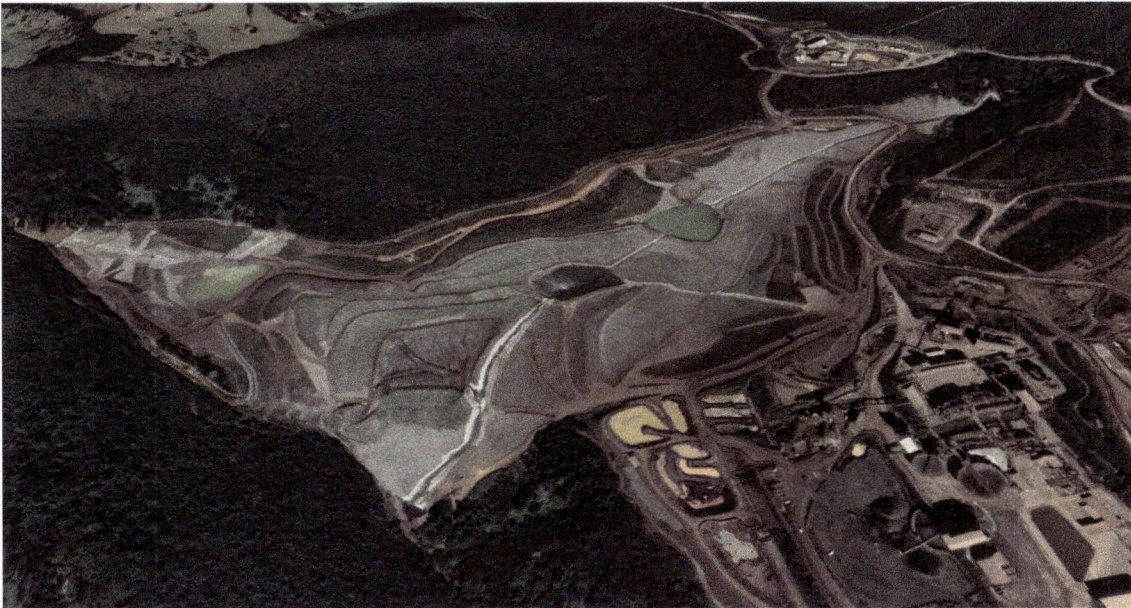

FIG 20 – Aerial view of the area now called Central Park, 2022.

FIG 21 – Aerial view of the area now called Central Park, March 2023.

ACKNOWLEDGEMENTS

The authors wish to thank Mineração Usiminas for allowing us to use this successful case, and the whole team effort that contributed in many ways to the conclusion of the reclamation in the old Central Dam, now known as Central Park.

REFERENCES

Projeto Conceitual De Fechamento Barragem Central Relatório Técnico Da Berma De Reforço [Conceptual design of central dam closure Technical report of the reinforcement berm], Walm Engenharia E Tecnologia Ambiental [Walm Engineering And Environmental Technology], report # Mj00b05000brtp0001, internal report.

Projeto Executivo De Descaracterização Barragem Central Relatório Técnico [Executive Project of Central Dam Decharacterization Technical Report], Walm Engenharia E Tecnologia Ambiental [Walm Engineering And Environmental Technology], report # Mj00b05000drgp0029, internal report.

Barragem Central Peer Review Dos Projetos De Descaracterização Da Barragem Central Relatório Técnico [Central Dam Peer Review of Central Dam Decharacterization Projects Technical Report], F&Z Consultoria E Projetos [F&Z Consulting and Projects], report # Mj20b05020drtp0034, internal report.

Conceptual study for the capping cover of a 10 ha in-pit tailings storage facility containing tailings with sub 10 kPa undrained shear strengths

A Shokouhi[1], M Llano-Serna[2] and A Marta[3]

1. Senior Geotechnical Engineer, Stantec, Fortitude Valley Qld 4006.
   Email: ali.shokouhi@stantec.com
2. Principal Geotechnical Engineer, Red Earth Engineering – a Geosyntec Company, Brisbane Qld 4000. Email: marcelo.llano@redearthengineering.com.au
3. Senior Principal; Director, Red Earth Engineering – a Geosyntec Company, Brisbane Qld 4000. Email: attila.marta@redearthengineering.com.au

ABSTRACT

In-pit tailings storage facility (TSF) design should consider the development of a sustainable post-closure landform. Furthermore, the facility should endeavour to achieve the closure design intent. Often, an encapsulating alternative is preferred. This paper describes some technical considerations encountered during the geotechnical stability assessment for a capping concept design. The concept design involved the use of geosynthetic reinforcement to improve constructability. This is due to the very low strengths, less than 10 kPa, of the tailings under the crust. The data presented in the paper comprises the geotechnical interpretation of a site investigation at a TSF below the existing ground surface in Australia. Polypropylene (PP) and Polyester (PET) are two common types of geotextiles used in engineering applications that require high tensile strength. However, their durability under long-term strain can vary depending on the pH of the service environment, exposure to ultraviolet radiation (UVR), and other factors. The paper aims to describe the technical considerations and methodology recommended for undertaking similar assessments in the future and provides some high-level discussion regarding geosynthetic reinforcement material selection in contrasting pH environments.

INTRODUCTION

A sustainable post-closure landform, often encapsulated, is a crucial step in mining life. This paper discusses a consolidation assessment and the constructability of capping alternatives with the outlook of completing closure activities of a mine pit that is backfilled with tailings materials. The site investigation was completed more than ten years after filling the mine pit. The geotechnical assessment aimed to understand the consolidation of the tailings material in the pit and how it would behave once a capping cover layer is installed. One representative sample of tailings was recovered and tested for geotechnical properties characterisation.

This paper presents interpretative results from a site and laboratory testing program. The field-testing program included CPTu, dissipation and Vane Shear testing (VST) at seven locations within the pit footprint. The testing and analysis results are presented herein, along with a set of numerical analyses, including trafficability assessment for the construction of capping alternatives and forward consolidation modelling to describe the long-term capping stability in terms of vertical settlement. Generally, the tailings material consists of granular fractions plus slimes, which are a by-product of the processing operations at the mining site. At present, no further information has been gathered regarding the chemical composition of the tailings. However, this is not expected to impact the reclamation procedure but may influence the final cover design. The reclamation area is around 200 m × 500 m, and the pit depth is a maximum of 26 m; the area covered is approximately 10 000 m².

PP and PET geotextiles are commonly used for different applications, depending on the required tensile strength. PP geotextiles are more durable in high pH environments than PET geotextiles, which are susceptible to alkaline hydrolysis at high pH. The composition of the sewing thread is also important in ensuring seam durability. The appropriate geotextile should be selected based on the pH of the service environment and loads applied over time (ie creep) to achieve the required design/operational life. It is possible to have geosynthetic reinforcement materials that range in initial tensile strengths from 30 kN/m to 5000 kN/m and strains at maximum load of 3 per cent to 15 per cent, hence the importance of selection based on a rational approach.

SITE CHARACTERISATION

Tailings

Seven CPTu probes were carried out in the tailings profile. Tailings maximum depths were interpreted to range between 15 m and 26 m.

Crust

A tailings crust of around 1 m was identified in the CPTu profiles, and the crust's undrained shear strength was variable, reaching values of less than 10 kPa and above 100 kPa. The crust has been classified mainly as a dilative material behaving between transitional soil and clay-like soil, according to Robertson (2016) SBTn plots.

Saturated soft tailings

The undrained shear strength of the tailings underlying the crust was consistently measured at strengths of less than 10 kPa between around 1 m and 10 m. An increasing undrained shear strength with depth was observed to reach around 30 kPa at 20 m. The strength in the bottom 5 m was observed to remain constant at around 30 kPa. The strength profile interpreted from CPTu probes is shown in Figure 1. Seven VST results were measured at three CPTu locations at depths of between 2 m and 5 m. VST was used to calibrate the undrained shear strength (S_u) profile from the CPTu results. The Nkt factor was observed to range between 11 and 15. Results are shown superimposed on the CPTu data presented in Figure 1. VST was completed in the softer region of the strength profile (ie CPTu test results at these depths are expected to have a decreased reliability due to the sensitivity of the strain gauge). The 80th percentile of the S_u estimated and calibrated from CPTu probes is also shown (see *Consolidation tests* section). The soft tailings materials are largely Clay-like contractive, according to Robertson (2016). CPTu results in Figure 1 were made semi-transparent; a denser (darker) scatter plot indicates regions of high data density.

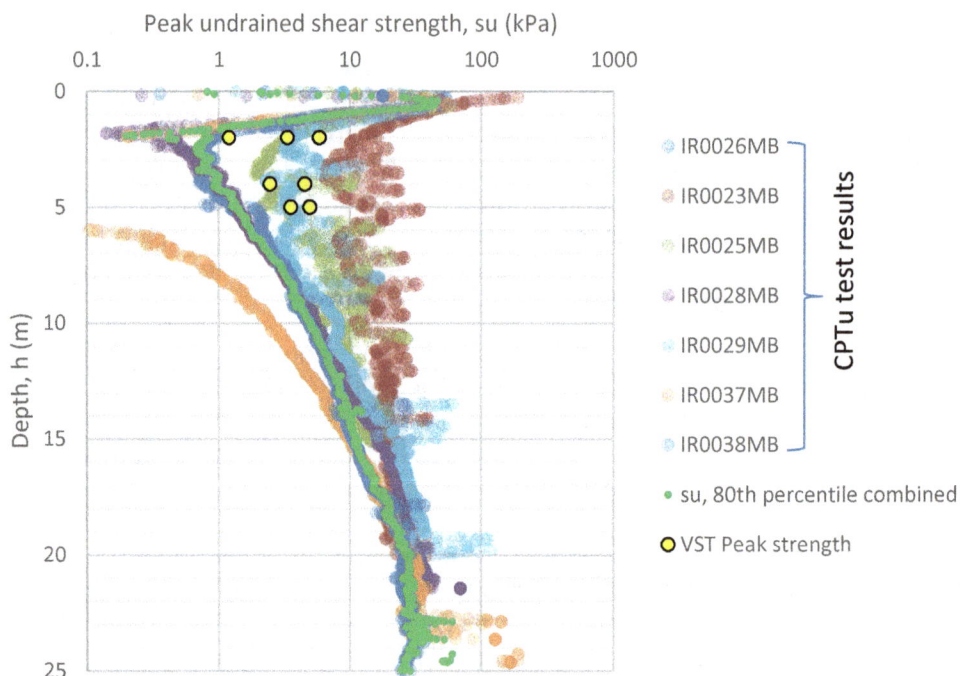

FIG 1 – Peak undrained shear strength profile interpreted from CPTu surveys.

LABORATORY TESTING PROGRAM

A laboratory program was designed to characterise the physical properties of the tailings sample and provide input parameters for the proposed analyses. Physical characterisation testing comprised settling test, particle size distribution, specific gravity, Atterberg limits (liquid limit obtained with the fall-cone test), and oedometric compression testing for consolidation assessment. Test results are compared with ICOLD 2017 classification framework (McLeod and Bjelkevik, 2017) with the following acronyms: CT, coarse tailings; HRT, hard rock tailings; ART, altered rock tailings; FT, fine tailings; and UFT, ultra-fine tailings.

Physical characterisation

Settling test

The tailings sample was mixed with tap water to form a 30 per cent solids content slurry and the height of the settled solids was measured over 23 days. It was noted that after 48 hours of settling, most of the coarse material had settled, but limited settlement of the finer fraction had occurred. At 23 days, the approximate settled solids content was estimated to be around 34 per cent by mass.

Particle size distribution, Atterberg limits and specific gravity

Wet sieve gradation tests (AS1289.3.6.1 (Standards Australia, 2009) and AS1289.3.6.3 (Standards Australia, 2020)) and Atterberg limits (AS1289.3.9.1 (Standards Australia, 2015), AS1289.3.2.1 (Standards Australia, 2009) and AS1289.3.3.2 (Standards Australia, 2009)) were completed on the bulk tailings. The characterisation tests indicate that the tailings are clayey sand, with sand materials of around 54 per cent by mass. The clay content (ie fraction finer than 0.002 mm) is reported to be about 29 per cent by mass. The plasticity index (PI = 13 per cent) is typical of low plastic clays. Index and gradation testing indicate that the tailings are classified as ART and CT, respectively. The specific gravity (AS1289.3.5.1 (Standards Australia, 2006)) of the sample was 2.65.

The gradation and index tests are shown in Figures 2 and 3.

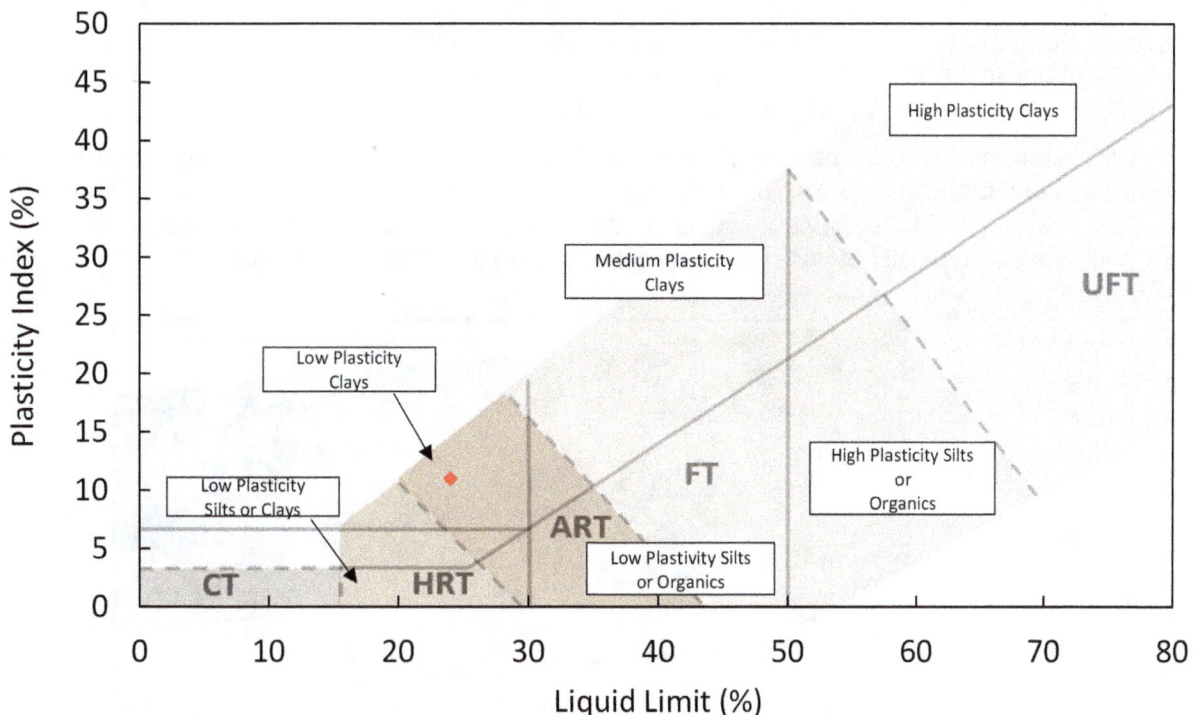

FIG 2 – Tailings plasticity chart compared with the ICOLD classification system.

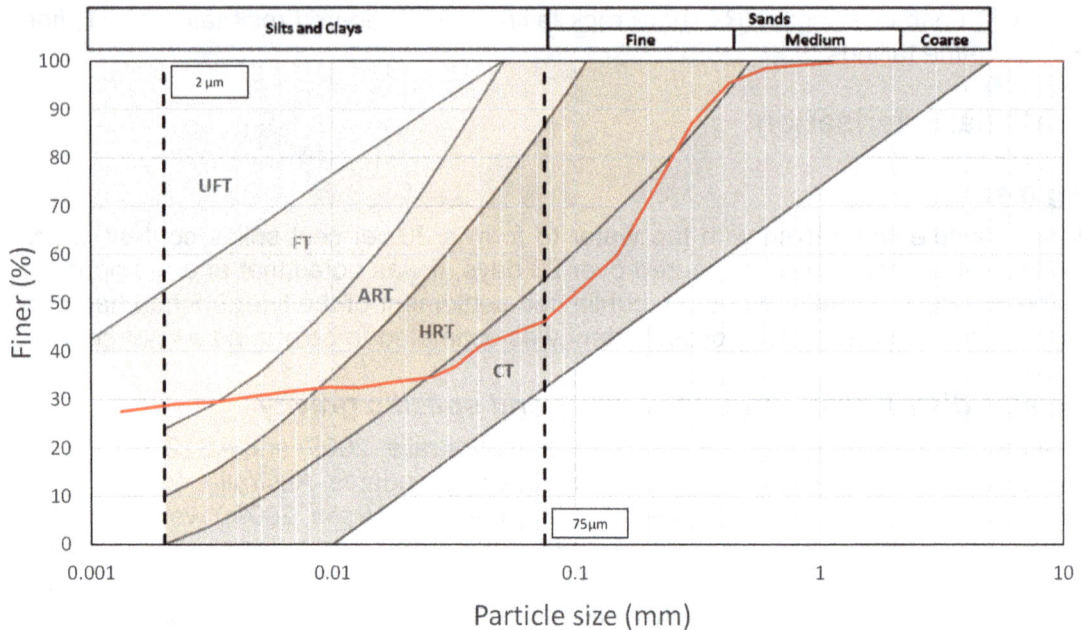

FIG 3 – Tailings particle size distribution compared with the ICOLD classification system.

Consolidation tests

Oedometric compression (AS1289.6.6.1 (Standards Australia, 2020)) was completed using a consolidometer apparatus. A sample was prepared at the liquid limit and loaded progressively to an effective vertical stress of 200 kPa; the sample was unloaded to an effective vertical stress of 50 kPa and re-loaded to an effective vertical stress of 1600 kPa.

As can be seen in Figure 4, the hydraulic conductivity values for the tailings calculated from the consolidation test range from 3.3×10^{-11} m/s and 4.0×10^{-10} m/s. The hydraulic conductivity of tailings measured indicates the behaviour of UFT according to ICOLD. The results suggest that consolidation times typically of very fine materials can be expected for the tailings.

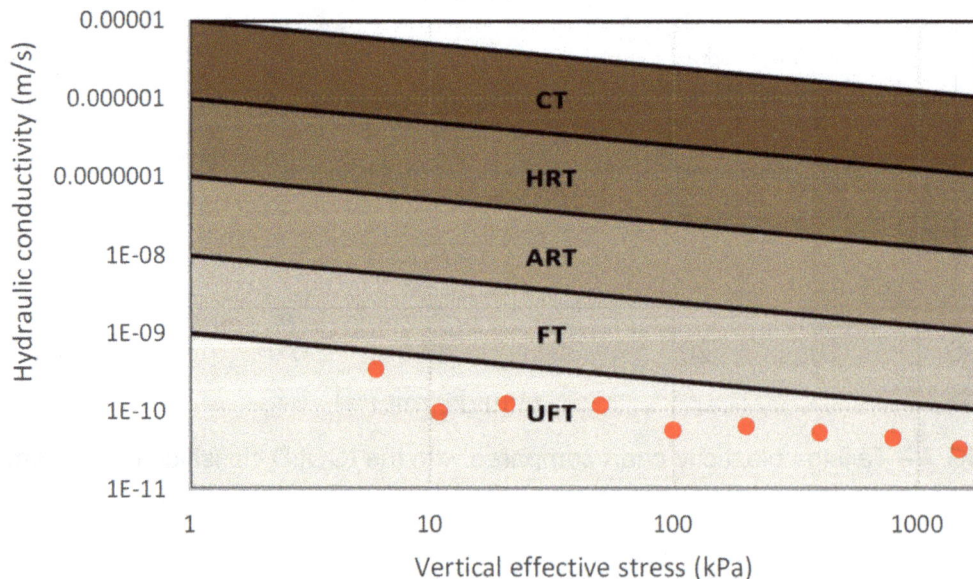

FIG 4 – Tailings hydraulic conductivity compared with the ICOLD classification system.

Figure 5 shows the interpreted compressibility plot. The figure shows the relationship between effective stress and the void ratio taken directly from the consolidation test results. Due to testing limitations, a sample had to be prepared at the liquid limit to test inside the oedometer apparatus. The authors acknowledge the limitations of this kind of testing for soft tailing, and budget constraints did not allow for a more appropriate testing technique. Due to the necessity to prepare a sample at the liquid limit, the initial portion of the compressibility plot was not captured. To offset this limitation, the test interpretation considered high stresses to capture the concave curvature typical of compressibility curves in tailings materials. CPTu data was interpreted to estimate the settled void ratio to accommodate for the lack of data at high void ratios. Estimates from CPTu correlations indicate that the settled void ratio for tailings deposited at the pit was around 2.9. The compressibility curve adopted for the study is shown as the red dotted line in Figure 5. The virgin compression line was determined by experience with this kind of material for other projects and the in situ void ratio.

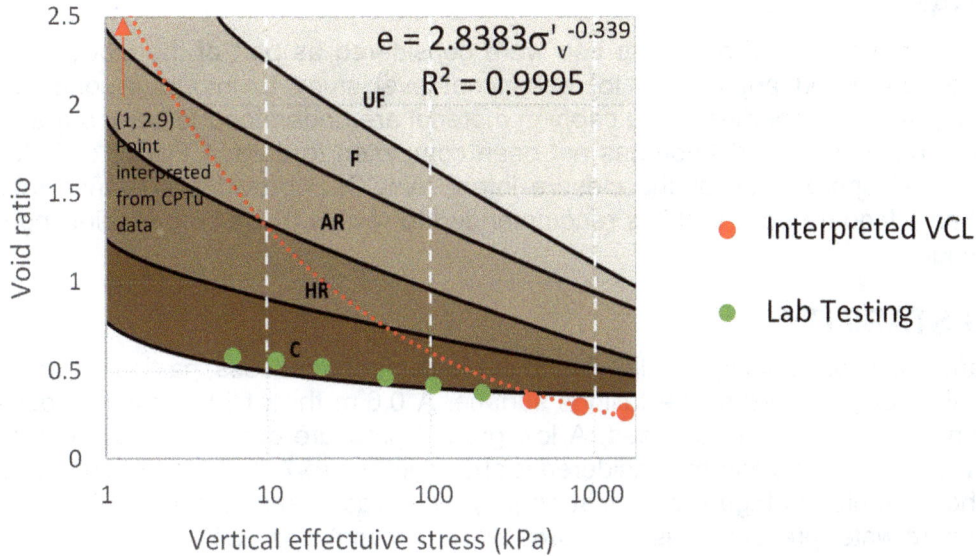

FIG 5 – Tailings compressibility compared with the ICOLD classification system.

It is recommended that modelling of one-dimensional consolidation of the tailings is undertaken using the large-strain consolidation theory (Fourie *et al*, 2022; Gibson, England and Hussey, 1967). The following equations express the theoretical relationship between the vertical effective stress, the void ratio, and hydraulic conductivity.

$$e = A.\sigma'^{B}_{v} \tag{1}$$

$$k = C.e^{D} \tag{2}$$

where, σ'_v is vertical effective stress, e is void ratio, k is hydraulic conductivity and A, B, C and D are empirical constants derived from testing.

Figure 5 presents the numerical regression to fit the parameters in Equation 1, whereas Figure 6 shows the numerical regression to fit the parameters shown in Equation 2. A summary of the fitted parameters is provided in Table 1.

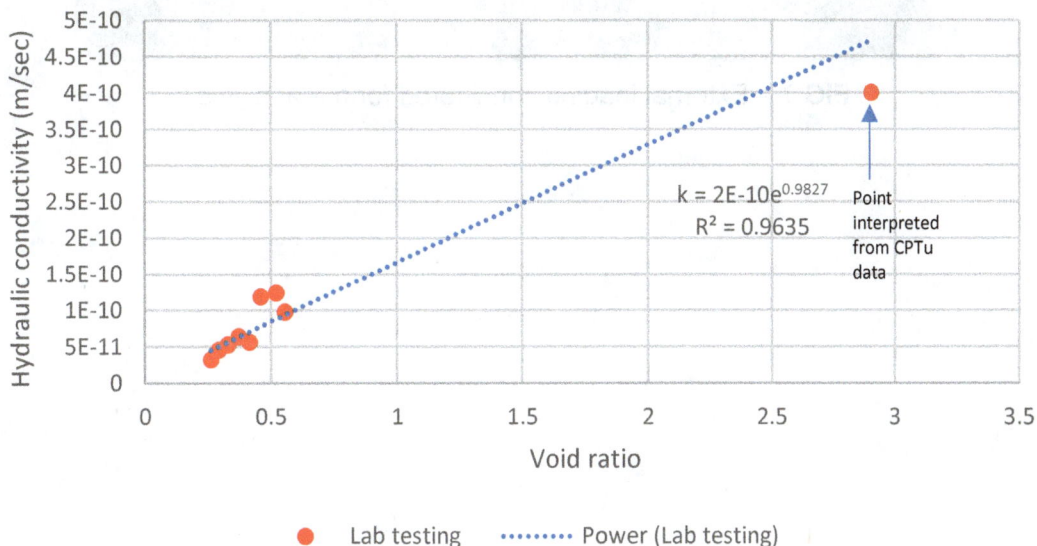

FIG 6 – Hydraulic conductivity change with void ratio changes.

TABLE 1

Summary of empirical Gibson's parameters for consolidation modelling.

A	B	C	D
2.84	-0.339	2.00×10^{-10}	0.983

DESIGN BASIS

Alternatives for the design of a closure cap were considered as part of the study. The proposed options were considered appropriate for a concept-level study to inform a conceptual closure strategy. The physical properties of the capping material are understood to be of a granular nature; however, laboratory characterisation has not been completed to date. A Factor of Safety (FoS) of 1.3 is considered appropriate for the closure intent. ANCOLD provides guidance tables for the selection of an adequate FoS, and it is recommended to review these before selecting one for the basis for design.

CAPPING STABILITY

It is important to assess capping stability during the reclamation process, particularly those resulting from external loading applied to the tailings surface. A 0.6 m thick fill layer was proposed as the minimum practical thickness to be used. A low ground pressure dozer LGP D6 was proposed to spread the fill, the loading diagram considered is shown in Figure 7. The strength profile adopted for design is shown in blue in Figure 8. The strength profile was based on the results from the CPTu (ie dynamic pore water pressure was measured during testing) and VST. The authors note that the data could be interpreted differently, but the design profile adopted herein was considered practical. The failure criteria adopted was Tresca for total stress conditions.

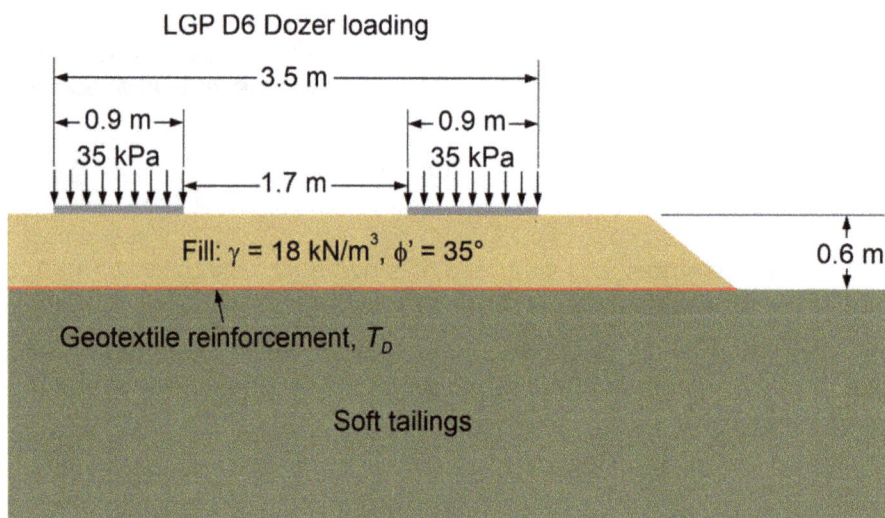

FIG 7 – External loading considered for the analysis.

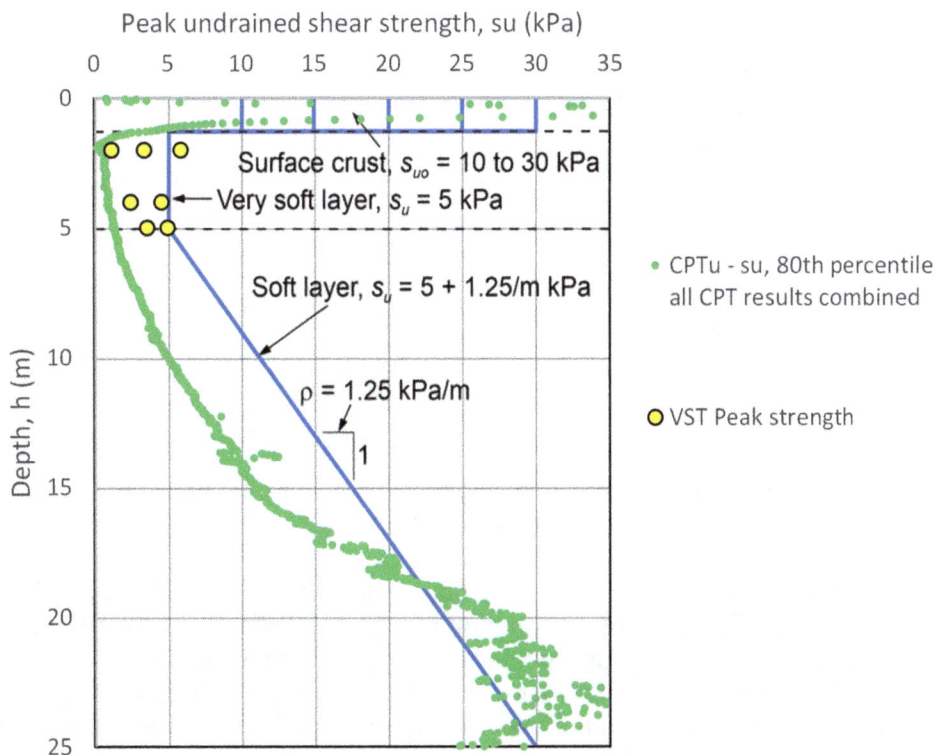

FIG 8 – Adopted strength profiles.

TECHNICAL ASSESSMENT

Geotechnical stability

The site has investigation has shown the presence of a crust with at an apparent over-consolidation state. The real undrained shear strength of the crust is spatially variable, consequently it is difficult to assess its contribution to the reclamation process. A sensitivity analysis approach aligned with the methods proposed by Lawson (2022) was adopted. Slip circle limit equilibrium analysis was performed using the design parameters from the strength profile shown in Figure 8 in the loading diagram provided in Figure 7. The very soft layer between 1 m to 5 m below the top of the tailings level is expected to govern the overall foundation stability. Critical slip surfaces passing through and near the base of the layer result in deep-seated failure surfaces. There are several methods to assess the geotechnical stability of a working platform such as Building Research Establishment (BRE), Temporary Works Forum (TWF) Guidance and Eurocode 7. However, each method has its own limitations, for example the BRE method does not apply for soft subgrades (Look and Honeyfield, 2016) and TWF and Eurocode 7 do not cover reinforced/stabilised soils (TWF, 2019).

The results following Lawson (2022) concepts are based on a sensitivity analysis carried out to evaluate geosynthetic reinforcement design tensile strengths required for different surface crust undrained shear strengths of 10, 15, 20, 25 and 30 kPa, see Figure 9. Slip circle limit equilibrium analysis to achieve a FoS > 1.5 is presented in Figure 9. The results indicate that for crusts strengths above 20 kPa geotextile reinforcement might not be required, although a geotextile separation layer may be recommended. For surface crust undrained shear strengths between 10 kPa and 20 kPa geotextile reinforcement is required.

It should be recognised that analytical models limit the reproduction of the complex interaction that takes place in a real platform under load (TWF, 2019). To overcome those limitations, a two-dimensional finite element (FE) method analysis using Optum G2 (Optum) software (Optum G2, 2016) was undertaken to assess the geotechnical stability of the capping platform. The geotextile was modelled using a geogrid element available in Optum. Geogrids are similar to Plates but cannot sustain uniaxial compression and offer no resistance to bending. As such, the material parameters concern only the tensile stiffness and strength.

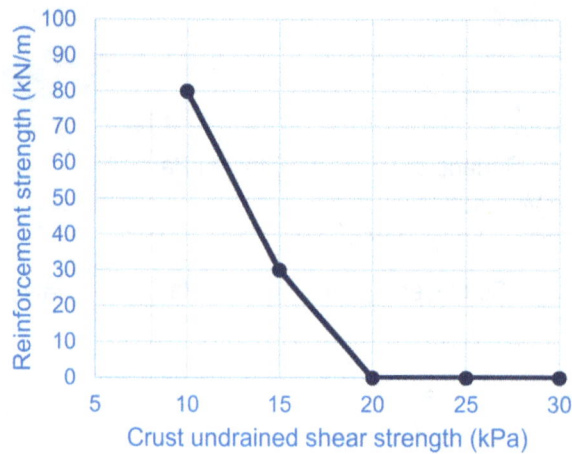

FIG 9 – Slip circle limit equilibrium sensitivity analysis results used to inform required reinforcement design tensile strength.

FE is a technique that allows discretising the groundmass (ie continuum) into a finite number of elements that form a mesh. Optum adopts triangular meshes to discretise the continuum. Each mesh element has a specific number of gauss points to compute stress and strain. Optum allows the development of the discretisation scheme using mesh adaptiveness. Based on strain control, up to five iterations were considered to refine the mesh based on strain outcomes automatically. The FoS calculation was undertaken using the plastic multiplier tool. The plastic multiplier analysis was completed under 2D conditions, ie plain strain using 6-node elements.

The FoS for construction was also calculated using the elastoplastic multiplier module from Optum. When using elastoplastic multipliers, a set of multiplier loads are incremented in a sequence of steps until collapse. The load multiplier is a numerical technique equivalent to the traditional FoS. The load multiplier method allows for assessing the stability of construction sequences with complex geometries and loadings that are not covered in conventional closed-form solutions found in design guidelines and codes (Fraldi *et al*, 2010). Tresca failure criterion was applied to model the shear strength of tailings. Tresca failure criterion, although simplistic, provides reasonable estimates of the mobilised shear strength if properly calibrated against field VST (Kouretzis *et al*, 2017). The material parameters adopted for numerical modelling are presented in Table 2.

TABLE 2

Material parameters adopted for numerical modelling.

Material type	Strength	Unit weight (kN/m³)	Interaction coefficient with geotextile
Tailings	Profile shown in Figure 7	14.5	0.7
Overburden (Platform)	c' = 4 kPa φ' = 35°	18.0	0.8
Geotextile	T_D of 30 kN/m	NA	NA

Figure 10 shows the deviatoric strains $|\varepsilon_1 - \varepsilon_3|$ for a sensitivity analysis case whereby the crust strength is 30 kPa. The loading condition is shown in Figure 7. The figure indicates that a perfect circular failure surface is a simplification. Instead, the numerical tool reveals a failure regions in red whereby upper bound strengths could be reached if a failure were to take place under the conditions modelled. Table 3 provides a summary of the FoS calculated for various scenarios for a combination of crust strengths and geosynthetic reinforcement. Results indicate that geosynthetic reinforcement might not be required for crusts with undrained shear strength of 15 kPa and above. Scenarios with crusts below 15 kPa undrained shear strength might see the benefit of a geosynthetic reinforcement layer.

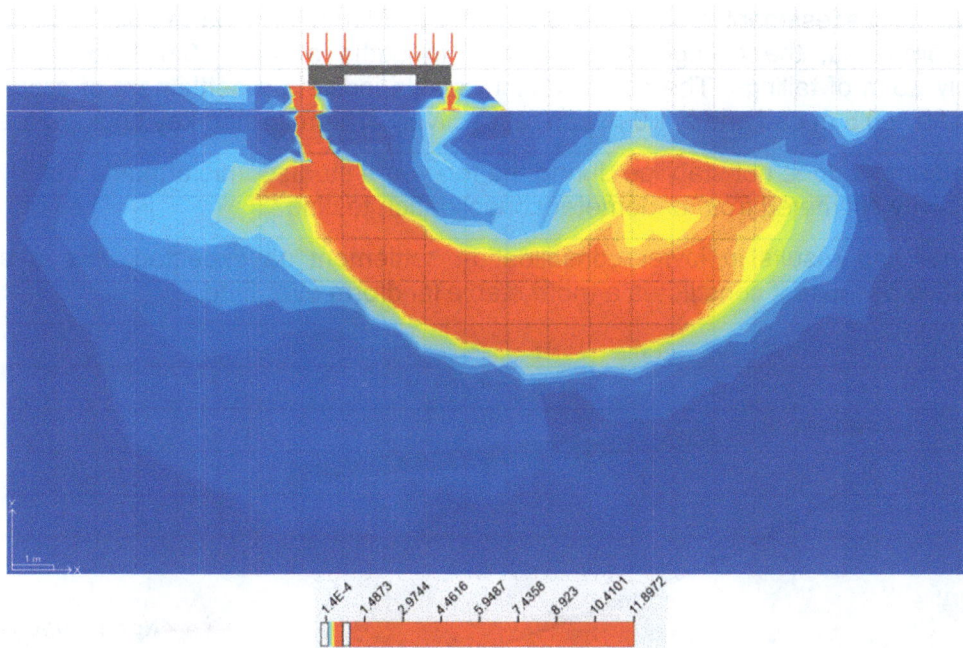

FIG 10 – Deviatoric strain colourmap distribution for a crust strength of 30 kPa, colourmap is exaggerated to indicate the location of the likely failure region. X and y axis with a scale are in the figure bottom left.

TABLE 3

FoS computed using Optum.

Crust strength based on Figure 8 profile	FoS	Reinforcement T_D (kN/m)
30 kPa	2.0	Not adopted
25 kPa	1.8	Not adopted
20 kPa	1.5	Not adopted
15 kPa	1.3	Not adopted
15 kPa	1.4	30
10 kPa	0.9	Not adopted
10 kPa	1.2	30

Long-term stability and consolidation settlement

Large strain consolidation modelling was undertaken using FSCA, finite strain consolidation analysis software (fscasoftware, 2023). FSCA is a numerical modelling package that allows large strain consolidation analysis. Large strain consolidation modelling was required because of the extremely soft tailings deposited in the pit. FSCA uses the properties of hydraulic conductivity and void ratio, interpreted in this report, with empirically derived parameter models, the tailings consolidation due to self-weight and the loading due to the capping layer. The model presented herein does not include the effect of vertical deformation due to the evaporation of water from the tailings surface that results in crusting.

Three scenarios were considered for the consolidation assessment:

1. No cap for ten years.

2. 0.5 m cap for ten years.

3. 1.0 m cap for ten years.

The consolidation assessment results are presented in Figure 11. The results indicate that if no capping is completed, the tailings are expected to settle around 50 mm for a thickness of approximately 25 m of tailings. The consolidation results indicate a settlement of around 250 mm and 350 mm for a 0.5 m cap and a 1.0 m cap respectively. The following key findings are noted:

- The settlement does not stabilise during the ten years studied; this attributed to low hydraulic conductivity estimated from oedometer consolidation testing.

- A thinner cap results in lower consolidation settlement as expected; this is a preferred alternative because it constitutes a more stable landform over time.

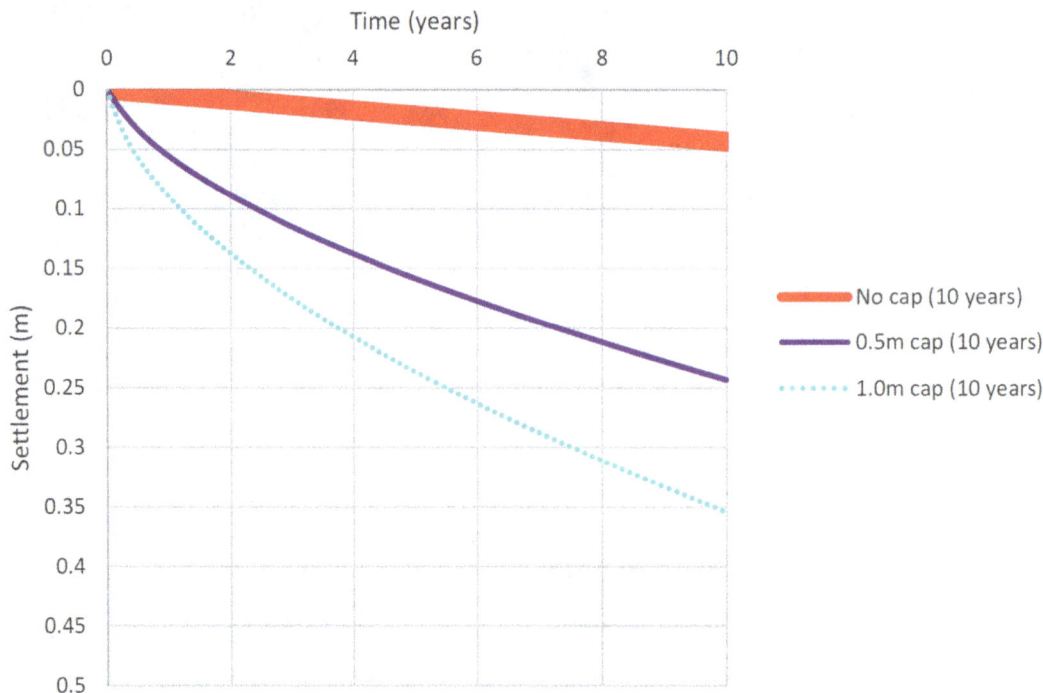

FIG 11 – Consolidation settlement modelling results of the tailings for ten years.

Geosynthetic reinforcement material selection

PP geotextiles are typically adopted where the required tensile strengths range from 80 to 200 kN/m, where PET geotextiles are used if higher tensile strengths in the range of 200 to 400 kN/m are required (TenCate, 2020). There are other geosynthetic reinforcements (eg HDPE geogrids), however, for very soft (S_u < 10 kPa) tailings deposits utilising a woven geotextile with biaxial strength in the two principal directions is preferred. The biaxial strength is created by using high strength sewn seams in the cross-direction of the rolls of geosynthetic reinforcement.

It is important to select the most appropriate geotextile for the given pH of the service environment to ensure its durability to achieve the required design intent and operational life. The durability of PP and PET geotextiles under long-term strain depends on factors such as molecular weight, degree of orientation, and crystallinity.

PP and PET degradation mechanisms

PET fibres can degrade in two independent mechanisms of hydrolysis. In acid or neutral environments, including water vapour, hydrolysis occurs throughout the cross-section of the PET products, leading to progressive rupture of the polymer chains and a consequent loss of strength of the fibres and bars. This type of hydrolysis, referred to as 'internal' hydrolysis, is related to the number of carboxyl end groups present at the end of polymer chains. The more carboxyl end groups present, the more likely the fibres are to undergo internal hydrolysis. This type of hydrolysis can be reduced by increasing the molecular weight, degree of orientation, and crystallinity of the polyester products.

In alkaline environments, the more aggressive hydroxyl ions (OH ions) cannot penetrate the bulk volume of the PET product and degradation takes place at the surface (Greenwood, Schroeder and

Voskamp, 2010). It is therefore known as 'external' hydrolysis. Material is removed from the surface leading to a loss in strength proportional to the reduction in cross-section. In this instance additives play a minor part. The sensitivity to oxidation can be reduced by controlling the structure of the polyester, limiting the concentration of diethylene glycol groups by appropriate conditions and catalysers during synthesis. Some crosslinking can also occur under harsh processing conditions. Hydrolysis stabilisers, mostly derivates of carbodiimides, may be introduced to reduce the number of effective carboxyl end groups which provide sites for the initiation of long-term hydrolytic degradation by hydrogen ions.

The impact of alkaline hydrolysis is strongly dependent on the transport by convection and diffusion of hydroxyl ions to the geosynthetic, which only happens in wet soil surroundings. Carbon dioxide, originating mainly from soil biology, also has to be regarded for its neutralising effects. External hydrolysis tends to occur around titanium oxide and antimony trioxide particles at the fibre surface, which can be present in PET fibres to control crystallisation. Tests have shown that at 50°C and below, the rate of reduction of strength in $Ca(OH)_2$ at pH 10 is similar to that in water. At 70°C and above, however, the strength diminishes faster in the alkaline solution (Greenwood, Schroeder and Voskamp, 2010).

According to Gibbs, Austin and Smith (2014), PET geotextiles are generally more resistant to photodegradation caused by ultraviolet radiation (UVR) than unstabilised polyolefin materials such as high density polyethylene (HDPE) and PP due to the inclusion of carbon-based benzene rings within the polymer structure. However, real-time outdoor exposure data collected over 3, 6, and 12 months for three sites around Australia revealed that the strength of the relationship between accelerated UVR data obtained through exposure in a xenon-arc weathering chamber for 500 hours and outdoor exposure data is heavily influenced by product variability, limitations of test methods, and abiotic factors concomitant with outdoor exposure. Therefore, it is important to carefully select the appropriate geotextile for a given pH of the service environment to ensure its durability and to achieve the required design and operational life.

Ehujuo *et al* (2022) found that the high pH value of leachates originating from landfills can chemically attack the material properties (mechanical and hydraulic) of non-woven geotextiles made of PP and PET polymers. By investigating the effect of chemical attack on the material properties of PP and PET non-woven geotextiles, Ehujuo *et al* was able to develop a system to categorise the environments suitable for PP and PET non-woven geotextiles based on their mechanical and hydraulic properties before and after application.

Jie *et al* (2022) conducted outdoor weathering tests at various exposure levels on woven and composite geotextiles to characterise their tensile behaviour. The study found that woven geotextiles underwent tensile strength degradation and became more brittle after five months of UV exposure, which showed the highest degradation of 69.55 per cent for strength retained. On the other hand, composite geotextiles only indicated a minimal degradation of 94.19 per cent for strength retained. Micro-scale approaches such as scanning electron microscopy (SEM) and Fourier-transform infrared (FTIR) spectroscopy were used to analyse the microstructure characteristics of the geotextiles and confirm the experimental results. Overall, the study provided clear evidence of the UV degradation of geotextiles in response to outdoor weathering.

CONSTRUCTABILITY

It should be noted that if the applied installation loads cause the crust to fail, the construction procedure becomes problematic. Dry surface crusts can present a relatively smooth surface when sands are present. Alternatively, crusts can undergo surface shrinkage, which results in a desiccated surface. Desiccated surfaces tend to be less stable and give rise to more significant construction difficulties. The presence of a dry surface crust greatly facilitates the fill cover procedure. Conversely, the presence of surface water can complicate the fill cover procedure and installation of the reinforcement geotextile.

Method for fill covering over reinforcement geotextile

Assuming that the surface crust is essentially dry, and the reinforcement geotextile can be placed conventionally, ie the geotextile reinforcement is laid out over the surface crust at right angles to the direction of the initial filling (finger dykes) the finger dykes can be constructed over the reinforcement geotextile at 8 to 10 m spacing, generating uniform tension over the surface, see Figures 12 and 13. Once the system is 'stressed', infilling between the finger dykes is carried out by completing the fill covering.

FIG 12 – Finger dyke construction (source: TenCate (2020)).

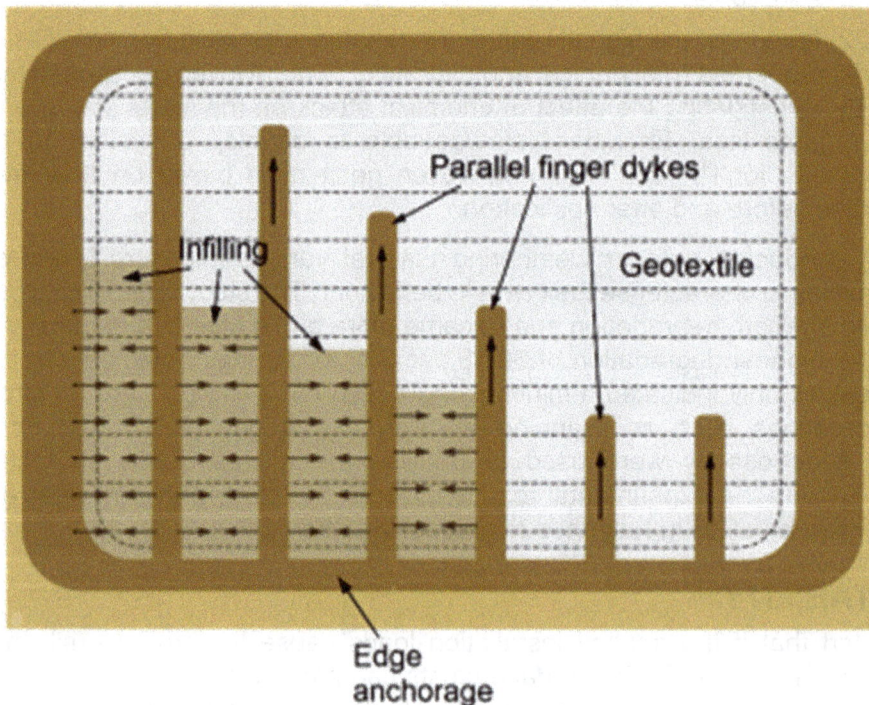

FIG 13 – Installing parallel finger dykes and infilling on geotextile panels (source: TenCate (2020)).

The current assessment does not consider the geotextile anchorage at the edges of the facility (see Figure 14), seaming requirements or a specific manufacturer/geotextile product. Furthermore, the current assessment does not provide technical guidance on techniques for measuring and quantifying the crust quality or determining when it is safe to proceed with construction. These technical assessments and recommendations must be carried out during a detailed design stage of the project.

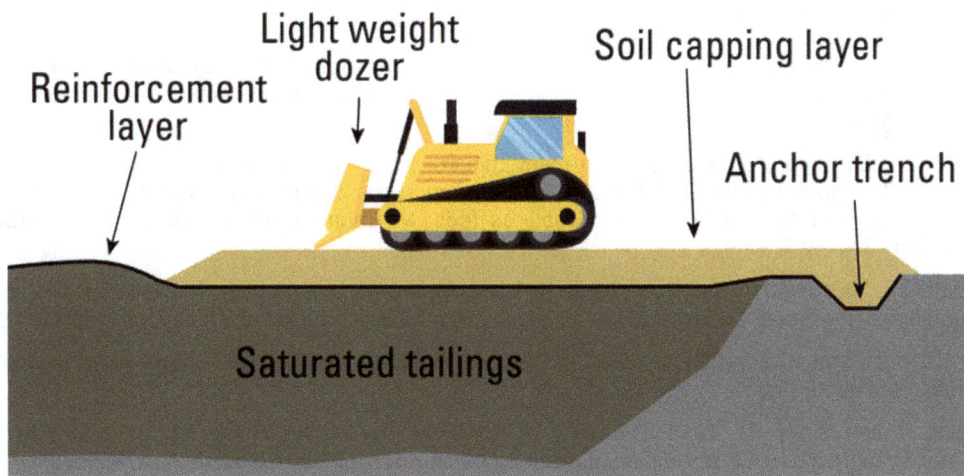

FIG 14 – Anchor trench schematic (source: TenCate (2020)).

CONCLUSIONS

A technical assessment to evaluate alternatives for constructing a capping layer over tailings deposited in a pit has been undertaken and presented in this paper. The technical assessment considered a series of sensitivity analyses to inform conceptual studies with appropriate construction equipment. The numerical modelling methods proposed are considered relevant, as they showed comparable results to the more traditional limit equilibrium approach. The methods allowed the representation of more complex failure regions. The numerical methods applied herein did not result in increased analysis time frames or numerical instabilities or divergences as is traditionally believed in the consulting industry for this kind of application. The methods proposed herein resolve limitations in traditional design guidelines such as BRE or Eurocode.

The alternatives analysed considered the construction of a capping layer with and without a reinforcement geotextile to aid constructability. The analysis revealed that a reinforcement geotextile is required to increase the FoS to acceptable levels under certain circumstances.

The proposed capping system comprised:

- A system featuring a linear layer, consisting of a reinforcement geotextile with characteristics to be selected.

- A protective and storage layer, likely comprising overburden material already existing in the vicinity of the pit.

- A topsoil layer to promote vegetation.

The adoption of a linear layer (ie reinforcement geotextile) was required due to very low FoS expected for some scenarios. The technical assessment indicates that a thinner capping layer is preferred because it imposes a lower load over the tailings while allowing a configuration to protect the reinforcement geotextile. The reinforcement geotextile allows trafficability and will later sustain the protective and storage layer.

The durability of geotextiles made of PP and PET polymers under long-term strain can vary depending on the pH of the service environment, exposure to UVR, and other factors. It is important to carefully select the appropriate geotextile for a given pH of the service environment to ensure its durability and to achieve the required design and operational life. Additionally, the composition of the sewing thread used in geotextile seams is also important to ensure the overall durability of the geotextile layer meets the design intent.

The assessment presented herein is limited to the cap structural stability during construction and over ten years of consolidation settlement. Other relevant evaluations such as surface drainage, anchor trenches, geochemical stability and non-load (ie durability) effects in order to select the most appropriate geosynthetic reinforcement material are recommended to progress the design further.

REFERENCES

Ehujuo, N, Kalumba, D, Sobhee-Beetul, L and Oriokot, J, 2022. A comparative study between the performances of polypropylene and polyester non-woven geotextiles in landfills, in 7th International Young Geotechnical Engineers Conference, Sydney, Australia.

Fourie, A, Verdugo, R, Bjelkevik, A, Torres-Cruz, L A and Znidarcic, D, 2022. Geotechnics of mine tailings: a 2022 State of the Art, in *Proceedings of the 20th International Conference on Soil Mechanics and Geotechnical Engineering (ICSMGE) – State of the Art and Invited Lectures* (eds: M Rahman and M Jaksa), pp 121–183 (Australian Geomechanics Society: Sydney).

Fraldi, M, Nunziante, L, Gesualdo, A and Guarracino, F, 2010. On the bounding of limit multipliers for combined loading, in *Proceedings of the Royal Society A: Mathematical, Physical and Engineering Sciences*, 466(2114):493–514.

Fscasoftware, 2023. Tailings Consolidation | FSCA Software [online]. Available at: <https://www.fscasoftware.com/> [Accessed: 13 March 2023].

Gibbs, D, Austin, R and Smith, D, 2014. Durability of Polyester Geotextiles Subjected to Australian Outdoor and Accelerated Weathering, in 7th International Conference on Environmental Geotechnics, Melbourne, Australia.

Gibson, R E, England, G L and Hussey, M J L, 1967. The theory of one-dimensional consolidation of saturated clays: 1. Finite non-linear consolidation of thin homogeneous layers, *Geotechnique*, 17(3):261–273.

Greenwood, J H, Schroeder, H F and Voskamp, W, 2010. Durability of Geosynthetics, Publication 243, CUR Building & Infrastructure, Netherlands.

Jie, S Y, Hoe, L Z, Paul, S C and Anggraini, V, 2022. Characterizing the Tensile Behaviour of Woven and Composite Fabrics under UV Exposure, *Appl Sci,* 12:11440.

Kouretzis, G, Pineda, J, Krabbenhøft, K and Wilson, L, 2017. Interpretation of vane shear tests for geotechnical stability calculations, *Canadian Geotechnical Journal*, 54(12):1775–1780.

Lawson, C, 2022. *Geosynthetic Reinforcements* (Lawson Publishing).

Look, B and Honeyfield, N, 2016. Working platforms to BRE or not to BRE is the question, *Aust Geomech Journal*, 51:11–21.

Robertson, P K, 2016. Cone penetration test (CPT)-based soil behaviour type (SBT) classification system—an update, *Canadian Geotechnical Journal*, 53(12):1910–1927.

TenCate, 2020. Soft site impoundment closure using geosynthetics [Presentation]. July 2020.

Standards Australia, 2009. AS1289.3.2.1:2009 Methods of testing soils for engineering purposes, Soil classification tests, Determination of the plastic limit of a soil, Standard method (Reconfirmed 2022). Available from: <https://infostore.saiglobal.com/en-au/standards/as-1289-3-2-1-2009-128264_saig_as_as_275148/>

Standards Australia, 2006. AS1289.3.5.1:2006 Methods of testing soils for engineering purposes, Soil classification tests, Determination of the soil particle density of a soil, Standard method. Available from: <https://infostore.saiglobal.com/en-au/standards/as-1289-3-5-1-2006-128260_saig_as_as_271107/>

Standards Australia, 2009. AS1289.3.6.1:2009 Methods of testing soils for engineering purposes, Soil classification tests, Determination of the particle size distribution of a soil, Standard method of analysis by sieving (Reconfirmed 2022). Available from: <https://infostore.saiglobal.com/en-au/standards/as-1289-3-6-1-2009-128258_saig_as_as_274937/>

Standards Australia, 2020. AS1289.6.6.1:2020 Methods of testing soils for engineering purposes, Soil strength and consolidation tests, Determination of the one-dimensional consolidation properties of a soil, Standard method. Available from: <https://infostore.saiglobal.com/en-au/standards/as-1289-6-6-1-2020-128224_saig_as_as_2873159/>

Standards Australia, 2015. AS1289.3.9.1:2015 Methods of testing soils for engineering purposes, Soil classification tests, Determination of the cone liquid limit of a soil. Available from: <https://infostore.saiglobal.com/en-au/standards/as-1289-3-9-1-2015-101148_saig_as_as_212507/>

Standards Australia, 2009. AS1289.3.3.2:2009 Methods of testing soils for engineering purposes, Soil classification tests, Calculation of the cone plasticity index of a soil (Reconfirmed 2022). Available from: <https://infostore.saiglobal.com/en-au/standards/as-1289-3-3-2-2009-128262_saig_as_as_275146/>

Standards Australia, 2020. AS1289.3.6.3:2020 Methods of testing soils for engineering purposes, Soil classification tests, Determination of the particle size distribution of a soil, Standard method of fine analysis using an hydrometer. Available from: <https://infostore.saiglobal.com/en-au/standards/as-1289-3-6-3-2020-128256_saig_as_as_2825523/>

Optum G2, 2016. Features, Optum Computational Engineering.

McLeod, H and Bjelkevik, A, 2017, July. Tailings Dam Design: Technology Update (ICOLD Bulletin), in Proceedings of the 85th Annual Meeting of International Commission on Large Dams.

Temporary Works Forum (TWF), 2019, April. Design of granular working platforms for construction plant A guide to good practice, Temporary Works Forum, c/o Institution of Civil Engineers, London.

Predicting the long-term erosional stability of regional cover designs using landscape evolution modelling

H G Vogler[1] and T K Rohde[2]

1. AAusIMM, Environmental Engineer, SGM Environmental, Windsor Qld 4030.
   Email: hvogler@sgmenvironmental.com
2. Director, SGM Environmental, Windsor Qld 4030. Email: trohde@sgmenvironmental.com

ABSTRACT

The Cobar region (the region) is located approximately ~700 km west of Sydney in central western New South Wales (NSW). The region has a semi-arid climate with low humidity, low rainfall and high evaporation. Annual rainfall is ~400 mm while evaporation averages 2000 mm. The Cobar basin is a significant metalliferous region in Australia, containing extensive base and precious metal deposits. During ore processing, potentially acid forming (PAF) tailings are generated and discharged to tailings storage facilities (TSF). PAF tailings require careful rehabilitation to minimise the risk of harm to the receiving environment.

Rehabilitation of a TSF usually requires the design of a cover to limit/prevent tailings interacting with oxygen and/or water to reduce the potential of acid mine drainage (AMD) that can be transported to the receiving environment. In semi-arid environments a store and release cover is typically used, in combination with other safe guards (ie a capillary break (CB) or reduced permeability layer (RPL)).

Due to material availability, cover designs regionally vary in thickness but are still considered suitable in regards to performance, often expressed as seepage. Notwithstanding, there are additional cover performance issues that need to be acknowledged.

Erosion plays an important part in Australian landscapes. Erosion (usually driven by water velocity) causes the cover surface to change with time. Long-term erosion stability is more complex than a simple laboratory analysis as material characteristics are a small part of understanding erosion stability. There are additional factors that need to be considered including; climate, slope and groundcover.

Typically, cover design does not consider erosion driven changes as a performance criterion. LAPSUS evolution modelling has been applied to three published cover designs for the Cobar region to examine performance changes over time in response to erosion driven change.

The LAPSUS modelling predicted that the cover design performances were affected by erosion after 100 years. All cover designs experienced localised erosion in the form of rills and gullies that penetrated the full thickness of the cover. In regards to overall soil loss, it was found that cover thickness is not a determining factor.

INTRODUCTION

The Cobar region (the region) is located ~700 km west of Sydney in central western New South Wales (NSW). The region extends from approximately 30 km north-west of Cobar to approximately 100 km south-east. The topography of the region is typically flat or gently undulating dotted with stony ridges and ranges. A large portion of the region is rangeland where vegetation consists of poplar box woodlands, mulga communities and cypress pine.

The region is one of the most significant metalliferous areas in Australia, containing extensive base and precious metal deposits. It has been a significant source of mineral wealth for 140 years since the discovery of the Great Cobar copper deposit in 1870.

During ore processing, potentially acid forming (PAF) tailings are generated and discharged to a tailings storage facility (TSF). PAF tailings created at mine sites (the Mines) in the region, contain sulfides, which when exposed to oxygen and water form acid mine drainage (AMD). The PAF tailings require careful management to minimise the risk of harm to the receiving environment. An analysis of the Mines identified that a cover is required to limit the potential for rainfall percolation into tailings and for AMD transport. Limiting percolation will decrease the potential for seepage and impact on the receiving environment. There are numerous examples of covers that have been built to limit interactions between tailings, oxygen and water in semi-arid Australia (Ayres *et al,* 2003; O'Kane and Walters, 2003; Rohde, Williams and Burton, 2011).

REVIEW

Regional geology

The region's geology is characterised by siliciclastic sediments locally intruded with felsic volcanics. The region has large, high-grade deposits that are hosted by marine sediments and consist of multiple lenses in steeply plunging, pipe-like clusters.

Regional soils

Soils are relatively uniform and relate closely to topographic position and local geology. Steeper slopes, ridges and crests tend to have sandy to earthy tenosols. Tenosols gradually grade down into red Dermosols, Kandosols and Calcarosols on the lower slopes, lowlands and flats with several variants existing. Deep alluvial and sandy soils commonly occur in the channels and creeks.

Land use

The predominant land uses in the region are agriculture and mining. Agriculture is dominated by sheep and some cattle grazing. Regionally, there are five operational mines that target a range of metals.

Cover design process

Climate and cover type

Among other factors, climate is an essential element in determining the type of cover most suited to TSFs in the region (International Network for Acid Prevention (INAP), 2009). The climate of the region is semi-arid with low humidity, low rainfall and high evaporation. Annual average rainfall is ~400 mm and is evenly distributed throughout the year. Rainfall is exceeded by evaporation and averages ~2000 mm per annum. The combined effect of evaporation and transpiration (evapotranspiration) at the mines is five times greater than rainfall. The Global Acid Rock Drainage (GARD) guide (INAP, 2009) suggests that a store-and-release cover (the cover) might be an effective tool, in combination with other safe guards (ie a capillary break (CB) or reduced permeability layer (RPL)) to reduce the potential for environmental harm after rehabilitation.

Determining preferred cover design

Cover design is a multi-stage process. Ideally, the process will include scaled trials that assess a range of potential cover options. This is usually done in combination with modelling in SVFlux (SoilVision, 2009), SEEP/W (GEO-SLOPE, 2012), or HYDRUS (PC-Progress, 2009). The models are used to simulate water movement for both saturated and unsaturated conditions.

Scaled trials may include either column trials or field trials. Field trials are used to test the cover over a period of time under natural atmospheric conditions. A limitation of field trials is that very long periods (greater than five years) is required to collect representative and reliable cover responses (Rohde, Defferrard and Lord, 2016). Column trials rely on artificial rainfall to speed up the data gathering process, and are comparatively faster than field trials. However, they are limited because they cannot be relied upon to evaluate cover performance. Rather the gathered data is used to calibrate a model which is used to predict future performance.

Regional cover designs

Based on previous work completed in the region by Jamson and Rohde (2019) a suitable cover (the monofil cover) should contain about 0.8 m of soil (the infiltration storage layer (ISL)). A monofil ISL of this thickness provides a suitable balance between infiltration, storage and evaporation resulting in low seepage. Notwithstanding, regional paucity of soil has made thinner (and more complex) covers more desirable. To achieve similar performance capabilities to a thick monofil cover, other safeguards including CBs and RPLs have been included in the design to counteract the loss of ISL infiltration storage capacity. The preferred cover thicknesses in the region vary from 0.8–1.55 m and contain multiple layers (from alternative sources), dependent on the physical and chemical properties of available cover material (Figure 1):

- Mine A:
  - 0.3 m RPL of dam sediment
  - 0.8 m ISL of non-acid forming (NAF) waste rock above the RPL
  - 0.15 m of soil above the ISL.
- Mine B:
  - 0.6 m CB of coarse NAF waste rock
  - 0.8 m ISL of NAF waste rock above the CB
  - 0.15 m of soil above the ISL.
- Mine C:
  - 0.8 m ISL of soil.

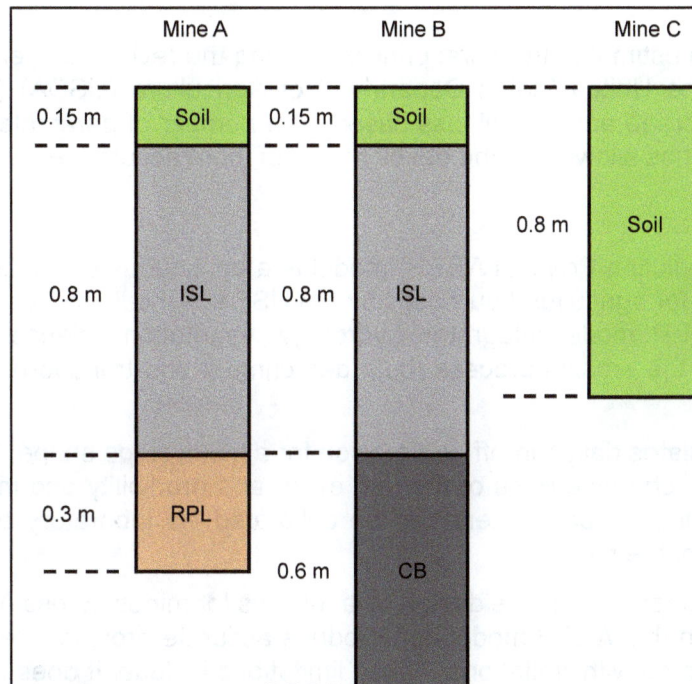

FIG 1 – Cover thicknesses of each mine.

TSF landscape

The purpose of a cover is to create a protective barrier on the landform that will reduce its impact to the receiving environment from AMD, neutral mine drainage (NMD) or saline drainage (SD). Therefore, successful cover design is often related to water movement through the cover (seepage). The process for determining if the cover is successful is based on how effectively it performs at limiting seepage. However, the cover will change over time, which can be expected to alter how the cover performs. Changes might include vegetation establishment and growth, grazing which may contribute to vegetation density changes, and consolidation which may change water flow pathways and velocities. All these changes make way for a major issue in cover performance; erosion.

Typically, physical and chemical properties of the cover is limited to laboratory analysis to estimate erosion potential. Long-term erosion stability is more complex, and poorly understood when only laboratory analysis is relied upon. There are additional factors that affect erosion including; climate, slope and groundcover. These factors cannot be measured in the laboratory (from inferred analysis) and require in-depth examination.

The outcome of erosion stability analysis is whether an acceptable rate of erosion can be maintained, and what impact this might have on cover performance. There is currently no broad agreement on what constitutes an acceptable rate of erosion on a rehabilitated landform. Research completed by Lu *et al* (2001) has shown that:

- The average Australian soil erosion rate is 6.3 tonnes per hectare per annum (t/ha/a)
- A low rate of erosion is defined as less than (<) 0.5 t/ha/a
- A high rate of erosion is defined as greater than (>) 10 t/ha/a.

Landloch Pty Ltd (2013) have shown that rehabilitated mine slopes have a low tendency to rill with an average erosion rate of <5 t/ha/a. The elimination of rilling will largely reduce the potential for channelised flow which can result in gully erosion. Therefore, avoiding rill formation will ensure landform stability by not exceeding an acceptable rate of erosion.

It is important to note that erosion rates can vary from year to year depending on climatic conditions. Therefore, the maximum annual erosion rate is also used as a secondary performance indicator. A target average and maximum erosion rate of 5–10 t/ha/a may therefore be appropriate for TSF landforms in the region.

Erosion modelling

Erosion stability can be estimated from first principals using the revised universal soil loss equation (RUSLE) created by the United States Department of Agriculture (USDA) (Renard *et al*, 1997). RUSLE predicts an average annual soil loss based on a number of parameters. Advancements in computer technologies has allowed for the development of more accurate erosion prediction models.

2D models

The Water Erosion Prediction Project (WEPP) model is a continuous simulation erosion prediction tool, initially developed for agricultural purposes by the USDA-Agriculture Research Centre (USDA-ARC, 1985). The WEPP model integrates hydrology, vegetation science, hydraulics and soil mechanics to simulate the erosion process (both detachment and transport) at a two-dimensional (2D) hillslope scale.

The WEPP model simulates daily run-off and erosion for a given slope shape, based on initial cover hydrological conditions, characteristics of the rain event and erodibility and infiltration parameters. The erodibility and infiltration parameters can be calibrated via laboratory or field studies which increase the accuracy of the model.

The WEPP model has been used in the design of landforms for mines across Australia (Howard and Roddy, 2012). Although the WEPP model can produce accurate erosion rates that represent field observations, it does come with limitations. These limitations include: it does not predict deposition patterns; it does not adjust hillslope elevation in response to erosion and it does not consider gully and tunnel erosion.

3D models (LEM)

Landform evolution models (LEMs) are used to simulate the three-dimensional (3D) development of landscapes over time and predict erosion and deposition processes on a landform scale. Each LEM requires a digital landscape which can be divided into two categories: digital elevation models (DEMs) and triangular irregular networks (TINs).

SIBERIA

SIBERIA is a LEM that was developed in Australia and has been widely applied since it was published in 1991. It predicts the long-term evolution of channels and hillslopes in a catchment. It does not use any actual rainfall or material erodibility parameters in its calculations, it instead relies on an activation threshold that is dependent on discharge and slope gradient. The parameters needed for SIBERIA are related to both run-off and soil erodibility and must be derived for each material, which is difficult and costly (Howard and Roddy, 2016).

CAESAR-Lisflood

CAESAR-Lisflood was originally developed in England to study long-term changes in river catchments, and is driven by fluvial processes (Howard and Roddy, 2016). Unlike SIBERIA, CAESAR-Lisflood processes rely on rainfall inputs, with input timesteps from minutes to days. The only material input required is particle size distribution (PSD); the other parameter inputs needed are associated with hydrological and water discharge processes. CAESAR-Lisflood is a computationally intensive model that prohibits large area and/or fine grids.

LAPSUS

LAPSUS is a multi-process LEM. It includes processes related to overland erosion, land sliding, tillage and tectonics (Schoorl, Veldkamp and Bouma, 2002). LAPSUS is a relatively simple cellular model using annual timesteps that has a hillslope focus, similar to SIBERIA. LAPSUS uses annual climate data (rainfall, infiltration and evaporation) to determine how much rainfall is converted to run-off. The main inputs related to the erosion processes are discharge, slope, sedimentation and erodibility.

METHODOLOGY

Without long-term field erosion data, an alternative approach to calibrating the LEM is required. The alternative approach to produce model parameters has been to use outputs from the WEPP model which can be calibrated by field or laboratory techniques. LAPSUS was chosen for the LEM as it is a relatively simple model that is not computationally intensive or requires a significant number of initial parameters.

Laboratory experiments (rainfall simulations)

The study of hydrological processes such as run-off, erosion and infiltration can be done using field plots under natural rainfall or field scale rainfall simulators. Laboratory scale rainfall simulators are a convenient and cost-effective method that has proven to be effective in predicting hillslope erosion rates and deriving site-specific erodibility parameters for mining landforms (Evans *et al,* 2000; Sheridan, Loch and So, 2000; Landloch Pty Ltd, 2010; So *et al,* 2018). The erodibility parameters are developed through a rainfall component and an overland flow component. Rainfall simulators are used to derive the materials interrill erosion factor and the effective hydraulic conductivity. Overland flow simulators derive the materials rill erodibility factor and critical shear stress.

These calibrated parameters can be used in conjunction with erosion models to assess the long-term stability of mining landforms.

WEPP modelling

WEPP model inputs can be broken up into four key components: climate, material properties, topography and surface conditions.

A 100 year synthetic climate file was generated using the CLIGEN stochastic weather generator. CLIGEN has been used and validated in Australia (Yu, 2003). The climate file needs daily temperature (minimum and maximum), rainfall and solar radiation data as well as pluviograph (six minute rainfall intensity) data.

The physical and chemical properties required for each cover material are: depth, cation exchange capacity (CEC), organic matter (OM), PSD and the calculated interrill and rill erodibility parameters (from the rainfall simulator experiment).

Slope parameters for the final landform are needed to set-up a linear slope cross-section in the WEPP model. Several slopes that are representative of the final landform were run in the WEPP model.

The WEPP model can also be run with different vegetative groundcovers. Vegetation groundcover was only varied if an initial model run predicted an erosion rate greater than 10 t/ha/a.

LEM with LAPSUS

LAPSUS requires several inputs before the model can be run successfully. The DEM/TIN created for the final landform should include the additional height of the preferred cover. A grid size should be used to give an appropriate resolution for the detail of the final landform. The grid size should be capable of showing relevant structures (eg buttressing, embankments and/or benches). The rainfall should be set to a value equivalent to the average rainfall of the region. Infiltration and evaporation are also required inputs, run-off is generated after infiltration and evaporation are subtracted from rainfall. The erosion parameters used are calibrated from the WEPP model outputs, creating a 3D model that reflects the predicted erosion rates of the WEPP model.

RESULTS

Laboratory experiments

Laboratory experiments were conducted on the following samples:

- Mine A: soil and NAF waste rock

- Mine B: soil and NAF waste rock

- Mine C: soil.

Table 1 gives the calculated interrill and rill erodibility parameters from the laboratory experiments.

TABLE 1

Interrill and rill erodibility parameters.

Parameter	Units	Mine A		Mine B		Mine C
		Soil	NAF waste rock	Soil	NAF waste rock	Soil
Interrill erosion factor	Kilogram seconds per quartic metre (kg s/m^4)	5.23×10^6	-	2.93×10^8	-	3.68×10^7
Effective hydraulic conductivity	Millimetre per hour (mm/hr)	100	100	100	100	100
Rill erodibility factor	Seconds per metre (s/m)	4.8×10^{-3}	-	9×10^{-4}	-	8.3×10^{-3}
Critical shear stress	Pascal (Pa)	65	-	16.6	-	8.9

It should be noted that the NAF waste rock has empty parameters, this is due to no run-off or rilling occurring during the laboratory experiments.

WEPP modelling

The outcome from the WEPP model produced results that indicated what the average and maximum erosion rates for the particular inputs would be. The erosion rates selected for modelling are shown in Table 2, these rates are to create 'conservative' erosion scenarios for each Mine TSF.

TABLE 2

Selected erosion rates used for LEM.

	Unit	Mine A	Mine B	Mine C
Average erosion rate	t/ha/a	3.1	6.16	4.72

Mine A

The WEPP model was run for multiple slope parameters (embankments and rehabilitated tailings beach) to determine a range of erosion rates that would be consistent with the final landform. The embankment and rehabilitated tailings beach were modelled separately as run-off is channelled to a drainage area; therefore, they are considered two separate systems (the run-off from the rehabilitated tailings beach does not affect the embankment). The rehabilitated tailings beach had negligible erosion which is to be expected for a low gradient. The embankments had increased erosion rates, which are comparatively steep.

The LAPSUS model was run using the average erosion rate predicted for the embankments (3.1 t/ha/a) to calibrate the erosion parameters. The erosion rate used is the highest average rate predicted by the WEPP model.

The LAPSUS modelling predicted that after 100 years the average soil loss depth of the landform would be 0.2 m with a maximum soil depth loss of 10.4 m. Figure 2 shows a 3D rendering of the final landform after 100 years.

FIG 2 – Mine A: 100 year erosion of final landform.

Mine B

The WEPP model was run for multiple slope parameters to determine a range of erosion rates that would be consistent with the final landform. Unlike Mine A, the final landform is free draining with no specific drainage areas.

The LAPSUS model was run using the average erosion rate predicted for the steepest slope (6.16 t/ha/a) to calibrate the erosion parameters.

The LAPSUS modelling predicted that after 100 years the average soil loss depth of the landform would be 0.4 m with a maximum soil depth of 20.1 m. Figure 3 shows a 3D rendering of the final landform after 100 years.

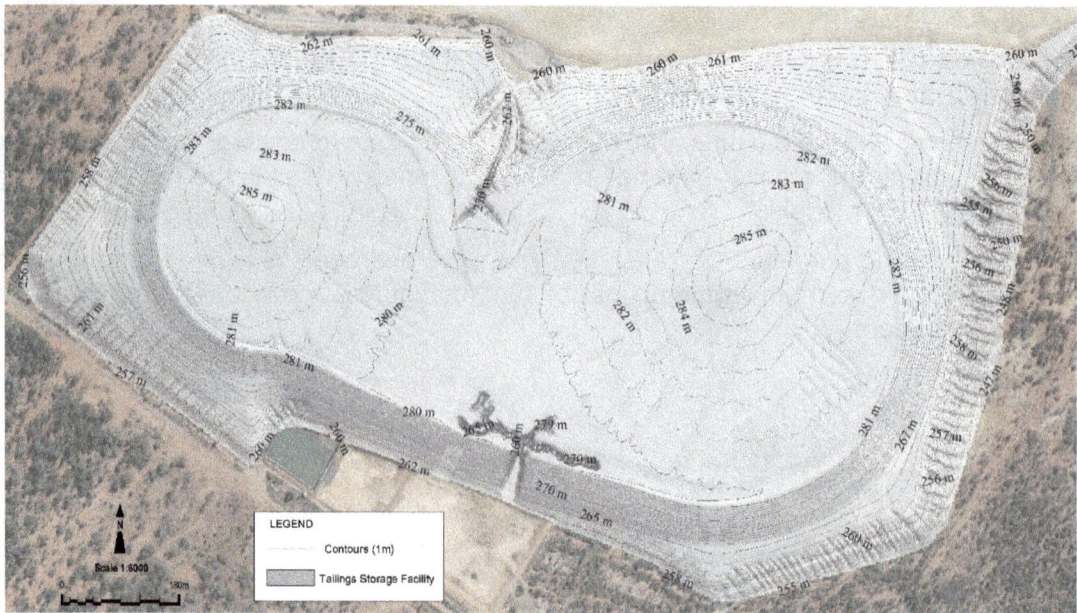

FIG 3 – Mine B: 100 year erosion of final landform.

Mine C

The WEPP model was run for multiple slope parameters to determine a range of erosion rates that represent the final landform. Like Mine B, the final landform is designed to be free draining with no preferred drainage areas.

The LAPSUS model was run using the average erosion rate predicted for the steepest slope (4.72 t/ha/a) to calibrate the erosion parameters, which is the highest average erosion rate predicted by the WEPP modelling.

The LAPSUS modelling predicted that after 100 years the average soil loss depth of the landform would be 0.1 m with a maximum soil depth loss of 10.7 m. Figure 4 shows a 3D rendering of the final landform after 100 years.

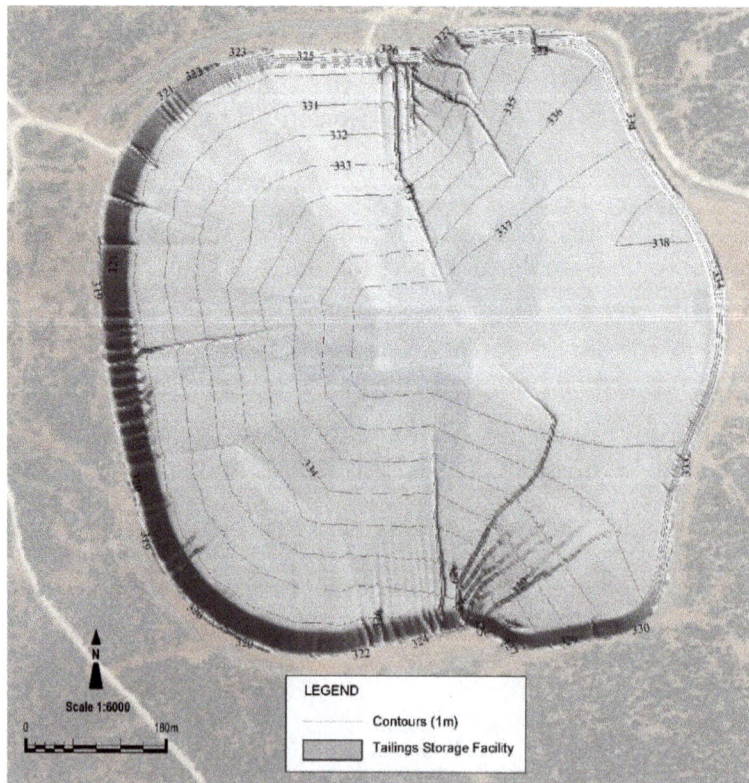

FIG 4 – Mine C: 100 year erosion of final landform.

DISCUSSION

LAPSUS indicates that the final landform of each Mine contains high-risk areas. Mine A modelling indicates a high-risk area is located along the south embankment, likely the drainage area. The 3D render also indicates that sections of the tailings beach is affected by rill erosion. Mine B modelling indicates that the north and south embankments between the two peaks are high-risk areas, because water flow is channelled by the relief. The south embankment is also impacted by rill erosion. Mine C modelling shows high-risk areas are associated at the north and south embankments (specifically where the landform fans out). The west section of the landform also appears to have significant rilling.

Figures 2 and 3 illustrate that erosion is more prevalent on the embankment sections than the tailings beach, because the steeper slopes induce greater flow velocity. The difference identified in Figure 1 is likely due to the channelled flow stopping water being directed over the embankments, hence the large erosion displayed at the drainage area.

Based on the average soil loss, each landform is considered stable after 100 years as the average depth of erosion does not exceed the depth of the preferred cover. Mine B had the most disturbed cover with about 26 per cent of the cover thickness eroded after 100 years, followed by Mine A and then Mine C (18 per cent and 14 per cent respectively).

Each model was run for 1000 years to determine when the landform is considered unstable (when the average depth of soil loss exceeds the cover thickness). As can be assumed from the 100 year results, Mine B failed the earliest with the cover thickness being completely deteriorated after 300 years. Mine A failed after 600 years and Mine C after 900 years. This suggests that determining a suitable cover thickness based on Mine specific requirements (eg the purpose of the cover, the available quantities of material and the landform design) does not heavily affect the outcome of landform erosion stability. It should be noted that all of the covers have localised and incised sections of erosion at 100 years, suggesting that failure has occurred at much shorter time scales.

CONCLUSION

The cover design process does not account for erosion potential of the cover, which is an integral part of creating a landform that is safe, stable and self-sustaining. Therefore, this paper examines the erosion potential of successful cover designs associated within the Cobar region, where these cover designs have proven effective for the purpose of a TSF cover (limited seepage and adequate water storage).

Erosion potential can be predicted through different modelling programs (2D and 3D). 2D modelling (WEPP) is a useful tool that can predict erosion rates for a number of different scenarios (climate, slope, material and vegetation). While this is helpful in understanding the general erosion potential of the final landform, it does not provide any potential landscape information (eg soil loss, areas of sheet and/or rill erosion). It can, however, be used to aid 3D modelling by providing a viable alternative to extensive field trials; thereby providing these landscape parameters. LAPSUS modelling showed that the TSF covers were stable up to 100 years, with the average depth of soil loss not exceeding the depth of the cover. However, the model also predicted that each cover was prone to localised erosion that would penetrate the cover thickness after 100 years. Modelling also indicated that each TSF cover would fail after 1000 years, when the average depth of soil loss exceeds the depth of the cover.

When determining erosion potential of a TSF, the thickness of the cover design plays a minimal role in the determination of erosion potential. Based on the three thicknesses modelled, the smallest cover (0.8 m) displayed the least amount of erosion potential, with the second highest erosion rate. The more important roles in erosion potential appear to be the overall landform design (slope) and water flow paths. The results present a conservative (worst-case) scenario; it is expected that these final landforms will have reduced erosion potential at a field scale, since the LAPSUS modelling assumes the entire TSF landform has the same erosion rate as the highest predicted WEPP outputs.

REFERENCES

Ayres, B, Silveria, C, Ellice, C, Christenson, D and O'Kane, M, 2003. Development of a Cover System Design for Potentially Acid-Forming Tailings at Peak Gold Mine, in *Proceedings of the Sixth International Conference on Acid Rock Drainage (ICARD)*, pp 957–963 (The Australasian Institute of Mining and Metallurgy: Melbourne).

Evans, K G, Saynor, M J, Willgoose, G R and Riley, S J, 2000. Post-Mining Landform Evolution Modelling: 1, Derivation of Sediment Transport Model and Rainfall-Runoff Model Parameters, *Earth Surface Processes and Landforms*, 25:743–763.

GEO-SLOPE International Ltd, 2012. *Seepage Modeling with SEEP/W* (GEO-SLOPE international Ltd: Canada).

Howard, E J and Roddy, B P, 2012. Evaluation of the water erosion prediction project model – validation data from sites in Western Australia, in *Mine Closure 2012: Proceedings of the Seventh International Conference on Mine Closure* (eds: A B Fourie and M Tibbett), pp 81–92 (Australian Centre for Geomechanics: Perth), https://doi.org/10.36487/ACG_rep/1208_09_Howard.

Howard, E J, and Roddy, B P, 2016. A comparison of landscape evolution models for waste landform closure designs, in Goldfields Environmental Management Group Workshop.

International Network for Acid Prevention (INAP), 2009. Global Acid Rock Drainage Guide, viewed 23 January 2023, www.gardguide.com.

Jamson, N P and Rohde, T K, 2019. Tailings storage facilities store-and-release cover design for the Cobar region, in *Mine Closure 2019: Proceedings of the 13th International Conference on Mine Closure* (eds: A B Fourie and M Tibbett), pp 621–634 (Australian Centre for Geomechanics: Perth), https://doi.org/10.36487/ACG_rep/1915_50_Jamson.

Landloch Pty Ltd, 2010. Rehabilitation of Waste Landforms: Woodie Woodie Manganese Operation, prepared for consolidated Minerals Limited by Landloch Pty Ltd.

Landloch Pty Ltd, 2013. Landform design study: Carmichael Coal Project, prepared for GHD Pty Ltd by Landloch Pty Ltd.

Lu, H, Gallant, J, Prosser, I P, Moran, C and Priestley, G, 2001. Prediction of Sheet and Rill Erosion Over the Australian Continent, Incorporating Monthly Soil Loss Distribution, Technical Report 13/01, CSIRO Land and Water, Canberra, Australia.

O'Kane, M and Walters, P, 2003. Dry trial covers at Mt Whaleback – A Summary of Overburden Storage Area Cover System Performance, in *Proceedings of the Sixth International Conference on Acid Rock Drainage (ICARD)*, pp 147–153 (The Australasian Institute of Mining and Metallurgy: Melbourne).

PC-Progress, 2009. The Hydrus-1D Model Description, Available from: <www.pc-progress.com/en/Default.aspx?h1d-description> [Accessed 20 March 2023].

Renard, K G, Foster, G R, Weesies, G A, McCool, D K and Yoder, D C, 1997. *Predicting soil erosion by water: a guide to conservation planning with the revised universal soil loss equation (RUSLE)*, United States Department of Agriculture (USDA), Agriculture Handbook No. 703, 404 p.

Rohde, T K, Defferrard, P L and Lord, M, 2016. Store and release cover water balance for the south waste rock dump at Century mine, in *Mine Closure 2016: Proceedings of the 11th International Conference on Mine Closure* (eds: A B Fourie and M Tibbett), pp 47–59 (Australian Centre for Geomechanics: Perth), https://doi.org/10.36487/ACG_rep/1608_0.6_Rohde.

Rohde, T K, Williams, D K and Burton, J, 2011. Store and release cover performance at Cadia Hill gold mine, Australia, in AB Fourie and M Tibbett, in *Mine Closure 2011: Proceedings of the Sixth International Conference on Mine Closure* (eds: A B Fourie and M Tibbett), pp 333–342 (Australian Centre for Geomechanics: Perth) https://doi.org/10.36487/ACG_rep/1152_36_Rohde.

Schoorl, J M, Veldkamp, A and Bouma, J, 2002. Modelling water and soil redistribution in a dynamic landscape context, *Soil Science Society of America Journal*, 66:1610–1619.

Sheridan, G, Loch, R and So, H, 2000. Use of laboratory-scale rill and interrill erodibility measurements for the prediction of hillslope-scale erosion on rehabilitated coal mine soils and overburdens, *Australian Journal of Soil Research*, 38:285–297.

So, H, Khalifa, A, Yu, B, Caroll, C, Burger, P and Mulligan, D, 2018. MINErosion 3: Using measurements on a tilting flume-rainfall simulator facility to predict erosion rates from post-mining landscapes in Central Queensland, Australia, PLoS ONE, 13(3).

SoilVision, 2009. *Introducing SVFlux,* Available from: <www.soilvision.com/products/svoffice-ge/svflux-ge> [Accessed: 20 March 2023].

USDA-ARC, 1985. *The Water Erosion Prediction Project (WEPP) Model*, United States Department of Agriculture, Agriculture Research Centre.

Yu, B, 2003. An assessment of uncalibrated CLIGEN in Australia, *Agricultural and Forest Meteorology*, 119 (3–4):131–148.

Numerical investigations of evaporation-driven salt uptake in store-release soil covers

W Zhang[1], C Zhang[2] and D J Williams[3]

1. Research Officer, The University of Queensland, Brisbane Qld 4072.
   Email: wenqiang.zhang@uq.edu.au
2. Research Fellow, The University of Queensland, Brisbane Qld 4072.
   Email: chenming.zhang@uq.edu.au
3. FAusIMM, Professor, The University of Queensland, Brisbane Qld 4072.
   Email: d.williams@uq.edu.au

ABSTRACT

In arid and semi-arid regions, evaporation dominates the evolution of water balance in store-release (SR) cover systems. It can facilitate diffusion-driven salt uptake from the hypersaline tailings into the covered soil by inducing upward advection. The performance of SR covers in limiting evaporation-driven salt migration fewerinvestigated needs to be investigated more. A one-dimensional numerical model that couples the transport of liquid water, water vapour, heat and solute was developed. The model was calibrated with the monitored data obtained from an instrumented column, which consisted of 0.6 m thick monolithic silt cover overlying 0.6 m thick compacted bauxite exploredons were conducted to explore the evaporation behaviour and salt distribution in monolithic covers using three representative cover materials over compacted and loose tailings. Under given atmospheric conditions, the maximum salinised depth in the cover was determined by the location where the upper boundary of the freshwater-saltwater mixing zone overlapped with the vapourisation plane. In one-year drying simulations, the fine sand cover was most vulnerable to salinisation as over two-thirds of the depths were affected by the uptaken salt when covering both types of tailings due to intense capillary effects. Although the clay cover could keep two types of tailings saturated in the one-year drying process, the silt cover was more recommended as the main component of SR covers due to its similar robust ability to limit salt uptake and moderate water-release capacity.

INTRODUCTION

Placing 'moisture store-release' (SR) soil covers over hypersaline tailings has been a promising way to rehabilitate closed tailings storage facilities (TSFs) and played an essential role in the sustainable development of the mining industry. In the soil cover system, dissolved salt from hypersaline tailings can migrate upwards under intrinsic molecular diffusion, evaporation-driven advection and mechanical dispersion. Due to the sharp solute concentration gradient between the cover soil and the underlying tailings, solute diffusion initiates when the cover-tailings interface is fully saturated (Kessler et al, 2010). Such a salt uptake process usually proceeds at extremely slow rates determined by the molecular diffusion of solute that depends on soil structures and water contents (Kelln, Barbour and Qualizza, 2008). As evaporation progresses, the unsaturated zone in the cover expands downwards, inducing an upward advective flux of pore water at lower depths and remarkably accelerating salt uptake. Although the water loss in the cover can weaken the solute diffusion, advection plays a predominant role in driving the upward salt migration, and its further development can lead to higher risks in salinisation of the cover soil and long-term degradation of the revegetated TSF (Dobchuk et al, 2013; Kelln, Barbour and Qualizza, 2008; Meiers et al, 2011). Due to the existence of solid particles, salt migration with porewater progresses in non-uniform velocities, and such tortuous movement is described as mechanical dispersion. Considering the mechanism of the abovementioned processes, salt migration is weak in the saturated zone due to the predominant molecular diffusion, while evaporation-induced advection and dispersion dominate in the unsaturated zone, allowing the salt migrates more steadily to the upper cover layer. In addition, evaporation-induced salt accumulation in the cover soil may cause precipitation of solid salt and formation of salt crust in the pore space, which impedes the vapourisation of pore water and hence mitigates the water-releasing capacity of the cover in the long run (Gran et al, 2011a; Zhang, Li and Lockington, 2014). Despite precipitation can somehow mitigate the upward salt migration by the downward flushing of fresh water (Meiers et al, 2011), it is impractical to rely only on natural water dilution in arid and semi-arid regions, where annual potential evaporation far exceeds rainfall. By virtue of the functionality of SR covers, maintaining a high water-saturated environment around the cover-tailings interface is critical to mitigating evaporation-driven advection and restricting salt uptake to progress under diffusion-only effect. Therefore, it is necessary to demonstrate that an SR cover design effectively controls the water balance of the cover system and limits upward salt migration under prolonged dry weather.

Previous studies demonstrated the performances of various SR covers on water balance control in response to the local climate variations, but were inadequate regarding salt transport and accumulation in the cover layers. This study aims to investigate evaporation-driven salt migration in monolithic SR numerically covers with various compositions over hypersaline tailings. A one-dimensional numerical model that couples liquid water flow, water vapour flow, heat and solute transport is introduced. The model is firstly calibrated against the monitoring results obtained from an instrumented column filed with 600-mm-thick monolithic soil cover over sea water-neutralised bauxite residue under a semi-arid climate. Infiltration is not considered in the model because negligible amounts of incidental rainwater could only moisturise a thin layer of the cover soil beneath the surface in dry seasons and get evaporated rapidly, as observed in the column test throughout the entire monitoring time. Besides, under a given weather condition, the moisture-releasing capacity of SR cover determines the intensity of internal water flux and influences the threshold of salt-affected depth in the cover. To gain better insights into how evaporation affects salt dynamics and distributions in SR covers with various hydrological parameters, several cover scenarios are discussed regarding their performances in limiting salt uptake under prolonged dry weather conditions.

METHODOLOGY

Numerical simulation

Numerical studies were carried out using SUTRAVAP, a modified version of an existing numerical model SUTRATM (developed by the United States Geological Survey) (Voss and Provost, 2010) to simulate one-dimensional water and salt transport between cover soil and hypersaline tailings with variable hydrological parameters. The model capability is extended to implement the evaporation process by considering surface resistance, salt resistance and aerodynamic resistance, water vapour flow in the unsaturated soil, salt precipitation and heat transport during evaporation. Major mathematical models and boundary conditions are summarised below. More details and applications of the model can refer to Zhang, Li and Lockington (2014).

Conceptual model, boundary conditions and initial conditions

A conceptual model simulating 1D water, salt and heat transport in a monolithic cover system was applied in the numerical simulation, as shown in Figure 1. The model domain **ABCD** encompasses one soil cover layer and an underlying hypersaline tailings layer. Evaporation takes place at the soil surface (**AD**) and is driven by the vapour density gradient between the soil surface and the atmospheric conditions (solar radiation level, relative humidity in the air, wind velocity and air temperature). Both liquid water flow and water vapour flow occur in unsaturated soil. No phase change of liquid water occurs below the water table. Vertical salt migration is induced by two processes: (i) solute advection and dispersion driven by a matric potential gradient due to evaporation; (ii) solute diffusion driven by the concentration gradient between fresh cover soil and hypersaline tailings. Energy exchange at the soil surface, downward heat transport to the underlying soil and lateral heat dissipation were also considered in the model.

The side boundaries **AB** and **CD** (Figure 1) were treated as no-flow boundaries because only vertical water movement was simulated in this study, but specific sensible heat flux was applied at these two sides to describe the heat exchange between the column wall and ambient environment. Seepage was not considered in the model, so boundary BC was also set as no-flow, similar to the situation that the instrumented column experienced. The upper boundary **AD** represents the soil surface, where evaporation and energy exchange takes place. Evaporation rate is determined by the soil moisture condition at the surface and the atmospheric condition. Constant heat source, wind flow and relative humidity were set at a reference point above **AD**, dictating the potential evaporation rate (PER) from the soil surface. As the degree of saturation at the soil surface decreases, water for vapourisation gradually becomes less available, leading to the increase of surface resistance and the consequent deviation of the actual evaporation rate (AER) from PER. In addition to the effects of surface resistance and atmospheric condition, precipitated salt in soil matrix can act as another resistance to surface evaporation and was considered in the computation of AER when solute concentration was beyond the solubility. Note that rainfall is not considered in this study as a dry season was predominant during the column test.

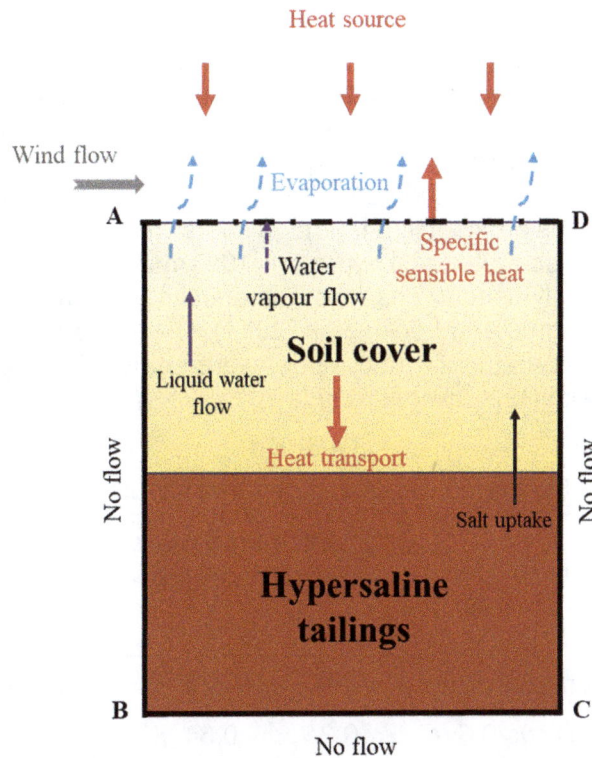

FIG 1 – Conceptual model of a monolithic soil cover over hypersaline tailings.

Model calibration was conducted by simulating a 50-day drying process of an initial water-saturated soil cover over compacted residue without rainfall event, aiming to reproduce the prolonged dry period (from 7 July to 25 August 2018) with negligible rainfall in an instrumented column test. Based on the weather station data near the column, the atmospheric condition, including solar radiation, wind velocity, air temperature and relative humidity, were assigned to be constant values representing their effective impacts. SWCC parameters, intrinsic permeabilities and thermal conductivities of the cover soil and residue were marginally adapted in the model to best fit the measured saturation, temperature profiles and calculated surface evaporation in the column test. The initial pore water salinities of soil cover and tailings were set to be 1×10^{-4} kg/kg and 3.5×10^{-2} kg/kg, respectively, similar to the measured values of the materials that were applied in the column test. The initial water pressure was set to be hydrostatic, considering the various fluid densities in the cover soil and tailings. The model domain was comprised of 1443 nodes and 960 elements, with a mesh size of 0.025 m. A time step of 5 seconds was assigned in the 50-day simulation.

SIMULATED COVER SCENARIOS

Based on the 50-day drying calibration, further simulation cases were set-up using conventional SR cover materials (ie fine sand, silt and clay) over compacted and loose tailings to explore the effect of cover compositions on evaporation and salt migration in monolithic cover systems. Table 1 summarises all the simulated cases, named with initials of material types (eg SCT indicates using silt cover over compacted tailings). The base case for cover scenarios was the extended calibration simulation of silty cover over compacted tailings for one year without extra water supply. The thicknesses of the cover, if considered in the case, were assigned to be 0.6 m. Such cover thickness was validated to prevent salt uptake through the two-year monitoring result from the instrumented column test. Below 0.6 m, tailings were set to be hypersaline with the same solute concentration as the calibration case. Cover and tailings types were characterised by assigning representative SWCCs and permeabilities, referring from the previous studies by Carsel and Parrish (1988) and Benson *et al* (2007). Different from the calibration case, a constant atmospheric condition with daily averaged net radiation (200 W/m^2), air temperature (25°C) and relative humidity (50 per cent) over one year was assigned at the reference point, as well as a constant aerodynamic resistance of 150 s/m, leading to a daily potential evaporation rate of 6 mm/day. Except for the calibration, each simulation was run for 365 days, representing a prolonged dry year with negligible rainfall. At the beginning of the 50-day desiccation, both the cover soil and tailings in the column were in relatively high degrees of saturation, mainly due to a series of rainfall events after the test commencement. Therefore, the nearly-full water-saturated state was set as the initial condition of the model. Only 3.8 mm of rainfall occurred in the first six days of the period, slightly moisturised the soil surface, and dry weather dominated the rest of the time. Such a drying dominated process contributes to calibrating hydrological parameters of the cover soil and tailings, as well as evaporation-driven salt migration process. The diffusion coefficient and dispersivity, which significantly influence salt transport in saturated and unsaturated soil, were 5 × 10^{-11} m^2/s and 0.01 m, respectively, similar to the values that were commonly used in previous numerical studies related to evaporation in saline soil (Gran *et al*, 2011b; Zhang, Li and Lockington, 2014). Given the column test was carried out in winter, initial air and soil temperatures in the model were set to 16°C, and evaporation at the soil surface proceeded slowly with PER being 2 mm/day.

TABLE 1

Simulated cases and cover configurations.

Simulation case	Cover configuration (material/thickness)	SWCC parameters (van Genuchten)			Intrinsic permeability (m^2)	Simulation period (days)
		S_{resw} (-)	α (m-1)	n (-)		
Calibration	Silt cover/0.6 m	0.14	0.98	1.35	1.8 × 10^{-15}	50
	Compacted tailings/0.6 m	0.2	0.34	1.2	5.6 × 10^{-17}	
SCT	Silt/0.3 m	0.14	0.98	1.35	1.8 × 10^{-15}	365
	Compacted tailings/0.3 m	0.2	0.34	1.2	5.6 × 10^{-17}	
SLT	Silt/0.6 m	0.14	0.98	1.35	1.8 × 10^{-15}	365
	Loose tailings/0.6 m	0.15	0.98	1.5	5.6 × 10^{-14}	
FSCT	Fine sand/0.6 m	0.12	1.47	6.5	2.8 × 10^{-12}	365
	Compacted tailings/0.6 m	0.2	0.34	1.2	5.6 × 10^{-17}	
FSLT	Fine sand/0.6 m	0.12	1.47	6.5	2.8 × 10^{-12}	365
	Loose tailings/0.6 m	0.15	0.98	1.5	5.6 × 10^{-14}	
CCT	Clay/0.6 m	0.18	0.98	1.2	5.6 × 10^{-16}	365
	Compacted tailings/0.6 m	0.2	0.34	1.2	5.6 × 10^{-17}	
CLT	Clay/0.6 m	0.18	0.98	1.2	5.6 × 10^{-16}	365
	Loose tailings/0.6 m	0.15	0.98	1.5	5.6 × 10^{-14}	

Note: cover soil types are characterised through SWCC parameters and intrinsic permeabilities.

RESULTS AND DISCUSSION

Numerical model calibration

Figure 2 shows the tested and simulated results during the calibration, which encompass temporal variations of water saturation, concentration, and temperature profiles with depth, as well as the actual evaporation rate. The simulated liquid water and vapour fluxes are also included to indicate the movement of the vapourisation plane. The model well captured the time trends of water, salt and heat transport in both the cover soil and tailings. Desaturation occurs predominantly near the soil surface, and a dry soil layer with a 50 mm thickness was formed below the surface after 50 days (Figure 2d). Upward salt migration from the underlying tailings to the cover soil was not evident as shown in both the simulation and the column measurement. As no upward water flow occurred near the cover-tailings interface (Figures 2n–2p), the salt movement was mainly driven by diffusion instead of evaporation-induced advection and dispersion. Despite distinct variations in diurnal air temperature onsite, the measured soil temperature gradients mainly existed to the depth of 200 mm. The simulated temperature profiles agree well with the measured ones, validating the assigned initial temperature and thermal conductivity used in the model. The evaporation rates derived from the column are slight higher than the simulated ones during the first six days while lower afterwards. Such discrepancies are attributed to: (i) occasional light rainfall events on the second and sixth days, resulting in partial resaturation at the surface and subsequent increase of the evaporation rate, (ii) the moisture sensor installed at the top cover layer might not well measure the residual water content of dry soil, leading to the slight underestimation of evaporation rate when the surface cover was dry.

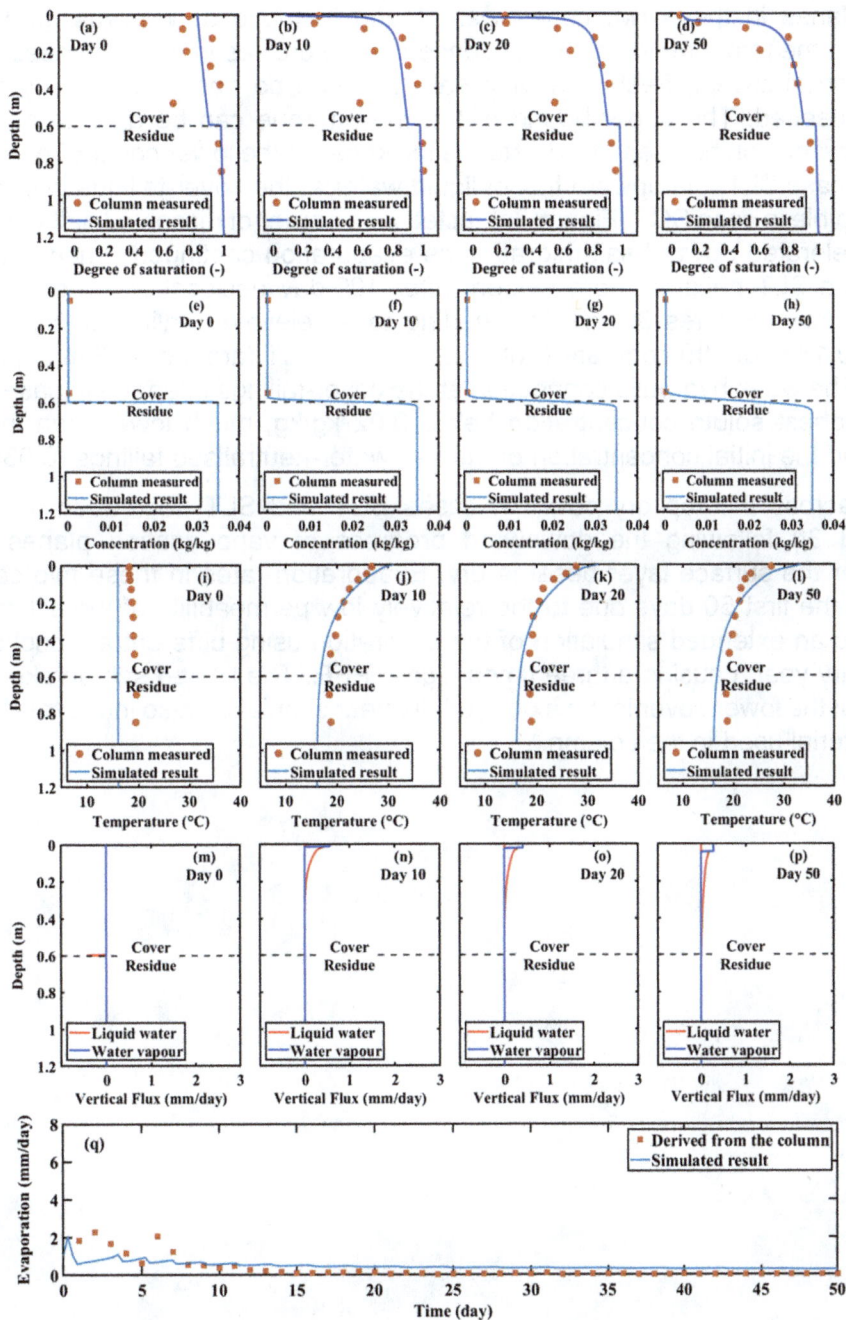

FIG 2 – Measured and simulated degree of saturation (a–d), concentration (e–h) and temperature (i–l) profiles; simulated liquid water and water vapour fluxes (m–p); and measured and simulated evaporation (q).

Soil cover scenarios

The performances of the three representative covers were assessed by their capabilities to release porewater and limit salt uptake. Figures 3–6 depict the one-year simulated results for the six cover scenarios, including saturation profiles, concentration profiles, liquid water and water vapour fluxes with depth, and temporal variations of evaporation rates. The comparison of the simulated cover scenarios can also serve as a sensitivity analysis on the soil hydrological parameters.

The cases that apply silt covers on two types of tailings (cases SCT and SLT) had similar evaporation rates over time (Figures 3e and 3j), while different saturation profiles in the tailings and salt distribution in the cover. In case SCT, the tailings remained nearly saturated over the 365-day simulation, with the lowest saturation at the tailings surface of 0.96, while the bottom of the cover experienced drying with saturation reduced from 0.92 to 0.79 (Figure 3a). Instead, in case SLT, more saline tailings water transport upward into the cover, causing the saturation at the bottom of the cover slightly higher (0.9) than that in case SCT. Such a strong upward liquid water flux at the cover-tailings interface in case SLT was attributed to the relatively high permeability of loose tailings. Both the vapourisation planes in these two cases were found to move downwards gradually with time. Similarly, through the temporal variations of concentration profiles in the two cases (Figures 3b and 3g), a mixing zone between freshwater and saltwater can be observed to progress upwards as evaporation proceeded. The upper bound of the mixing zone can be defined as the salinisation plane, and its location can be used to indicate the thickness of the lower cover affected by salt uptake from tailings. In case SLT, the upward flux of liquid water at the cover-tailings interface (Figure 3h), which was negligible in case SCT (Figure 3c), acted as a carrier of dissolved salt from the underlying tailings and accelerated upward salt migration as evaporation continued. Such evaporation-driven advection in case SLT became more evident after 100-day desiccation, when the bottom cover started to desaturate (Figures 3f and 3g), leading to an elevated salinisation plane in the cover (approximately 0.18 m depth) compared with that in case SCT (around 0.15 m depth) at the end of the simulation. The weak hydraulic connection at the cover-tailings interface in case SCT effectively restricted the highest solute concentration below 0.02 kg/kg, much lower than that of case SLT (0.028 kg/kg) and the initial concentration of the sea water-neutralised tailings (0.035 kg/kg).

Vapour fluxes across the top dry cover in cases SCT and SLT decreased in the same trend (Figures 3d and 3f) following the downward progress of vapourisation planes and sustained evaporation after the surface layer became dry. Evaporation rates in these two cases decreased progressively in the first 50 days due to the relatively low permeability of the silt cover. It is noted that case SCT is an extended simulation of the calibration using different atmospheric parameters representing a dry year, causing a three times higher PER. The limited salt uptake and maintained high saturation at the lower cover in the extended dry year simulation also indicates the effectiveness of the cover material used in the column.

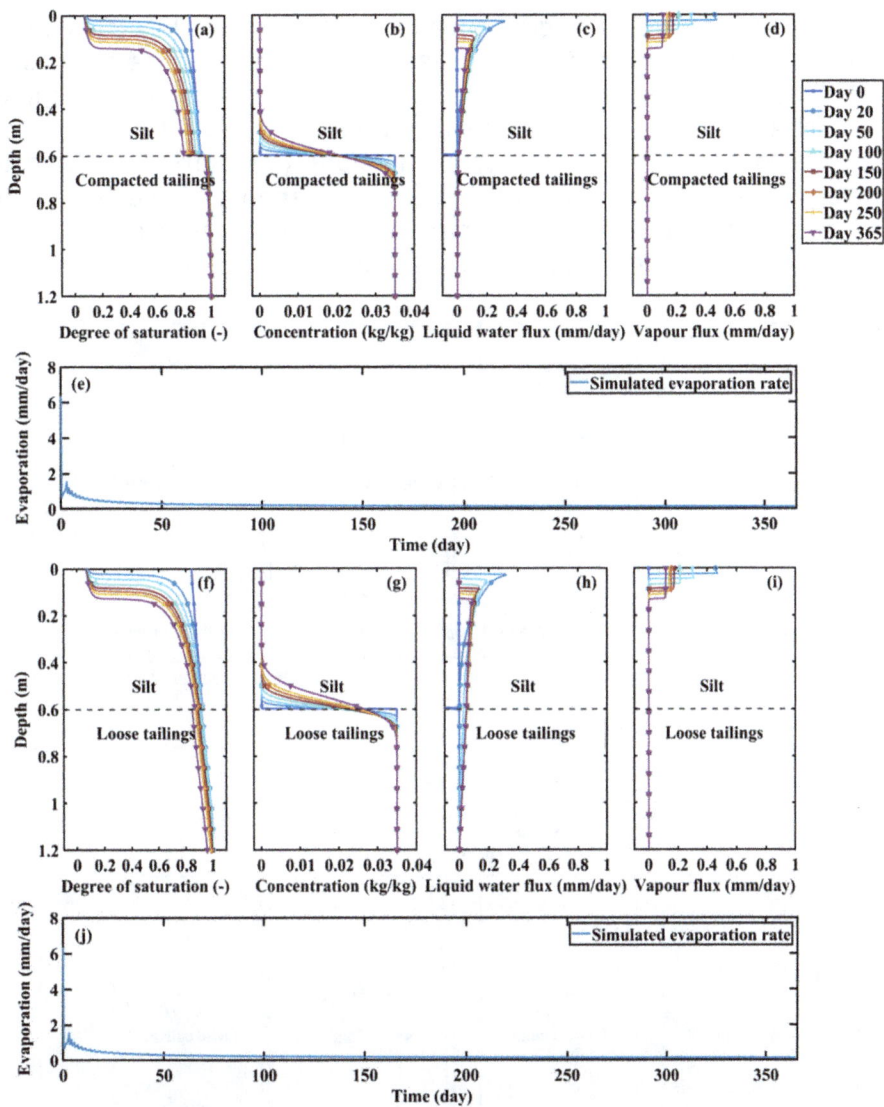

FIG 3 – Simulated results for cases SCT and SLT, including (a) (f) degree of saturation; (b) (g) concentration, (c) (h) liquid water flux, (d) (i) water vapour flux, and (e) (j) temporal variations of evaporation.

Figure 4 displays the simulated results for cases FSCT and FSLT. The overall time trends of the degree of saturation, concentration, liquid water and vapour fluxes in the two cases varied in wider ranges compared to cases SCT and SCT, especially for the highest solute concentration and liquid water flux, which were one order of magnitude higher than silt cover cases. Both of the fine sand covers released the majority of porewater at high rates within the first 50 days and became extremely dry with the depth-averaged degree of saturation below 0.2 at the end of the simulation (Figures 4a and 4f). Despite considerable water loss in the cover, the overall degrees of saturation of the compacted and loose tailing were still at high levels (over 0.98 and 0.85, respectively). This is mainly due to different SWCC behaviours of the fine sand and tailings, which enables the latter to keep high water content at the same suction where the former was in low saturation levels.

The major differences between cases FSCT and FSLT lie in the saturation profiles in the tailings and salt distribution in the covers. The compacted tailings were not affected by evaporative flux and remained near-saturated across the depth after 365 days, while the loose tailings in case FSLT desiccated evidently with a saturation drop of 0.16 occurring at the tailings surface. The stronger hydraulic connection between the cover and tailings in case FSLT supplied saline tailings water to the cover, leading to a relatively thin dry soil layer in the cover (0.2 m thick) and pronounced upward saltwater intrusion (from 0.6 m to 0.1 m depth, average solute concentration close to the tailings level of 0.035 kg/kg) with a hypersaline zone (solute concentration of 0.088 kg/kg) formed at 0.2 m depth. On the contrary, the salt-affected depth in case FSCT was about 0.4 m from the cover bottom to the surface, with a steep concentration gradient and a mild salinity spike at 0.28 m depth. Although salt accumulation was most pronounced near the vapourisation plane in the two cases, the elevated solute concentration was still below the solubility limit. Note that the dry soil layer formed below the cover surface can act as a barrier to salt uptake as it breaks the hydraulic connection between the salinised soil and the overlying non-saline soil. Therefore, the maximum depth that salt could upwards migrate to is determined by the location where the vapourisation plane and the salinisation plane overlap, and salt accumulation would then proceed beneath the overlapped plane instead of further moving towards the soil surface (eg Figure 4b after day 250 and Figure 4g after day 150).

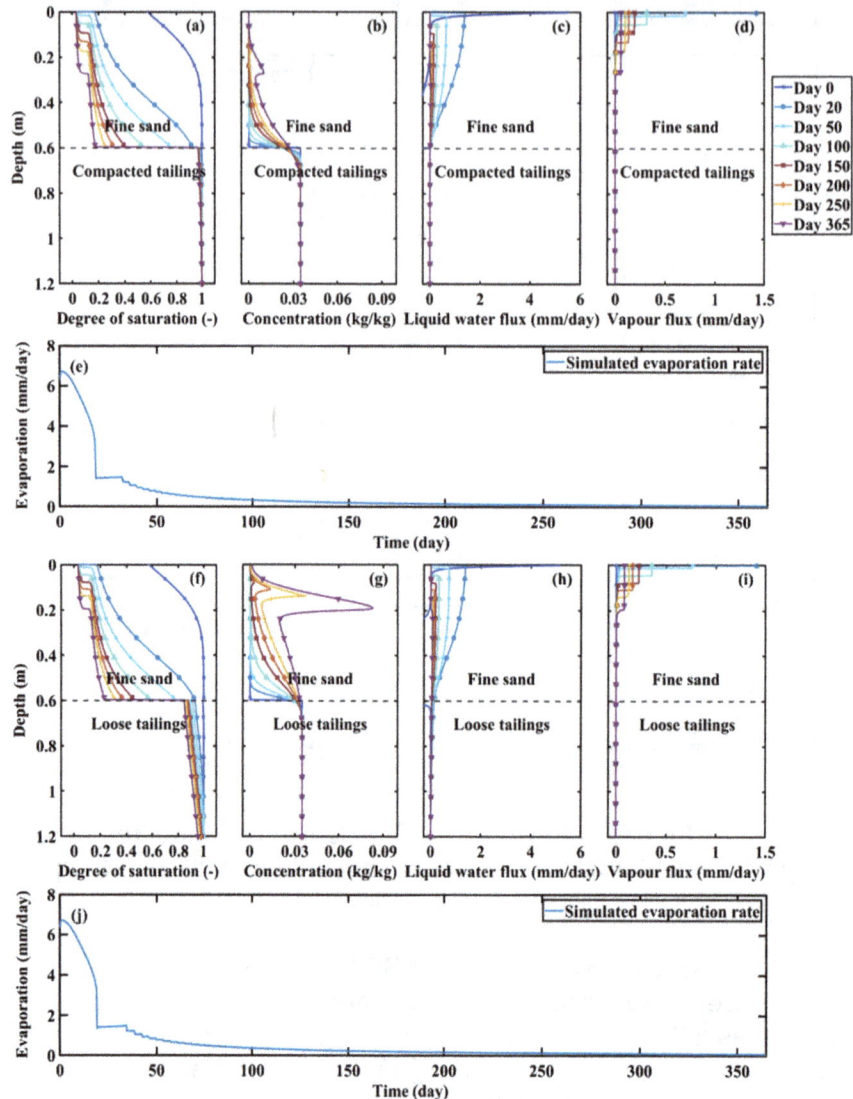

FIG 4 – Simulated results for cases FSCT and FSLT, including (a) (f) degree of saturation; (b) (g) concentration, (c) (h) liquid water flux, (d) (i) water vapour flux, and (e) (j) temporal variations of evaporation.

The simulations of FSCT and FSLT indicate that using fine sand as the main component of a 0.6 m-thick monolithic SR cover has poor performance in limiting salt uptake from tailings and storing sufficient water under dry weather. However, a thinner layer (eg 0.2 m) of fine sand may serve as a hydraulic barrier between a silt cover and tailings to restrict the salinisation plane at a lower depth.

Figure 5 shows the simulated results for cases CCT and CLT. The general variations of all parameters in the two simulations were almost the same and proceeded more slowly than the cases using silt and fine sand as covers. Throughout the 365-day simulation, desaturation only occurred in the clay covers following the similar pattern as in case SCT, and the unsaturated zone just expanded to 0.4 m depth in both two cases (Figures 5a and 5f). A dry soil layer was formed on the first day of desiccation, and vapour flow was found to dominate the rest of the drying process with an average rate below 0.2 mm/day (Figures 5d and 5f). Due to the low permeability and weak water-releasing capacity of the clay cover, the compacted tailings and loose tailings were not affected by surface evaporation and remained near-saturated all the time. Such a highly-saturated condition around the cover-tailings interface indicated that upward salt migration was only driven by the slow diffusion process and not influenced by evaporation-driven advection and dispersion. It is noted that the vapourisation plane moved downwards at a similar rate to the elevation of the salinisation plane in both cases, as the final salt-affected thicknesses in the cover were restricted to 0.18 m from the bottom towards the surface, and the dry tailings layers formed were also around 0.18 m below the surface. Since the salt uptake is still ongoing slowly, the maximum salt-affected depths (ie the location where the vapourisation plane and the salinisation plane overlaps) in cases CCT and CLT are assumed to be around 0.3 m below the cover soil surface. Overall, the 0.6 m-thick clay cover is robust in insulating the compacted and loose tailings from desiccation and hence limiting salt uptake in a dry year. But it may not be suitable to serve as a main component of SR cover given its poor water-releasing capacity.

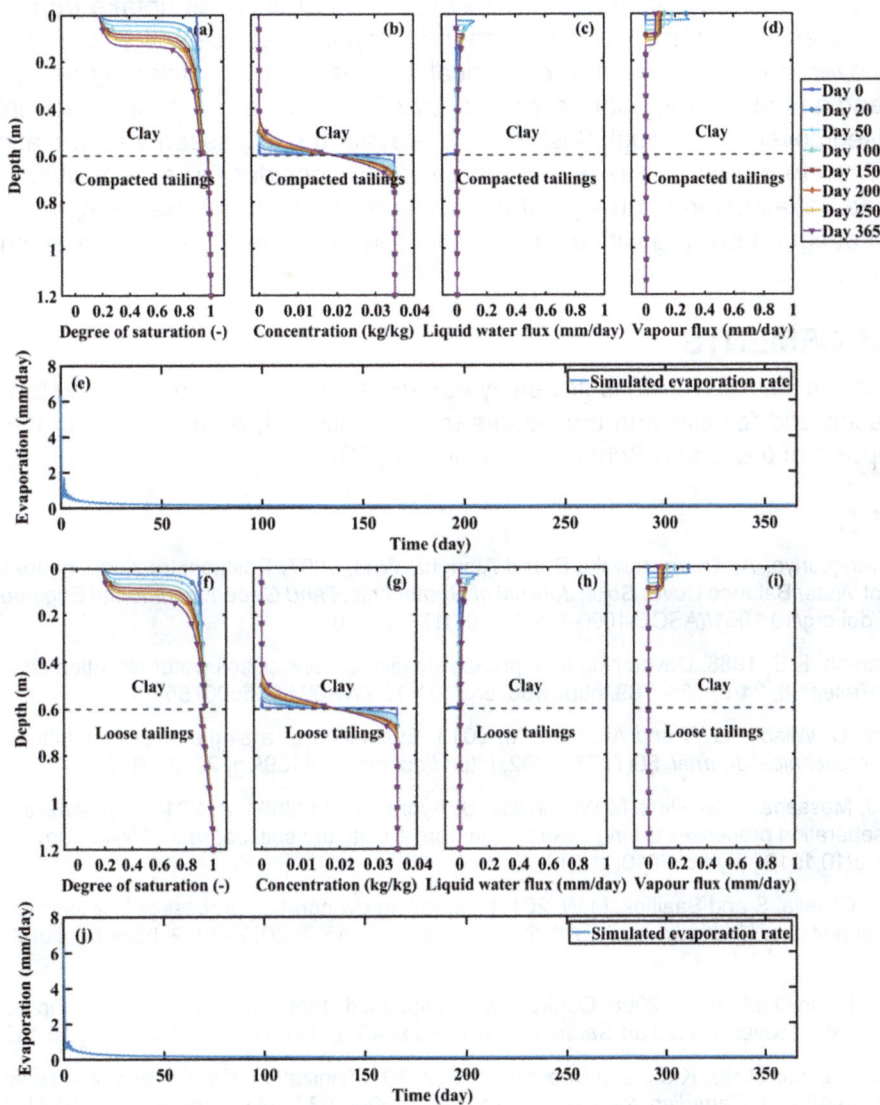

FIG 5 – Simulated results for cases CCT and CLT, including (a) (f) degree of saturation; (b) (g) concentration, (c) (h) liquid water flux, (d) (i) water vapour flux, and (e) (j) temporal variations of evaporation.

CONCLUSIONS

This paper investigates the effect of evaporation on salt uptake in various monolithic covers over hypersaline tailings. A one-dimensional numerical model coupled heat, liquid water, water vapour and solute transport was used to simulate evaporation processes in uncovered tailings and six representative cover scenarios. The model was first calibrated with the monitored results from an instrumented column filled with 0.6 m-thick silt cover over 0.6 m-thick compacted bauxite residue under a semi-arid climatic condition. Good agreements were obtained between the measured and simulated degree of saturation profile, solute concentration profile, temperature profile and daily evaporation rate throughout 50-day desiccation without rainfall. We further examined the performance of three representative cover materials over two types of tailings with respect to limiting evaporation-driven salt uptake in a dry year. Through the simulated results, we found that: (i) the 0.6 m-thick fine sand cover is most vulnerable when applied over loose tailings as almost the entire cover layer was salinised in a dry-year simulation; (ii) a salinisation plane was proposed to describe the upper bound of the freshwater-saltwater mixing zone in the cover during the salt uptake process, and the maximum salinised depth can be dictated by the overlap of the vapourisation plane and the salinisation plane; (iii) the evaporation impacts on salt migration in silt and clay cover cases were not evident due to their capabilities of maintaining a highly-saturated state at the cover-tailings interface to avoid advection and dispersion effect, but silt cover is more recommended to act as the main component of an SR cover for its higher release capacity.

These findings reveal the mechanisms of how evaporation affects salt uptake from the hypersaline tailings into soil cover and highlight the importance of appropriate selection in cover material for a monolithic SR cover. It is noted that the numerical studies were simplified by using homogeneous cover material and annual average atmospheric data without rainfall. Meteoric water infiltration would dilute the salinised cover soil through flushing and elevate the ingressed vapourisation plane in the cover system. Therefore, future work is recommended to consider the effects of long-term wetting-drying cycles and variations in hydrological parameters of cover soil. Besides, the performance of other SR cover configurations (eg with a capillary barrier, multilayer) can also be examined to extend the model capability.

ACKNOWLEDGEMENTS

Queensland Alumina Limited (QAL) is gratefully acknowledged for commissioning the work on which this paper is based and for allowing the results to be published. Also, special gratitude is given to the financial support of the China Scholarship Council (CSC).

REFERENCES

Benson, C H, Sawangsuriya, A, Trzebiatowski, B and Albright, W H, 2007. Postconstruction Changes in the Hydraulic Properties of Water Balance Cover Soils, *Journal of Geotechnical and Geoenvironmental Engineering*, 133(4):349–359. https://doi.org/10.1061/(ASCE)1090–0241(2007)133:4(349)

Carsel, R F and Parrish, R S, 1988. Developing joint probability distributions of soil water retention characteristics, *Water Resources Research*, 24(5):755–769. https://doi.org/10.1029/WR024i005p00755

Dobchuk, B, Nichol, C, Wilson, G W and Aubertin, M, 2013. Evaluation of a single-layer desulphurized tailings cover, *Canadian Geotechnical Journal*, 50(7):777–792. https://doi.org/10.1139/cgj-2012–0119

Gran, M, Carrera, J, Massana, J, Saaltink, M W, Olivella, S, Ayora, C and Lloret, A, 2011a. Dynamics of water vapor flux and water separation processes during evaporation from a salty dry soil, *Journal of Hydrology*, 396(3–4):215–220. https://doi.org/10.1016/j.jhydrol.2010.11.011

Gran, M, Carrera, J, Olivella, S and Saaltink, M W, 2011b. Modeling evaporation processes in a saline soil from saturation to oven dry conditions, *Hydrology and Earth System Sciences*, 15(7):2077–2089. https://doi.org/10.5194/hess-15–2077–2011

Kelln, C, Barbour, S L and Qualizza, C, 2008. Controls on the spatial distribution of soil moisture and solute transport in a sloping reclamation cover, *Canadian Geotechnical Journal*, 45(3):351–366. https://doi.org/10.1139/T07–099

Kessler, S, Barbour, S L, van Rees, K C J and Dobchuk, B S, 2010. Salinization of soil over saline-sodic overburden from the oil sands in Alberta, *Canadian Journal of Soil Science*, 90(4):637–647. https://doi.org/10.4141/cjss10019

Meiers, G P, Barbour, S L, Qualizza, C V and Dobchuk, B S, 2011. Evolution of the Hydraulic Conductivity of Reclamation Covers over Sodic/Saline Mining Overburden, *Journal of Geotechnical and Geoenvironmental Engineering*, 137(10):968–976. https://doi.org/10.1061/(ASCE)GT.1943–5606.0000523

Voss, C I and Provost, A, 2010. SUTRA: A model for saturated-unsaturated, variable-density ground-water flow with solute or energy transport, in US Geological Survey Water-Resources Investigations Report 02–4231, vol 2010. http://pysot.readthedocs.io/en/latest/index.html

Zhang, C, Li, L and Lockington, D, 2014. Numerical study of evaporation-induced salt accumulation and precipitation in bare saline soils: Mechanism and feedback, *Water Resources Research*, 50(10):8084–8106. https://doi.org/10.1002/2013WR015127

Monitoring and surveillance

Use of drone video for weekly surveillance at remote site tailings storage facilities

D Brett[1], F Daliri[2], J Belen[3], E Lim[4] and R Medalla[5]

1. Senior Technical Director, GHD, Hobart 7000 Tas. Email: david.brett@ghd.com
2. Senior Tailings Engineer, GHD, Vancouver, Canada. Email: farzad.daliri@ghd.com
3. Civil Engineer, GHD, Manila, Philippines. Email: jeffreygerald.belen@ghd.com
4. Civil Engineer, GHD, Manila, Philippines. Email: ernestgabriel.lim@ghd.com
5. Civil Engineer, GHD, Manila, Philippines. Email: ryan.medalla@ghd.com

ABSTRACT

Drone video has been used successfully for the past two years for weekly surveillance of tailings storage facilities (TSFs) under construction/operation at remote sites in the Philippines as part of an overall technical review role for the projects. This role includes full-time technical support by in-country personnel with an annual in-person field inspection by senior personnel from Australia and/or Canada. The videos are taken by mine technicians under the direction of site engineers. Technical details of the TSF construction/monitoring are then reviewed by the site team with off-site senior engineers in a weekly Teams meeting.

The videos have been found to be very useful and provide the reviewers with more information than still images. The drone route typically follows an established pattern that gives an overview of the site while more detailed close-up views can be requested by the reviewers if a particular issue arises. Most of the time, the reviewers view the videos before the Teams meeting and can then fast-forward to critical viewpoints during meeting discussion.

The video meeting format has been instrumental in the identification of issues such as potential instability of surrounding cuttings and waste rock dumps in the steep topography, poor drainage control leading to potential severe erosion, borrow area management, tailings discharge management and access road planning. The drone video also allowed a high level of technical support to site supervisory staff at the remote site that would otherwise be impossible.

INTRODUCTION

During the COVID-19 lockdown period and even in normal times, it has not been practical for a high level of detailed field inspection of construction projects in remote areas by experienced technical engineering staff. This was the case for a tailings storage facility (TSF) construction project in the Philippines. This could equally apply in developed countries, where the travel time to site covers more than a few hours. At the Philippine site, this issue was overcome using on-site, in-country technical engineering staff working on a rolling shift basis, supported by senior engineering personnel in Australia and Canada undertaking weekly Teams conferencing and an annual site inspection by the ex-pat team.

The weekly Teams conference reviewed video footage of the site. The videos have been found to be very useful and more informative than still images. The drone route typically follows an established pattern that gives an overview of the site while more detailed close-up views can be requested by the reviewers if a particular issue arises. Most of the time, the reviewers view the videos before the Teams meeting and then fast – forwarded to critical viewpoints during meeting discussion.

DRONE FEATURES

Description of drone

A drone is a flying robot or unmanned aerial vehicle (UAV) which is controlled remotely and collects data, such as survey points, video footage and photographs. From a military background, drones have evolved to be used in a range of civilian roles. Typical commercial-use drones are compact in design but can vary in size to aircraft scale. For this project, the drone used was a DJI Mavic 2 Pro (Figure 1), which features omnidirectional vision systems, infrared sensing systems, and a fully stabilised 3-axis gimbal with a 1" CMOS sensor camera that shoots 4k video and 20-megapixel photos. It is capable of connecting to a smart mobile phone for remote screen viewing and to pilot the drone.

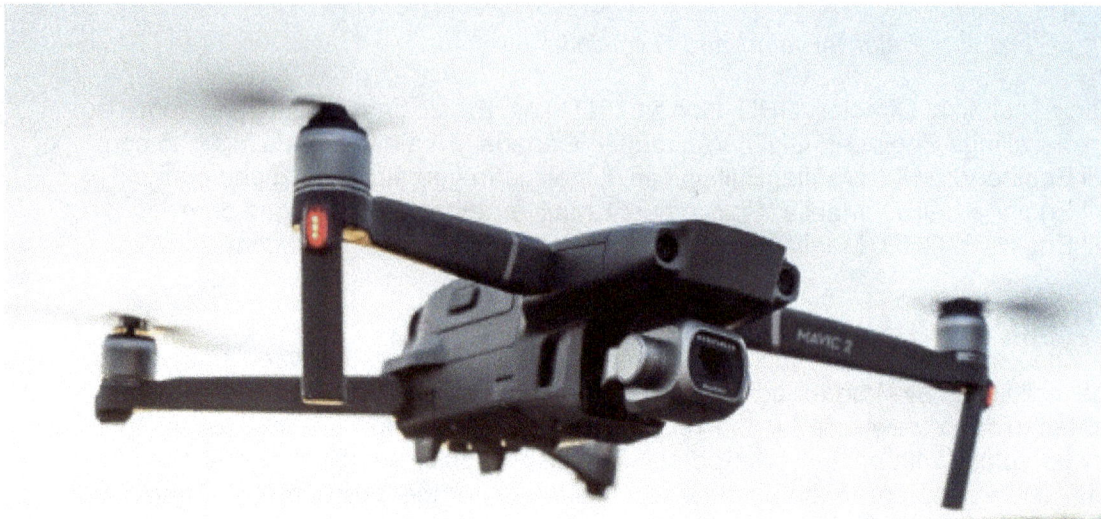

FIG 1 – DJI Mavic 2 Pro drone.

This drone is also capable of flight plan mode which can be configured through its device setting depending on the specified location, area and pre-defined pathway.

A Mavic 2 Pro or similar drone retails for approximately AU$3000.

Comments on practical aspects

Resolution/file size/time

The usual output from the drone comprises a 300–500 megabytes (mb) MP4 file at 4k resolution. This provides three to five minutes of viewing and is sufficient for a flying route around the TSF to give good coverage of the site from different viewpoints. Additionally, the drone allows full high definition (FHD) resolution this but was found to be impractical due to file size. The 4k resolution has been adequate for the purpose of review.

Weather effects

It is crucial to have a detailed plan and a review of the weather forecast prior to drone operations. Wet and windy conditions will complicate the drone control and should be avoided.

Different weather elements that impact drone operations are as follows.

Precipitation/rain event

Some drones are equipped to be waterproof, however, most cannot tolerate water and moisture. Rainwater can damage electrical components and render a drone unusable. If rain occurs during the scheduled date of drone activity, the flight should be rescheduled.

Strong winds

Difficulty arises from flying drones during windy conditions as wind limits of drones are based on their models as indicated in the operation manual. The drone model used on-site could withstand a level-5 wind resistance, which is 29–38 kilometres per hour (kph). Before flying a drone in wind, the battery should be fully-charged as more power is consumed during strong winds. Otherwise, the drone will slow down due to radio receiver issues because of power drainage. Consequently, if the wind speed is above the threshold, the drone operator might lose control due to momentum changes. The crucial decision on whether to continue or delay drone operations during the prevalence of strong wind should be taken into account by site engineers and drone operators.

Visibility

Fog, mist, and haze affects visibility and can block drones from view. The drone operator must always keep the drone in line of sight. Part of the drone flying activity is the flight plan, which is provided by the site engineer to the drone operator for initial assessment and overview. Any area, which has poor visibility that could potentially block the line of sight, must be double checked and rerouted or delayed as necessary.

Cold temperatures

Low temperature can reduce battery performance but this has not been a problem at the Philippine site.

Natural hazards

Conducting drone activities in rural and remote areas entails careful planning. Planning includes identifying external and internal factors that could affect the flight. Aside from drones being operator-dependent as a remote-controlled device, external factors such as environmental hazards require strict planning and execution.

Topography

The limitation of flying drones in remote areas is grounded on the fact that terrain is often obstructed with dense flora. To ensure safe drone flight activities, the site engineer and the drone operator must use cautious judgement and planning to consider the difference in elevation between way-points. Obtaining survey maps of the area will provide an overview of the pre-set maximum altitude of the drone, preventing the loss of signal and possible crash.

Avians

The behavioural adaptation of birds reflects their responses to environmental situations, which includes territorial instinct. Issues with bird attack can be addressed through careful and well-thought countermeasures which include setting up a flight plan limiting interactions with their breeding and brooding patterns.

On-site drone activities are typically scheduled in the morning as large species of birds use thermals for soaring and gaining altitude. Planning drone activities early would also limit avian interactions as their flight patterns are typically observed on warmer parts of the day as the ground heats up from the solar radiation. Arguably, drones can be flown at any time, but using this measure would minimise associated risks. Bird interaction has not been a problem at the Philippine site but has been reported at several Australian sites.

Humans

Part of the risk assessment in drone activities includes the surveillance on security and safety risks to people. Surveillance should provide information dissemination on the drone flight plan and precautionary announcements to prevent potential hazards/injuries.

PROCEDURE FOR VIDEO CAPTURE

A flight plan is prepared by the site engineer using way-points plotted on a drawing plan, taking into account the key areas that the site engineer aims to capture on video. The flight plan is then discussed with the drone operator, along with instructions such as having to zoom in on specific areas, to ensure that all the necessary details are clearly shown to the viewers and allow for a more effective discussion. A copy of the drone footage is then obtained and sent out prior to the meeting to allow the reviewers to view the footage and note all the concerns that need to be raised to the site engineer and addressed by the site team.

PROCEDURE FOR TECHNICAL REVIEW

The drone videos were typically taken at the start of each week (Monday), unless weather conditions or other issues prevented this, in which case the flight would be undertaken at the earliest time. The videos were then emailed to Canada and Australian staff for review the following day in a meeting between the on-site engineer and the off-site senior engineers. The meetings using Teams were typically 30 minutes duration. The findings made during the review were then discussed by the site engineer with the client at a weekly meeting to ensure that the client construction supervisors understood the issues prior to giving out instructions to personnel in the field. The schedule was designed to capture the construction progress for the past week and give ample time for the site team to address all the concerns that were raised during the drone review meeting. The availability of video also simplified the client discussions.

An important aspect of the system of video capture and review meetings has been the responsiveness of the client and construction team in addressing the concerns of the review team.

EXAMPLE FINDINGS

The video meeting format was instrumental in the identification of issues such as potential instability of surrounding cuttings and waste rock dumps in the steep topography, poor drainage control leading to potential severe erosion, borrow area management, tailings discharge management and access road planning. The drone video allowed a high level of technical support to site supervisory staff at the remote site that would otherwise be impossible.

Below are some findings discussed from drone footage weekly reviews.

Unsafe slope upstream of the TSF

The drone image shown in Figure 2, shows a cut slope located on the right abutment of the impoundment area. This was noted by reviewers as potentially unstable with any failure potentially causing waves that could threaten the embankment. Dam design engineers notified the client of the risks and impact if not corrected immediately.

FIG 2 – Upstream batter noted to be potentially unstable.

Remediation works allowed stabilisation of the batter as shown in Figure 3.

FIG 3 – Batter after stabilisation works..

Overtopping risk (ongoing gabion spillway construction)

The drone image in Figure 4 shows the progress of the gabion spillway construction. The gabion spillway was built in stages. The reviewers noted that the raising of the spillway walls was progressing at a lower rate than the spillway crest. This meant that if there was a rise in the pond level leading to overtopping the spillway, there was a potential for unplanned spill to both sides of the spillway leading to damage and potential failure of the spillway. The dam engineers advised the client on the spillway construction sequence, with the mid-section serving as the low point and the gabion sidewall being built in advance resulting in the correct geometry shown in Figure 5 and the completed structure in Figure 6.

FIG 4 – Gabion spillway walls lower than crest.

FIG 5 – Properly constructed spillway.

FIG 6 – Completed spillway.

Figure 5 shows the sequence of the gabion spillway construction which has been modified based on the recommendations of the dam engineers. This includes the construction of the gabion side walls before the gabion steps to ensure that the flow is directed into the middle portion of the spillway in case of an overtopping event.

Erosion along the abutment

The drone image in Figure 7 shows erosion along the right abutment caused by surface run-off. Dam designers informed the client of the potential risk for instability if the erosion progresses further.

Figure 8 shows the remediated scouring along the South abutment. Surface run-off was also diverted away from the abutment to prevent erosion from subsequent rainfall events.

FIG 7 – Significant erosion on right abutment.

FIG 8 – Corrected erosion.

CONCLUSION

Drones, which allow data collection quickly and remotely, are one of the advanced technologies that can be used for on-site dam inspections. They allow site engineers, dam designers, and owners to interact during online discussions and conduct technical online reviews and assessments of site construction progress using video footage.

Drones can serve multiple functions in construction. They are capable of capturing a clear overview of site progress for technical and non-technical viewers, providing flexibility for inspection covering specific areas of concern particularly on matters that require immediate resolution and updates, and serving as a tool to identify potential dam hazards and risks that could lead to harm, damage, and failures. Importantly they have proven an excellent tool for communication between designers and constructors to allow timely correction of issues noted by external reviewers.

Monitoring tailings storage facilities with muon tomography

M A McLean[1], F Scutti[2], R Seikel[3], S Krishnan[4], C Webster[5], S Collins[6], C Goss[7], S Palanisamy[8], P H Panchal[9], J Donovan[10] and A R Duffy[11]

1. Fellow in Applied Geophysics, School of Geography, Earth and Atmospheric Sciences, The University of Melbourne, Parkville Vic 3010. Email: m.mclean@unimelb.edu.au
2. Postdoctoral Research Fellow, Swinburne University of Technology, Hawthorn Vic 3122. Email: fscutti@swin.edu.au
3. Senior Software Engineer, Swinburne University of Technology, Hawthorn Vic 3122. Email: rseikel@swin.edu.au
4. Senior Lecturer, Swinburne University of Technology, Hawthorn Vic 3122; mDetect Pty Ltd, South Yarra Vic 3141. Email: skrishnan@swin.edu.au
5. Senior Technical Officer, Swinburne University of Technology, Hawthorn Vic 3122; mDetect Pty Ltd, South Yarra Vic 3141. Email: cwebster@swin.edu.au
6. Research Associate, Swinburne University of Technology, Hawthorn Vic 3122. Email: scottcollins@swin.edu.au
7. Tailings Metallurgist, Prominent Hill, BHP, Adelaide Airport SA 5950. Email: craig.goss@bhp.com
8. Director, Manufacturing Futures Research Platform, Swinburne University of Technology, Hawthorn Vic 3122. Email: spalanisamy@swin.edu.au
9. Research Associate, Swinburne University of Technology, Hawthorn Vic 3122. Email: ppanchal@swin.edu.au
10. Entrepreneurship and Innovation Associate Professor, Swinburne University of Technology, Hawthorn Vic 3122; mDetect Pty Ltd, South Yarra Vic 3141. Email: jdonovan@swin.edu.au
11. AAusIMM, Pro Vice-Chancellor Flagship Initiatives, Swinburne University of Technology, Hawthorn Vic 3122; mDetect Pty Ltd, South Yarra Vic 3141. Email: aduffy@swin.edu.au

ABSTRACT

Tailings storage facilities present a global and growing challenge for mining companies. Tailings, the by-products of the processing plant, need to be carefully managed, as the tailings storage facility could present a potential risk to lives and livelihoods in the event of a collapse. Such collapses can lead to environmental impacts of significant and permanent damage to regions downstream. Consequently, real-time monitoring of the structural stability of such facilities is a significant opportunity for the safety and sustainability of active mine sites and legacy mines. To our knowledge, this paper explores the world's first application of muon tomography to analyse tailings dam structures by mDetect Pty Ltd and Prominent Hill at the Prominent Hill Mine site in South Australia. We present a novel method and technology to investigate the inner structure of tailings storage facilities performed using muons, a natural source of radiation that can be used to take X-ray-like scans of large structures. Muon technology, therefore, offers a unique opportunity to provide deep structural insights into tailings storage facilities and potentially a platform from which future research can investigate the development of early warning systems for change detection.

INTRODUCTION

Large structures, such as Tailings Storage Facilities (TSF), represent a unique challenge to existing technology utilised within the mining sector due to their extensive physical scale and slow evolution over time. Therefore it is desirable for a TSF to be the subject of long-term monitoring with surface and intrusive testing for stability and liquefaction assessments, ideally over significant volumes of the site. Remote monitoring capability for tailings structures is currently available with detection systems such as extensometers, LIDAR, total station, taseometers, and seismic sensors, to name but a few, but their range and coverage of detection area are limited. Furthermore, contemporary techniques might also come with compounding issues such as background noise that inhibits detection ability, radiation licences that create barriers to deployment, and significant energy consumption of the systems which increase cost and restrict the longevity and flexibility of deployment.

Utilising naturally occurring particles from space, known as muons, that randomly but with well-characterised frequency and direction impact a site and penetrate through hundreds of metres of rock has long been considered a potential means to scan large volumes of matter. The rate at which a muon loses energy as it interacts with matter is proportional to the density of the material it travels through. Shadowing from dense structures is a property that muons share with X-rays, but unlike X-rays, muons are not easily created in the lab. However, muons are continually being created by high-energy cosmic rays arriving from (mostly) interstellar space and colliding with Earth's atmosphere at altitudes of about 15 km. The average muon flux at sea level is about 1 muon per square centimetre per minute.

Initially explored by E P George in 1955 to estimate tunnel overburden in the Snowy Mountains in Australia (George, 1955); detector capabilities have greatly improved permitting mineral exploration (Bryman, Bueno and Jansen, 2015), volcano monitoring (Nagamine et al, 1995), tide/tidal wave detection (Tanaka et al, 2022) and even mapping of corridors inside pyramids (eg Procureur et al 2023).

In this paper, to the best of our knowledge, we demonstrate a world-first use case of muon tomography in the context of a Tailings Storage Facility. Exploiting the natural occurrence of muons means that no radiation license is required, their highly penetrating nature permits large volumes to be surveyed, and the robust but low-cost detectors offered by mDetect Pty Ltd allow long-term deployments for monitoring changes over time.

METHODOLOGY

We will introduce the site where the trial was undertaken, discuss the detectors deployed and then explain the reconstruction methodology for the inversion of recorded muons to inferred densities.

Deployment context

The Prominent Hill copper-gold mine is located approximately 640 km north-west of Adelaide in northern South Australia. Prominent Hill has one TSF impounded within the Integrated Waste Landform (IWL, shown in Figure 1). The TSF is a singular, circular facility containing tailings generated from processing ore from the open pit (safely closed in 2018) and underground mines. It has a diameter of 1750 m covering an area of about 2.4 km². Construction of the TSF has been ongoing, with five embankment lifts occurring since commissioning in 2009. The TSF embankment is a minimum 60 m wide waste rock dump rising 25 m high. On the tailings side of the embankment, a filter zone made up of finer waste rock, typically 3 m wide, was built concurrently with a 6 m clay liner adjacent to the tailings.

Instrumentation

A unique and patented solution incorporating smart sensor-based coincidence detectors (Krishnan et al, 2019) has been developed by mDetect Pty Ltd to monitor tailings dams' structural integrity using multiple telescopic muon detectors. This innovative technology can provide complementary and unique data insights to existing methods, offering a more comprehensive early warning capability for site stability.

Detector unit design

The core detection unit is a 'telescopic' muon detector with several plastic scintillators within a unit that emit light when struck by a muon. This light is then received by mounted single-pixel silicon photomultipliers (SiPMs; Krishnan et al, 2020). By utilising optically isolated sub-detectors within a panel plane, the arrival direction of intersecting muons can be estimated geometrically. As shown in Figure 2 (top panels) the sub-detectors provide angular resolutions of better than five degrees within a field-of-view of order 17 × 17 degrees per pointing.

Each detector observed multiple pointings over the course of nearly ten days, collecting 231 hours of data to determine the muon flux as a function of angle across the TSF embankment.

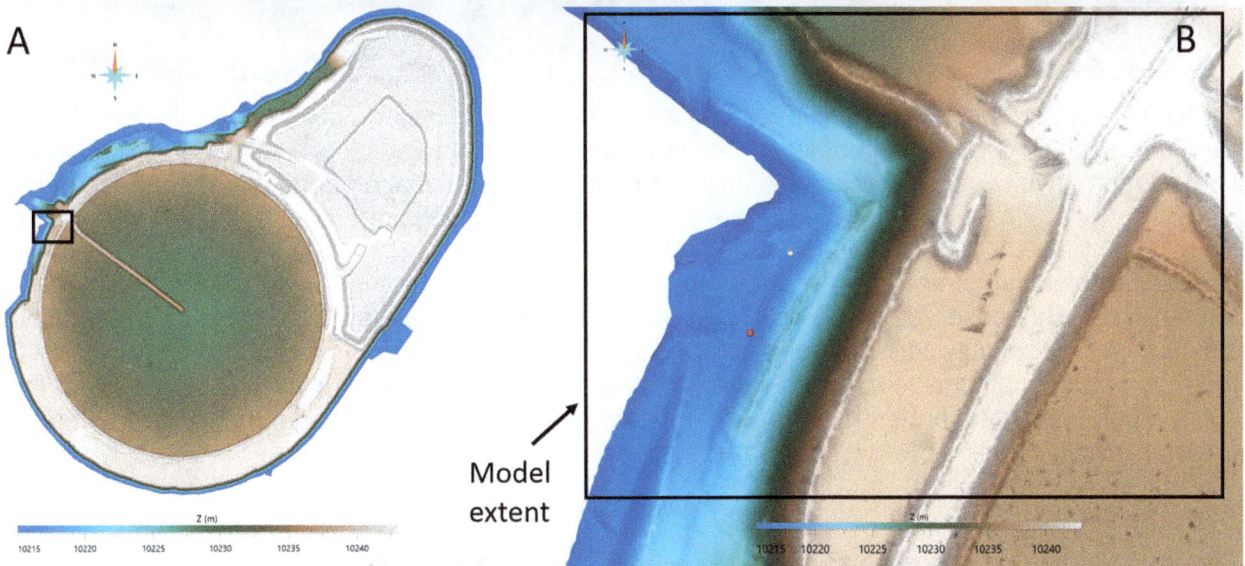

FIG 1 – Top, aerial view of Prominent Hill tailings storage facility with white box showing modelled region. Bottom, topography surfaces generated from LiDAR data for: (a) the entire Prominent Hill tailings storage facility (black box shows the location of the scanned region), and (b) the scanned region zoom in, with two telescopic detector deployments (discussed later) as the white and red dots. These detectors are placed approximately 10 m from the beginning of the rise, spaced 50 m from each other.

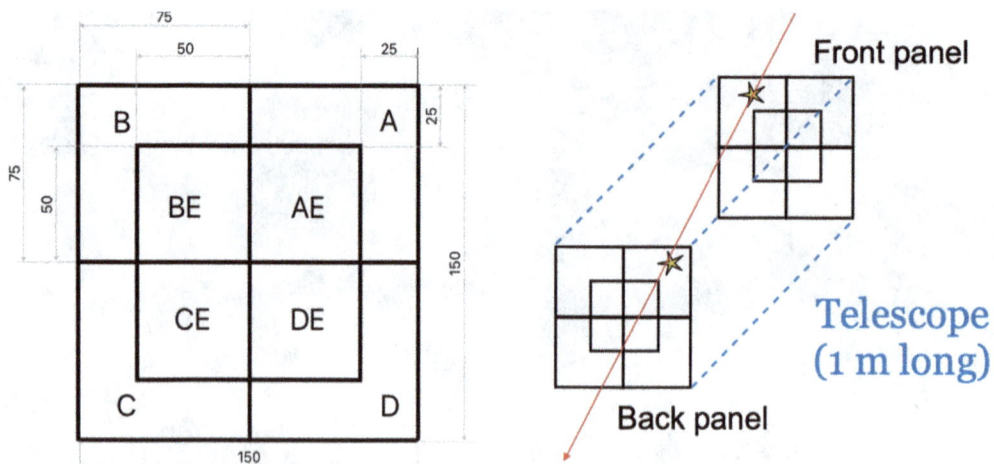

FIG 2 – Top, showing a qualitative sketch representing, left, optically isolated scintillation slabs and, right, which allow for a geometric reconstruction of a muon trajectory (indicated by a star/red line). Bottom, two telescopes mid-installation (note not their final placements), with reference muon counter as the small black box at its base.

Integrating this scan over longer periods provides a robust angular muon arrival rate (equivalently termed flux). This observed muon flux depends, in addition to the elevation angle of observation, only on another quantity called *opacity* (or projected surface density) which is directly proportional to the muon path length inside the material and to the density of the material itself. Using at least two observation points (separated by 50 m) then provides a stereoscopic 3D constraint on the material within overlapping fields of view, as discussed later.

System design

The stand-alone system is solar-powered, with data transferred over 4G and the ability to work independently of the power and comms mine infrastructure, if required. Some of the issues that commonly occur with remote data handling include data access control management, encryption, authentication, and other privacy issues. mDetect Pty Ltd operates as an autonomous system independent of site infrastructure, easing many IT-related overheads and permissions from the host site.

Reliable performance in a harsh field environment is possible as the onboard data storage and solar-powered operation mitigate the potential data loss due to network and power disruptions. Furthermore, asset monitoring can be performed remotely over 4G. The system can support additional sensors (ie pressure sensors, crack sensors, and inclinometers), which could lead to new insights and innovations when time-correlated with the density data.

As shown in Figure 3, three mDetect detectors (seen in Figure 2) were used to monitor the region of interest identified in Figure 1. The primary and secondary detectors were telescopes used to scan the region dynamically. The third reference detector was a simple muon counter recording background rates. The primary and secondary detectors are powered by an independent solar array with a single communication link used for data transfer and power distribution. Data acquisition and 4G gateway then arbitrates the data from all three detectors and forwards data to the cloud. The system allows robust assessments of risks associated with any structure changes, with a high degree of sensitivity and confidence. It is a unique, low-cost, low-maintenance system that can support companies in monitoring the stability of structures such as tailings dams.

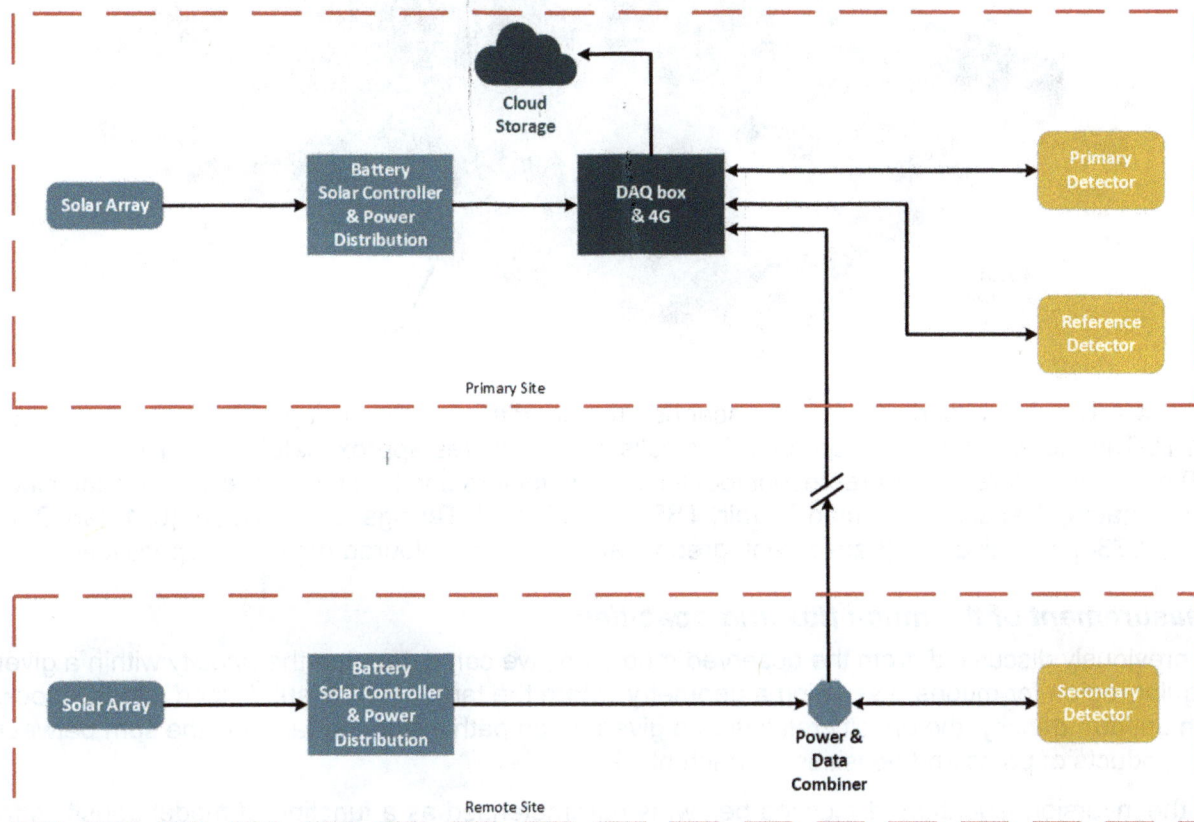

FIG 3 – mDetect Pty Ltd System Block Diagram for the deployment to the Prominent Hill TSF, indicating the two telescope muon detectors as schematically shown in Figure 1 and a simple reference muon counter.

Software analysis

We first introduce the geophysical modelling, leveraging site data implemented with SKUA-GOCAD, and then a proprietary reconstruction algorithm to invert recorded events to physical densities using open-source software PUMAS.

Geological modelling

We performed three-dimensional (3D) modelling using AspenTech's SKUA-GOCAD (version 21). The SKUA-GOCAD modelling system is a data-based, 3D geometrical modelling scheme that integrates information through a discrete mathematical approach (Mallet, 1992). SKUA-GOCAD was used to visualise the topographic surfaces of the tailings facility, discretise models into a 3D block model and visualise results.

The approach applied for the Prominent Hill TSF model was to use existing company LiDAR data obtained to build a 3D surface (shown in Figure 1). The LiDAR surface accurately characterised; the shape of the top of the TSF, the top of the tailings inside the facility, and the surrounding topography outside the TSF. The base of the TSF was well constrained using a prior LiDAR survey carried out before the tailings facility was built. These surfaces were used to divide a discretised block model into regions intended to represent geological domains (Figure 4). Four geological domains were built to characterise the geometries of the TSF. A density and rock identifier was attributed to each domain (see caption Figure 4).

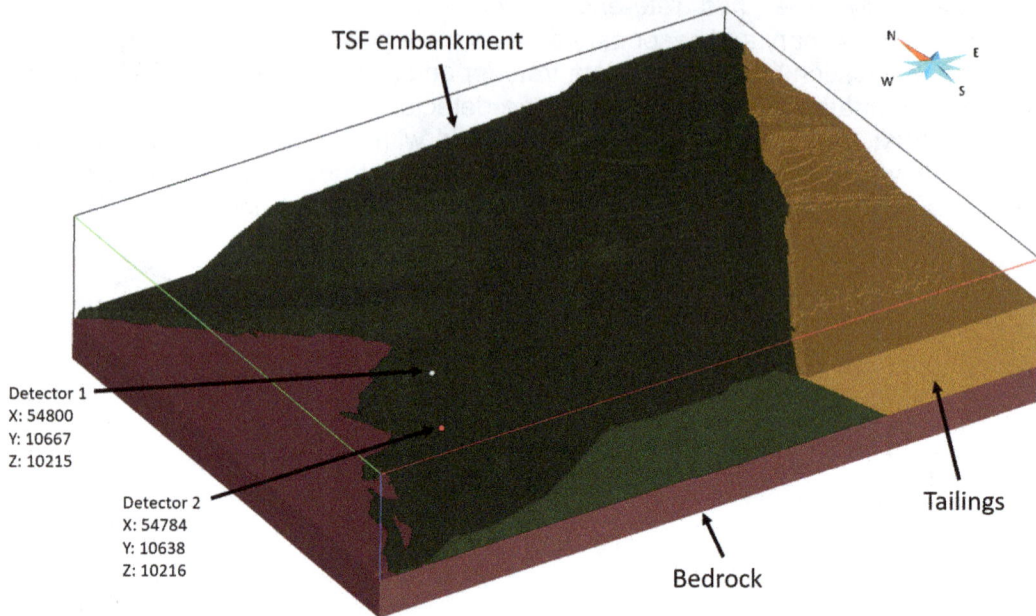

FIG 4 – The discretised model of the scanned region of the tailings storage facility at Prominent Hill. The model is characterised by 0.5 m cells and measures approximately 244 m × 179 m × 50 m. White and red points represent locations for detectors one and two (specified in local mine coordinates). Densities assumed for [air, TSF embankment, Tailings, Bedrock] are [0, 1.495, 2.3, 1.734] g/cc given by [transparent, green, yellow, brown] coloured regions, respectively.

Measurement of the muon flux and opacities

As previously discussed, from the observed muon flux, we can determine the *opacity* within a given angular region for muons. Assuming a geometry where the target can be subdivided into 3D blocks with uniform density, the opacity relative to a given muon path inside the target is the sum between the products of paths and densities for each block.

As the inversion algorithm, discussed below, is parameterised as a function of model densities for an ideal model of the target, these need to be compared with measured opacity values proportional to the true target densities to provide a constraint on the idealised parameters. It is then necessary to convert measured muon fluxes into measured opacity values.

First, the measurement of the muon flux is performed. Detector panels, segmented in sub-panels, provide digital information regarding the two-dimensional traversal coordinate of the muon trajectory through the panel, referred to as a *muon hit* in the following. Therefore, the panel's geometry and segmentation dictate the spatial resolution of the hit. The trajectory of the muon, as qualitatively depicted in Figure 2, is reconstructed using a randomisation procedure, which uses the combination of two sub-panels, front and back, that have been activated by a muon event. The position of the muon hit is generated randomly within each sub-panel in the combination. A straight line is assumed to pass through the two-hit combination. By knowing the orientation of the detector, the absolute values of the azimuthal and elevation coordinates are assigned to the muon trajectory.

When counting muon events, the limited geometrical acceptance of the detector must be taken into account. This acceptance depends on the muon incident angle with respect to the detector's longitudinal axis. A detector simulation is performed to account for this effect, where particle trajectories are simulated with uniform orientation and fixed intensity, and the detector incident angle is varied. The acceptance is estimated as the fractional flux variation with respect to a vertical configuration and is parameterised as a function of the incident angle. The inverse of the acceptance is defined as the acceptance correction factor, which is used as a multiplicative weight to each muon count.

For a given elevation angle of observation, known opacities lead to the same flux and vice versa. Simulations have been used where the opacity is predetermined to establish a relation between flux and opacity. This relation is used to convert the measured flux. The simulation utilised a half-sphere model where a detector is placed in the centre of the half-sphere base, providing uniform line-of-sight lengths in all directions. It is conducted with the PUMAS software library (Niess, 2022). Flux versus opacity relations are obtained for different elevation values in one-degree intervals.

The measured flux is taken in every angular bin of the measured muon flux maps. It is associated with an elevation value corresponding to the centre of the bin, which is used to select the simulated flux-vs-opacity relation. The opacity is then measured by interpolating the measured flux, comprising 231 hours of data captured over 9.6 days, and the uncertainty on the opacity is assigned via uncertainty propagation. In Figure 5, we see in the top panel recorded muon flux from detector one (two) in the left (right) column and on the bottom rows their respective inferred opacities. The data analysis for measuring muon fluxes and opacities has been conducted with the Pyrate software package (Scutti, 2022).

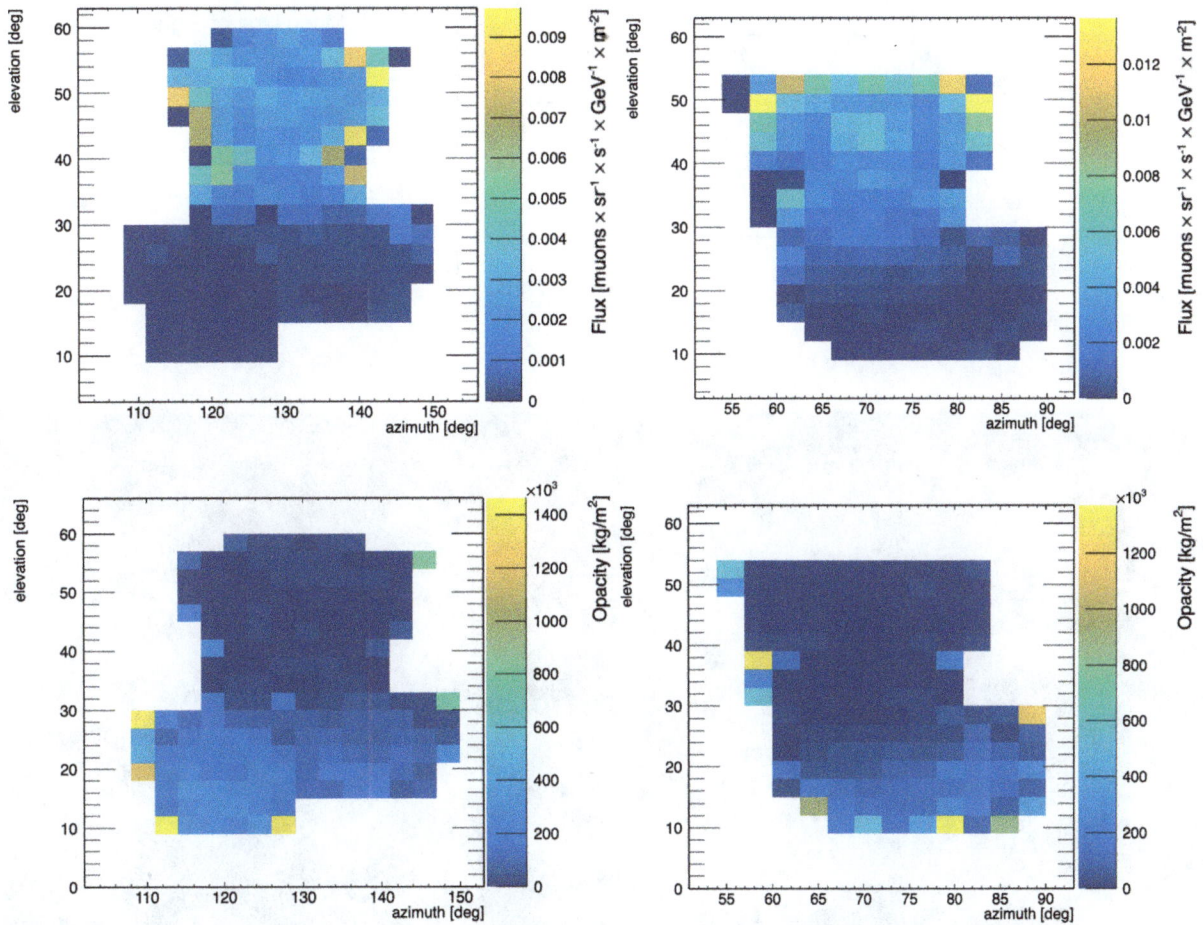

FIG 5 – Top, recorded muon fluxes for detector one (two) in the left (right) column. Bottom, inferred opacities using derived flux-opacity relationships from PUMAS (Niess, 2022).

Muon tomography inversion

The basis of the algorithm for 3D muon tomography, as implemented in this case study, is that the lines of sight (LOS) of different detectors intersect common cells in the 3D block model space. They are diagrammatically shown in Figure 6 (top). In this case, we define the behaviour of muons as our action (A) as they travel through the medium of the dam wall as they are then detected to produce our projection. We call this the Forward Projection, given by Equation 1 as:

$$A.x = b \qquad (1)$$

where A is the action, *x* is the image, and *b* *is* the resulting (or observed) projection. Simplifying assumptions about the interactions of muons with the medium of the tailings embankment are as follows. The first is that muons travel in straight lines at the energy ranges that predominate the ambient muon flux. The second, muons are attenuated at a rate (Λ) proportional only to the density of the medium traversed (ρ_j) and distance traversed (Δl_j) given by Equation 2 as:

$$\Lambda = \sum_{j=1}^{N} \rho_j \, \Delta l_j$$

(2)

The aim of tomography is to find the inverse of the Forward Projection equation, termed the Backward Projection. The basis of our adopted algorithm for 3D Muon Tomography is that of comparing reconstructed LOS opacities observed against those computed by Equation 2. The computed LOS opacity uses a 3D regular grid traversal algorithm (Figure 6, top) through an *a priori* discretised geological model of densities (termed the block model).

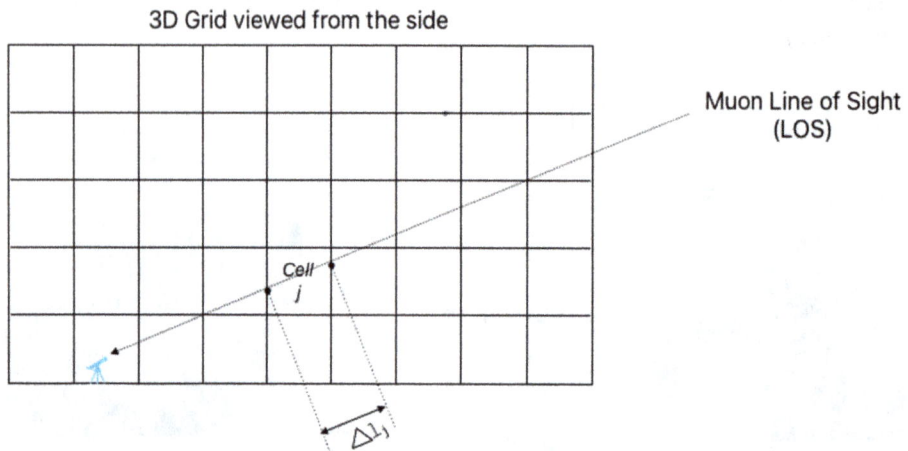

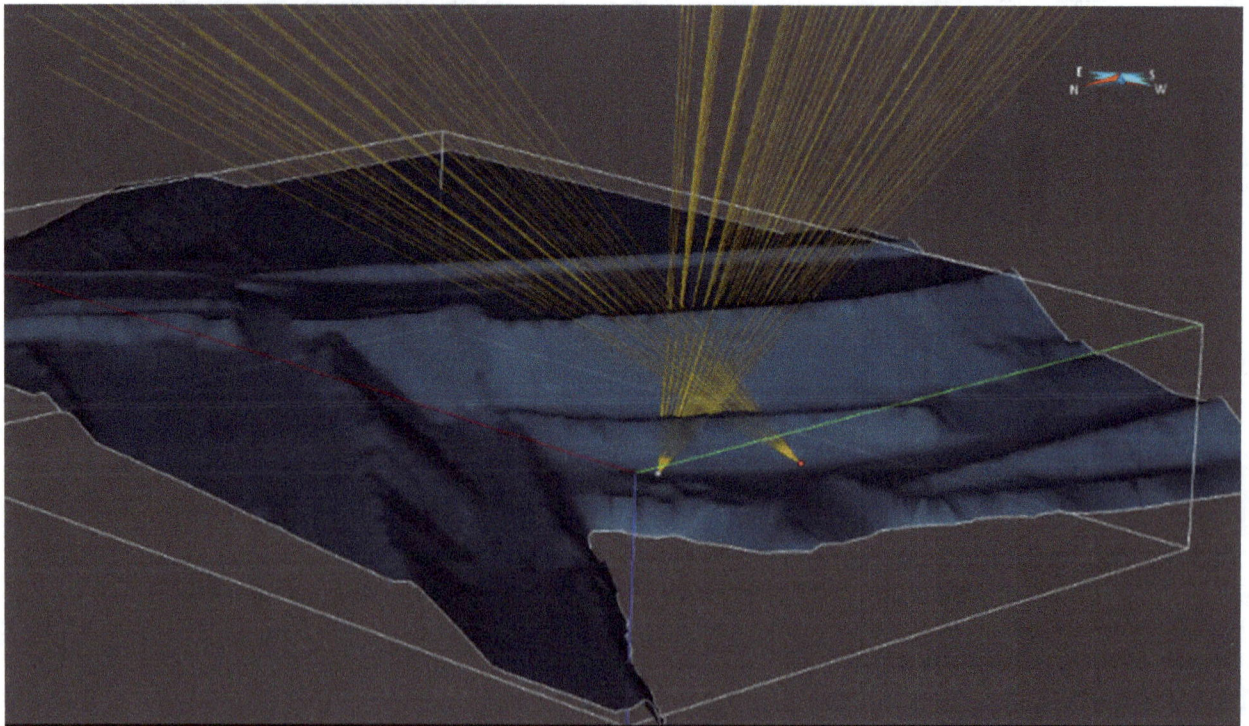

FIG 6 – Top, 3D Regular Grid Traversal Algorithm; with the opacity along a muon's line of sight (LOS) computed using Equation 2. Bottom, two detectors with intersecting LOS through the embankment.

3D muon tomography – the MCMC method

The simplifying assumptions allow for the action A of Equation 1 to be represented as a matrix where the rows represent the multiple LOS that make up the observations (Figure 6, bottom). The columns of each row represent the distance of muons travel through each of the cells of the 3D regular grid of the discretised geological model where cells not traversed are set to zero. Furthermore, the projection b of Equation 1 is a column matrix representing the observed opacity for each LOS (row). Finally, the image x of Equation 1 is a column matrix representing the density for each cell of the 3D regular grid of the discretised geological model and, as such, is the unknown that the algorithm aims to solve for. In other words, the algorithm aims to find x such that Equation 3 is minimised.

$$| b - A.x | \tag{3}$$

The algorithm developed to solve Equation 1 is based on a Monte Carlo Markov Chain (MCMC) approach. Starting with an *a priori* discretised geological model of densities to populate the initial guess for values of the image x, precomputed segment lengths of muon trajectories for the action A, and reconstructed opacities derived from the observed muon events as the projection b, an iterative method stochastically perturbs the image one randomly selected cell at a time looking to minimise Equation 3.

This approach allows for geological constraints to be applied to the evolution of the model during its convergence (Guillen *et al*, 2004). The ability to use geological constraints is an advantage the MCMC approach offers over matrix methods such as SIRT (eg Ren and Kalscheuer, 2020).

The process that reconstructs the muon flux and, ultimately, the muon opacity along each line of sight from the raw field muon event observations also propagates the expected uncertainty and provides this as an input to the MCMC inversion. The MCMC inversion makes use of these uncertainties in the way it weighs contributions from different detectors towards the calculation of the global misfits. The result is an apparent density within each cell, as well as propagated errors from the detector measurement and computational algorithms.

RESULTS

The MCMC-derived post-inversion apparent densities can only be distributed along lines of sight. Consequently, there are more points closer to the muon telescope where the lines of sight converge and fewer points distal to the muon telescope (Figure 7a). It is therefore not possible to generate a 'solid volume' of apparent densities from muon tomography, making it necessary to 'fill in the gaps' between apparent density estimates. We used an 'ordinary kriging' process to accomplish this.

Semi-variograms were calculated to determine the correlation characteristics of the inverted density data and determine how far densities could be statistically interpolated. The result of this process was a search ellipse which facilitated the distribution of the data. Once a search ellipse was created, the inverted densities could be distributed throughout the discretised 3D block model. Although there were increasingly larger gaps between inverted apparent densities further away from the muon telescope, there still was a sufficient volume of points to enable the search ellipse to completely fill the scanned volume.

On the margins of the scanned volume (further away from the inverted apparent densities), kriged values were not representative of the expected densities. This is expected as the kriging process is less robust where there are limited data points (or no data points at all). This was demonstrated by relatively high variance values of kriged data points at the margins of the scanned volume. To help with visualisation, these spurious values were clipped off the density distribution. The variance was used as a measure of where to clip the kriged volume. A variance cut-off value was arbitrarily chosen based on the quantity of spurious apparent density values. Lastly, the kriged data was clipped by the topography surface (Figure 7b).

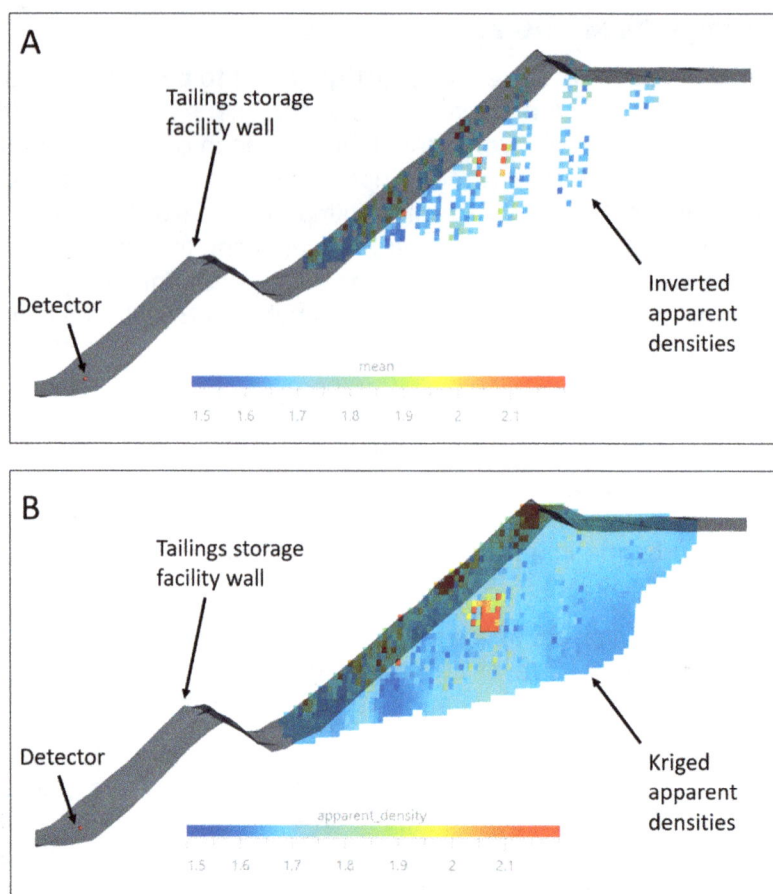

FIG 7 – Apparent densities, ranges represented by colour, imaged in the Prominent Hill TSF (a) post inversion and (b) post kriging, for a single slice through the wall.

GOCAD integrated interpretation

The kriged apparent densities show mostly consistent values throughout the scanned volume. Many of the apparent densities have a relatively low density as demonstrated by the median value of 1.68 g/cc. A large proportion (87.8 per cent) of values fall between 1.46 g/cc and 1.80 g/cc (light-blue range in Figure 8). Correspondingly a small proportion (9.4 per cent) of the scanned volume has a density above this range between 1.80 g/cc and 2.2 g/cc (orange-red range in Figure 8). These higher values are typically characterised by discrete, localised anomalies which often reside close to the western margin of the TSF wall (Figure 8c and 8d).

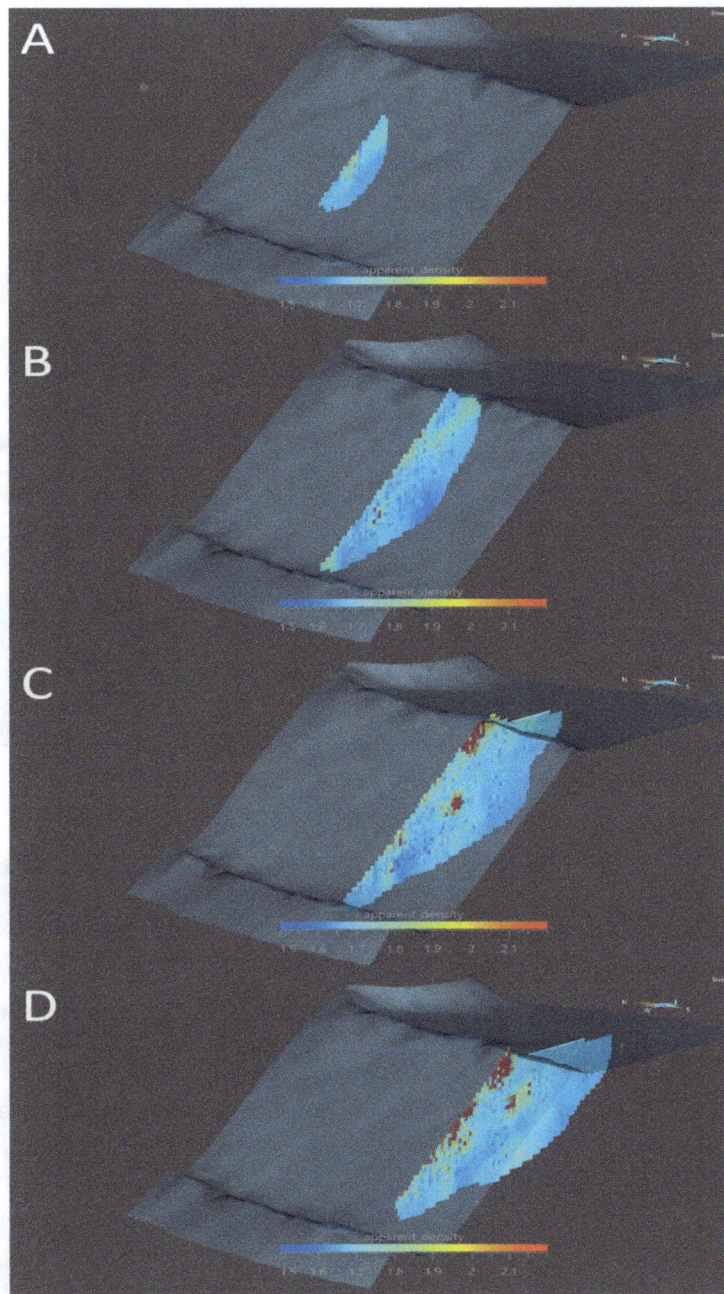

FIG 8 – Apparent densities kriged throughout the scanned volume, shown as a progressive slice from North (a) to South (d). The transparent blue surface is the topography of the TSF.

DISCUSSION

Due to the short duration of the observational deployment, only a relatively limited extent of one wall of the TSF was imaged using muon telescopes. A robust and proprietary algorithm was developed to invert the recorded muon flux and resulting inferred opacities into apparent 3D physical densities.

The results of the muon-derived apparent densities, post-kriging, are found to be a heavily skewed distribution, with the 10th and 90th percentile found at 1.64 g/cc and 1.82 g/cc, respectively. While there are some outliers with impossibly high apparent densities (over 10 g/cc), these make up a very small proportion of the scanned volume (0.3 per cent of the scanned volume falls between 3 g/cc and 10.2 g/cc) and are considered to be numerical artefacts.

As shown in Figure 9, relatively high apparent densities (between 1.8 g/cc and 2.2 g/cc) are robustly detected although their cause remains unclear. We suspect these values correspond to cemented backfill and basement waste which has been confirmed to have been used to build the TSF in the observed direction of the telescope. Apparent densities estimated from muon tomography have tended to be slightly higher (median of 1.68 g/cc) compared with the measured value (1.495 g/cc).

We suggest the most likely explanation for this is that the median value has been influenced by these high values which would not have affected the laboratory-measured value.

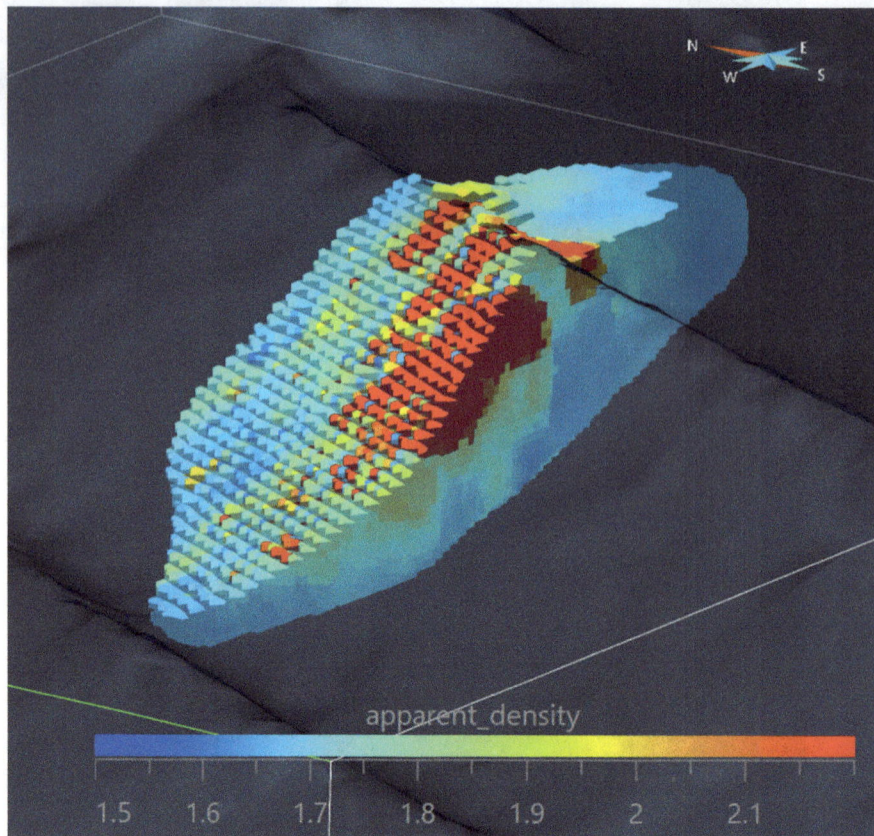

FIG 9 – The scanned region of the TSF coloured by the apparent density (g/cc). There are significant detections of enhanced densities relative to the background, and assumed, uniform distribution of materials that aligned with waste materials known to be embedded in the site.

In general, however, there is little variability in the inferred values for the apparent density within the volume imaged, reflecting that essentially only one 'rise' was scanned that would reasonably exhibit a similar density throughout construction.

CONCLUSIONS

This work represents a world-first use of imaging a Tailing Storage Facility (TSF) with muon detectors. A short-duration deployment was undertaken at the Prominent Hill TSF, with several days of muon events captured by two telescopic muon detectors arranged to provide a limited 3D stereoscopic resolution.

The muon rate, or flux, detected across the extent of the embankment was converted to an opacity using an open-source muon physics transport engine, PUMAS (Niess, 2022). Regions in which opacities, or equivalently projected surface densities, had intersecting sightlines were then inverted to produce a 3D apparent density using a proprietary Monte Carlo Markov Chain algorithm. These discrete points were then kriged using the standard industry package SKUA-GOCAD to create a 3D volumetric apparent density.

Within that resulting 3D volumetric region, a range of apparent densities was inferred, heavily skewed to the expected wall material density (10th and 90th percentile found at 1.64 g/cc and 1.82 g/cc respectively). Relatively higher density regions (between 1.8 g/cc and 2.2 g/cc) are corresponded to known areas of cemented backfill and basement waste in the wall, suggesting a real cause of these outliers.

Although only a limited area of the Tailings Storage Facility was scanned in this short deployment the utilisation and efficacy of muons for cheaply scanning large volumes, with no site disruption or installation requirements is clear.

ACKNOWLEDGEMENTS

The authors gratefully acknowledge the support of the Prominent Hill mine in facilitating the trials, in particular Antoinette Stryk. mDetect recognises the valuable ongoing support of the Swinburne University of Technology, with special thanks to David Smoors for technical efforts. This work was made possible by the Commonwealth Government's Advanced Manufacturing Growth Centre, AMGC, grant number C259.

REFERENCES

Bryman, D, Bueno, J and Jansen, J, 2015. Blind Test of Muon Geotomography for Mineral Exploration, *ASEG Extended Abstracts*, pp 1–3. https://doi.org/10.1071/ASEG2015ab054

George, E P, 1955. Cosmic rays measure overburden of tunnel, Commonwealth Engineer, pp 455–457 (Tait Publishing Co.: Melbourne).

Guillen, A, Courrioux, G, Calcagno, P, Lane, R, Lees, T and McInerney, P, 2004. Constrained gravity 3D litho-inversion applied to Broken Hill, in ASEG 17th Geophysical Conference and Exhibition, August 2004, Sydney.

Krishnan, S, Webster, C, Duffy, A R, Brooks, G, Clay, R and Mould, J, 2019. SiPM Smart Sensor Based Portable Coincidence System for Distributed Field Applications, *JINST*, 14(9):P09026.

Krishnan, S, Webster, C, Duffy, A R, Brooks, G, Clay, R and Mould, J, 2020. Improving the energy resolution while mitigating the effects of dark-noise, for a microcontroller based SiPM sensor, *JINST*, 15(9):P09028.

Mallet, J L, 1992. GOCAD: a computer aided design program for geological applications, in *Three-Dimensional Modelling with Geoscientific Information Systems* (ed: K Turner), NATO ASI Series, pp 123–142.

Nagamine, K, Iwasaki, M, Shimomura, K and Ishida, K, 1995. Method of probing inner-structure of geophysical substance with the horizontal cosmic-ray muons and possible application to volcanic eruption prediction, *Nucl Instrum Methods Phys Res Section A*, 356:585–595.

Niess, V, 2022. The PUMAS Library, *Comp Phys Comms*, 279:108438. https://doi.org/10.1016/j.cpc.2022.108438

Procureur, S, Morishima, K, Kuno, M, Manabe, Y, Kitagawa, N, Nishio, A, Gomez, H, Attié, D, Sakakibara, A, Hikata, K, Moto, M, Mandjavidze, I, Magnier, P, Lehuraux, M, Benoit, T, Calvet, D, Coppolani, X, Kebbiri, M, Mas, P, Helal, H, Tayoubi, M, Marini, B, Serikoff, N, Anwar, H, Steiger, V, Takasaki, F, Fujii, H, Satoh, K, Kodama, H, Hayashi, K, Gable, P, Guerriero, E, Mouret, J-B, Elnady, T, Elshayeb, Y and Elkarmoty, M, 2023. Precise characterization of a corridor-shaped structure in Khufu's Pyramid by observation of cosmic-ray muons, *Nature Communications*, 14:1144. https://doi.org/10.1038/s41467-023-36351-0

Ren, Z and Kalscheuer, T, 2020. Uncertainty and Resolution Analysis of 2D and 3D Inversion Models Computed from Geophysical Electromagnetic Data, *Surv Geophys*, 41:47–112. https://doi.org/10.1007/s10712-019-09567-3

Scutti, F, 2022. Pyrate: a novel system for data transformations, reconstruction and analysis, *Journal of Physics: Conference Series*, 2438:012061. https://iopscience.iop.org/article/10.1088/1742-6596/2438/1/012061

Tanaka, H K M, Aichi, M, Balogh, S J, Bozza, C, Coniglione, R, Gluyas, J, Hayashi, N, Holma, M, Joutsenvaara, J, Kamoshida, O, Kato, Y, Kin, T, Kuusiniemi, P, Leone, G, Lo Presti, D, Matsushima, J, Miyamoto, H, Mori, H, Nomura, Y, Okamoto, N, Oláh, L, Steigerwald, S, Shimazoe, K, Sumiya, K, Takahashi, H, Thompson, L F, Tokunaga, T, Yokota, Y, Paling, S and Varga, D, 2022. Periodic sea-level oscillation in Tokyo Bay detected with the Tokyo-Bay seafloor hyper-kilometric submarine deep detector (TS-HKMSDD), *Scientific Reports*, 12:6097. https://doi.org/10.1038/s41598-022-10078-2

Instrumentation and monitoring systems in tailings dams' construction

M Ghamgosar[1], G Style[2] and K Tehrani[3]

1. AAusIMM, Senior Associate Geotechnical Engineer, ATC Williams, Brisbane Qld 4051
   Email: mortezag@atcwilliams.com.au
2. Senior Principal Engineer, ATC Williams, Brisbane Qld 4051.
   Email: gideons@atcwilliams.com.au
3. Senior Tailings Engineer, ATC Williams, Brisbane Qld 4051. Email: kathyt@atcwilliams.com.au

ABSTRACT

The design process and implementation of effective construction plans are critical aspects of high-risk geotechnical structures such as tailings dams. Apart from legal, financial, environmental, and political considerations in the construction or upgrading of tailings dams, comprehensive design approaches should comprise possible adverse impacts on the facility's structural integrity. This is based on the detailed geotechnical assessments and possible structural responses before, during and after the construction stages. According to the ANCOLD consequence guidelines (2012), active earthquake zone design parameters and operational requirements should consider the severity level from minor to major catastrophic damages. Subsequently, the detailed design and management system could identify mitigating strategies for high seismic zones to achieve high-quality operation outcomes. Therefore, real-time monitoring techniques will add to the risk management system to alert operators to any changes in the tailings dam's external or internal structural components to assess new or residual risks influenced by earthquake loadings. Real-time monitoring data was collected and studied from 13 Vibrating Wire piezometers (VWPs) and five Shape Array inclinometer devices installed at the dam foundation and crest alignment. Geotechnical conditions and accumulative excess water pore pressure development within the dam foundation were studied in conjunction with construction and seismic loadings. The outcomes of this study provided appropriate instructions to manage the construction stages. It included the accommodation of acceptable levels of pore pressure fluctuation to avoid geo-structural instability due to the expected large settlement within clayey deposits in the dam foundation at depth. In addition, the required time to dissipate excess pore pressure was determined and considered to prevent unfavourable deformations imposed by external earthquake loadings.

INTRODUCTION

Tailings storage facilities (TSF) are classified as high-risk engineered structures. Due to recent failures in South America and Australia, ANCOLD risk management and Global Industry Standard on Tailings Management (GISTM; ICMM, 2020) guidelines provide a systematic approach to identify and evaluate potential risk factors to causes of failure and practical mitigation measurement to be implemented for long-term stability. In TSF projects, associated risks are pore pressure build-up during construction and uncertainty of the geotechnical information related to foundation, embankment, and tailings material. This can be expanded into internal and external factors that impact the TSF, resulting in a dynamic system that requires proactive dam safety management programs from the cradle to the grave.

In addition to the geotechnical data and technical elements set out in the guidelines for TSF assessments, extra technical analyses are required to assess the performance during its life cycle. The TSF performance assessment increases complexity when tailings are located in highly seismic areas. This paper described and discussed two earthquake impacts on two TSFs. Critical in the analysis are the practical consequences and the monitoring system's role in the safe operation of the TSF. Earthquake data was recorded with the first example located in PNG approximately 65 km east of the mine site and at a depth of 90 km with a moment magnitude of 7.6 in 2022. The second data set consisted of two small earthquakes occurring in Australia close to a TSF with a moment magnitude of 2.7 in 2018.

As part of the safety management system, site operations at these two mines were temporarily suspended on the date of the earthquakes. In Australia, the wall located south of the TSF had a breach that formed, with some sections of the northern embankment collapsing into the southern tailings dam (Jefferies *et al*, 2019). In contrast, at the PNG earthquake on 11 September 2022, with a magnitude of 7.6, minor damage was recorded at the TSF. This earthquake resulted in 21 people losing their lives with 30 people injured. Due to structural failure, 389 homes were lost, predominantly focused in the Eastern Highlands, PNG.

Outcomes at the PNG and Australian TSF are different, as the hydro-mechanical properties, construction methods and foundation conditions differ; principal concepts of risk assessment can be applied to evaluate earthquake consequences during the operating or post-seismic situations. In addition to the detailed field and laboratory studies linked with third-party and independent reviewers for such high-risk incidences, the critical finding is to utilise comprehensive monitoring systems. These systems provide data to all involved in monitoring the performance of a TSF, resulting in timely responses to mitigate the extensive impact on people, the environment and a facility's production. The following sections outline and discuss the elements of a successful monitoring plan and interpretation of data to understand the TSF geotechnical behaviour under the internal or external forces applied.

MONITORING AND SURVEILLANCE TSF

Monitoring management is the objective of maintaining flexibility in developing a TSF. Different surveying techniques can be utilised depending on the construction or operational stages for an Operation, Maintenance and Surveillance (OMS) action plan. Implementation of the monitoring system, type and required quantities are defined in the preliminary design due to the consequence classification criteria assessment of the TSF.

Geotechnical critical cross-section selection for monitoring

International Council on Mining and Metals (ICMM) states in the tailings management good practice guideline that collecting data and monitoring dams' performance shall be undertaken with the confidence to address the basic elements of the monitoring plan (Morrison, 2022). This includes the purpose, indicator, baseline, target, data collection, instrumentation frequency, responsibility, reporting protocol and quality control. Therefore, the most significant subject for the monitoring plan is to identify the most critical section of a dam and foundation prominently characterised upon the initial and detailed design by complying with all historical and recent geotechnical information.

Among the other critical components in all dam structures, the foundation of a tailings dam is the most complex zone to interpret, as it relies on the *in situ* foundation condition, key geotechnical parameters of material types and construction methods employed. Critical geotechnical sections for monitoring purposes can be selected along the embankment crest. Installation of sensors should occur prior to the construction phase of the embankment, with instruments operating until the post-operation and closure stages.

As a part of additional raises in one of the tailings dams in PNG, uncertainties regarding the extended foundation were studied in detail and resolved through a comprehensive geotechnical investigation. Figure 1 shows a cross-section with a possible sudden consolidation zone identified that can occur in the clay lenses or softer seams. These zones are susceptible to high compressibility and low permeability conditions and could result in geotechnical instability.

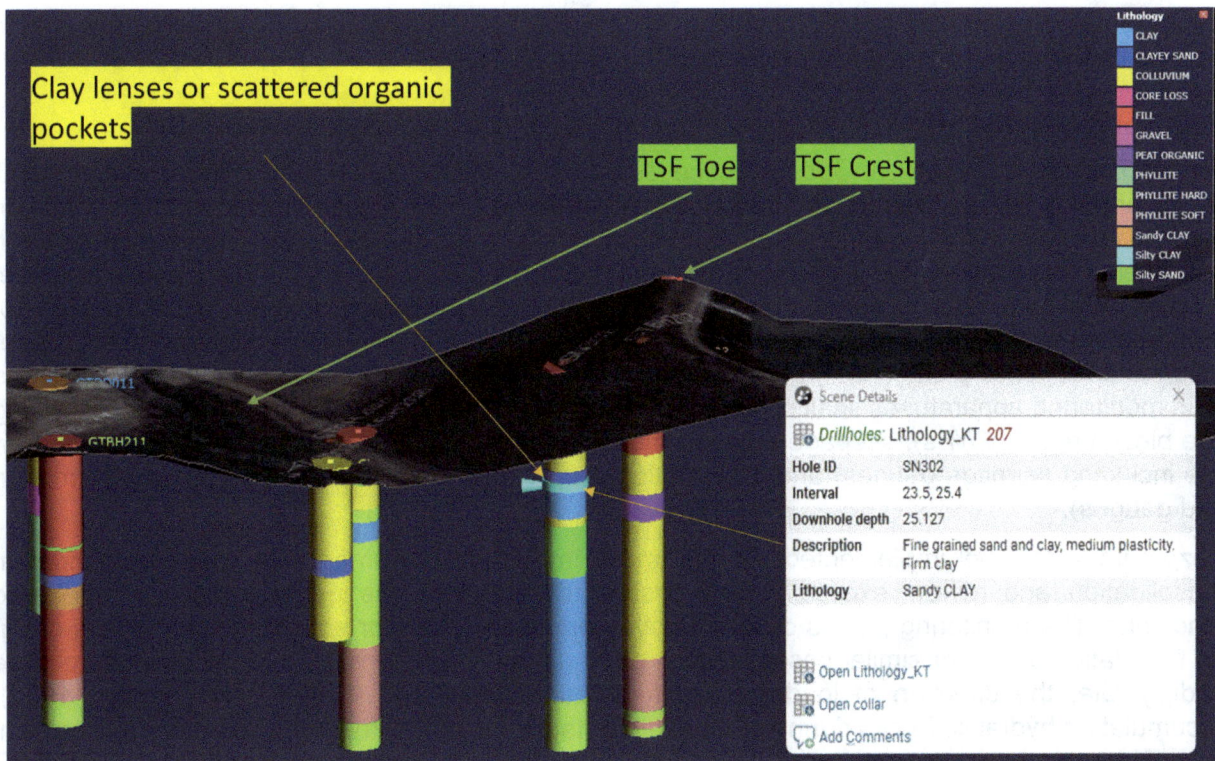

FIG 1 – Clay lenses and organic depositions in the dam foundation.

Based on Figure 1, unexpected vertical and horizontal consolidation responses were recorded by deformation sensor arrays. It should be noted that although it did not cause a significant impact on the stability of the structure, it was disregarded from the initial geotechnical assessment and not considered important in the ground model for the site. In conclusion, selected cross-sections should reflect the critical lithology, such as high plasticity clay formation and over-consolidated clays in which strains may localise during the construction or operational stage under static loading (rapid rise) or earthquake forces (rapid dynamic).

Analysis methodology

In a TSF, tailings are discharged mostly in a horizontal direction through the spigots, thus, utilising simplified deformation assessment, elastic-perfect plastic, and isotropic constitutive models to predict deformation optimistically compared to the anisotropic models. Nimbalkar, Annapareddy and Pain (2018) showed that the horizontal slice method (HSM) formulation to compute the internal stability of reinforced soil retaining structure could be successfully used for the deformation analysis in tailings.

Recently reviewed numerical tools for the deformation analysis of tailings dams by (Geppetti, Facciorusso and Madiai, 2022) showed that the zones with the highest liquefaction risk are near the surface of the settling basin, for which the linear method (LM) or the equivalent linear method (ELM) could underestimate the tailings' non-linear behaviour. Thus, the degree of robustness in numerical assessment is mainly related to tailings characteristics studies undertaken in the field or laboratory studies. Based on these studies, the appropriate non-linear stress-strain constitutive models are selected and calibrated based on available monitoring data or laboratory analysis. Adopting the 1D consolidation theory or Newmark's rigid block analogy (Bergado *et al*, 1991; Newmark, 1965) to the tailing material can compromise this analysis. Pore-water pressures would be underestimated for the tailings consolidation, and this theory is not recommended for the liquefaction and deformation analysis. Reports for the Australian TSF failure emphasised that deformation analysis must employ advanced numerical methods with 2D (plane strain) and 3D analysis with appropriate stress-strain relations such as NorSand, UBCSand, or PM4Sand (Jefferies *et al*, 2019). NorSand (Jefferies, Shuttle and Been, 2015) could accurately model Static and deformation analysis. Regarding the three or two-dimensional analysis, ANCOLD provided a hierarchy of numerical modelling requirements for static and dynamic analysis. However, there are technical debates about whether results obtained from 3D or 2D modelling for the tailings material are conservative and which represents the most conservative estimate of the Factor of Safety.

Monitoring systems

Per the Tailings Management Good Practice Guide (GPG; ICMM, 2021), a monitoring system shall be implanted to assess the required geotechnical parameters. The design and implementation of an installed instrumentation and monitoring plan shall include a pre-defined acquisition period for each instrument. This includes the instrument type, maintenance plan, accountable responsible engineer (for the data acquisition), methods for data acquisition (cloud-based or manual), processes for documenting and sharing to designer and third-party reviewer, and ultimately integrated to the trigger action and response plan (TAPR) and emergency preparedness and response plan (EPRP). Three main monitoring apparatuses shall be included in all tailings designs, which are automated vibrating wire piezometers (VWPs), seepage measurement (automated V-notch for conductivity and outflow rate including seepage gauges) and deformation arrays (fully automated shape array deformation rigid gauges).

Piezometers can be implemented as the most important instruments to monitor foundation consolidation and review excess pore pressure to re-assess the general stability of altered conditions. The monitoring plan is critical if clayey organic pockets or softer lithology lenses are in the foundation area. For similar cases, excess pore pressure can be related to the quick surcharge loading rate, the direction of loading, and foundation anisotropy (Chang and Duncan, 1983). Accumulating hydraulic forces during embankment rise could cause the induction and development of considerable pore pressure within the dam's core (Hosseini, Mohammad and Fard, 2003). 3D laser, light detection and ranging (LiDAR) and interferometric synthetic aperture radar (InSAR) scanning tools could benefit the open pit and slope stability assessment or as-built quality surveillance plan. However, due to the tailings dam brittleness (Olalla and Cuellar, 2001) and outflow speeds modelled by mudflow and other numerical programs for the Brumadinho dam, it was indicated that LiDAR scanning might not be effective due to the delay in receiving updated data (Lumbroso et al, 2021). Therefore, standalone data logger systems integrated with VWPs, and deformation array gauges are versatile and reliable tools to collect and transmit live data for quick review and action. All data are seamlessly uploaded to the server's cloud and graphically demonstrate results by custom-made software or interactive dashboards, which automatically process data, plot, and share with appropriate technical and responsible parties (Figure 2).

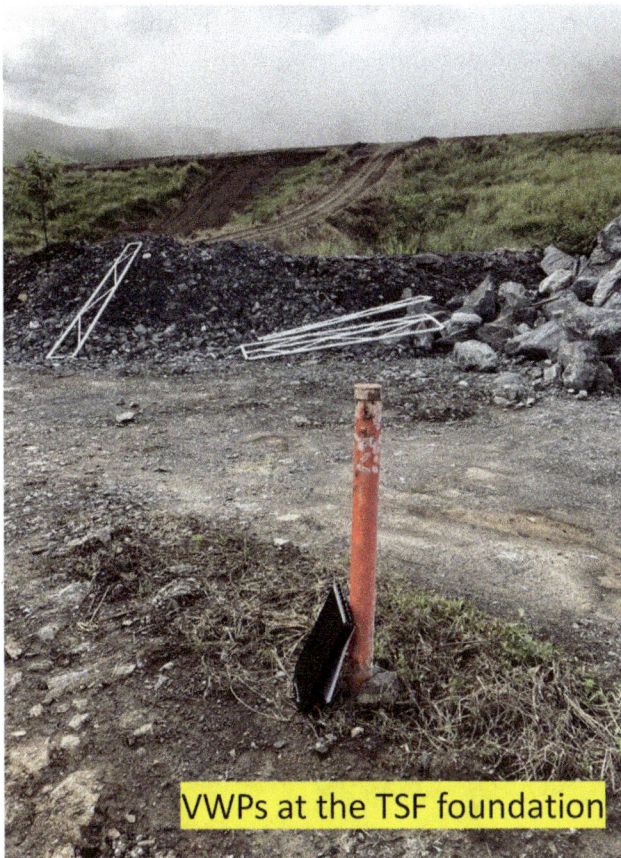

VWPs at the TSF foundation

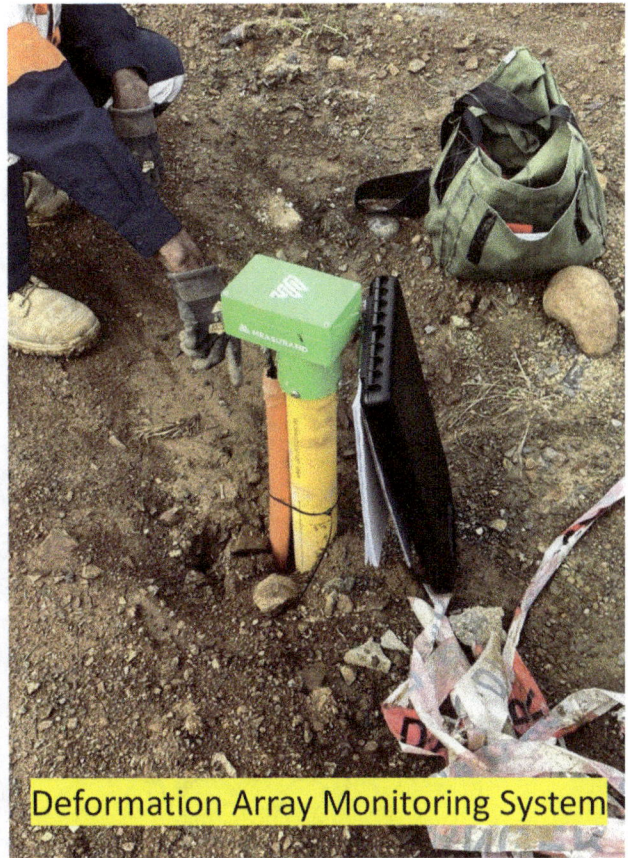

Deformation Array Monitoring System

FIG 2 – Installed online stream monitoring systems at the TSF Foundation (left) and dam crest (right).

Another critical monitoring system component is seepage management, which was measured manually in the last few decades; however, recently, automated outflow apparatuses have been employed in large tailings facilities to control run-off waters during construction and post-closure stages. An example of a fully automated V-notched seepage system versus a manual approach is shown in Figure 3.

V-Notched automated seepage system

Manually seepage measurement system

FIG 3 – Automated seepage system (left) versus manual method (right).

Despite the requirement of seepage data for the verification and environmental studies, flow rate, electrical resistivity, and Electrical Conductivity (EC) provide early warning of potential leakage to the rapid response of tailings embankment and foundation subject to external or internal dynamic forces. An example of a sudden increase in EC is presented in Figure 4.

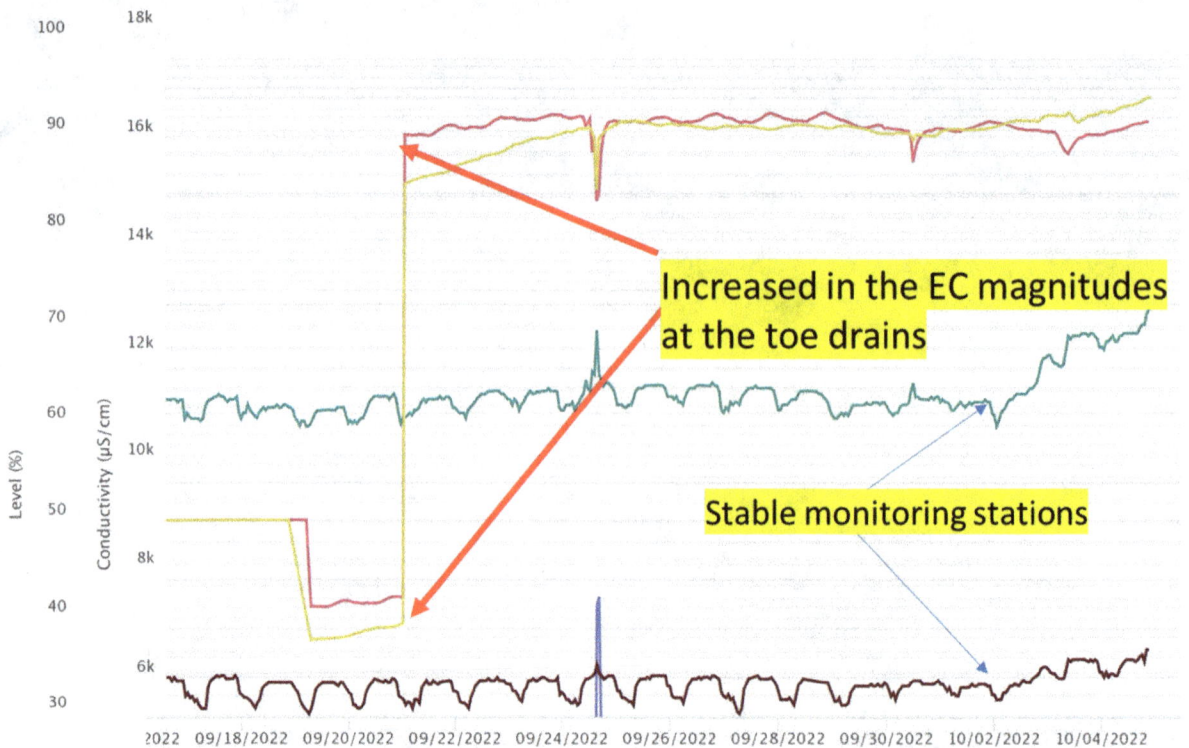

FIG 4 – Sudden change of the conductivity rate at the seepage outlet.

The EC results imply that the seepage emission could be impacted by unusual seepage activities, influenced by an earthquake, liner system damages or internal erosion. In this case, the conductivity level raised from 7.5 K μs/cm to approximately 15.3 K μs/cm within a short period and the water level at the V-notch reached 90 per cent of the maximum height. Further monitoring and stability analysis indicated possible root causes and adequate remedial actions to mitigate risks. Tracers such as fluorescent leak detection dyes were used in an array of settings to identify potential leakage spots along the embankment. At the same time, the outflow was continuously monitored for a breakthrough in the dye. Analysis of seepage rates, concentration curve of dye in outflow and location of dye expression assisted in defining the scope of the event and remedial actions required.

Monitoring data interpretation

Despite the accuracy and simplicity of commercially available surveillance systems, interpretation of recorded data requires some level of engineering judgment, geotechnical understanding of site conditions and a conceptual model of failure mechanism. A successful monitoring management plan engages the active participation of the engineer accountable for the site, together with engineering staff collecting data and sharing it with other parties to fulfil the emergency preparedness and response plan requirements.

In a particular case, increased pore pressure of the TSF embankment is expected during a new rise, resulting in short-term construction loads that decrease the shear strength of key zones. However, if clay lenses in the dam foundation experience sudden anisotropic consolidations, the impact is a sudden spike of pore pressure causing a geotechnical instability. Deformation-monitored data could reflect sudden volume changes as a spike and raise concerns regarding the associated risks.

Accumulating excess pore water pressure and dissipation rates could indicate localised displacements, which do not necessarily lead to an embankment's upstream or downstream side instability. Figure 5 illustrates three different responses of a TSF foundation to rapid pore pressure response for three distinct lithologies, ie silty gravel, sensitive organic clay, and a lump of highly plastic clay.

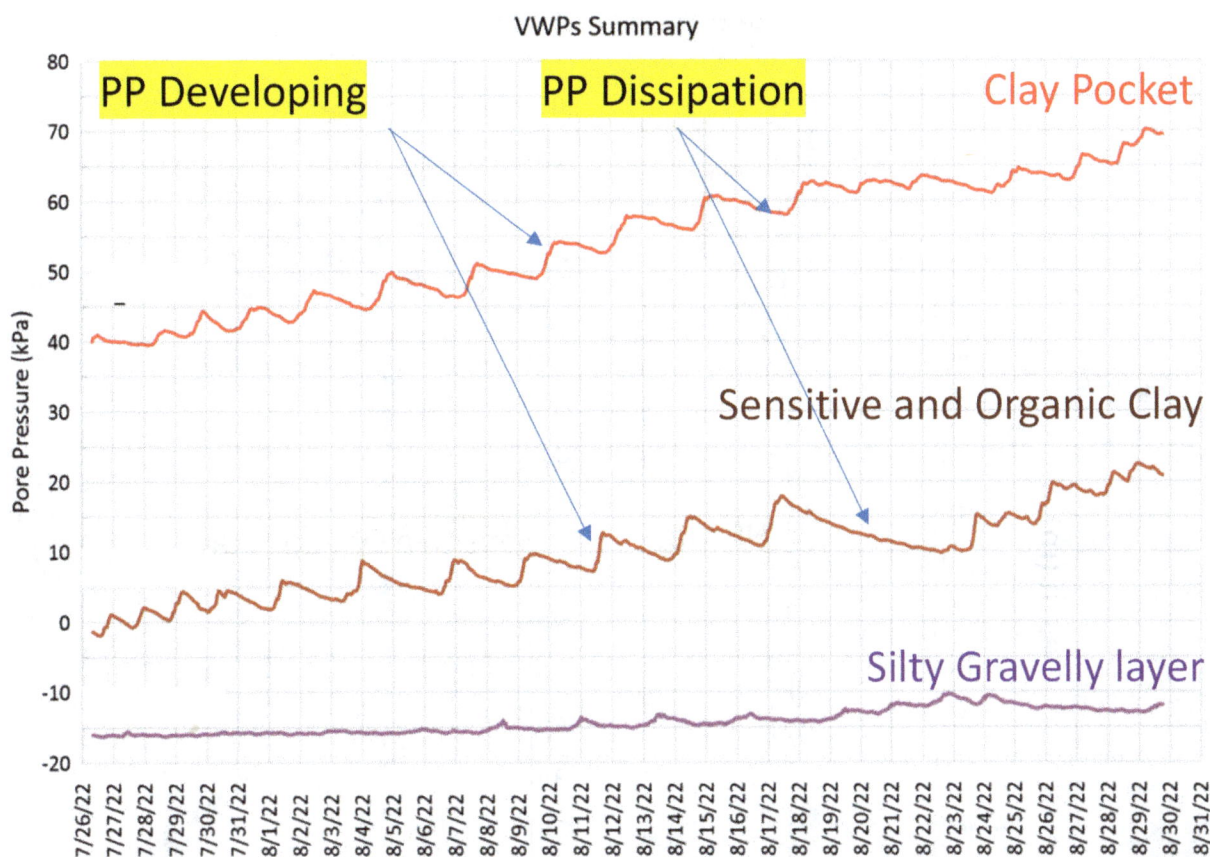

FIG 5 – Developing and dissipation of pore water pressure for silty gravelly, organic, and normal clays during the rising construction.

In Figure 5, a piezometric response to the quick rise in sensitive clay (PI>50) is almost 1.46 times greater than the response to highly plastic clay (PI>35) and almost no changes to highly permeable silty gravel deposits. A trial was undertaken to measure the dissipation time for excess pore water pressure. The results showed that 2.5 days are required for the piezometric level to return to the previous level for highly plastic clay. In contrast, for the sensitive organic clay, it took 5.5 days. Vertical cumulative displacement results (Figure 6) indicated aggressive responses to sensitive clay compared to high plasticity clay. The maximum vertical consolidation settlement was measured at less than 5 mm for the high-plasticity clay and 36 mm for the organic clay. A negative vertical deformation is observed with lateral depressions and might result in the rotation of the soft soil lenses.

The unexpected deformation or sudden pore pressure increase is critical to managing a TSF. It should be addressed in the operational manual system (OMS) to record all occurrences of events with documented details to allow for further analysis. As a result of observations from Figures 5 and 6, construction sequences was adjusted to allow built-up pore pressures to dissipate to acceptable levels. Excess pore water pressure could increase during the operational stage due to various factors such as rapid loading, changes in groundwater levels, or the consolidation process. Excess pore pressure build-up can lead to decreased shear strength, increased potential for liquefaction, and reduced bearing capacity of the foundation. On the other hand, the dissipation of excess pore pressure is also crucial. Proper drainage and dissipation of excess pore pressure help maintain the stability and integrity of the TSF. Design considerations for TSFs must account for the development and dissipation of excess pore pressure to ensure the safe and reliable operation of the facility.

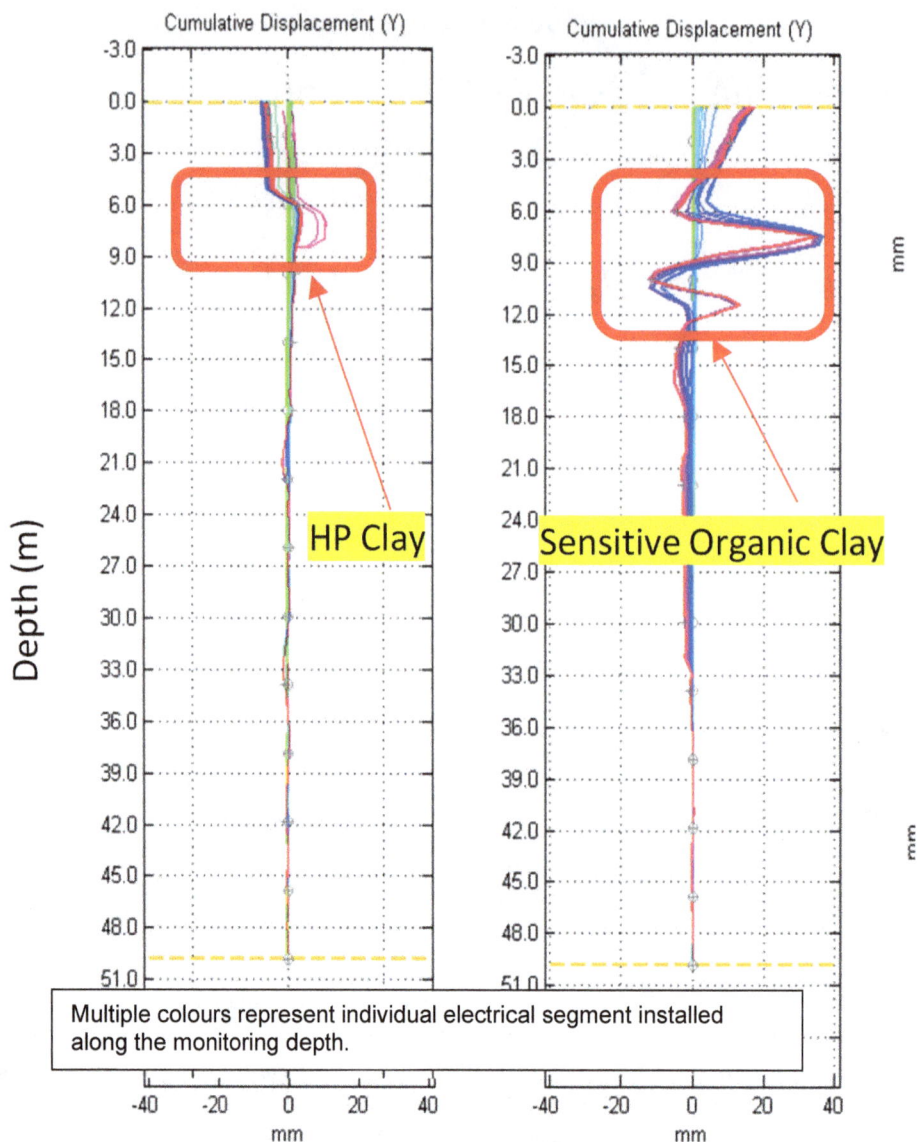

Cumulative Displacement (Y)

Cumulative Displacement (Y)

HP Clay

Sensitive Organic Clay

Depth (m)

Multiple colours represent individual electrical segment installed along the monitoring depth.

FIG 6 – Cumulative vertical deformation along the monitoring depth for high plasticity clay (left) and organic clay (right).

The response to events described in the preceding section should be filed in a well-documented database that includes surveys and monitoring data. It will facilitate the effective transfer between responsible engineers for the TSF. Consequently, it can optimise a rapid construction methodology and assess mitigation measurement for severe events.

Earthquake impacts

Dynamic forces acting on the TSF could increase internal loading at the foundation and embankment, impacting the stability of the TSF. Large dams such as Fundão Tailings Dam experienced a fragile state (strain stress curves). They quickly could be transferred to yielding and plastic envelope, whilst kinematics of the emerging failure factors is less critical for small dams. According to the GISTM guideline, monitoring information after any earthquake event shall be detailed, recorded, and shared with the project engineer of record (EoR) and the Responsible Tailings Facility Engineer (RTFE) to be reviewed and signed off if the earthquake incident is minor. Factors such as crack propagation patterns, size, width, and direction are key to visually assessing in association with the live monitoring data. In the example illustrated in Figure 7, after an earthquake event, shallow transitional depth and narrow scattered cracks were identified and recorded with detailed surveying data to assess the earthquake's impact. The crack propagation was not continuous or extensive in the vertical plane (depth less than 0.6 m), but pore pressure data increased during the earthquake. The increase in pore pressure was almost 4.5 times higher for the clayey type of material than the regular increase-decrease rate under normal construction activities.

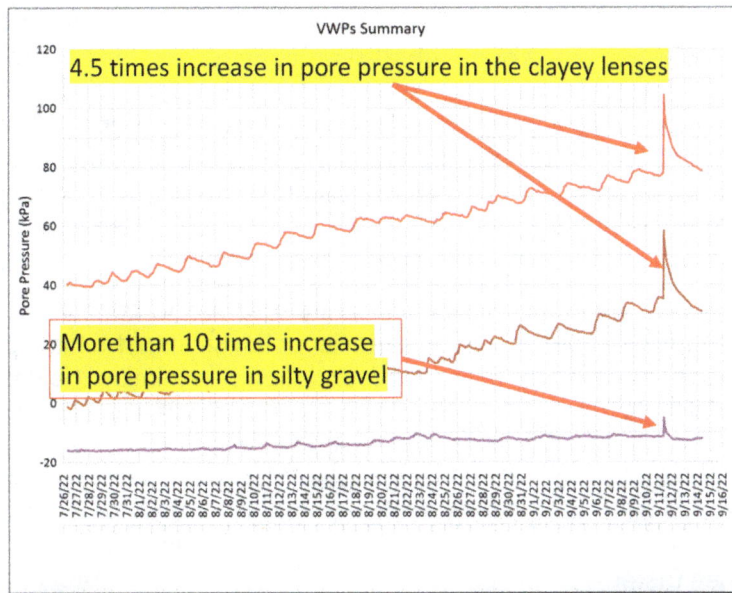

FIG 7 – Pore pressure sudden increase in the earthquake event (top) and irregulated scatter and transitional cracks along the dam crest (bottom).

Meanwhile, for the silty gravelly material (with a higher void ratio), the change of excess pore water pressure was measured about ten times higher. However, the excess pore pressure dissipated quickly for the silty gravel layer compared to the clayey layers. The latter required about two days to return to previous pore pressure levels.

In this case (Figure 7), silty layers are not restricted between the low permeability clays and based on the thin thickness, it was noted that excess pore pressure quickly dissipated. In unfavourable conditions (thick silty deposits or the tailings material itself), a fully dynamic assessment (coupled hydromechanical analysis) is required. Recorded seismographs from the nearby stations will be used to reassess the TARP levels against the design probabilistic and deterministic assumptions (in addition to the extreme event case Maximum Credible Earthquake (MCE)) to produce appropriate risk assessment outcomes and possible mitigation plans.

CONCLUSION

In this study, multiple data sources were used to emphasise the role of monitoring and surveillance management in a TSF. According to the approaches and principles outlined in the GISTM guideline, monitoring conformance protocols shall be developed and strictly adhered to the design basis and current situation of a TSF. The effect of the short-term construction loading impacts was discussed by focusing on dam foundation studies. It was shown that VWPs, SAVV and hydraulic measurements tools are effective instruments to improve the design and management of TSF dams by providing valuable data and insights into the tailings dam behaviour and performance. Utilising adequate and appropriate measurement systems and reviewing data by engineers, operators, and stakeholders can provide a clear understanding of the performance of the TSF dam, identify potential problems or risks, and take suitable actions to mitigate operational and post-closure hazards.

It was pointed out how *in situ* materials can differently respond to construction raises. It was also shown that the VWPs information could be used to optimise construction programs to facilitate construction loadings over the critical area in the foundation to minimise the risk to the TSF integrity. In addition, a vertical deformation array system could effectively reflect the quick consolidation in the foundation or the embankment and provide a reliable indication tool to mitigate any instability risks. Seepage information and electrical conductivity results were also discussed as warning signs regarding dam leakage or internal hydraulic erosions. Finally, it was shown that earthquakes could increase pore water pressure in silty layers compared to clay layers, but it was dissipated approximately 2.5 times faster.

Therefore, developing a robust design integrates all collected knowledge and monitoring information in a data management system. Data collected should be leveraged to minimise the risk of a TSF failure that limits negative impacts on the community and the environment.

ACKNOWLEDGEMENTS

The authors acknowledge that ATC Williams provided the opportunity to prepare and present this paper at the Mine Waste and Tailings Conference 2023 using multiple pieces of information collected from various projects.

REFERENCES

ANCOLD, 2012. *Guidelines on the Consequence Categories for Dams,* Hobart, Tasmania: Australian National Committee on Large Dams.

Bergado, D T, Asakami, H, Alfaro, M C and Balasubramaniam, A, 1991. Smear effects of vertical drains on soft Bangkok clay, *Journal of Geotechnical Engineering*, 117(10):1509–1530.

Chang, C S and Duncan, J M, 1983. Consolidation analysis for partly saturated clay using an elastic–plastic effective stress–strain model, *International Journal for Numerical and Analytical Methods in Geomechanics*, 7(1):39–55.

Geppetti, A, Facciorusso, J and Madiai, C, 2022. Tailings Dams Numerical Models: A Review, in *Proceedings of the 7th World Congress on Civil, Structural and Environmental Engineering (CSEE'22)*, pp 241–247.

Hosseini, S, Mohammad, M and Fard, R A, 2003. Pore pressure development in the core of earth dams during simultaneous construction and impounding, *Electronic Journal of Geotechnical Engineering*, 8(1).

International Council of Mining and Metals (ICMM), 2020. Global Industry Standard on Tailings Management (GISTM), ICMM.

International Council of Mining and Metals (ICMM), 2021. Tailings Management: Good Practice Guide, ICMM.

Jefferies, M, Morgenstern, N, Van Zyl, D and Wates, J, 2019. Report on NTSF embankment failure, Cadia Valley Operations, for Ashurst Australia.

Jefferies, M, Shuttle, D and Been, K, 2015. Principal stress rotation as cause of cyclic mobility, *Geotechnical Research,* 2(2):66–96.

Lumbroso, D, Davison, M, Body, R and Petkovšek, G, 2021. Modelling the Brumadinho tailings dam failure, the subsequent loss of life and how it could have been reduced, *Natural Hazards and Earth System Sciences*, 21(1):21–37.

Morrison, K F, 2022. Tailings Management Handbook: A LifeCycle Approach, Society for Mining, Metallurgy and Exploration, Incorporated.

Newmark, N M, 1965. Effects of earthquakes on dams and embankments, *Geotechnique*, 15(2):139–160.

Nimbalkar, S, Annapareddy, V R and Pain, A, 2018. A simplified approach to assess seismic stability of tailings dams, *Journal of Rock Mechanics and Geotechnical Engineering*, 10(6):1082–1090.

Olalla, C and Cuellar, V, 2001. Failure mechanism of the Aznalcóllar dam, Seville, Spain, *Géotechnique*, 51(5):399–406.

A noise reduction algorithm for automatic interpretation of periodical survey data for tailings dam performance monitoring

C Han[1], K F Wong[2], W Ludlow[3] and Z R Yung[4]

1. Senior Engineer, Red Earth Engineering – a Geosyntec Company, Brisbane Qld 4000. Email: chao.han@redearthengineering.com.au
2. Data Scientist, Scovano, Pty Ltd, Brisbane Qld 4000. Email: eric.wong@scovano.com
3. Principal Engineer, Red Earth Engineering – a Geosyntec Company, Brisbane Qld 4000. Email: wade.ludlow@redearthengineering.com.au
4. Tailings Engineer, Red Earth Engineering – a Geosyntec Company, Brisbane Qld 4000. Email: george.yung@redearthengineering.com.au

ABSTRACT

The implementation of the Global Industry Standard on Tailings Management (GISTM) requires regular review of technical monitoring data to assess the performance of tailings dams against design expectations. These monitoring programs often include survey monuments/prisms at discrete dam locations to detect embankment movements based on interpretations of recorded data at prescribed intervals. A timely and reliable interpretation of these movements is crucial as it forms part of a site-wide emergency response system. However, there are often significant survey tolerance issues that exceed movement trigger limits and limitations on the regularity and frequency of the survey data collection that affect the reliability and accuracy of the readings.

This paper discusses limitations that are common within the industry and presents an automatic interpretation algorithm of the survey data that can be implemented in a web-based monitoring application to address some of these challenges. The proposed algorithm provides an estimation using a weighted moving average of survey gradients. The algorithm has proven to be unbiased for linear movements, even when time intervals between observations are randomised. Optimal window size selection using a dynamic k-nearest neighbours (d-kNN) algorithm for non-linear movement is also provided to control the accuracy of predicted movement trend informed by real-time monitoring data.

The implementation of the automated algorithm can provide a tool that increases confidence when reviewing poor quality survey data with poor tolerance.

INTRODUCTION

An increasing number of international mining companies have endeavoured to achieve compliance with the Global Industry Standard on Tailings Management (GISTM) (International Council on Mining & Metals (ICMM), 2020). GISTM Requirements 7.3 and 7.4 stipulate the importance of regularly reviewing tailings facility performance against design intents with the use of monitoring and surveillance data. An integrated dam surveillance system typically involves the use of sensors, monitoring instruments, and other advanced technologies to detect potential dam issues such as embankment settlement, seepage, and erosion, all of which could compromise the integrity of a dam. In recent years, the development of new sensors, data analytics, and remote monitoring technologies has provided the industry with access to real-time information, enabling early detection of potential issues and allowing proactive measures/mitigations to be implemented to address the concern. However, the reliability and timely data interpretation for real-time systems is critical since the prescribed thresholds are typically associated with a corresponding Trigger Action and Response Plan (TARP).

Ground surveying is a commonly used monitoring technique in tailings facilities to detect movements in the embankment by periodically measuring fixed points (often survey monuments with mounted prisms) on the facility. The ground survey method utilises ground-based equipment, such as a total station and a global positioning satellite (GPS) receiver, as shown in Figure 1, to measure changes in horizontal and vertical coordinates. GPS receivers are prevalent over total stations in tailings dam applications due to the following key advantages:

- Mobility to access difficult terrain and ease of set-up without requiring stationing the instrument.

- Lower equipment cost and higher robustness in outdoor environments.

- Minimal risks of equipment damage due to surrounding construction activities.

FIG 1 – Photographs of total station and GPS rover (source: Hanrahan and Hanmore, 2021).

The main issue with using the above system is the limitation on survey accuracy which depends on the number and duration of satellites used for the survey, experience of the surveyor and duration available to complete the survey, location of the site and sophistication of the survey system used. The author's experiences are that survey tolerance can be fairly poor using this equipment (say up to 60 mm) and the survey data contains a lot of noise that needs to be filtered out to establish data trends.

This paper discusses the current practices of using GPS receivers in a TSF located in northern Australia and highlights some of the key challenges and limitations encountered when using the standard GPS system to monitor dam movements. These challenges were resolved by introducing an interpretation algorithm implemented via a web application that reduces data noise by adopting a weighted moving average of the movement gradient. The basis for the proposed algorithm is presented in this paper through an application example to demonstrate its functionality.

COMMON PRACTICES AND CHALLENGES

Survey monuments or markers are normally installed on the dam surface at critical locations nominated by the Design Engineer. A TARP is then developed for each survey monument and included in the facility Operational Maintenance and Surveillance (OMS) Manual. The OMS and TARPs provide dam operators a reading frequency and the corresponding quantitative trigger(s) which when exceeded, yields a corresponding actionable response.

The TSF discussed in this paper requires completion of monthly surveys on monuments concreted into the dam crest and has a trigger tolerance of 25 mm for the horizontal and/or vertical embankment displacements per month. Exceeding this trigger level, results in heightened alertness and a response action to be implemented such as increasing the reading and inspection frequencies, notifying the Responsible Tailings Facility Engineer (RTFE), the Engineer of Record (EoR), and establishing risk mitigation resources immediately after the trigger exceedance.

Survey of the dam crest monuments and data collection is carried out manually by surveyors and will typically involve three repeated readings of the GPS coordinates, including Northing (N), Easting (E), and Elevation (H). The average of the repeated readings is considered the representative coordinates for each of the locations. The monthly horizontal and vertical displacements is interpreted using Equation 1:

$$\Delta h_n = \sqrt{(N_n - N_{n-1})^2 + (E_n - E_{n-1})^2};$$
$$\Delta v_n = H_n - H_{n-1}$$

(1)

where Δh_n and Δv_n are interpreted monthly horizontal and vertical displacements based on the latest readings; N_n, E_n, and H_n are the latest coordinates; N_{n-1}, E_{n-1}, and H_{n-1} are the coordinates from the previous month readings.

Figure 2 depicts the cumulative changes in recorded movements from the original coordinates at a surveyed location over a 6.5 year monitoring period. It is evident that no obvious trend of movement can be reasonably quantified from this figure, except for a significant level of noise (ie survey tolerance) that was recorded. The corresponding monthly horizontal and vertical displacements calculated from Equation 1 are plotted in Figure 3 and compared with the 25 mm and -25 mm triggers (represented by red dashed lines) for the respective monthly horizontal displacement and settlement. The figure shows that the triggers were frequently exceeded throughout the reported period despite the absence of an observed movement trend. The 'false alarm' events can be attributed to the following main reasons:

- The 25 mm trigger is within the data noise limits of up to 45 mm for vertical displacement, and hence the trigger can be exceeded without any observational indications.

- Variations in reading frequency can make Equation 1 invalid since it assumes that the interval between two consecutive readings is approximately 30 days. For instance, inaccessibility of survey locations, unavailability of survey personnel, and other makeshift arrangements can result in the interval being significantly longer or shorter than 30 days.

- Additionally, bespoke manual interpolations to overcome the above limitations can result in inconsistent approaches being used each time the data are reviewed, causing delays in interpretation, which may result in missing the best time window to react.

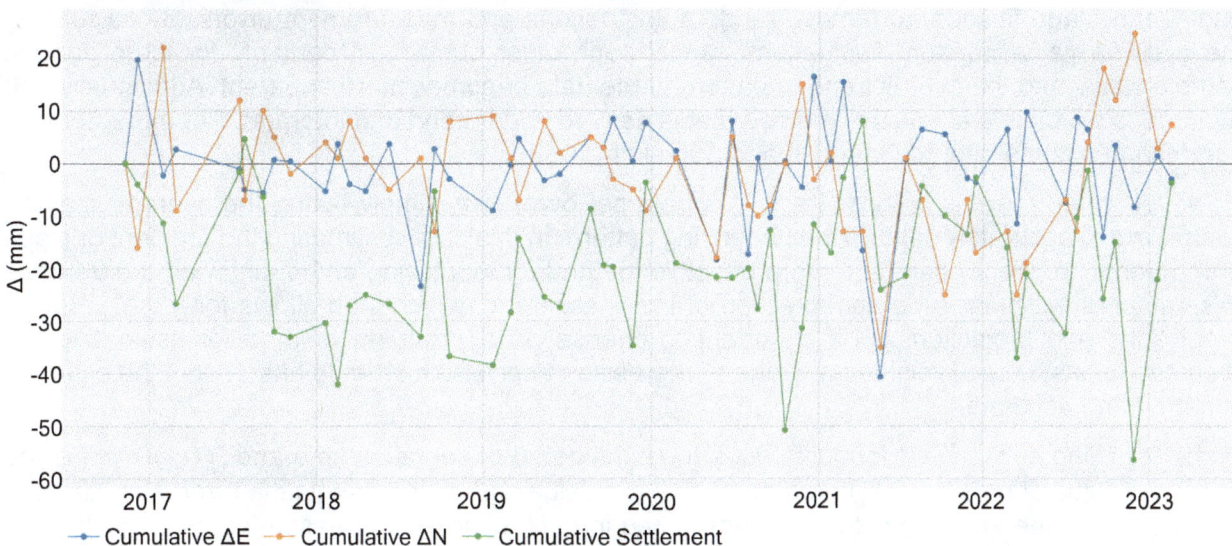

FIG 2 – Coordinate change from the origin at a survey location.

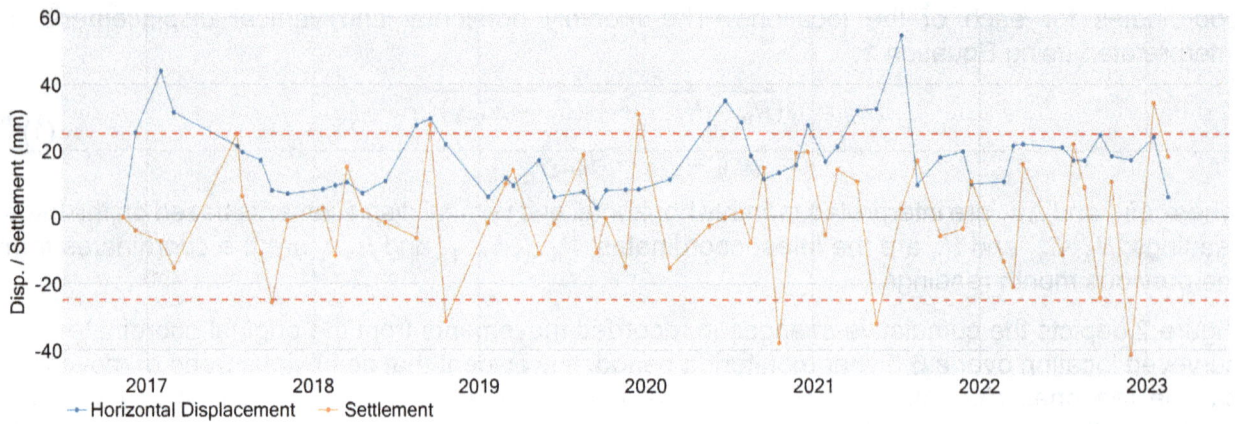

FIG 3 – Inferred monthly displacements from Equation 1 compared with ±25 mm trigger limits.

The practicality and reliability of responses to the survey readings have been constrained by the above challenges and hence a new algorithm was proposed to:

- establish movement trends and reduce 'noise' in the survey record

- account for irregular reading frequency

- implement a consistent approach.

PROPOSED METHOD

Noise reduction

The purpose of reducing data noise is to eliminate unwanted or irrelevant information from a data set so that meaningful insights and informed decisions can be made without bias. The most widely used noise reduction method is taking the moving average of a homogeneous data set to improve the signal-to-noise ratio. This method has been used in a variety of scientific fields, including physics, engineering, and finance, to remove random fluctuations and measurement errors. By calculating the average value, random fluctuations cancel each other out almost certainly for large enough sample sizes, and the overall trend or pattern of the data becomes more apparent. Additionally, the average is a simple and intuitive method that is easy to understand and interpret. For more detailed information, please refer to Grimmett and Stirzaker (1992).

However, when there is a linear trend in a data set over time, simply taking the average may be insufficient to accurately capture the underlying pattern. In this case, estimating the gradient or slope of the trendline directly can be a more effective method for identifying and quantifying the trend as this captures the magnitude and direction of the linear trend, which then allows for more accurate predictions and projections. It is important to choose an appropriate method for estimating the gradient, for instance a weighted average or regression analysis, in order to ensure that the estimate is robust and accurate.

In the following subsequent formulae, t_n is the considered observation time and $[f(t_n) + e_n]$ is the observed value at time t_n. $f(t_n)$ represents the true value and e_n represents the error. Note that the true value and the error cannot be de-aggregated in each observation. Assuming that $t_{n-1}, \ldots, t_{n-k}$ are the k nearest observation times in which the true values have the same gradient to that at the time t_n, the proposed estimator of the gradient at time t_n is given by Equation 2:

$$\hat{g}(t_n) = \sum_{i=1}^{k} \frac{[f(t_n) + e_n] - [f(t_{n-i}) + e_{n-i}]}{t_n - t_{n-i}} \times \frac{t_n - t_{n-i}}{\sum_{j=1}^{k}(t_n - t_{n-j})}$$

$$= \frac{\sum_{i=1}^{k}[f(t_n) + e_n] - [f(t_{n-i}) + e_{n-i}]}{\sum_{j=1}^{k}(t_n - t_{n-j})}$$

(2)

The gradient is computed as the total change in observed values divided by the total time difference. Additionally, $\hat{g}(t_n)$ is an unbiased estimator of the gradient at time t_n, which gives an equal weight to all the error terms within the k nearest observation times. Consequently, the change that occurred over a certain time interval can be estimated by multiplying the estimated gradient by the length of the interval.

Given that the data set may exhibit piecewise linear trends over a prolonged monitoring period and that the estimated gradient may only be valid within the most recent observations, the dynamic k-nearest neighbours (d-kNN) method introduced by Zhong *et al* (2017) and Wu, Cai and Gao (2010) was adopted. The d-kNN is a non-parametric method that estimates the value of a new observation by utilising the values of its k nearest neighbours in the data set. This approach is flexible and adaptable to various data types and patterns. Additionally, it is worth noting that piecewise linear trends can serve as a good approximation to non-linear trends. For more detail, please refer to Dunham (1986).

Automated algorithm

The above noise reduction approach, which uses a weighted moving average of gradients complemented by the d-KNN algorithm, was implemented into a customised web application to interpret the monitoring data. The application allows for instant uploading of raw data, automatic processing, and real-time visualisation. The automatic interpretation of uploaded data is achieved through the following steps:

Step 1: Merging the survey measurements into daily averages.

Step 2: Selecting potential values for k.

Step 3: Compute $\hat{g}(t_n)$ using k=2 and k=3, denoted as $\widehat{g_2}(t_n)$ and $\widehat{g_3}(t_n)$, respectively.

Step 4: Calculate the cumulative movement, $\widehat{h_k}(t_n)$, using k=2 and 3, respectively, based on the formula $\widehat{h_k}(t_n) = \widehat{h_k}(t_{n-1}) + (t_n - t_{n-1}) * \widehat{g_k}(t_n)$.

Step 5: Select the value of k (ie k=2 or k=3) based on the smallest difference between the observed cumulative movement and $\widehat{h_k}(t_n)$.

Step 6: Calculate the 30-day (ie monthly) increment by multiplying $\widehat{g_k}(t_n)$ by 30, where k is the value selected in step 5.

The proposed algorithm has the following key merits:

- Reduce potential human errors associated with data entry and manipulation.

- Consistent calculation method used for large data sets from all instruments throughout a site to avoid *ad hoc* and/or bias subjective decisions.

- Minimise the time taken to review and interpret the results of the latest survey readings to address any potential issues from the survey trends.

APPLICATION EXAMPLE

The example application focuses on interpreting the movement of the embankments over a 30-day period using survey data with excessive noise/large tolerance. The selection of the d-kNN method in the proposed application has enabled a linear approximation of the actual non-linear trends. However, this approach is inevitably biased, and the magnitude of the bias can vary with the data window size, ie the k value. A smaller k value means that the true movements are observed within a shorter time interval, assuming that the linear trend remains unchanged over the time window when its true value changes. If this assumption is incorrect, the detection of displacement risks may be delayed. Based on the typical observation interval of 30 days, an adopted k value of less than or equal to three was carried forward as this reflected a time window no longer than three months.

When k is equal to 1, the 30-day incremental displacements are calculated by linearly interpolating or extrapolating the differences between two consecutive readings, depending on the observation interval. The solid line in Figures 4 and 5 shows the horizontal movements calculated from Equation 1, and the dotted line shows the estimated 30-day increments using Equation 2. If the interval between measurements is less than 30 days, the dotted line will be further from zero than the solid line, and vice versa. It should be noted that selecting k=1 does not reduce noise as intended.

Additionally, with k=1, the estimated cumulative embankment movement from the origin is equal to that of the observed value.

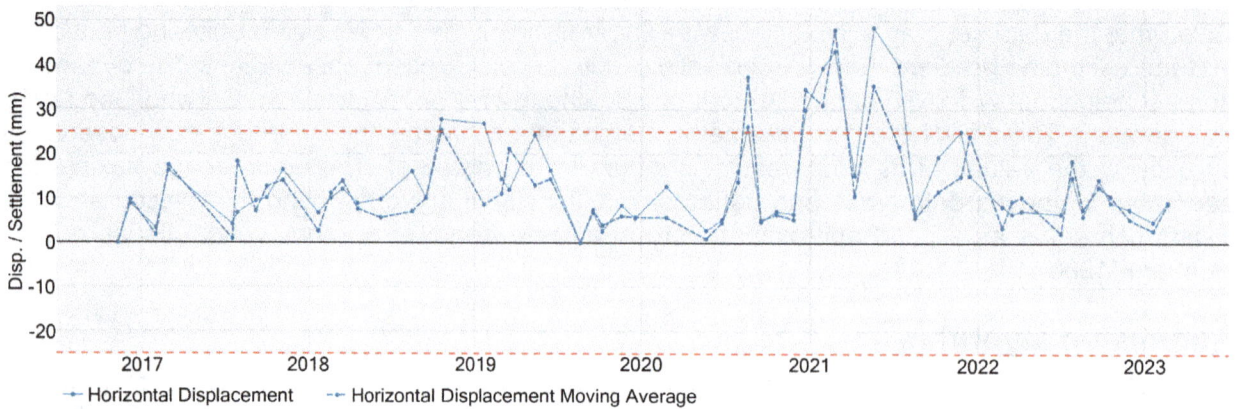

FIG 4 – Observed and estimated 30-day horizontal incremental displacement when k=1.

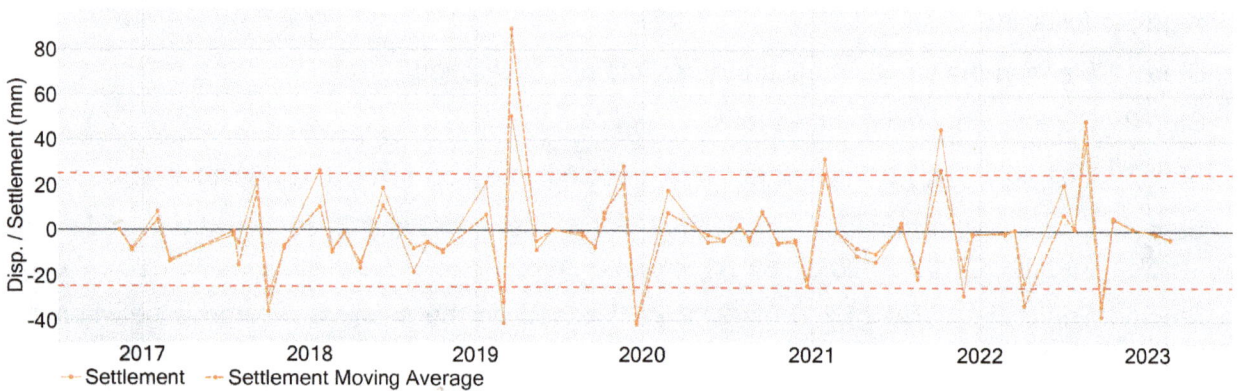

FIG 5 – Observed and estimated 30-day settlement when k=1.

When k is increased to 2 or 3, the estimated cumulative movement from the origin using Equation 2 tends to exhibit biases, especially when applied to settlement interpretation (ie vertical movement), as illustrated in Figures 6 and 7. The bias becomes more prominent over time, which may lead to misleading trend interpretation.

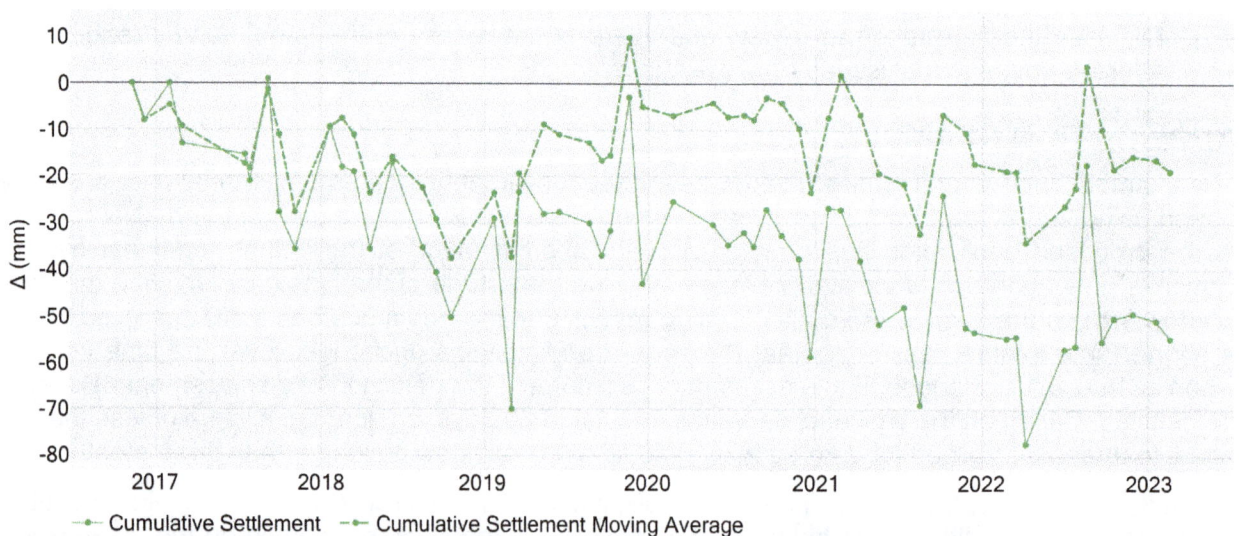

FIG 6 – Observed and estimated cumulative settlement over time when k=2.

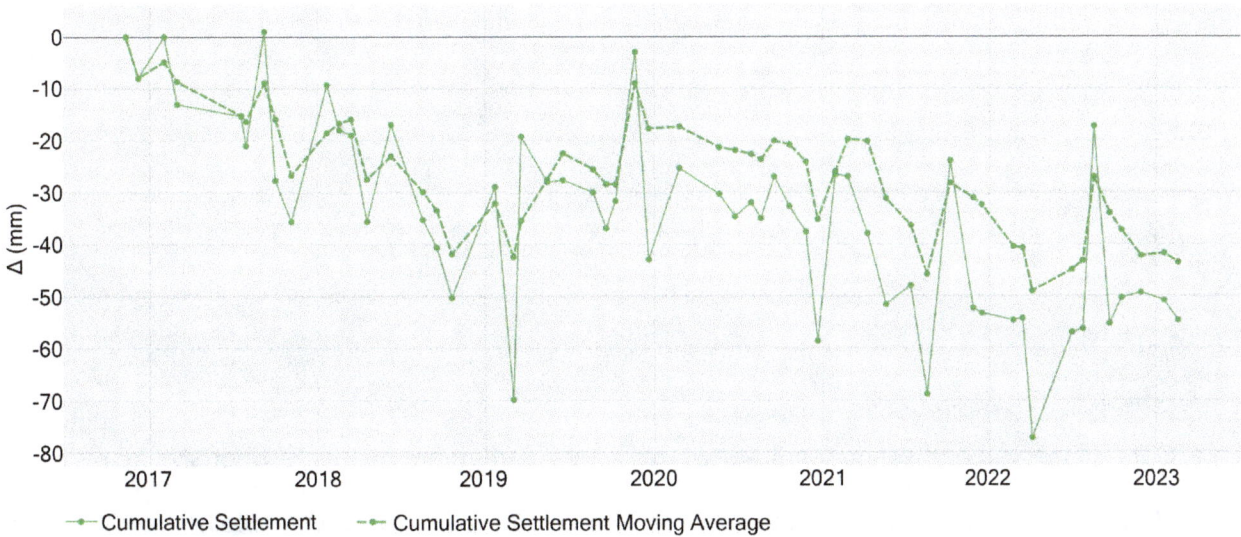

FIG 7 – Observed and estimated cumulative settlement over time when k=3.

Balancing noise reduction and bias is often a trade-off in data analysis. Based on the analysis of historical data, it was found that dynamically selecting a value of k between 2 and 3 yields the most optimal results considering the monthly observation frequency requirement. The principle for selecting k is based on the minimal difference between the measured cumulative movement from the origin and the estimated value using the weighted moving average of gradients method. The smaller the difference, the smaller the bias, indicating that a linear trend is more appropriate within the selected interval. Figure 8 illustrates the improvement in data consistency when k is determined dynamically for each data point. It is noticeable that the observed cumulative settlement trend (dotted line) almost overlaps with the estimated trend (solid line).

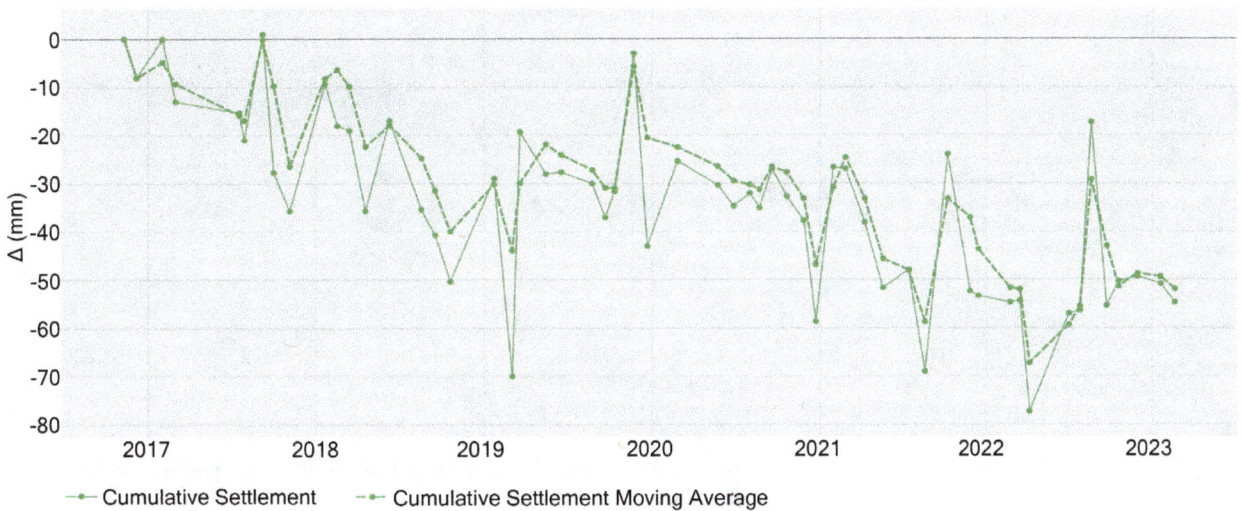

FIG 8 – Observed and estimated cumulative settlement over time when 2 <= k <= 3.

Figure 9 illustrates the corresponding incremental settlement over a period of 30 days, calculated using the proposed d-KNN algorithm. It is shown that the false alarms which would have been triggered by the observed data can be prevented by the estimated results from the proposed method.

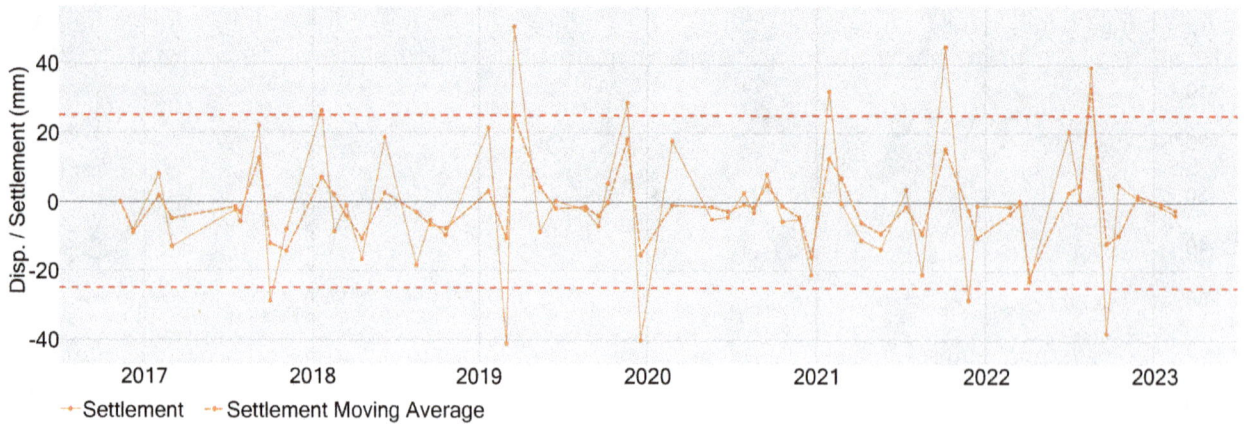

FIG 9 – Observed and estimated 30-day incremental settlement over time when 2 <= k <= 3.

Apart from settlement measurements, the proposed method is also valuable in reducing the occurrence of false alarms for horizontal movement. Figures 10 and 11 show the graphs of cumulative horizontal movement from the origin and incremental movement over time, respectively. Again, the solid line represents the observed results, while the dotted line represents the estimated results. Based on the subsequent cumulative movement, numerous false alarms were observed during the previous movement. However, the proposed method dramatically reduces the occurrence of these false alarms. Additionally, the estimated cumulative movement fits well with the observed cumulative movement.

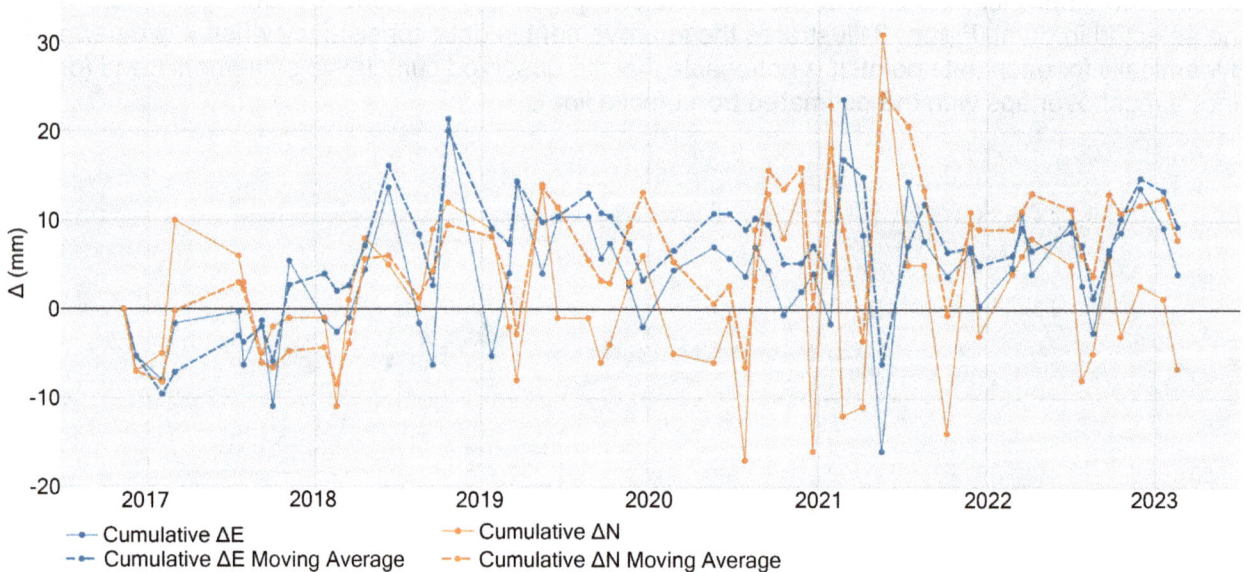

FIG 10 – Observed and estimated cumulative movements, Easting and Northing respectively, over time when 2 <= k <= 3.

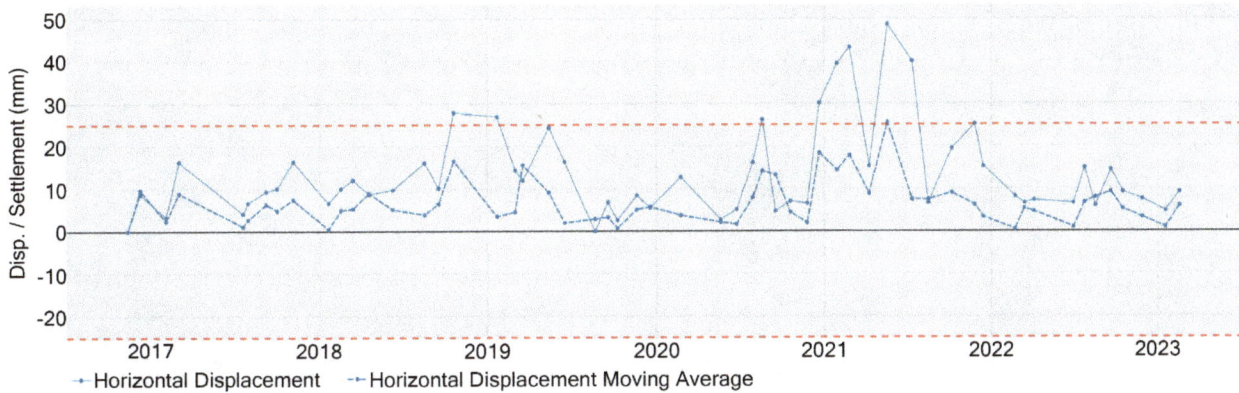

FIG 11 – Observed and estimated 30-day incremental horizontal movement over time when 2 <= k <= 3.

CONCLUSIONS

This paper highlights the survey accuracy limitations of some of the current practices using GPS receivers for monitoring tailings dam movement and has subsequently introduced an alternative interpretation algorithm to address these challenges. The proposed algorithm reduced the noise in the survey and reduced the bias in interpreting the data trends.

The method presented offers a robust and reliable way to interpret periodic survey data that suffers from poor tolerance. By quantifying and reducing data noises, engineers and dam operators will be better suited and informed in making decision throughout the life of the facility.

REFERENCES

Dunham, J G, 1986. Optimum Uniform Piecewise Linear Approximation of Planar Curves, *IEEE Transactions on Pattern Analysis and Machine Intelligence*, pp 67–75.

Grimmett, G R and Stirzaker, D R, 1992. *Probability and Random Processes*, 2nd edition (Oxford Clarendon Press).

Hanrahan, P and Hanmore, R, 2021. GNSS Rover vs Total Station – Pros & Cons [online]. Available from: <https://hitechniques.ie/blog/gnss-rover-vs-total-station-pros-cons/> [Accessed: 27 March 2023].

International Council on Mining & Metals (ICMM), 2020. *Global Industry Standard on Tailings Management*. International Council on Mining & Metals (ICMM).

Wu, J, Cai, Z and Gao, Z, 2010. Dynamic K-Nearest-Neighbor with Distance and attribute weighted for classification, in *Proceedings of the 2010 International Conference on Electronics and Information Engineering*, vol 1, pp 356–360 (Institute of Electrical and Electronics Engineers: Chengdu).

Zhong, X, Guo, Z, Gao, L, Shan, H and Zheng, J, 2017. An Improved k-NN Classification with Dynamic k, in *Proceedings of the 9th International Conference on Machine Learning and Computing*, pp 211–216 (Association for Computing Machinery: Singapore).

Monitoring the impact of mining-induced seismicity on LaRonde main tailings storage facility

E Masengo[1], E-P Ingabire[2], M R Julien[3], J Huza[4], M James[5] and T Lépine[6]

1. Engineer of Record, Agnico Eagle Mine Limited, Toronto ON M5C 2Y7, Canada. Email: edouard.masengo@agnicoeagle.com
2. Responsible Tailings Facility Engineer, ArcelorMittal Mining Canada GP, Longueuil QC J4K 5G7, Canada. Email: edouardinepascale.ingabire@arcelormittal.com
3. Vice-President Environment and Critical Infrastructures, Agnico Eagle Mine Limited, Toronto ON M5C 2Y7, Canada. Email: michel.julien@agnicoeagle.com
4. Director Environment and Operational Risks, Agnico Eagle Mine Limited, Toronto ON M5C 2Y7, Canada. Email: jessica.huza@agnicoeagle.com
5. Engineer of Record, Agnico Eagle Mine Limited, Toronto ON M5C 2Y7, Canada. Email: michael.james@agnicoeagle.com
6. Engineer of Record, Agnico Eagle Mine Limited, Toronto ON M5C 2Y7, Canada. Email: thomas.lepine@agnicoeagle.com

ABSTRACT

LaRonde Mine (LaRonde) is an operating gold mine currently exploiting deposits more than 3 km below ground surface under conditions of mining-induced seismicity. In 2014, LaRonde commissioned a microseismic system to record induced seismic events continuously to evaluate their impact on underground mining operations. Subsequently, there were concerns that mining-induced seismicity could potentially create conditions to trigger tailings storage facility (TSF) failure. Therefore, in July 2020, a vibration triaxial sensor (accelerometer) was installed at the Main TSF to record the peak ground acceleration (PGA) and peak particle velocity (PPV) of the seismic events induced by underground mining activities. In addition, seven vibrating wire piezometers (VWP) are set to record excess porewater pressures at 0.1 second intervals rather than the normal recording interval of one to four hours for the other VWP within the TSF. In case of a seismic event with a magnitude, $M_R \geq 2$, high frequency signal acquisition is automatically triggered, and the seven selected VWP generate an automated report covering a two-minute period, one minute before and one minute after the seismic event. The data collected from the VWP installed at the TSF are used to understand the behaviour and response induced by the seismic loading on the tailings and the cohesive soils in TSF foundation after each underground mining seismic event. Since July 2020, two significant seismic events have been recorded; the first one was recorded 28 December 2021, with a magnitude of $M_R = 3.8$ and a PGA = 0.127 g and the second one was recorded on 29 November 2022, with a magnitude $M_R = 3.5$ and a PGA = 0.422 g. This paper presents the results of a case study of an instrumentation program that has been implemented by LaRonde to monitor the TSF behaviour with respect to mine-induced seismicity activity.

INTRODUCTION

LaRonde Mine is owned and operated by Agnico Eagle Mines Limited (AEM) and is located in the Abitibi region of north-western Québec, Canada. The mine site is approximately 60 km west of the City of Val d'Or. In production since 1988, LaRonde mine currently operates 3.2 km below the surface. LaRonde is currently considered the deepest underground gold mine in the western hemisphere. The objective of this paper is to present the results of a case study of an instrumentation program that has been implemented by LaRonde to monitor the TSF behaviour subject to mine-induced seismicity activities in order to understand the behaviour and response induced by the seismic loading on the tailings and the cohesive soils in TSF foundation after each underground mining induced seismic event.

GEOLOGICAL SETTING

LaRonde is located near the southern boundary of the Archean-age (2.7 billion years old) Abitibi Subprovince and the Pontiac Subprovince within the Superior Geological Province of the Canadian Shield. The most important regional structure is the Cadillac-Larder Lake fault zone, marking the contact between the Abitibi and Pontiac Sub-provinces, located approximately two kilometres to the south of LaRonde.

The geology that underlies LaRonde consists of three east–west trending, steeply south-dipping and generally south-facing regional groups of rock formations. From north to south, they are:

- 400 m (approximate true thickness) of the Kewagama Group, which is made up of a thick band of interbedded wacke.

- 1500 m of the Blake River Group, a volcanic assemblage that hosts all the known economic mineralisation on the property.

- 500 m of the Cadillac Group, made up of a thick band of wacke interbedded with pelitic schist and minor iron formation (Mercier-Langevin *et al*, 2017).

MINING-INDUCED SEISMICITY

Due to a relatively high incidence of mining-induced seismic events, in 2014, LaRonde commissioned a microseismic monitoring system to record induced seismic events continuously to be able to manage the associated risks through a combination of control measures, including a ground-control-driven mining sequence and level design, dynamic ground support systems, and procedures to limit workforce exposure.

For the mining operations, it is always important to have an accurate event source location. This accuracy depends on several factors such as the number of sensors used, seismic system array, velocity model, etc. There are generally two types of seismic monitoring systems for recording seismic events in the mining area, geophones, and accelerometers. Two types of geophones, 4.5 Hz and 15 Hz, are commonly used for mine seismicity monitoring. The 4.5 Hz geophones are used for recording events with larger moment magnitudes ($0.5 < M_w < 3$) while the 15 Hz geophones record smaller events ($-1.5 < M_w < 1.5$). Accelerometers record very small events ($-1.5 < M_w < -3.0$). Thus, to have a decent coverage of different seismic event magnitudes, it is necessary to have a combination of geophones and accelerometers. Each of the geophones or accelerometers can be uniaxial or triaxial. Triaxial sensors consist of three orthogonally mounted uniaxial sensors. The primary function of uniaxial and triaxial sensors is different as uniaxial sensors provide accurate event source location, whereas the triaxial sensors determine the seismic source parameters. Uniaxial sensors have lower costs but to have a full record of seismic wave energy (leading to accuracy in source parameter and mechanism), a more accurate S-wave recognition, and an ideal seismic array, it is necessary to have a good balance between triaxial and uniaxial sensors. As a rule of thumb, for every three uniaxials, one triaxial is needed.

The different sensors present the followings applications and advantages (Queen's University, 2014):

- Geophone:
  - Uniaxial:
    - suitable for lower frequencies
    - accurate detection of larger seismic events in soft rocks.
  - Triaxial:
    - three-dimensional data recording (full energy)
    - direction vector of an event
    - seismic events that might be clipped on higher frequency can be detected.
- Accelerometer:
  - Uniaxial:
    - economical
    - easy to install
    - lower noise
    - optimal for locating seismic events in hard rocks.

- o Triaxial:
  - three-dimensionally data recording (full energy)
  - direction vector of an event
  - ideal for locating an event
  - event magnitude and many other source parameters can be calculated.

MONITORING MINING-INDUCED SEISMICITY

Since 2003, the LaRonde Mine has used seismic monitoring. The current monitoring system comprises 121 accelerometers with a frequency of 50 Hz (25 triaxial and 96 uniaxial sensors), as well as 39 triaxial geophones with a frequency of 15 Hz.

When large seismic events (greater than 2 M_w) occur at LaRonde, they tend to saturate the underground seismic system. In these situations, the location of the large seismic event can be accurately determined, but the magnitude is misrepresented. To compensate for the low dynamic range of the underground seismic system, 13 geophones of 4.5 Hz are installed at the surface in communities close to the mine site as part of a regional network (Figure 2). These geophones are located far from the source of seismic events and provide a more reliable magnitude, expressed as a Local Richter magnitude (M_R).

With these instruments, PGA and PPV of the seismic events induced by the underground mining activities are recorded and the seismic event can be accurately located. The monitoring system is capable to quickly (30 sec) and accurately (± 5 m) locate the seismic event (LaRonde Engineering Department, 2019).

For all events of M_R higher than 2.5, the system sends a warning by text message to LaRonde personnel and the surrounding communities.

Figure 1 presents the location and arrangement of the underground and surface sensors while Figure 2 presents the plan view of all the seismic events of $M_R \geq 2$ since 2019. These seismic events occurred at 1.0 to 1.8 km from the triaxial sensor installed at the Main TSF.

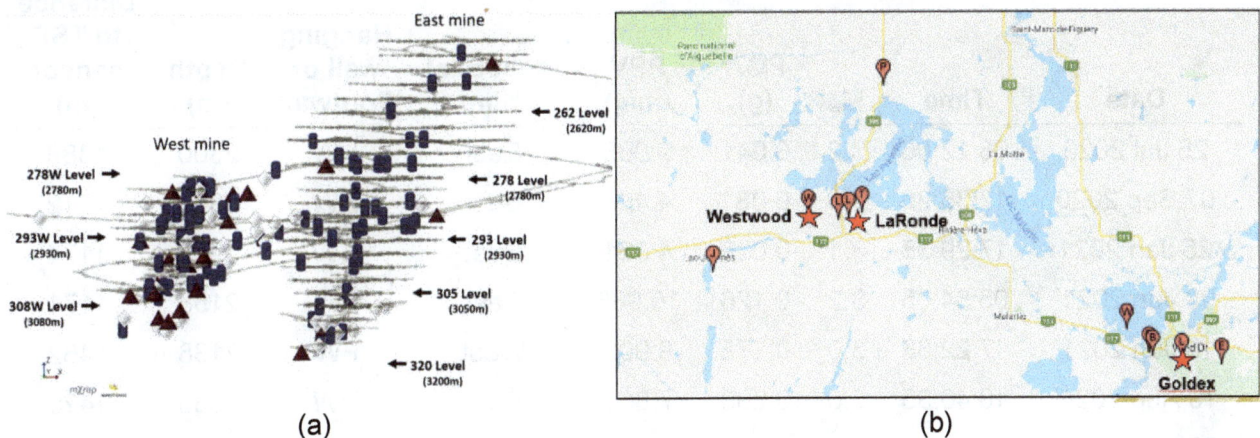

(a)   (b)

FIG 1 – Location and arrangement of sensors: (a) underground and (b) 4.5 Hz surface geophone sensors (Sasseville, Turcotte and Falmagne, 2022).

FIG 2 – Plan view of the location of all seismic events with a magnitude of $M_R \geq 2$ since 2019 (LaRonde Environment and Tailings Management, 2021; LaRonde Engineering Department, 2019).

Table 1 provides the details for the mining-induced seismic events with a magnitude $M_R \geq 3$ since July 2022. The results indicate that there were seven of 13 seismic events of $M_R \geq 3.0$ that exceeded the PGA design value of 0.094 g (corresponding to a region seismic event with a recurrence interval of 2475 years and an of M_w 6 to 6.5) indicating that not all high magnitude seismic events are inducing significant PGA and that the same magnitude events do not necessarily induce the same PGA. The underground mining-induced seismic event with the greatest PGA, 0.422 g, was recorded on 29 November 2022, and had a M_R of 3.5. No trends or patterns related to the magnitude, PGA, location, distance, or depth of the seismic events were identified.

TABLE 1

Major induced seismic events exceeding $M_R \geq 3.0$ since July 2020.

Date	Time	M_R	PGA (g)	PPV (m/s)	Location West/East Mine	Hanging wall or Footwall	Depth (m)	Distance to TSF sensor (m)
25 Jul 2020	08:12:05	3.2	0.041	5.005	East	HW	2300	1389
07 Sep 2020	17:36:20	3.0	0.081	4.463	East	HW	1975	1212
25 Jan 2021	17:29:35	3.1	0.029	4.321	East	HW	1978	1174
08 Jun 2021*	05:54:05	3.2	0.126	10.966	East	FW	2158	1171
18 Jun 2021*	17:22:38	3.0	0.175	9.905	West	FW	2138	1462
10 Aug 2021*	18:46:35	3.0	0.230	7.861	East	HW	2085	1429
29 Nov 2021*	17:45:46	3.0	0.159	7.725	West	HW	2032	1770
28 Dec 2021*	18:22:41	3.8	0.127	17.350	East	FW	2098	951
03 May 2022*	19:03:05	3.0	0.148	5.819	West	HW	2222	1511
02 Sep 2022	19:38:13	3.0	0.066	4.331	East	HW	2088	1429
29 Nov 2022*	00:30:57	3.5	0.422	17.011	West	FW	2268	1459
08 Dec 2022	19:30:57	3.2	0.062	7.249	East	HW	2129	1323
16 Jan 2023	17:27:05	3.0	0.026	3.486	West	HW	2042	1692

* Exceeding the PGA design value of 0.094 g.

TSF DESCRIPTION AND CHARACTERISATION

LaRonde's Main TSF was initially constructed in 1988 and was expanded to the east in 1997. Slurry-deposited tailings are primarily contained within the central portion of the facility (Main TSF), confined by Dikes 1 and 7. The Main TSF is divided into two parts (original TSF to the west and the East Extension) with a central internal embankment (Dike 2). The initial perimeter dykes utilised zoned earth fill and were built to elevation 337 m (a maximum height of 10 m). The Main TSF was incrementally raised in the upstream direction between 2000 and 2019 at an average rate of 1.0 m/a. The perimeter embankments were subsequently raised using the central method between 2000 and 2002 to reach elevation 343 m and then by upstream raise method to its final elevation 358 m in 2019. A typical cross-section of the Main TSF is presented in Figure 3 (LaRonde Environment and Tailings Management Department, 2021).

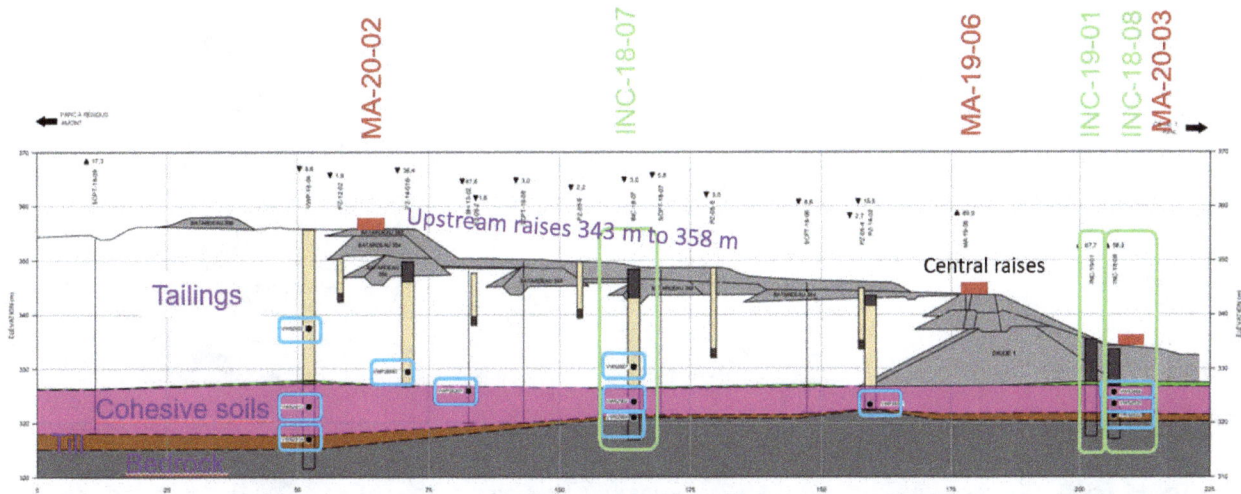

FIG 3 – Typical cross-section of Dike 1 on the Main TSF.

Stratigraphy

The Main TSF is underlain by thin layers of topsoil and peat and then by a layer of silty clay/clayey silt up to 8 m-thick, dense till (sandy silt to silty sand) to 7 m in thickness and, ultimately, bedrock composed of schist and andesite. Bedrock lays beneath the TSF at depths ranging from near surface to the north of the basin to greater than 15 m depth near the south-west corner of the TSF.

Characterisation of the tailings

An extensive geotechnical investigation was carried out that included boreholes, CPTu/SCPTu, field vane tests, and an advanced laboratory testing program to evaluate the undrained shear resistance (S_u) for peak and post-liquefaction conditions and to confirm the contractive or dilative state of tailings using the critical state soils mechanics approach.

The results of the investigation indicate that the tailings are contractive with characteristic state parameter $\psi_k > -0.05$ (Figure 4). However, based on CPTu/SCPTu tests and field vane testing the tailings were found to be partially drained (Figure 5). The hydraulic conductivity ranges from 1×10^{-7} to 8×10^{-7} m/s and then tailings are non-plastic.

The Chandler (1988) method was applied to the tailings field vane test (FVT) results to verify the drainage condition during the tests. The time factor (T) was estimated with the horizontal consolidation coefficient (C_{vh}) based on porewater pressure dissipation (PPD) tests. The results of the analysis presented on Figure 5 suggest that the shear occurred during the FVT was in drained or partially drained conditions.

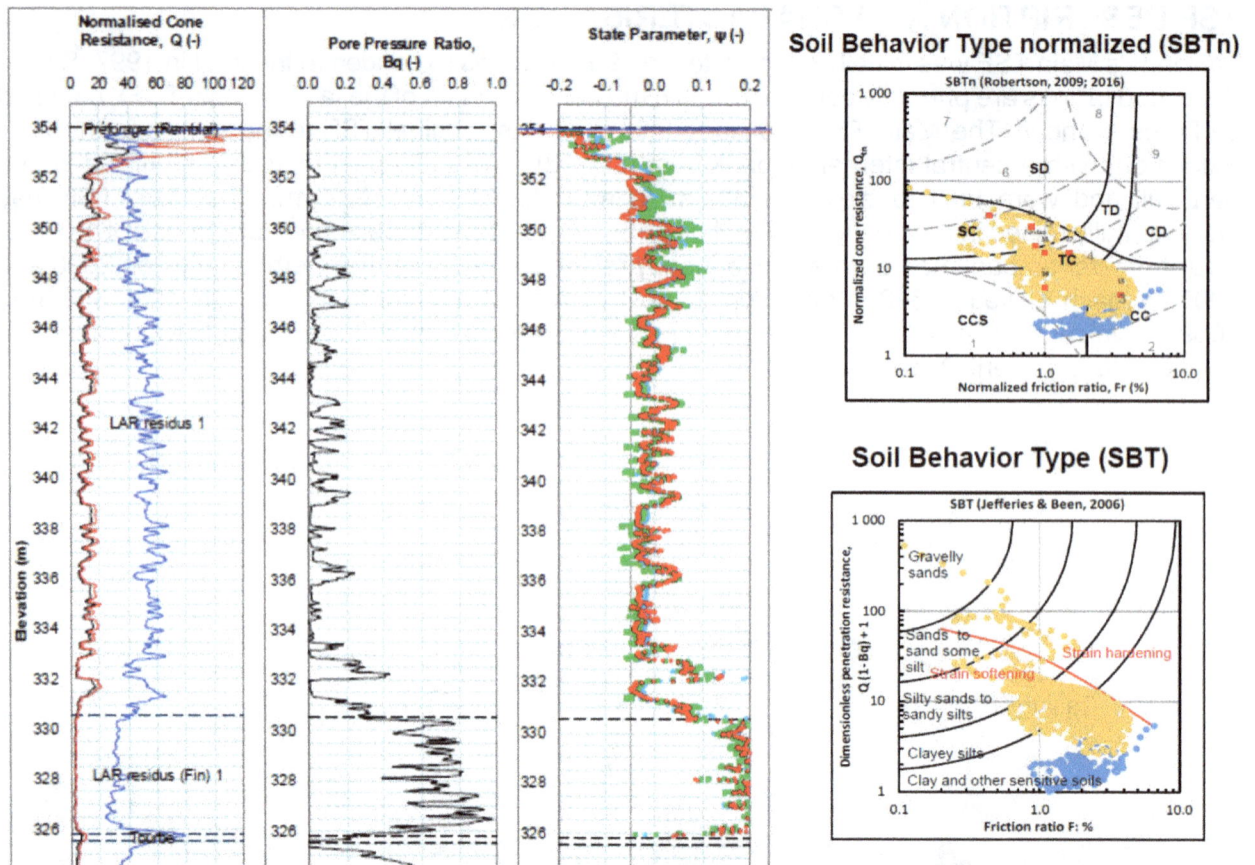

FIG 4 – Contractive behaviour of the tailings in the Main TSF (Golder Associates Ltd, 2021b).

Characterisation of the cohesive soils in the TSF foundation

During the geotechnical investigation, high quality samples of cohesive material were collected for laboratory testing using Shelby tubes and a large-diameter Laval sampler. Standard and advanced laboratory tests (particle size distribution, Atterberg Limits, liquidity Index (LI), CIU, Ck_0U, DSS, CDSS) were carried out on the cohesive soils of the foundation. The results indicate that these soils are prone to strain-softening, strain-softening after 4 to 8 per cent shear strain (Figure 6), which is common for cohesive material with an LI higher than 1.0 and ranging between 0.05 to 1.5.

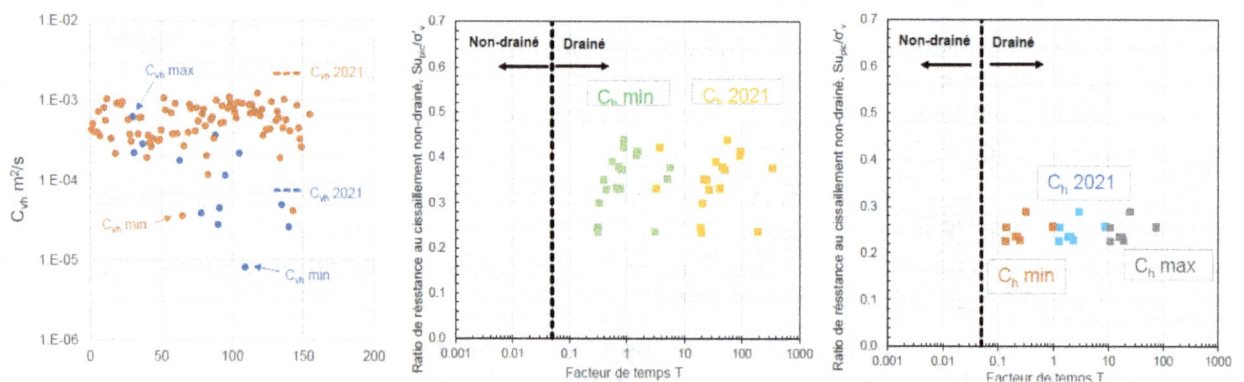

FIG 5 – Drained or partially drained behaviour of the tailings during field vane tests (Golder Associates Ltd, 2021b).

In addition, cyclic direct simple shear tests (CDSS) were carried out to assess potential for cyclic softening of silts and clays using the method proposed by Boulanger and Idriss (2007). The results presented on Figure 7 indicate that the cohesive material is prone to cyclic softening with respect to the 2475-year design earthquake.

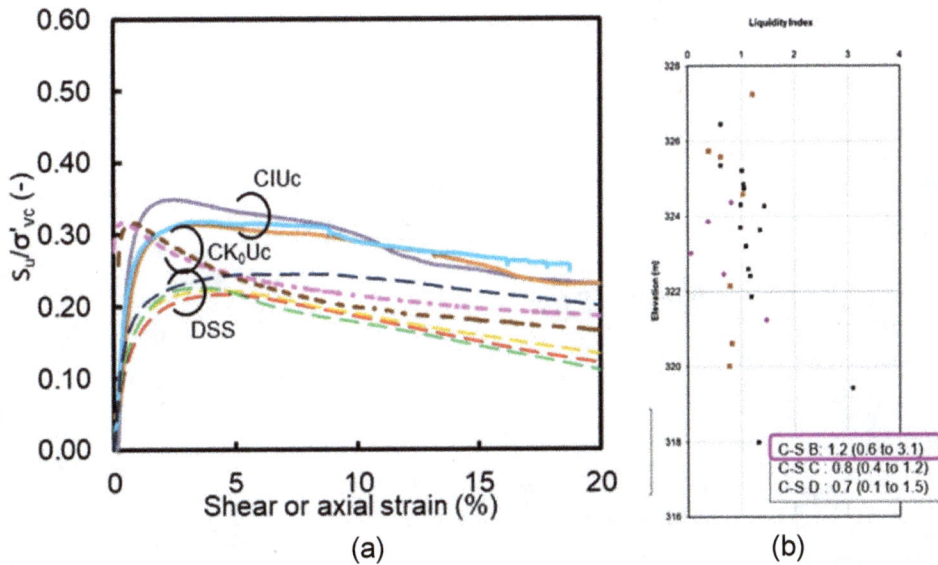

FIG 6 – Assessment of the potential of strain weakening based on laboratory test results: (a) CIU, Ck₀U, DSS test results and (b) Liquidity Index profile (Golder Associates Ltd, 2021a).

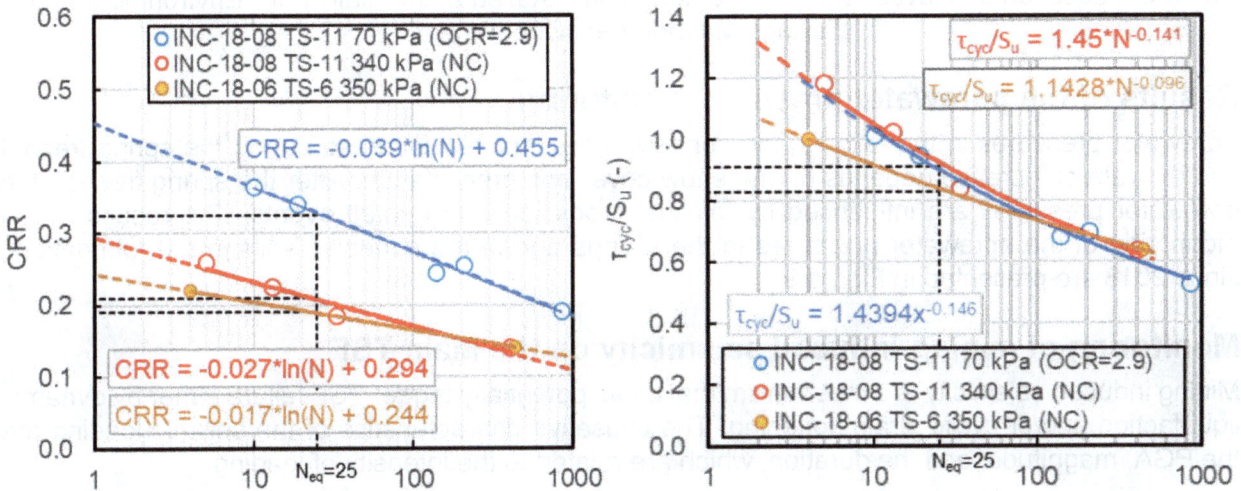

FIG 7 – Assessment of the potential of cyclic softening using CDSS Test Results (Golder Associates Ltd, 2021a).

INSTRUMENTATION AND MONITORING

A comprehensive instrumentation monitoring program connected to Wonderware via LTE (Wonderware database is operational since June 2019) with appropriate coverage was implemented on-site. This program includes Trigger Action Response Plan (TARP) with levels and alarms and is monitored by qualified professionals to detect significant deviations that could substantially affect the performance of the TSF.

The Main TSF instrumentation monitoring program includes:

- 39 monitoring wells to monitor water levels in the tailings.

- 103 vibrating wire piezometers (VWP) to monitor porewater pressure in the tailings and the clayey foundation.

- 30 survey monuments to monitor surface displacements at the crest on the starter dams.

- 18 inclinometers to monitor displacements at depth below the surface in the tailings and the clayey foundation.

Figure 8 presents the location and layout of all the instruments installed at the Main TSF.

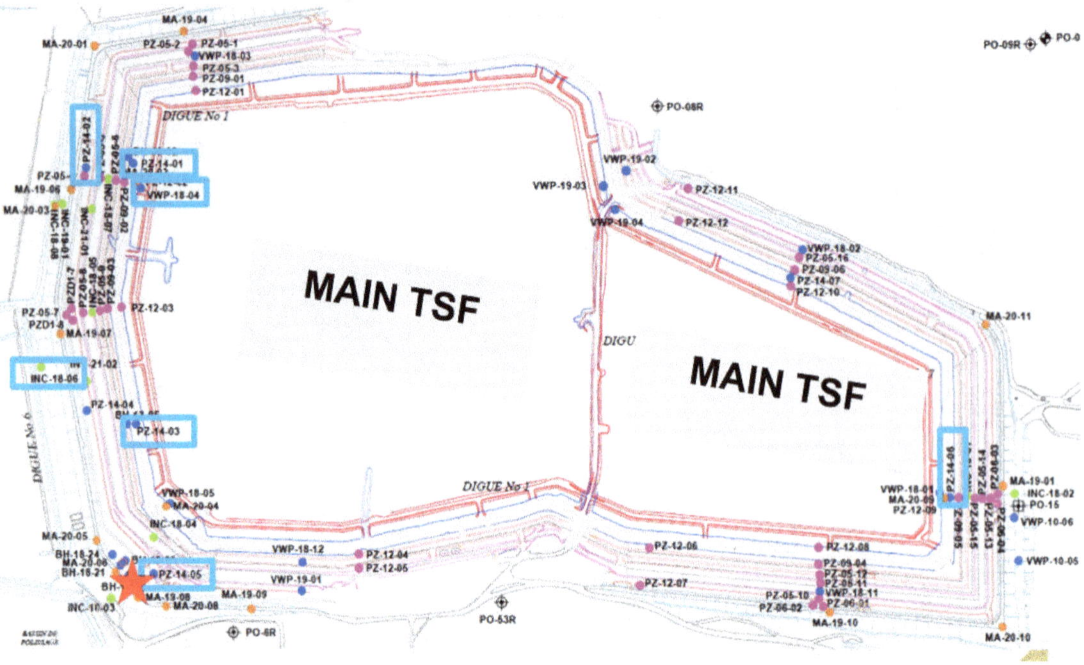

FIG 8 – Location and layout of the instrumentation installed at the Main TSF (Environment and Tailings Management Department).

Results of the porewater pressure monitoring

Porewater pressures in the tailings and cohesive soils typically increase during the spring freshet and the rate of increase depends on the snow cover and precipitation. After the spring freshet, the porewater pressures are influenced by tailings deposition and rainfall events. The results of the monitoring of the porewater pressures in the tailings and in the cohesive soils in the foundations since 2018 are presented in Figure 9.

Monitoring of mining-induced seismicity on the Main TSF

Mining induced seismicity is a mechanism that could potentially trigger TSF failure either by dynamic liquefaction and/or cyclic strain softening. The causative characteristics of the seismic loading are the PGA, magnitude, and the duration, which are related to the intensity of loading.

Therefore, in July 2020, a series of instruments were installed and/or dedicated to the monitoring of the impacts of mining induced seismic events on the Main TSF to collect the magnitude, PGA, and duration of the induced ground motions.

The monitoring system installed at the Main TSF for the mining induced seismicity consists of:

- Seven VWP, which are set to record excess porewater pressures at 0.1 second intervals. In case of a seismic event with a magnitude of $M_R \geq 2.0$, a high frequency signal acquisition is triggered and these VWP generate a report from one minute before until one minute after the event.

- One vibration triaxial sensor (accelerometer) recording the peak ground accelerations (PGA) and the peak particle velocities (PPV). The PGA value is used to check if the design PGA value is not exceeded and the PPV value is used to check if the regulatory requirements are not exceeded.

These reports are automatically generated and sent to all the members of the Environment and Tailings Management (ETM) department.

The location of these instruments is presented on Figure 5. The seven VWP are surrounded by a blue rectangular and the accelerometer is represented by the red star (see also Figure 2 for the location of the accelerometer).

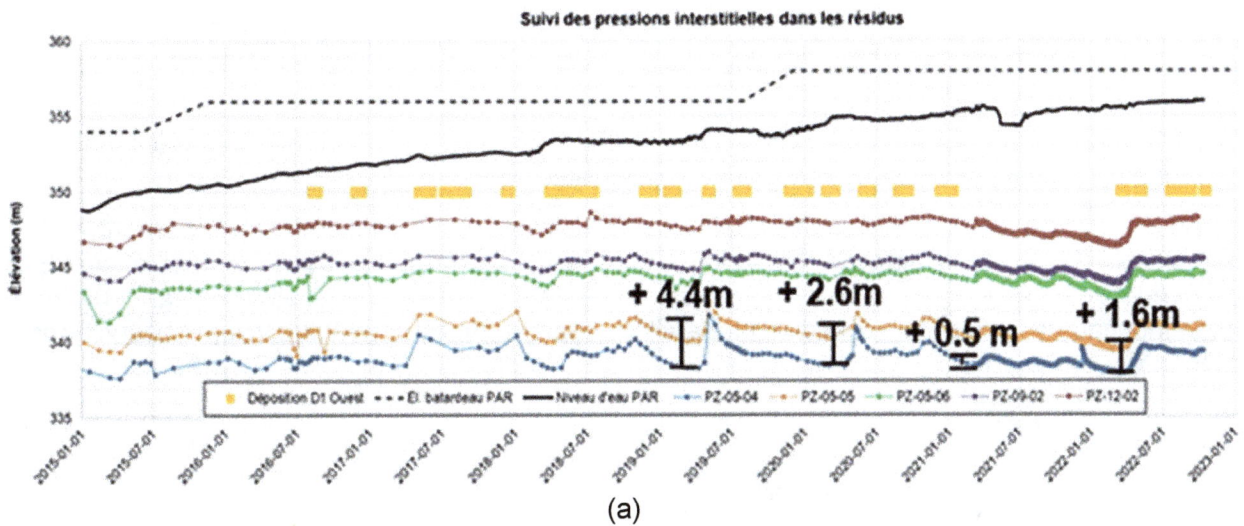

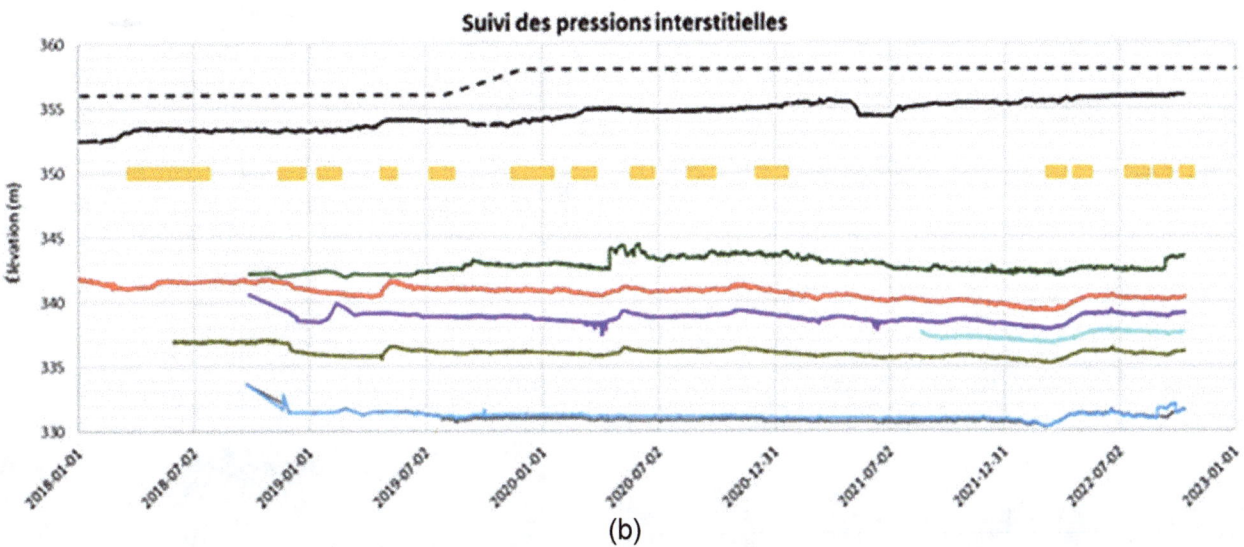

FIG 9 – Results of the porewater pressure monitoring since 2018: (a) in the tailings; (b) in the cohesive soils in the TSF foundation (Environment and Tailings Management Department).

Details of the installation of the vibrating wire piezometers

Out the seven VWP dedicated to the induced seismicity monitoring, three are installed in the tailings (PZ-14–01 VW28573, PZ-14–03 VW28568, PZ-14–06 VW28572), three are installed in the cohesive foundation soils (PZ-14–02 VW28571, PZ-14–05 VW28570, WP-18–04 VW52913), and one in the lower crust of the cohesive foundation soils (Inc-18–06 VW52928). This lower VWP has been excluded from the current analysis because it doesn't reflect the behaviour of the cohesive soils. Details about the VWP installation (depth and stratigraphic layer) are provided on Figure 3.

Automated generated reports

Following each seismic event, two types of automated reports are generated and sent to the TSF management team:

1. One minute after the seismic event an initial automated report is generated and issued, which includes:

   o the date, the time and the duration of the event

   o the acceleration time history

   o the PGA and PPV values recorded by the triaxial sensor

   o for each vibrating wire piezometer, a graph showing the excess porewater pressures for a two minute interval centred on the event.

In the graph, there is also an embedded a table with the mean, maximum and minimum values of the porewater pressure during the two minute interval. A typical automated report is presented in Figure 10.

2. One hour after the event, a graph showing the evolution of the excess porewater pressures during the last 48 hours is automatically generated.

In addition to the automated generated report, following each induced seismic event with a magnitude of $M_R \geq 3$, a detailed visual inspection is also carried out by a member of the ETM department to identify if there are any cracks, settlements, and/or displacements that could have been induced by the seismic event and a detailed report summarising all the visual observations is prepared.

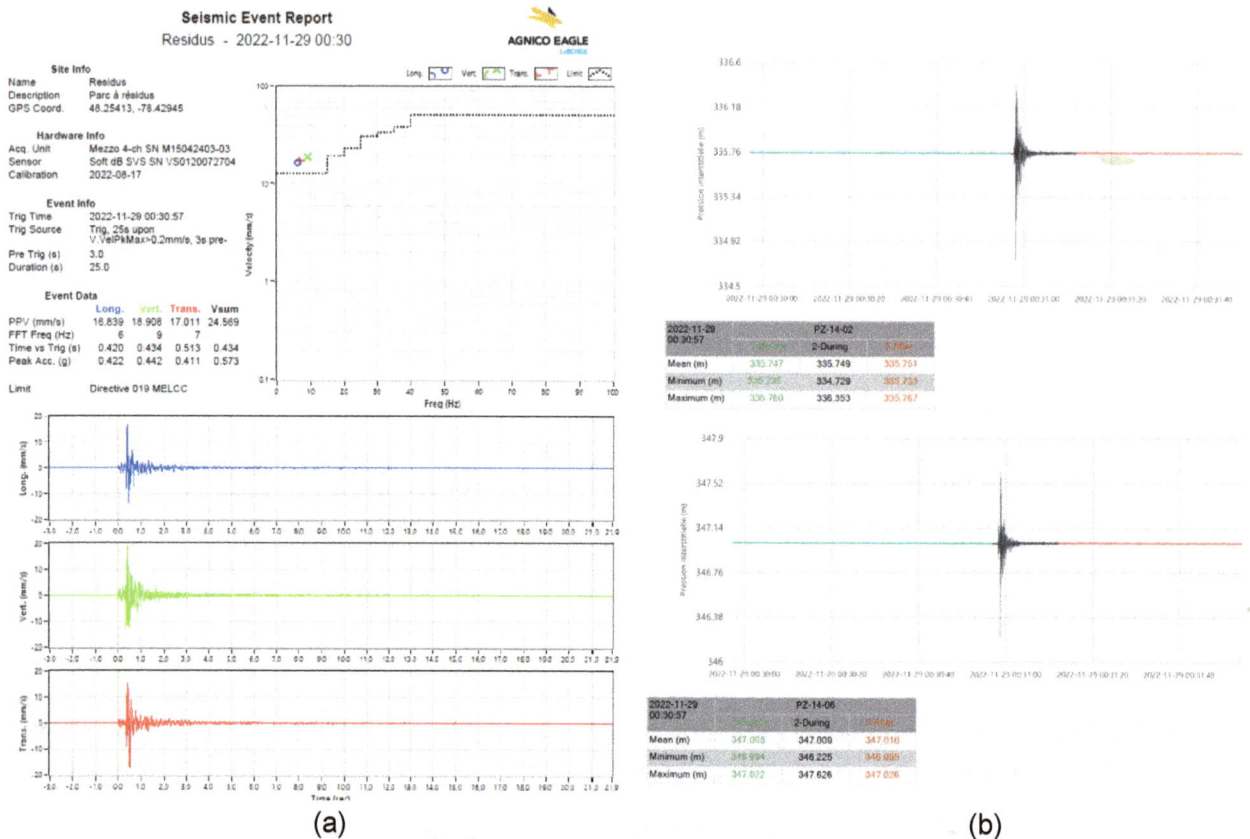

(a)   (b)

FIG 10 – Typical automated reports generated after each significant event: (a) PGA and PPV; (b) excess porewater pressures one minute before, during and one minute after the seismic event.

Analysis of PGA values induced by seismic events with a magnitude of $2.0 \leq M_R < 3.0$ in 2021 and 2022

A review of all the seismic events with a magnitude of $2 \leq M_R < 3$ that occurred in 2021 and 2022 were carried out to analyse if lower magnitude seismic events with a magnitude $M_R < 3.0$ could have induced PGA values that exceed the PGA design value of 0.094 g.

In 2021, there were six seismic events with a magnitude of $2 < M_R < 3$ that exceeded the PGA design value, while in 2022, there are seven such events. These events are summarised in Table 2.

Overall, in 2021 and 2022, a total of 20 mining induced seismic events with a magnitude $M_R \geq 2$ exceeded the PGA design value, which would have triggered a visual inspection if the threshold for a visual inspection was the PGA design value compared to the 14 visual inspections that were carried out based on the magnitude threshold of $M_R = 3$.

RESULTS AND DISCUSSIONS

The design earthquake for the TSF has a recurrence interval of 2475 years and the corresponding parameters are a PGA of 0.094 g and an M_w of 6 to 6.5. Initially, a Trigger Action Response Plan based on a seismic event with a M_R at least 3 was arbitrary chosen; the protocol specified that a visual inspection by a member of the ETM department after each induced seismic event with a $M_R \geq 3$.

However, as microseismic information was evaluated it was found that the PGA values of the events were relatively independent of M_R. Subsequently, only a PGA of 0.094 g with an M_R of at least 2 was retained. Therefore, it has been recommended to carry out a visual inspection based on the induced PGA value every time that it is exceeding the PGA design value.

The difference between the design and TARP loadings is essentially the number of equivalent cycles associated with the magnitudes, so the TARP event is lower than the design event and provides for an alert before design values associated with the capacity of the TSF are attained.

It was also noted that the other instrumentation in the TSF, such as inclinometers and survey monitoring points, were not affected by the mining-inducted seismic activity. In addition, visual inspections carried out after each event with an $M_R \geq 3$ didn't indicate any abnormal findings.

TABLE 2

Seismic events with a magnitude of $2 \leq M_R < 3$ that exceeded PGA design value of 0.094 g in 2021 and 2022.

Date	Time	M_R	PGA (g)	PPV (m/s)	West/East Mine	Hanging wall or Footwall	Depth (m)	Distance to TSF sensor (m)
25 Apr 2021	13:12:43	2.6	0.130	4.824	West	HW	2164	1601
09 Jun 2021	06:30:28	2.3	0.136	4.381	East	FW	2157	1052
18 Jun 2021	17:35:44	2.4	0.202	5.868	West	FW	2130	1401
15 Aug 2021	23:57:57	2.7	0.191	6.783	East	HW	2217	1146
27 Sep 2021	17:25:51	2.9	0.109	6.330	East	HW	1887	1360
14 Oct 2021	17:27:11	2.2	0.153	4.942	West	FW	2250	1463
05 Apr 2022	11:04:01	2.4	0.120	4.938	East	FW	2209	1045
11 Apr 2022	18:42:07	2.8	0.190	6.668	East	FW	2183	987
03 May 2022	17:26:59	2.6	0.097	4.685	West	HW	2224	1568
28 May 2022	20:26:33	2.1	0.098	3.579	East	FW	1909	1115
26 Jul 2022	00:31:54	2.7	0.209	8.242	West	HW	2125	1771
02 Sep 2022	17:25:19	2.5	0.101	3.275	East	FW	1999	1328
29 Nov 2022	00:12:14	2.6	0.208	6.468	West	FW	2144	1432

Following the review and analysis of all the 2021 and 2022 seismic events with a magnitude of $2 \leq M_R < 3$, 13 mining induced seismic events that exceeded the PGA design value were identified and no visual inspection was carried out after these seismic events.

Of the 20 mining induced seismic events with a magnitude $M_R \geq 2$ that exceeded the PGA design value, ten were in the West mine and ten in the East mine, eight were in the hanging wall and 12 were in the footwall.

Finally, the results presented in Tables 1 and 2 indicate that the seismic events with the highest magnitude did not necessarily inducing the highest PGA. The following six examples indicate that the three first seismic events with a magnitude $M_R \geq 3$ induced lower PGA values compared to the last three seismic events with a magnitude $M_R < 3$ that produced higher PGA values:

- The event of 25 July 2020 with a magnitude M_R of 3.2 induced a PGA of 0.041 g.

- The event of 28 December 2021 with a magnitude M_R of 3.8 induced a PGA of 0.128 g.

- The event of 16 January 2023 with a magnitude M_R of 3.0 induced a PGA of 0.026 g.

- The event of 18 June 2021 with a magnitude M_R of 2.4 induced a PGA of 0.202 g.

- The event of 26 July 2022 with a magnitude M_R of 2.7 induced a PGA of 0.209 g.

- The event of 29 November 2022 with a magnitude M_R of 2.6 induced a PGA of 0.208 g.

The observed ground motions at the TSF are much higher than expected for seismic events smaller than M < 4. Ground motions at these levels could produce structural damage which has not been observed on-site. Atkinson (2020) estimated that induced earthquakes must be at least M 4 in size to have damage potential within 5 km of the hypocentre.

Induced excess porewater pressures in the cohesive soils and in the tailings

The analysis of the results of the generated excess porewater pressures indicates that the seismic events with the highest magnitude induced higher excess porewater pressures, even though they didn't induce the highest PGA values.

Following the review of excess porewater pressure data induced by the seismic events presented in Table 1 and 2, it was found that the following three seismic events generated the largest excess porewater pressures:

- The event of 29 November 2022 with a magnitude M_R of 3.5 and a PGA of 0.422 g generated excess porewater pressures of +60.4/-102.0 cm in the clayey soils in the foundation at PZ-14–02 and +61.7/-78.4 cm in the tailings at PZ-14–06.

- The event of 28 December 2021 with a magnitude M_R of 3.8 and a PGA of 0.127 g generated excess porewater pressures of +73.9/-82.5 cm in the clayey soils in the foundation at PZ-14–02 and +61.7/-78.4 cm in the tailings at PZ-14–06.

- The event of 8 June 2021 with a magnitude M_R of 3.2 and a PGA of 0.126 g generated excess porewater pressures of +29.0/-42.0 cm in the clayey soils in the foundation at PZ-14–02 and +49.0/-58.6 cm in the tailings at PZ-14–06.

Normally, during significant seismic loading characterised by such PGA values, the generation of significant excess porewater pressures in contractive tailings and cohesive soil is expected and their dissipation after shaking is relatively slow. However, the results presented on Figure 11 for the 29 November 2022 seismic event indicate that the excess porewater pressures in the tailings and the cohesive soils in the foundations were quickly and completely dissipated after each seismic event. The same behaviour was also observed for the 28 December 2021 and 8 June 2021 seismic events. This behaviour can be attributed to the relatively low magnitudes of the events, which were not sufficient to induce liquefaction. As shown by Seed and Idriss (1971), when the seismic loading is significantly below the level required to induce liquefaction (excess pore water pressure << vertical effective stress), excess pore water pressures tend to dissipate during or shortly after shaking.

Figure 11 also presents the acceleration time history of the seismic event. On the acceleration time history, we can observe that there are two types of general waveforms produced from a seismic event; P-waves and S-waves. P-waves are compressional waves, and travel at the greatest velocity, making them the first-arrival waveforms. The S-Waves are shear waves, travel at a velocity slower than the P-waves making them the second arrival waveforms. When comparing the waveforms of the acceleration time histories and to the excess porewater pressure development, it appears that the P-waves, that are the first to reach the VWPs, do not induce excess porewater pressures; it is only when the S-waves hit the VWPs that excess porewater pressures develop.

In general, the duration of seismic events varies between three and five seconds and the S-waveforms of the accelerograms are quite similar to the excess porewater pressure waveforms, this has been observed for all the seismic events.

In addition, the results indicate that for each seismic event, the mean values of the excess porewater pressures one minute before, during, and one minute after the seismic event are quite similar, which could be an indication that the tailings and the cohesive soils in the foundations are behaving as a drained or at least a partially drained material under the significant seismic loadings induced by the seismic event.

This observed behaviour is somehow unexpected or counter-intuitive when dealing with contractive tailings and cohesive soils prone to cyclic softening.

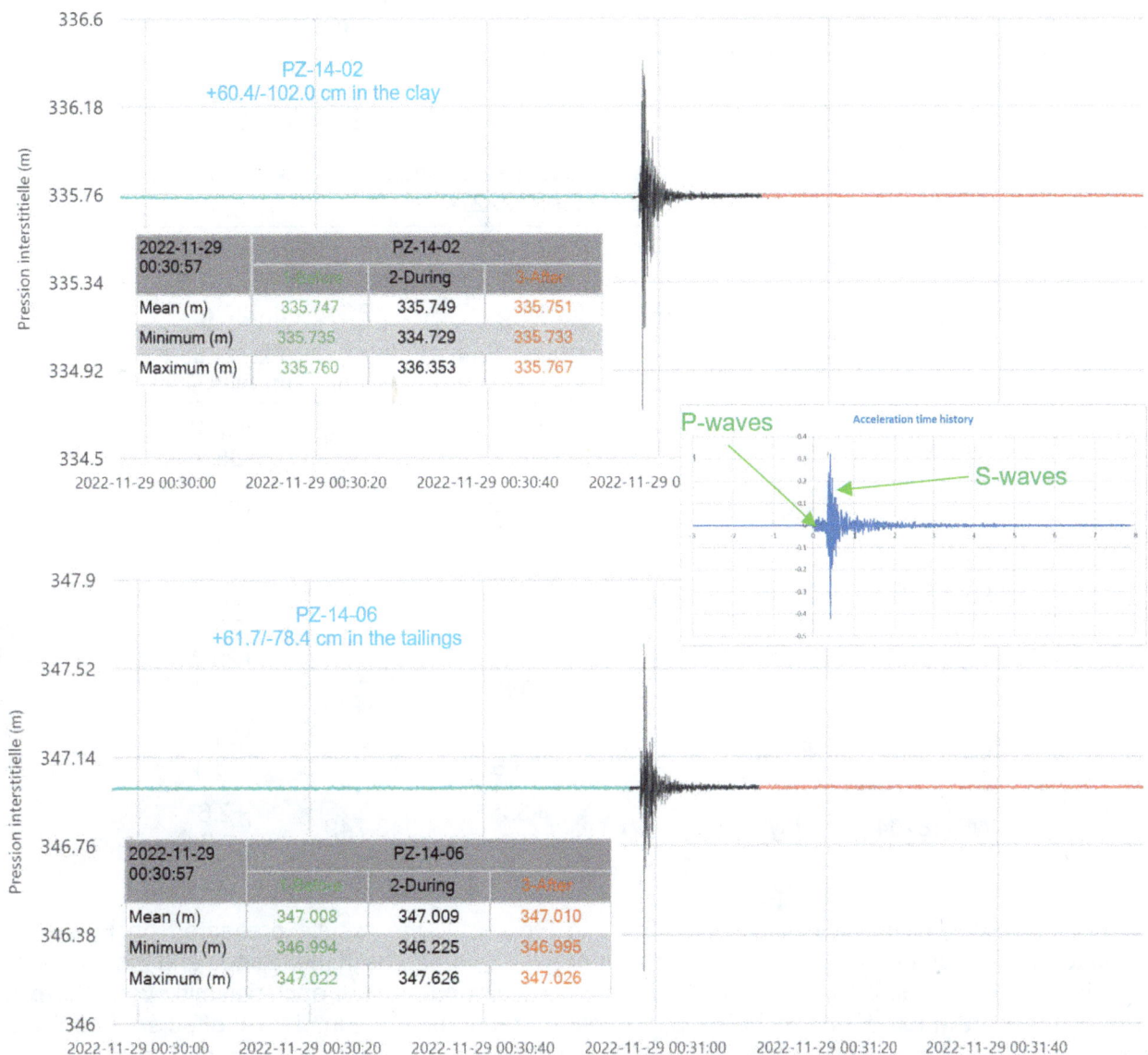

FIG 11 – Excess porewater pressure generated by 29 November 2022 Seismic Event M_R = 3.5 and PGA = 0.422 g.

Cumulative effects of multiple consecutive seismic events

Factors that could affect the dynamic tailings liquefaction and the cyclic strain-weakening of the cohesive soils in the TSF foundations are magnitude and duration of the ground shaking caused by an underground mining-induced seismic event. To analyse the impact of magnitude and duration of the seismic events on the potential dynamic tailings liquefaction and cyclic strain-softening of the cohesive soils in the foundations, we have reviewed the excess porewater pressures induced by two significant consecutive seismic events that occurred on 29 November 2022 within 18 minutes; the first one occurred at 00:12:14 with a magnitude M_R = 2.6 and a PGA = 0.208 g while the second one occurred at 00:30:57 with a magnitude M_R = 3.5 and a PGA = 0.422 g. Both seismic events induced a PGA exceeding the design PGA value of 0.094 g.

The detailed information of these consecutive events including the induced excess porewater pressures is summarised in Table 4.

The purpose of the review of the excess porewater pressures induced by the two events was to verify if there was any cumulative excess porewater pressure build-up that could result in the dynamic tailings liquefaction and cyclic strain-weakening of the cohesive soils in the TSF foundations.

Normally, for contractive tailings and cohesive soils prone to strain weakening and cyclic softening, following consecutive seismic events that have induced significant peak ground accelerations, it should be expected to observe a significant excess porewater pressure build-up that would take time to dissipate.

TABLE 4

Excess porewater pressure following two consecutive seismic events within 18 minutes on 29 November 2022.

Piezometer	Layer	Event 29 Nov 2022, 00:12:14, M_R = 2.6 PGA = 0.208 g		Event 29 Nov 2022, 00:30:57, M_R = 3.5 PGA = 0.422 g	
		Mean value (m)	Excess pore pressure (cm)	Mean value (m)	Excess pore pressure (cm)
PZ-14–01	Tailings	346.710	+21.7 -21.1	346.698	+32.0 -35.1
PZ-14–02	Clay	335.747	+31.8 -18.6	335.749	+60.4 -102.0
PZ-14–03	Tailings	346.340	+10.3 -8.2	346.344	+31.4 -27.7
PZ-14–05	Clayey Silt	330.239	+10.1 -9.8	330.228	+26.2 -23.7
PZ-14–06	Tailings	347.009	+21.2 -33.9	347.009	+61.7 -78.4
WP-18–04	Clay	343.750	+5.2 -4.2	343.749	+2.6 -7.7

As already observed on the results presented in Figure 11, the results presented on Figure 12 indicate that for each seismic event, the excess porewater pressures in the tailings and the cohesive soils in the TSF foundations are quickly and completely dissipated after each seismic event; there is no excess porewater pressure build-up; therefore, there is no cumulative effects of multiple consecutive seismic events that could lead to dynamic tailings liquefaction and cyclic strain-weakening of the cohesive soils in the TSF foundations. In addition, the mean values of the excess porewater pressures one minute before, during, and one minute after the second seismic event are the same compared to the mean values of the first seismic event, 18 minutes earlier.

These results confirm that the tailings and the cohesive soils in the TSF foundations are probably behaving as a drained or at least a partially drained material under the seismic loading induced by the seismic event.

Once again, the results indicate that, the excess porewater pressures in the tailings and the cohesive material in the TSF foundations are quickly and completely dissipated after each seismic event; there is no excess porewater pressure build-up; therefore, there is no cumulative effects of multiple consecutive seismic events.

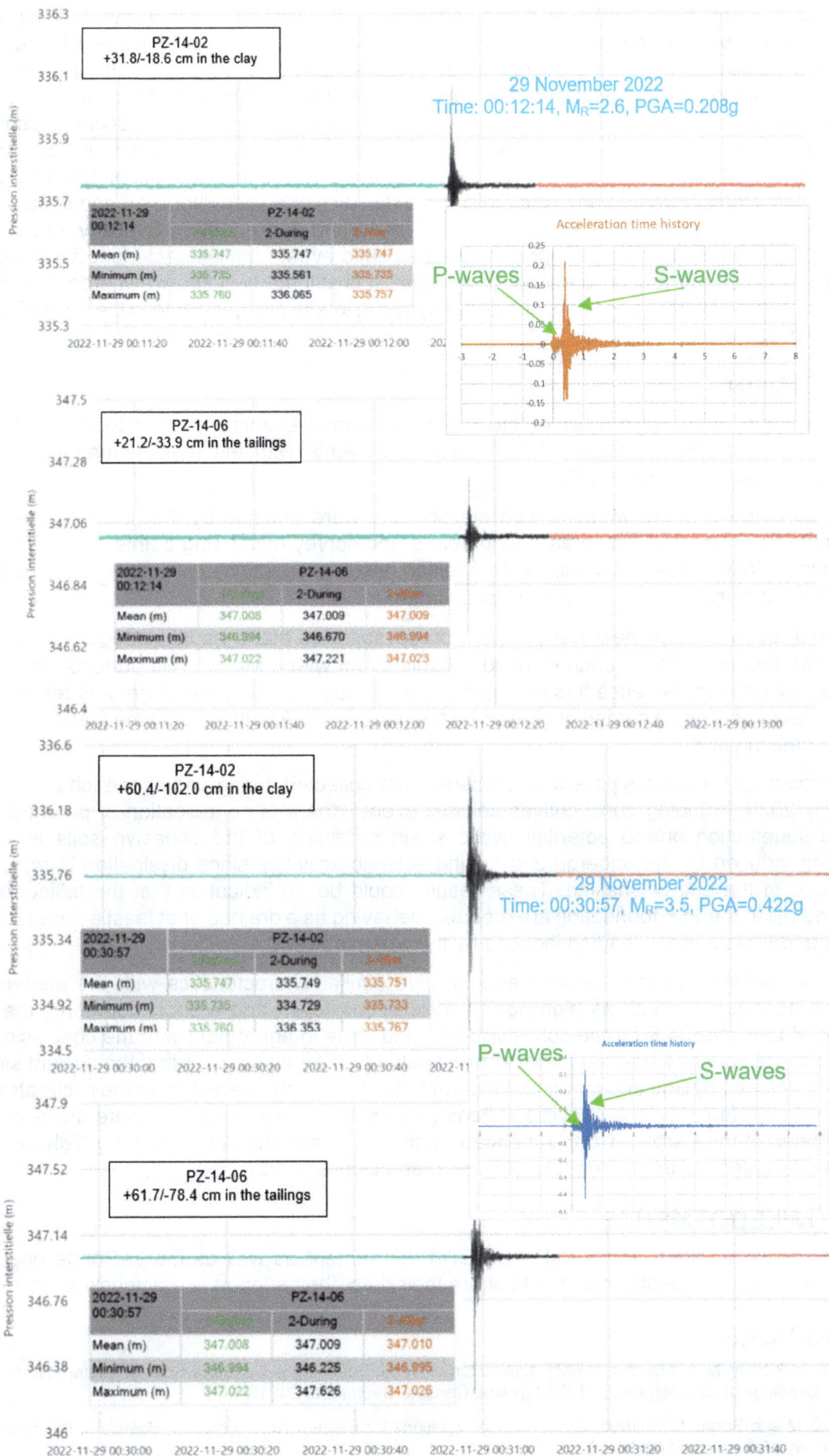

FIG 12 – Excess porewater pressures induced by two significant consecutive induced seismic events.

This observation on the potential of cyclic softening in the cohesive soils in the TSF foundations seems to agree with the results of research carried out at Sherbrooke University by Abdellaziz, Karray and Lashin (2021) on the evaluation of cyclic softening in cohesive soils of eastern North America. The main characteristics of these induced seismic events is that the acceleration time histories are characterised by a high peak acceleration (a_{max}) and short durations. According to Abdellaziz, Karray and Lashin (2021), the current simplified method developed by Boulanger and Idriss (2007) could overestimate the cyclic stress ratio (CSR) and then the risk of cyclic softening in eastern North America where one of the main characteristics of eastern North American earthquakes is the fact that they are rich in high frequency (Atkinson and Boore, 1995). Similarly, to the eastern North American earthquakes, LaRonde induced seismic events are characterised by a high peak acceleration (a_{max}) and a short duration. In addition, it is still not well known whether the standard evaluation method of clayey soil deposits under dynamic loading is applicable in the case of sensitive eastern Canada clays.

CONCLUSION

When comparing the magnitudes of the different seismic events to the induced peak ground accelerations, the results indicate that the seismic events with the highest magnitude are not necessarily inducing the highest PGA.

It was also noted only the vibrating wire piezometers were affected by the seismic events, other instrumentation in the TSF, such as inclinometers and survey monitoring points, were not affected by the mining-inducted seismic activity. In addition, visual inspections carried out after each event with an $M_R \geq 3$ didn't indicate any abnormal findings.

On an operational point of view, the current protocol specifies that a visual inspection by a member of the ETM department after each induced seismic event with a $M_R \geq 3$, this protocol was found to be somehow arbitrary because it is not linked to any design criteria; therefore, it is recommended that a visual inspection be carried out based on the induced PGA value every time that it is exceeding the PGA design value.

After review of all the excess porewater pressure data collected during and after each seismic event since July 2020, including consecutives seismic events, there is no indication of potential tailings dynamic liquefaction or/and potential cyclic strain softening of the cohesive soils in the TSF foundation induced by the underground mining seismic activities since dissipation is rapid, within one minute of the end of shaking. These results could be an indication that the tailings and the cohesive soils in the TSF foundation are probably behaving as a drained or at least a partially drained material under the seismic loading induced by the seismic event.

Given that LaRonde induced seismic events have similar characteristics with the eastern North American earthquakes such as high peak accelerations (a_{max}) and a short durations, the results collected at LaRonde TSF for the cohesive soils tend to be in agreement with the conclusion of the research carried out by Abdellaziz, Karray and Lashin (2021) pointing out that the current simplified method developed by Boulanger and Idriss (2007) could probably overestimate the cyclic stress ratio (CSR) and therefore, the risk of cyclic softening in Eastern North America where the acceleration time histories of the Eastern North American earthquakes are characterised by a high peak (a_{max}) and a short duration of the earthquake (Atkinson and Boore, 1995).

ACKNOWLEDGEMENTS

The authors would like to thank LaRonde ETM department as well as the LaRonde engineering department who generously accepted to share their data, their support is graciously acknowledged.

REFERENCES

Abdellaziz, M, Karray, M and Lashin, I, 2021. Evaluation of Cyclic Softening in Cohesive Soils of Eastern North America, Proceedings of GeoNiagara 2021 Conference, Canada, September 2021.

Atkinson, G M and Boore, D M, 1995. Ground-motion relations for Eastern North America, *Bulletin of the Seismological Society of America*, 85(1):17–30.

Atkinson, G M, 2020. The Intensity of Ground Motions from Induced earthquakes with implications for damage potential, *Bulletin of the Seismological Society of America*, 110(5):2366–2379.

Boulanger, R W and Idriss, I M, 2007. Evaluation of cyclic softening in silts and clays, *Journal of Geotechnical and Geoenvironmental Engineering*, 133(6):641–652.

Chandler, R J, 1988. The in-situ measurement of the undrained shear strength of clays using the field vane, *Vane Shear Strength Testing in Soils: Field and Laboratory Studies*, pp 13–44.

Golder Associates Ltd, 2021a. Cohesive soils characterization at the Main TSF Presentation during the Independent Review Board in October 2021.

Golder Associates Ltd, 2021b. Tailings characterization at the Main TSF Presentation during the Independent Review Board in October 2021.

LaRonde Engineering Department, 2019. Sismicité à la mine LaRonde [Seismicity at the LaRonde mine]. Presentation during the Independent Review Board meeting in October 2019.

LaRonde Environment and Tailings Management Department, 2021. Geotechnical Monitoring and surveillance activities, Presentation during the Independent Review Board in October 2021.

Mercier-Langevin, P, Dubé, B, Blanchet, F, Pitre, D and Laberge, A, 2017. The LaRonde Penna Au-rich volcanogenic massive sulfide deposit, *Reviews in Economic Geology*, 19:225–245.

Queen's University, 2014. Mining induced seismicity, QueensMineDesignWiki [online]. Available from: <https://minewiki.engineering.queensu.ca/mediawiki/index.php/Mining_induced_seismicity>

Sasseville, G, Turcotte, P and Falmagne, V, 2022. Control Measures to Manage Seismic Risk at the LaRonde Mine, a Deep and Seismically Active Operation, Proceedings of the 56th US Rock Mechanics/Geomechanics Symposium, New Mexico.

Seed, H B and Idriss, I M, 1971. Simplified procedure for evaluating soil liquefaction potential, *Journal of Soil Mechanics and Foundations Division, ASCE*, 97:1249–1273.

Historical topographic surveys of Jagersfontein tailings dam produced from archive satellite images

S Rivet[1]

1. Solutions Advisor, PhotoSat, Vancouver BC V5N4E8, Canada. Email: sam.rivet@photosat.ca

ABSTRACT

This paper is an illustrative example of how available archive satellite imagery can be used to produce topographic and reconstructed surfaces and then draw valuable insights from them for the purpose of tailings storage facility (TSF) management.

A search of commercial and government archives yielded hundreds of images covering the Jagersfontein TSF. This satellite imagery dates back to the 1970s and includes both mono and stereo images from Keyhole, Landsat, Sentinel, and WorldView satellites. Where stereo pairs are available, it is possible to directly produce a topographic surface (DEM). When stereo imagery is unavailable, mono images can be used to produce a reconstructed surface of the dyke structure and deposited tailings.

The orthophotos and surfaces (DEM) produced show the geotechnical features of the tailings facility in detail. Spectral analysis of the images makes it possible to determine the distribution of tailings and high-moisture tailings. The surfaces (DEM) for each year can be used to directly measure dam crest height, operational freeboard, environmental containment freeboard, storage capacity, and beach profiles.

Lastly, comparison of successive surfaces provides measurements of cut-and-fill volumes. Historical surfaces are valuable for verifying existing survey data and filling in gaps in records. This type of information can help engineers understand how legacy assets were constructed, operated, and maintained. This type of information may also be valuable for determining total tailings volume contained in the TSF, total tailings volume released in the event of a failure, and measure movement in the dyke structure.

This type of study has potential applications for managing and monitoring legacy assets, particularly those where as-built or operational survey data is unavailable or a change of ownership has occurred.

INTRODUCTION

This case study is an illustrative example of how available satellite imagery from different archives can be processed to produce high quality surveys from different years and derive additional data insights from the surveys. The study was independently produced following the Jagersfontein Dam Collapse on September 11th 2022. No site visit was required to produce this study.

The topographic surveys shown in this study are digital elevation models (DEM). Depending on the source satellite imagery, some DEMs may be equivalent to a digital terrain model (DTM) where ground features such as trees or buildings are removed during processing. Other DEMs may be equivalent to a digital surface model (DSM) as surface features are not removed during processing.

This topographic data produced in this study was used to measure:

- Outflow volume resulting from the September 11th dam failure.
- Volumes placed on or within the TSF on an iterative surface to surface basis.
- Design Inflow retaining capacity from each topographic survey.
- Beach freeboard from each topographic survey.
- Dam crest elevation from each topographic survey.

For material placed within the TSFs footprint prior to March 9th 2011, the outflow from failure was studied.

Through spectral analysis of mono images, water bodies and high moisture tailings were mapped throughout the TSF. Precision orthorectified satellite imagery was utilised to measure downstream dyke movement starting in 2019. Finally, 11 cross-section locations were selected in order to visualise the evolution of the tailings storage facility from 1977 to 2022 through the topographic surveys produced directly and reconstructed.

METHODS

To produce this study, a search of satellite archives was conducted. From the available mono and stereo satellite imagery, a variety of data was produced, including but not limited to topographic surfaces, boundaries of geotechnical features, mapping of water and high moisture tailings, volume changes, and measurements of capacity and freeboard.

Satellite archive search

A search of government and commercial archives yielded dozens of satellite images, taken between the 1960s and present day, which cover the Jagersfontein tailings storage facility (TSF) area.

Stereo imagery

When a satellite captures two or more images of the same area from different angles, the images are called stereo pairs. It is possible to process stereo pairs from high-resolution optical satellites to produce highly-accuracy ground elevation measurements. It is ideal when images that form stereo pairs are captured on the same date, however, given low levels of TSF construction and tailings placement activity in 2011 and in 2022 post-failure, utilising scenes that are 1–2 weeks apart, should yield high-level survey data.

As listed in Table 1, three different sets of stereo pairs were utilised to produce topographic surfaces (DEM/DTM). In addition, Keyhole stereo imagery was processed to produce a 1977 topographic surface (DEM). Additional information on the technical specification of the Worldview-1 and Worldview-2 satellites can be found in the data sheets (DigitalGlobe, 2020a, 2020b).

TABLE 1

Stereo imagery date, satellite, and resolution.

Image 1 date	Image 2 date	Satellite(s)	Resolution (cm)
February 4, 1977	February 4, 1977	Keyhole	60–120
February 26, 2011	March 9, 2011	WV-2 and WV-1	50
October 19, 2017	October 19, 2017	WV-1	50
September 16, 2022	September 24, 2022	WV-2	50

Mono imagery

When images are captured from one angle, they are referred to as mono images. Over 20 mono images from satellites – Keyhole, WorldView, LandSat 5, LandSat 8, and Sentinel-2A – were used to observe activity, measure structure movement, and generate geotechnical boundaries of the Jagersfontein TSF embankment structure, tailings, high-moisture tailings, and water bodies.

Composition mapping

Within the footprint of the topographic surfaces produced, geotechnical boundaries of the TSF's embankment structure, tailings, high-moisture tailings, and water bodies were determined and utilised to segment the TSF into modelled components. The boundaries were created by observing features in the high-resolution WorldView imagery and then cross-referencing them with a multispectral analysis of the LandSat and Sentinel data bands.

Embankments structure

The embankment structure (Dyke/Wall) boundaries were primarily determined through visual observation of the high-resolution WorldView imagery.

The general location of tailings material was determined through observation. However, the tailings material was segregated into tailings and high-moisture tailings. This distinction is significant given the difference in behaviour between tailings material that is relatively dry versus tailings material with an elevated water content.

Normalised soil moisture index

Soil water content is expressed on a gravimetric or volumetric basis. Gravimetric water content is the mass of water per mass of dry soil. It is measured by weighing a soil sample, drying the sample to remove the water, then weighing the dried soil.

The measurement of soil water content using optical satellites is performed by determining the Normalised Soil Moisture Index (NSMI). This is based on the work done by Alonso-Jimenz *et al* (2019) working with Sentinel-2A and initially and more conceptually by Haubrock *et al* (2008). NSMI is a normalised difference of bands 5 and 8 from Landsat-5 (US Department of the Interior, US Geological Survey (USGS), 2013), or bands 11 and 12 from Sentinel-2A, and it is shown to correlate with the moisture content of soil. The areas of high NSMI values observed in either the Landsat-5 or Sentinel-2A band analysis were utilised to establish the footprint of high-moisture tailings material within the TSF.

Normalised difference water index

Similarly, the Normalised Difference Water Index (NDWI) was used to determine the boundaries of water bodies within and south of the TSF. The NDWI is calculated using the GREEN-NIR (visible green and near-infrared) combination, which allows the detection of subtle changes in water content of water bodies. This makes it possible to identify them against surrounding soil and vegetation.

All boundaries determined through the spectral analysis described above were cross-referenced with what could be observed in the higher resolution WorldView imagery to verify the results.

Topographic surfaces from stereo satellite imagery

From dates between 2011 and 2022, three different sets of stereo pairs were utilised to produce topographic surfaces (DEM/DTM). In addition, Keyhole stereo imagery was processed to produce a 1977 topographic surface (DEM). It is ideal when the images used to form a stereo pair are captured on the same date. However, given the minimal construction and tailings placement activity in 2011 and in 2022 post-failure, combining scenes from dates that are 1–2 weeks apart still allows for the production of a high-quality survey.

Reconstructed 2022 pre-failure surface

The most recent high resolution stereo imagery prior to the dam failure is from 2021. As a result, to produce a pre-failure topographic surface a mono image from Sentinel-2A from September 6th 2022 was used in combination with the post-failure topographic surface data from the September 16th and 24th survey. The September 6th Sentinel-2A mono image can be seen in Figure 1, and the post failure Worldview Satellite image from September 24th 2022 can be seen in Figure 2.

To reconstitute a pre-failure surface, the post-failure topographic surface was used as a starting point. This post-failure topographic surface, born from the September 16th and September 24th Worldview stereo imagery can be seen in Figure 3. The surface is represented as it appears, in plan view, in our in-house VTW software. The format shown is colour, shaded and highlighted (CSH) which communicates slope, aspect, and elevation. Warm colours represent higher elevations and cold colours lower elevations.

The corresponding legend displaying the range of the elevations within the topographic on the post-failure September 16th/September 24th surface can also be seen in Figure 3.

FIG 1 – Sentinel 2A Satellite Mono Image: September 6th 2022.

FIG 2 – September 24th 2022 Worldview Satellite Image.

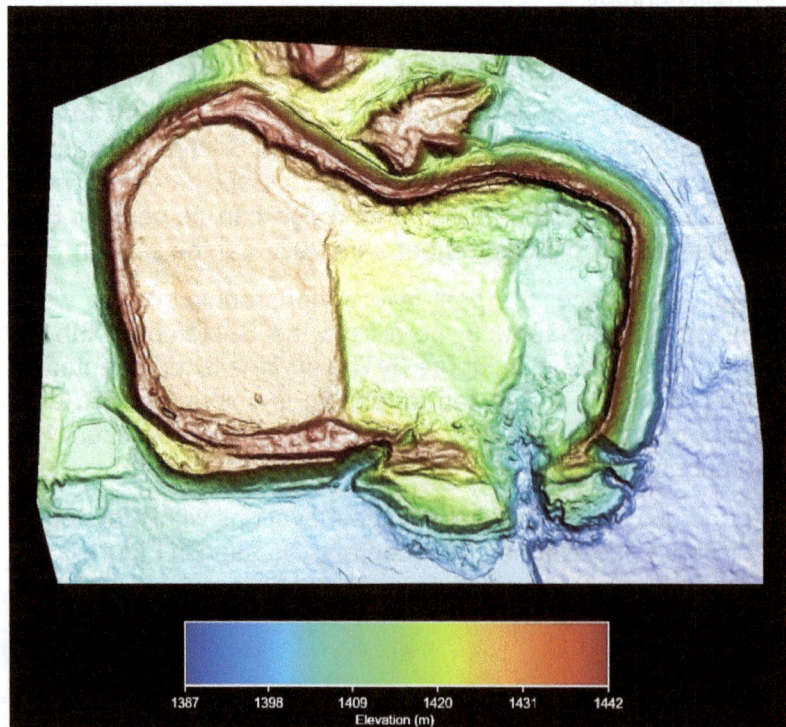

FIG 3 – September 16th/24th 2022 post-failure Topographic Surface.

The 2022 post-failure surface was modified to re-construct the TSF embankment structure in areas visibly damaged by the dam failure:

- The three breached areas to the south-east.
- The damaged upstream portion of the Dyke crest in most of the entire eastern TSF.

'Patch' surfaces were designed and merged into the pre-failure surface to address the three breached areas to the south-east. It was important to utilise the existing survey data to maintain the correct crest height and footprint in the breached areas.

Based on the observations of the imagery, it was assumed that the width of the dyke crest and wall slopes were consistent throughout the TSF. The western portion of the TSF crest and upstream wall of the dyke were not damaged in the failure; therefore, based on the existing September 16th/24th topographic surface (Figure 3), the width of the dyke crest was estimated to be ~40 m based and the upstream slope of 1.5:1 to 2:1.

Although substantial damage was observed in the upstream portion of the crest in most of the eastern 'cell', a portion of the downstream crest remained intact, excluding the previously mentioned breached areas. Therefore, it was possible to extract the height of the damaged dyke crest from the existing survey.

As a result, the pre-failure dam crest's width, height and upstream slope could be reconstructed to a reasonable accuracy. The September 6th 2022 Sentinel-2A imagery (Figure 1) was then used to establish the eastings and northings of the tailings footprint within the TSF prior to failure. The elevation of that boundary could then be determined wherever it contacted the re-constructed upstream dyke wall.

In order to visualise the re-constructed pre-failure surface versus the September 16th/24th post failure surface shown in Figure 3, two cross-sections were create showing both surfaces section. Figure 4 displays two cross-section planes overlaid over the September 24th 2022 Worldview satellite image and Figure 5 displays the two surface in section along those planes. Section 1 running west to east within the TSF and section 2 running north to south within eastern and damaged portion of the TSF.

FIG 4 – September 24th 2022 Worldview satellite image with cross-section planes overlaid.

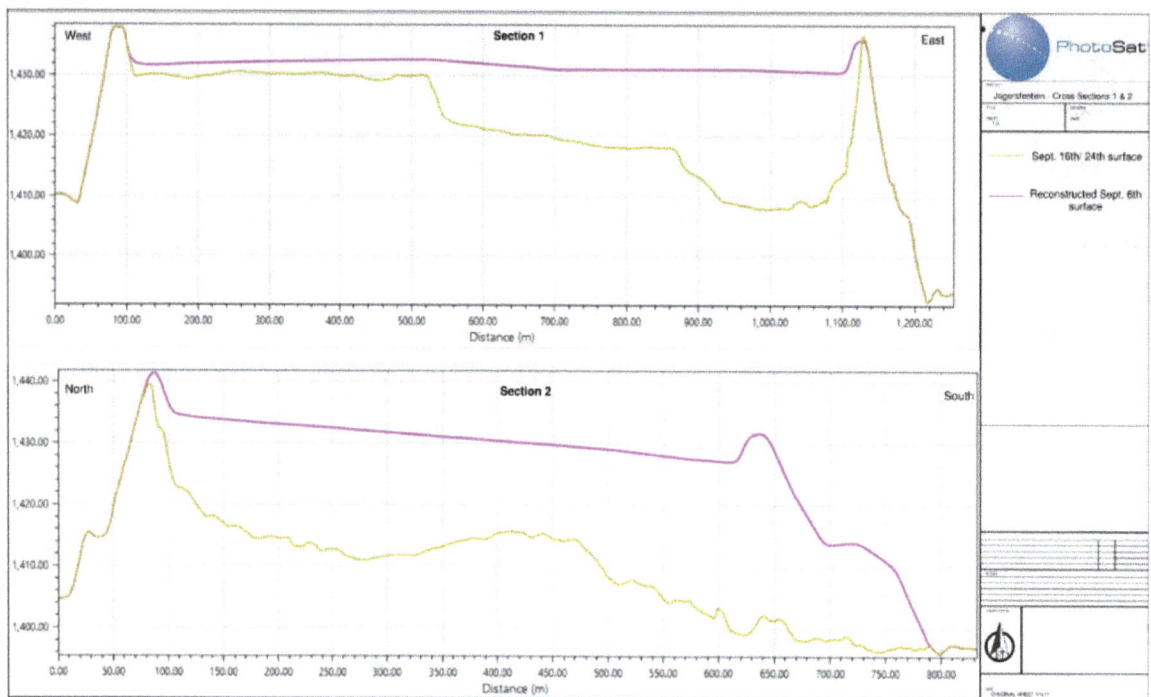

FIG 5 – Cross-sections 1 and 2 of September 6th 2022 pre-failure reconstructed surface and September 16th/24th 2022 post-failure surface.

To additionally visualise the reconstruction sequence, from left to right, in Figure 6, we see, in plan view and CSH format:

- The September 16th/24th 2022 post-failure topographic surface.

- A partially reconstructed September 6th 2022 pre-failure topographic surface where the dyke embankment structure has been rebuilt.

- A completed September 6th 2022 pre-failure surface where the dyke embankment structure has been rebuilt and the tailings surface overlaid.

FIG 6 – Topographic surfaces for September 16th/24th 2022 (left), partially reconstructed September 6th 2022 (centre), and fully reconstructed September 6th 2022 (right).

RESULTS

Topographic surfaces and component mapping by year

As shown in the following sections, the topographic surfaces display the segregated components, 1 m contours, and 2022 TSF post-failure footprint overlayed. The topographic surfaces were either produced from stereo pairs or reconstructed from mono imagery.

1977 topographic surface and component map

This topographic surface (Figure 7) was produced from Keyhole stereo satellite imagery. The 'dyke/wall', 'tailings material', 'waterbodies' and 'unknown material' boundaries, listed in the Figure 7 legend, were determined through visual observation of the satellite imagery.

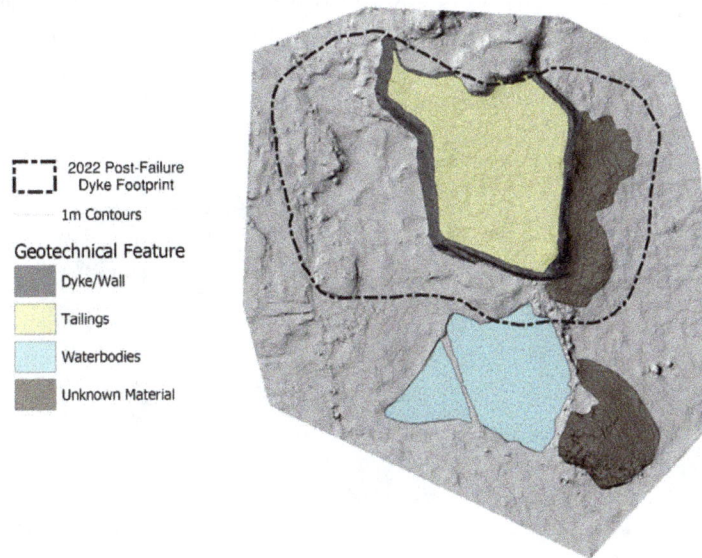

FIG 7 – Component map for 1977 surface.

2011 topographic surface and component map

This topographic surface (Figure 8) was produced from WV-1 and WV-2 stereo satellite imagery from February 26th 2011 and March 9th 2011. The 'dyke/wall' and 'unknown material' boundaries, listed in the Figure 8 legend, were determined through visual observation of the satellite imagery.

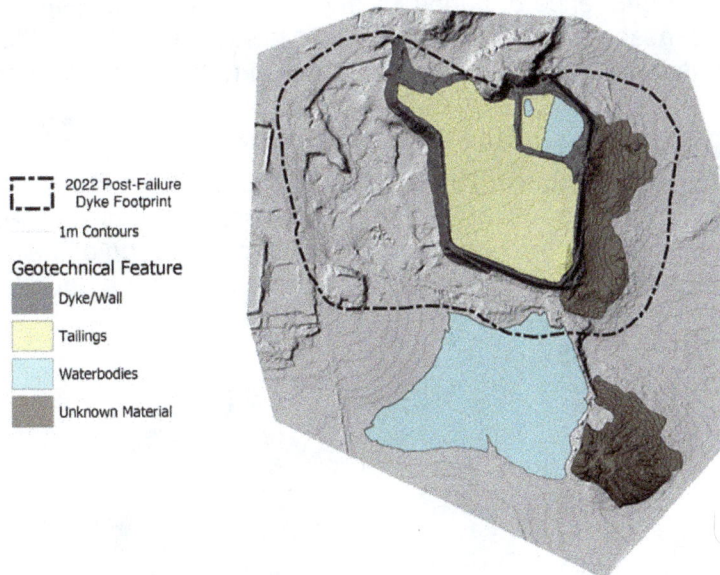

FIG 8 – Component map 2011 surface.

The Normalised Difference Water Index (NDWI) was used to determine the boundaries of water bodies within and south of the TSF. The NDWI is calculated using the GREEN-NIR (visible green and near-infrared) combination, which allows the detection of subtle changes in water content of water bodies.

The tailings boundary, generally, was determined through visual observation of the satellite imagery. Additionally, a normalised difference of bands 5 and 8 from Landsat-5 imagery dated February 16th

2011 was performed to assess the Normalised Soil Moisture Index and therefore the moisture content within the tailings area. That analysis yielded no high-moisture tailings' areas.

2017 topographic surface and component map

This topographic surface (Figure 9) was produced from WV-1 stereo satellite imagery from October 19th 2017. The 'dyke/wall' boundary was determined through visual observation of the satellite imagery. The Normalised Difference Water Index (NDWI) was used to determine the boundaries of water bodies within and south of the TSF.

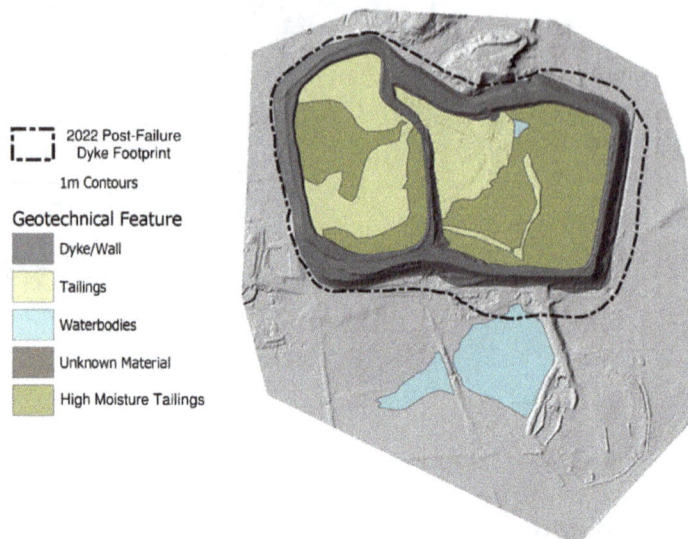

FIG 9 – Component map of 2017 surface.

The tailings boundary, generally, was determined through visual observation of the satellite imagery. Additionally, a normalised difference of bands 11 and 12 from multiple Sentinel 2-A images within 2017 was performed to assess the Normalised Soil Moisture Index and therefore the moisture content within the tailings area. That analysis yielded a distinct area of high moisture tailings.

2022 pre-failure topographic surface and component map

This surface (Figure 10) was produced by utilising the September 16th/24th 2022 post-failure topographic surface and September 6th 2022 Sentinel 2-A imagery, which is detailed in Figures 4 to 6. The 'dyke/wall' boundary was determined through visual observation of the satellite imagery. The Normalised Difference Water Index (NDWI) was used to determine the boundaries of water bodies south of the TSF.

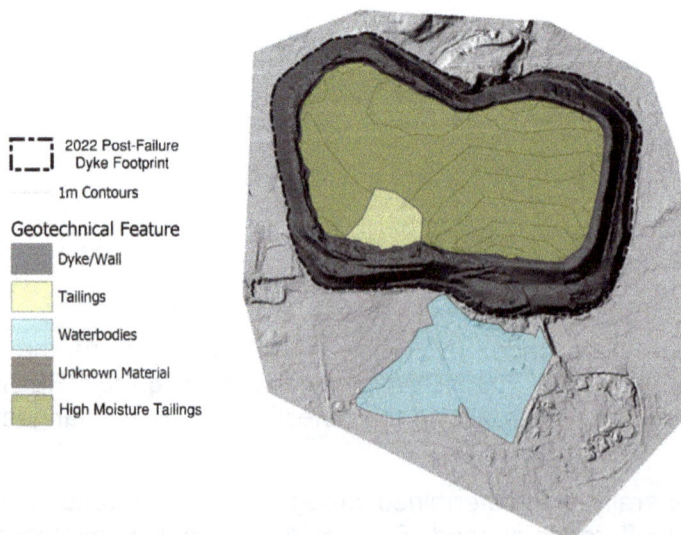

FIG 10 – Component map of 2022 pre-failure surface.

The tailings boundary, generally, was determined through visual observation of the satellite imagery. Additionally, a normalised difference of bands 11 and 12 from the Sentinel 2-A imagery dated September 6th 2022 (Figure 1) was performed to assess the Normalised Soil Moisture Index and therefore the moisture content within the tailings area. That analysis yielded a distinct and extensive area of high moisture tailings that can be seen in Figure 10.

2022 post-failure topographic surface and component map

This topographic surface (Figure 11) was produced from WV-2 stereo satellite from September 16th and 24th 2022 Figure 2).The components were mapped through spectral analysis of Landsat and Sentinel-2A imagery.

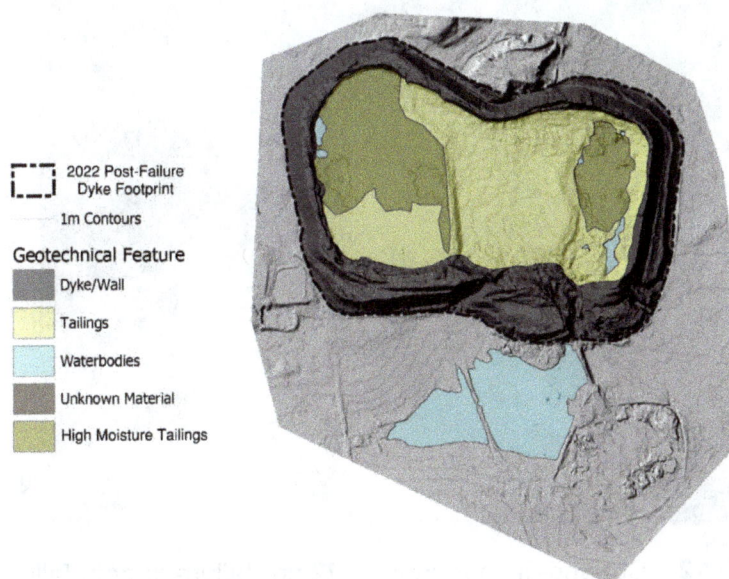

FIG 11 – Component map of 2022 post-failure surface.

The 'dyke/wall', tailings and high moisture tailings boundaries were determined through visual observation of the satellite imagery.

The Normalised Difference Water Index (NDWI) was used to determine the boundaries of water bodies within and south of the TSF.

General observations from topographic surfaces and component mapping

As can be observed in Figures 7 and 8 displaying the composition and layout of the TSF on February 4th 1977 and February 26th/March 9th 2011, there was minimal footprint expansion between 1977 and 2011. Figures 7 and 8 also display some type of fill material appears loosely amassed outside eastern wall of TSF in February of 1977 and February/March of 2011. The Construction of north-eastern tailings cell can also be observed in Figure 8.

Figure 9 display the composition and layout of the TSF on October 19th 2017. In Figure 9, a distinct dividing west–east wall can be seen. As can be see in Figure 10, showing the composition and layout of the TSF on September 6th 2022, the dividing wall is submerged in tailings material.

As can be seen in the composition maps of Figures 7, 8 and 9, displaying the TSF in 1977, 2011 and 2017, the water body to the south of the TSF within 2022 dyke post-failure footprint in the area where southern wall failed.

The composition maps of the TSF from October 19th 2017 (Figure 9) and September 6th 2022 (Figure 10), we can observe a significant amount of high moisture tailings within the TSF. In contrast, the composition maps from February 4th 1977 and February 26th/March 9th 2011, Figures 7 and 8 respectively, display no areas of high moisture tailings. Construction of a significant dyke feature, possibly a buttress, can be observed downstream of TSF in Figures 9 and 10 as well.

Comparison of 2022 pre-failure and post-failure surface

To determine the volume of material released from the dam collapse on September 11th 2022, the September 16th/24th 2022 post-failure surface (Figure 11) and the reconstructed September 6th 2022 pre-failure surface (Figure 10) were compared.

In Figure 12, a heat map shows the vertical change between the Figures 11 and 10 surfaces. The total and segregated embankment structure and the tailings cut-and-fill volumes between those periods are listed in Table 2.

FIG 12 – Cut-and-fill map from 2022 pre-failure to post-failure.

TABLE 2

Cut-and-fill volumes from 2022 pre-failure to post-failure.

Zone	Area (m²)	Fill (m³)	Cut (m³)	Net (m³)
Embankment structure	80 000	10 000	660 000	-650 000
Tailings	475 000	-	4 390 000	-4 390 000
Total*	555 000	10 000	5 050 000	-5 040 000

As listed in Table 2, it is estimated that over five million cubic metres of material exited the TSF during failure, with approximately 13 per cent of this material originating from the embankment structure. Most of the material originating from the eastern portion of the TSF with thicknesses as high as 25 m.

Volume and origin of discharged material

To locate and quantify the origin of the discharged material, the 2011 topographic surface (Figure 8) was compared against the post-failure 2022 topography (Figure 11) and produced a 'Cut' map within the footprint of the 2011 TSF.

Volume of discharged material

The 'Cut' map comparing the 2011 topographic surface (Figure 8) surface with the 2022 post-failure surface can be seen. This map provides insight into the location, height and quantity of material placed, prior to February 26th/March 9th 2011, that left the TSF during the failure. The volume of material highlighted in Figure 13 is 208 000 m³.

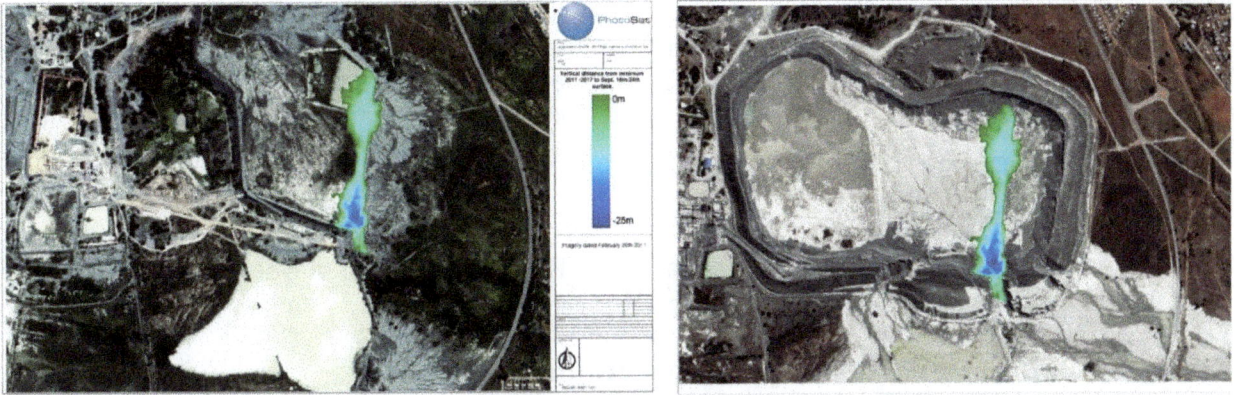

FIG 13 – Cut map overlaid on 2011 and 2022 post-failure orthophoto.

Figure 14 shows the same cut area from Figure 13 overlayed over a October 1st 2014 WV-2 satellite image. Figure 15 shows the area in October 2014 with the 2011 dyke footprint overlaid. In the imagery, there appears to be significant earthwork being conducted in the eastern wall area of the 2011 dyke.

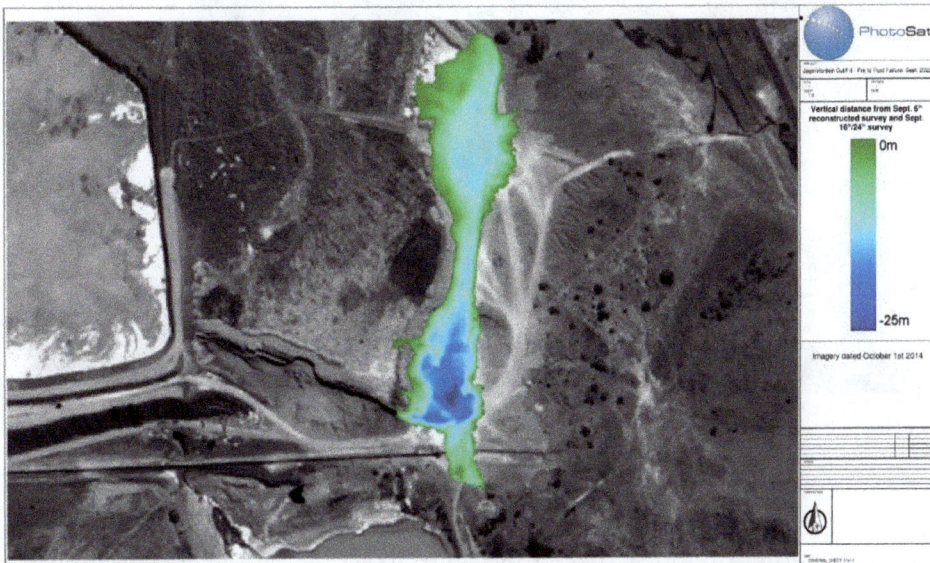

FIG 14 – Cut map overlaid on the October 1st 2014 orthophoto.

FIG 15 – 2011 Dyke footprint overlaid over October 1st 2014 WV-2 satellite image.

Furthermore, it appears that the 2011 eastern dyke wall has been excavated and pushed back in a configuration that matches the location of the material, from the 2011 Dyke footprint, that left the TSF during the failure. The configuration of the excavated 2011 eastern wall, as shown in the 2014 imagery, resembles the September 24th 2022 post-failure configuration (*Beach freeboard* section).

These observations suggest that:

- Material that exited the Dyke during the failure was not emplaced prior to 2011.

- Either the failure material was outside the 2011 TSF footprint or what was within the 2011 TSF footprint, was excavated after 2011.

Cut-and-fill maps with corresponding volumes

The 'Cut-and-fill' map comparing the 2011 topographic surface (Figure 8) with the 2022 post-failure surface (Figure 9) can be seen in Figure 16. This map provides insight into the location, height and quantity of material placed between February 26th 2011/March 9th 2011 and October 19th 2017.

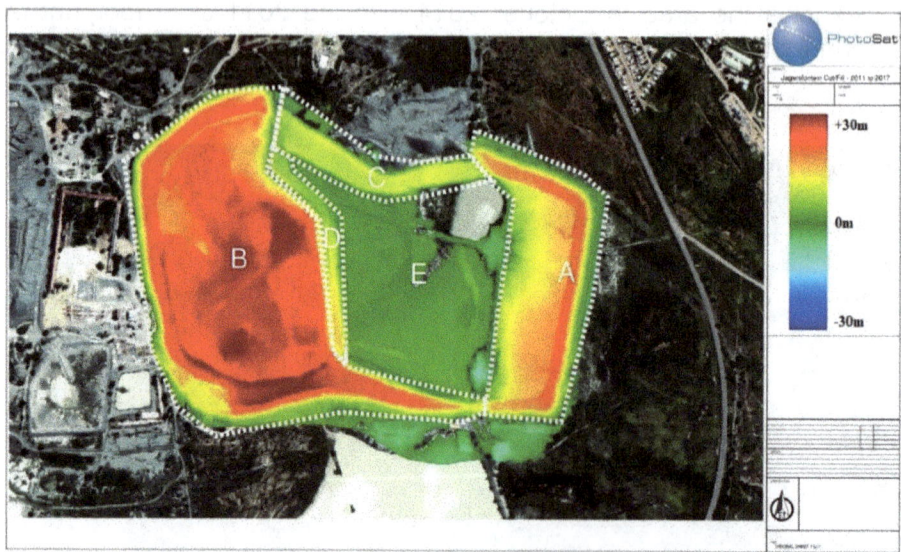

FIG 16 – Cut-and-fill map from 2011 to 2017.

The 'Cut-and-fill' map comparing the 2017 topographic surface (Figure 9) with the pre-failure September 6th 2022 surface (Figure 10) can be seen in Figure 17. This map provides insight into the location, height and quantity of material placed between October 19th 2017 and September 6th 2022.

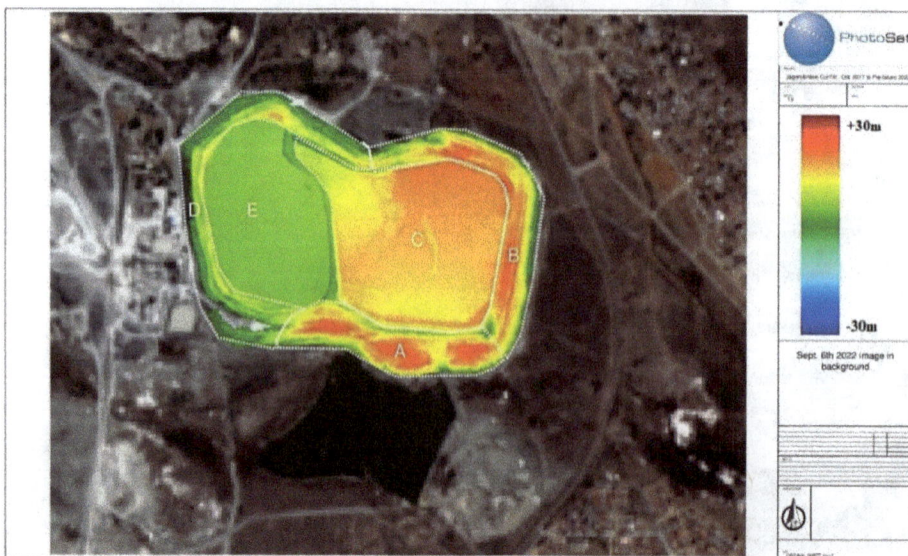

FIG 17 – Cut-and-fill map from 2017 to 2022 pre-failure.

The total and zone segregated cut-and-fill volumes for those periods are listed in Tables 3 and 4. These maps (Figures 16 and 17) display vertical surface to surface changes greater than 50 cm. However, total volume changes of all values are accounted for in the volume tables.

TABLE 3

Cut-and-fill volume from 2011 to 2017 surface.

Zone	Area (m²)	Fill (m³)	Cut (m³)	Net (m³)
A	132 000	1 641 000	0	1 641 000
B	337 000	7 114 000	2 000	7 112 000
C	37 000	290 000	0	290 000
D	18 000	145 000	0	145 000
E	192 000	236 000	43 000	193 000
Total*	848 000	9 426 000	45 000	9 381 000

*The total volumes row corresponds to the material movement within the entire 2022 Dyke footprint, and it is not the sum of the zone rows above.

TABLE 4

Cut-and-fill volumes from 2017 to 2022 pre-failure surface.

Zone	Area (m²)	Fill (m³)	Cut (m³)	Net (m³)
A	229 000	1 284 000	0	1 284 000
B	264 000	3 997 000	0	3 997 000
C	136 000	450 000	28 000	422 000
D	117 000	1 416 000	4 000	1 412 000
E	103 000	1 387 000	0	1 387 000
Total*	848 000	8 534 000	32 000	8 502 000

*The total volumes row corresponds to the material movement within the entire 2022 Dyke footprint, and it is not the sum of the zone rows above.

Design inflow retaining capacity

Figure 18 defines the Design Inflow Retaining Capacity. It is the vertical distance between the surface of the pond/tailings and the low point of the dam crest. It is based on the definition of freeboard within *DSC3F Tailings Dams Guidance Sheet* developed by New South Wales Dams Safety Committee (NSW DSC, 2012). What this paper is defining as the 'Design Inflow Capacity' was referred to as the 'Environmental Containment Freeboard' in the DSC3F Tailings Dams Guidance Sheet. The decision to update the nomenclature was made following discussions with experienced industry members.

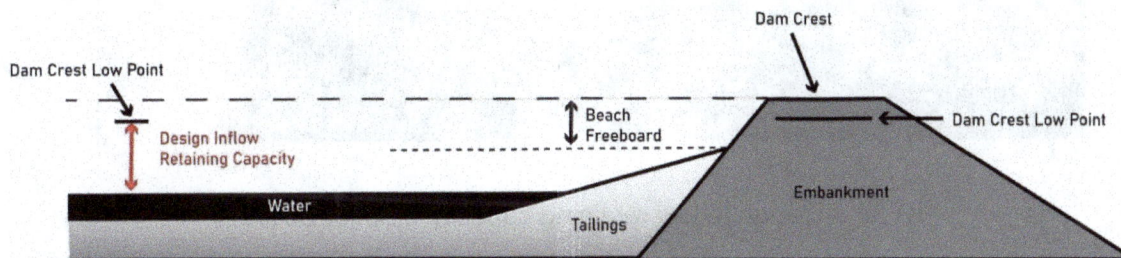

FIG 18 – Definition of design inflow retaining capacity.

Design inflow retaining capacity for 2011

Figure 19 displays the design inflow retaining capacity based on the February 26th/March 9th 2011 surface (Figure 8). It was measured at 205 000 m³. The low point of the TSF is in the south-west corner of the TSF. Freeboard in the south-east corner of the dyke was considered 0 m as material is loosely amassed above that elevation but not in a 'constructed' fashion where it could be considered a retaining wall. There is no capacity in the north of TSF where tailing's elevation is above the 1415.25 m low point.

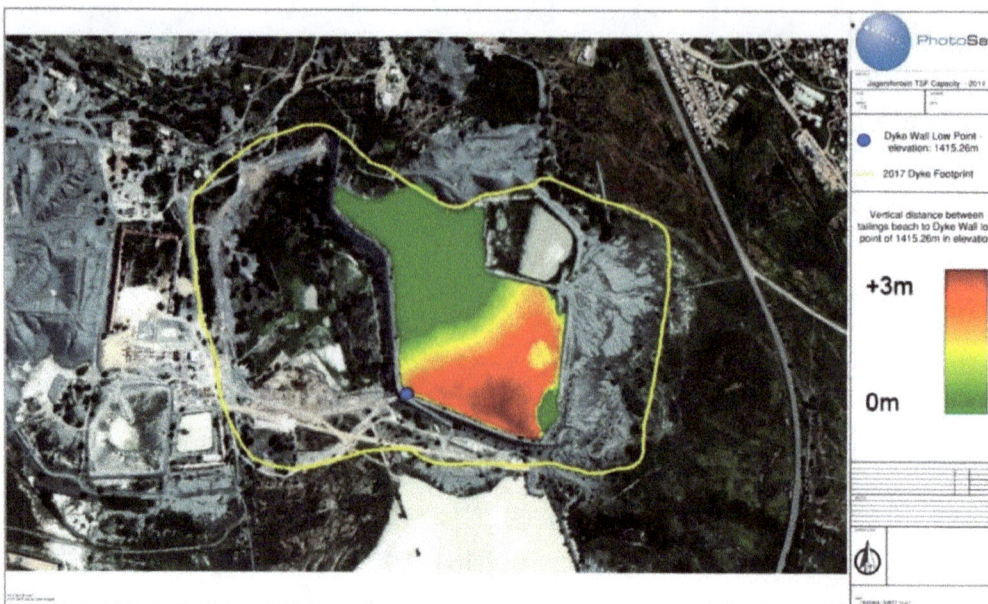

FIG 19 – Design inflow retaining capacity map for 2011.

Design inflow retaining capacity for 2017

Figure 20 displays the design inflow retaining capacity based on the October 19th 2017 surface (Figure 9). It was measured at 632 000 m³. The low point of the TSF wall is in the south-east corner but along the southern wall in proximity to where failure occurred in 2022. There is no design inflow capacity in the west cell of the TSF where tailing's elevation is above the 1417.71 m throughout.

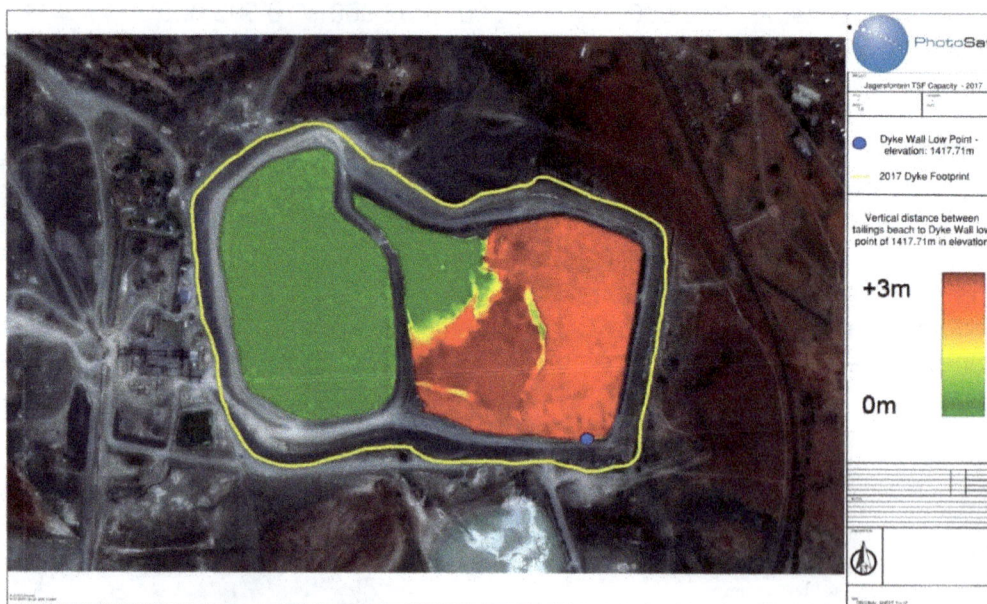

FIG 20 – Design inflow retaining capacity map for 2017.

Design inflow retaining capacity for 2022 pre-failure

Figure 21 displays the design inflow retaining capacity based on the September 6th 2022 surface (Figure 10). It was measured at 333 000 m³. The dyke wall low point of 1431.13 m is located in the south area and in proximity to where failure two of the failures occurred.

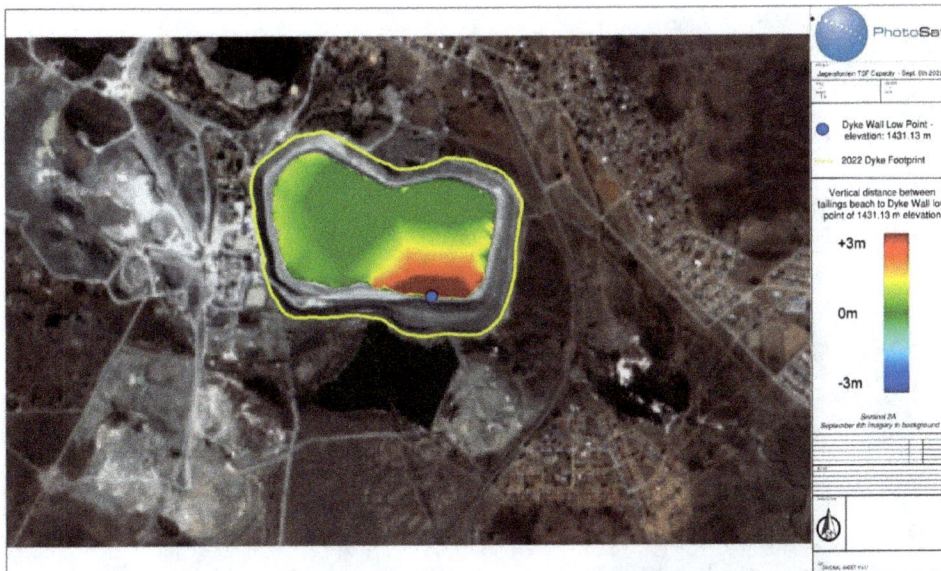

FIG 21 – Design inflow retaining capacity for 2022 pre-failure.

Dam crest elevation

Figure 22 defines the Dam Crest Elevation. It is the highest point along the embankment throughout a TSF. Figure 22 is based on the definition of freeboard in the *DSC3F Tailings Dams Guidance Sheet* developed by New South Wales Dams Safety Committee (NSW DSC, 2012).

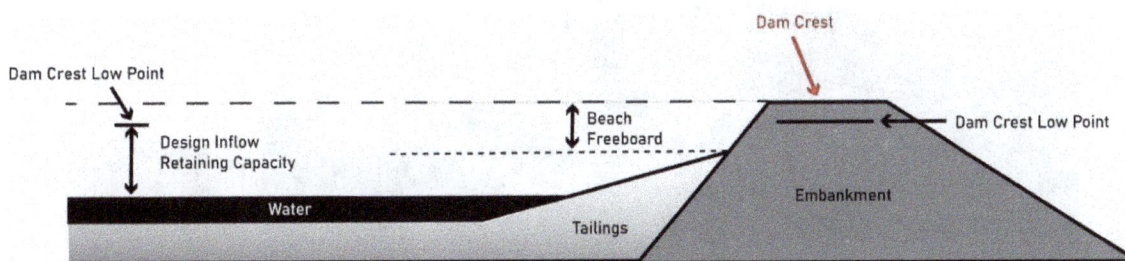

FIG 22 – Definition of dam crest height.

Dam crest elevation for 2011

Figure 23 shows the range in the wall elevation along the entire TSF based on the February 26th/March 9th 2011 surface (Figure 8). In some areas of the TSF, no embankment was observed, and for these areas, no dam crest measurement is shown.

FIG 23 – Dam crest height for 2011.

Dam crest elevation for 2017

Figure 24 shows the range in the wall elevation along the entire TSF based on the October 19th 2017 surface (Figure 8).

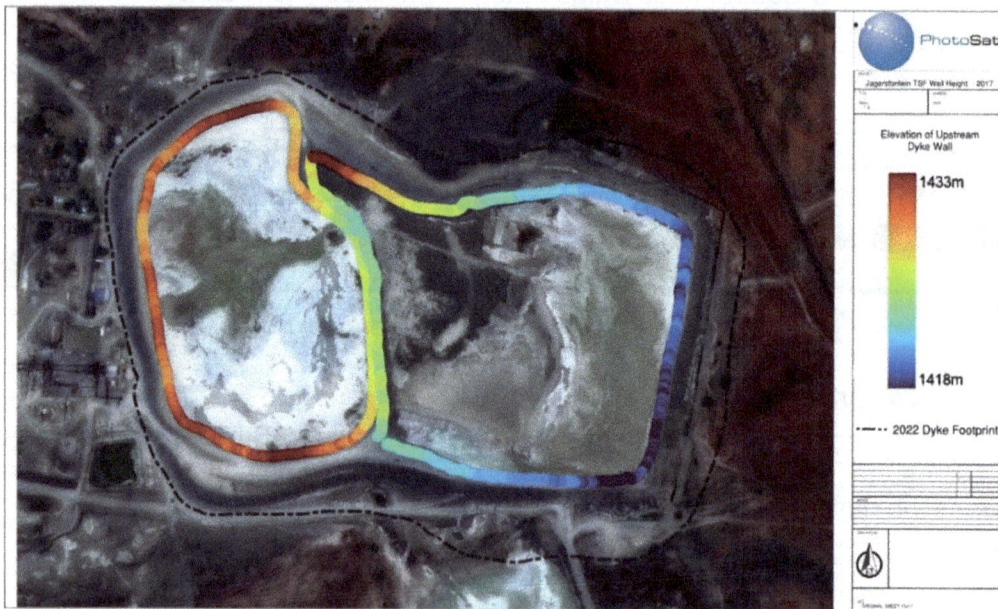

FIG 24 – Dam crest height for 2017.

Dam crest elevation for pre-failure 2022

Figures 25 shows the range in the wall elevation along the entire TSF based on the September 6th 2022 surface (Figure 20).

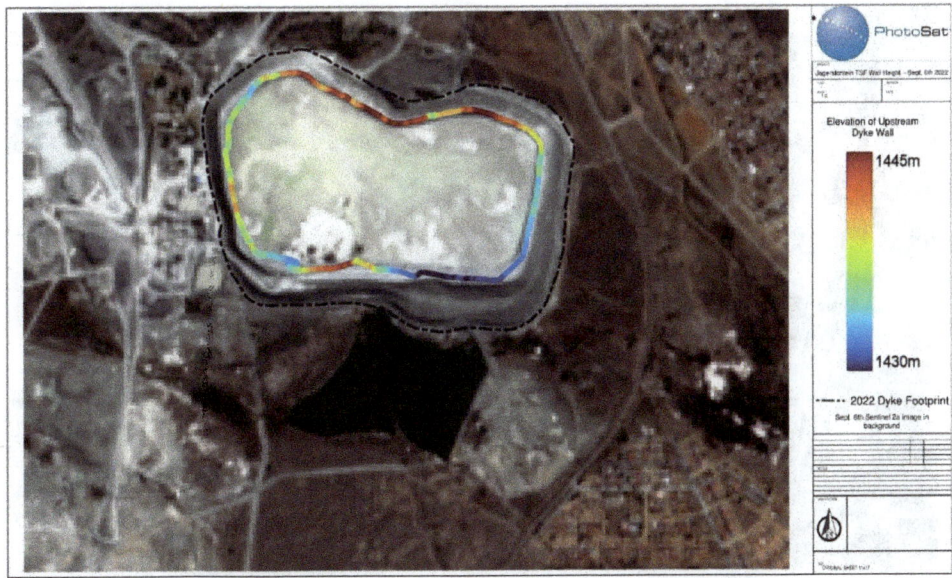

FIG 25 – Dam crest height for 2022 pre-failure.

Beach freeboard

Figure 26 defines the Beach Freeboard. It is the vertical distance between the point where tailings contact the dyke wall and the upstream shoulder of the dyke wall at that location. It is based on the definition of freeboard within *DSC3F Tailings Dams Guidance Sheet* developed by New South Wales Dams Safety Committee (NSW DSC, 2012). What this paper is defining as the 'Beach Freeboard' was referred to as the 'Beach Freeboard' in the DSC3F Tailings Dams Guidance Sheet. The decision to update the nomenclature was made following discussions with experienced industry members.

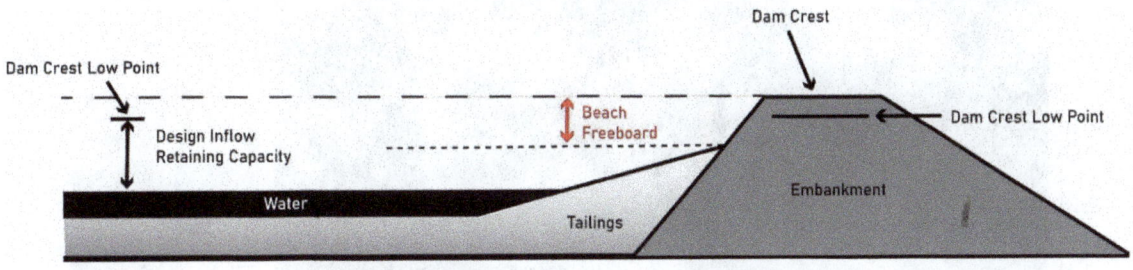

FIG 26 – Definition of beach freeboard.

Beach freeboard for 2011

Figure 27 shows the beach freeboard along the entire TSF based on the February 26th/March 9th 2011 surface (Figure 8). In some areas of the TSF wall either do not appear constructed or some other material submerged the wall. In those area, beach freeboard was not measured.

FIG 27 – Beach freeboard for 2011.

Beach freeboard for 2017

Figure 28 shows the beach freeboard along the entire TSF based on the October 19th 2017 surface (Figure 9).

FIG 28 – Beach freeboard for 2017.

Beach freeboard 2022 pre-failure

Figure 29 shows the beach freeboard along the entire TSF based on the September 6th 2022 surface (Figure 10). It is important to note that the September 6th 2022 surface is a reconstructed surface and measurement derived from this surface are estimates with more uncertainty than the topographic surface data produced directly from stereo imagery in 2011, 2017 and post-failure in 2022 (Figures 8, 9 and 11 respectively).

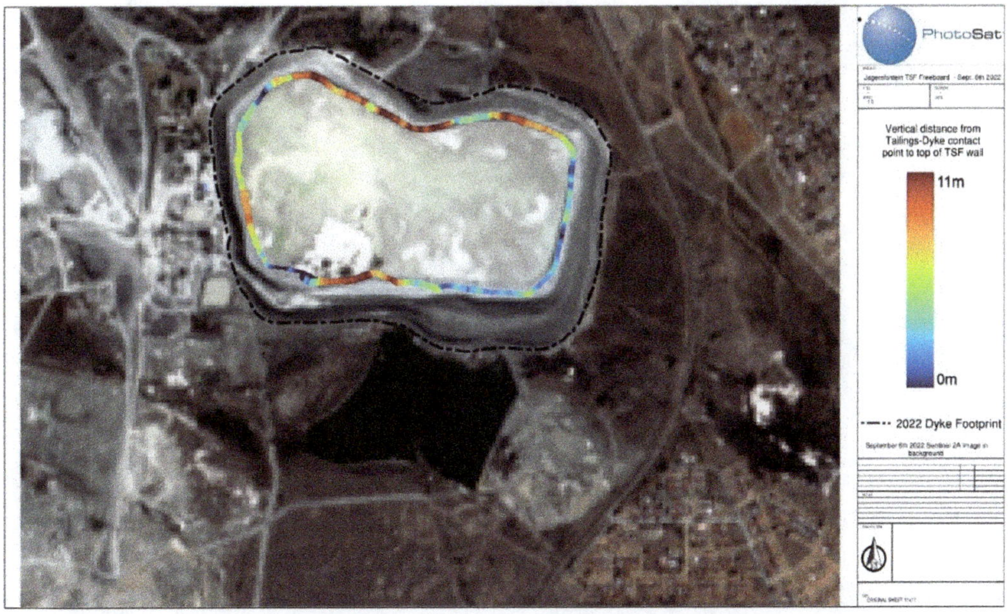

FIG 29 – Beach freeboard for 2022 pre-failure.

CONCLUSION

This paper shows the potential for leveraging existing satellite data to produce tailored and retroactive insights into TSF monitoring/investigations.

The satellite data provides the ability to confirm construction and/or placement timelines (Figure 16 Figure 17), determine total and iterative placed volumes (Tables 3 and 4), embankment heights (Figure 23–25), measure the TSF capacity (Figure 20–22) freeboard (Figure 27–29) and visualise the evolution of the composition of the facility itself over time (Figure 7–11). As well, specific period investigations can be performed to provide insight on the facilities construction and operational management (Figure 13–15).

Furthermore, the ability to reconstruct topography, such as was performed in this study for the September 6th 2022 pre-failure surface (Figure 10), provides a unique and powerful tool that can be utilised to gain insight in situations where limited information exists.

The satellite data and subsequent topographic surface data studied in this report display a facility that grew substantially between 2011 and 2022 (Figure 16–17), with substantial embankment and tailings volumes being placed (Tables 3 and 4). As well, the composition mapping (Figure 7–11) show a progressive proliferation of high moisture tailings that eventually.

REFERENCES

Alonso-Jimenz, C, López, P, Benito, R and Tarquis, A M, 2019. Correlation between vegetation index and solid moisture index using Sentinel-2, University de Madrid, pp 75–78.

DigitalGlobe, 2020a. WorldView-1 Data Sheet. Available at: https://resources.maxar.com/data-sheets/worldview-2//worldview-1.pdf

DigitalGlobe, 2020b. WorldView-2 Data Sheet. Available at: https://resources.maxar.com/data-sheets/worldview-2//worldview-2.pdf

Haubrock, S-N, Chabrillat, S, Lemmnitz, C and Kaufman, H, 2008. Surface soil moisture quantification models from reflectance data under field condition, January 2008, Helmholtz-Zentrum Potsdam, Deutsches GeoForschungs Zentrum GFZ.

New South Wales Dams Safety Committee (NSW DSC), 2012. DSC3F Tailings Dams Guidance Sheet, New South Government, Australia.

US Department of the Interior, US Geological Survey (USGS), 2013. Landsat – A Global Land-Imaging Mission, Fact Sheet 2012–3072, revised May 2013. Available at: http://pubs.usgs.gov/fs/2012/3072/fs2012–3072.pdf (Accessed: 21 August 2015).

Operational aspects and case histories

Tailings deposition management plan – design and implementation including operational aspects and case history

B Batchelder[1] and A Uz[2]

1. Senior Technical Director, Geotechnical, Tailings and Mine Water Management Dams, GHD, St Leonards NSW 2120. Email: bob.batchelder@ghd.com
2. Senior Geotechnical Tailings Engineer, GHD, Sydney NSW 2065. Email: attila.uz@ghd.com

ABSTRACT

The tailings deposition management plan (TDMP) provides the methodology for optimal deposition of tailings slurry. It forms an integral part of the design of a tailings storage facility (TSF). This paper focuses on subaerial (air-dried) tailings deposition, which is extensively utilised within Australia.

The correct implementation of the TDMP determines the condition of the tailings deposit in terms of trapped moisture, deposited dry density (which directly affects storage capacity), saturation/excess pore pressure (which influences static liquefaction), strength/stability (risk), the deposited beach slope and the difficulty/cost of rehabilitation. Correct design and implementation of the TDMP is necessary to achieve the design requirements, to reduce risk, and to enable cost-effective rehabilitation.

Once implemented, the TDMP is validated by observation and adjustment of spigot discharge duration (per spigot in some cases) and the number/location of spigots to provide the required beach shape to minimise the pond area in the TSF. This in turn maximises deposited dry density and hence TSF storage, increases flood routing capability, reduces risk of static liquefaction, non-compliant discharges/overtopping and high rehabilitation costs.

Other aspects of the deposition which are recorded during this process include channel erosion and incorrect beach slope which are often due to spigot type, the number of spigots open and/or decrease in percent solids of the tailings. The first two aspects are handled by operator training, the third will require a revision of the TDMP as a 'change management' measure.

This paper presents the testing conducted and typical results for a metalliferous mine in western NSW (site unnamed) showing the most recent implementation of the TDMP design. This TDMP is about to be implemented. The paper also presents a case study for a coal tailings TSF in the Western Coalfields of NSW west of Lithgow (site unnamed), which has successfully been implemented and achieved deposited subaerial beaching meeting the required (estimated density) as per the TDMP design.

INTRODUCTION

The tailings properties utilised in the design of the tailings deposition management plan (TDMP) are determined by careful laboratory medium to large scale simulation testing, comprising column/permeability/consolidation undrained settling tests combined with air drying tests (in buckets under lights with a reference water sample) for a range of deposited layer thicknesses. These results, informed by the Australian Bureau of Meteorology (BOM) data for a site, are used to model the tailings deposition for the TSF geometry under the site-specific climatic conditions, in order to optimise the deposition layer thickness and cycle time for the emplacement of tailings at the maximum achievable dry density for the site.

THE REQUIREMENT FOR TDMP

Tailings storage facilities (TSF) employ subaerial discharge under a carefully prepared TDMP in order to achieve significant gains in deposited dry density and strength in a cost-effective manner with acceptable risk for the planned subaerial discharge operations. In particular, the TDMP is used to inform and assess the potential for upstream/centreline raising that can be achieved by optimising the combination of 'layer' thickness and air-drying time, in combination with managing the tailings geometry and deposition flow management, in order to: maximise storage, assist in upstream staging, reduce operational and long-term risks; and to reduce rehabilitation costs.

Once implemented, the TDMP will be conducted using the 'Observational Approach' (OA), whereby aspects of the deposition are recorded and adjusted as necessary during the deposition process, including:

- Avoiding beach channel erosion by adequate number of spigots open, addressing dysfunctional discharge from the spigots, and keeping the percent solids as high as is reasonably practical (minimum 44 per cent as per the supplied tailings sample).

- Achieving the correct beach shape to minimise the decant pond, by keeping it to a minimum area centred around the gravity decant inlet. This is achieved through positioning and varying the length of discharge time across the spigots, for each deposition segment.

- Ensuring proper operation of the decant by adding short riser pipes (150 mm to 300 mm length) to prevent tailings ingress whilst keeping the length of riser pipe/pipes to a minimum. In order to minimise the decant pond depth. Noting that any areas covered by water will remain saturated and thus not gain density increase through air drying, thus subaqueous not subaerial deposition will be occurring with corresponding increase in risk and much lower strength and deposited density.

PREPARATION OF THE TDMP

The TDMP is prepared based on large scale laboratory simulation testing to model the settling/consolidation behaviour of the deposited tailings layer using typically at least two different layer thicknesses. The assessment of multiple layer thickness is required in order to model supernatant production and the time to achieve end of primary consolidation (start of the air drying portion of the slurry sample in the 'cycle time).

Large scale air drying tests are commenced on the 150 mm and 300 mm slurry thickness samples after they have reached the end normal consolidation. The air drying simulation test is conducted in order to model the density and strength gains achieved by over-consolidation through the air drying process ie achieving an unsaturated state. The site climatic data is used in combination with the results of this test, to estimate site evaporation potential.

Laboratory simulation testing

The laboratory simulation testing conducted includes:

- **Settling and Consolidation Test Phase 1** – settling and consolidation tests – top and bottom undrained.

- **Settling and Consolidation Test Phase 2** – top undrained, base drained.

- **Air drying under lights** – with a water container for correlation to BOM Class A Pan Factor.

Soil Classification results are also conducted in order to classify the tailings and to estimate the specific gravity (SG) of the slurry sample. An SG of 2.79 was adopted as part of the example TDMP. The tailings test samples are taken from the throughput to the TSF.

Slurry settling and consolidation tests

Large scale undrained deposition settling and consolidation tests (see Figure 1) are carried out to evaluate the rate of settling and to estimate the initial deposited dry density likely to be achieved in the layer deposition. An in-house laboratory test procedure is used to assess the settling and consolidation behaviour for tailings deposition. This test simulates placement of a tailings layer over underlying low permeability tailings deposits, prior to air drying.

Duplicate sets of samples (Samples A and B) are tested for repeatability, for each of deposited tailings slurry layer thicknesses assessed (150 mm and 300 mm). These samples are placed into cylinders at the solids concentration of mine throughput.

The method for optimising the deposited dry density includes division of the TSF depositional area into a number of portions, and assessing the cycle time for the sequential deposition over each portion ie how thick a slurry layer is placed in the portion and how long the drying time is for each layer until a recurrence of deposition (end of cycle time) occurs. The implementation of this technique is conducted in conjunction with the OA, as described above.

The times for settling and consolidation for the above test samples are recorded in the settling and consolidation test shown in Figure 1. This is achieved by recording the decrease in slurry height at suitable time intervals, from layer placement until secondary consolidation (creep) is established. This enables the time for effective completion of the primary consolidation to be established for each sample.

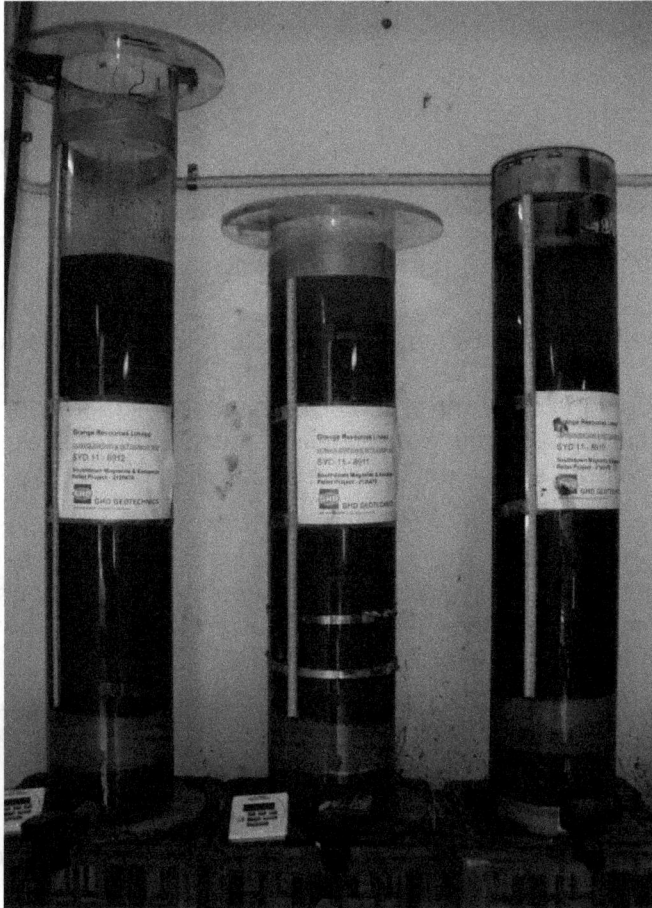

FIG 1 – Column settlement and consolidation test.

Following completion of Phase 1, the sand filter drainage layer beneath the sample is drained via a drainage tap, located near the base of the test cylinder, thereby decreasing the pore pressure at the base of the sample and enabling a second stage of consolidation testing for the sample to be assessed from the data collected (if desired).

The sample permeability is recorded during this second stage. The seepage and supernatant water quality can be assessed during the test, if desired. The interpreted tests results for the settling and consolidation tests on the subject samples are presented graphically in Figure 2.

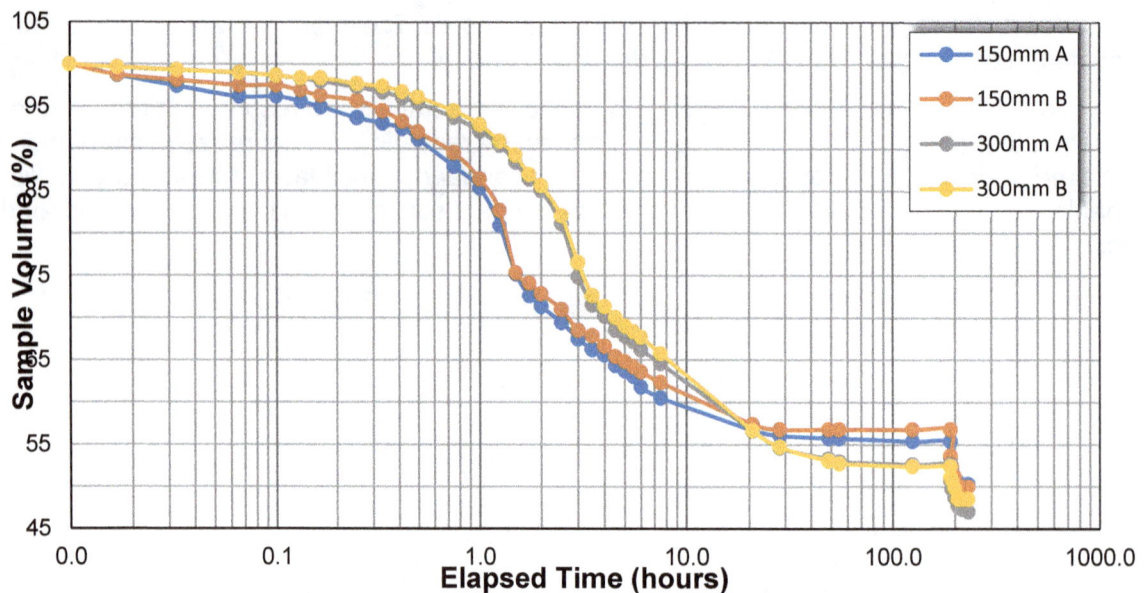

FIG 2 – TSF tailings settling test – 150 and 300 mm slurry layer thickness (Samples A and B).

The settlement testing demonstrates three phases: mixed sedimentation and consolidation (the steeper portion of the curve); primary consolidation (less steep curve); then flattening out to secondary consolidation (creep) settlement until the termination of the test.

The steep section at the curve following completion of Phase 1 of the test, is the test Phase 2 where only permeability was calculated for the subject samples. Consolidation data, while able to be obtained from the recorded settlement and the small effective stress decrease from the draining (release of pore pressure) from the base, is relevant for subaqueous deposition and hence is not calculated for subaerial modelling.

The testing results, as included below for the two slurry thickness samples, provide performance data for the slurry layers as follows:

150 mm slurry layer

- A percent solids of 63–64 per cent (γ_d = 1.07–1.08 t/m³) was recorded for the 150 mm thickness slurry layer samples at the end of primary consolidation, which took approximately one day.

- A percent solids of 63–65 per cent (γ_d = 1.07–1.11 t/m³) was recorded at the end of testing (190 hrs) for the 150 mm thick layer sample.

- The permeability recorded in the Phase 2 testing ranged from 3.7 to 7.4 × 10⁻⁷ m/s.

300 mm slurry layer

- A percent solids of 55–57 per cent (γ_d = 0.86–0.89 t/m³) was recorded for the 300 mm thickness slurry layer samples at the end of primary consolidation. Primary consolidation took approximately 1.5 days.

- A percent solids of 67 per cent (γ_d = 1.16–1.17 t/m³) was recorded at the end testing for the 300 mm thick layer sample (190 hrs).

- The permeability recorded in the Phase 2 testing ranged from 4.7 × 10⁻⁷ m/s to 1.1 × 10⁻⁶ m/s for the 300 mm thick slurry samples.

Air drying tests

Large-scale air drying tests are conducted to simulate the air drying behaviour of the tailings in areas not covered by water eg beneath the (minimum sized) decant pond or other ponded water on the tailings surface (not recommended). Where water cover exists, only the NC density will be achieved.

Samples of tailings are placed into buckets and allowed to settle, with supernatant water being removed, until NC conditions are reached, prior to conducting the air-drying test.

In the air-drying test (Figure 3), the evaporative loss of water and settlement/shrinkage of the tailings sample is recorded, and used to calculate the dry density increase in the tailings sample. The evaporative water loss from the tailings surface is compared to that from a similar bucket of free water, in order to calculate the Class A Pan evaporation Factor for the tailings, by correlation to evaporation rate from the free water surface.

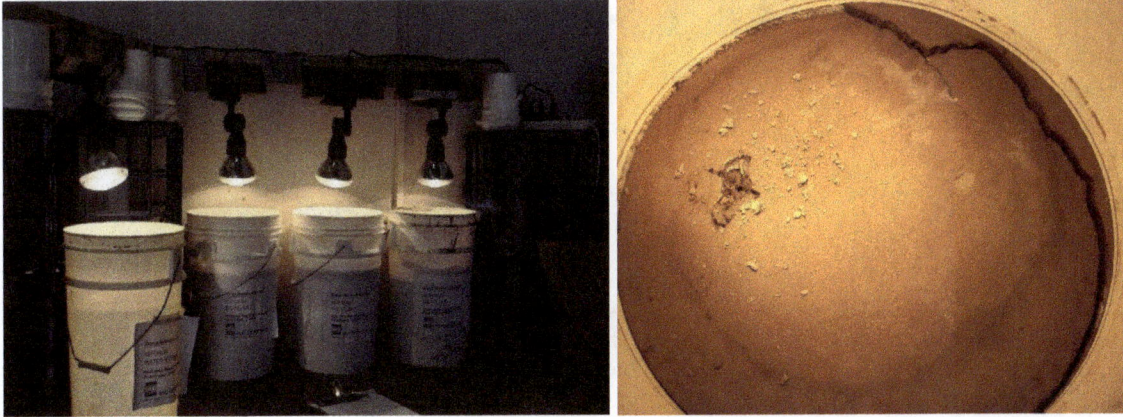

FIG 3 – Air drying test.

The tailings samples are periodically weighed and measured, in order to estimate moisture content, dry density and incremental evaporation losses. The water sample is periodically weighed at the same time as the tailings samples. From this data it is possible to calculate the variation of Pan Factor within the tailings versus the sample moisture content.

The Class A Pan Factor for the soil typically reduces with the drying time of the tailings sample, due to increased soil suction.

This test models the air drying performance potential for the slurry deposition with varying slurry layer thickness and cycle time. The expected dry density and rate of rise of the tailings deposit can be calculated based on the tailings throughput and subaerial deposition area. This also enables the assessment of 'change management' for differing throughput values or an extended periods of ponding (other than that the minimum decant pond area). and provides understanding as to whether the tailings are likely to achieve sufficient drying to sustain upstream or centreline construction.

From this data it was possible to calculate the variation of pan factor with the tailings sample moisture content, as illustrated in Figure 4, for 150 mm and 300 mm thickness slurry deposition layers.

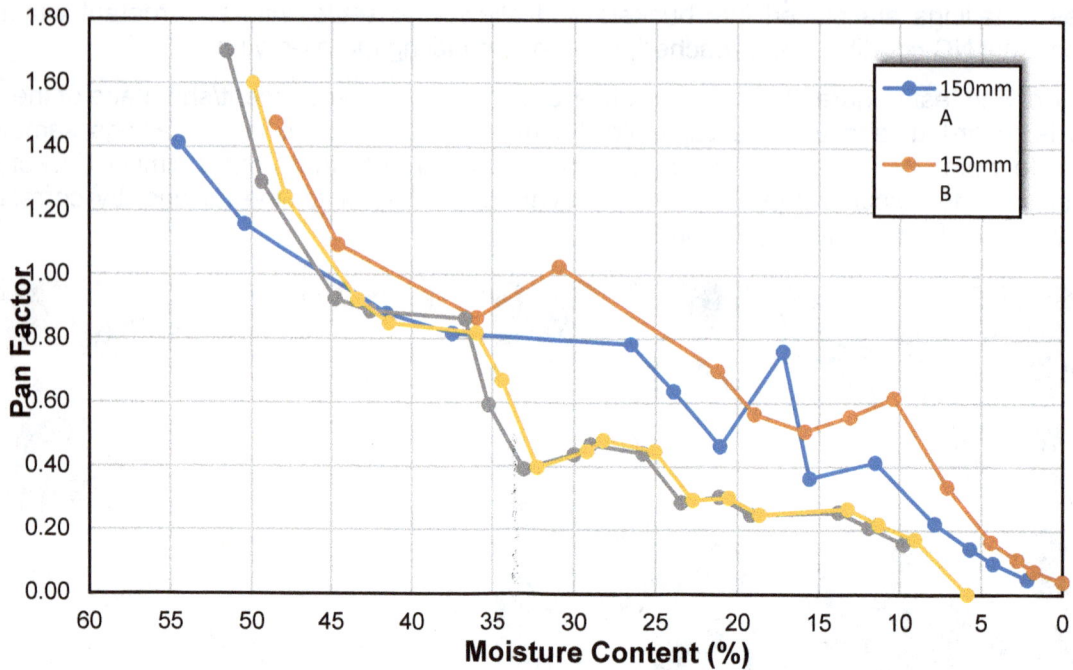

FIG 4 – Tailings air drying tests – pan factor versus moisture content.

The results of these tests inform the estimation of the tailings air dried density that will be attained through the use of site based evaporation data and hence the expected rate of rise of the tailings beach deposit for the throughput. It also provides an understanding as to whether the tailings are likely to achieve sufficient strength to sustain upstream or centreline construction.

The quantified evaporative loss, using the site evaporation data, results in a revised moisture content with corresponding dry density for the slurry layer, which allows choice of the optimum cycle time and layer thickness combination for the planned deposition. The results as plotted in Figure 5 show the advantage of the 150 mm layer for air drying, even though the 300 mm layer has a marginally higher density at end of primary consolidation.

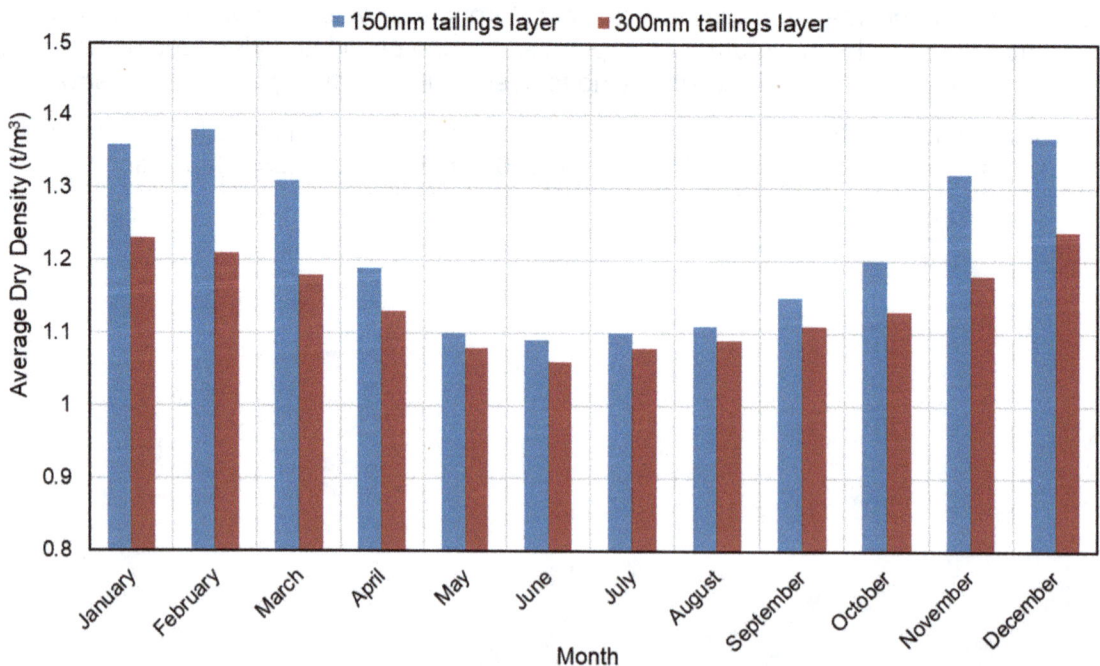

FIG 5 – Tailings air drying tests – average monthly dry density calculated for the 150 mm and 300 mm slurry layer thicknesses based on the air drying test data.

Mine Waste and Tailings Conference 2023 | Brisbane, Australia | 13–14 July 2023

CASE STUDY

The above illustration of the preparation and implementation of a TDMP relates to a metalliferous mine site in Central Western NSW. It is the latest refinement of presentation for the TDMP, but is as yet to be implemented at the site.

The case study was chosen to illustrate the advantages of this methodology, relates to an earlier application of the methodology, for an unnamed coal tailings site west of Lithgow.

The coal tailings site was using subaqueous deposition into an old Open Cut (OC) named 'A' Pit, which lies to the north of the current TSF, and which was at the time rapidly becoming unsustainable, due to the site' available area constraints and the rate of rise of the tailings deposit.

It was assessed from TDMP design for the site, that the use of the maximum possible site area was needed for the subaerial deposition, given the throughput combined with the site rainfall and evaporation characteristics resulted in a rate of rise of some 3 m per annum. It was considered unlikely that the site would support upstream construction and potentially problematic, at least in the early stages, to support centreline construction.

Based on the TDMP testing, a design chosen for what was for a co-disposal of Coarse Coal Reject (CCR) to be used as material for by downstream construction of the TSF 'turkeys nest' embankment. With subaerially deposited coal fines placed under a detailed TDMP into the TSF. The TSF deposition area ranged from some 10 ha initial footprint, to approximately 13 ha footprint at the final TSF height. The previous TSF, 'A' Pit, had a maximum 4 ha available.

Using the methodology described in the simulation testing above, a cycle time and 'slurry' layer thickness were obtained, to target the optimum number and positions of spigots, in order to maximise the deposited dry density for the slurry type, throughput and site characteristics. This comprised varying the number of '12 hour' shifts per deposition point, and the sequence of deposition over 11 deposition points, while maintaining a beach that sloped towards the decant (in the west), without low points or ponding away from the decant occurring. The deposition was modelled as a series of deposition cones emanating from each spigot point, the diameter of which were dependent on the number of shifts per spigot point. This deposition pattern is shown in plan on Figure 6.

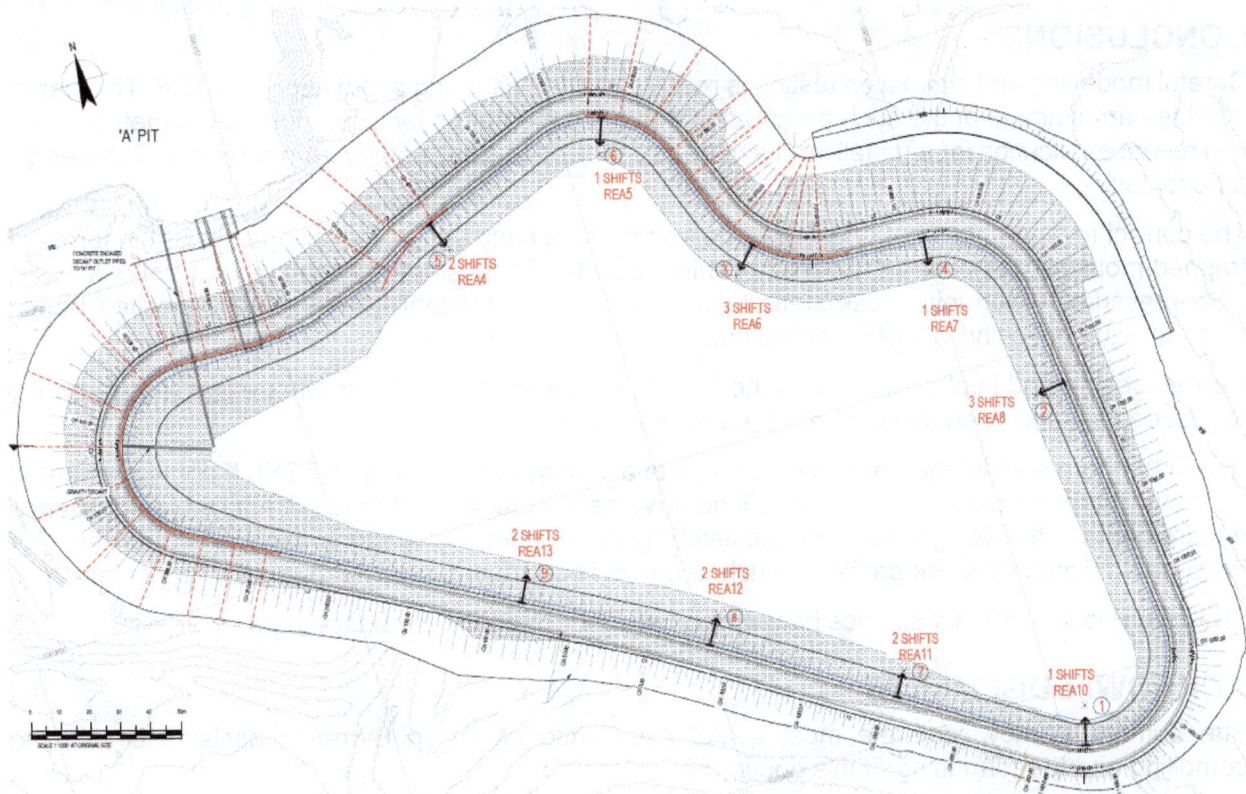

FIG 6 – Tailings deposition plan spigot arrangement.

Modelling in the TDMP showed that if managed correctly under the TDMP, there would be approximately 60 per cent to 70 per cent increase in the deposited dry density (from 0.6 t/m^3 to 1 t/m^3) in the placed coal tailings, after being over-consolidated by air drying rather than placed under water cover in the OC. Achievement of this air dried density was borne out by subsequent LIDAR and in situ testing.

The spigot shift times, sequencing, variation of spigot positions (addition/subtraction) was conducted using daily observation of the deposition.

During the deposition, a significant increase in coal fines production occurred for an extended period of time. The anticipated deposited density (with higher throughput) was modelled in the TDMP for contingency action planning which informed the 'change management' contingency action requirements. One such contingency action was to deposit back into the old OC (now settled/shrunk by about 2 m by consolidation and air drying) by including a further three spigot points, and thereafter to include it in the TDMP sequence to further improve the achieved deposited dry density.

FIG 7 – Case study – typical deposition condition, winter.

CONCLUSIONS

Careful modelling and simulation testing is required for the planning and design of a TSF. This paper provides an example of the such preparation and implementation by illustration of the methodology in a recent application for a metalliferous mine and by way of a case study from an earlier TSF design for coal tailings.

The correct implementation of the TDMP determines the condition of the tailings deposit in terms of trapped moisture, deposited dry density (which directly affects storage capacity), saturation/excess pore pressure (which influences static liquefaction), strength/stability (risk), the deposited beach slope and the difficulty/cost of rehabilitation.

Correct design and implementation of the TDMP is necessary to achieve the design requirements, to reduce risk, and to enable cost-effective rehabilitation.

The TDMP forms an integral part of the design of a tailings storage facility (TSF). If not implemented correctly, the performance of the TSF will be adversely impacted ie it may not fulfil the performance requirements of the design, nor have the flexibility to respond to changes in tailings composition nor produce an acceptable risk profile for such items as flood routing and static liquefaction potential.

The TDMP is in summary science backed modelling for deposition management planning.

ACKNOWLEDGEMENTS

The authors acknowledge the most useful assistance of Vivian Byrne in assistance with the compiling and the formatting of this document.

Using a quantitative risk assessment to demonstrate opportunity for reduced buttressing for an upstream raised tailings storage facility

H Lewis[1], B Tiver[2], D Pham[3] and S Sharma[4]

1. Project Tailings Engineer, WSP, Adelaide SA 5000. Email: harrison.lewis@wsp.com
2. Principal Tailings Engineer, WSP, Adelaide SA 5000. Email: brad.tiver@wsp.com
3. Tailings Engineer, WSP, Adelaide SA 5000. Email: daniel.pham@wsp.com
4. Senior Project Manager, Whyalla SA 5600.

ABSTRACT

Slope stability analyses using the 'Factor of Safety' (FoS) approach have been considered as a main method in the industry when assessing the stability of tailings storage facilities (TSFs). Typically, risk mitigation measures and actions are recommended where the FoS for a slope is less than the values recommended by (ANCOLD, 2019), based on deterministic limit equilibrium (LE) modelling.

The deterministic LE modelling undertaken for the TSF considered in this paper indicated additional buttressing was required to meet minimum FoS requirements for future embankment raises. This paper presents a case study where a quantitative risk assessment (QRA) approach was used to estimate the probability of failure (p[F]) for the same TSF slope, with the intent to demonstrate that the risks are acceptable without additional buttressing.

Various triggers for slope instability were identified and fault trees developed to support estimates of the likelihood of each of these events occurring. The p[F] for each of these triggering events was estimated based on the product of the probability the event occurring, the probability slope instability would trigger given the trigger occurs and the p[F] from stability modelling for each condition. The potential loss of life (PLL) was assessed based on a time-based risk exposure based on discussions with the site owner.

The assessment indicated the TSF would remain in the zone where risks are tolerable only if they satisfy the 'As Low As Reasonably Practicable' (ALARP) principle, as per the f-N chart from the ANCOLD risk guidelines (2022) for 'New Dams'. The QRA results indicated that no additional buttressing was required for the next embankment raise as the TSF remained in the broadly acceptable zone indicated by (ANCOLD, 2022). Additional studies, including a numerical deformation model, were recommended to further assess the likelihood of catastrophic release of tailings under the triggers considered in the QRA and to identify if other reasonably practicable risk reduction measures could be installed.

INTRODUCTION

This paper describes the quantitative risk assessment (QRA) method adopted for a case study where slope stability compliance was assessed using the ANCOLD risk guidelines for a TSF, instead of the typical minimum factor of safety (FoS) approach outlined by (ANCOLD, 2019). The TSF is a 40 m high upstream raised facility that has been in operation since 2007 and is shown in Figure 1. Due to a change in configuration of the TSF, upstream raising of the embankments from the starter dam started in 2014 and has continued with generally 3 m high raises every few years. The next embankment raise, Raise 8, is due for construction in 2023.

FIG 1 – Locations of cross-sections considered.

Following the history of significant TSF failures around the world in recent years and evolving standards, more detailed geotechnical investigations were proactively undertaken on the TSF in 2019, and to support the design of Raise 7. Arising from the investigations and subsequent analyses, a buttress was designed to improve the FoS. A key issue that was discovered was the presence of contractive soils in the foundation of the TSF, which resulted in deep seated, sliding type failures being indicated by limit equilibrium (LE) modelling, with FoS values that were below those recommended by ANCOLD under peak strength conditions (ie FoS values <1.5). A staged approach to buttress design and construction was undertaken to return the TSF slope to compliance with the minimum FoS requirements outlined by ANCOLD for the Raise 7 TSF embankment height.

Detailed design of the Raise 8 embankments was undertaken using the 'traditional' FoS approach to guide the slope geometry, as well as a QRA approach. The trade-off has indicated to the site owner that if the QRA approach is accepted and the 'As Low As Reasonably Practicable' (ALARP) position can be supported, it could save placing about 800 000 m^3 of compacted fill for Raise 8 operation that is required to maintain the ANCOLD FoS. Therefore, there is significant benefit at this site in adopting a QRA approach. This paper presents how the QRA was undertaken and the recommendations that were made.

GEOTECHNICAL INVESTIGATIONS AND BACKGROUND

A detailed geotechnical investigation was undertaken to support the design of the Raise 7 embankments in October 2019. The outcomes of this investigation triggered the initial stages of the buttress design. The analyses indicated uncertainties, and therefore large buttresses, which necessitated further investigation to design and construct the buttressing in a stage approach. The second phase of investigation was undertaken between November 2020 and March 2021. The investigations included the following:

- A cone penetration test with pressure measurement (CPTu) investigation was undertaken in October 2019 targeting both the deposited tailings and foundation soils for the western flank (highest embankment) of the TSF. The investigation was accompanied with classification and strength testing on both tailings and foundation soil samples.

- A borehole investigation was undertaken at the toe of the first stage of buttressing in November 2020. The investigation included collection of samples that were subjected to direct simple shear testing (DSS) and classification testing. This testing was undertaken to highlight the natural variability in the foundation soils. In the design of the first stage of buttressing, the potentially contractive foundation soils were only interpreted from the CPTu, and were categorised as a single unit. This additional investigation indicated the presence of eight units in the foundation, which were represented by two foundation layers in slope stability analyses. These layers were referred to as 'sand' and 'clay' layers, both of which indicated contractive behaviour on the DSS test work.

- A cross-hole geophysics investigation was undertaken at two locations in March 2021. The investigation was undertaken to inform a seismic ground response analysis (SGRA) to assess the potential for cyclic liquefaction to occur.

- Vibrating wire piezometers (VWPs) were installed in the foundation underlying the starter dam and samples were collected for subsequent laboratory testing. The investigation provided an opportunity to undertake additional testing to characterise the contractive foundation soils.

In summary, the TSF has been well investigated and characterised, and the supporting slope stability analysis completed is thorough and have been subject to external review by a Senior Independent Technical Reviewer (SITR).

INSTABILITY TRIGGERS

Sequence leading to failure

The probability of failure p[F] associated with each slope instability trigger comprises the following components:

1. Triggering event needs to occur.
2. Slope instability needs to be triggered due to the event.
3. p[F] should be estimated given the triggering event has occurred.

A fault tree was developed, based on a series of logical events that was used to estimate the probability of slope instability triggering. Probabilities were assigned to each logical event identified with reference to (Barneich et al, 1996) and in consultation with the Engineer of Record (EOR) and internal subject matter experts (SMEs). The fault trees were dependent on the outcomes from probabilistic stability analyses, which each assumed an instability trigger had occurred.

Triggering events

The probability of failure for the embankment was estimated assuming failure was caused by the following triggers:

1. A rise in the phreatic surface (within the tailings, foundation, or both).
2. The occurrence of an earthquake equivalent to either of the design earthquakes. Probabilities were assigned to the following earthquake scenarios based on the outcomes of a liquefaction assessment and SGRA:
   - Tailings deposited above RL 156 m liquefy and all other materials strain soften, reducing in strength by 20 per cent.
   - Tailings deposited above RL 156 m liquefy.
   - Tailings deposited above RL 156 m liquefy and only tailings below RL 156 m strain soften, reducing in strength by 20 per cent, ie the foundation materials and embankment materials do not strain soften.
3. The embankment toe is undercut.

4. Scour erodes significant portions of the reinforcing buttress.

5. Future embankment construction works trigger slope instability via static liquefaction.

6. Lateral extrusion of the starter embankment or toe buttress.

STABILITY ANALYSES

General

Stability analyses were undertaken on three representative cross-sections through the western embankment to assess the individual risk associated with each instability trigger, at the locations shown in Figure 1. Laboratory testing, geotechnical investigation data and compaction test results collected during construction quality assurance (CQA) services for the buttress construction were used to develop probability density functions (PDFs) for the following material parameters:

1. The bulk unit weight of the buttress.

2. The undrained strength ratio (USR) for contractive zones within the foundation.

3. The USR for tailings deposited into the TSF under both triaxial compression and direct shear load conditions.

The PDFs developed to inform the analyses considered both normal and beta distributions. Slope stability modelling was undertaken using a Monte Carlo simulation inbuilt into the software used to undertake the analyses. A total of 10 000 scenarios were considered for each instability triggering case.

Buttress and tailings density functions

Deterministic assessment

The anisotropic strength function presented in Table 1 was developed to vary the strength of the tailings depending on the angle of the critical failure surface for the deterministic assessment. Slope failure through the upper embankments is shown in Figure 2, with annotations indicating where the different load conditions govern.

TABLE 1

Tailings anisotropy strength function.

Strength conditions	Shear angle of inclination		USR adopted
	From	To	
Triaxial Extension (TXE)	-90	-45	0.15
Direct Simple Shear (DSS)	-45	45	0.19
Triaxial compression (TXC)	45	90	0.25

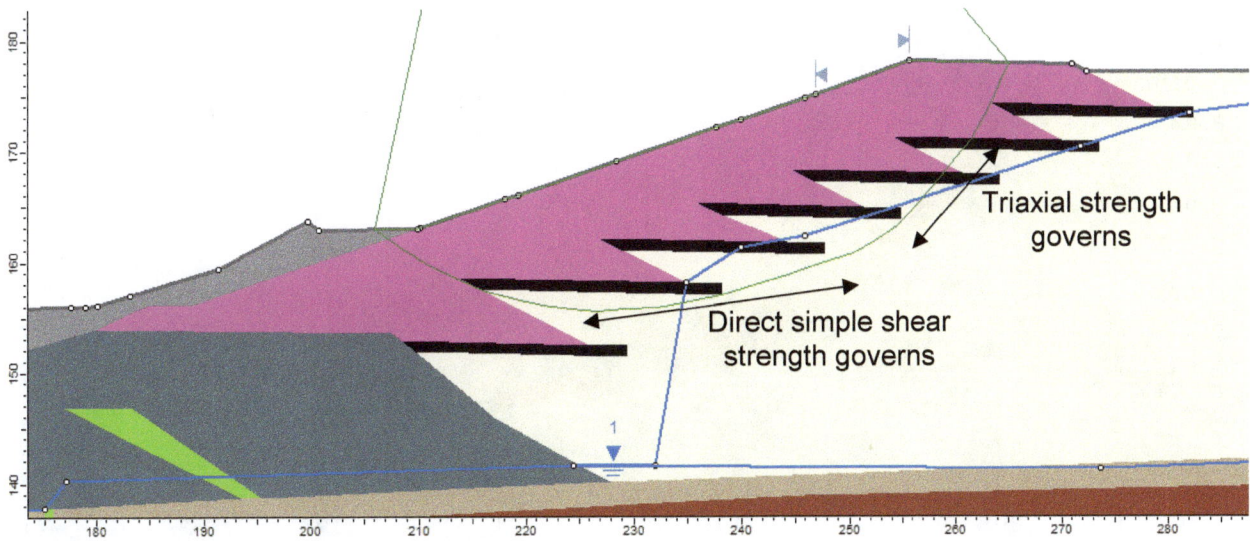

FIG 2 – Combination of triaxial and direct simple shear strength applied to tailings for local failure.

Tailings static conditions

The state parameter (Ψ) (Been and Jefferies, 1985) defines the difference between the void ratio of an element of soil, and the void ratio on the critical state line (CSL) at the same mean effective stress. Generally, soils with Ψ less than -0.05 are considered to be dilative, and not susceptible to brittle strength loss or softening behaviour post-peak strength during rapid or cyclic shearing. When Ψ is greater than -0.05 there is a risk of strength loss under these conditions. The likelihood of occurrence, and the severity of strength loss increases as Ψ increases. Equation 1 is proposed by (Jefferies and Been, 2015) to estimate Ψ:

$$\psi = \frac{\ln\left(\frac{Qp(1-Bq)+1}{k\prime}\right)}{-m\prime} \tag{1}$$

Where, QP is tip resistance normalised by mean effective stress and B_q is normalised excess pore pressure.

Interpretation parameters k' and m' are based on the rigidity of the soil and are presented in Equations 2 and 3. The direct input method proposed by (Jefferies and Been, 2015) was used in to inform this assessment and estimates λ_{10} using the slope of the CSL from critical state testing. CSL testing undertaken in 2019 suggested 1.21 and 0.2 were appropriate estimates for M_{tc} and λ_{10} respectively.

$$k' = M_{tc} \times (3 + \frac{0.85}{\lambda_{10}}) \tag{2}$$

$$m' = 11.9 - 13.3\lambda_{10} \tag{3}$$

Relationships can be developed between state parameter and undrained strength ratio (USR), as described by (Been and Jefferies, 1985). Laboratory tests results for DSS and TXC testing were plotted in Figures 3 and 4, for the respective test methods and relationships were developed between Ψ and USR. The N_{kt} value used to inform USR estimates for the tailings was calibrated to fit the relationships developed for DSS and TXC testing. The variability in tailings strength estimated using the CPTu data was then used to develop PDFs for the tailings under each strength condition.

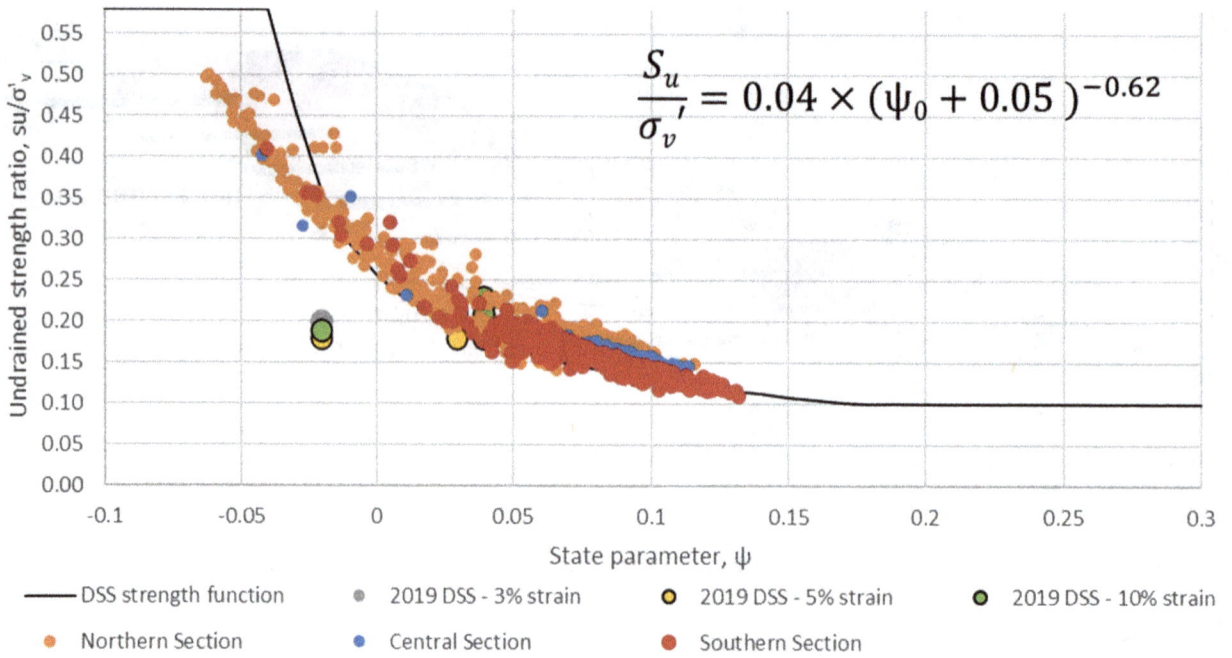

$$\frac{S_u}{\sigma_v'} = 0.04 \times (\psi_0 + 0.05)^{-0.62}$$

FIG 3 – CPTu data fitted to DSS test results.

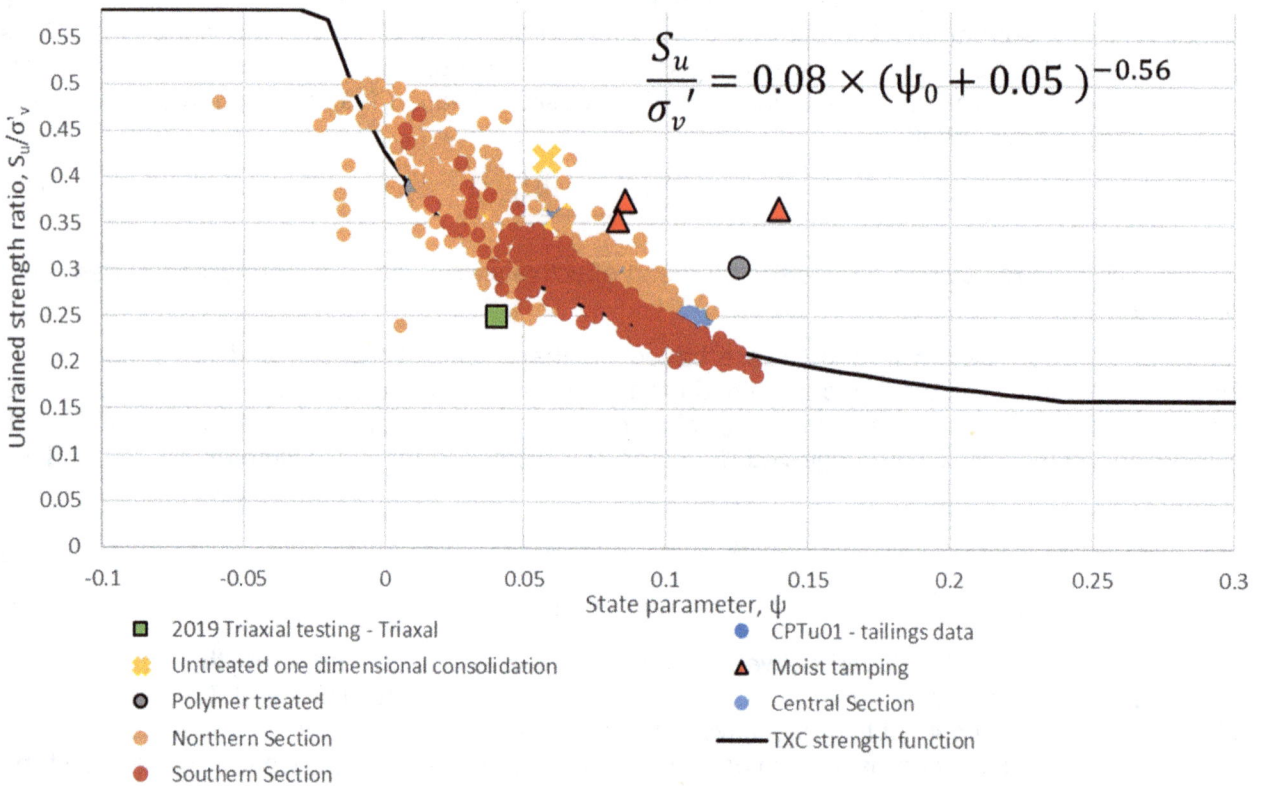

$$\frac{S_u}{\sigma_v'} = 0.08 \times (\psi_0 + 0.05)^{-0.56}$$

FIG 4 – CPTu data fitted to TXC test results.

Tailings post-seismic conditions

The SGRA undertaken in prior assessments indicated that tailings deposited above RL 156 m would be susceptible to cyclic liquefaction under the design earthquake. This assessment was undertaken using earthquake acceleration time histories derived from the site-specific probabilistic seismic hazard assessment undertaken for the TSF.

Based on the outcomes of the SGRA and the probabilities assigned to events by the SMEs, the earthquake scenario where tailings deposited above RL 156 m liquefied was critical to the assessment. Accordingly, no strain softening parameters were adopted in the assessment and have therefore not been presented in this report.

The residual tailings strength ($\frac{s_r}{\sigma'_{vo}}$) was calculated using relationships presented by (Jefferies and Been, 2015), assuming a geostatic stress ratio (K_0) of 0.5. The post-liquefaction tailings strength for tailings assumed to liquefy (ie tailings deposited from 16 m and above) was calculated using Equation 4, developed from case histories as a 'best practice' approach.

$$\frac{s_r}{\sigma'_{vo}} = \frac{1+2 \cdot K_0}{3} \cdot 0.051 \cdot M_{tc} + 0.8 \cdot (\lambda_{10} - \psi_o) \tag{4}$$

A PDF was developed for residual tailings strength, based on the variability of the inferred *in situ* characteristic state parameter from CPTu testing and are summarised in Table 2.

TABLE 2
Distribution statistics for tailings USR.

Loading condition	Distribution	Minimum	Maximum	Mean	Standard deviation
DSS	Normal	0.11	0.50	0.19	0.07
TXC	Beta	0.19	0.50	0.29	0.06
Residual strength	Normal	0.11	0.30	0.16	0.03
TXC (strain softening)	Normal	0.15	0.40	0.24	0.05

Distribution statistics

The distribution statistics associated with these PDFs are summarised in Table 2. The same PDF was adopted for each cross-section under static conditions, even though tailings deposited at the Southern Section typically has a higher ψ, and therefore a marginally lower *in situ* strength.

Buttress bulk density

Reinforcing buttresses have been progressively designed and constructed near the downstream toe of the western embankment for the TSF. Field density testing was undertaken to check the bulk unit weight of fill, once placed and compacted, to confirm that the minimum unit weight adopted in the slope stability assessments were being throughout construction.

The field density test data indicates there is variability associated with the bulk unit weight achieved during buttress construction. The statistics presented in Table 3 summarise the available compaction data from all phases of buttress construction, for each material type.

TABLE 3
Distribution parameters for buttress bulk unit weight of the buttress (kN/m^3).

Loading condition	Distribution	Minimum	Maximum	Mean	Standard deviation
Coarse Tailings	Beta	24.4	28.3	26.3	0.6
Natural Sand	Beta	17.0	22.6	18.3	0.9

Foundation density functions

Empirical N_{kt} method

An assessment of the peak undrained shear strength of the foundation sand below the phreatic surface and foundation clay has been made using the CPTu data over the respective material zones. The empirical 'N_{kt}' method was used to estimate the undrained strength ratio (USR) for each material Equation 5, as proposed by (Robertson, 2009).

$$S_u(peak) = \frac{q_t - \sigma_{vo}}{N_{kt}} = \frac{q_{net}}{N_{kt}} \tag{5}$$

Where q_t is corrected tip resistance, σ_{vo} is vertical overburden pressure, and N_{kt} is an empirical cone factor.

Contractive strength functions

Samples collected during the geotechnical investigations for both foundation sand and clay were subjected to DSS testing for effective stresses ranging between 100 kPa and 600 kPa. A strength relationship between effective stress and undrained strength ratio (USR) was developed based on the outcomes of this testing for clay zones and saturated sand layers within the foundation, as shown in Figures 5 and 6 respectively. The N_{kt} value adopted to estimate the mean USR for CPTu data was calibrated to approximately fit the strength functions developed for each material. The range of USR values from CPTu testing is also presented on Figures 5 and 6 for the respective soils.

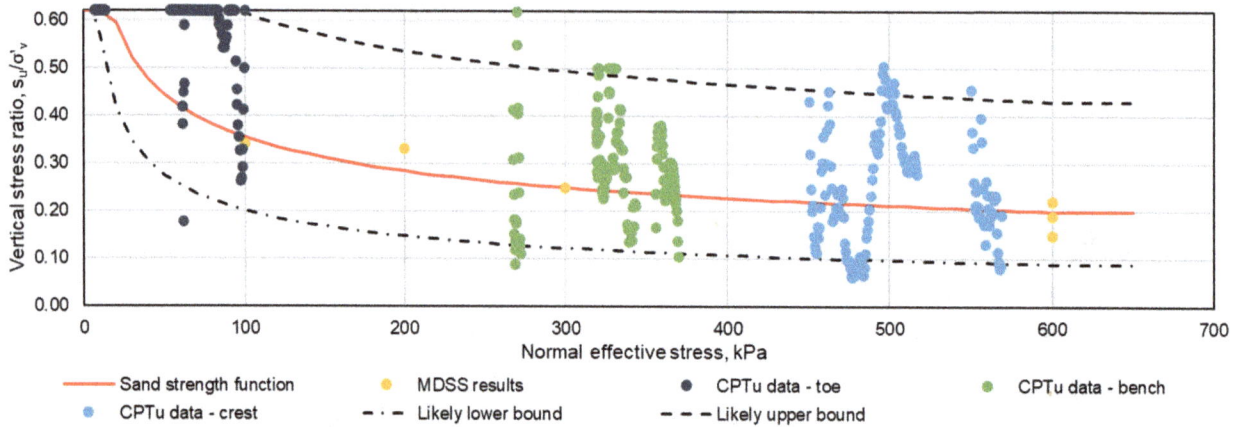

FIG 5 – Sand strength function and USRs from 2019 CPTu investigation data.

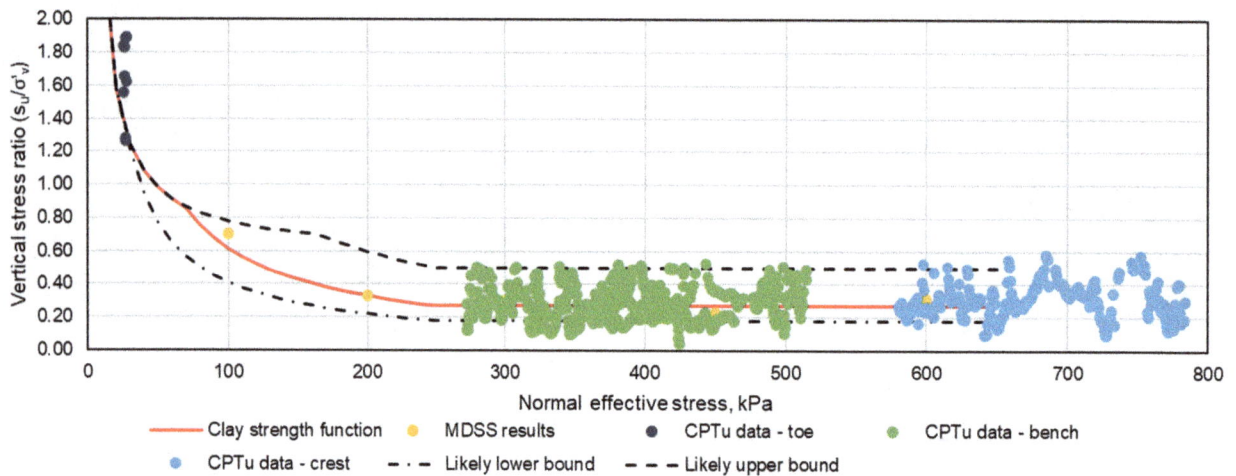

FIG 6 – Clay strength function and USRs from 2019 CPTu investigation data.

Effective stress translation

The effective stresses applied to the foundation changed since the CPTu investigation was undertaken when buttress fill was placed and compacted to the west of the TSF. An effective stress model was developed to replicate conditions at the time of the investigation. The TSF geometry was then updated to replicate the effective stresses applied to the foundation during Raise 8 operation, as shown in Figure 7.

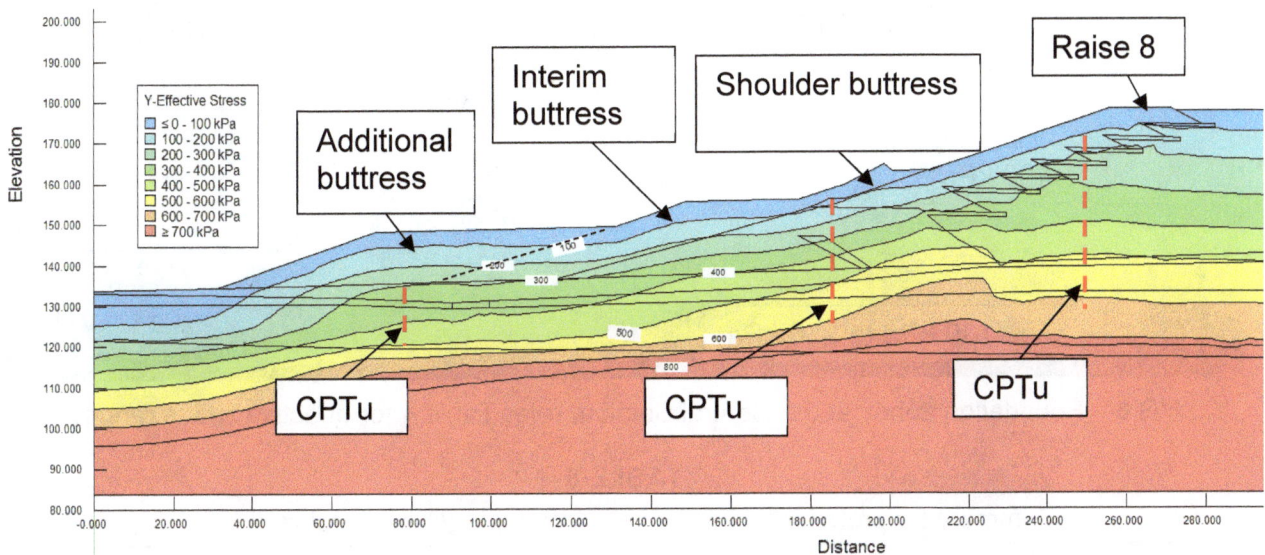

FIG 7 – Modelled effective stresses once Raise 8 construction is complete at the Central Section.

The strength functions presented in Figures 5 and 6 were used to estimate the mean USR for zones within the foundation. The standard deviation of the CPTu data fitted to the strength functions was then used to estimate the strength variability for stress ranges within the foundation. The soils within the foundation were split into the effective stress zones presented in Table 4.

TABLE 4
Distribution parameters foundation USR.

| Material type | Effective stress range (kPa) | | Colour | Effective stress (kPa) | Zoned USR parameters | | | |
	From	To			Mean	Standard deviation	Min	Max
Sand	<100			Constant USR of 0.62 adopted (Note)				
	100	400		325	0.24	0.12	0.07	0.48
	>400			600	0.20	0.12	0.06	0.41
Clay	<100			100	0.60	0.51	0.44	20
	100	250		200	0.33	0.24	0.08	1.40
	>250			300	0.27	0.11	0.04	0.54

Note, this is consistent with the unsaturated, drained strength adopted during deterministic LE modelling.

The extent of each of these zones adopted in the stability model was dependent on the stresses modelled in Figure 7 and are shown in Figure 8 for the base Raise 8 model (no slope instability trigger incorporated). A stability model was then set-up for the respective instability triggers and the PoF was estimated using the Monte Carlo simulation built into the stability model software. The PoF assuming each trigger occurred is summarised in Table 5. Note, the risk associated each failure was also dependent on the probabilities assigned in the fault trees.

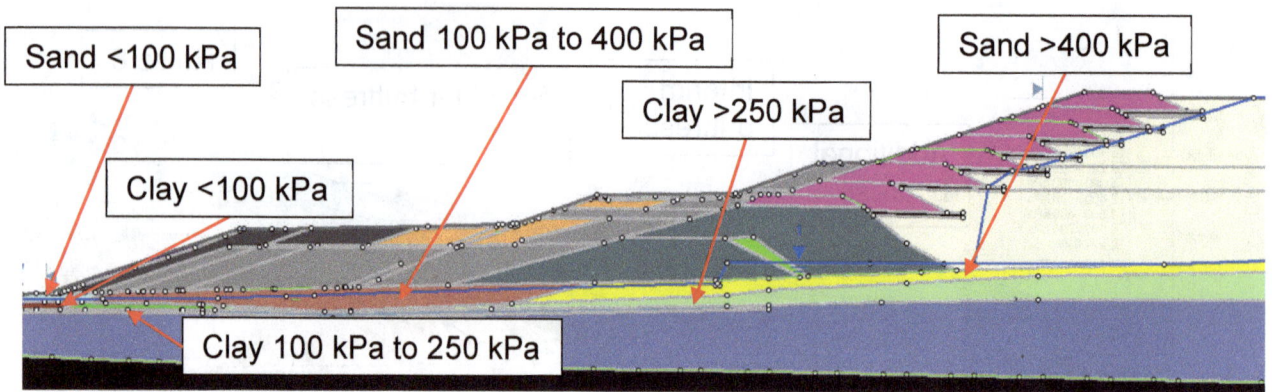

FIG 8 – Foundation zones at the Central Section, refer Table 4 for effective stress ranges.

TABLE 5

Summary for probability of failure assuming trigger has occurred (%).

Instability trigger	Upper embankment failure			Foundation failure		
	Northern Section	Central Section	Southern Section	Northern Section	Central Section	Southern Section
Rising phreatic surface	5.8	0	6.3	14.6	8.3	9.9
Earthquake	3.7	0.9	5	3.6	8.8	7.6
Geometry change due to undercutting of the toe or scour	5.5	5.8	7.7	15.8	9.7	5.2
Construction	5.5	2	4.2	4	9.6	6.5
Lateral extrusion	0	6.9	1.3	-	4	36.3

Foundation sensitivity

A likely lower bound strength function was considered for both sand and clay materials, shown in Figures 5 and 6, to check suitably conservative parameter combinations were being considered using the Monte Carlo simulations. The FoS achieved for the base model ranged between 0.59 and 2.28. The FoS comparatively ranged between 0.76 and 0.77 when adopting the lower bound strength function for both contractive foundation soils. As the zoned approach achieved FoS values less than the lower bound strength function, the parameter combination from the zoned assessment was considered appropriate. Additionally, the relatively close range of FoS outputs for the lower bound functions indicate most variability is associated with the strength adopted for the foundation soils.

QUANTIFYING RISK

Individual risk

The risk an individual is exposed to by entering the inundation zone downstream of the TSF is summarised in Table 6. The table indicates the risk attributed to each instability trigger for both failure modes considered. The individual risks are based on the PoF estimated from the stability analyses and the probabilities assigned to fault tree events in the QRA.

TABLE 6

Breakdown of individual risk for each instability trigger.

Instability trigger	Individual risk	
	Upper embankment failure	Foundation failure
Rising phreatic surface	1.27×10^{-5}	6.65×10^{-6}
Earthquake	1.80×10^{-5}	1.85×10^{-5}
Geometry change due to undercutting of the toe or scour	1.58×10^{-13}	1.06×10^{-9}
Construction	4.00×10^{-5}	5.50×10^{-5}
Lateral extrusion	$-$ *	3.38×10^{-9}
Individual risk – each failure	7.07×10^{-5}	8.02×10^{-5}
Combined – TSF failure	1.51×10^{-4}	

* – Lateral extrusion was not considered through the upper embankments due to compaction achieved during construction for the upstream embankment raises.

Population at risk

A dam break assessment (DBA) was previously undertaken for the TSF to consider the extent of the inundation zone if breach was to occur. That assessment was published in June 2020 and was based on a range of assumptions that are now superseded with the improved knowledge and planning for the TSF. Notwithstanding, it was well-understood in the QRA that the population at risk (PAR) was not a broader societal risk and was confined to mine site staff. In that context, the f-N charts are presented herein with the PLL being that of the mine site as the societal PLL is zero. Based on the estimates provided by owner, the PLL was estimated to be 0.31, ie there is no permanent population in the inundation zone, only itinerant mine site personnel for inspections and maintenance.

The highest annual probability of failure of the three cross-sections examined was estimated to be 2.26×10^{-5}. The embankment of the TSF that was examined, the highest confining embankment, combined with the PLL of the mine site workers, plots in the tolerable zone, as shown in Figure 9.

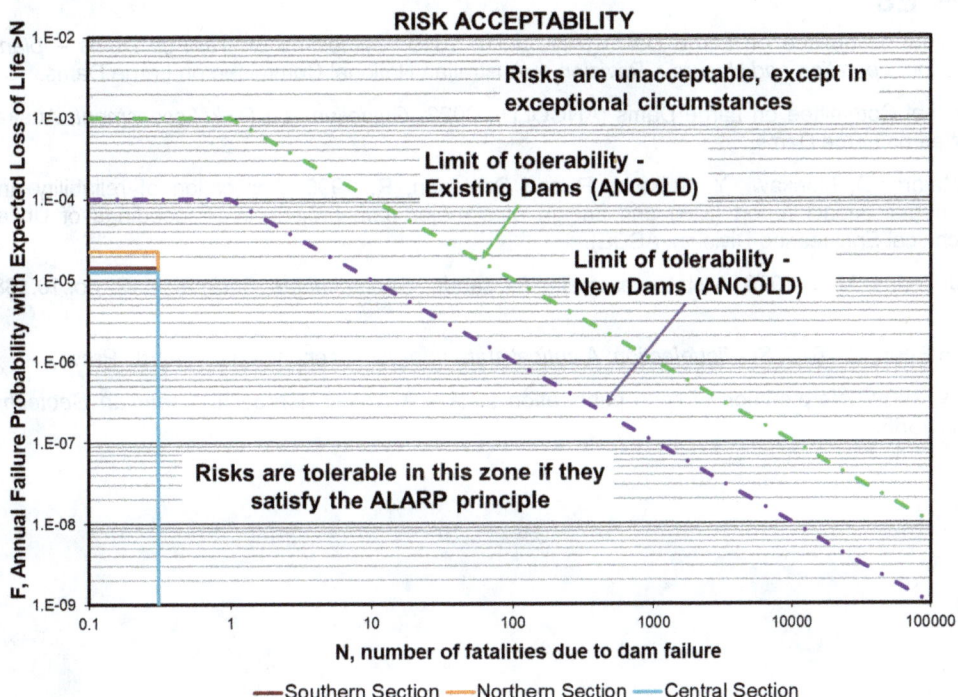

FIG 9 – Societal risk for catastrophic failure of the western embankment.

SATISFYING ALARP

The authors acknowledges that the risk profile for the western embankment of the TSF is in within the tolerable risk zone of the f-N charts, so long as they satisfy the ALARP principle. On the basis of the assessment undertaken, the EOR recommended that the Raise 8 embankments be constructed and operated without any further buttressing. However, to continue to satisfy the ALARP principle, further studies and/or improvements were recommended to be undertaken to improve the understanding and risk assessment of the western slope of the TSF, as follows:

1. A numerical deformation model should be developed to indicate the deformations of the slope under the static and cyclic liquefaction triggers considered in the QRA. This could potentially indicate that additional buttressing is required based on the model performance. This assessment is being undertaken by the owner/EOR at the time of writing.

2. The DBA should be revisited for the proposed geometry of the Raise 8 embankments to assess if there is any change to the societal PLL. Note that this assessment is being undertaken by the owner/EOR at the time of writing.

3. The QRA could be expanded to consider other potential failure modes such as overtopping and piping erosion. The TSF has storage capacity for more than the probable maximum flood (PMF) event and therefore, overtopping is considered low risk. However, quantifying the risk associated with failure due to internal erosion is an opportunity for improvement.

CONCLUSION

The outcomes of the QRA presented in this paper indicates the benefit of using a risk-based approach to the design of the performance of the slope of a TSF. There is more work to be undertaken, such as the DBA and the deformation model, which are considered to be continuations of satisfying the ALARP principle. It strongly indicates that investment of millions of dollars in buttressing just to satisfy FoS would be better invested in the supplementary studies to improve the knowledge base of the TSF and better characterise the risk, which would also help to inform future raise designs and closure of the TSF.

ACKNOWLEDGEMENTS

The authors acknowledge the support of the TSF owners to undertake the work and share the information documented in this paper.

REFERENCES

Australian National Committee on Large Dams (ANCOLD), 2019. *Guidelines on Tailings Dams – planning, design, construction, operation and closure – Revision 1*, Australian National Committee on Large Dams.

Australian National Committee on Large Dams (ANCOLD), 2022. *Guidelines on Risk Assessment,* Australian National Committee on Large Dams.

Barneich, J, Majors, D, Morikawi, Y, Kulkami, R and Davidson, R, 1996. Application of reliability analysis in the environmental impact report (EIR) and design of a major dam project, in Proceedings of Uncertainty 1996, (Geotechnical Engineering Division: ASCE).

Been, K and Jefferies, M G, 1985. A state parameter for sands, *Géotechnique*, 35(2):99–112. doi:10.1680/geot.1985.35.2.99

Jefferies, M and Been, K, 2015. *Soil liquefaction: A critical state approach*, second edition (CRC Press).

Robertson, P K, 2009. Interpretation of cone penetration tests – A unified approach, *Canadian Geotechnical Journal*, 46:1137–1355.

Investigation of a coalmine spoil pile failure

J E Wilson[1], M Llano-Serna[2], L Diaz Amaya[3] and D J Williams[4]

1. MAusIMM(CP), Principal Geotechnical Engineer, BHP Mitsubishi Alliance, Goonyella Riverside Mine, Moranbah Qld 4744. Email: jordan.wilson@bhp.com
2. Principal Geotechnical Engineer, Red Earth Engineering – a Geosyntec Company, Brisbane Qld 4000. Email: marcelo.llano@redearthengineering.com.au
3. Masters Student, The University of Queensland, St Lucia Qld 4072. Email: l.diazamaya@uq.net.au
4. FAusIMM, Professor of Geotechnical Engineering, The University of Queensland, St Lucia Qld 4072. Email: d.williams@uq.edu.au

ABSTRACT

A significant spoil pile failure occurred at Goonyella Riverside Mine in the Bowen Basin Coalfields of Queensland, Australia, between 1 and 5 July 2022. The spoil pile consisted primarily of Tertiary silts and clays excavated from the adjacent pit and occurred during a rainfall event that was preceded by several large rainfall events over the 24 months prior to the failure. These rainfall events led to the wetting up and degradation of the silty and clayey spoil, forming soft layers. The failure was captured using various instruments and field monitoring techniques. It was concluded that wet spoil layers with high clay mineral content behaving undrained during shearing played a key role in the failure. The failure has been the subject of back-analysis using the finite element method, sampling and laboratory testing, and a piezocone investigation, aimed at understanding the degradation and strength loss on wetting up of the clay mineral-rich spoil and preventing similar failures in the future. The paper describes the construction and composition of the spoil pile, the impact of rainfall events, the failure, and the results of the investigations. The investigations have led to an improved understanding of the degradation and strength loss of the clay mineral-rich spoil. Recommendations are made to prevent similar failures occurring in the future.

INTRODUCTION

Goonyella Riverside Mine (GRM) is situated in the Moranbah coal measures within the Bowen Basin of Queensland, Australia. The mine produces hard, low volatile, metallurgical coal for the seaborne market. The overburden consists primarily of Tertiary age, unconsolidated clays, silts, and sands overlaying Permian age sedimentary rock. At the northern extent of the mine, a large paleochannel feature cuts through the end wall, thickening the Tertiary unit from an average of 30 m from the natural surface to up to 90 m. The mining process involves both dragline and truck/shovel operations to remove overburden, with the trucks dumping in spoil piles that progressively follow the dragline strip excavation.

The stability of these spoil piles has direct consequences, from the safety of the workforce and the financial feasibility of the operation to long-term environmental and reputational damage.

A large spoil pile failure occurred in the GRM R16 spoil on 1 July 2022 during a rainfall event that had locked out access to the pit (Figure 1). The failure was captured with a combination of CCTV, a Slope Stability Radar and periodic UAV photogrammetry as the movement progressed over the next five days. The failure slowed to negligible levels on 5 July 2022, after which recovery activities commenced. The failure involved 15 million cubic metres of spoil and ran out 200 m from the toe of the RL240 level (Relative Datum +280 m).

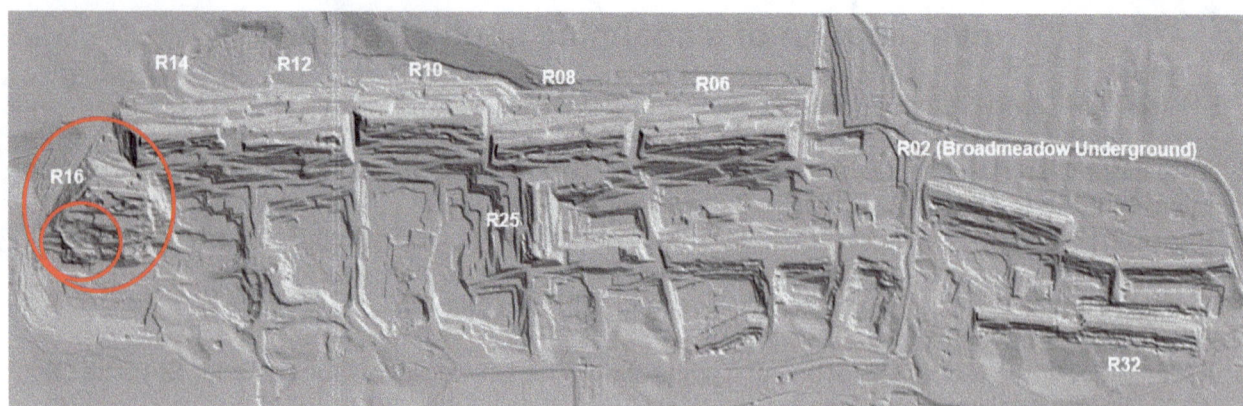

FIG 1 – Aerial relief map of the main excavation of GRM, with operational pits labelled. Ramp 16 and spoil failure are marked by red circles.

Despite the significant volume involved, the failure did not have material consequences to the site. Due to a combination of its location distant from the pit, pit progression and a mature hazard management process, no people were put at risk during the failure, and operations resumed within days of the failure.

There remains a significant concern, however, as the spoil pile was not considered to be a 'high-risk' design. With slight changes to the location, velocity or runout of the failure, harm to people and/or significant financial impact could arise from a similar failure. Therefore, it is critical to understand the causes of this failure to ensure that there are no other 'low-risk' or future spoil piles comprising similar materials with a potential to fail.

TIMELINE OF EVENT

27 June 2022 (t-3 days, Figure 2):

Cracking at RL280 was observed by Open Cut Examiners and site-based Geotechnical Services were called to inspect. The geotechnical assessment at the time was based on previous monthly LiDAR, indicating movement of ~2 m between early April and 27 June 2022, equating to less than 1 mm/hr of movement.

Action was taken to include the spoil piles in the weekly LiDAR capture area to enable weekly monitoring of slope movement and to inform whether closer monitoring with a slope stability radar was warranted. In addition, the observed cracking was recommended to be filled before the predicted wet weather arrived in the coming days.

30 June 2022 (t-12 hrs):

Cracking reappeared at the same location on the RL380 level and a GroundProbe Slope Stability Radar was to be installed on the adjacent highwall. The radar data indicated spoil pile movement of ~10 mm/hr between the RL380 and RL310 levels. An inverse velocity plot pointed to failure within 24 hrs, and pit operations were stopped.

1 July 2022 (failure, Figure 3):

The spoil velocity at the RL310 level progressed to >45 mm/hr at 12:05 am 1 July 2022, exceeding the range of the radar. At this time, movement began to accelerate at the RL240 level.

2 July 2022 (t+24 hrs, Figure 3):

Approximately 24 hours later, at 11:30 pm 1 July 2022, spoil velocity at the RL240 level exceeded the 45 mm/hr radar detection limit, setting in motion broad-scale movement laterally. Radar and survey data indicated a progressive failure, in which lower-level soil was triggered to failure. During daylight hours on 2 July 2022, the failure was captured on mine CCTV and later analysed to estimate its velocity to be constant at 5 m/hr.

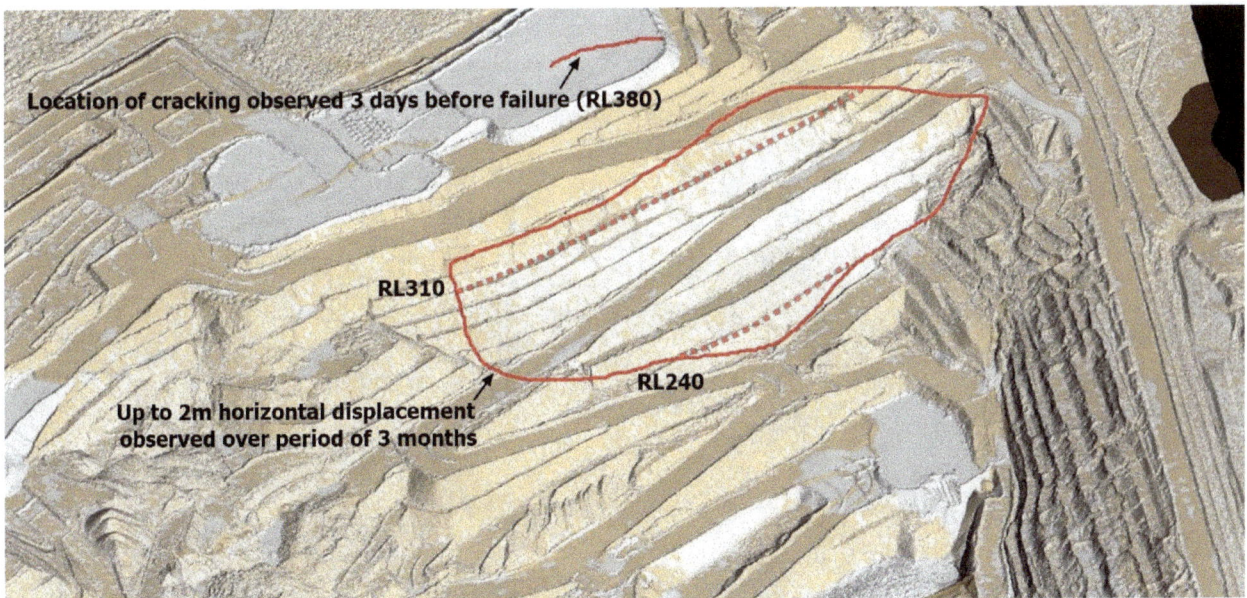

FIG 2 – Spoil creep observed after reported cracking on 27 June 2022. The white surface indicates June LiDAR topography, and the beige surface indicates April LiDAR.

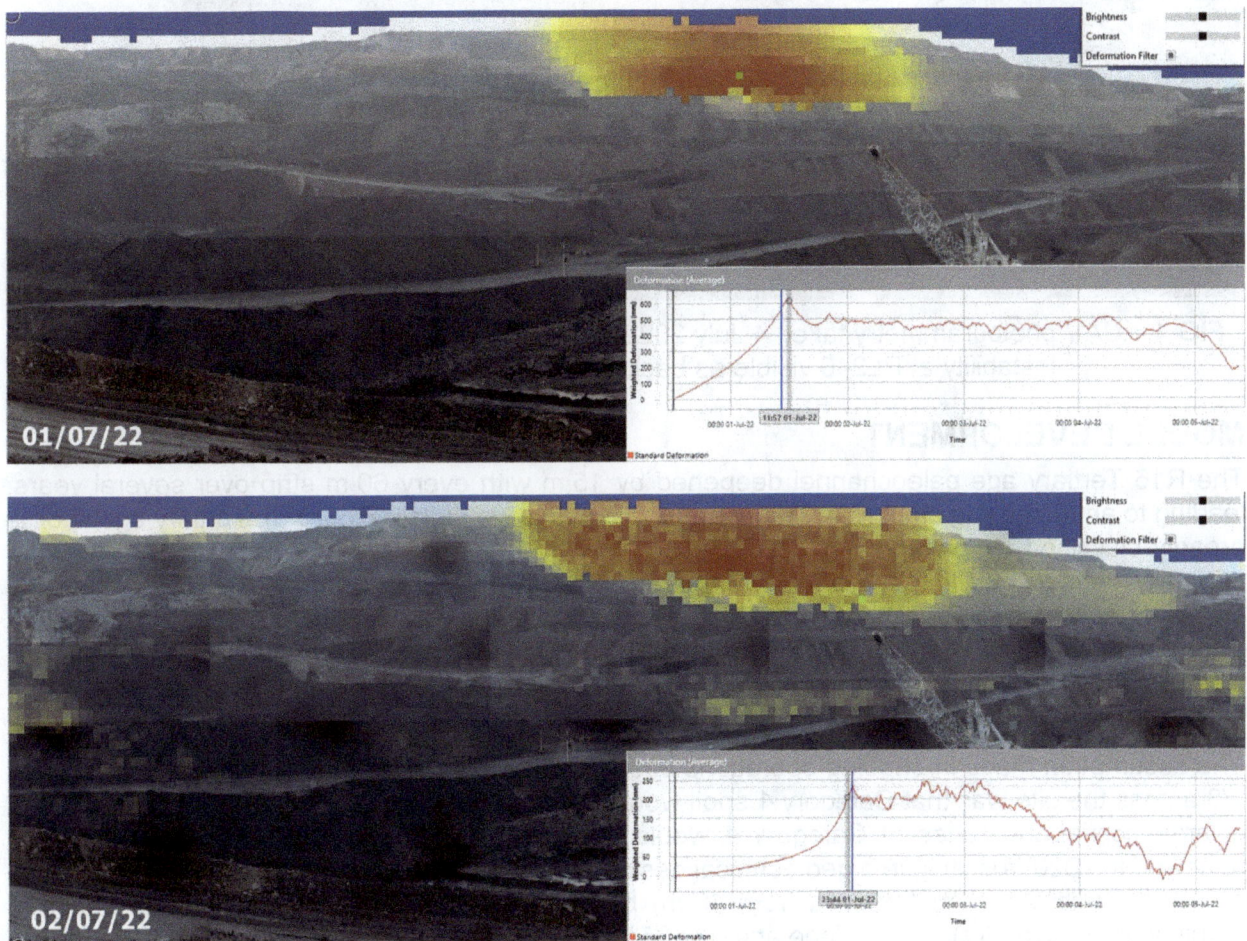

FIG 3 – Radar output on 1 and 2 July 2022 showing the failure at RL310 and RL240, respectively. Characteristic hyperbolic deformation is evident before failure at both locations.

4 July 2022 (t+72 hrs, Figure 4):

Movement of the failure mass ceased by 4 July 2022, as the rainfall abated, allowing access into the pit to assess recovery requirements. UAV photogrammetry was flown to determine the final failure geometry.

FIG 4 – UAV photogrammetry taken 4 July 2022, with initial instability at RL310 and subsequent instability at RL240 visible as back scarps within overall failure mass.

MODEL DEVELOPMENT

The R16 Tertiary age paleochannel deepened by 15 m with every 60 m strip over several years, leading to an increasing amount of Tertiary age silty and clayey spoil (Figure 5). Tertiary age spoil is broadly classified as 'Category 1' according to the BMA Material Spoil Classification Framework originally defined by Simmons and McManus (2004). Permian age sedimentary rock is typically classified as Category 2 or 3 based on other metrics within the Framework. It should be noted that a key assumption in applying the BMA Material Spoil Classification Framework is that the spoil remains largely unsaturated and hence that excess pore water pressures are not generated when sheared.

The assumption made in ongoing operations based on the relative ratio of Category 1 to Category 2/3 across the site was that Category 1 spoil would be distributed with Category 2/3 spoil such that it large contiguous zones of Category 1 would not be possible. This assumption for R16 was reviewed in 2020, resulting in a geotechnical design update requiring a lower overall spoil pile slope angle. Specifically, 15 m (reduced from 10 m) benches between 20 m spoil lifts were adopted, resulting in an overall spoil pile slope angle of 25°.

The overall spoil pile slope angle as-constructed was in fact significantly flatter than 25° due to the addition of haul roads on the slope, resulting in an overall slope angle at the time of failure of only 16°.

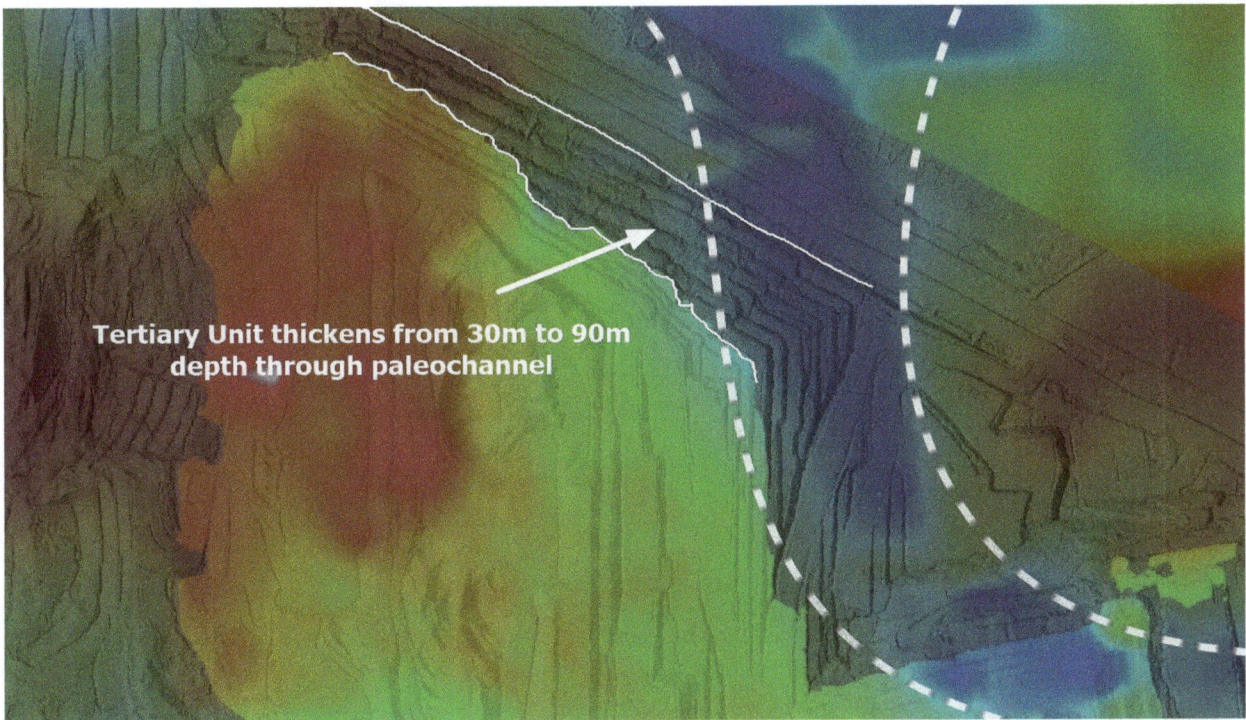

FIG 5 – The base of the Tertiary in the pit has progressively become deeper strip by strip (traced in the solid white, along with the natural surface). Base of Tertiary (BUTE) is coloured by the z-axis; red is the shallowest, and blue is the deepest.

The specific location of where the Category 1, 2 and 3 spoil is dumped within a spoil pile is not explicitly tracked on-site, as designs have historically been material invariant. However, to back analyse the failure, a model of the spoil pile was generated using historic dig/dump coordinates, sourced from truck GPS beacons, and stored in the GRM fleet management database (Figures 6 and 7). These data were combined with rainfall from the Australian Bureau of Meteorology rain gauge at the Moranbah (MOV) Airport, about 30 km south of GRM.

FIG 6 – Cross-section of the pre-failure topography, with dump points coloured by material type. Tertiary age spoil is displayed in orange and Permian age spoil in grey.

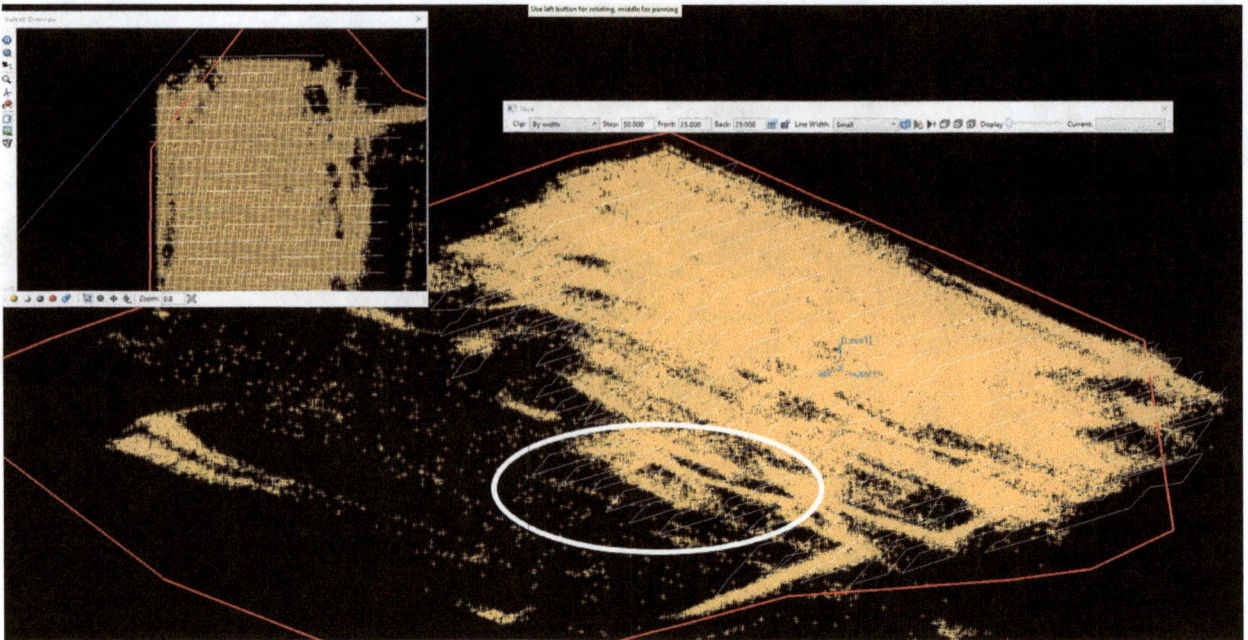

FIG 7 – Isometric view of the Tertiary age dump point cloud of R16. Circled in white is the location of the failure primarily confined to zones with Tertiary age spoil dumped down to RL280.

RAINFALL ANALYSIS

Tertiary age spoil is distinct from Permian age spoil in its inability to drain water rapidly. Therefore, layers of Tertiary age spoil are likely to host a series of perched water tables or regions of high saturation. The spoil failure toe was 100 m above the basal seam in this pit, sitting atop well-draining Permian spoil. As a result, vibrating wire piezometers would not have provided meaningful data in the areas of concern unless the location and hazard posed by the wet layers was already known.

Buried clayey spoil is likely to retain its as-deposited or wet-up moisture content for decades. To model locations of acute high saturation in the spoil pile, historical daily rainfall data from MOV was analysed on rolling 30 and 365-day bases (Figure 8).

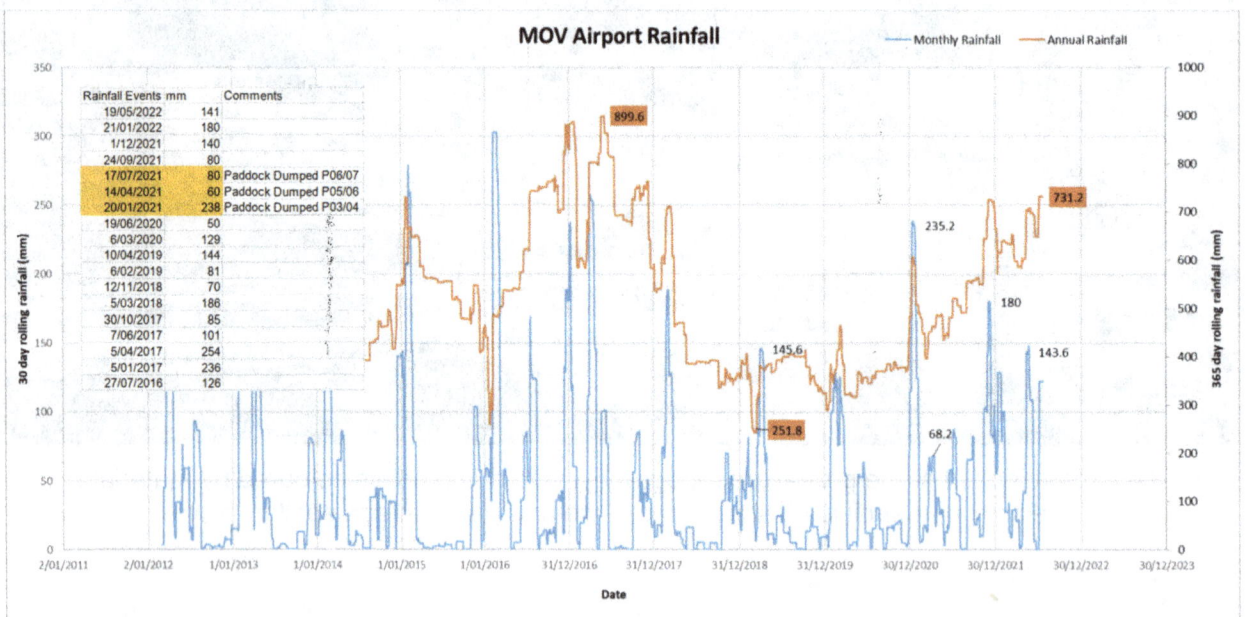

FIG 8 – Daily rainfall data aggregated to rolling 30 and 365-day totals for Moranbah Airport. Source: Australian Bureau of Meteorology.

While monthly rainfall events have been higher in previous years, the 2020 to 2022 rainfall was characterised by many smaller and unseasonal events through the winter (which is normally the dry season), leading to a near decade-high annual rainfall of 731 mm at the time of the failure, and frequently wetting-up exposed Tertiary age spoil.

The corresponding monthly LiDAR topography was inspected for each rainfall event as per the table in Figure 8. Flat areas with poor drainage potential and paddock dumps were marked with polygons for each wet month and extrapolated vertically by 5 m to create a 'wet material' model that could be coupled with the spoil category model of the R16 spoil pile.

Aerial photography was compared with the LiDAR topography to validate this approach, as shown in Figures 9, 10 and 11.

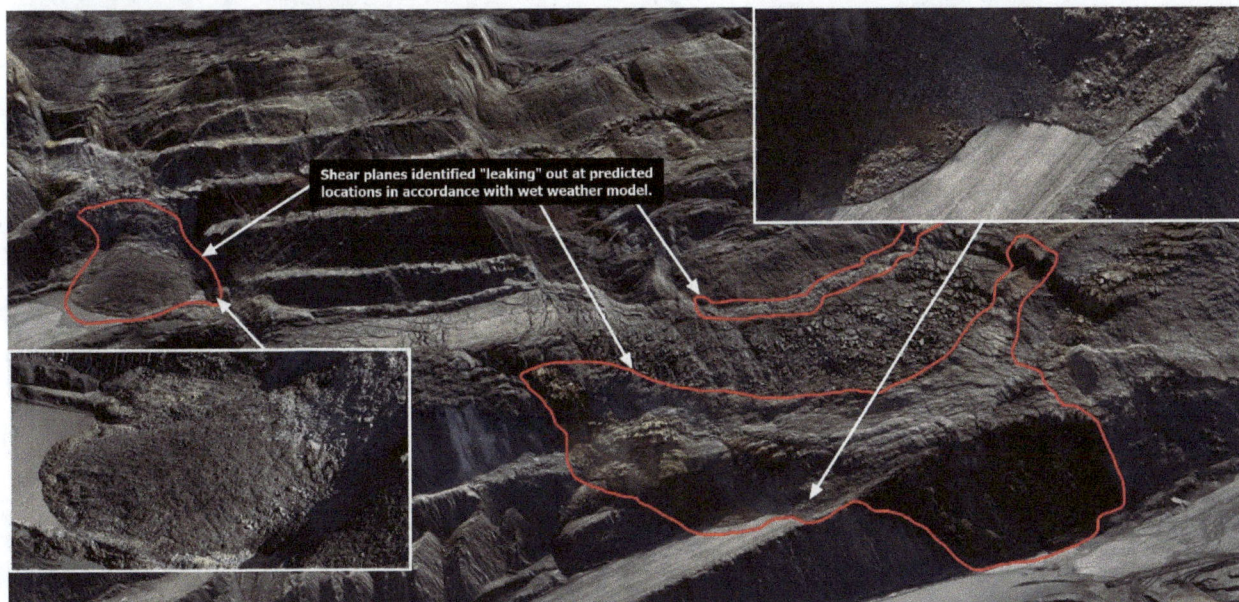

FIG 9 – Post-failure photography showing locations where the spoil flowed out of the broader failure mass. These locations correspond to historically identified wet zones in Figure 11.

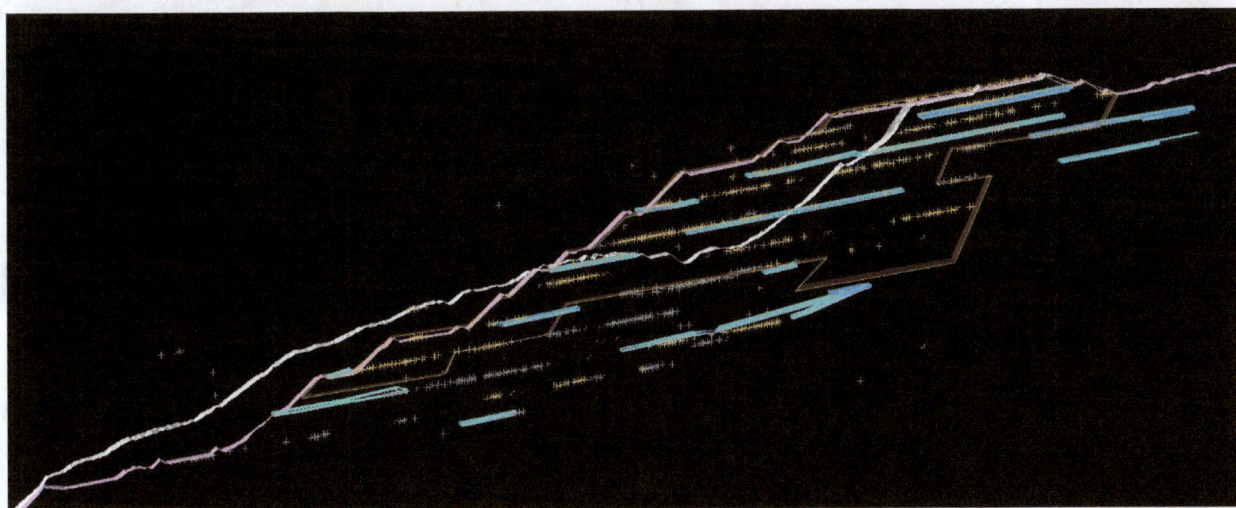

FIG 10 – Cross-section through the centreline of the R16 spoil pile failure used to generate 2D models. The wet layer model is coloured in Cyan, the pre-failure surface in pink, and the post-failure surface in white. The extent of the Tertiary age spoil is displayed in beige.

FIG 11 – 2021 topography of the RL310 level of the R16 spoil pile, comprising paddock-dumped Tertiary age spoil. Monthly rainfall totalled 250 mm, and water is seen pooling between the paddock-dumps, preventing dozing.

LABORATORY TESTING

Eleven representative spoil bulk samples from several lithologies dumped in the R16 spoil pile were collected from two strategic locations that allowed safe access to the failure during a site visit on 22 and 23 August 2022. The collection of disturbed samples followed conventional geotechnical engineering practice, see Figures 12 and 13.

FIG 12 – Sample collection of Grey Tertiary age spoil (left) and breaking up of one of the Grey spoil samples (right).

FIG 13 – Approximate location of spoil sample collection locations at levels ~RL220 and ~RL310.

Six clayey spoil types were collected. The extensive and ongoing laboratory testing program conducted at The University of Queensland Geotechnical Engineering Laboratories included characterisation testing, and material strength testing using a Simple Shear Test (SST) apparatus. Characterisation test results for the spoil samples collected are shown in Table 1. After reviewing the characterisation test results, it was decided to conduct strength testing on the Grey-1 and the Mottled White Yellow (MWY) samples. The criteria for selecting the Grey-1 and MWY samples was: (i) Grey-1 had the lowest *in situ* Gravimetric Moisture Content (GMC) and the lowest Liquid Limit (LL), and (ii) MWY had the highest LL, highest clay content and lowest GMC, representing the range of samples collected.

The GDS Electro-Mechanical Cyclic Simple Shear Device (GDS-Instruments, 2023) used to conduct the strength tests restrains specimens in a rubber membrane to allow control of drainage, and thin horizontal steel rings to support the specimen. The top cap is moved horizontally relative to each the base to apply shear loading.

The SST undrained strength testing was conducted using applied vertical stresses at 50 kPa, 100 kPa, 200 kPa, 400 kPa, and 800 kPa to simulate nominal overburden depths of approximately 2.5 m, 5 m, 10 m, 20 m, and 40 m, respectively. Various GMCs were tested to simulate wetting-up by rainfall after paddock dumping. Each specimen was consolidated overnight at the applied vertical stress and sheared undrained at a strain rate of 0.7 mm/min or 1 per cent/min.

TABLE 1

Summary of the characterisation tests performed on the 11 samples collected at the R16 spoil dump failure.

Bucket No.	Material	GMC (%)	LL (%)	PL (%)	PI (%)	G_s	PSD			
							Clay Content (%)	Fines Content (%)	Sand Content (%)	Gravel content (%)
1	Yellow-1	5	55	20	34	2.71	33	82	18	0
2	RB-1	28	74	26	48	2.66	27	77	21	2
3	MWY	5	97	26	71	2.69	62	99	1	0
4	Grey-1	16	36	17	19	2.65	15	43	37	20
5	Grey-2	22	38	14	24	2.65	16	42	38	20
6	RB-2	31	72	31	41	2.63	27	76	21	3
7	Grey-3	20	36	15	21	2.66	18	41	39	20
8	Yellow-2	5	84	23	60	2.69	61	92	8	0
9	WCS	6	91	23	68	2.62	50	97	3	0
10	Olivine-1	29	63	34	28	2.63	17	45	29	26
11	Olivine-2	21	60	34	26	2.66	8	18	70	12

The Grey-1 sample (Table 1) was tested at its as-sampled GMC=16 per cent and at the LL, which equates to GMC=36 per cent. For the MWY sample, the GMC was increased from as-sampled (5 per cent) to GMCs of 20 per cent, 40 per cent and 60 per cent. Testing at higher GMCs was not possible due to the slurry-like specimen squeezing between the ring and the loading piston, see Figure 14. The Grey-1 sample was tested 'as sampled' and wetted up to the LL, whilst the MWY sample was ground and wetted up to the abovementioned GMC. Samples were moist tamped within the testing ring prior to testing.

FIG 14 – Squeezing and excessive deformation of MWY specimens at GMC=80 per cent (left) and GMC=97 per cent (LL) under only 50 kPa applied vertical stress.

For MWY at GMC=40 per cent and GMC=60 per cent, the specimen also experienced slurry-like squeezing behaviour similar to that seen in Figure 14 when the applied vertical stress was above 400 kPa for the former and 200 kPa for the latter, limiting the applied vertical stresses accordingly.

The undrained SST results shown in Figure 15 include the shear stress-shear strain plots for MWY prepared at GMC=20 per cent and tested at an applied vertical stress of between 100 kPa and 800 kPa, giving peak undrained shear strengths of between about 95 kPa and 185 kPa, respectively, and minimal softening post-peak. Similar behaviour was observed for Grey-1.

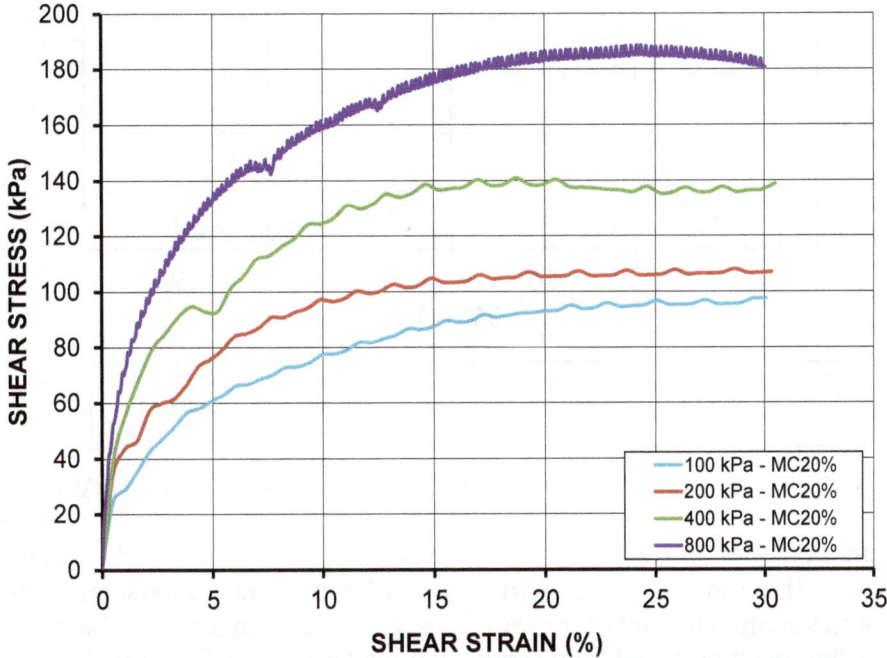

FIG 15 – Stress-Strain curves for MWY at GMC=20 per cent.

Figure 16 shows the interpreted strength envelopes obtained from SSTs at different GMC for Grey-1 and MWY, indicating that the undrained shear strength ratio (s_u/σ'_v) and the undrained shear strength (s_u) at $\sigma'_v=0$ decreased with increasing GMC. For MWY, s_u/σ'_v ranged between 0.01 and 0.28 for GMC between 60 per cent and 5 per cent, respectively. For Grey-1, s_u/σ'_v ranged between 0.19 and 0.25 for GMC between 36 per cent and 16 per cent, respectively.

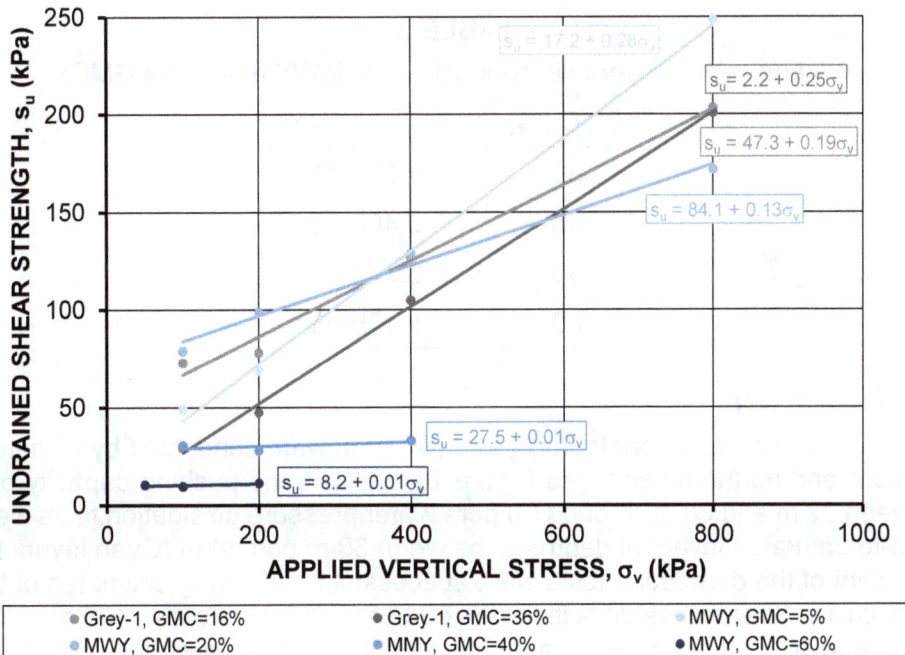

FIG 16 – Strength envelope for Grey-1 and MWY at different GMC.

Figure 17 shows the relationship between s_u/σ'_v and GMC and between the undrained shear stress at no confinement stress ($s_{u\ \sigma'v=0}$) and GMC. Suction values of about 57 MPa for the MWY at its as-sampled GMC were estimated using a Dew Point Potentiometer, making sample preparation in the SST difficult and accounting for its high apparent strength.

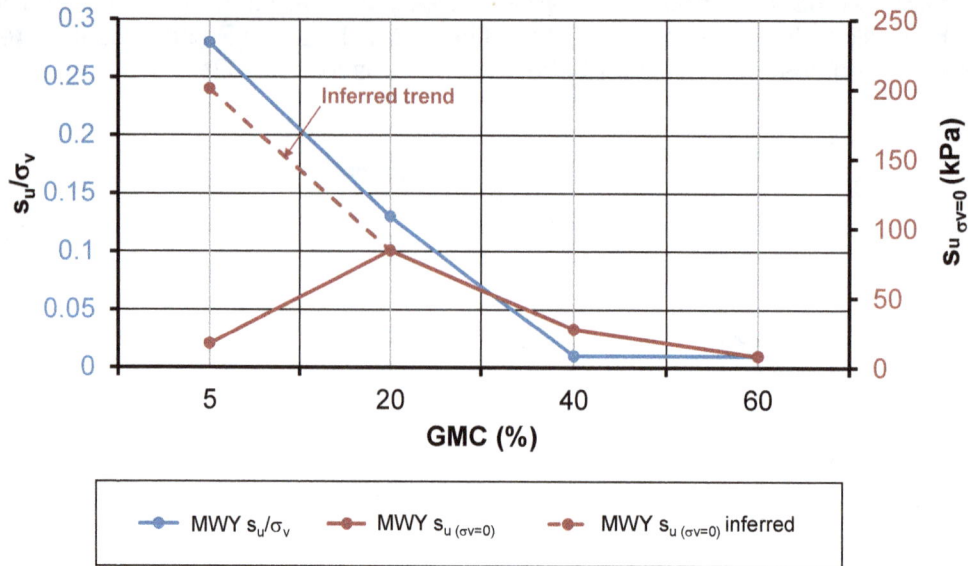

FIG 17 – Relationship between s_u/σ'_v and $s_{u\ \sigma'v=0}$ for different GMC.

For the SST tests at higher GMCs, the increase of water reduced the suction making the sample preparation easier. The red dashed line in Figure 17 represents an estimate of the $s_{u\ \sigma'v=0}$ at GMC=5 per cent (as-sampled), consistent with the $s_{u\ \sigma'v=0}$ values obtained for higher GMC. The solid red line indicates the result obtained from regression analysis of the SST results.

Four oedometer tests were conducted on MWY to estimate vertical hydraulic conductivity. Applied vertical stresses were 22.5 kPa, 50 kPa, 100 kPa, 250 kPa, 500 kPa and 1000 kPa. Specimens were prepared at initial GMCs of between 30 per cent and 60 per cent. The specimen prepared at GMC=30 per cent underwent swelling under 22.5 kPa, due to residual suction being higher than this applied vertical stress.

Table 2 summarises the average vertical hydraulic conductivity (k_v) values obtained for each GMC, covering the narrow range from 1.70E-10 to 5.46E-10 m/s, equating to approximately 5 mm/yr.

TABLE 2

Average vertical hydraulic conductivity of MWY at various GMCs.

GMC (%)	$k_{v\ average}$ (m/s)
30	1.70E-10
35	5.46E-10
40	3.77E-10
60	5.13E-10

SITE INVESTIGATION

Six 'post mortem' Seismic Cone Penetration Tests (SCPTu) were conducted by Conetec around the failure main scarp and northern flank, see Figure 18, with a cone-pushing capacity of 150 MPa to depths of between 22 m and 68 m. Around 50 pore water pressure dissipation tests were attempted targeting close-to-saturated layers at depths of between 30 m and 50 m (Cyan layers in Figure 10). Around 30 per cent of the dissipation tests were successful in providing estimates of the horizontal hydraulic conductivity. Shear wave velocities ranged between 200 m/s and 400 m/s, irrespective of the degree of saturation. Compression wave velocities ranged between 500 m/s and 900 m/s for essentially dry spoil and were higher than 1000 m/s and up to 1400 m/s in near-saturated fine-grained spoil. Vane shear strength testing was attempted in wet spoil layers but with no success.

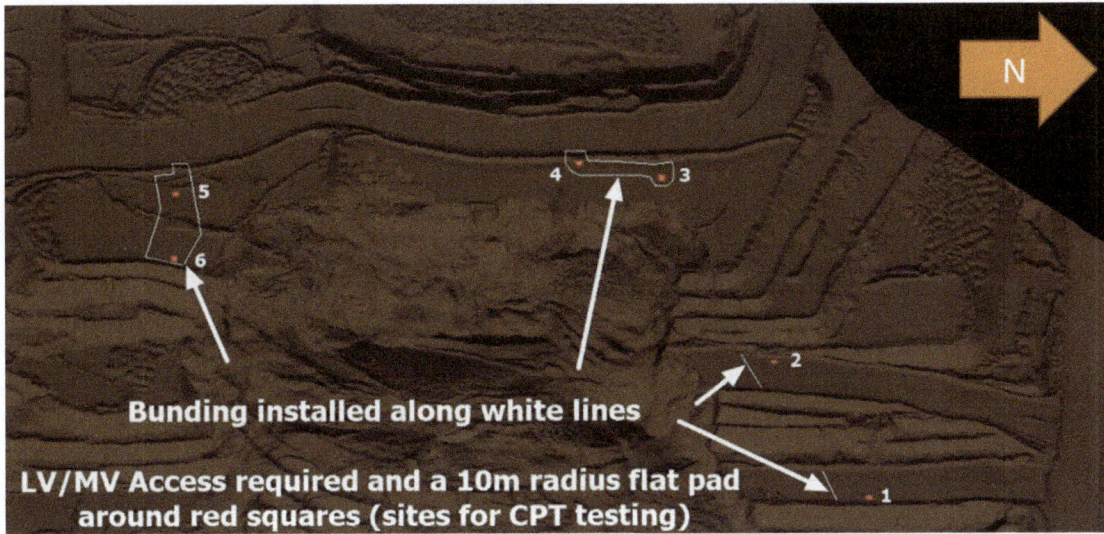

FIG 18 – SCPTu investigation locations surrounding the R16 Dump failure scarp.

High pore water pressures were measured in fine-grained spoil, that had been wet-up by incident rainfall on paddock-dumping, see Figure 19.

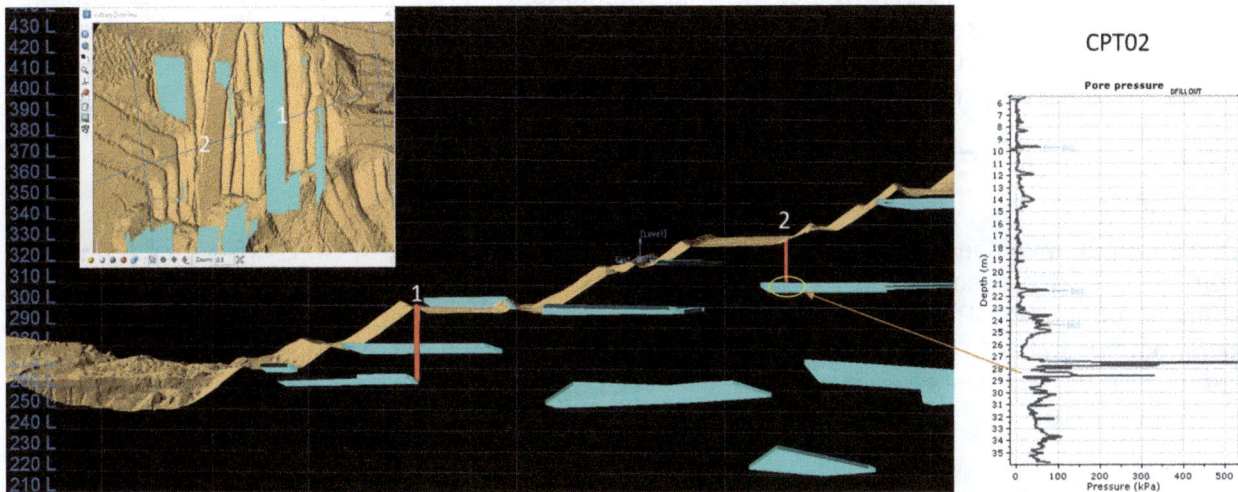

FIG 19 – Peaks in pore water pressure were measured in paddock-dumped spoil wet-up by incident rainfall as shown at the CPT02 location.

Normalised Soil Behaviour Type (SBTn) characterisation from the CPTu results (Robertson, 2016) was of little value because it does not consider the pore water pressure response observed in the wet-up spoil. Instead, the excess dynamic pore pressure based SBT chart proposed by Schneider *et al* (2008) was applied, see Figure 20, which indicates that the wet-up spoil can mainly be classified as a transitional soil.

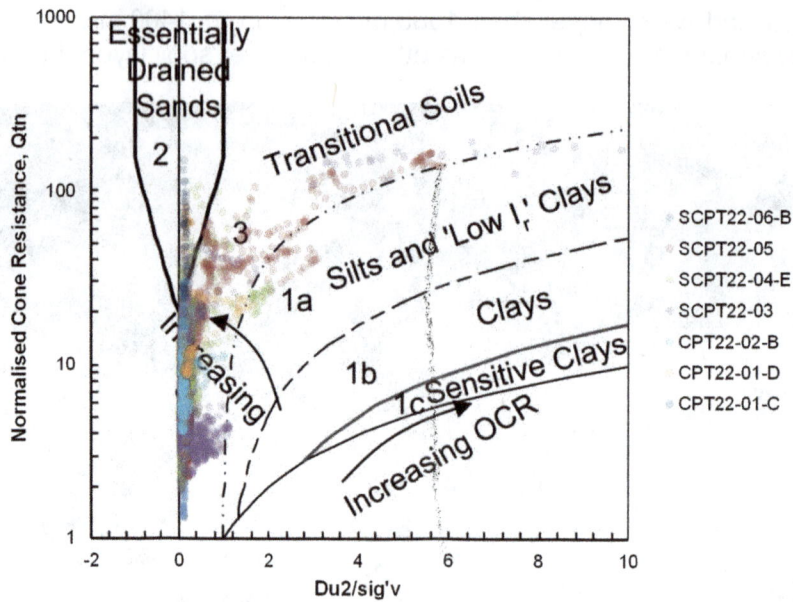

FIG 20 – Application of CPTu classification proposed by Schneider *et al* (2008).

The N_{kt} factor is required to estimate the undrained shear strength from CPTu cone resistance. However, the failure of vane shear testing prevented an estimate of the N_{kt} factor. Therefore, a sensitivity analysis was undertaken using N_{kt} factors of between 5 and 20. N_{kt}=20 resulted in s_u values close to the CPTu sleeve friction, which Monforte, Arroyo and Gens (2022) suggested may conservatively be taken as the residual strength. Figure 18 shows that site investigation was undertaken outside but nearby to the footprint of the failure, where it is assumed that soils encountered were not fully impacted by the failure.

COMPARISON BETWEEN SITE INVESTIGATION AND LABORATORY TESTING

Two strength envelopes from MWY at GMC=20 per cent and Grey-1 at GMC=16 per cent obtained from Figure 16 are compared in Figure 21 with the sleeve friction values (residual undrained strengths) from the CPTu tests in clayey spoil. The laboratory strength envelopes are approximate lower bounds to the CPTu sleeve friction data. The CPTu data suggest a drop in residual undrained strength at vertical stresses above 800 kPa, as has been highlighted by Bradfield, Fityus and Simmons (2019).

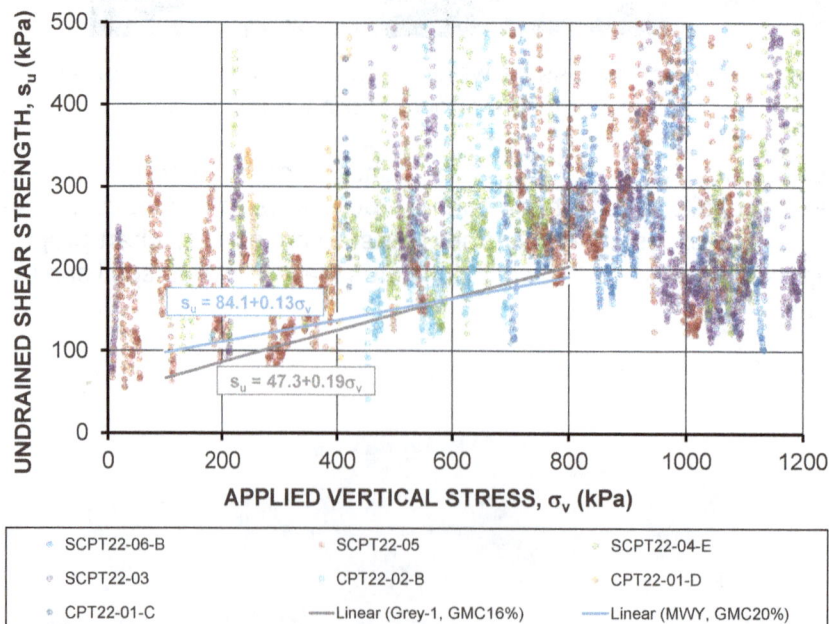

FIG 21 – Comparison between CPT investigation results and UQ-GEC Simple Shear Testing.

Sixteen pore water pressure dissipation tests indicated horizontal hydraulic conductivity (k_h) ranging between 5E-10 m/s and 9E-8 m/s, which are compared with estimates of k_v from the oedometer testing in Figure 22. k_v generally decreases with increasing vertical stress, by around 1.5 orders of magnitudes for an increase in vertical stress of about 2 orders of magnitude. The field k_h values are higher than the laboratory k_v values and increasingly so above a vertical stress of 200 kPa, by up to 4 orders of magnitude. This is attributed layering of spoil in the field and to three-dimensional effects in the field compared with one-dimensional in the laboratory. Chapman (2008) recorded k_v/k_h ratios in mine spoil and reported two data sets. The first data set included glacial tills and clays, and a k_v/k_h of 10:1 was reported. The second data set included cobblestones and boulders (siltstones), and results reported a k_v/k_h of approximately 100:1. The k_v/k_h value for this project was estimated to range between 25:1 to 50:1.

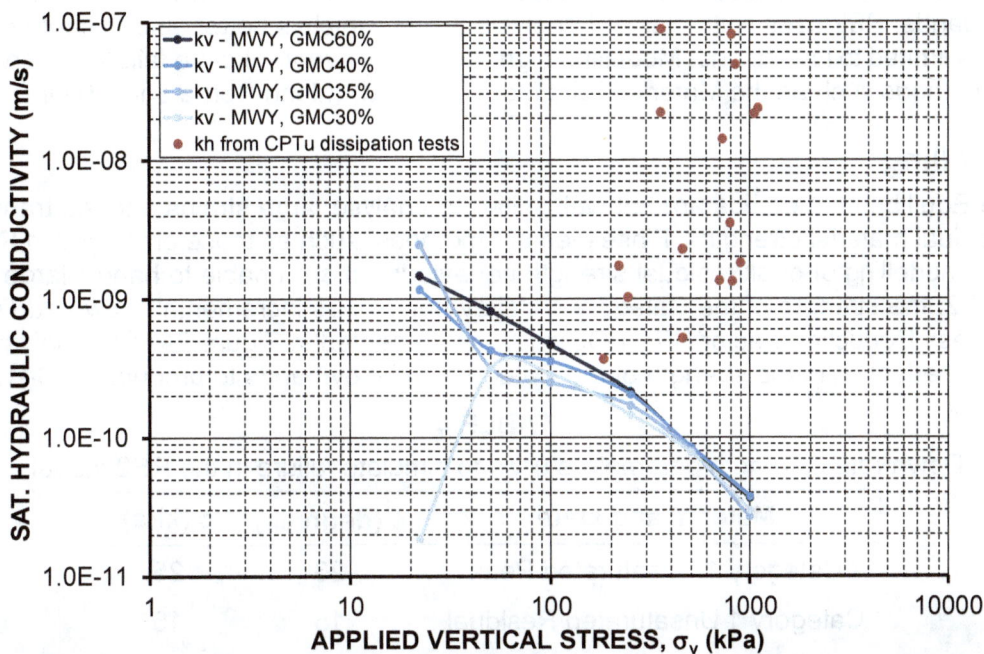

FIG 22 – Comparison between vertical saturated hydraulic conductivity at different GMC from laboratory oedometer testing and horizontal hydraulic conductivity from the CPTu dissipation tests.

BACK ANALYSIS

Slide2 modelling

Slide2 and RS2 models (Figure 23) were generated with the Vulcan data given in Figure 10. BMA spoil category strengths were adopted as a preliminary base case (Table 3) and the Coulomb failure criterion was applied in both Slide2 and RS2. Zones affected by high rainfall were assumed to be an ~5 m band of completely saturated material; however, a water table was not employed due to the expected slow-draining nature of the clayey spoil.

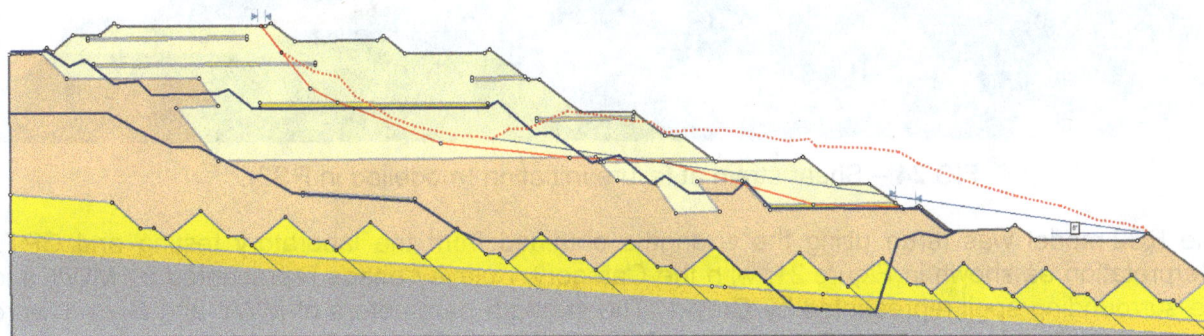

FIG 23 – R16 Slide2 Model generated (later fed into RS2).

TABLE 3

Strength parameters adopted as a preliminary base case.

Material	φ (degrees)	c (kPa)
Category 1 Unsaturated	23	25
Category 1 Saturated	18	0
Category 2 Unsaturated	28	30

The base case results from Slide2 gave a factor of safety of 1.64. This high value is attributed to the limit equilibrium method on which Slide2 is based being unable to capture the destabilising effect of thin, weak bands. RS2 uses finite element analysis (FEA), which is better able to capture this effect although it is limited to low-strains. Modelling high-strain flow events like the failure observed require specialist methods such as the Material Point Method (MPM) (Llano-Serna and Williams, 2018).

RS2 modelling

Limit State Equilibrium methods are not well suited to analyse large strains and the transition from peak to residual material strengths. Finite element methods, such as those employed in Rocscience RS2 allows modelling of peak/residual strength behaviour but are unable to handle large strains. In applying RS2, the strength of the Category 1 clayey bands was incrementally lowered (Table 4) to obtain a Stress Reduction Factor SRF(FOS)=1.0, as presented in Figure 24. The analysis required a substantial reduction in the strength of the Category 1 clayey bands to produce FOS=1.0.

TABLE 4

Down-rated strength parameters used to simulate failure in the RS2 model.

Material and state	φ (degrees)	c (kPa)
Category 1 Unsaturated Peak	23	25
Category 1 Unsaturated Residual	15	15
Category 1 Saturated Peak	10	0
Category 1 Saturated Residual	9	0
Category 2 Unsaturated	28	30

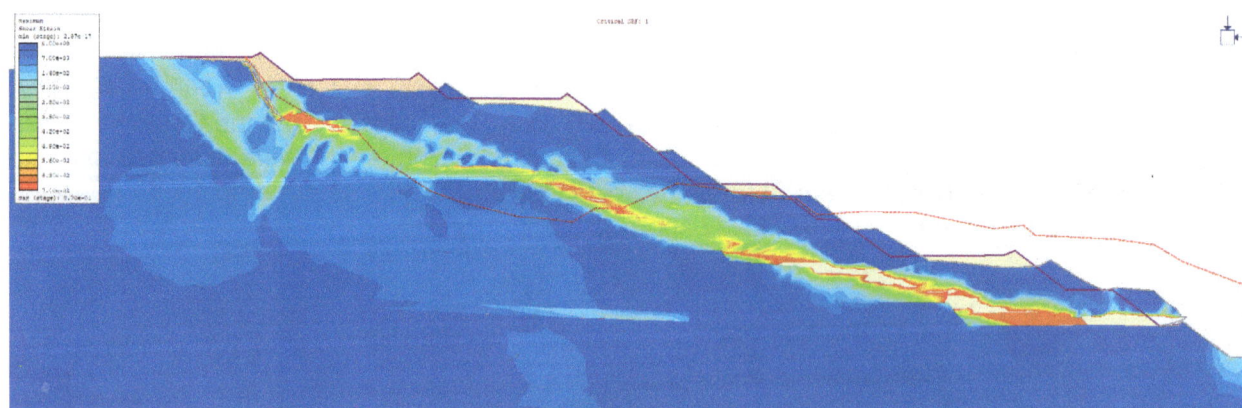

FIG 24 – Shear strain at failure initiation (modelled in RS2).

The RS2 model was rerun using the strengths obtained from the laboratory testing and CPTu interpretation as shown in Figure 21, with the Category 1 clayey bands represented by MWY and the surrounding spoil represented by Grey-1. The strength parameters of MWY and Grey-1 were again manually adjusted (Table 5) to yield an SRF (FOS)=1.0 (Figure 25).

TABLE 5

Down-rated strength parameters based on testing used to simulate failure in the RS2 model.

Material and state	Assigned material	φ (degrees)	c (kPa)
Category 1 Unsaturated Peak	Grey-1	13.5	110
Category 1 Unsaturated Residual	Grey-1	9	110
Category 1 Saturated Peak	MWY	4	55
Category 1 Saturated Residual	MWY	0	46
Category 2 Unsaturated	-	28	30

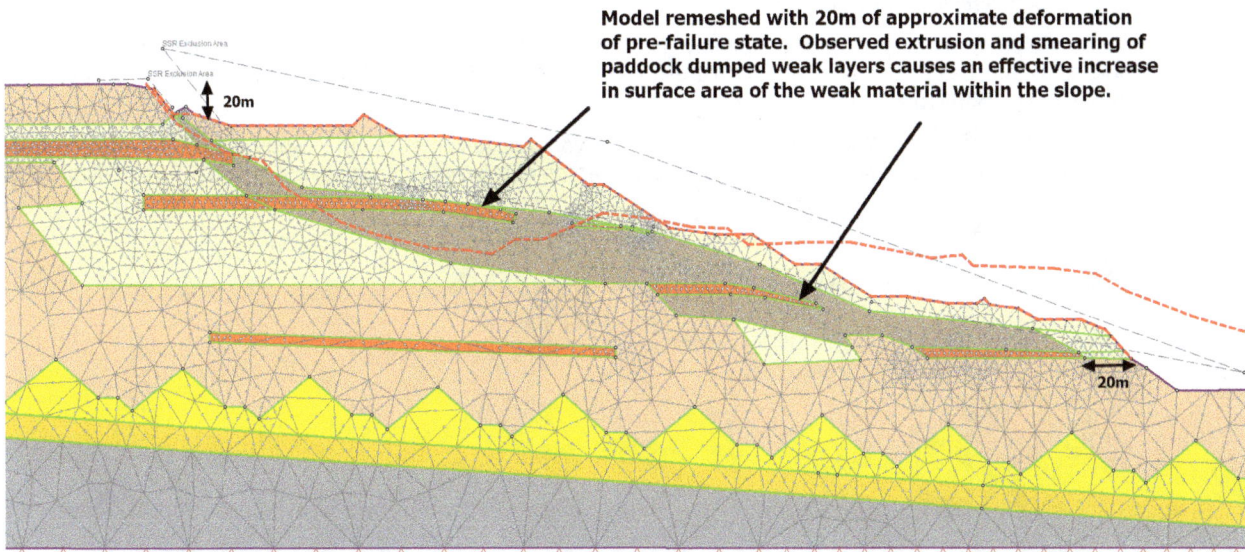

FIG 25 – Remeshed RS2 model with an imposed deformation step of ~20 m.

Sensitivity to inclusion of wet layers

A check on the sensitivity to the inclusion of wet layers was made, with one or several wet layers being removed and the model rerun. When the lower two wet layers (deposited in April and July 2021) were removed, the SRF(FOS) rose to 1.09, indicating that without the series of rainfall events in succession throughout 2021 it is unlikely that the failure would have occurred.

CONCLUSIONS

The R16 spoil pile failure was the result of paddock-dumped clayey spoil layers that were impacted by a series of rainfall events. The BMA spoil category strength parameters do not represent the low strengths that these wet clayey spoil layers developed. The results of the laboratory and field testing and back-analyses of the failure provide a basis for estimating the low strengths of such wet clayey spoil layers, which led to the failure. To avoid such failures, large areas of paddock-dumped, clay-rich spoil should not be exposed to rainfall.

Further modelling using more advanced methods, such as the MPM, is proposed to understand better the observed deformations and failure run-out.

ACKNOWLEDGEMENTS

The authors acknowledge the BHP – Mitsubishi Alliance (BMA) and the Goonyella Riverside Mine (GRM) team for assistance provided during the site visit, for the Scholarship granted to support the work of the third author towards a Master of Engineering – Research Thesis, and for their approval to publish this important case study. The authors thank Sebastian Quintero, Manager of the Geotechnical Laboratories at The University of Queensland for his ongoing assistance in undertaking the extensive laboratory testing associated with this study.

REFERENCES

Bradfield, L, Fityus, S and Simmons, J, 2019. Shear strength characterisation for very-high coal mine spoil dumps: Applicability of an industry-accepted framework, *ARMA, American Rock Mechanics Association*, pp 1–15.

Chapman, D, 2008. Hydrogeologic characterization of a newly constructed saline-sodic clay overburden hill, Master Degree Thesis, Saskatoon: University of Saskatchewan.

GDS-Instruments, 2023. GDS Electro-Mechanical Dynamic Cyclic Simple Shear. Available from: https://www.gdsinstruments.com/gds-products/gds-electro-mechanical-dynamic-cyclic-simple-shear

Llano-Serna, M and Williams, D, 2018. The Application of Run-out Analysis to Mine Site Landslides, in *Proceedings of the XIV Congreso Internacional de Energia y Recursos Minerales: Slope Stability 2018*, 7 p.

Monforte, L, Arroyo, M and Gens, A, 2022. Undrained strength from CPTu in brittle soils: A numerical perspective, Cone Penetration Testing 2022, pp 591–597.

Robertson, P, 2016. Cone penetration test (CPT)-based soil behaviour type (SBT) classification system—an update, *Canadian Geotechnical Journal*, pp 1910–1927.

Schneider, J, Randolph, M F, Mayne, P and Ramsey, N, 2008. Analysis of factors influencing soil classification using normalized piezocone tip resistance and pore pressure parameters, *Journal of Geotechnical and Geoenvironmental Engineering*, pp 1569–1586.

Simmons, J V and McManus, D A, 2004. Shear strength framework for design of dumped spoil slopes for open pit coal mines, *Advances in geotechnical engineering: The Skempton Conference Proceedings*, pp 981–991 (Institution of Civil Engineers).

Practical considerations for tailings infrastructure planning and the impact on project costs

J T Wylie[1], D W Wall[2] and P G McDonald[3]

1. MAusIMM, Project Engineer, GPA Engineering, Brisbane Qld 4000.
   Email: jason.wylie@gpaeng.com.au
2. Lead Process Metallurgical Engineer, GPA Engineering, Melbourne Vic 3000.
   Email: dean.wall@gpaeng.com.au
3. Lead Mechanical Engineer, GPA Engineering, Brisbane Qld 4000.
   Email: patrick.mcdonald@gpaeng.com.au

ABSTRACT

The value of good planning is undeniable; however, it can be challenging to know exactly what to plan for. In the case of a tailings facility, the list of potential changes that may occur over the life of a project can be long and convoluted. Deposition rates, wall raises, alternate flow paths, changes in ore mineralogy, and technology advances can all significantly impact operations.

The design of tailings infrastructure, such as pumps, piping, power supply and control systems, does not often receive the same level of rigour as the design of the dams or upstream processing plants. Often, these systems are designed to manage a specific set of conditions with limited flexibility. This can lead to cases where significant investment in retrospective upgrades is required.

However, decisions can be made in the design phase to improve the flexibility of tailings infrastructure systems and assist with future-proofing the operation. This paper outlines common pitfalls encountered in tailings and decant return infrastructure design and presents mitigation strategies that can be used to avoid these. Examples of this include the installation of multiple tailings pipelines, selection of piping materials, sizes and rating, pump selection, and capacity requirements for electrical and control system infrastructure.

The paper also presents the economic case for strategic investment in key areas by considering different upgrade scenarios across the project's life. These scenarios and the costs shown are based on the authors' experience across several projects of this type.

INTRODUCTION

Developing a new mining project includes a significant design focus on the process plant and the tailings storage facility (TSF). However, the interfaces between these facilities (tailings transfer and decant) do not receive the same attention.

The same is true for brownfields optimisation and expansion projects of tailings disposal and decant infrastructure. There is often a push to defer capital expenditure and do the 'bare minimum' to sustain operations. This leads to decisions that may not consider the future needs of the operation, particularly in the lower cost, interstitial aspects.

Designing for the total life cycle of a system means that the operator can make early investments in capacity-enabling infrastructure and understand the timing of required capital upgrades. As per the Australian/New Zealand Guide on Life Cycle Costing, AS/NZS 4536:1999 (Standards Australia, 1999), it is important to note that the total life cycle cost should not be considered in isolation and that there are many other qualitative factors. Some of these factors include; technology redundancy (will the technology need to be updated during the mine life), system complexity, operability, and maintainability.

This paper will assess two hypothetical scenarios, intending to highlight the qualitative and quantitative benefits in the early planning of tailings disposal and decant infrastructure.

The purpose of this paper is to specifically focus on the design of a tailings disposal system rather than the tailings storage facilities themselves. Please note that the following items have not been considered as part of this investigation:

- TSF design and construction costs, including wall raises and associated minor piping changes.

- Project closure and planning costs.

- Depreciation and amortisation, time value of money and detailed financial analysis.

- Pump type selection (positive displacement versus centrifugal).

- Operational and maintenance requirements and costs.

- Other tailings process types, such as dry stacking or paste.

METHODOLOGY

Example application

A theoretical example application was selected as the best way to highlight the many options and decisions associated with the design and installation of tailings disposal and decant systems. It also shows that these decisions are all related and are not completed in isolation.

The example application referenced in this paper is theoretical; however, the design is based on a combination of several operating tailings facilities across several mines and clients, primarily involved in base and precious metals extraction. It is common for sites to install multiple TSFs over the life-of-mine, with multiple stages of crest raises to prolong the life of the facilities. The dimensions/distances used have been selected to simplify the system and avoid some additional design decisions (centrifugal versus positive displacement pumps). Figure 1 shows the example application layout.

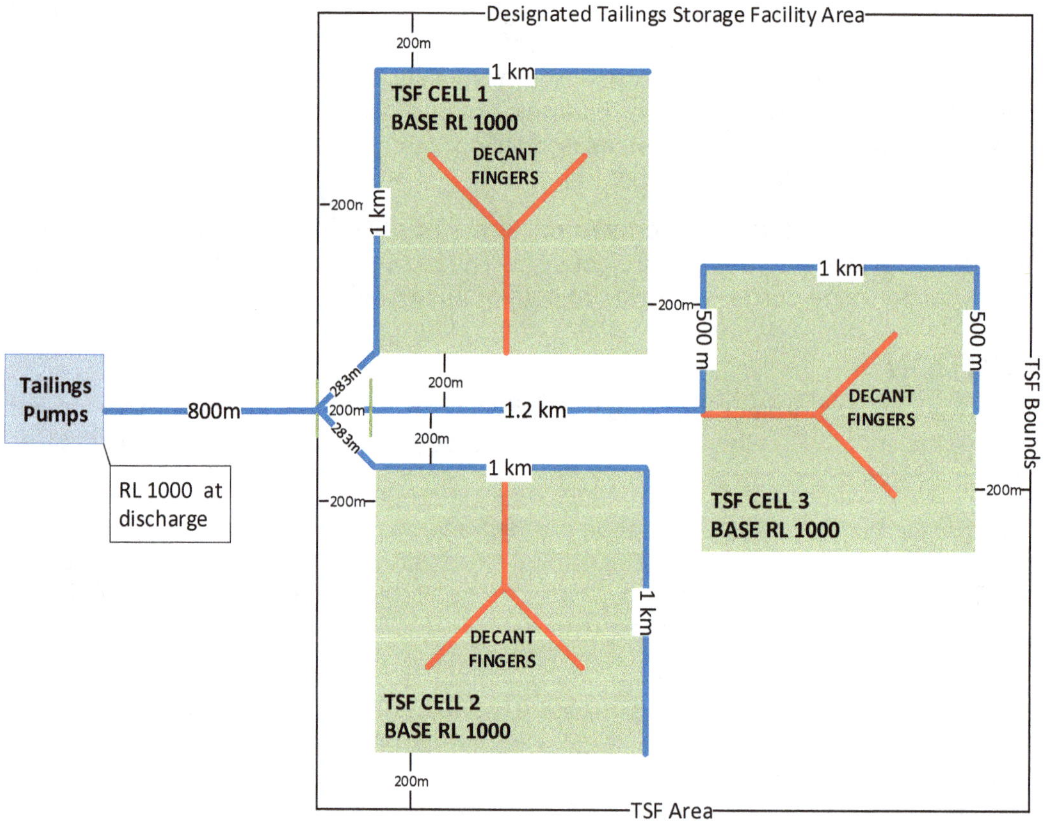

FIG 1 – Example application TSF layout.

Key items to note:

- 800 m from the tailings pump discharge to the TSF area.

- Three 1 km × 1 km TSFs.

- Spigots spaced every 50 m.

- 200 m between the initial TSF cell walls and the area boundaries, pipelines and other cells.

The tailings facility is assumed to be based on the following key design parameters:

- 25-year mine life.

- Only one active TSF at a time.

- Newtonian settling slurry with water as the carrier fluid:

  o Solids SG = 3.0.

  o Particle size P_{50} = 50 μm.

  o Maximum 55 w/w% solids.

  o No appreciable changes to mineralogy during the mine life.

- Tailings solids are considered non-abrasive, meaning that piping materials are unlined carbon steel or high-density polyethylene (HDPE).

- Maximum temperature of HDPE piping is 50°C.

Table 1 outlines the changes that are expected to occur around tailings disposal and the TSFs during the life of the mine representing the wide range of conditions for which the tailings deposition system must be designed.

TABLE 1

TSF lifetime operating summary.

Case	Active facility	Operating years	Crest height (m)	Discharge rate (dry t/h)
1		1–5	10	500
2	TSF 1 (Stage 1)	6–7	14	865
3		8	18	936
4		9–13	10	962
5	TSF 2 (Stage 2)	14–15	14	1000
6		16	18	962
7		17–22	10	936
8	TSF 3 (Stage 3)	23–24	14	647
9		25	18	500

Each TSF has two planned wall raises included in the operating plan. The first occurs after five years, followed by a second raise after the seventh year.

The tailings discharge rate presented in Table 1 refers to the dry tonnes of solid tailings material leaving the process plant. This paper does not consider the reasons for these flow rate changes; however, typical reasons include ore feed rate increases, changes in ore mineralogy and plant optimisation (throughput and recovery).

Design variables

Each pump and pipeline system was designed to meet three primary targets:

1. Fluid velocity above the Durand solids settling velocity (ideally 1.1–1.2 times the settling velocity).

2. Fluid remaining below the piping system's Maximum Allowable Operating Pressure (MAOP).

3. Pumps operating near Best Efficiency Point (BEP).

Each of these three targets is affected by three set operating factors:

1. Solids flow rate.

2. TSF location.

3. Crest height.

The following parameters were altered to achieve these operating factors:

- Pipeline size (nominal diameter).

- Pipeline material selection.

- Pipeline pressure rating.

- Solids concentration.

- Pump selection and number of pumps used.

Each of these design parameters affects more than one of the three design targets above, creating an iterative design process to achieve the most appropriate parameters. Furthermore, each case builds on the installed pipeline and equipment of the previous case.

Solids settling velocity

The slurry in this design is a settling slurry. Consequently, if the fluid velocity in the pipeline decreases below the slurry's settling velocity, solids will settle out of the slurry and cause pipe blockages (Grzina, Roudnev and Burgess, 2002). The Durand method was used to calculate solids settling velocity, as it yields higher pumping velocities than other methods and has more built-in safety against pipe blockages.

In the hydraulic design of each case, the actual pumping velocity was kept at approximately 110 per cent to 120 per cent of the Durand settling velocity to cater for uncertainties while not using an excessive pumping velocity which would cause unnecessary frictional losses or erosion.

This example application varies the solids throughput, which affects the calculated Durand settling velocity. Where the velocity was required to increase, the slurry was diluted from 55 w/w% solids to no less than 40 w/w% solids to increase the flow rate and prevent settling.

Maximum allowable operating pressure of piping

Two materials were used for the tailings pipelines; high-density polyethylene (HDPE) and carbon steel. HDPE is the preferred material, as it is more cost-effective and has a roughness of 30 times less than carbon steel (The American Society of Mechanical Engineers, 2010; Standards Australia, 2009), reducing frictional loss. HDPE is limited to an MAOP of 2.5 MPa (PN25 PE100) before de-rating; fittings are often further de-rated due to their fabrication methodology. Consequently, HDPE was used where possible, and carbon steel was used in sections where the desired MAOP could not be reached when using HDPE. Many other construction materials may be used for tailings pipelines, including various lining options for carbon steel; however, the above materials were chosen for simplicity.

HDPE de-rating

The HDPE used in the design was de-rated, as the nominal pressure rating (PN) assigned to the piping equates to a performance at 20°C; however, if the wall temperature is higher than this, the MAOP of the pipe reduces (Plastics Industry Pipe Association of Australia Limited, 2019). The HDPE used in this design was de-rated to 50°C, the maximum HDPE piping temperature, which gives an MAOP of approximately 70 per cent of the HDPE's MAOP at 20°C.

Carbon steel rating

The Schedule 20 carbon steel used in the hydraulic design has an MAOP of 5.7 MPa at up to 204°C (The American Society of Mechanical Engineers, 2010). This is significantly more than is required by the pipeline. The rating of the carbon steel would consequently be determined by the MAOP of the valves chosen for the carbon steel line sections. Either 2 MPa or 4 MPa-rated valves were used depending on the MAOP needed.

Pump selection

For simplicity, an industry-standard horizontal centrifugal slurry pump was used for each design. Each pump has the best efficiency point for a given produced head and volumetric flow rate. The approximate duty point for each case is known, allowing for the selection of an appropriate pump, operating within 60 per cent to 120 per cent of the pump's Best Efficiency Point (BEP). Head and efficiency de-rating was not considered.

Scenarios

Two project execution scenarios were considered:

Scenario 1 – reactive design.

Scenario 2 – proactive design.

In Scenario 1, an individual solution was designed for each case and considered only the requirements of the case and previously installed equipment. In Scenario 2, the needs of future cases were also considered (crest height, solids flow rate, and TSF location).

Scenario 1 design

Each case in Scenario 1 was designed iteratively. The first step was to size the pipeline. The model iterated the HDPE pipe's pressure rating and diameter to optimise these two factors while achieving the fluid velocity constraints. If the pressure exceeded the rating of HDPE, then carbon steel was specified. The final design produced a pipeline where the pressure is close to, but not exceeding, the line's MAOP ensuring the line is not overdesigned. The selection procedure was aided graphically by a hydraulic grade line. This hydraulic grade line shows the system pressure at each point along the pipeline remaining below the MAOP of each pipeline section.

Once the pump duty (flow rate and head) was calculated, the pumping configuration was determined (pump model, stages and number of trains).

Subsequent cases were then calculated by evaluating the validity of the five design parameters and implementing changes as needed to meet the new requirements of the system. This methodology was used for each case, giving nine different designs.

Scenario 2 design

In Scenario 2, the design was split into the needs of the three TSF cells, reducing the nine cases to three. This approach minimises project complexity while considering the plan over the entire life-of-mine.

Each TSF was designed iteratively, as in Scenario 1, with the addition of each of the three pump and pipeline designs needing to meet the requirements of all three cases. The only changes inside each design were limited to very short sections of pipeline changes due to wall raises and slurry dilution to remain above the settling velocity.

Scenario 2 also considered, for all cases together, the following items:

- Future space demand for pipelines and new equipment
- Peak motor size for each pump to prevent constant upgrading
- Peak power demand, which affects the high voltage supply line, the transformers, the switchboards/MCC, and the power cables to each pump motor
- Switchroom size for additional future equipment.

Economic assessment methodology

The economic assessment completed in this investigation is preliminary and designed to facilitate a comparison between the two execution scenarios. The specific figures presented have accuracy in line with the low level of definition completed.

This paper's equipment and material cost data was sourced from historical vendor cost information sought for recent, relevant projects. Installation costs were based on benchmarked figures provided by installation contractors for several projects across Australia.

All costs presented are in 2023 Australian dollars, with adjustments for time and exchange rates applied where relevant. Indirect costs, such as design/engineering and owner's costs, were included as industry-standard factors, common to all scenarios. No allowance for site-specific requirements, project contingency or risk provisions has been made. The cost assessment also avoids the excluded items listed in the Introduction.

RESULTS

Scenario 1 – reactive design

Scenario 1 was designed as per the Design Methodology. Figure 2 shows the completed pump and pipeline design for each case. Note that the discharge header around the TSF indicates the number of pipes operating; two pipes indicates both the left and right side of the TSF are operating not duplicate pipes.

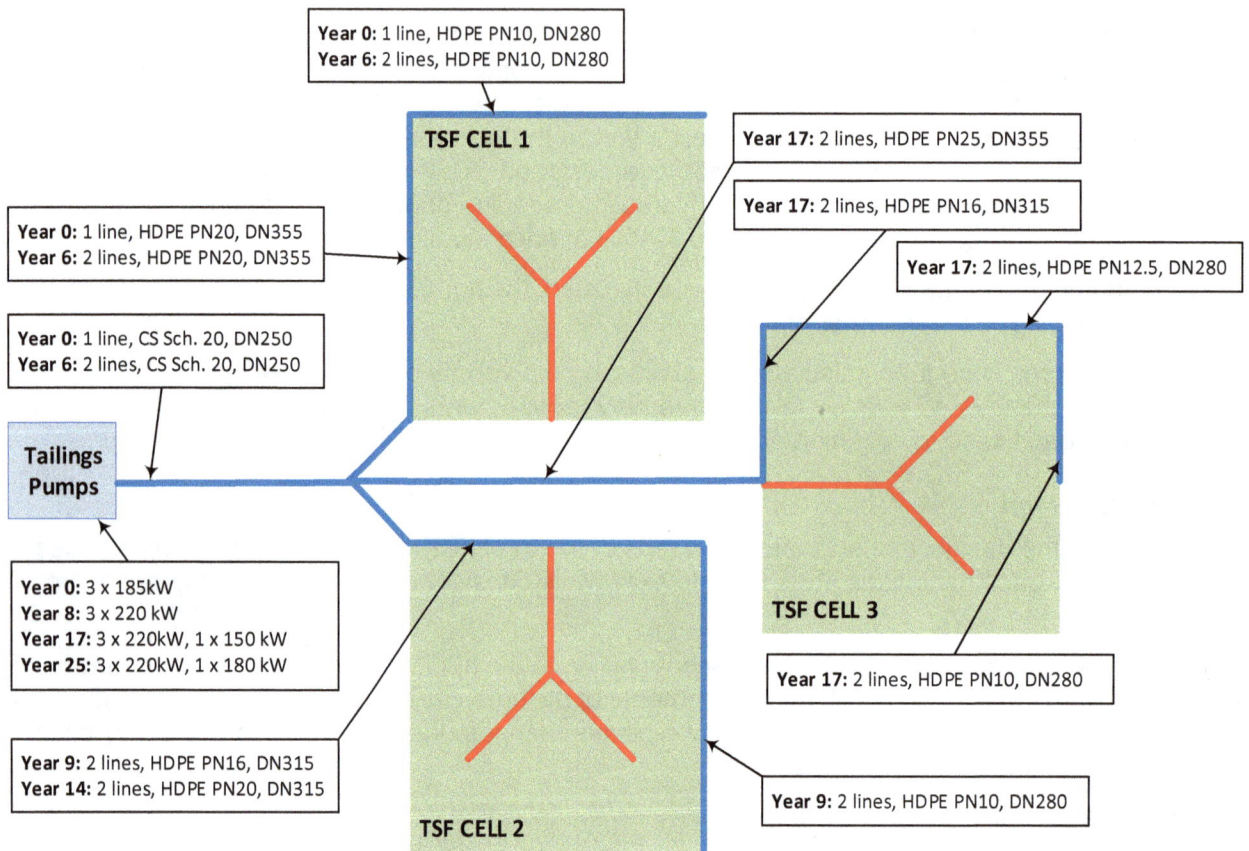

FIG 2 – Scenario 1 tailings pump and pipeline design.

The significant increase in flow rate into Case 2 (see Table 1) necessitated the pipeline to be twinned. A series of upgrades/changes were then required throughout the life of the project.

Scenario 1 requires the completion of nine separate projects over the life-of-mine – the initial installation, followed by eight expansion or upgrade projects. These projects, their estimated cost and their expected timing, are summarised in Table 2. Please note that projects less than $100 000 in value have not been included in this table, as these would be considered minor upgrades completed by site operations and would not necessarily require a dedicated project.

TABLE 2

Scenario 1 project summary.

Case	Year	Cost	TSF change	Description
1	0	$6.3 M	New TSF	Original installation (1 pump train and pipeline) with discharge header to TSF 1
2	5	$5.3 M	Flow increase Wall raise	Pump station and line duplication to discharge headers
3	7	$3.1 M	Flow increase Wall raise	Pump motor replacement and associated electrical upgrades Extension of carbon steel sections in twinned pipeline
4	8	$1.7 M	New TSF Flow increase	Tie-in and extension of twinned carbon steel pipeline New discharge headers for TSF 2 Upgrade valves to 4MPa
5	13	$0.8 M	Wall raise Flow increase	Extension of twinned carbon steel pipeline Upgrade of HDPE spigot headers to higher pressure rating
6	15	$0.2 M	Wall raise	Extension of twinned carbon steel pipeline
7	16	$6.2 M	New TSF	Tie-in and extension of twinned carbon steel pipeline New discharge headers for TSF 3
9	24	$0.4 M	Wall raise Flow rate decrease	Extension of twinned carbon steel pipe Second pipeline redundant/changed to standby operation

$24.1 M

Scenario 2 – proactive design

With the benefit of the future information developed for Scenario 1, Scenario 2 was designed in a proactive manner, where the electrical, instrumentation and control (EIC) infrastructure, pumps, and pipelines require as few changes as possible during each stage.

The design in Figure 3 utilises this information, giving a design that only required three major construction periods, and minor piping changes commensurate with wall raises.

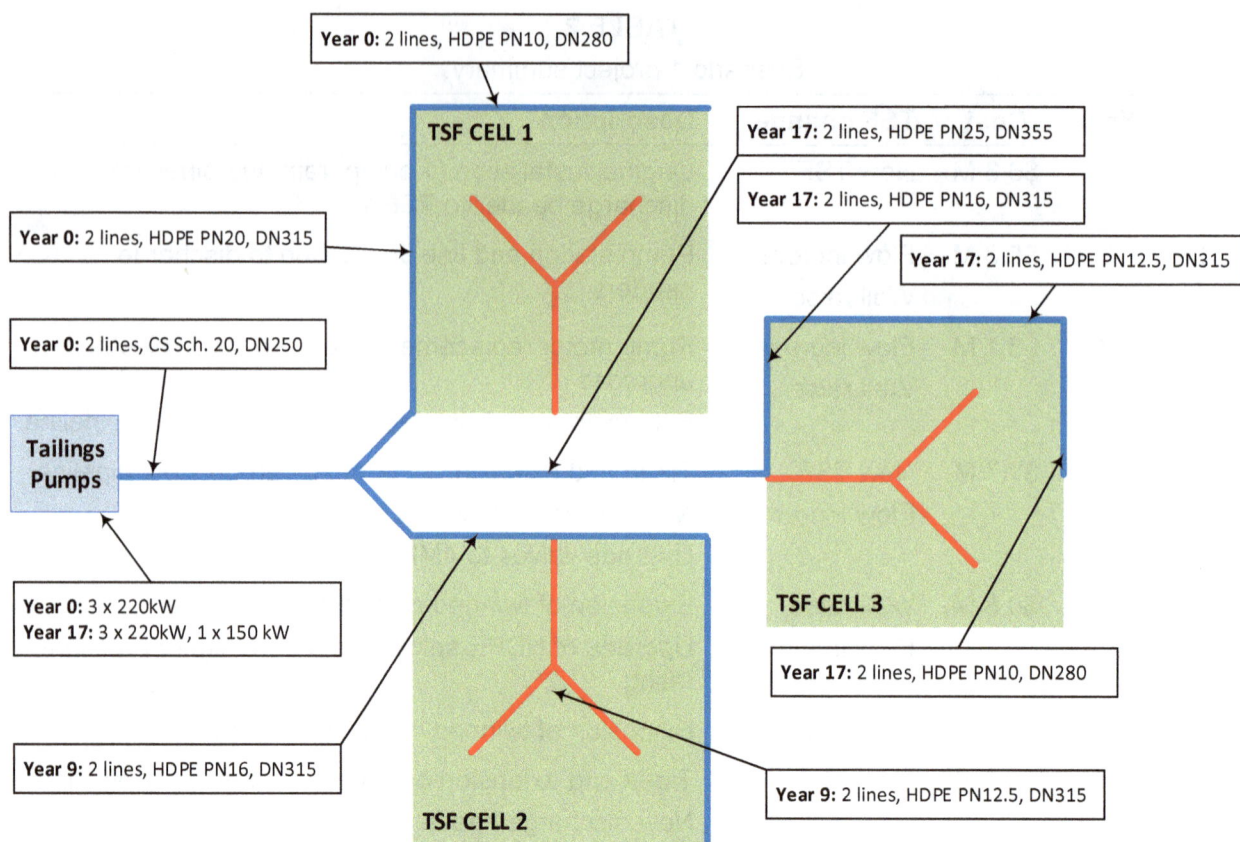

FIG 3 – Scenario 2 tailings pump and pipeline configuration.

Scenario 2 requires the completion of four separate projects over the life-of-mine – the initial installation, followed by three expansion or upgrade projects. These projects, their estimated cost and expected timing are summarised in Table 3.

TABLE 3

Scenario 2 project summary.

Case	Year	Cost	TSF change	Description
1	0	$8.4 M	New TSF	Original installation (2 pump trains and single pipeline) with discharge headers to TSF 1
2	5	$2.0 M	Flow increase Wall raise	Line duplication to discharge headers Minor piping changes for wall raise
3	8	$2.1 M	New TSF Flow increase	Tie-in and extension of twinned carbon steel pipelin New discharge headers for TSF 2
4	16	$3.8 M	New TSF	Tie-in and extension of twinned carbon steel pipeline New discharge headers for TSF 3
		$16.4 M		

DISCUSSION

The example applications presented in this paper represent two extreme cases of tailings disposal system planning. Very few mine sites will be able to plan their life-of-mine tailings system with complete certainty; however, this paper will argue that forward planning based on reasonable assumptions will allow stakeholders to make informed decisions and prevent capital regret while also reducing the complexity of future brownfields upgrades and optimisation projects.

Project costs

Figure 4 shows that Scenario 1 has a reduced initial capital expenditure, but the required capital for each subsequent upgrade exceeds Scenario 2 for most other cases. The cumulative cost trend indicates that the initial capital investment of Scenario 1 starts to pay-off after the fifth year of operation.

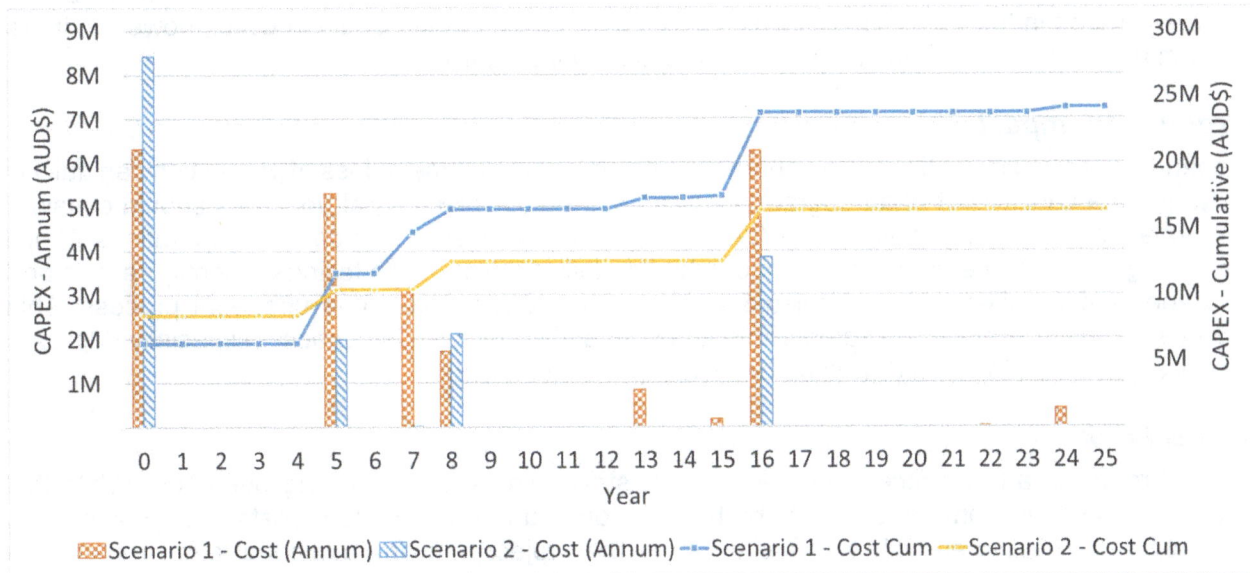

FIG 4 – Project costs over life-of-mine (2023 AUD).

The two scenarios have also been compared using a 9 per cent discount factor; Scenario 1 – reactive design – $14.2 M, Scenario 2 – proactive design – $11.7 M.

While this is not a complete Net Present Valve calculation, it does show that, in this case, even with the time-cost of money considered, early expenditure and good planning can prevent spending more capital later.

Figure 5 compares the two scenarios separated into key functional areas. The graph reinforces that the main difference between the scenarios is the supporting infrastructure (electrical), an increase in civil works, and the indirect costs associated with executing a higher number of projects.

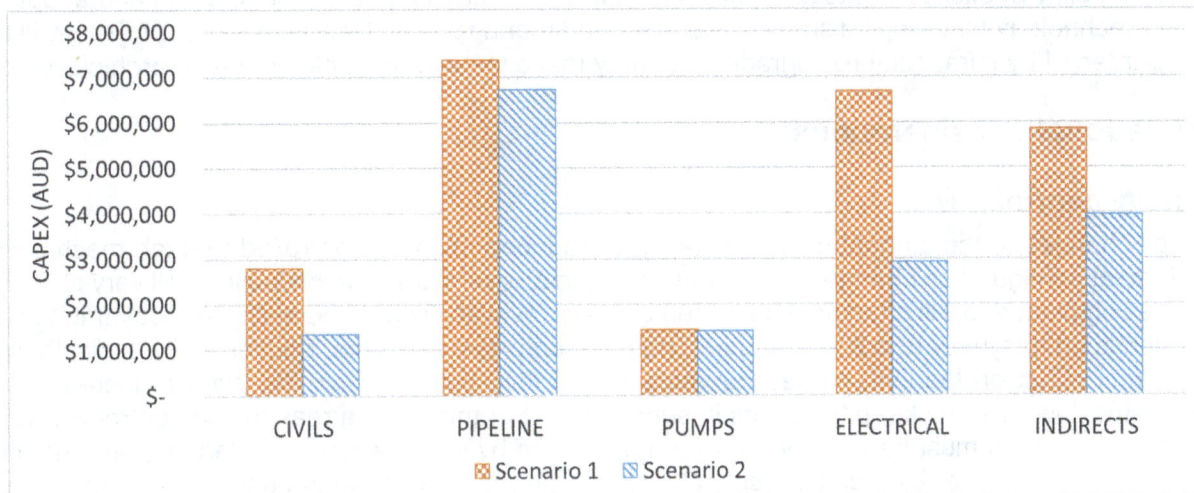

FIG 5 – Total area costs (2023 AUD).

Project considerations

Mobilisation/demobilisation

Contractor and equipment mobilisation and demobilisation can be a significant cost to a project, particularly on remote sites without permanent, large-scale contractor presence. When works are completed in combined packages, these costs can be minimised. Mobilisation and demobilisation were estimated in the example applications based on a percentage of direct costs; however, there is often a fixed portion that impacts smaller projects more significantly.

Production impact

Depending on the size of the mining operation, the cost of downtime or loss of production can quickly outweigh the capital cost of an upgrade/modification. Future planning of upgrades should consider the impact on operations where no redundant systems are available to continue operations. While some upgrades can be completed during scheduled shutdowns, often tailings reconfigurations and upgrades are required to be completed without any interruptions to the upstream process. This increases the value of early, strategic investment in additional tailings capacity, especially duplicate disposal pipelines, to avoid downtime or throughput constraints.

Project risk

Capital regret is a key concern for the financial stakeholders in any mining operation. While this paper offers suggestions for capital to be brought forward to reduce future costs and complexity, it is noted that this comes with a financial risk to the project and any additional effort required to approve the allocation of capital. This investigation does not consider the value of delaying capital expenditure, but it is a key driver behind many of these and other similar decisions.

When considering the expedition of works, the design team must consider the following:

- What is the likelihood that the assumptions impacting this decision are incorrect?
- What is the project's cost now compared to if it needs to be completed later?
- Is there a cheaper alternative that will provide a provision for future installation?

Whatever the outcome, this process should be documented as part of the project decision process.

Technology advances

The availability of excess capacity in critical infrastructure may help to enable the trial and adoption of new technology for tailings disposal, treatment and management. New technology projects will be less burdened by infrastructure upgrades that may make the cost of implementation prohibitive.

Operational considerations

Density control

Not all changes to the amount or type of tailings produced must be controlled through mechanical equipment configuration changes. The process variable 'levers' that can be 'pulled' will vary for each processing plant and will often be determined by water availability and the ability to recover it. Water addition is an easy way to adjust the operating point on a system curve. Dilution alone typically has a minimal effect on the Durand settling velocity of a fluid but will allow the plant to increase the volumetric flow rate and keep the solids in suspension. For this to be a reasonable control variable, the tailings system must be designed with a mechanism by which water is added, ie pump hopper with a controlled water inlet and consideration must be made for tailings beach requirements and consolidation targets. The capacity of the decant water system must also be sufficient such that the additional water can be returned to the concentrator/water storage facility.

Another requirement for density control is the installation of a density transmitter for online control loop feedback. These elements are typically nucleonic instruments, which are expensive and require close control by a Radiation Safety Officer (RSO) on-site.

Pigging

A pipeline intervention gadget (pig) is a tool inserted into a pipeline and used to remove scale, settled solids and/or air from a pipe. The requirement for pigging is dependent on the nature of the fluid transported. Creating a piggable slurry pipeline comes with several challenges that can be minimised if considered early. The key challenge with creating a 'piggable' line is the variable internal diameter of different piping materials and pressure ratings. In the example application, the pipeline's internal diameter generally has a variance of approximately 10–20 mm, allowing for a foam pig to be used. The downstream spigot header has a much smaller internal diameter and would need to be maintained separately through either replacement or a separate pigging run.

Design considerations

Mechanical

Piping

In both scenarios, the piping configuration becomes unsuitable within the first five years of operation. It is common for mines to ramp up and down and, in some situations, do so multiple times throughout their operational life.

The initial piping design in Scenario 1 was optimised for a start-up operation but quickly became undersized for the nameplate throughput of the plant. Sites with a long ramp-up time of several years often decide to delay capital expenditure by installing the equipment required for start-up and upgrading when the throughput is increased.

In the second scenario, the design incorporates the second train of pipework ('twinning') and pumps as standby pumps at the initial construction stage. While there is not necessarily anything wrong with delaying the capital expenditure, the planning for this installation must be considered in the early phase of the project to ensure the correct provisions are made for the installation across all areas.

The rheology of tailings will significantly affect the choices made during throughput variations. The example application presented in this paper focuses on a settling slurry with a very limiting velocity constraint. Many tailings slurries, particularly slimes, exhibit non-settling and/or non-Newtonian behaviours and do not need to be operated within the same boundaries. These slurries tend to have greater frictional losses. They may benefit from a larger pipe diameter/higher rated pipe installed during the early operation phases to prevent overpressure/capacity constraints.

Designers should also consider the practicality and cost of installing/realigning hard pipelines (like carbon steel). These pipelines are typically installed above ground on plinths and may require thrust/anchor blocks to retain them. Wall raises may also require additional buttressing/ramp realignment, impacting pipeline alignment. This should be considered when using a hard pipeline up to the crest of a TSF.

Valves

In an HDPE system, valves are rarely the pressure constraint, with most slurry valve styles available in pressure classes aligned with HDPE pressure ratings. In a carbon steel system, the pipeline will likely be rated to a much greater pressure than the valves. In these cases, the MAOP of the system will be defined by the pressure rating of the specified valves; this is often defined within a site standard pipe specification. Early phase design of the tailings system will identify whether an increased valve pressure rating should be selected to prevent MAOP limitations within the carbon steel section of a tailings disposal line. The replacement of valves can be a costly operational change, particularly if there are many valves or an exotic material of construction is specified due to the composition of the tailings fluid.

Pumps

Pumps need to be designed for the entire expected operating range of flow rates and pressures. The capacity and efficiency of a centrifugal pump quickly diminish as backpressure increases and the system curve moves further away from the best efficiency point. Pumps must be designed alongside the pipeline to ensure the optimum number of pump stages, and trains are selected in addition to pump model and motor size. Consideration should be given to availability requirements – installation of standby pump trains and whether a standby train can be repurposed as a future duty train if the throughput increases. Additional pumps will incur an increased operational cost due to maintenance, although the increased availability of the system may offset this.

Motors should be selected with a significant design margin and future-proofed to ensure they are sufficiently sized. The cost difference between various sizes of motors tends to be insignificant compared to the cost of replacing a motor and the associated infrastructure.

Any pumping system chosen should be easily maintainable with the ability to be flushed during upset and shutdown conditions.

Hoppers and tanks

Hoppers and tanks in slurry service are often rubber lined and cannot be readily modified. Modifying these vessels usually requires replacement or fabrication in a workshop which can result in significant cost and time impacts. A relatively low-cost solution is to allow spare/additional flanges/nozzles for equipment that may be required. This is particularly pertinent for pump hoppers where only a single train has been installed but may require a standby train/second-duty train installed in the future.

Electrical, instrumentation and controls

One of the significant challenges that is often presented during brownfields projects is power capacity of existing transformers and switchrooms. It is common practice in the industry to allow for a nominal spare capacity, which adds to the physical space within EIC infrastructure, based on the maximum demand for the equipment. However, this philosophy often leads to shortcomings during upgrades, with few options to physically locate new infrastructure in overcrowded switchrooms.

There are many options for the sizing of this infrastructure; however, selecting the correct option can only be informed through the future planning of the powered equipment. Larger or additional transformers, switchrooms and/or panels do not necessarily need to be installed at the start of operations, but the provision of physical space or electrical capacity may prevent an expensive modification in the future.

Civil and structural

TSF layout

Final landform planning is necessary when defining the life-of-mine piping and road infrastructure. Changes to dam standards, measured phreatic levels, and crest raises beyond design all have the potential to impact the footprint of a TSF. Conservatively planning for these eventualities by placing permanent infrastructure outside the potential long-term footprint may prevent spatial clashes.

Sufficient width must also be allowed for in pipe traces to allow for any future expansion requirements, in addition to the snaking width of HDPE pipe (expansion/contraction associated with temperature changes).

Concrete

During the initial construction of a minerals processing facility, a concrete batch plant is typically available, and slabs, bunds, plinths etc can generally be formed and poured in large quantities at a significantly reduced cost compared to brownfields projects. If the additional infrastructure has been identified for future installation, it may be pragmatic to consider expanding the bunded areas and including additional plinths as a provision. This becomes increasingly important in facilities with tight spatial constraints, where additional equipment would be difficult to retrofit if not designed to be installed at the outset.

Structural

The structural considerations of this system are more likely to be focused on practicality, and ensuring that supports are not placed in areas where future expansion is required. Early design drawings showing provisions for future equipment can assist with planning for these issues.

CONCLUSIONS

This paper has investigated a comparative example application to examine the impact of forward planning on tailings disposal system design and cost. Two scenarios were presented and modelled over a 25-year mine life to determine both the cost implications and the opportunities to optimise the design to minimise the number of upgrades.

Several opportunities were identified across a range of engineering disciplines. Some of these opportunities which provide high value for future-proofing a design include:

- Complete a TSF scenario mapping exercise across the lifespan of the project, similar to that presented in this paper.

- Specifying increased motor rating on pumps sufficient to operate during the highest throughput and farthest discharge location.

- Allowing sufficient physical and electrical capacity in transformers, switchrooms and MCCs for pump station expansion.

- Condensation of project phases to facilitate forward planning during operations:

  o For example, planning the maximum possible crest elevation when designing the next TSF, not just the next period.

  o This example achieved this by condensing nine design stages into three for Scenario 2.

- Allowing space within bunded areas for planned pumping infrastructure expansions and concrete pouring.

- Adding mechanical provisions for future expansion (ie nozzles/flanges) on equipment, particularly rubber-lined equipment.

- Allocating future space for additional pipelines in pipe traces/culverts to reduce impacts on road design.

Some other items which should be considered and identified but may be completed later include:

- Concrete and steel structures (plinths, supports etc).

- Additional pumps, motors, drives and VSDs.

While the above items may not be installed, any provisions that need to be made for their future implementation should be considered.

ACKNOWLEDGEMENTS

Adam Freeling provided hydraulic modelling and calculation assistance for this investigation.

Process data for this paper came from a range of Australian mining operations, primarily base and precious metals. This data was then anonymised to form the basis of the example application presented.

Cost data was based on pricing received for recent, similar projects, as provided by reputable vendors.

REFERENCES

Grzina, A, Roudnev, A and Burgess, K E, 2002. *Slurry Pumping Manual* (Warman International Ltd).

Plastics Industry Pipe Association of Australia Limited (PIPA), 2019. Industry Guidelines: Temperature Rerating of PE Pipes, PIPA.

Standards Australia, 1999. AS/NZS 4536:1999 – Life cycle costing – An application guide, Australian/New Zealand Standard.

Standards Australia, 2009. AS/NZS 4130 Polyethylene (PE) pipes for pressure applications, Australian/New Zealand Standard.

The American Society of Mechanical Engineers, 2010. ASME B36.10M Welded and Seamless Wrough Steel Pipe, ASME.

Selection of parameters for TSF and WRD design

Advanced numerical approach for investigation of rock fill mechanical behaviour

R Asadi[1,5], B Ghahreman-Nejad[2,3] and M M Disfani[4]

1. PhD Candidate in Geotechnical Engineering, The University of Melbourne, Parkville Vic 3052.
   Email: rasadi@student.unimelb.edu.au
2. Chief Technical Officer, ATC Williams, Mordialloc Vic 3195.
   Email: behroozg@atcwilliams.com.au
3. Senior Honorary Fellow, The University of Melbourne, Parkville Vic 3052.
   Email: behrooz.g@unimelb.edu.au
4. Associate Professor, The University of Melbourne, Parkville Vic 3052.
   Email: mahdi.miri@unimelb.edu.au
5. Geotechnical Engineer, ATC Williams, Mordialloc Vic 3195. Email: rezaa@atcwilliams.com.au

ABSTRACT

Considering the large particle size of rock fill materials (ie -1 cm to 1 m), conducting laboratory tests is prohibitively expensive and time-consuming. Consequently, current design methods and analysis of rock fill structures rely heavily on empirical correlations developed based on historical large-scale experimental studies. Extensive use of rock fill in embankment dams, waste rock dumps, offshore structures etc, on the one hand, and lack of a clear understanding of its mechanical behaviour, on the other hand, highlight a clear need for further investigation. To seek an alternative approach, many studies have been conducted employing parallel particle size grading techniques in which a scaled-down assembly with parallel particle size distribution (PSD) is analysed. It should be considered that rock fill behaviour is significantly scale dependent which predominantly stems from the effect of particle breakage phenomenon. Hence, developing a robust technique (experimental or numerical) that can assimilate effective parameters, including particle shape and breakage, ie allowing simulation and study of large-scale assemblies with realistic boundaries, is considered necessary. Particle breakage is a fundamental phenomenon in the mechanical behaviour of rock fill which substantially influences shear strength, deformability, porosity etc, at different stress levels. The effect of particle breakage can be understood by rock mineralogy, particle shape, surface roughness, PSD, and confining stresses. In this study, a computationally efficient breakage method based on the Discrete Element Modelling (DEM) technique with the use of a flexible membrane is investigated. The results of some large-scale triaxial tests (100 × 180 cm), along with UCS tests are analysed and employed to improve and validate the developed DEM model for large-scale triaxial testing.

INTRODUCTION

Rock fill refers to an assembly of rock aggregates ranging from below 1 cm to over 1 m in size with the range generally limited to between 10 cm and 80 cm in most geotechnical applications. Rock fill is extensively used as the main construction material in embankment dams, railway subgrades, offshore structures etc. Despite this, the current understanding of the mechanical behaviour of rock fill and its effective parameters is limited. Considering the large particle size of rock fill materials, conducting laboratory tests is prohibitively time consuming and expensive. Consequently, current design methods and analysis of rock fill structures heavily rely on limited historical large-scale experimental studies (Leps, 1970; Marsal, 1967). Due to the scarcity of large-scale triaxial machines globally, several studies (Alonso, Tapias and Gili, 2012; Frossard et al, 2012; Jiang, Xie and Liu, 2018) have conducted testing using a parallel particle size grading technique, in which a downscaled assembly of rock fill particles with parallel particle size distribution (PSD) to the actual rock fill PSD is tested and analysed. The parallel PSD procedure is illustrated in Figure 1 with the formulation summarised in Equation 1 and Equation 2. In these equations, the stress-strength behaviour of the bigger assembly (material B) can be evaluated based on the test results of the smaller assembly (material A) using the Weibull modulus extracted from particle breakage test results and the PSD scaling factor (D_B/D_A). The principal stress ratio (eg σ_1/σ_3) are considered to be the same in samples A and B, which is expected to result in the same mobilised internal friction angle (Frossard et al, 2012). These underlying assumptions for the parallel grading technique may not be generalised given the rock fill behaviour is significantly scale-dependent (Alonso, Tapias and Gili, 2012; Marsal, 1973), predominantly, due to rock particle breakage (Alonso, Tapias and Gili, 2012). Hence, developing alternative analyses techniques which can adequately assimilate the effective parameters (ie particle shape, breakage etc) on the mechanical behaviour of rock fill are considered necessary.

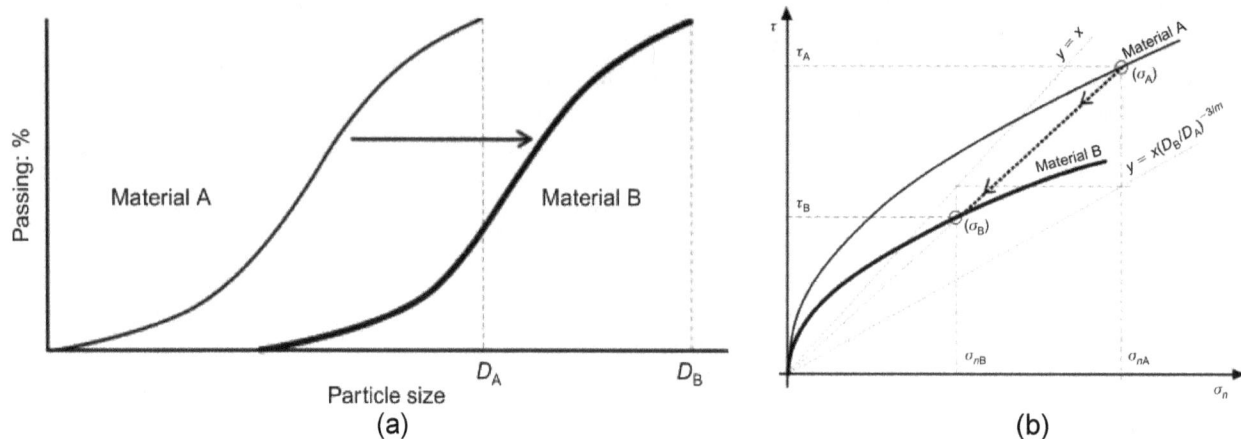

FIG 1 – Parallel grading scaling method: (a) parallel PSDs for Sample A and B, (b) scaled failure envelopes (Frossard *et al*, 2012).

$$\sigma_B = \left(\frac{D_B}{D_A}\right)^{-3/m} \times \sigma_A \qquad (1)$$

$$\tau_B = \left(\frac{D_B}{D_A}\right)^{-3/m} \times f \left\langle \sigma_{nB} \times \left(\frac{D_B}{D_A}\right)^{-3/m} \right\rangle \qquad (2)$$

Particle breakage is a fundamental phenomenon, which substantially influences the rock fill shear strength, deformability, porosity etc under different stress conditions (Lade, Yamamuro and Bopp, 1996; Marsal, 1967, 1973). This phenomenon is directly influenced (to varying degrees) by rock mineralogy, particle shape, surface roughness, PSD, and confining stresses (Tapias, Alonso and Gili, 2015; Yang *et al*, 2019). Large settlements in rock fill dams in excess of 1 per cent of the dam height (Oldecop and Alonso Pérez de Agreda, 2007) generally arise from rock particle breakage and rearrangements. The breakage phenomenon results in an alteration to the rock fill void ratio and PSD which can be quantitatively assessed through Relative Breakage Index (Br) (Einav, 2007) considering the initial, current and ultimate PSDs for a given sample. It has been well established that particle breakage affects the shear strength of angular particles more than rounded particles. For the latter larger rock particles lead to more interlocking and an increase in shear strength (Varadarajan *et al*, 2003). Particle breakage has been studied through different methods including single particle crushing and X-ray tomography (Zhao *et al*, 2015), and laboratory tests such as oedometer (Oldecop and Alonso Pérez de Agreda, 2001) and triaxial (Xiao *et al*, 2014; Yu, 2017) tests. At a glance, the focus of most experimental studies to date has been the particle breakage and associated parameters at the end of testing (Jia *et al*, 2017). Numerical methods have allowed researchers to investigate the evolution of breakage and its effect on the mechanical behaviour of rock particle assemblies.

DEM has been widely used to study the behaviour of granular materials. The method is based on the simulation of individual particles which interact with each other through simple equations of motion. The ability of DEM to capture the complex interactions between distinct elements makes it an ideal tool for modelling the behaviour of granular materials such as rock fill. Different particle shapes including spherical, elliptical, and complex angular assemblies (eg overlapped, bonded non-overlapped sub-particles, edged rigid blocks) have been employed for DEM modelling, as shown in Figure 2. More complex shapes can be obtained by sacrificing computational efficiency and using numerous finer sub-particles with different methods introduced to simulate particle breakage in DEM. As a complex and computationally intensive method, the coupled DEM with Finite Elements Methods (FEM) can be considered to accurately capture the breakage surface (Bagherzadeh-Khalkhali, Mirghasemi and Mohammadi, 2008). In this coupled DEM-FEM method, the particles interaction is modelled by DEM whilst particles are individually discretised and analysed in FEM.

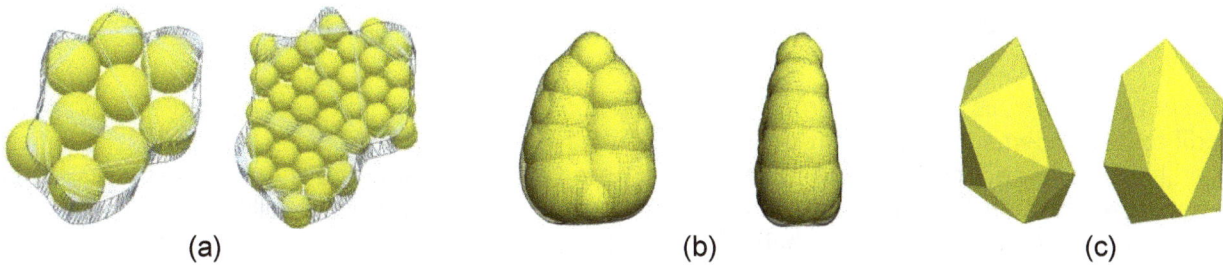

FIG 2 – Illustration of generating angular particles using: (a) bonded non-overlapped sub-particles; (b) overlapped sub-particles; (c) edged rigid blocks.

A more computationally efficient approach is employing bonded particles in which the particle breakage is modelled by changing bond strength between the sub-particles (Xu, Hong and Song, 2018). In this method, each mother particle is represented by bonded sub-particles. Although the simulation of granular assembly using the bonded particles is faster than the coupled DEM-FEM approach, it is still computationally intensive for modelling of a reasonable size particle assembly. Although these methods offer a thorough understanding of the micro-mechanics of particle breakage (Ciantia *et al*, 2016), other researchers have devised a particle replacement technique (Bruchmüller *et al*, 2011; Lobo-Guerrero and Vallejo, 2005) as an alternative with acceptable accuracy and remarkably improved computational demand. In the replacement method, the broken particle is recognised using different breakage criteria (de Bono and McDowell, 2016; Russell and Muir Wood, 2009) and replaced by an assembly of finer particles. The particles in this method are mainly simulated as spherical or circular elements. A more sophisticated model may be developed by considering different breakage patterns and particle sizes for the newly substituted smaller particles based on the stress condition of the mother particle (Zhou *et al*, 2020). The two principal aspects of the replacement method that require improvement are: (i) shape of the mother particle, and (ii) mass loss due to the replacement of the mother particle by an assembly of daughter particles with smaller volume (mass). To address these two aspects, the authors have introduced an advanced replacement method (Asadi, Disfani and Ghahreman Nejad, 2021) with different effective parameters to study the mechanical behaviour of rock fill material. The effect of boundary conditions in triaxial testing, has also been considered and analysed.

For DEM modelling of laboratory tests, two different techniques can be employed for the simulation of physical boundary condition: (i) assembly of spheres/balls located around the sample, and (ii) the rigid walls that are the most common boundary condition. The former approach may be based on either bonded marginal particles that will be kept at the sample perimeter and treated as the membrane during the simulation or considering the marginal particles in each cycle of simulation as the membrane without any bond. For the unbonded marginal elements method, the boundary particles may move into a loose assembly and the computational efficiency deteriorates to search for peripheral elements in each given cycle. Although researchers have made an effort to provide strategies for the calibration of microparameters especially related to the bonded particles of the membrane (Zhang *et al*, 2022), sophistications may be added to the model due to: (i) calibration of membrane bonds in different stress conditions and for various particles size and shape, (ii) necessary modifications to the bonded particles due to large deformations, and (iii) a large number of marginal particles to sufficiently address the membrane performance (de Bono, McDowell and Wanatowski, 2012). Hence, rigid walls have been most commonly used in DEM modelling.

For the triaxial test modelling, a rigid cylindrical wall composed of a different number of rectangular platens (depending on the resolution) may be used. In this method, the sample not only gets consolidated symmetrically but also cannot locally bulge. Liu, Yang and Dong (2022) studied the effect of the resolution of segmentation of cylindrical walls into rigid vertical segments also deformable articulated wall segments composed of triangular parts on compaction and mechanical behaviour of rock fill. Alternatively, segmentation could be conducted in a horizontal direction as small cylindrical rings (Zhao and Evans, 2009) to achieve a flexible boundary, however, in this method, the sample will be confined symmetrically in a diagonal direction since the independent rings may contract or expand and its centroid is fixed. Therefore, to benefit from the simplicity and computational efficiency of the rigid wall along with a fully deformable membrane in all directions, a combination of horizontal and vertical segmentation may be used as tile theory (Khoubani and Evans, 2018).

Consequently, in this study, the mechanical behaviour of rock fill particles is studied using the Modified Replacement Method previously introduced by the authors as the breakage simulation algorithm also the effect of flexible boundary on the particle breakage and mechanical behaviour is investigated through the tile theory. Independent small rigid rectangular walls will account for a cylindrical deformable membrane that can freely move inward and outward to achieve the target confining stress.

NUMERICAL MODELLING

Modified Particle Replacement Method (MPRM)

As mentioned earlier, the modified particle replacement method has been employed to efficiently capture the particle breakage phenomenon. The detail of this method is presented and discussed with details in a separate research paper (Asadi, Disfani and Ghahreman Nejad, 2021). The MPRM model stands out from other methods by its ability to simulate the particle shape, which has not been considered by others. Any particle shape is simulated using non-overlapped spherical particles and apollonian packings which is one of the useful advantages of this method. The second key advantage of this method is its computational efficiency, which is achieved by modelling the particles as rigid clumps from the outset. Due to the rigidity of the clump particles, the computational demand is significantly reduced since there is no need to analyse internal contacts between sub-particles. Throughout the simulation, the stress condition of each particle is periodically monitored at specified intervals using the contact forces between particles. Several breakage criteria have been proposed and utilised in DEM simulations for studying particle breakage such as: (i) Octahedral shear stress, (ii) Mean particle stress, (iii) Major principal stress, and (iv) Particle stress derived from the maximum contacts force and its diameter ($\sigma = F_{max}/d^2$).

In a study by de Bono and McDowell (2016), the above criteria were evaluated through DEM simulation of one-dimensional consolidation tests. Their findings showed that the maximum contact force criterion exhibited the best agreement with experimental results and observed behaviour. The maximum contact force at breakage (F_{max}) can be derived from the particle's characteristic strength ($\sigma_{0|d_0}$) at the reference diameter (d_0) as shown in Equation 3.

$$F_{max} = \left[\sigma_{0|d_0} \times \left(\frac{d}{d_0}\right)^{-3/m} \times \pi \left(\frac{3r_{avg}}{4E_{avg}}\right)^{2/3}\right]^3 \qquad (3)$$

In this equation, the contact area is calculated using the average contact radius (r_{avg}) and average modulus of elasticity (E_{avg}), and the characteristic strength may be scaled for different particle sizes using Weibull's modulus (Weibull, 1951). The characteristic strength of particles can be evaluated using the compressive (σ^c) and tensile (σ^t) strengths of the particle based on Russell and Muir Wood's 2009 study.

As shown in Figure 3, during the simulation cycling, if a particle is diagnosed in critical stress (ie 90 per cent of its characteristic strength), it will be replaced by an assembly of bonded balls of the same size at the same position which is able to breakdown and provide the model with breakage simulation capability. In this way, the model will only investigate the internal contacts of a particle that is near breakage, so the simulation will be fast and computationally efficient.

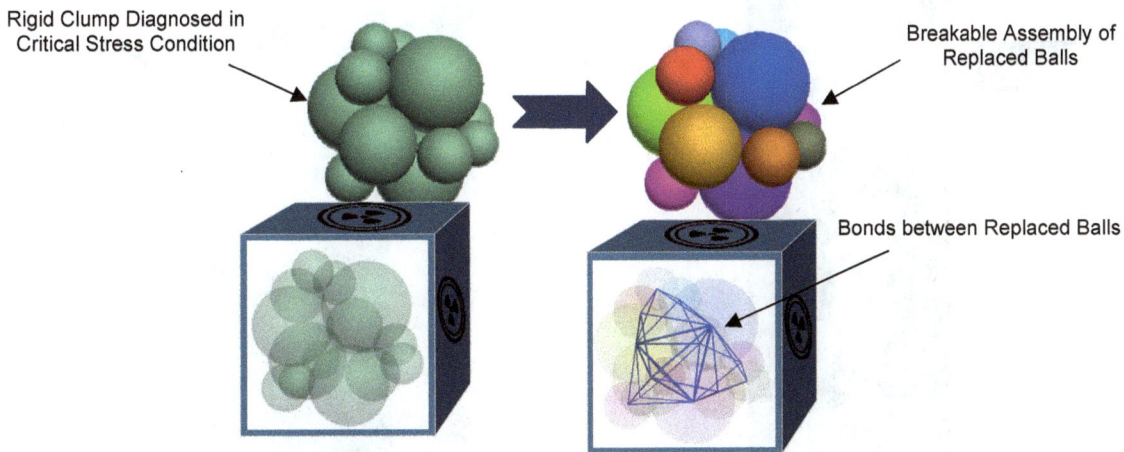

FIG 3 – Schematic illustration of Modified Replacement Method.

Tile-Based Flexible Membrane (TBFM)

Deformable boundary conditions and flexible membrane in this study are simulated using a group of rectangular segments known as Tiles. ITASCA PFC³ᴰ software has been used for the DEM simulation in this study. Considering the difficulties with: (i) calibration, and (ii) the limited confining pressure available through bonded circumferential individuals as a deformable boundary, on one hand, and non-interactive and independent walls in the PFC platform, on the other hand, TBFM provide an optimum and novel approach for simulation of the flexible membrane in triaxial testing.

The cylindrical flexible membrane in triaxial testing has been traditionally simulated using a semi-rigid membrane. As shown in Figure 4a, the membrane is divided into rectangular vertical strips (walls) which move forward and backward together to apply target confining stresses. A preliminary assessment of particle breakage using a semi-rigid membrane is presented in a paper at the 2021 ANCOLD conference (Asadi, Disfani and Ghahreman Nejad, 2021).

A fully flexible membrane is achieved by a series of overlapped tiles generating rings around the sample. These rings together form a flexible cylinder, as illustrated in Figure 4b. In contrast to a semi-rigid cylinder in which all segments must move together, each tile can independently move forward or backward to reach any stress levels. Hence, bulging and block shearing of the sample can be adequately captured in this method. The top and bottom caps are simulated as a platen wall as shown in Figure 4c.

The number of tiles and their size should be adjusted to have a balance between the computational demand and deformability of the membrane. The minimum dimension of a tile should also be larger than the smallest particle size so it remains in contact with the particles and the target confining stress could be calculated. Each tile is overlapped (Figure 4c) with its neighbours in a ring of tiles so that no gap could form to backward movement of the tile. The number of stacked rings vertically should be enough to accurately capture the sample's volumetric deformation (bulging or shearing) of the assembly. The circular cross-section of the membrane can be estimated by the different number of tiles in each ring (H_{part}). The area difference with the circumscribed circle for different number of tiles is illustrated in Figure 4d, in which it is clear that for $H_{part} > 20$, the area difference is below 1 per cent. Hence, in this study the number of tiles in each ring is limited to 20 with different number of rings (V_{sec}) investigated.

The servo program controls the stress condition of each tile independently, so, considering the contact forces acting on each tile, it will move forward or backward to achieve a given stress. Loading rate and the velocity of walls, including top and bottom platens, and tiles, are limited in a way that the confining stress remains constant, and a quasi-static condition is attained. Hence, the maximum inertial number (I) in the numerical simulations is limited to 10^{-4} which is strictly below the recommended value of 10^{-3} (Khoubani and Evans, 2018).

In Equation 4, I is the inertial number, $\dot{\varepsilon}$ stands for strain rate, d_{50} is average particle diameter, p' is the mean effective stress, and ρ_s is the particle density:

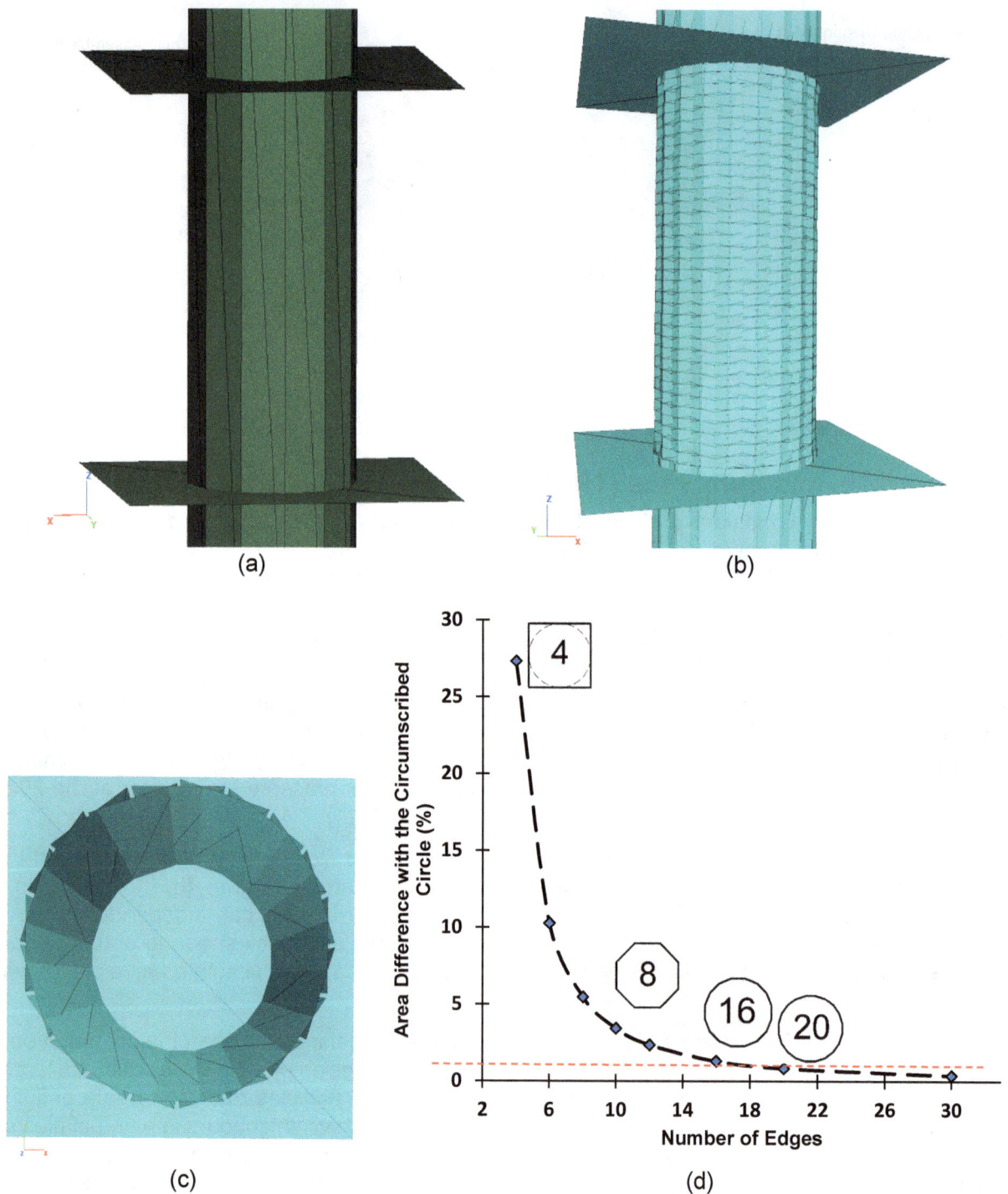

(a)

(b)

(c)

(d)

FIG 4 – Triaxial membrane model: (a) Semi-rigid (Solid) vessel composed of vertical strips, (b) Cylindrical membrane generated by independent rectangular tiles (V_{sec} =36), (c) Plan view of the top platen and first row of overlapped tiles (H_{part} = 20), (d) Area difference with the circumscribed circle for different number of tiles.

$$I = \frac{\dot{\varepsilon} \times d_{50}}{\sqrt{p'/\rho_s}} \tag{4}$$

Triaxial testing simulations

In this study, two different particle shapes composed of 14 and 4 balls have been employed in the numerical simulations as shown in Figure 5. For the parametric study of the numerical model, a series of triaxial tests have been modelled on a cylindrical sample, 30 cm diameter and 70 cm tall with a porosity of 58 per cent. The adopted DEM parameters are listed in Table 1. The triaxial testing simulations have been conducted using semi-rigid and tile-based flexible membranes with H_{part} = 20, and V_{sec} = 12, 24, and 36. The linear contact model is employed for the interaction of particles, with the parallel bonding adopted between the sub-particles of replaced critical particles.

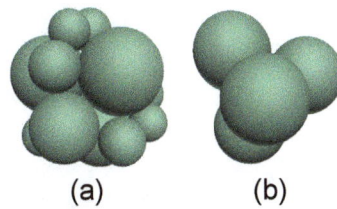

(a)   (b)

FIG 5 – Particle shapes employed for numerical simulations: (a) 14 balls, (b) 4 balls.

TABLE 1

DEM parameters.

Parameters	Value		
	Weak	Medium	Strong
Characteristic strength of particle (MPa)	23.2	43.2	350.7
Linear parallel bond tensile strength (N/m^2)	0.3×10^8	0.5×10^8	4.2×10^8
Linear parallel bond cohesion (N/m^2)	0.6×10^8	1.0×10^8	8.4×10^8
Young's modulus, E (MPa)	46.8		
Poisson ratio, v	0.3		
Linear normal stiffness (N/m)	5.8×10^6		
Linear shear stiffness (N/m)	2.9×10^6		
Linear parallel bond normal stiffness (N/m)	9.6×10^8		
Linear parallel bond shear stiffness (N/m)	4.8×10^8		
Sample initial porosity	58%		

To prepare the samples, non-overlapping particles are assembled in a larger vessel/mould for each sample. The vessel is then contracted to its actual size of 30 cm × 70 cm. It is worth noting that a uniform particle size of 2 cm is used in the parametric study. Once the samples are prepared, the target confining pressure will be achieved by servo-controlled movement of the walls, and finally, the shearing load will be applied by the top and bottom platen. In addition to the inertial number criteria, the shearing rate and the velocity of all the walls are limited to a very slow rate so that the target confining stress is maintained at a constant level (tolerance <0.1 per cent) throughout the simulation.

TBFM resolution

Figure 6 shows the numerical Triaxial test results for the 14-ball shape rock particle using flexible membrane with different number of tile rows (V_{sec}) and solid (semi-rigid) membrane at 2 MPa confining stress. For these models the number of tiles in each row/ring (H_{part}) is 20, resulting in total tile numbers of 720, 480, and 120 for the 36, 24, and 12 row membranes, respectively. The results indicate that the samples in the solid membrane exhibit a more dilative behaviour. Additionally, even weak particles show less contraction in the solid membrane as compared to the flexible membrane.

The predicted deviatoric stress in the solid vessel is slightly higher than that in the flexible membrane for assemblies with Medium to Strong particles (ie limited broken particles). Notwithstanding, for weak particles, where breakage is expected to be prevalent (Figure 6c), the peak and residual mobilised strengths are higher for the analyses with flexible membranes.

The difference in the predicted stress-strain behaviour between the flexible and solid membranes is relatively small with the difference becomes more pronounced for weak particles, where significant particle breakage is observed. It can be concluded that the (20 × 36) tile arrangement produces the highest accuracy for simulation of the breakable rock fill particles.

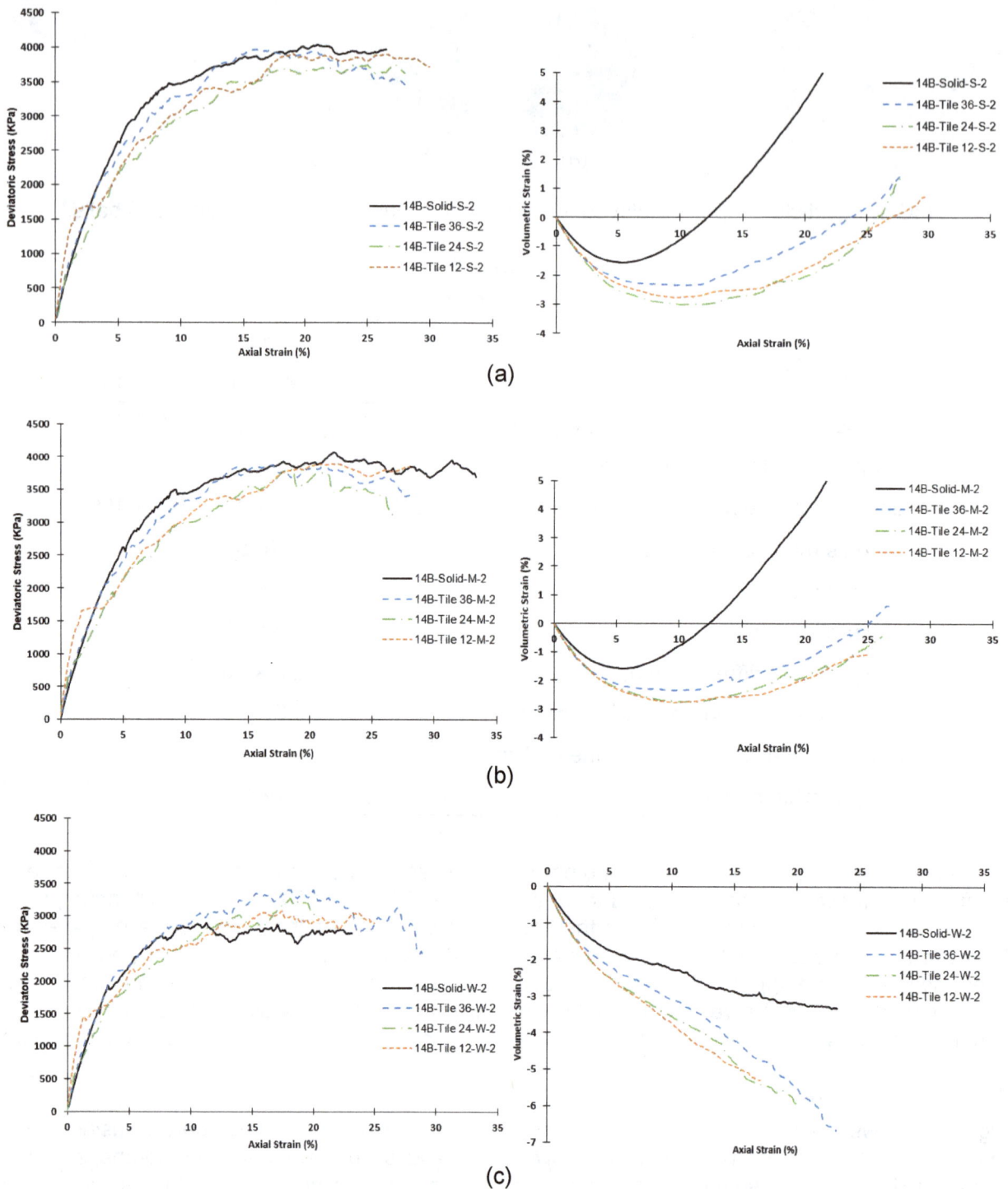

(a)

(b)

(c)

FIG 6 – Numerical triaxial test results for different number of tiles' rows (V_{sec}) and solid (semi-rigid) vessel for 2 MPa confining stress and: (a) strong, (b) medium, (c) weak particles (the legend is title base on: Particle Shape – Tile Resolution – Particle Strength – Confining Stress).

Particle shape effect

As previously stated, MPRM offers the advantage of incorporating particle shape in the modelling process with optimal computational efficiency from the beginning of modelling. To examine the stress-strain behaviour of samples comprised of various particle shapes, Figure 7 presents the results of triaxial modelling using two distinct particle shapes composed of 14 and 4 balls (shown in

Figure 5) at two different confining stresses of 1.5 MPa and 2 MPa, using uniform particle sizes of 2 cm.

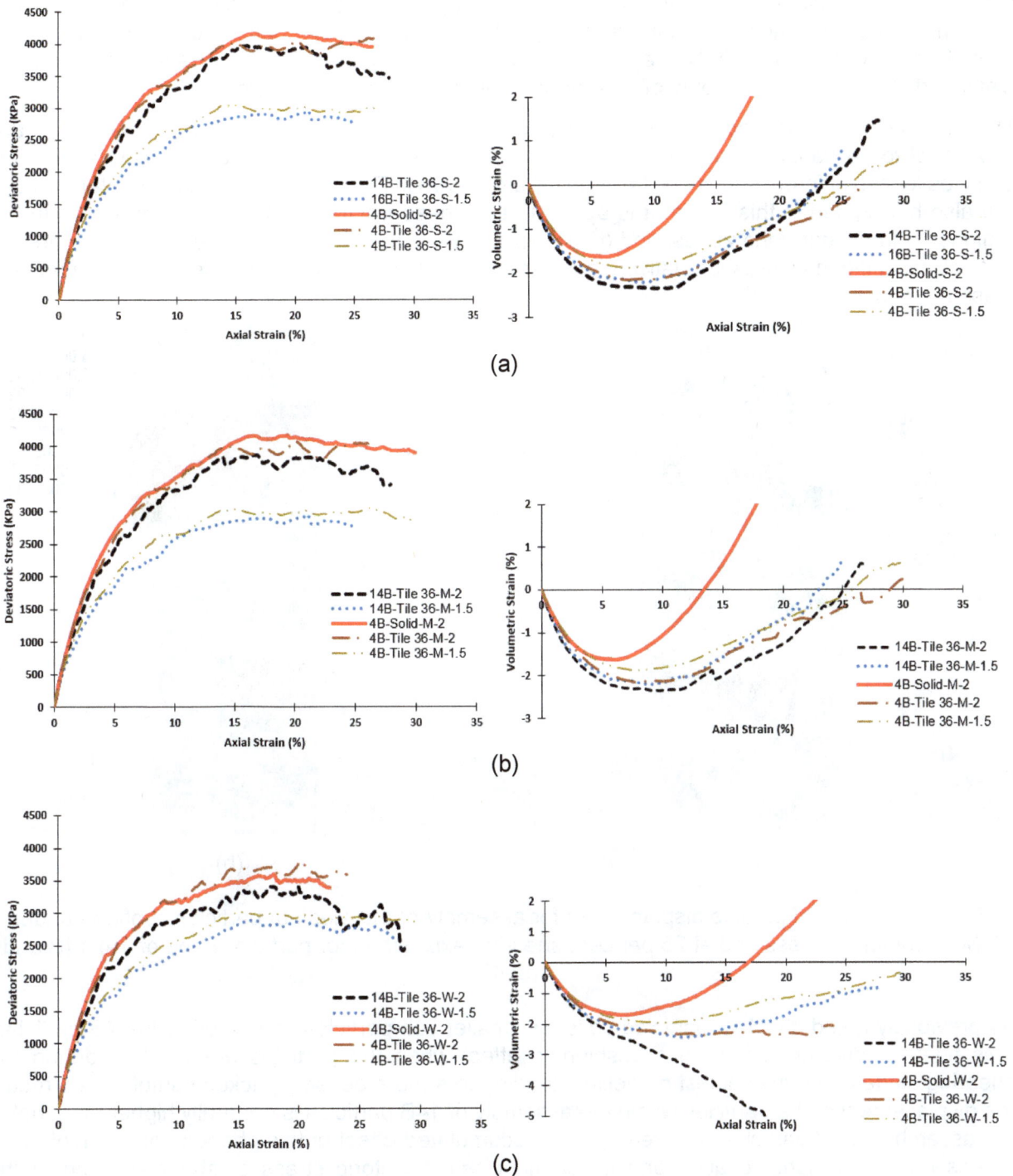

FIG 7 – Triaxial numerical test results for different particle shapes composed of 14 balls and 4 balls for: (a) strong, (b) medium, and (c) weak particles.

It is evident from the comparison of Figures 6 and 7 that the difference in mobilised deviatoric stress between the solid and flexible membranes is lower for particle shape 4B than for particle shape 14B. However, the trend in volumetric behaviour for both shapes is almost identical. The assembly of 4B particles shows a higher strength in comparison to that of 14B particles due to higher angularity. This effect is particularly evident at lower confining stresses (1.5 MPa), where fewer particles are broken. The increased angularity of 4B particles causes greater dilation at lower axial strains. However, as axial strain increases, the breakage phenomenon overcomes the effect of angularity, resulting in lower dilation at the residual stage compared to 14B particles. This trend is more obvious for the strong and medium particles confined under 1.5 MPa stress.

In the case of 14B particles, the emergence of finer daughter sub-particles following the replacement of the mother particle under critical stress conditions leads to greater compaction of the assembly. This phenomenon is particularly exaggerated at high confining stresses of 2 MPa and in the case of weak particles where the majority of mother particles are broken, as shown in Figure 7c.

The Diagram of particle displacement confined under 2 MPa stress using TBFM with H_{part} of 20, and V_{sec} equal to 36 (labelled as 'Tile 36' in the triaxial models' results) for the assembly of Medium particles for two different shapes of 14B and 4B are shown in Figure 8. The critical daughter particles are also highlighted in this figure. It is apparent that these fractured particles are clustered in two regions: (i) the shear band zones, and (ii) the areas where the particle displacements are markedly distinct, resulting in increased relative displacement and higher applied stresses on the particles in these regions.

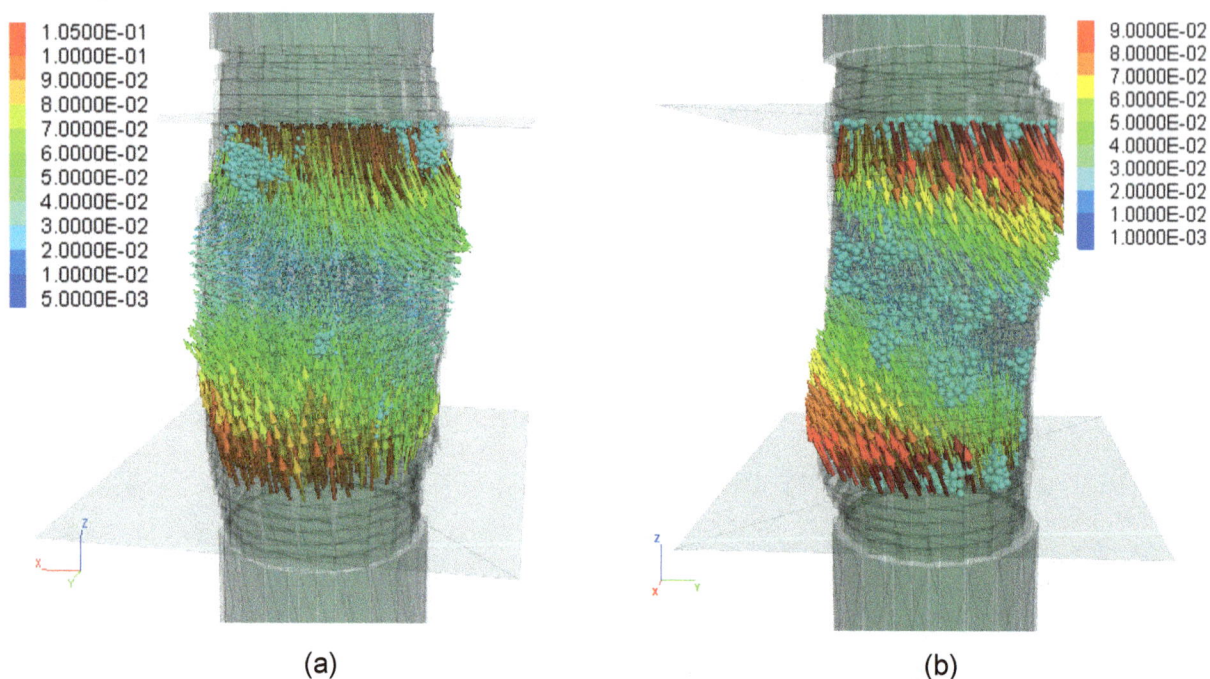

(a)   (b)

FIG 8 – Diagram of particle displacement for assembly of the Medium particles confined under 2 MPa confining stress and at 25 per cent shearing axial strain for particle shape of: (a) 14B and (b) 4B.

As previously noted, the 14B particles possess smaller sub-particles compared to the 4B particles. This finer subdivision results in a cushioning effect when the particles are crushed during the shearing stage of the triaxial test modelling, leading to a more densely packed sample. As a result, the displacement of the particles within the assembly of 14B particles is generally higher than that in the assembly of 4B particles (Figure 8a). The accumulated effect of this phenomenon at high axial strains results in higher dilation and lower mobilised deviatoric stress or shear strength of the assembly in comparison to the assembly of 4B particles as demonstrated in Figure 7b.

Moreover, the assembly of 14B particles is in the bulging condition which stems from the smoother stress distribution due to the rearrangement of daughter sub-particles. It is worth to be noted that the cushioning effect and redistribution of stress resulted in the fewer number of broken particles. On the other hand, the assembly of the 4B particles is showing two distinct shear blocks sliding against each other, as shown in Figure 8b. Also, the number of broken particles in this assembly is higher than the 14B particles' assembly which is due to: (i) the lower intrinsic strength of bigger pebbles, and (ii) bigger daughter sub-particles of this particle shape and concentrated stress on the critical particles.

LARGE-SCALE TRIAXIAL EXPERIMENTS

ATC Williams has designed and commissioned a series of large-scale triaxial tests on rock fill material from a dam located in Peru. The tests were conducted on representative samples of rock fill at the University of Chile. The PSD's of the samples tested experimentally and numerically are illustrated in Figure 9. To seek computational efficiency, the minimum particle size in the numerical model is limited to 1 cm. The sample bulk density is 2.1 t/m³ with a particles specific gravity of 2.65. The actual rock fill samples are composed of 30 per cent intrusive and 70 per cent calcitic rocks, hence, in the numerical model, two different intrinsic strengths (ie Medium and Weak) have been employed.

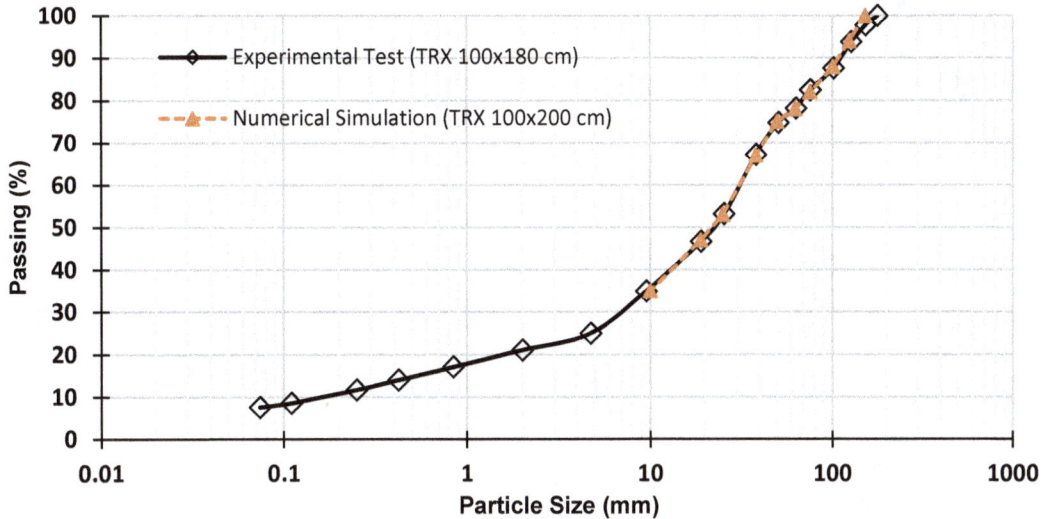

FIG 9 – Grain size distribution curve for the experimental and numerical tests.

The results for the actual large-scale triaxial and numerical model tests using coupled MPRM-TBFM methods are presented in Figure 10. The particles' intrinsic strength and micro parameters including parallel bond tensile and cohesive strength are calibrated using the Unconfined Compression Strength (UCS) tests from the laboratory testing of the rock fill particles. For model calibration, the results of the sample confined under 2 MPa have been employed, and the model performance is evaluated for the samples under 1 MPa and 1.5 MPa effective confining stresses.

Reference to Figure 10, the sample under lower confining stress (1 MPa) shows a dilative behaviour, and as the confining stress increases, the dilation moves toward a more contractive behaviour. Obviously, the samples under higher confining stresses show higher strength and mobilised deviatoric stress. The particle breakage phenomenon becomes more dominant for higher axial strains for the sample under 1 MPa confinement, and rearrangement of finer broken particles along with lower confining stress makes the stress condition prone to dilation after a given amount of compaction at higher shear strength.

Figure 11 illustrates the variation of mobilised friction angle by the axial strain for different confining stresses. In the case of granular particles, the mobilised friction angle can be directly interpreted as the stress ratio between deviatoric stress to the mean effective stress ($\eta = {}^q/_p$). The sample exhibits a higher peak friction angle under lower confinement. However, after the predominance of particle breakage and dilation, the residual stress ratio and friction angle decrease below those observed at higher confining stresses of 1.5 MPa and 2 MPa.

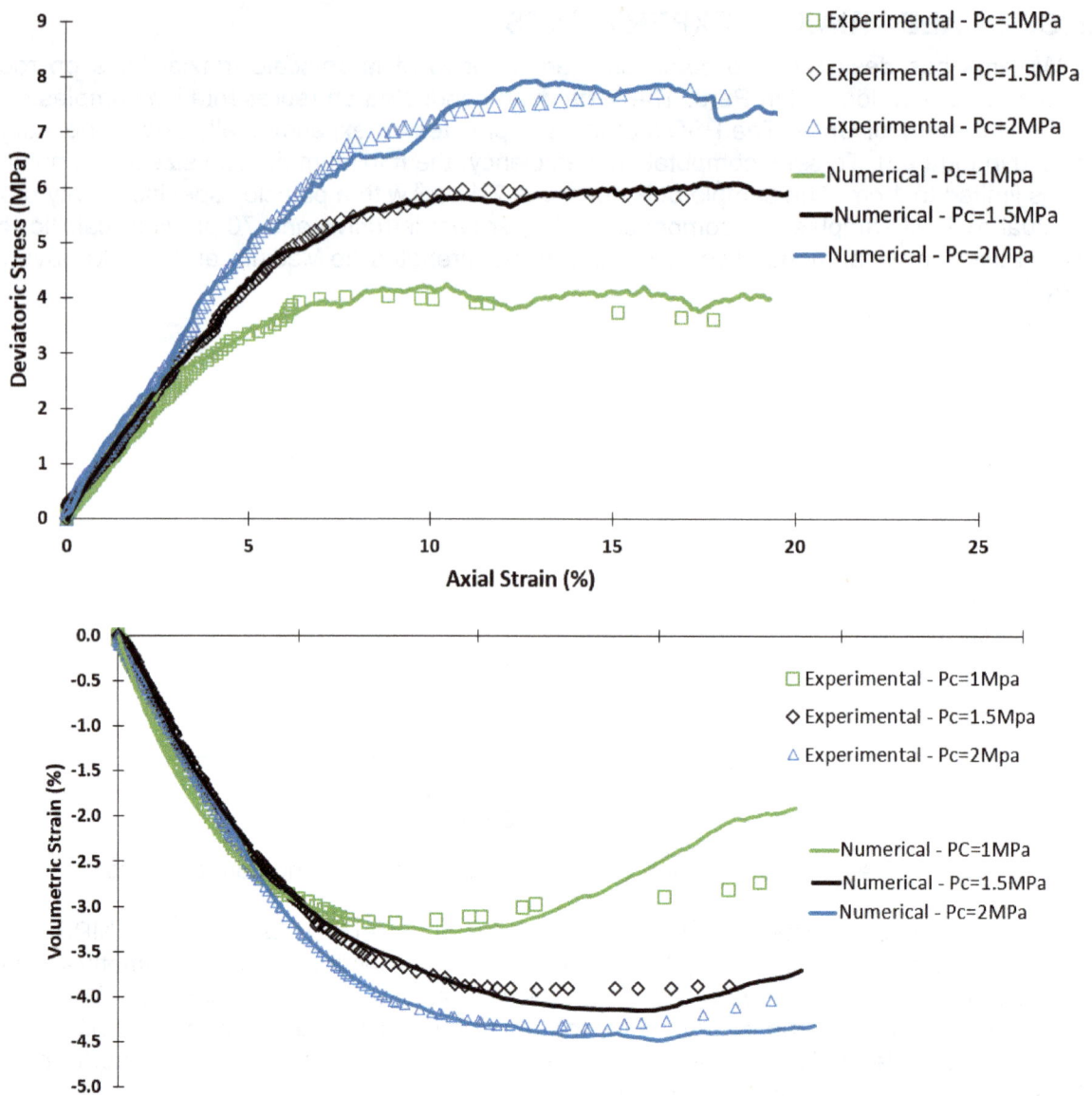

FIG 10 – Large-scale triaxial result for experimental tests and numerical models using coupled MPRM-TBFM.

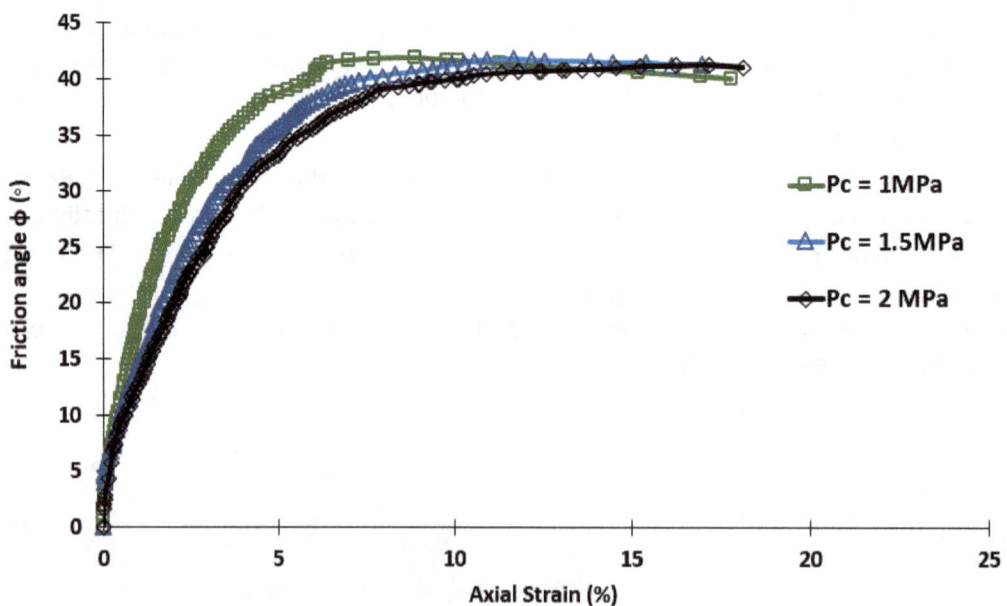

FIG 11 – Variation of mobilised friction angle by the axial strain for different confining stresses.

CONCLUSIONS

In this paper, the effect of particle shape, confining stress, breakage phenomenon, and boundary condition on the mechanical behaviour of rock fill particles is studied using the Modified Particle Replacement Method (MPRM) previously introduced by the authors as the breakage simulation algorithm. Furthermore, the effect of a flexible membrane in the triaxial test modelling on the particle breakage and mechanical behaviour is investigated through the Tile-Based Flexible Membrane (TBFM) theory in which independent small rigid rectangular walls form a cylindrical deformable membrane that can freely move inward and outward to achieve the target confining stress.

Apart from the numerical modelling and parametric study on the granular assemblies, the large-scale triaxial results on the material of a dam in Peru which have been conducted at the IDIEM laboratory of the University of Chile are encompassed in this study. The samples tested at the IDIEM laboratory have a diameter of 1 m and a height of 1.8 m, which have been modelled as 1 m (diameter) × 2 m (height) samples.

In this study, two different particle shapes composed of 14 and 4 balls have been employed in the numerical simulations. For the parametric study of the numerical model, a series of triaxial tests have been conducted on a sample of 30 (diameter) × 70 cm (height) using semi-rigid (solid) and flexible membranes with H_{part} = 20, and V_{sec} = 12, 24, and 36.

The deviatoric stress in the samples with solid membrane is slightly higher than that in the samples with flexible membrane for assemblies of Strong and Medium particles. However, the peak and residual mobilised strength are higher in the flexible membrane samples with weak particles' assembly. The flexible membrane mainly affects volumetric behaviour. This becomes more pronounced for weak particles, where a large number of breakages are observed.

The assembly of particles with higher angularity shows a higher strength and greater dilation at lower axial strains. This effect is particularly evident at lower confining stresses. Nonetheless, as the axial strain increases, the effect of angularity is subdued by the particle breakage, resulting in lower dilation at the residual strength level.

The Diagram of particle displacement using coupled MPRM and TBFM shows that the fractured particles are clustered in two regions: (i) the shear band zones, and (ii) the areas where the particle displacements are markedly distinct, resulting in increased relative displacement. The finer daughter sub-particles result in a cushioning effect when the particles are crushed during the shearing stage, leading to a more densely packed sample and higher dilation and lower shear strength.

Moreover, the assembly of 14B particles is in the bulging condition which stems from the smoother stress distribution due to the rearrangement of finer daughter sub-particles. On the other hand, the assembly of the 4B particles is showing two distinct shear blocks sliding against each other. Also, the number of broken particles in this assembly is higher than that in the 14B particles assembly which is due to: (i) lower intrinsic strength of bigger pebbles, and (ii) bigger daughter sub-particles of this particle shape and concentrated stress on the critical particles.

The developed Discrete Element Model has been calibrated and validated using the large-scale triaxial test results. A good agreement in stress-strain behaviour is captured between the modelled and experimental results.

ACKNOWLEDGEMENTS

This research project has been initiated by ATC Williams and carried out at the University of Melbourne. The authors would like to express their appreciation for the sponsorship and ongoing support of this study.

The authors are also grateful to the ITASCA group for their willingness to share information and provide support through ITASCA Educational Partnership (IEP) program.

Furthermore, the authors would like to express their sincere gratitude to Dr Matteo Ciantia, the Senior Lecturer of Soil Mechanics and Geotechnical Engineering at the University of Dundee for guidance and insightful feedback that greatly enhanced the quality of this work.

REFERENCES

Alonso, E E, Tapias, M and Gili, J, 2012. Scale effects in rock fill behaviour, *Géotechnique Letters*, 2(3):155–160.

Asadi, R, Disfani, M M and Ghahreman Nejad, B, 2021. An Overview to Mechanical Behaviour of Rock fill, Focus on Numerical Modelling using DEM, ANCOLD 2021 Conference – Dams: Perceptions and Realities of Risk.

Bagherzadeh-Khalkhali, A, Mirghasemi, A A and Mohammadi, S, 2008. Micromechanics of breakage in sharp-edge particles using combined DEM and FEM, *Particuology*, 6(5):347–361.

Bruchmüller, J, van Wachem, B G M, Gu, S and Luo, K H, 2011. Modelling discrete fragmentation of brittle particles, *Powder Technology*, 208(3):731–739.

Ciantia, M O, Arroyo, M, Butlanska, J and Gens, A, 2016. DEM modelling of cone penetration tests in a double-porosity crushable granular material, *Computers and Geotechnics*, 73:109–127.

de Bono, J and McDowell, G, 2016. Particle breakage criteria in discrete-element modelling, *Géotechnique*, 66(12):1014–1027.

de Bono, J, McDowell, G and Wanatowski, D, 2012. Discrete element modelling of a flexible membrane for triaxial testing of granular material at high pressures, *Géotechnique Letters*, 2(4):199–203.

Einav, I, 2007. Breakage mechanics–Part I: Theory, *Journal of the Mechanics and Physics of Solids*, 55(6):1274–1297.

Frossard, E, Hu, W, Dano, C and Hicher, P-Y, 2012. Rock fill shear strength evaluation: a rational method based on size effects, *Géotechnique*, 62(5):415–427.

Jia, Y, Xu, B, Chi, S, Xiang, B and Zhou, Y, 2017. Research on the Particle Breakage of Rock fill Materials during Triaxial Tests, *International Journal of Geomechanics*, 17(10).

Jiang, M, Xie, Y and Liu, J, 2018. DEM Investigation on the Scale Effect on Macroscopic and Microscopic Mechanical Behavior of Rock fills, GeoShanghai International Conference (Springer).

Khoubani, A and Evans, T M, 2018. An efficient flexible membrane boundary condition for DEM simulation of axisymmetric element tests, *International Journal for Numerical and Analytical Methods in Geomechanics*, 42(4):694–715.

Lade, P V, Yamamuro, J A and Bopp, P A, 1996. Significance of particle crushing in granular materials, *Journal of Geotechnical Engineering*, 122(4).

Leps, T M, 1970. Review of shearing strength of rock fill, *Journal of Soil Mechanics and Foundations Div*.

Liu, D, Yang, J and Dong, B, 2022. Discrete element analysis of the influence of compaction quality on mechanical properties of rock fill materials, *Computers and Geotechnics*, 151.

Lobo-Guerrero, S and Vallejo, L, 2005. DEM analysis of crushing around driven piles in granular materials, *Géotechnique*, 55(8):617–623.

Marsal, R J, 1967. Large-scale testing of rock fill materials, *Journal of the Soil Mechanics and Foundations Division*, 93(2):27–43.

Marsal, R J, 1973. *Mechanical properties of rock fill* (John Wiley and Sons, Incorporated).

Oldecop, L A and Alonso Pérez de Agreda, E, 2001. *A model for rockill compressibility*.

Oldecop, L A and Alonso Pérez de Agreda, E, 2007. *Theoretical investigation of the time-dependent behaviour of rock fill*.

Russell, A R and Muir Wood, D, 2009. Point load tests and strength measurements for brittle spheres, *International Journal of Rock Mechanics and Mining Sciences*, 46(2):272–280.

Tapias, M, Alonso, E E and Gili, J, 2015. A particle model for rock fill behaviour, *Géotechnique*, pp 1–20.

Varadarajan, A, Sharma, K, Venkatachalam, K and Gupta, A, 2003. Testing and modeling two rock fill materials, *Journal of Geotechnical and Geoenvironmental Engineering*, 129(3):206–218.

Weibull, W, 1951. A statistical distribution function of wide applicability, *Journal of Applied Mechanics*, 18(3):293–297.

Xiao, Y, Liu, H, Chen, Y and Jiang, J, 2014. Strength and Deformation of Rock fill Material Based on Large-Scale Triaxial Compression Tests, II: Influence of Particle Breakage, *Journal of Geotechnical and Geoenvironmental Engineering*, 140(12).

Xu, M, Hong, J and Song, E, 2018. DEM study on the macro- and micro-responses of granular materials subjected to creep and stress relaxation, *Computers and Geotechnics*, 102:111–124.

Yang, G, Yan, X, Nimbalkar, S and Xu, J, 2019. Effect of Particle Shape and Confining Pressure on Breakage and Deformation of Artificial Rock fill, *International Journal of Geosynthetics and Ground Engineering*, 5(2).

Yu, F W, 2017. Particle breakage and the critical state of sands, *Géotechnique*, 67(8):713–719.

Zhang, P, Sun, X, Zhou, X and Zhang, Y, 2022. Experimental simulation and a reliable calibration method of rock fill microscopic parameters by considering flexible boundary, *Powder Technology*, 396:279–290.

Zhao, B, Wang, J, Coop, M, Viggiani, G and Jiang, M, 2015. An investigation of single sand particle fracture using X-ray micro-tomography, *Géotechnique*, 65(8):625–641.

Zhao, X and Evans, T M, 2009. Discrete simulations of laboratory loading conditions, *International Journal of Geomechanics*, 9(4):169–178.

Zhou, W, Wang, D, Ma, G, Cao, X, Hu, C and Wu, W, 2020. Discrete element modeling of particle breakage considering different fragment replacement modes, *Powder Technology*, 360:312–323.

Hollow cylinder testing systems – the case for their use in tailings engineering practice

R Fanni[1], D Reid[2] and A B Fourie[3]

1. Principal Tailings Engineer, Golder Associates, Perth WA 6008. Email: riccardo.fanni@wsp.com
2. Research Fellow, Department of Civil, Environmental and Mining Engineering (CEME), University of Western Australia, Crawley WA 6009. Email: david.reid@uwa.edu.au
3. MAusIMM, Professor, Department of Civil, Environmental and Mining Engineering (CEME), University of Western Australia, Crawley WA 6009. Email: andy.fourie@uwa.edu.au

ABSTRACT

The quality and frequency of detailed laboratory test programs to characterise tailings have greatly improved over the past five years. However, one area where limitations in current commercial engineering practice remains is in accounting for below-slope stress conditions – specifically plane strain conditions with a non-vertical principal stress angle. Although commercially available SGI and NGI direct simple shear testing devices offer some features relevant to such conditions, they also suffer a number of limitations, particularly with respect to an inability to measure principal stresses. This paper presents an approach for the use of a Torsional Shear Hollow Cylinder (TSHC) device in high-level tailings commercial laboratory programs on the basis of an examination of typical below-stress conditions, a review from the literature on the effects of a non-vertical principal stress angle and a critical appraisal of testing options that currently exist for commercial laboratory practice. Finally, some examples of the additional insight provided by the TSHC, and the potentially lower values of strengths (compared to triaxial compression) from various stress-paths relevant to static liquefaction failures, are provided for further illustration.

INTRODUCTION

Over the past five years, the amount and quality of the laboratory characterisation of tailings has greatly increased owing largely to a number of prominent tailings storage facility (TSF) failures. In many cases, the increased quality and selection of better approaches to tailings testing is simply adopting what was clearly laid out decades earlier – for example the seminal work of Ladd (1991), or the detailed guidance outlined by Jefferies and Been (2006) and Shuttle and Cunning (2007) on tailings testing to determine the critical state line (CSL) and infer in situ state parameter Ψ (Been and Jefferies, 1985). However, prior to the impetus of recent failures and the clear illustration of their use in a series of failure investigations, these approaches were not adopted as widely as perhaps would have been advisable.

While the increased quality of testing is clear, a major limitation that remains is that the two readily-available devices for commercial testing – the triaxial and the direct simple shear (DSS) – each suffer limitations, some of which are probably not sufficiently appreciated by the industry. First, evidence of the ubiquity of these two devices in tailings practice can be seen by a review of recent TSF failure investigations provided in Table 1. The triaxial test has been the key tool used to characterise tailings shearing behaviour in these studies, with the DSS applied in some cases where cyclic triggering required investigation. Importantly, the triaxial test, from which most of the information was derived in these investigations, has a vertical principal stress angle – a condition that differs significantly from conditions below much of the slope.

TABLE 1

Test types used to characterise tailings in recent failure investigations.

Failure investigation	Triaxial tests	DSS tests
Fundão (Morgenstern *et al*, 2016)	CSL and NorSand calibrations Lateral extrusion triggering tests	Cyclic liquefaction triggering assessment
Cadia (Jefferies *et al*, 2019)	CSL and NorSand calibrations Stress path triggering tests	Static and cyclic liquefaction triggering assessment
Feijão (Robertson *et al*, 2019)	CSL and NorSand calibrations Strength versus state parameter profile Creep triggering tests	No DSS tests on tailings

The purpose of this paper is to argue that with the ongoing efforts to improve tailings characterisation, we have reached a point as an industry where use of more sophisticated tests has become both practical and, indeed, advisable. In particular, we suggest that the well-established Torsional Shear Hollow Cylinder (TSHC) device provides the ideal combination of control and capabilities to enable a richer and more relevant tailings test program to be carried out. This argument is outlined first by reviewing below-slope stress conditions, then drawing the reader's attention to the wide library of data that highlights the importance of principal stress direction on soil behaviour, followed by an appraisal of available test devices and the factors that, in our view, make the TSHC the best available tool to be applied in current practice.

BELOW-SLOPE PLANE STRAIN STRESS CONDITIONS

To frame the discussion of this work, a review of below-slope plane strain stress conditions, in contrast to axisymmetric conditions (ie below level ground) is useful. The differences are highlighted schematically in Figure 1, with the following aspects of particular note below slopes:

- Intermediate principal stress σ'_2 and minor principal stress σ'_3 are no longer equal.

- The direction of the principal stress σ'_1 angle α rotates away from the vertical, with the amount of rotation being dependent on the position within the slope.

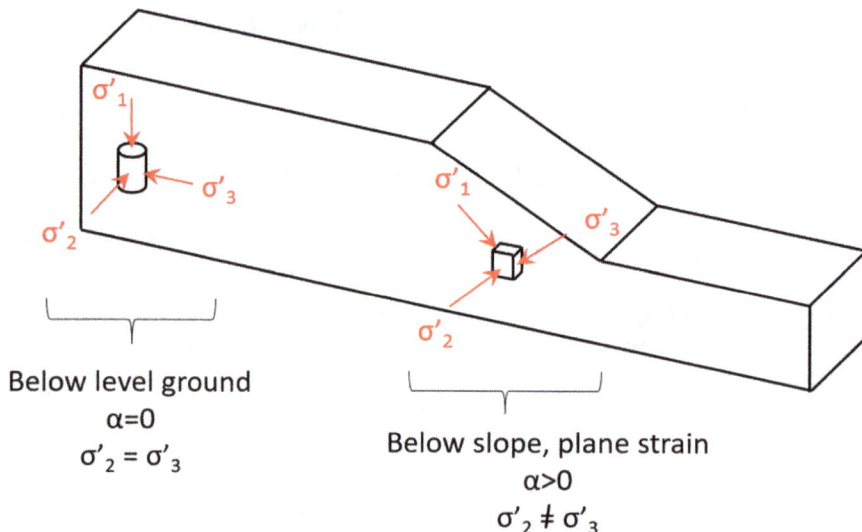

FIG 1 – Schematic contrast of stress conditions below level ground and a slope.

For review, in order to characterise the differences between the two conditions shown in Figure 1, in addition to the principal stress angle from vertical α, the parameter b as defined by Bishop (1966) is used:

$$b = \frac{\sigma'_2 - \sigma'_3}{\sigma'_1 - \sigma'_3}$$

Where b = 0 represents triaxial compression, b = 1 represents triaxial extension, and a value ranging from 0.2 to 0.4 is generally produced in plane strain testing in different types of test devices and on different soils (Shibuya and Hight, 1987; Wanatowski and Chu, 2007).

While schematic representations such as Figure 1 are common in the geotechnical literature, further elaboration and demonstration of these conditions using a 'real' example is useful. To this end, an idealised TSF numerical model was developed using the finite difference code FLAC v8.1, as follows:

- The geometry is shown in Figure 2, having been selected to represent a reasonable analogue of many upstream-raised TSFs. A final TSF height of 50 m and slope of 1V:2.5H are the most notable features.

- The TSF is assumed to be located above a strong foundation such that these materials may be dilative if sheared.

- The tailings were modelled using the NorSand constitutive model, with the parameters developed by the Cadia Panel (Table 2; Jefferies *et al*, 2019) adopted, as these are arguably the most robustly-determined publicly available data set for a silty tailings.

- The tailings were incrementally deposited in 1 m layers, with the phreatic surface increased proportionally with each raise consistent with the final phreatic surface profile shown in Figure 2.

- After completion of the model development using NorSand, a script was implemented to 'switch' the tailings to the Mohr Coulomb constitutive model with an undrained strength ratio s_u/σ'_{vc} of 0.25 adopting Tresca approach, ie implemented in Mohr–Coulomb by imposing friction and dilatation equal to zero with strength given as cohesion calculated from the consolidated vertical effective stress. This process of using NorSand to develop the stresses within a model while applying Mohr Coulomb to carry out subsequent discrete analysis is consistent with both the Cadia and Feijão modelling approaches (Jefferies *et al*, 2019; Robertson *et al*, 2019).

- The factor of safety (FoS) for the slope under these conditions was found using the shear strength reduction technique within FLAC. This indicated a FoS of 1.44 for the slope. Clearly, the idealised TSF represents a scenario for which further detailed information on the tailings' behaviour would be sought, including by means of laboratory characterisation, due to the factor of safety not meeting globally accepted values for static conditions.

- A script was implemented to extract the stress conditions along the critical failure surface identical to the previous step.

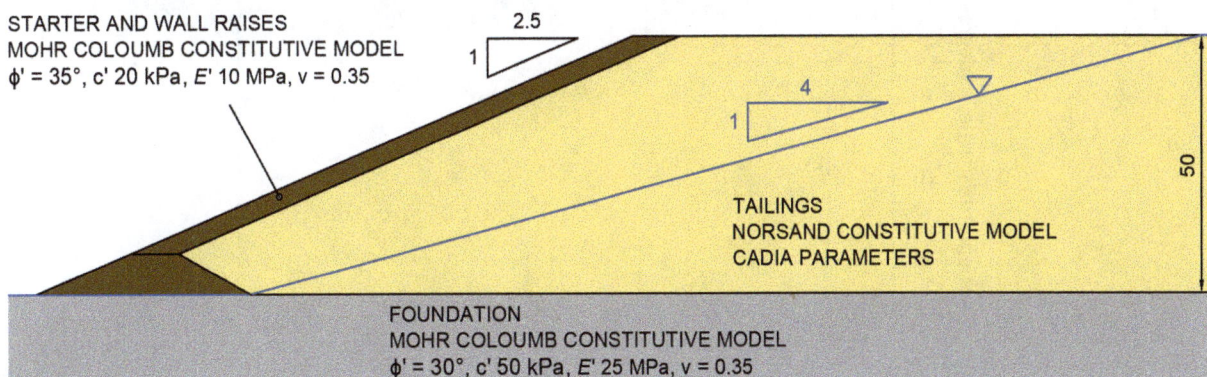

FIG 2 – Idealised TSF geometry for demonstration purposes.

TABLE 2

Cadia NorSand parameters used for tailings model development (Jefferies *et al*, 2019).

Parameter	Value
CSL definition	$e_c = 0.906\text{-}0.355 * (p'/100)^{0.119}$
M_{tc}	1.5
N	0.3
X	8
Plastic hardening	50-450Ψ
Elasticity	17 MPa $* (p'/100)^{0.76}$
ν	0.2
Seed Ψ	+0.06

The critical potential failure surface with a FoS of 1.44 identified in this process is shown in Figure 3. Also shown in Figure 3 are the principal stress angles along the identified critical slip surface within the tailings. These transition from zero near-surface (ie outside the area of influence of the slope) to up to about 50° near the toe, just upstream of the starter embankment. The value of b along the slip surface is generally close to 0.3. In both cases, clearly, conditions along much of the failure surface are quite distinct from axisymmetric with a vertical principal stress angle (ie the triaxial compression test). The potential implications of these stress conditions are discussed in more detail subsequently.

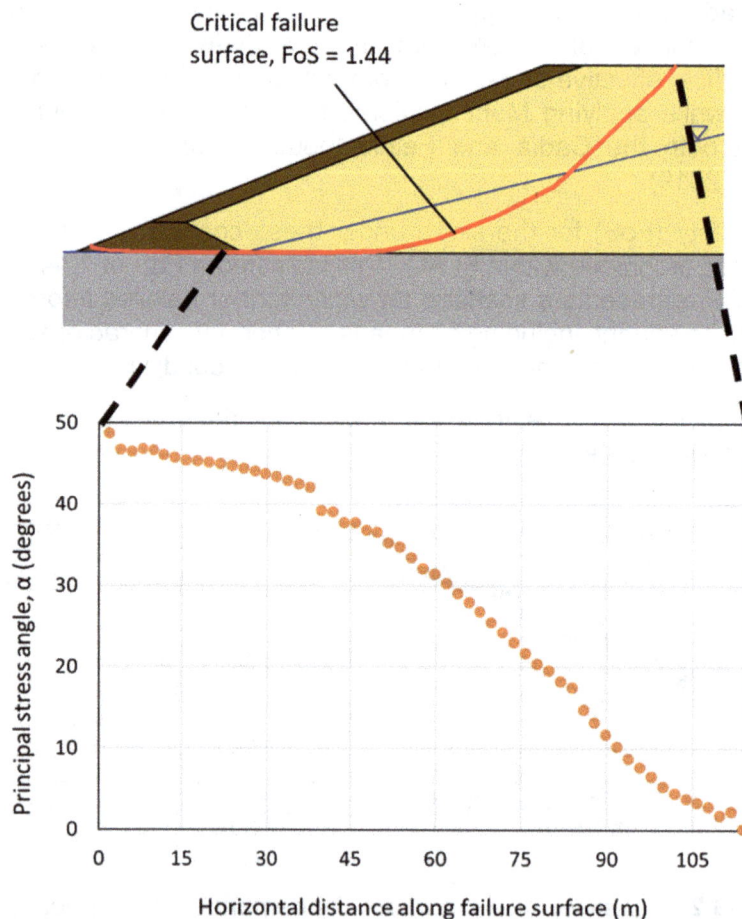

FIG 3 – Critical failure surface and principal stress angle along the failure surface within the tailings.

EFFECT ON SHEARING BEHAVIOUR

General

Of the two main differences between conditions below level ground and below a slope – principal stress direction and intermediate stress conditions – the focus in this work is primarily on principal stress direction. To appreciate how a non-vertical principal stress angle can affect shearing behaviour, it is instructive to consider the stress-history of an element of hydraulically-deposited tailings. When first deposited as a slurry, the tailings settle out of suspension and begin to develop effective stress under a vertical gravitational force. This means that the particle to particle contacts that develop are, in general, best at resisting stresses applied in a vertical direction. As vertical stresses increase so too does the cross anisotropy. Over the years and decades that follow, the slope geometry that develops means that some elements of tailings are now exposed to a significantly rotated principal stress direction (for example, refer to Figure 3). The rotation occurs gradually and is different throughout the slope. The main question then is – what effects is this rotation likely to have?

Undrained shearing

The clearest demonstration of the effect of a rotated principal stress is probably the detailed TSHC experimental work outlined by Sivathayalan and Vaid (2002). They water-pluviated samples of Fraser River sand, then consolidated the sand under a range of principal stress angles with a stress ratio K_c (σ'_1/σ'_3) of 1.5 – in other words, a very relevant consolidated state to the below-slope conditions of, for example, a hydraulically deposited structure built up over a period of years. Through careful preparation, each sample tested was at an essentially identical void ratio. Given that they were also consolidated to the same mean effective stress, this implies the same Ψ for each test discussed herein.

The results of Sivathayalan and Vaid's (2002) testing is synthesised in Figure 4, showing the significant reduction in strength with increasing principal stress angle. Importantly, while peak strengths reduce, the reduction in the minimum post-peak strength is even greater. In other words, the material is both weaker and more brittle in below-slope conditions, compared to what might be inferred from a triaxial test with a principal stress angle of zero. The increased brittleness is particularly noteworthy in light of attempts to characterise brittleness using a triaxial test, given how new tailings standards make particular note of the importance of brittleness and minimum post-peak strength (ICMM, 2020). While others have shown similar general effects from a rotated principal stress direction (Yoshimine, Ishihara and Vargas, 1998; Shibuya, Hight and Jardine, 2003), the work of Sivathayalan and Vaid is arguably the most important, given the water pluviated fabric used and the selection of consolidation and shearing procedures that simulate development of stresses below a slope.

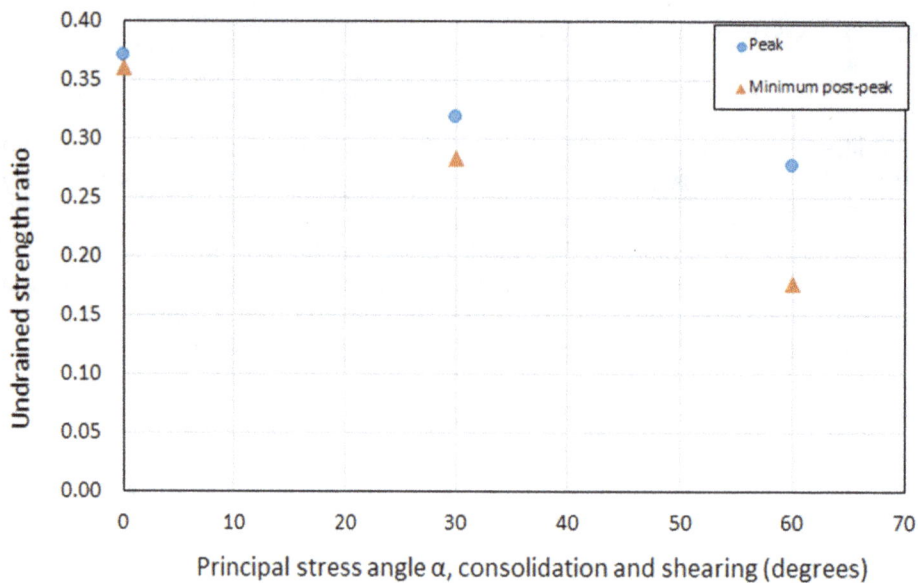

FIG 4 – Effects of anisotropy on undrained shearing behaviour of Fraser River sand (Sivathayalan and Vaid, 2002).

Liquefied shear strengths

While laboratory element tests are commonly used in geotechnical and tailings practice to refine in situ tests and develop a synthesis of strengths for use in design, the measurement of liquefied shear strengths has been an area where it has proved particularly challenging to obtain reasonable values in the laboratory. Specifically, many flow liquefaction case histories of sandy fills (eg Nerlerk, Fundão) had in situ states at, or perhaps even slightly dense of, the CSL (Jefferies and Been, 2015; Morgenstern *et al*, 2016). Element tests sheared from such a condition will produce dilative behaviour at large strains, contrary to the full-scale behaviour seen. Reid *et al* (2022b) outlines an argument, largely based on a review of historical simple shear tests carried out using the TSHC, that anisotropy appears to be a likely candidate to explain the discrepancy seen between triaxial compression element tests and field-scale behaviour for lightly dilative fills. This work draws on the trends seen for minimum post-peak strengths with increasing principal stress angle from Sivathayalan and Vaid (2002) (refer Figure 4) but is particularly focused on sands tested in the TSHC that can be referenced to a CSL.

The results synthesised by Reid *et al* (2022b) are presented in Figure 5 for TSHC simple shear testing on Toyoura (Yoshimine, Ishihara and Vargas, 1998) and M31 (Georgiannou, Konstadinou and Triantafyllos, 2018) sands. As regularly-used standard laboratory sands, the reference CSL for both is available (Verdugo and Ishihara, 1996; Vranna and Tika, 2015), thus allowing the presentation of the TSHC programs in the context of Ψ. These data are shown within the presentation framework developed by Jefferies and Been (2015), where the x-axis is the inferred characteristic in situ state (ψ_k) while the y-axis is the estimated liquefied shear strength from the failure ($s_{u\,(LIQ)}/\sigma'_{vc}$). As noted previously, many of these flow events are from lightly dilative states, where triaxial testing with a vertical principal stress angle will not provide appropriate strengths – rather, the samples will dilate at high strain, contrary to field-scale behaviour. However, the simple shear tests carried out in the TSHC align favourably with the case histories. This strongly points to the benefits of TSHC testing in obtaining more representative values of post-peak shear strengths from element tests, while also being able to directly calculate the Ψ value for the tests (unlike the DSS, where it is not possible to determine all principal stresses).

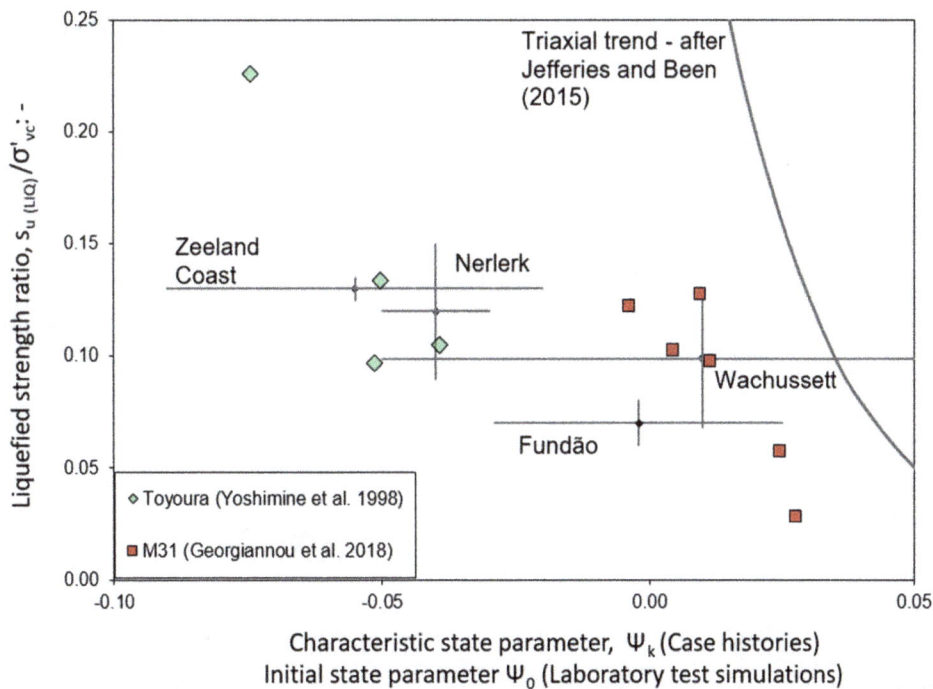

FIG 5 – Results of simple shear tests carried out within the TSHC for Toyoura and M31 sands (after Reid *et al*, 2022b). TSHC test data from Yoshimine, Ishihara and Vargas (1998) and Georgiannou, Konstadinou and Triantafyllos (2018), with general presentation framework after Jefferies and Been (2015).

Constant shear drained triggering

Triggering of static liquefaction under a rising phreatic surface – typically referred to as constant shear drained (CSD) triggering – is arguably one of the most dangerous triggering mechanisms as it is often preceded by little to not visible warning and is underappreciated by many practising engineers. Notwithstanding these issues, a significant number of studies have been carried out since the early, 1990s to prove the CSD trigger and provide useful data for model calibration and analyses (Eckersley, 1990; Sasitharan *et al*, 1993; Skopek *et al*, 1994; Chu *et al*, 2012).

As noted by Reid (2020) and Reid *et al* (2022a), all the publicly-available examples of static liquefaction triggering under CSD loading are based upon triaxial compression tests or, in one case, plane strain tests (Wanatowski and Chu, 2007). Importantly, all of these tests have a vertical principal stress direction, and thus if the type of anisotropy presented previously (eg Figure 4) applies to CSD triggering, then the instability stress ratios ($\eta_{IL} = q/p'$) developed with triaxial or plane strain tests will be unconservative. While forms of a DSS have been developed to enable CSD testing (eg Reid and Fourie, 2019), such an approach suffers from the lack of detailed stress information on the sample, and the DSS device is thus better-suited for demonstration purposes. The limitations of the DSS in the context of stress measurement are discussed later.

To address CSD triggering with a rotated principal stress direction, Fanni, Reid and Fourie (2022a) carried out an TSHC laboratory program where samples of silica fine sand were prepared at a range of densities adopting the moist tamping (MT) and dry pluviation (DP) preparation methods. The specimens were then consolidated to different principal stress angles (0, 22 and 45°), followed by a CSD stress path under the same principal stress angle. As the testing was carried out in a highly responsive TSHC capable of dynamic testing, the initiation of instability was easily detected by a rapid acceleration of displacements and subsequent liquefaction. The results of this test program are summarised in Figure 6. Significant reductions in the stress ratio η_{IL} are seen, with a rotated principal stress angle – up to 30 per cent, for example, at $\alpha = 45°$. Interesting to note is also the difference in η_{IL} between the MT and DP specimens, wherein for the same ψ_0 the DP indicates a much more pronounced reduction in η_{IL} than MT, indicating that fabric anisotropy (ie particle arrangement) plays a significant role in the liquefaction behaviour of soils and tailings. Furthermore, at $\alpha = 0°$ DP exhibited dilative behaviour, while at $\alpha = 45°$ DP indicated behaviour consistent with liquefaction up to $\psi 0 = -0.04$. As noted previously in Figure 3, a 45° principal stress angle is not unreasonable near the toe of many slopes.

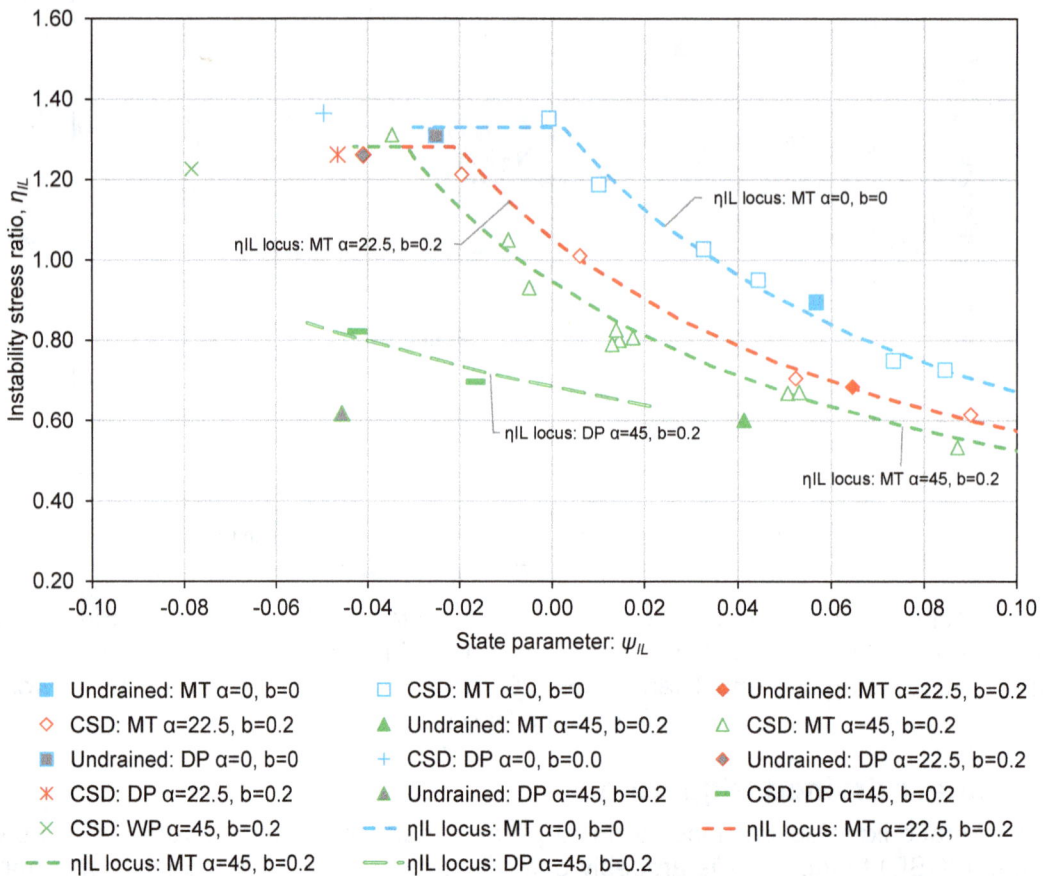

FIG 6 – Results of CSD stress path carried out under different principal stress angles (after Fanni, Reid and Fourie, 2022a).

OPTIONS TO SIMULATE BELOW-SLOPE CONDITIONS

General

The previous discussion provided evidence of the importance of principal stress direction, generally derived from the TSHC. However, other devices can test with a principal stress angle rotated from vertical, and it is important therefore to discuss the principles of each and their merits and limitations in the context of this paper's thesis regarding the benefits of the TSHC. A summary of such devices is provided as Table 3.

TABLE 3

Summary of test devices capable of testing with a non-vertical principal stress angle and/or plane strain conditions.

Device type	Plane strain or stress controlled	Principal stress direction	Advantages / disadvantages	Select references
Plane strain	Plane strain	Vertical	Unable to simulate below-slope conditions with alpha>0	(Cornforth, 1964; Wanatowski and Chu, 2007)
DSS (NGI or SGI type)	Plane strain, but unknown value	Rotates during shear	Unknown stress conditions, potential issues with non-uniformity at high strains	(Kjellman, 1951; Bjerrum and Landva, 1966)
Berkeley-type simple shear	Stress controlled	Rotates during shear	Allow to undertake testing at different K_0, hence give more freedom to assess the implication of different stress conditions on material behaviour. Lack of standardisation on test methods, limited commercial availability	(Boulanger et al, 1993)
Instrumented, rectangular simple shear	Plane strain	Rotates during shear	Mechanic difficulties, limited commercial availability	(Roscoe, 1953; Den Haan and Grognet, 2014)
TSHC	Plane strain or stress controlled	Can be directly controlled, or allowed to rotate during simple shear loading	Limited commercial availability. Non-uniformity in some stress combinations	(Hight, Gens and Symes, 1983; Shibuya and Hight, 1987; Sivathayalan and Vaid, 2002)
Directional shear cell	Stress controlled	Directly controlled	No commercial availability, limited consolidation stresses	(Arthur, Chua and Dunstan, 1977)

As can be seen from this summary, while a range of options exist, the TSHC combines the benefits of:

1. Current commercial accessibility.

2. A wide range of testing options.

3. Measurement of the full stress state of the sample.

This third benefit does not require a significant amount of material and testing/preparation time is similar to conventional triaxial testing.

It is for these reasons that we suggest in this work that the TSHC forms the most practical tool to further advance tailings engineering practice in the context of below-slope stress conditions. However, as the current leading practice tool is the DSS, further consideration of that device's merits and disadvantages is warranted.

On the DSS

The DSS, of either the NGI (Bjerrum and Landva, 1966) or SGI (Kjellman, 1951) type, has found an increasingly wide range of application to the measurement of undrained shear strengths (Ladd, 1991) and the cyclic response of a wide range of soils. Indeed, its use to characterise tailings has been promoted for these reasons (Reid, 2018), and the DSS has formed the key tool to infer the strength of critical foundation layers below TSFs (eg Morgenstern, Vick and Van Zyl, 2015). The device is a well-established tool that enables rapid assessment of strength, benefits from decades of testing experience and procedures, and utilises the constant volume technique to simulate undrained conditions, a method with practical advantages for test duration and labour time. Drained static bias can also be readily applied, particularly useful for simulating below-slope conditions where a significant amount of shear stress on the horizontal plane will have developed owing to the development of the structure over preceding decades.

The clear benefits of the DSS aside, one well established issue is related to the lack of knowledge of the stress conditions of the tested specimen. Only vertical stress and shear stress on the horizontal plane are known – insufficient to enable principal stresses to be inferred, and thus deviator or mean effective stress. This produces uncertainty in the actual shear stress experienced by the sample, and requires significant assumptions in estimating Ψ. Many of these issues have previously been discussed (eg Jefferies, Shuttle and Been, 2015), and are difficulties not suffered by the TSHC.

Another potential limitation of the DSS is an inability to reliably measure liquefied shear strengths and/or the brittleness of loose soils at high strains. This has previously been suggested on the basis of comparing DSS laboratory element behaviour to quick clay flowslides (eg Locat *et al*, 2016), and was then further investigated by Fanni, Reid and Fourie (2022b) by carrying out a laboratory study using the DSS and TSHC on loose samples of gold tailings for which the CSL was already available (Reid *et al*, 2022b).

The results of the work of Fanni, Reid and Fourie (2022b) are summarised in Figures 7 and 8, as shear stress versus shear strain and as a state diagram. The results clearly indicate that while the TSHC shows a brittle post-peak response and are tending towards the CSL inferred using as a reference the triaxial compression CSL (Reid *et al*, 2022a), the DSS tests show far less brittleness and do not tend to the same CSL. This is despite the samples all being sheared from the same range of void ratio and mean effective stress.

The non-uniform shearing behaviour recognised in literature for DSS testing (eg Saada and Townsend, 1981; Airey and Wood, 1987; Budhu, 1984; Menkiti, 1995) may have contributed to the higher post-peak horizontal shear stress recorded in the DSS testing compared to TSHC testing in this study. We note that the stress conditions erasing in the TSHC testing under simple shear conditions are such that small non-uniformities in stresses and strains develop in the TSHC (eg Fanni, Reid and Fourie, 2022b). This makes the TSHC an ideal device to assess the behaviour of material under simple shear conditions at small and large strains.

Given the increased focus on understanding the brittleness of tailings materials (ICMM, 2020), this post-peak behaviour of the DSS represents a considerable potential concern as it likely overpredicts post-peak strengths. Therefore, while the DSS will always represent a practical tool for cyclic triggering assessments and inferring peak strengths for foundation soils (for example), it seems clear that the TSHC is needed to better quantify brittleness of tailings under principal stress angles such as those present below slopes – consistent with observations made by Sivathayalan and Vaid (2002) almost two decades ago.

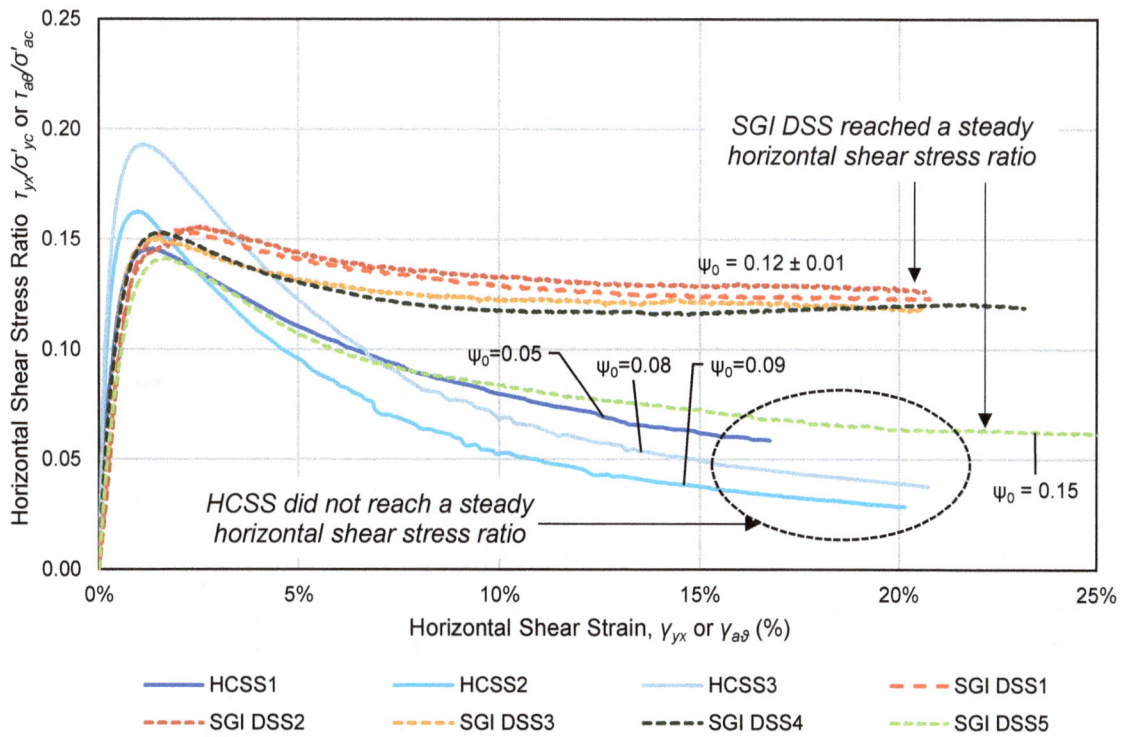

FIG 7 – Stress-strain data for comparison TSHC and DSS tests – after Fanni, Reid and Fourie (2022b).

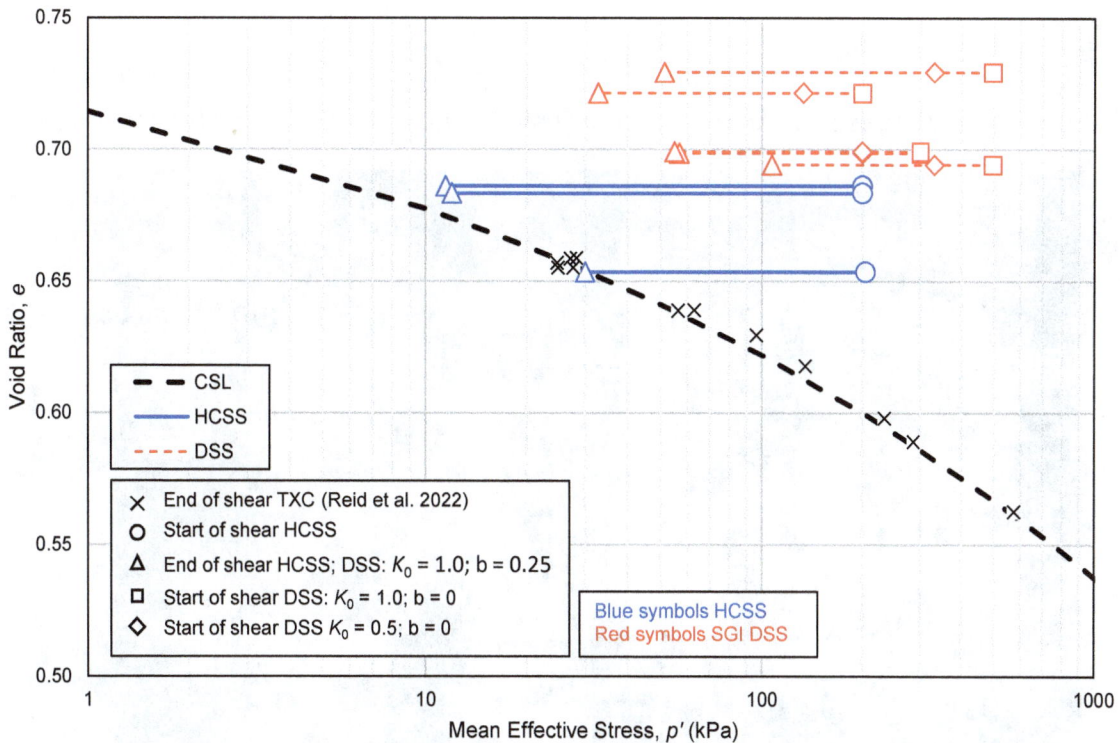

FIG 8 – State diagram comparison of TSHC and DSS tests – after Fanni, Reid and Fourie (2022b).

OVERVIEW OF TSHC APPARATUS

As the TSHC is less familiar to most tailings' practitioners than the triaxial or DSS, a review of the device is now provided for reference. The TSHC comprises a hollow cylinder mounted, similar to a triaxial, on top and bottom platens and separated from external water pressure by means of a membrane. A schematic describing the stress conditions acting on the TSHC and an image of a current system are provided as Figure 9, while Figure 10 shows a picture of a specimen prepared in a modern TSHC device developed by GDS (left picture) along with a sheared specimen (right picture).

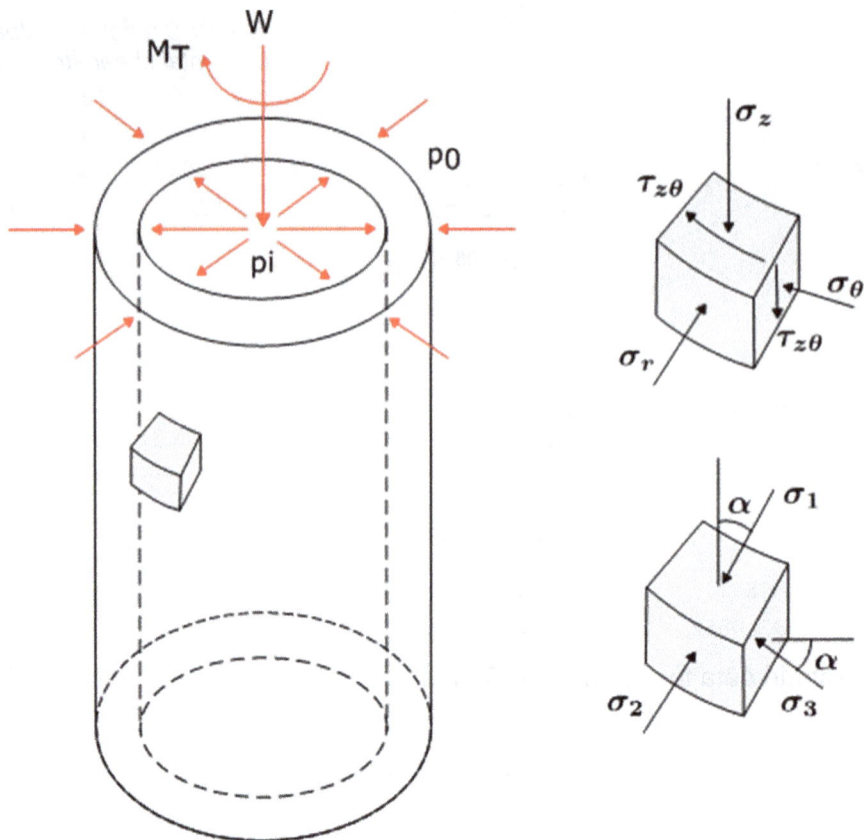

FIG 9 – TSHC stress conditions (figure adjusted from Hight, Gens and Symes (1983).

FIG 10 – Image of a modern TSHC developed by GDS Instruments with specimen inside cell (left) and after being sheared (right).

The major difference between the TSHC and a triaxial specimen is the ability to vary the inner and outer cell pressures, and to apply a torque. The combination of these additional degrees of control enable almost any stress combination to be achieved – barring some limits on non-uniformity (Hight, Gens and Symes, 1983) – or for various specialised forms of testing to be carried out such as enforcing plane strain conditions under either drained or undrained loading. Simple shear loading can easily be carried out by rotating the specimen under undrained and constant height conditions, thus providing data on a loading condition critical to the stability of slopes while measuring all principal stresses throughout the test.

Importantly, the form of modern TSHC device outlined in Figure 10 is readily incorporated into a modern tailings testing laboratory. Many aspects of the device – such as computer-controlled pressure pumps – differ little from examples that are now commonplace for tailings triaxial testing. Testing of low plasticity silt materials (most tailings) undergoing large saturation collapse strains if moist tamped can be carried out using simple methods familiar to most commercial laboratories with high accuracy in determination of ψ_0, eg Fanni, Reid and Fourie (2023). As such, we suggest that increased application of the TSHC device is both appropriate and practical given ongoing tailings characterisation efforts, particularly with respect to brittleness.

CONCLUSIONS

The quantity and quality of the testing of tailings has dramatically increased over the past five years, with current test procedures now regularly employing the excellent guidance that was provided in the geotechnical literature between ~1990 and 2006. However, one area where current commercial techniques are still limited is accounting for the effects of principal stress direction on peak strengths, liquefied strengths, brittleness, and instability stress ratio. A review of data from the literature carried out showed the clear effects of principal stress direction on shearing behaviour for soils and tailings – importantly, indicating that the use of triaxial compression tests with a vertical principal stress angle would be unconservative in this context.

REFERENCES

Airey, D W and Wood, D M, 1987. An evaluation of direct simple shear tests on clay, *Géotechnique*, 37(1):25–35.

Arthur, J R F, Chua, K S and Dunstan, T, 1977. Induced anisotropy in a sand, *Géotechnique*, 27:13–30.

Been, K and Jefferies, M G, 1985. A state parameter for sands, *Géotechnique*, 35:99–112.

Bishop, A W, 1966. Sixth Rankine Lecture: The strength of soils as engineering materials, *Géotechnique*, 16:89–130.

Bjerrum, L and Landva, A, 1966. Direct Simple-Shear Tests on a Norwegian Quick Clay, *Géotechnique*, 16:1–20.

Boulanger, R W, Chan, C K, Seed, H B, Seed, R B and Sousa, J B, 1993. A low-compliance bi-directional cyclic simple shear apparatus, *Geotechnical Testing Journal*, 16(1):36–45.

Budhu, M, 1984. Nonuniformities imposed by simple shear apparatus, *Canadian Geotechnical Journal*, 21(1):125–137. doi:10.1139/t84–010.

Chu, J, Leong, W K, Loke, W L and Wanatowski, D, 2012. Instability of Loose Sand under Drained Conditions, *Journal of Geotechnical and Geoenvironmental Engineering*, 138:207–216.

Cornforth, D H, 1964. Some Experiments on the Influence of Strain conditions on the Strength of Sand, *Géotechnique*, 14:143–167.

Den Haan, E J and Grognet, M, 2014. A large direct simple shear device for the testing of peat at low stresses, *Géotechnique Letters*, 4:283–288.

Eckersley, D, 1990. Instrumented laboratory flowslides, *Géotechnique*, 40:489–502.

Fanni, R, Reid, D and Fourie, A, 2023. A simple method to calculate void ratio of very loose silts and silty sands in torsional shear hollow cylinder testing, *Geotechnical Testing Journal* (in press).

Fanni, R, Reid, D and Fourie, A, 2022a. Effect of principal stress direction on the instability of sand under the constant shear drained stress path, *Géotechnique*.

Fanni, R, Reid, D and Fourie, A, 2022b. On reliability of inferring liquefied shear strengths from simple shear testing, *Soils and Foundations*, 62(3):101151. The Japanese Geotechnical Society.

Georgiannou, V N, Konstadinou, M and Triantafyllos, P, 2018. Sand Behavior under Stress States Involving Principal Stress Rotation, *Journal of Geotechnical and Geoenvironmental Engineering*, 144:04018028.

Hight, D W, Gens, A and Symes, M J, 1983. The development of a new hollow cylinder apparatus for investigating the effects of principal stress rotation in soils, *Géotechnique*, 33:355–383.

International Council on Mining and Metals (ICMM), 2020. Global Industry Standard on Tailings Management (GISTM), ICMM.

Jefferies, M and Been, K, 2006. *Soil Liquefaction.*

Jefferies, M, Morgenstern, N R, Van Zyl, D V and Wates, J, 2019. Report on NTSF Embankment Failure, Cadia Valley Operations for Ashurst Australia, 119 p.

Jefferies, M, Shuttle, D and Been, K, 2015. Principal stress rotation as cause of cyclic mobility, *Geotechnical Research,* 2:66–96.

Jefferies, M G and Been, K, 2015. *Soil Liquefaction: A Critical State Approach*, (CRC Press: Boca Raton).

Kjellman, W, 1951. Testing the shear strength of clay in Sweden, *Géotechnique,* 2:225–232.

Ladd, C, 1991. Stability Evaluation during Staged Construction, *Journal of geotechnical engineering*, 117:540–615.

Locat, A, Leroueil, S, Fortin, A, Demers, D and Jostad, H P, 2016. The 1994 landslide at Sainte-Monique, Quebec: geotechnical investigation and application of progressive failure analysis, *Can Geotech J Editors' Choice,* 1:490–504.

Menkiti, O C, 1995. Behaviour of clay and clayey-sand, with particular reference to principal stress rotation, Doctor of Engineering Thesis, Imperial Collage.

Morgenstern, N R, Vick, S G and Van Zyl, D V, 2015. Report on Mount Polley tailings storage facility breach. Report of independent expert engineering investigation and review panel.

Morgenstern, N R, Vick, S G, Viotti, C B and Watts, B D, 2016. Fundão Tailings Dam Review Panel: Report in the immediate causes of the failure of the Fundão Dam.

Reid, D and Fourie, A, 2019. A direct simple shear device for static liquefaction triggering under constant shear drained loading, *Géotechnique Letters,* 9:142–146.

Reid, D, 2018. Importance of the simple shear loading direction to the stability of tailings storage facilities, in *Proceedings of the Mine Waste and Tailings Conference 2018*, pp 274–284 (The Australasian Institute of Mining and Metallurgy: Melbourne).

Reid, D, 2020. On the effect of anisotropy on drained static liquefaction triggering, *Géotechnique Letters,* 10(3):393–397.

Reid, D, Chiaro, G, Fanni, R and Fourie, A, 2022b. Arguments for increasing the use of hollow cylinder testing in the characterization of tailings, in Proceedings of Tailings and Mine Waste, Civil & Environmental Engineering Department, Colorado State University.

Reid, D, Dickinson, S, Mital, U, Fanni, R and Fourie, A, 2022a. On some uncertainties related to static liquefaction triggering assessments, in *Proceedings of the Institution of Civil Engineers-Geotechnical Engineering*, 175(2), pp 181–199.

Robertson, P K, de Melo, L, Williams, D J and Wilson, G W, 2019. Report of the Expert Panel on the technical causes of the Failure of Feijão Dam I.

Roscoe, K H, 1953. An apparatus for the application of simple shear to soil samples, in *Proceedings of the 3rd International Conference on Soil Mechanics and Foundation/Geotechnical Engineering (ICSMFE)*, 1:186–191.

Saada, A S and Townsend, F C, 1981. State of the art: laboratory strength testing of soils, Laboratory shear strength of soil, ASTM STP, 740, pp 7–77.

Sasitharan, S, Robertson, P K, Sego, D C and Morgenstern, N R, 1993. Collapse behavior of sand, *Canadian Geotechnical Journal,* 30:569–577.

Shibuya, S, Hight, D W and Jardine, R J, 2003. Four-Dimensional Local Boundary Surfaces of an Isotropically Consolidated Loose Sand, *Soils and Foundations,* 43:89–103.

Shibuya, S and Hight, D W, 1987. On the stress path in simple shear, *Géotechnique,* 37:511–515.

Shuttle, D A and Cunning, J, 2007. Liquefaction potential of silts from CPTu, *Canadian Geotechnical Journal,* 44:1–19.

Sivathayalan, S and Vaid, Y P, 2002. Influence of generalized initial state and principal stress rotation on the undrained response of sands, *Canadian Geotechnical Journal,* 39:63–76.

Skopek, P, Morgenstern, N R, Robertson, P K and Sego, D C, 1994. Collapse of dry sand, *Canadian Geotechnical Journal,* 31:1008–1014.

Verdugo, R and Ishihara, K, 1996. The steady state of sandy soils, *Soils and Foundations,* 36:81–91.

Vranna, A and Tika, T, 2015. The liquefaction resistance of weakly cemented sands, *Geomechanics from Micro to Macro,* pp 1225–1229.

Wanatowski, D and Chu, J, 2007. Static liquefaction of sand in plane strain, *Canadian Geotechnical Journal,* 44:299–313.

Yoshimine, M, Ishihara, K and Vargas, W, 1998. Effects of principal stress direction and intermediate principal stress on undrained shear behavior of sand, *Soils and Foundations,* 38:179–188.

A performance-based approach for calibration of fine tailings parameters used for closure design

C Han[1], W M Teh[2], M Llano-Serna[3], W Ludlow[4] and Z R Yung[5]

1. Senior Engineer, Red Earth Engineering – a Geosyntec Company, Brisbane Qld 4000. Email: chao.han@redearthengineering.com.au
2. Principal Engineer, Red Earth Engineering – a Geosyntec Company, Perth WA 6000. Email: ming.teh@redearthengineering.com.au
3. Principal Geotechnical Engineer, Red Earth Engineering – a Geosyntec Company, Brisbane Qld 4000. Email: marcelo.llano@redearthengineering.com.au
4. Principal Engineer, Red Earth Engineering – a Geosyntec Company, Brisbane Qld 4000. Email: wade.ludlow@redearthengineering.com.au
5. Tailings Engineer, Red Earth Engineering – a Geosyntec Company, Brisbane Qld 4000. Email: george.yung@redearthengineering.com.au

ABSTRACT

The adoption of a performance-based design approach is gaining momentum among tailings engineers, as recommended by the Global Industry Standard on Tailings Management (GISTM). Performance-based design generally refers to utilising surveillance data throughout the life cycle of the facility to re-calibrate the existing design parameters and forecast future performance.

This paper presents a performance-based approach used to re-calibrate the design parameters of a fine tailings material with high plasticity to inform subsequent closure works. The calibration has utilised the data collected from multiple instruments during a closure capping trial on the surface of the fine tailings, including vibrating wire piezometers (VWPs), settlement plates and topographic survey. A Modified Cam Clay model was developed, incorporating the laboratory testing results from Rowe Cell, oedometers and triaxial tests. The predicted strengths are compared to the results from *in situ* seismic cone penetration tests and vane shear tests. The calibrated model was then used to further indicate the performance of the fine tailings and the adjacent dam structure under the design and construction loading conditions, using the finite element software, Optum. The calibration process has demonstrated good consistency between the predicted and measured deformation and strength values, which provided confidence in the inputs for the subsequent risk-informed option study of closure design.

The approach presented in this paper demonstrates a feasible and reliable way to undertake the performance-based design method. Using performance-based design in tailings management can result in more cost-effective and sustainable solutions. It allows designers to consider a wider range of options and optimise the design based on site-specific conditions.

INTRODUCTION

Performance-based design is an engineering design approach that focuses on achieving desired performance outcomes rather than simply meeting minimum code requirements. It has been recommended by the Global Industry Standard on Tailings Management (GISTM) (ICMM, 2020) forming part of a tailings management system for Performance-Based Risk-Informed, Safe Design, Construction, Operation and Closure of tailings facilities (PBRISD) (Morgenstern, 2018). This approach typically involves iterative evaluations of the design performance of tailings facilities utilising analysis and simulation results from the anticipated performance parameters calibrated by subsequent surveillance data throughout the facility life cycles. The evaluation outcomes are compared against the prescribed performance objectives to validate or further optimise the design solution regularly.

The performance-based approach offers several advantages compared to the conventional design approach that relies on prescriptive criteria such as Factor of Safety (FoS), including:

- The use of monitoring instruments to collect data at targeted locations for updating analyses can improve knowledge of local shear and pore water properties of materials near the shear plane.

- The calibrated simulations can predict the potential static and seismic deformations under extreme events, which provides a quantitative reference to freeboard management. It can also provide information at the onset of localisation of deformations, so progressive failure can be determined. This is critical for strain-weakening materials susceptible to progressive failure due to strain incompatibility and/or brittle materials susceptible to static liquefaction.

- The improved simulation and calibration with time also provides a more authoritative record to demonstrate that the overall response of the facility based on constructed conditions is in accordance with the design intent. Such records can be used to support regulators in addressing safety concerns.

Based upon the analysis of past failures, the performance-based approach can be particularly valuable in achieving a robust design and maintaining integrity to prevent the collapse of tailings facilities. It is particularly relevant when conditions are spatially complex and variable, with problematic (compressible, strain weakening and/or liquefiable) materials in the structural zones. These problematic materials are sensitive to pore pressure increase due to loading from the tailings facility (ICMM, 2021).

The approach presented in this paper is schematically shown in Figure 1. Two key concepts are developed: i) the Trial Domain and ii) The Application Domain. The Trial Domain corresponds to a benchmarking exercise that allows the prediction of future performance in the Application Domain. Each domain forms part of an independent and mutually calibrated performance-based design cycle(s). The data is used to calibrate and update the numerical model by iterative cycles through the loop shown in Figure 1.

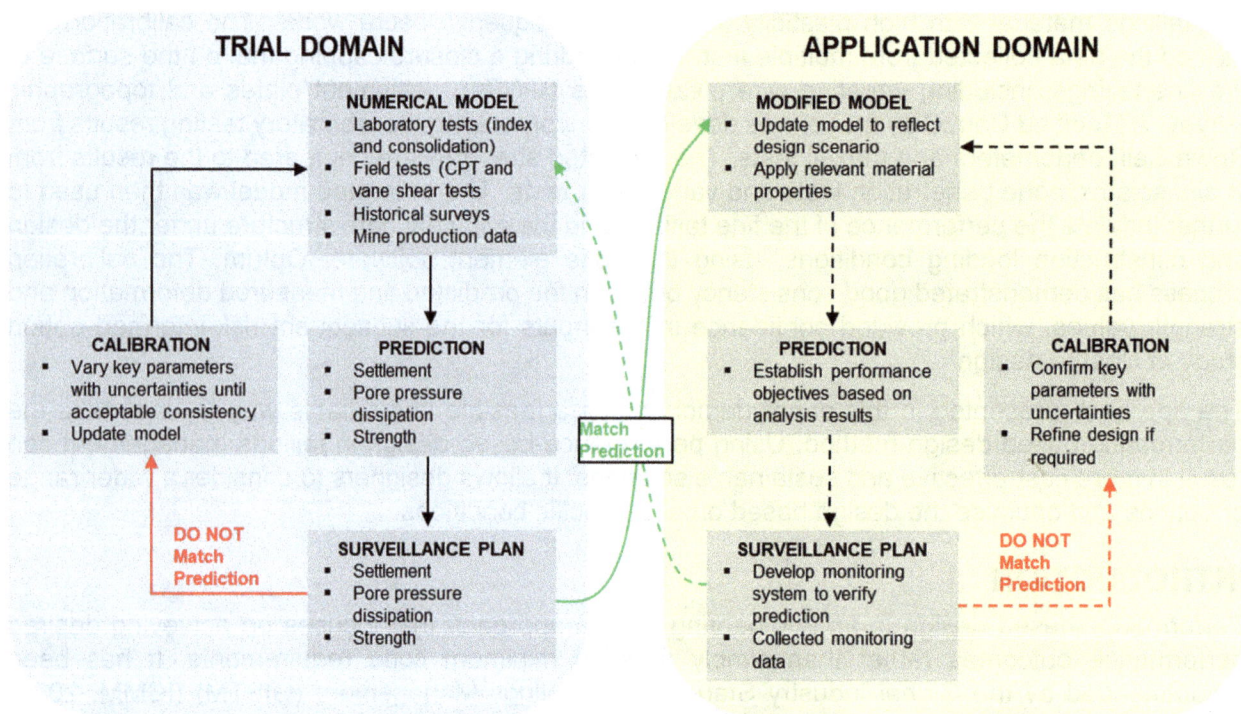

FIG 1 – Performance-based design approach presented in this paper.

To note:

- The Trial Domain begins with the development of a geotechnical numerical model based on available historical data. The results are used to inform a monitoring and instrumentation plan for the trial works and the collected monitoring data is then compared to the model to calibrate the key parameters with uncertainties. The works in Trial Domain conclude once an acceptable consistency between the predicted and monitored data is established.

- The Application Domain begins with the modified numerical model based on the future design concept. The results are used to inform a tailored surveillance plan for the design, accompanied by performance objectives and the collected monitoring data is then compared to the design predictions for parameter calibration and design refinement. The works in Application Domain will continue throughout the life cycle of the facility.

- Convergence between predictions and measurements in each domain will improve model reliability and hence, facility safety.

The paper structure includes a summary of the closure trial. The field trial section is followed by a detailed description of the numerical model used to calibrate the instrumentation data (Trial Domain). The validated numerical model in the Application Domain is used to inform future works. A summary of conclusions and key lessons learnt are provided in the last section.

TRIAL SET-UP

The closure trial comprised the construction of a 1.5 m thick capping layer on an active fine tailings facility located in northern Australia. Tailings deposition ceased on 7 November 2018, with intermittent deposition into the facility between May 2019 and August 2019. The construction of the capping layer commenced on 11 July 2020 and lasted for ten days. The capping layer comprised unsaturated sand tailings excavated from a nearby tailings facility.

The objective of the trial was to evaluate the current tailings conditions and the trafficability of the available equipment, comprising a CAT 740 Ejector Haul Truck and a skid steer loader. Data was progressively collected from the trial over ten weeks through four VWPs and settlement plates (SP). The original locations of the instruments were informed by preliminary analyses based on a concept-level design landform, as shown in Figure 2. However, as the work front progressed, bow waves and cracks began to develop at the south-eastern extents of the capping layer, as marked in Figure 2. Personnel and equipment were removed from the area immediately due to safety concerns, and the construction portion of the trial ceased before the planned footprint was achieved.

FIG 2 – Capping trial plan.

Two of the planned VWPs were installed 2 m to 3 m below the ground surface spaced at 15 m, and they were connected to a data logger to record pore pressures every 30 seconds during the trial and every 30 minutes post-trial. Data was captured over ten weeks. Settlement plates were fabricated and positioned at 15 m spacing adjacent to the VWP locations and were surveyed pre-trial, mid-trial and post-trial to record the progressive settlement. Cross-sectional details of the installed instruments are shown in Figure 3.

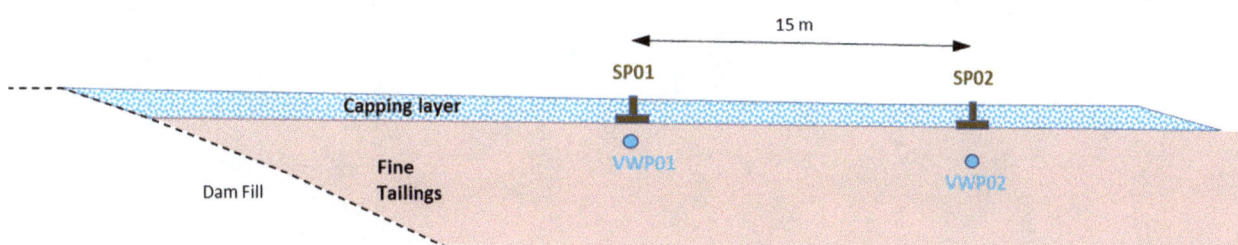

FIG 3 – Cross-sectional details of installed instruments.

TRIAL DOMAIN

The numerical validation was carried out as per the steps below:

1. Develop a numerical model based on the trial site configuration and construction sequence.

2. Initial characterisation of soils using available laboratory and *in situ* testing data.

3. Compare predicted pore pressures and settlements with trial instrument measurements, and adjust the key parameters with uncertainties in the numerical model if required.

4. Repeat steps 2–3 until acceptably consistent results are obtained between the predicted and measured values.

Model configuration

Two-dimensional finite element analyses were conducted using the commercial software Optum G2 (Krabbenhoft, Lymain and Krabbenhoft, 2021) to assist with the interpretation of the trial measurements. The site configuration during the trial was modelled in accordance with historical survey data, with the subsurface stratigraphy and the instrumentation locations shown in Figure 4. The soil regions were discretised with 6-node triangular elements. The horizontal and vertical extents of the soil mesh were selected to be sufficient to eliminate boundary effects, as confirmed through initial numerical trials with varying horizontal and vertical extents. The phreatic conditions within the model were informed by site-wide piezometer monitoring data coincident with the trial period. The trial stages were modelled in three phases:

1. The initial phase representative of the pre-trial condition was modelled using the 'Initial Stress' analysis option;

2. The capping layer placement was modelled using the 'Elastoplastic' module combined with the 'Short-term' time scope to reproduce the undrained response of the tailings; and

3. Post-trial consolidation was modelled using the 'Consolidation' analysis for a 10-week period.

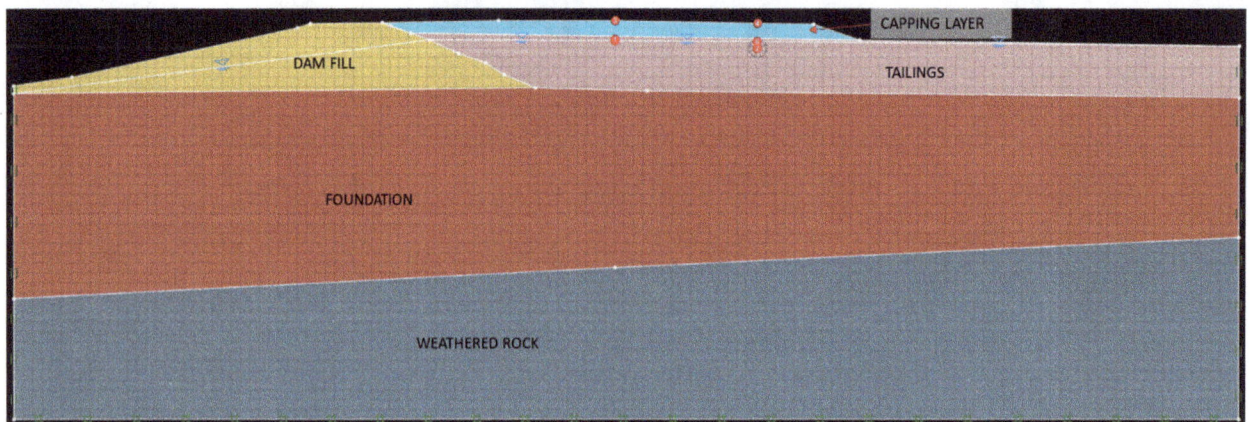

FIG 4 – Subsurface stratigraphy at trial location modelled in Optum G2.

Material characterisation

Table 1 summarises the key parameters used for the numerical models, the selection of which is outlined in the following subsections.

TABLE 1

Geotechnical parameters for numerical models.

Model type	Parameters	Unit	Tailings	Dam fill	Foundation	Sand cap
General	Unit weight, γ_{sat}	kN/m^3	14	20	18	16
	Permeability, k	m/s	3.5×10^{-8}	1.0×10^{-7}	1.2×10^{-7}	Free draining
	Poisson's ratio, υ	-		0.2	0.2	0.2	0.2
Modified Cam Clay	Slope of virgin consolidation line, λ	-	0.35	-	-	-
	Initial void ratio, e_0	-	2.0	-	-	-
	Slope of swelling line, κ	-	0.05	-	-	-
	Critical state friction angle, ϕ'_{CS}	Degree	15	-	-	-
Mohr Coulomb	Internal friction angle, ϕ'	Degree	-	30	26	-
	Cohesion, c'	kPa	-	5	0	-
Elasticity	Young's modulus, E	MPa	-	30	15	25

Tailings

Geotechnical index tests, including Particle Size Distribution and Atterberg Limits indicated that the tailings are predominantly high plasticity CLAY with a plasticity index of approximately 30 per cent. The Modified Cam-Clay (MCC) model was used to represent the constitutive behaviour of the tailings resembling soft clay, and the consolidation and swelling parameters were inferred from 1D oedometric compression testing (ie Oedometer and Rowe Cell) on tailings samples. Based on the production tonnages and the corresponding topographic surveys, the tailings unit weight was inferred from the historical average dry density reconciliation results. The average dry density coincident with the trial period is approximately 1.2 t/m^3, which is equivalent to a saturated unit weight of 14 kN/m^3 and an average void ratio of 2.0, incorporating the measured specific gravity of 3.5. The permeability of tailings has significant uncertainties when tested in a laboratory, and CPT dissipation results at discrete locations suggest values ranging from 1×10^{-7} m/s to 1×10^{-9} m/s. The value presented in Table 1 is the calibrated result from the trial VWP measurements.

Site-specific critical state testing results were unavailable for the tailings. Interpreted undrained shear strength profiles s_u, from nearby CPT tests close to the trial date were used to back-calculate the critical state friction angle, using the Equation 1 (Wood, 1991). Equation 1 assumes a triaxial compression loading condition typical of the stress state below the cap layer. A 1.5 m layer of crust with a minimum undrained strength of 2 kPa was also considered. Figure 5 provides a comparison of the CPT-interpreted undrained shear strength, s_u, profiles and interpreted profile for Optum.

$$s_u = \frac{Mp'_0}{2^{1+\Lambda}}\left[\left(\frac{q_0}{Mp'_0}\right)^2 + 1\right]^{\Lambda} \tag{1}$$

Where p'_0 and q_0 are the pre-trial mean effective stress and deviatoric stress, respectively, assuming $K_0 = 1 - \sin(\phi'_{CS})$; $\Lambda = 1 - \kappa/\lambda$.

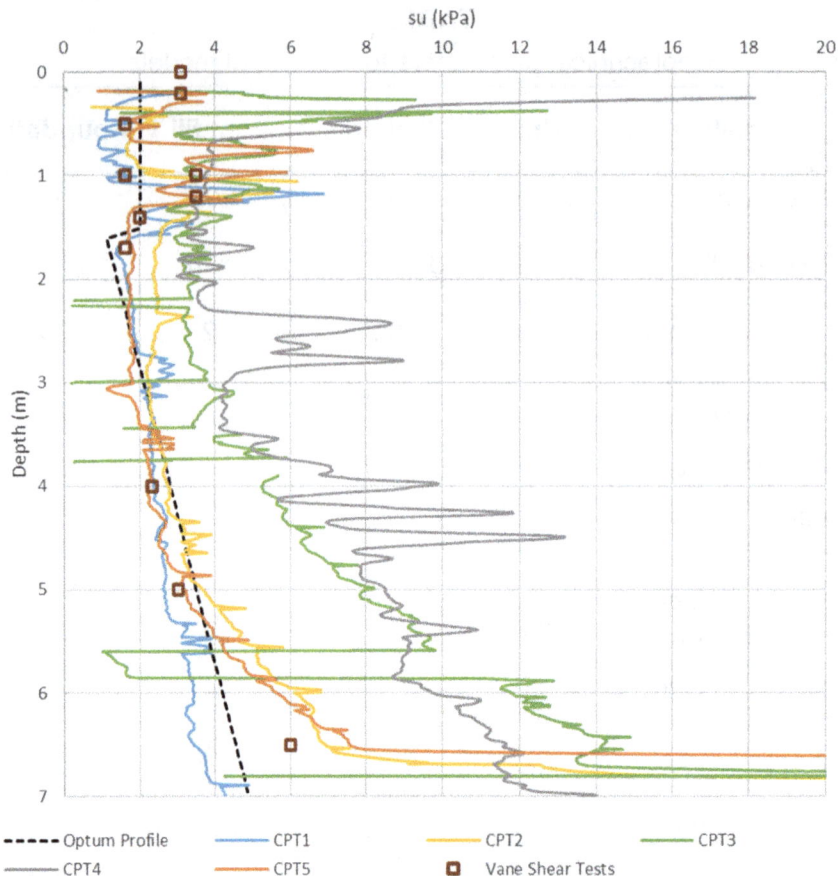

FIG 5 – Comparison of CPT interpreted undrained shear strength profiles and equivalent profile in Optum.

Dam fill and foundation

Dam Fill and Foundation are predominately low to medium plasticity CLAY. The Dam Fill was compacted to at least 95 per cent maximum dry density at the optimal moisture content, while Foundation material was loosely placed historically to backfill mining voids. Mohr–Coulomb models were used for these materials, and the parameters were adopted based on the results of laboratory (triaxial tests on compacted samples) and field (multiple CPT tests completed throughout the facility life cycle) tests. Preliminary sensitivity analyses suggested that the variation in material properties has minimal impact on the results at the instrument locations due to their distance to the zones of influence. Therefore, the properties of these materials did not form part of the calibration process.

Sand cap and weathered rock

Sand Cap material comprises fine to medium grain SAND with a fines content of less than 10 per cent. It was modelled as an elastic material to avoid non-convergence issues due to local failure in the capping layer. The existing testing results informed the properties in Table 1.

Weathered Rock was modelled as a non-permeable rigid body.

Calibration results

An equivalent B-bar can be defined as the ratio of excess pore pressure (Δu) over the cap surcharge pressure ($\Delta \sigma_v$). Figure 6 shows B-bar calculated using the instrument data over the 10-week trial monitoring duration. Calculated data is shown in red, while monitoring data is shown in black. A general consistency between the data sets can be observed, indicating i) the shallower monitoring location (VWP01) has a lower B-bar response than VWP02 due to the relatively shorter drainage path length; ii) a 50 per cent pore pressure dissipation requires a one to two week waiting period. Slight differences between the calculated and observed data are explained due to the drainage paths on the field being 3D and the numerical model abstraction undertaken in 2D plain strain. Despite the slight differences, the model captures the field observations remarkably well.

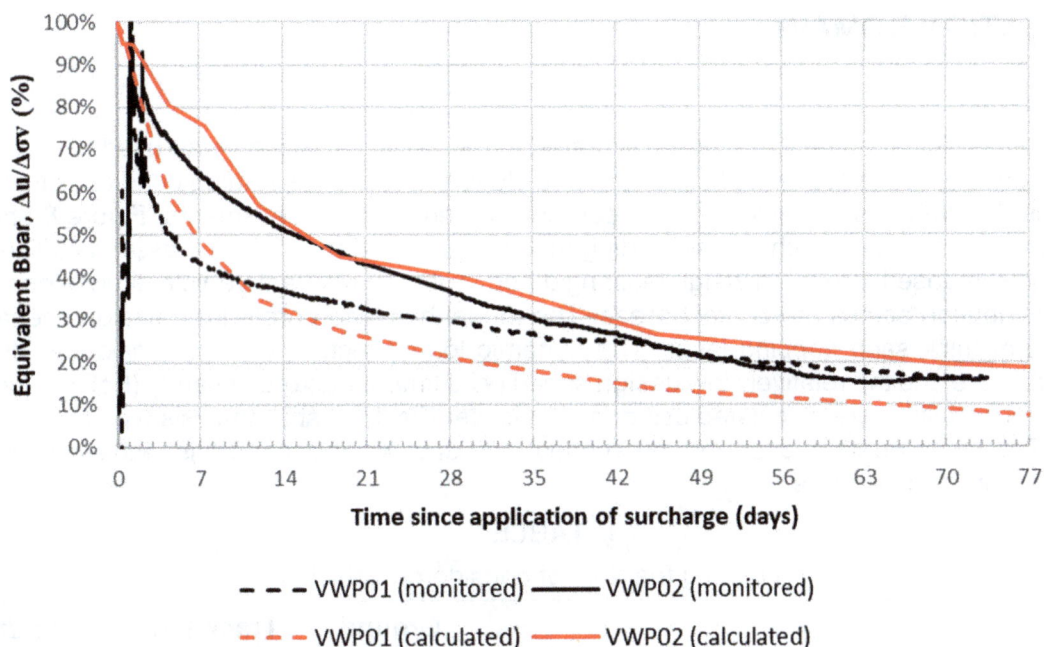

FIG 6 – Comparison of measured and calculated equivalent B-bar.

A comparison of the settlement development between the calculated and the measured data over the 10-week trial monitoring duration is shown Figure 7. It is shown that the calculated trends of settlements at different locations agree with the discrete monitoring data, albeit up to 30 mm discrepancy is observed.

Based on the above calibration outcomes, the numerical model was considered to be capable of reasonably predicting future performance that completes the Trial Domain component schematised in Figure 1.

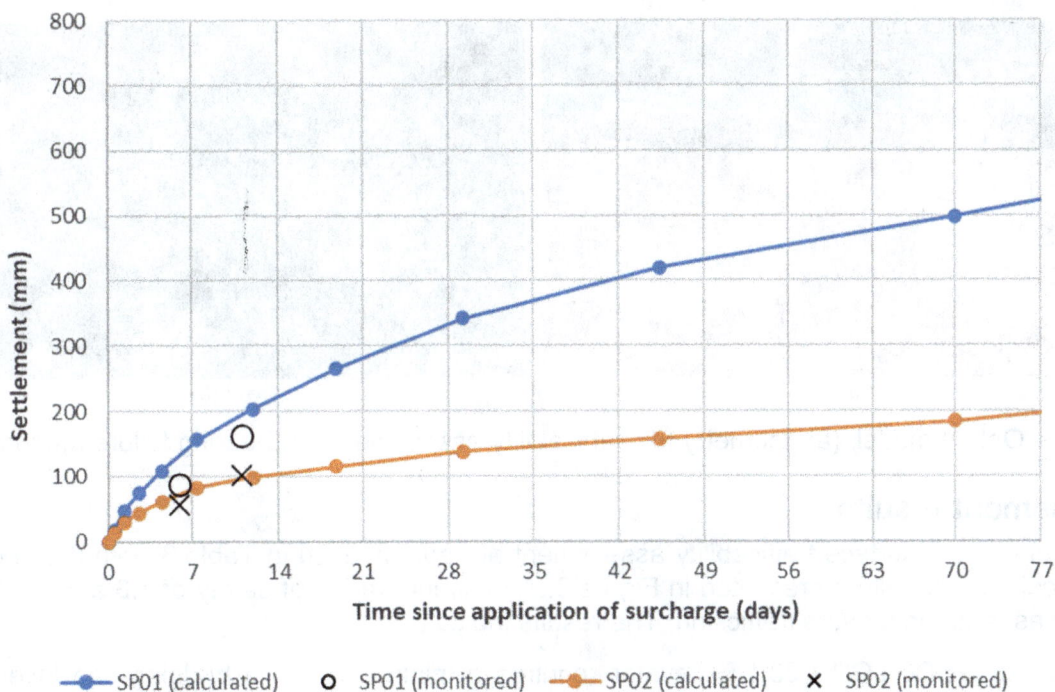

FIG 7 – Comparison of measured and calculated settlements.

APPLICATION DOMAIN

Model input

The two-dimensional numerical model calibrated in the previous section was used for subsequent surface trafficability assessments to inform future closure works. The loading conditions imposed by the nominated vehicles in Table 2 have been considered in the assessments. Figure 8 shows the modified model geometry with corresponding material parameters in Table 1 assigned. The vehicle loading was imposed through a weightless, rigid body with the relevant vehicle dimensions to allow load redistribution across the contact areas. The model has incorporated the design scenario with up to a 3 m thick sand capping layer. The dynamic load effects were not considered due to its transient nature and the relatively low design speed (<20 km/h). Factors of safety (FoS) against track or tyre bearing failure are calculated using the 'Multiplier Elastoplastic' analysis type incorporated in Optum G2. Consolidation over the prescribed 100-day waiting period is achieved using the 'Consolidation' analysis type.

TABLE 2

Loading features of considered vehicles.

Vehicle model	Purpose	Ground pressure (kPa)	Track/tyre contact footprint	Track/tyre spacing (m)
Caterpillar D6 LGP – 36 inch shoes	Material spreading	36	0.9 m (W) × 3.2 m (L)	2.3
20 t Excavator (JCB JS220– 900 mm shoes)	Spigot handling	28	0.9 m (W) × 3.7 m (L)	2.4
Caterpillar 777 Haul Truck	Material transportation	310	1.5 m (2W) × 1.2 m (L)*	3.6

Note: * assumed based on the provided loaded radius from the manufacturer's specifications.

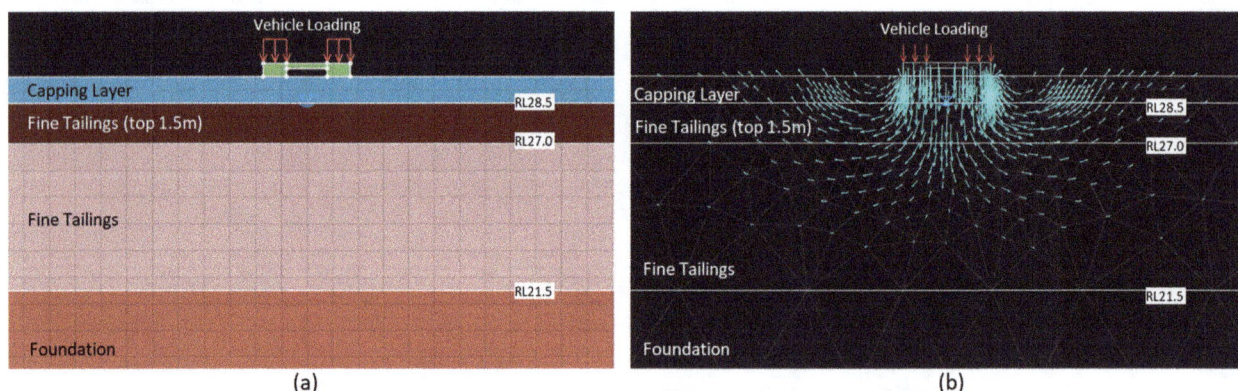

FIG 8 – Optum model: (a) geometry for trafficability assessment; (b) bearing failure mechanism.

Assessment results

The results of the surface trafficability assessment are summarised in Table 3, with the vectors of the typical bearing failure presented in Figure 8. A minimum factor of safety of 1.3 against bearing failure was targeted for safe trafficking. The results indicate:

- Caterpillar D6 LGP / 20 t Excavator requires a minimum of 1 m bridging cap layer and a subsequent 100-day waiting time prior to achieving a minimum FoS = 1.3.

- Caterpillar CAT 777 trafficking on up to 3 m thick of flat sands tailings beach cannot be safely achieved (ie FoS < 1.0), and options to improve the current tailings strength further are recommended, such as mud-farming, dewatering or geogrid bridging.

These findings and the associated recommendations are planned to be incorporated into subsequent risk-informed decision-making with the support of quantitative risk assessments.

TABLE 3

Results summary of trafficability assessment.

Vehicle model	Capping layer thickness	FoS after various waiting period		
		0-day	100-day	Infinite
Caterpillar D6 LGP / 20 t Excavator	0.5	0.68	-	0.70
	1.0	1.07	1.31	1.33
Caterpillar 777 Haul Truck	3.0	0.28	-	0.56

CONCLUSIONS

This paper presents a performance-based approach that was used to calibrate the numerical model developed for the closure capping works of a fine tailings facility. The parameters of the numerical model were developed based on the existing laboratory and field testing data, topographic survey records and mine production data. The calibration was undertaken by varying the key uncertainties in design parameters against the monitoring data collected from VWPs and settlement plates during and after the trial works. The calibration outcome has demonstrated acceptable consistency between the predicted and measured deformations, pore pressure response and undrained shear strengths, which provided confidence in the inputs for the subsequent closure design.

The use of performance-based design in tailings management can result in more cost-effective and sustainable solutions by:

- Overcoming several limitations of the conventional prescriptive approach provided in a number of industry design guiding documents.

- Allowing designers to utilise piecewise information gathered at different phases of the life cycle of the facility and continuously challenge the initial design assumptions via staged simulation of facility behaviour and incremental forward projection of future performance.

- The numerical methods applied herein were developed within a short time frame (two days) and further updates were executed rapidly. Additionally, the models appeared to be stable and robust in various application scenarios and capable of providing reliable output.

- The consistency between prediction and measurements can also further improve confidence in dam safety.

REFERENCES

International Council on Mining and Metals (ICMM), 2020. *Global Industry Standard on Tailings Management*, International Council on Mining and Metals.

International Council on Mining and Metals (ICMM), 2021. *Tailings Management Good Practice Guideline,* International Council on Mining and Metals.

Krabbenhoft, K, Lymain, A and Krabbenhoft, J, 2021. *Optum Computational Engineering Optum G2 Analysis* [computer software]. Retrieved from https://optumce.com/products/brochure-and-datasheet.

Morgenstern, N, 2018. Geotechnical Risk, Regulation, and Public Policy (Victor de Mello Lecture), *Soils and Rocks*, 41:107–129.

Wood, D M, 1991. *Soil Behaviour and Critical State Soil Mechanics* (Cambridge University Press).

CPT interpretation for tailings dam stability analyses – considerations for defining tailings' undrained strength

J Rola[1], J Eldridge[2] and P Barbieri[3]

1. Principal Consultant, SRK Consulting (Australasia) Pty Ltd, West Perth WA 6005.
   Email: jrola@srk.com.au
2. Consultant, SRK Consulting (Australasia) Pty Ltd, West Perth WA 6005.
   Email: jeldridge@srk.com.au
3. PhD candidate, The University of Western Ontario, London Ontario N6A 3K7, Canada.
   Email: pbarbie2@uwo.ca

ABSTRACT

Undrained shear strength ratios (USRs) are commonly adopted within the industry to characterise tailings strength, allowing for straightforward statistical analysis. *In situ*, the undrained shear strength (s_u) is typically assessed using cone penetration test (CPT) data, that is normalised by the inferred effective vertical stress (σ'_v) at the time of testing to define the USR. A strong understanding of σ'_v, and consequently piezometric pore water pressure (PWP) profile, seepage and pore fluid density, is key to developing representative USRs. However, from CPT, s_u is a function of only corrected cone resistance (q_t), total vertical stress (σ_v) and cone factor (N_{kt}). As such s_u is estimated with reasonable confidence from the CPT data, independent of the *in situ* pore water understanding.

This paper presents considerations for defining undrained strengths within tailings from CPT data, discussing different approaches to characterising strength using statistical methods. This paper highlights the sensitivity of USR to varying interpretations of piezometric PWP. It presents an equivalent approach to represent the tailings' undrained strength using total stresses, that removes the need to understand the PWP profile/regime. The assessment applies a stochastic model fitting process to the total stress approach to develop a statistical representation of the tailings' strength, such that a single characteristic strength can be adopted in the stability analyses, similar to the USR approach. The process followed to develop a statistical representation of tailings strength using the total stress approach is presented. This procedure is not intended to replace the USR approach but to provide an alternative screening method when uncertainty surrounds the *in situ* PWP conditions within a tailings deposit.

INTRODUCTION

Conventional slurry tailings depositional practices commonly lead to loose contractive tailings deposits. An understanding of the tailings strength properties is critical to assessing the safety of the storage facility when contractive tailings are part of its structural component. The CPT has become the premier tool within the tailings industry for estimating *in situ* strength of deposited tailings. Using the CPT, tailings strength properties are commonly represented as USRs for use in stability analyses to assess the safety of the storage facility.

The USR approach is a consolidated undrained strength representation and when used in conjunction with the CPT, requires a strong understanding of the piezometric PWP profile across the investigated storage facility, as this impacts the estimated *in situ* effective vertical stress (σ'_v). A simplified definition of PWP may be adopted through an assumption of the prevailing hydrostatic conditions and an estimate of standing water table depth from the inference of the CPT dynamic PWP response and/or nearby instrument measurements (ie vibrating wire piezometers and standpipe piezometers). Depending on the available information, a more realistic representation of PWP can be adopted through use of static PWP measured at various depth intervals within the investigated profile. Typically, measurements are from vibrating wire piezometers (VWPs) and/or CPT dissipation tests run to static PWP. However, in many instances, there are no VWPs near a CPT and running dissipation tests to static PWP is not feasible as test durations can vary significantly from seconds in coarse materials to days in finer-grained materials. As such, a simplified estimate of PWP is commonly adopted by practitioners.

This paper presents considerations for defining undrained strengths within tailings from CPT data, discussing different approaches to characterising strength using statistical methods. This paper highlights the sensitivity of USR to varying interpretations of piezometric PWP and presents an equivalent approach to represent the tailings' undrained strength using total stresses, removing the need for an understanding of the PWP profile.

The procedure presented in this paper is not intended to replace the USR approach but to provide an alternative screening method when uncertainty surrounds the *in situ* PWP conditions within a tailings storage facility (TSF).

BACKGROUND

Undrained strength from CPT

The conventional CPT used within the tailings industry provides the following three key measurements:

1. Cone resistance (q_c), which is converted to corrected cone resistance (q_t) to consider pore water effects.

2. PWP, most commonly located behind the cone (u_2).

3. Sleeve resistance (f_s).

Undrained shear strength (s_u) from the CPT is most commonly calculated using a relationship between s_u and q_t that was developed through theoretical solutions and having the following form (Robertson and Cabal, 2022):

$$s_u = \frac{q_t - \sigma_v}{N_{kt}}$$ (1)

where, σ_v is the total vertical stress and N_{kt} is the cone factor. As s_u is a total stress calculation, the PWP does not affect the assessed strength. With reference to Equation 1, q_t is measured by the CPT, σ_v is calculated based on a known depth and measured/estimated bulk unit weight and N_{kt} is calibrated with companion testing or calculated with empirical relationships. Therefore, it can generally be concluded that s_u is assessed from the CPT with reasonable confidence, which is why it is currently considered the premier tool within the tailings industry for estimating *in situ* strength of deposited tailings.

USR from CPT

USRs are calculated from CPT by dividing the above s_u by the *in situ* effective vertical stress (σ'_v) at the time of testing, as shown in Equation 2:

$$USR = \frac{s_u}{\sigma'_v} = \left(\frac{q_t - \sigma_v}{N_{kt}}\right)\left(\frac{1}{\sigma'_v}\right) = \left(\frac{q_t - \sigma_v}{N_{kt}}\right)\left(\frac{1}{(\sigma_v - u)}\right)$$ (2)

where u is the *in situ* static PWP at the time of testing. Given s_u is calculated from the CPT as shown in Equation 2, it is important to highlight that USR from CPT is wholly dependent on an inversely proportional relationship to σ'_v. Therefore, the calculated USR can vary significantly based on the interpretation of PWP, which is influenced by piezometric level, seepage condition and pore fluid density. As such, a strong understanding of the PWP profile at each CPT location is necessary for the calculated USRs to be representative of the actual undrained strength of the tailings. Where there is uncertainty in the PWP profile, care must be taken to understand how this impacts the USR calculation.

PWP condition within TSFs

TSFs are typically earth retaining structures that may have complex PWP conditions that are often difficult to characterise. Martin (1999) discusses the following PWP regimes in upstream-raised TSFs:

- hydrostatic
- over hydrostatic

- over hydrostatic with bottom drainage

- positive pressure below hydrostatic (ie under hydrostatic)

- pore pressure near zero

- unsaturated.

These conditions can lead to uncertainty in even identifying the saturation level/depth with confidence, let along the PWP profile. It is the authors' experience that for some scenarios uncertainty may remain in defining the saturation level and PWP profile with depth, even with a large degree of monitoring instrumentation (VWPs and standpipes), non-standard *in situ* measurements (seismic P-wave velocity and resistivity) and extensive dissipation testing.

In addition to the above, it is noted that the interpreted PWP conditions may also be impacted (although to a lesser extent) by potential variability in the pore fluid density from dissolved solids and/or geochemical reaction occurring within the tailings mass.

USR SENSITIVITY TO THE ADOPTED PWP PROFILE

General

As noted above, the calculation of USR from CPT is dependent on interpretation of the PWP profile. The sensitivity of this interpretation in relation to USR is further emphasised by displaying the variability in calculated USR for the following variations in the adopted PWP profile:

- depth to saturation level for hydrostatic conditions

- varying PWP regimes for a consistent saturation level.

Varying depth to saturation level

The variation of calculated USR for different interpretation of saturation level under hydrostatic conditions is displayed in Figure 1. It is noted that for the CPT data presented in the first three figures, the upper 2.2 m of the data set is removed as this material was drilled out.

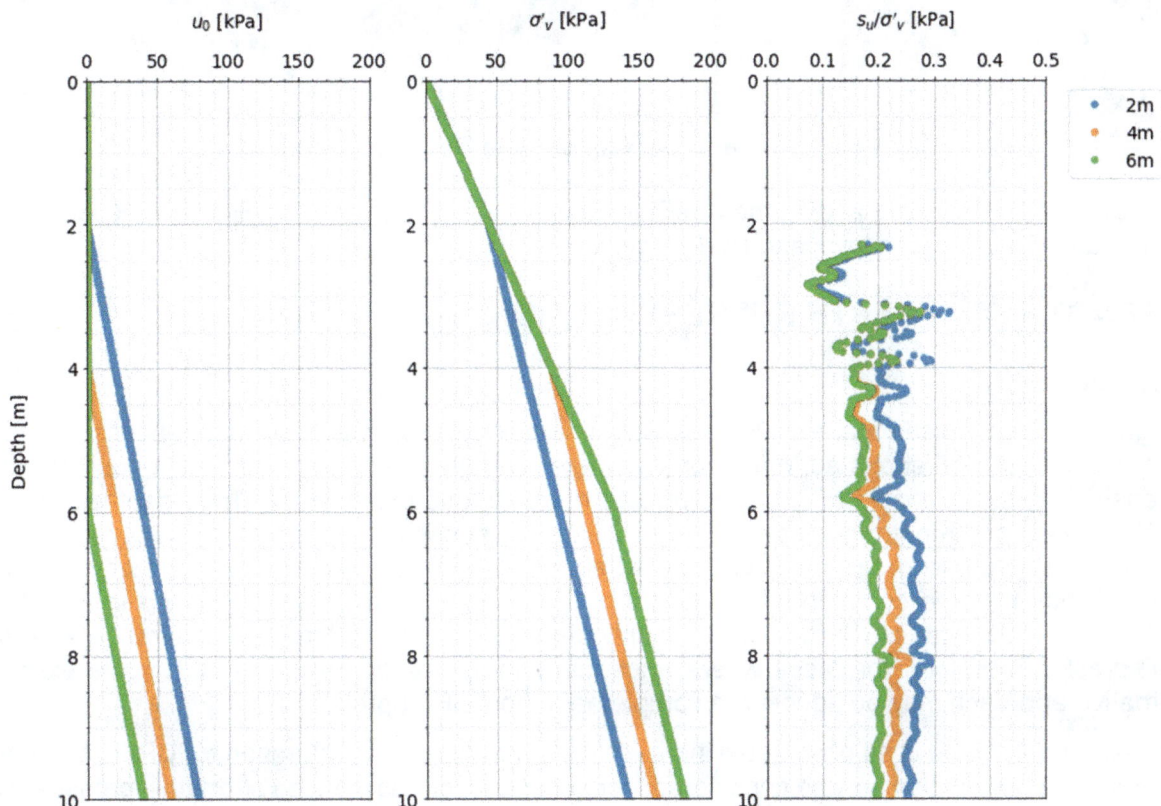

FIG 1 – σ'_v and USR for varying depth to saturation.

Varying PWP regimes

The variation in calculated USR under the following PWP regimes is displayed in Figure 2.

- hydrostatic

- over hydrostatic

- over hydrostatic with bottom drainage

- positive pressure below hydrostatic (ie under hydrostatic)

- pore pressure near zero.

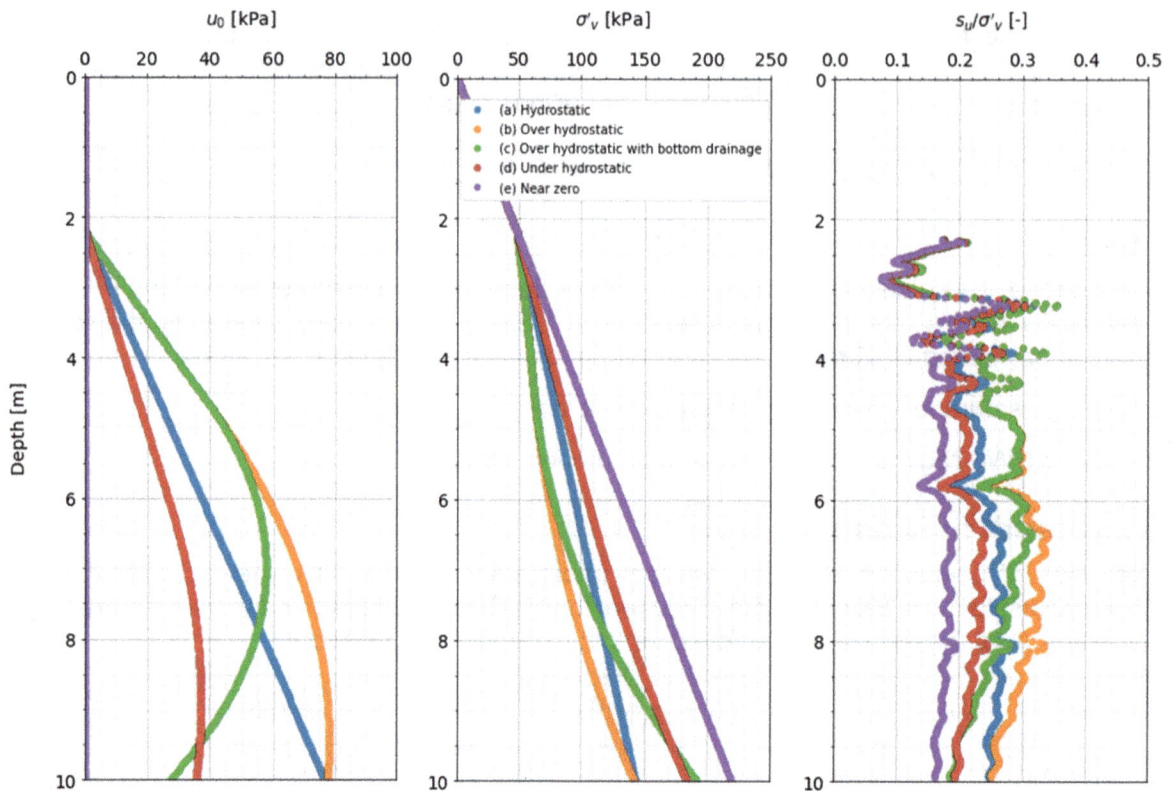

FIG 2 – σ'_v and USR for varying PWP regimes.

As is evident in Figures 1 and 2, there is substantial variation in the calculated USR for varying interpretation of saturation level and PWP regime.

STRENGTH WITH DEPTH REPRESENTATION

General

It is generally recognised that s_u increases with increasing stress within tailings deposits. As such, the conventional total stress approach for s_u, where strength may be adopted as a constant for a material/layer, is not considered an appropriate representation for tailings. This together with the ability to forecast strength increases due to changes in stress as well as its simplicity in which a statistical representation can be developed when large data sets are available, is why the USR approach has become common within the tailings industry. However, as discussed in the above sections, the calculated USR from CPT is highly dependent on the interpreted saturation level and PWP profile. Therefore, for cases where uncertainty in saturation level or PWP profile exist, an alternative approach may be adopted to complement the USR approach.

For cases where there is an observable increase in s_u with depth of penetration for CPTs, a strength with depth representation may be considered, accounting for stress induced strength increases. This representation is not new and many of the available limit equilibrium slope stability packages (GeoStudio 2022.1 SLOPE/W, version 11.4.2.250 by GEOSLOPE International) allow input of this type of strength representation. In fact, it likely provides the most realistic representation of s_u at the time of testing. However, uncertainty may arise when considering this type of strength representation for changed stress conditions (ie unknown future conditions), but it will be shown that forecasting can be achieved where increasing stress conditions are consistent across the period of interest.

Stress increases

Stress conditions within a tailings deposit vary with time, but when considering uniform embankment raising or typical tailings deposition, a generally consistent stress increase can be considered. This consistency in stress increase may allow adoption of a strength with depth representation for known future conditions. However, it is advised that the validity of adopting this type of representation is checked specifically for each site and the associated stress increases. This can be done by comparing CPT strength with depth profiles at adjacent locations but completed at different time periods. For example, Figure 3 displays both a strength with elevation and strength with depth profile for CPTs conducted approximately one year apart and within about 10 m of each other on a tailings beach.

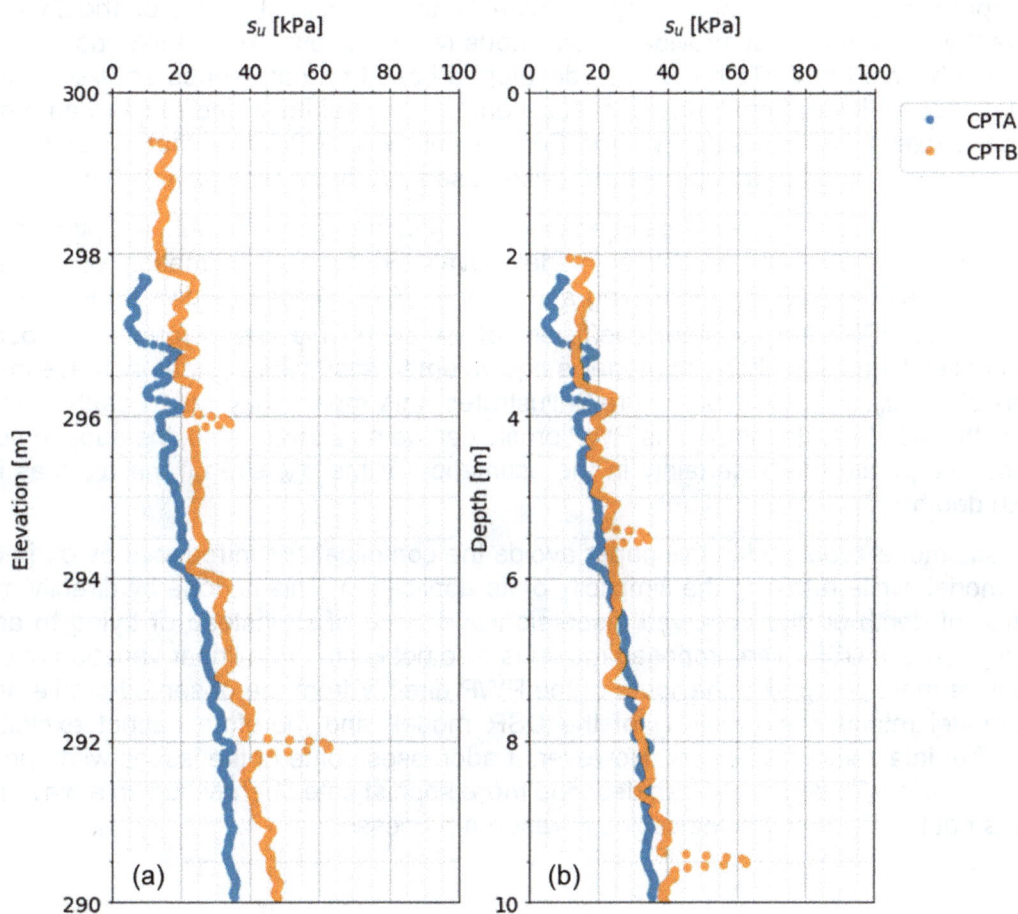

FIG 3 – (a) s_u with elevation; (b) s_u with depth.

When considering the s_u with elevation plot (Figure 3a), there is an observable offset in the strength data for the two CPTs for the same elevation. However, when s_u is plotted with depth of penetration (ie Figure 3b), the data generally overlap and the s_u offset shown in Figure 3a can be attributed to the strength increase of the tailings due to continued deposition (ie increased stress from the overlying tailings).

When the above condition exists for a tailings deposit, it provides the opportunity to adopt a strength with depth representation for future conditions provided the stress increases are consistent across the data set. Moreover, it allows grouping of historical CPT data together to create a data set for statistical representation.

STATISTICAL APPROACH

Background

According to Phoon and Kulhawy (1999) there are three distinct components of uncertainty in geotechnical parameters: the inherent soil variability, the measurement variability, and the transformation model. The inherent variability refers to the spatial and temporal distribution of a parameter in the soil mass and can be mathematically modelled as a random field (Vanmarcke, 2010). The measurement variability arises as only a limited sample of the spatial distribution of the geotechnical parameter is measured, and the measuring process introduces random errors. The transformation model is the uncertainty introduced when a model is applied to estimate a soil property. Equation 1 is an example of such a model where s_u is derived from the measured q_c and depth. The uncertainty in the transformation method can only be assessed by comparing multiple measurements of the same parameter across sites and different estimation models. The transformation model statistics are predetermined and not a function of the site, whereas inherent variability and measurement variability are site specific.

CPT tests provide a unique opportunity to quantify the combined effect of the inherent and measurement variability as they provide a continuous record across depth. Methods of analysis of the variability observed in CPTs have been developed based on data series analysis (Cami *et al*, 2020). A data series differs from typical statistical analysis in that the samples have an order given by depth that is relevant to the sample values. For example, Figure 3 shows that s_u has a clear trend increasing with depth, and the variability can be expressed as a fluctuation deviating from such trend.

A simple way of addressing the dependence of s_u with depth is through adoption of a USR representation, as is done by statistical models available for the estimation of s_u based on parameters such as the OCR, I_p, etc (Ching and Phoon, 2012). These models adopt a log-normal distribution of the USR with constant coefficient of variation. If a linear variation of pore water pressures applies to the site, then the model is equivalent to assuming a linear increase in-depth of the median of the s_u. However, as Figure 2 illustrates, this model may fail to capture the actual behaviour within a tailings deposit as the PWP profile can be non-linear. Also, the adoption of a USR model must incorporate the uncertainty in the estimation of the σ'$_v$, even if the σ'$_v$ does increase linearly with depth.

The stochastic model proposed in this paper avoids the complications introduced by σ'$_v$ in the USR stochastic model, while retaining the simplicity of its application. This is done by directly modelling the influence of depth on the observed dispersion and trend of s_u instead of trying to address it through adoption of a USR. The proposed model is also not limited to a linear variation of σ'$_v$, being able to capture more complex behaviours of the PWP directly from the observed profile of s_u. The proposed model retains the simplicity of the USR model, and therefore cannot exploit the full potential of the data series framework. However, it addresses some of the issues with applying the USR model to tailings deposits and can describe more sophisticated behaviours in a way amenable to engineers not familiar with the intricacies of random processes.

Method

The model proposed assumes that at any given depth the s_u follows a log-normal distribution. Log-normal distributions are commonly observed in mechanical properties at large and geotechnical parameters in particular (Baecher and Christian, 2005), they cannot assume negative values and have a positive eccentricity. The probability distribution of a log-normal distribution is given by (Kottegoda and Rosso, 2008):

$$F_X(x) = P[X \leq x] = \Phi\left(\frac{\ln(x) - \mu_{lnX}}{\sigma_{lnX}}\right) \tag{3}$$

where Φ is the normal standard distribution and μ_{lnX} and σ_{lnX} are the log-normal distribution parameters. A log-normal variable is characterised by the fact that its natural logarithm $\ln(x)$ has a normal distribution with mean μ_{lnX} and standard deviation σ_{lnX}. These are however not the mean μ_X and standard deviation σ_X of the variable itself, but are related to them through:

$$\sigma_{lnX} = \sqrt{\ln(1 + \delta_X^2)} \tag{4}$$

$$\mu_{lnX} = \ln(\mu_X) - \sigma_{lnX}^2/2 \tag{5}$$

where $\delta_X = \sigma_X/\mu_X$ is the coefficient of variation (COV) of the random variable. The stochastic model proposed for the s_u assumes that the parameters of the log-normal distribution vary with depth:

$$F_{S_u}(s_u|z) = P[S_u \le s_u|z] = \Phi\left(\frac{\ln(s_u) - \mu_{lnSu}(z)}{\sigma_{lnSu}(z)}\right) \tag{6}$$

Different models can be fitted by adopting different functions for $\mu_{lnSu}(z)$ and $\sigma_{lnSu}(z)$. The simplest case, from a mathematical standpoint, would be to assume that these parameters are constant with depth. This would result in the same distribution of s_u for any given depth and fail to capture the depth dependency. More complex behaviours of the overall trend and the dispersion around it can be modelled by choosing more complex functions for $\mu_{lnSu}(z)$ and $\sigma_{lnSu}(z)$, like polynomials of increasing order. There is a trade-off between complexity and predictive capabilities. On one extreme, a model constant with depth defeats the purpose of capturing the general trend of s_u, while an overly-complex model results in unnecessary parameters to be calibrated. Statistical methods such as the Akaike Information Criterion (Akaike, 1998, 1974) could be applied to compare the trade-off between predicting capacity and simplicity, but are not considered as part of this paper.

In this case, it is assumed that the coefficient of variation of the undrained shear strength (δ_{Su}) is constant, which implies that σ_{lnSu} is also independent of depth. The resulting stochastic model is therefore given by:

$$F_{S_u}(s_u|z) = \Phi\left(\frac{\ln(s_u) - (a + b\,z)}{\sigma_{lnSu}}\right) \tag{7}$$

The maximum likelihood method is applied to fit the model to the observed data set. The parameters are chosen such that the likelihood (L) of the observed samples is maximised. The likelihood is the product of the probability density function $f_{S_u}(s_u|z)$ evaluated at every sample:

$$L = \prod_{j=1}^{N} f_{Su}\left(s_{u_j}|z_j\right) = \prod_{j=1}^{N} \frac{1}{\sqrt{2\pi}\,\sigma_{lnSu}\,s_{u_j}} \exp\left[-\left(\frac{\ln\left(s_{u_j}\right) - a - b\,z_j}{\sigma_{lnSu}}\right)^2\right] \tag{8}$$

This is equivalent to maximising the log-likelihood (LL) given by:

$$LL = \ln(L) = \sum_{j=1}^{N} \ln\left(\frac{1}{\sqrt{2\pi}\,\sigma_{lnSu}\,s_{u_j}}\right) - \sum_{j=1}^{N}\left(\frac{\ln\left(s_{u_j}\right) - a - b\,z_j}{\sigma_{lnSu}}\right)^2 \tag{9}$$

The values of a, b and σ_{lnSu} are thus obtained by maximising LL. An example of the fitting process can be seen in Figure 4, where the fitted model is represented by the distribution percentiles across depth. As expected, the mean value of the expected s_u increases with depth. Given the definition of the log-normal distribution, assuming a constant COV also implies that the range of s_u for any confidence level also increases with depth. Thus, the model predicts a larger variability as the depth increases.

A common issue when fitting a stochastic model is the elimination of outliers from a data set. An outlier is defined as a sample whose value is not the product of the stochastic process that generates the rest of the data set. In a CPT, an outlier can be attributed to an intrusion of a different material in an otherwise homogenous soil mass, due to equipment malfunction or perturbations during the recording procedure. Outliers can be identified and removed from the data set by the very fact that they present an abnormal stochastic behaviour relative to the overall model. There are several methods that can be applied to achieve this, with a wide range of mathematical complexity (Kottegoda and Rosso, 2008). Once more, a simpler method is preferred due to the robustness of its application.

Values that are known to correspond to an abnormality in the sample *a priori* (instrumentation error, a thin layer of material that perturbates an otherwise contiguous reading across a given material etc) should be removed before the analysis. For example, Figure 4 shows sudden peaks corresponding with thin layers of coarser tailings. Since this is not the same material whose resistance is being characterised by the model, the samples are not considered in the calibration. Given that this type of outlier results in concentrated zones of higher strength, its exclusion from the model is on the conservative side in a structural reliability analysis.

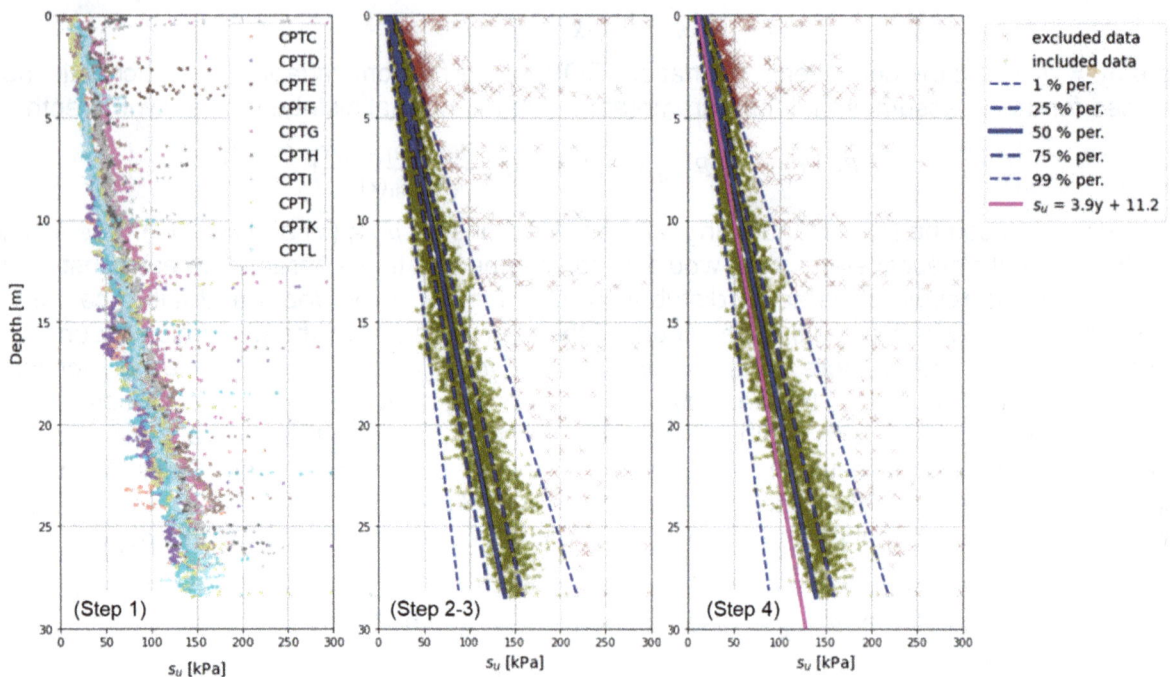

FIG 4 – Example strength with depth model application.

However, not all outliers are due to causes known *a priori*, and thus easily removable. A simple method to detect outliers is to remove samples that exceed a given percentile of the distribution. Samples above the chosen exceedance probability are removed from the data set and the model is fitted to the remaining samples. A statistical test can then be performed based on the exceedance probability associated with the removed samples in the model fitted to the filtered data set. Or a simpler evaluation can be conducted by simply assessing the impact on the model parameters after removing the suspected outlier. In a large enough data set, no individual sample should be so predominant that it controls the value of the model's parameters. If removing the suspected outlier results in a significant change in the model parameters, then it is likely that the sample did not belong to the same random process that produced the rest of the data set and is indeed an outlier.

Application

The application of the above method for a strength with depth representation is demonstrated below using the following steps:

- Compile the derived strength with depth data set from CPTs where there is consistency in the stress condition (ie data from CPTs conducted on the tailings beach).

- Remove samples that are known to be outliers *a priori*.

- Use the maximum likelihood method to fit the model parameters.

- Adopt a strength with depth profile based on a confidence level (eg the 25th percentile of the model).

CONCLUSIONS

A USR approach is commonly adopted within the industry to characterise tailings strength using CPT data, but requires a detailed understanding of the saturation level and PWP profile/regime. It is shown through sensitivity analyses that there is substantial variation in the calculated USR for varying interpretation of saturation level/PWP regime and that caution should be given when uncertainty surrounds these items. Given s_u is estimated with reasonable confidence from the CPT data, an equivalent approach to represent the tailings' undrained strength using total stresses is presented removing the need to understand the PWP conditions. The approach applies a stochastic model fitting process to develop a statistical representation of the tailings' strength, such that a single characteristic strength can be adopted in stability analyses, similar to the USR approach.

This procedure is not intended to replace the USR approach but to provide an alternative screening method when uncertainty surrounds the *in situ* PWP conditions within a TSF.

REFERENCES

Akaike, H, 1974. A new look at the statistical model, in *IEEE transactions on automatic control*, 19(6):716–723.

Akaike, H, 1998. Information theory and an extension of the maximum likelihood principle, selected papers of Hirotugu Akaike, pp 199–213.

Baecher, G B and Christian, J T, 2005. *Reliability and statistics in geotechnical engineering* (John Wiley & Sons Ltd: Chichester).

Cami, B, Javankhoshdel, S, Phoon, K K and Ching, J, 2020. Scale of fluctuation for spatially varying soils: estimation methods and values, *ASCE-ASME Journal of Risk and Uncertainty in Engineering Systems, Part A: Civil Engineering*, 6(4):03120002.

Ching, J and Phoon, K K, 2012. Modeling parameters of structured clays as a multivariate normal distribution, *Canadian Geotechnical Journal*, 49(5):522–545.

Kottegoda, N T and Rosso, R, 2008. *Applied statistics for civil and environmental engineers* (Blackwell Publishing: Malden).

Martin, T E, 1999. Characterization of pore pressure conditions in upstream tailings dams, in *Proceedings of the Sixth International Conference on Tailings and Mine Waste 1999*, pp 303–314 (AA Balkema: Rotterdam).

Phoon, K K and Kulhawy, F H, 1999. Characterization of geotechnical variability, *Canadian Geotechnical Journal*, 36(4):612–624.

Robertson, P K and Cabal, K, 2022. *Guide to cone penetration testing* (Gregg Drilling LLC: Signal Hill).

Vanmarcke, E, 2010. *Random fields: analysis and synthesis* (World Scientific Publishing Company).

Selection of soil shear strength parameters based on integrated *in situ* tests, lab tests and numerical calibration approach

Z Song[1], R Truby[2] and S Luo[3]

1. Senior Geotechnical Engineer, Knight Piésold, East Perth WA 6004.
   Email: zsong@knightpiesold.com
2. Senior Engineer, Knight Piésold, East Perth WA 6004. Email: rtruby@knightpiesold.com
3. Principal Geotechnical Engineer, Knight Piésold, East Perth WA 6004.
   Email: jluo@knightpiesold.com

ABSTRACT

Site investigations were conducted to obtain soil parameters for stability and deformation assessment for an upstream tailings storage facility. Three material types that control the tailings dam stability were identified during the site investigations, comprising coarse-grained silty/sandy tailings, fine-grained silty/clayey tailings and residual soil. Advanced *in situ* tests, including Seismic Cone Penetrometer Tests (SCPT) and Vane Shear Tests (VST) were conducted. A comprehensive laboratory testing program was undertaken, which included direct simple shear (DSS) tests and triaxial tests to assess the Critical State Line (CSL).

An integrated approach was adopted to determine soil parameters, takings advantage of both *in situ* and laboratory test results. The peak undrained shear strengths ratios (s_u/σ'_v) of the fine-grained silty/clayey tailings were determined from the DSS test results. The CPT cone factor (N_{kt}) was calibrated using *in situ* s_u/σ'_v strengths of the materials in relatively thin layers. Subsequently, the *in situ* strengths of all other layers were calculated using the N_{kt} factor. Due to partial drainage in the coarse-grained tailings, its peak and residual undrained shear strengths were assessed using triaxial and non-standard VST test results.

The yield stress of the residual soil was determined using laboratory tests, Self-boring Pressuremeter (PMT) and FLAC™ numerical modelling results. The undrained shear strength of the residual soil was based on the calculated yield stress, PMT undrained shear strength and DSS test results.

INTRODUCTION

A comprehensive site investigation was carried out to characterise the undrained shear strengths for deposited tailings and foundation soil, due to the relatively high phreatic surface in an upstream tailings storage facility. Three types of materials that control the stability of the facility were identified during the site investigations, comprising:

- coarse-grained silty/sandy tailings
- fine-grained silty/clayey tailings
- foundation residual soil.

To carry out detailed stability assessments, the tailings and residual soil shear strengths need to be characterised. An integrated approach was selected, using *in situ* tests, laboratory tests and numerical modelling to select the soil strength parameters.

The site investigations comprised drilling and sampling to obtain high-quality undisturbed samples, laboratory testing, and the following *in situ* tests:

- Seismic Cone Penetration Tests (SCPTu), and non-standard Vane Shear Test (VST) in coarse-grained tailings.
- SCPTu and standard VST in fine-grained tailings.
- Pressuremeter (PMT) and standard VST in residual soil.

The methodologies to select the undrained shear strength parameters are described in this paper.

INTEGRATED APPROACH

Mayne (2005) recommended a hybrid of empirical, analytical, experimental, and/or numerical methods to interpret ground behaviour due to the inherent benefits and limitations of the various approaches. Table 1 summaries the benefits and limitations of *in situ* and laboratory test. It is evident that they are, in most cases, complementary. Consequently, the reliability of each method is greatly enhanced when used in conjunction with the other methods.

TABLE 1

Advantaged and disadvantages of *in situ* and laboratory tests.

	In situ tests	Laboratory tests
Pros	• Continuous Profile (CPT). • Tests are carried out in place, in the natural environment, without sampling disturbance. • Testing is usually fast and repeatable. • A large soil volume can be tested.	• Samples are reconsolidated for testing at the proposed stress level. • Tests are performed in a controlled strain-rate, drainage and temperature environment. • Soil behaviour under cyclic loading conditions can be assessed.
Cons	• Sample cannot be tested at an increased stress level. • Some engineering properties are not measured directly and are based on correlations. • Some loading conditions cannot be tested eg cyclic loading.	• Samples may not represent the *in situ* condition. • A small soil volume is tested.

In addition to *in situ* and laboratory tests, numerical modelling was undertaken to calibrate soil parameters. One of the primary benefits of numerical modelling is the ability to incorporate advanced soil constitutive models, enabling the simulation of soil behaviour under complex loading conditions. Shape Acceleration Arrays (SAAs) were installed in the tailings storage facility, enabling the comparison of numerical results and SAA monitoring results to calibrate the soil parameters.

TYPICAL DAM SECTION

The typical tailing dam section is shown in Figure 1, illustrating that the coarse tailings is underlain by a layer of fine-grained tailings and residual soil at depth. The relevant undrained shear strength parameters were determined based on the potential failure mode, as follows:

- Triaxial compression for the coarse-grained tailings.

- Direct simple shear for the fine-grained tailings and residual soil.

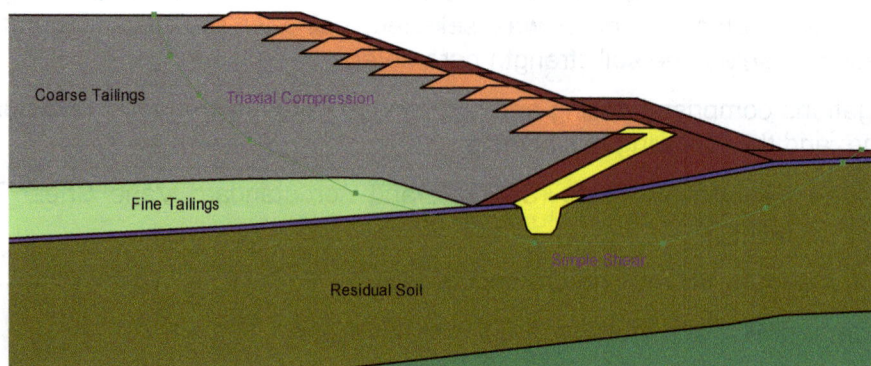

FIG 1 – Typical cross-section of tailings storge facility.

Based on the CPT results and the soil behaviour types by Robertson (1990, 2009, 2016), the following materials classification were made:

- Coarse tailings can be classified as 'Sand – Clean sand to silty sand', and 'Sand mixture – Silty sand to sandy silt'. Negative dynamic pore pressures were often recorded in dilative tailing layers and positive pore pressures in contractive tailings layers.

- Fine tailings, if encountered, can be classified as 'Silt mixtures – clay silt to silty clay'. Positive dynamic pressures were recorded in the tailings due to the undrained contractive condition.

- Residual soil, if encountered, can be classified as 'Silt mixtures – clay silt to silty clay'.

Figure 2 shows typical CPT results and material types from the CPT interpretation.

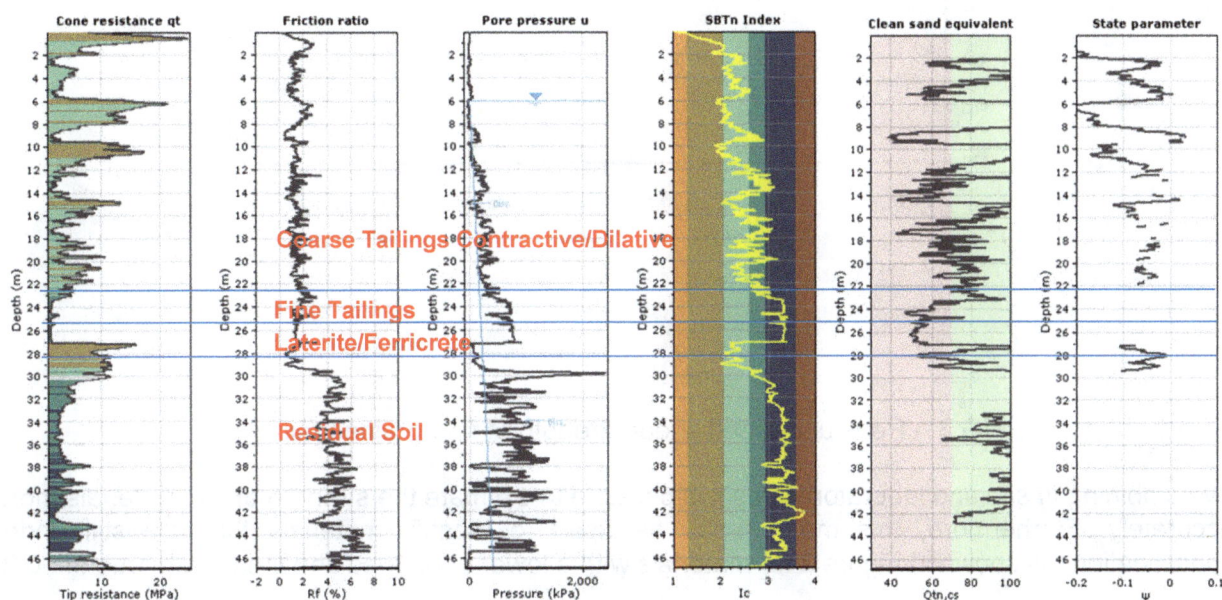

FIG 2 – Typical CPT results.

GEOTECHNICAL CHARACTERISATION

The characterisation of each material type is presented below.

Fine-grained tailings

The fines contents of the fine-grained tailings are typically 60 per cent to 90 per cent. The material exhibits typical clay-like soil behaviour.

The undrained shear strengths of the fine-grained tailings can be estimated using the total cone resistance (q_t) and a cone factor (N_{kt}). While a cone factor of 14 is typically adopted, literature indicates values ranging from 8 to 16.

The fine-grained tailings undrained shear strength was determined as follows:

- Determine the peak undrained shear strength ratio of the material under triaxial compression or direct simple shear using laboratory test results.

- Use the strength of the thin *in situ* material layers, which should be normally consolidated, to calibrate the N_{kt} factor.

- Use the calibrated N_{kt} factor to determine the shear strength of materials at other locations, including relatively thick layers.

Figure 2 shows the interpreted soil shear strength ratios:

- The s_u/σ'_v ratio under triaxial compression is approximately 0.31.

- The s_u/σ'_v ratio under direct simple shear is approximately 0.25.

Based on a s_u/σ'_v ratio of 0.25 and the thin fine-grained materials with a thickness of less than 5 m, a N_{kt} factor of 12 was assessed for material under simple shear. Figure 3 shows the interpretation of the undrained shear strength with a N_{kt} factor of 12 in comparison with laboratory test results.

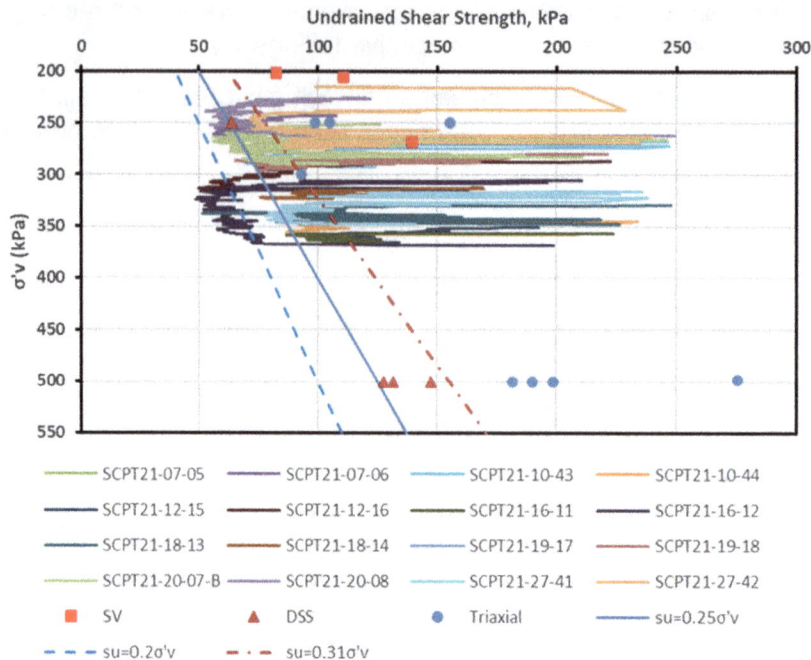

FIG 3 – Peak undrained shear strength of fine-grained tailings.

By combining *in situ* and laboratory tests it is possible to estimate the strength of the materials more accurately. Furthermore, this method can be used to identify materials that are still under consolidation, as shown in Figure 3 for materials with a lower undrained shear strength ratio of ~0.2.

Coarse-grained tailings

Undrained shear strength

The fines content for coarse tailings ranges from 40 per cent to 80 per cent, with low plasticity. The 50 per cent CPT dissipation time (t_{50}) is usually less than 10 seconds, suggesting partial drainage (Robertson, 2012; DeJong and Randolph, 2012). The vane shear test was therefore conducted at a non-standard rate of 240°/min to ensure undrained behaviour (Reid, 2016).

The peak undrained shear strength was proposed as a function of the state parameters based on the CSL triaxial test results and 240°/min vane shear test results. Figure 4 shows the relationship. Some of the VST tests undertaken during the early project stages were carried out at the standard shear rate of 12°/min. Due to partial drainage, these tests indicated higher undrained shear strengths.

Figure 5 shows the residual undrained shear strength, based on vane shear tests results and CSL triaxial test results. The relationship correlates well to the Jefferies and Been (2016)'s best practice line.

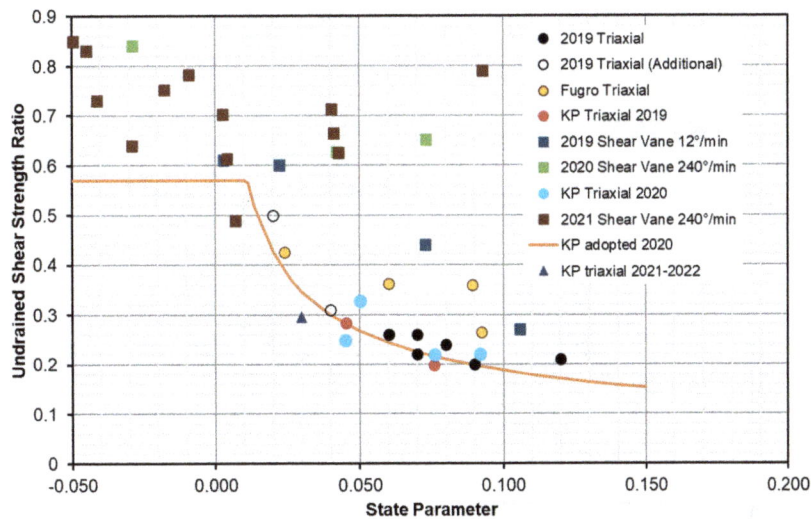

FIG 4 – Peak undrained shear strength of coarse-grained tailings.

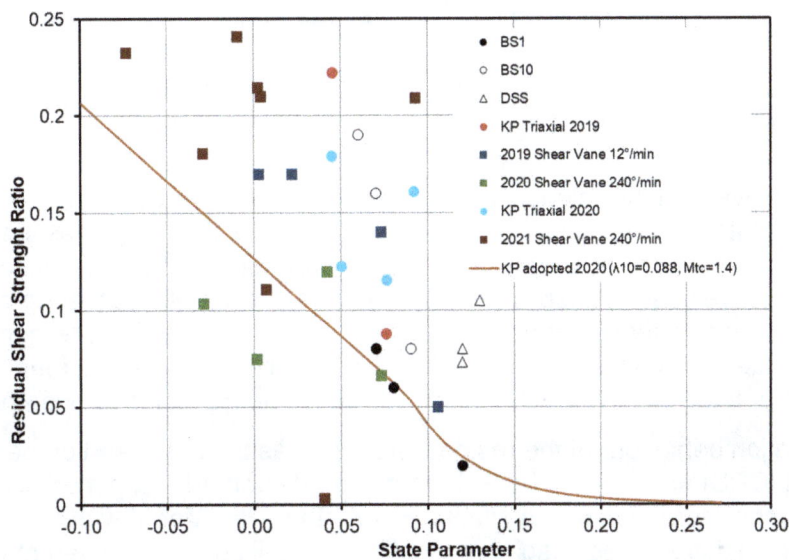

FIG 5 – Residual undrained shear strength of coarse-grained tailings.

State parameter

The state parameter is an important parameter that should be determined accurately for the coarse-grained tailings. The interpretations of the state parameters were based on the following methods: Robertson and Cabal (2015) and Plewes, Davies and Jefferies (1992).

The CPT 'widget' (Shuttle, 2019), which is a finite element cavity expansion program, further improved CPT data interpretation.

All methods yielded similar results as shown in Figure 6. Therefore, either of Robertson and Cabal (2015) or Plewes, Davies and Jefferies (1992) methods can be used for the state parameter interpretation. Robertson updated the Robertson and Cabal method in 2022 (Robertson, 2022). The updated state parameters using the Robertson (2022) method are typically ~0.02 smaller than those from the Robertson and Cabal (2015) method. However, the residual undrained shear strength would be similar if the following methods are adopted:

- Jefferies and Been (2016) residual strength method using either Plewes, Davies and Jefferies (1992) or Robertson and Cabal (2015) state parameters.

- Robertson (2022) residual strength with Robertson (2022) state parameters or $Q_{tn,cs}$.

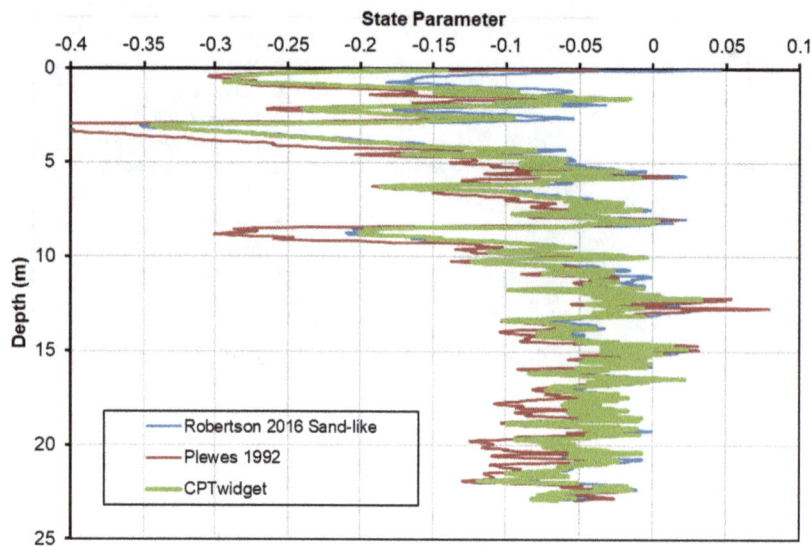

FIG 6 – State parameters by Robertson and Cabal (2015), Plewes, Davies and Jefferies (1992) and CPT 'widget' method.

Residual soil

Yield stress

All residual soils behave as if over-consolidated to some degree (Leroueil and Barbosa, 2000). Their compressibility is relatively low at low stress levels. Once a threshold has been exceeded, the compressibility increases. Traditionally, the threshold was termed as 'pre-consolidation stress' for transported soil. However, residual soils are formed as the result of weathering and decomposition of rocks and thus stress history is of no relevance (Wesley, 2010). A more general term of 'yield stress' is recommended for residual soils (Wesley, 2010; Mayne, 2017) as a measure of the strength of the interparticle, or inter-mineral crystal bonds remaining in the soil after weathering.

The over-consolidation behaviour of the residual soils was assessed based on oedometer, Constant Rate of Strain Cell (CRS) and self-boring PMT tests. In addition, FLAC™ modelling was undertaken to predict embankment deformations. A Plastic Hardening (PH) Model with memory of yield stress (equivalent pre-consolidation) was used to model the residual soil. A representative yield stress was determined through calibration of the FLAC™ model against inclinometer monitoring results through a trial-and-error approach. The stiffness of the residual soil was determined by the PMT tests, laboratory tests and Standard Penetration Tests (SPT). In addition, in order to accurately model the brittle coarse tailings behaviour, a NORSAND model was adopted for the coarse tailings. The calibration indicates a yield stress of between 500 kPa and 550 kPa.

Table 2 summarises the determined yield stress through the laboratory tests, *in situ* tests and numerical modelling. Figures 7 and 8 show the predicted embankment deformation and the calibration of the FLAC™ modelling against the inclinometer monitoring results.

TABLE 2

Yield stress of residual soil.

Methods	Yield stress (kPa)
Oedometer	290–1230
CRS	730–1080
Self-Boring PMT	340–660
Numerical Calibration	500–550

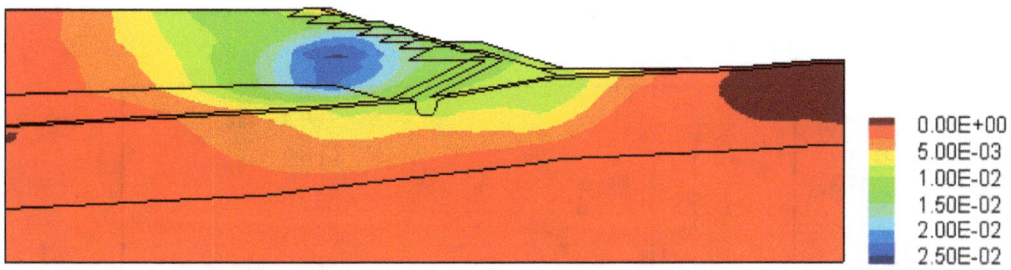

FIG 7 – Predicted FLAC horizontal displacement results (Unit: m).

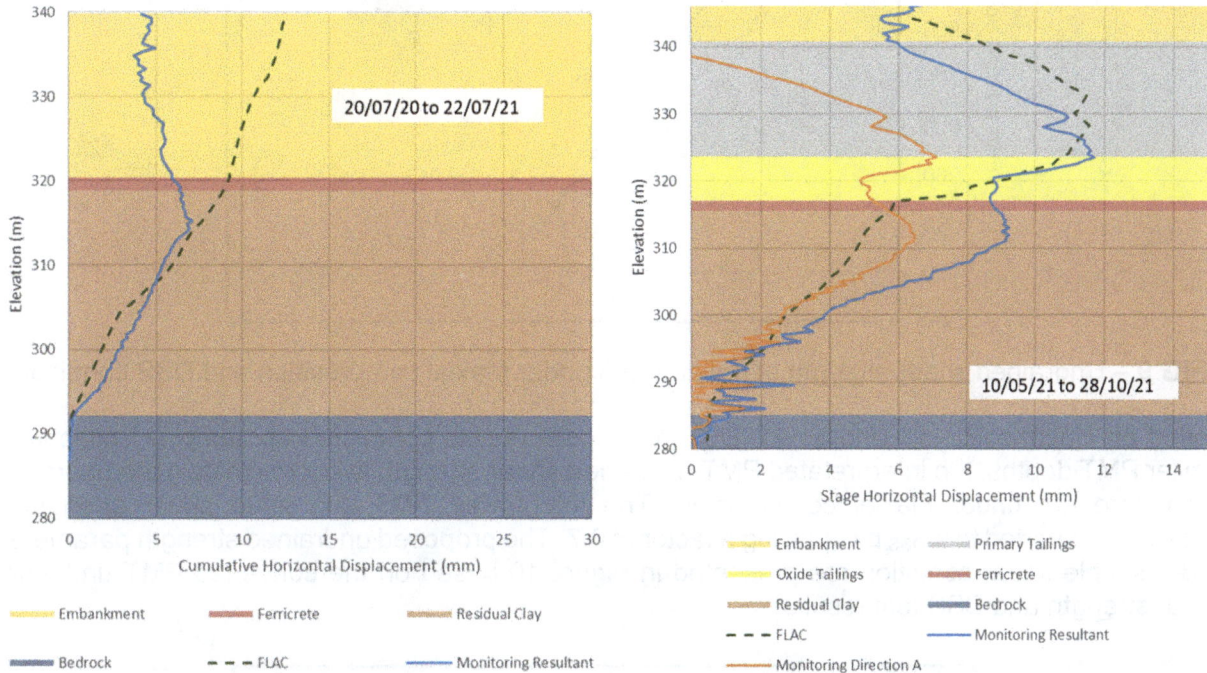

FIG 8 – Comparison of predicted deformations versus site monitoring data.

Undrained shear strength

Figure 9 illustrates all triaxial and DSS test results conducted on undisturbed residual soil samples. It shows a clear overall trend with the triaxial compression strength exceeding the DSS strength. The s_u/σ'_v ratio under direct simple shear is approximately 0.26. The factor of $s_{uDSS}/s_{uTriaxial}$ is approximately 0.7, which is consistent with Mayne (1985).

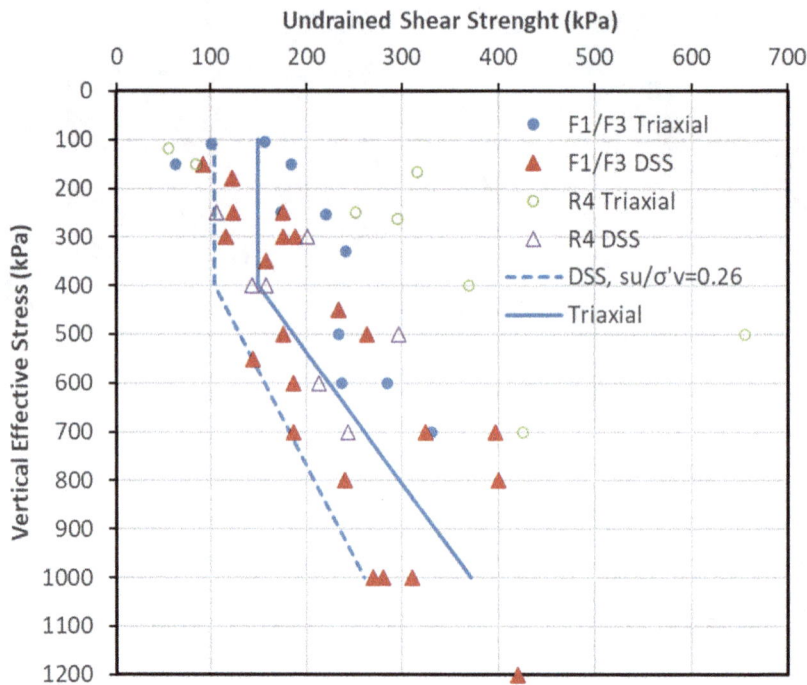

FIG 9 – Undrained shear strength of residual soil under triaxial compression and DSS condition.

Based on Unconsolidated Undrained (UU) triaxial test results conducted on samples collected at similar PMT depths, the interpretated PMT undrained shear strength is approximately the undrained shear strength under triaxial compression. The interpreted PMT undrained shear strength is therefore converted to s_{uDSS} by applying a factor of 0.7. The proposed undrained strength parameters under simple shear condition are presented in Figure 10 based on the converted PMT undrained shear strength and DSS test results.

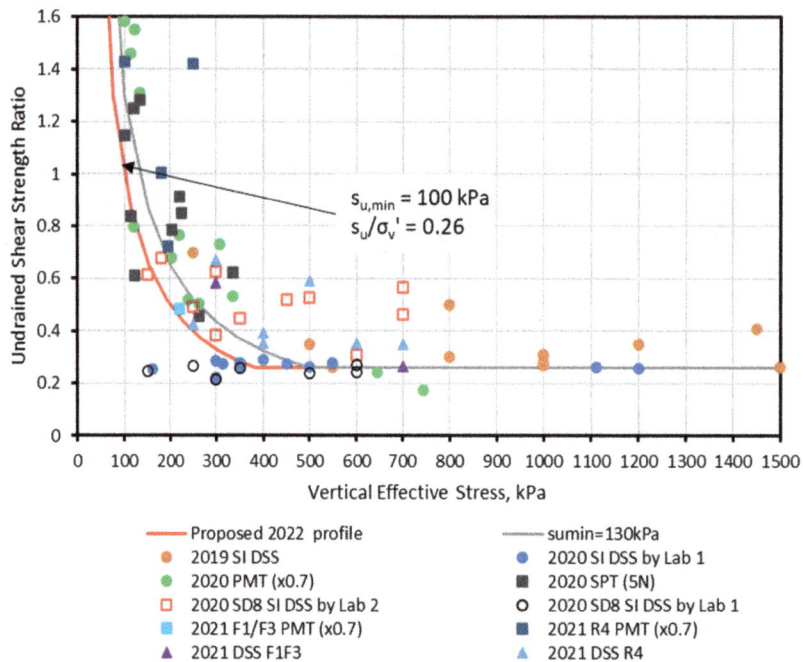

FIG 10 – Undrained shear strength of residual soil under triaxial compression and DSS condition.

Some of the early DSS tests were conducted independently in two different labs and they showed different results, with one indicating over consolidated behaviour and the other nearly normally consolidated behaviour. Upon comparison with the PMT tests and FLAC numerical modelling results, it was determined that the normally consolidated behaviour results may be erroneous, and they were excluded from the interpretations.

The following undrained shear strength was recommended for the stability assessments:

- Minimum undrained shear strength of 100 kPa (corresponding to a yield stress about 400 kPa).

- An undrained shear strength ratio of 0.26.

The recommended minimum undrained shear strength of 100 kPa is approximately 80 per cent confidence level of the materials with vertical effective stress less than 300 kPa. The mean minus half standard deviation value is around 130 kPa, which corresponds to a yield stress about 500 kPa. This is consistent with FLAC numerical modelling calibration.

CONCLUSIONS

Both *in situ* and laboratory tests have their inherent benefits and limitations. By conducting both types of tests, it is possible to calibrate the *in situ* tests and identify any improperly conducted laboratory tests.

This study has illustrated that:

- Numerical modelling can be used to calibrate undrained shear strengths and residual soil yield stresses.

- It is possible to simulate soil behaviour under complex loading conditions using advanced soil constitutive models.

By utilising the benefits of *in situ* tests, laboratory tests and numerical modelling, the undrained strengths of tailings and residual soil could be more accurately determined for different stress conditions.

REFERENCES

DeJong, J T and Randolph, M, 2012. Influence of partial consolidation during cone penetration on estimated soil behaviour type and pore pressure dissipation measurements, *Journal of Geotechnical and Geoenvironmental Engineering*, 138(7):777–788.

Jefferies, M and Been, K, 2016. *Soil liquefaction: a critical state approach,* 2nd edition, pp 291–292 (CRC Press).

Leroueil, S and Barbosa, P S de A, 2000. Combined effect of fabric, bonding, and partial saturation on yielding of soils, in *Proceedings of the Unsaturated Soils for Asia*, pp 527–532 (Balkema: Rotterdam).

Mayne, P W, 2005. Integrated ground behavior: *In situ* and labs tests, *Deformation characteristics of geomaterials*, pp 162–185 (CRC Press).

Mayne, P W, 2017. Stress history of soils from cone penetration tests, *Soils Rocks*, 40:203–218.

Mayne, P W, 1985. A Review of Undrained Strength in Direct Simple Shear, *Soils and Foundations,* 25(3):64–72.

Plewes, H D, Davies, M P and Jefferies, M G, 1992. CPT based screening procedure for evaluating liquefaction susceptibility, Proceedings of the 45th Canadian Geotechnical Conference.

Reid, D, 2016. Effect of rotation rate on shear vane results in a silty tailings, 5th International Conference on Geotechnical and Geophysical Site Characterisation (ISC'5).

Robertson, P K and Cabal, K L, 2015. *Guide to Cone Penetration Testing for Geotechnical Engineering* (6th edition).

Robertson, P K, 1990. Soil classification using the cone penetration test, *Canadian Geotechnical Journal*, 27(1):151–158.

Robertson, P K, 2012. Interpretation of *in situ* tests–some insights, *Mitchell Lecture-ISC*, 4:1–22.

Robertson, P K, 2016. Cone penetration test (CPT)-based soil behaviour type (SBT) classification system—an update, *Canadian Geotechnical Journal*, 53(12):1910–1927.

Robertson, P K, 2022. Evaluation of flow liquefaction and liquefied strength using the cone penetration test: an update, *Canadian Geotechnical Journal*, 59(4):620–624.

Robertson, P K, 2009. Interpretation of cone penetration tests—a unified approach, *Canadian Geotechnical Journal*, 46(11):1337–1355.

Shuttle, D, 2019. CPTwidget: a finite element program for soil-specific calibration of the CPT.

Wesley, L D, 2010. *Geotechnical engineering in residual soils* (John Wiley and Sons).

Application of Bayesian methods to the estimation of shear strength parameters for slope stability analysis of a TSF

I Vernengo Lezica[1] and L F Contreras[2]

1. Senior Consultant, SRK Consulting, Brisbane Qld 4000. Email: ivernengo@srk.com.au
2. Associate Corporate Consultant, SRK Consulting, Brisbane Qld 4000.
   Email: lfcontreras@srk.com.au

ABSTRACT

This paper covers the estimation of shear strength parameter values for input into slope stability analysis of tailings storage facilities. The paper describes the current industry practice of selecting a single (sometimes arbitrary) conservative value for each parameter and proposes to use Bayesian statistics to obtain a meaningful probabilistic evaluation of the strength envelope based on the available data and knowledge of the parameters. The concept of prediction intervals was used to define the strength parameters with selected levels of confidence for the analysis. The paper describes the advantages of the Bayesian approach, including mainly the quantification of uncertainty, traceability of the results, incorporation of prior knowledge and update with new information. The paper uses data from a real project to show the shear strength parameter estimation results for both methods and undertakes a 2D limit equilibrium slope stability analysis, highlighting the differences in the results obtained. The authors believe that the proposed method is robust and systematic and aligns with the risk approach to tailings management proposed by the Global Industry Standard on Tailings Management (GISTM). The use of the proposed Bayesian approach provides the required inputs to estimate the probability of failure for an embankment, which is considered a preferred outcome when compared with a factor of safety, as it is usually the practice in the industry. A discussion including further work and the potential to use these techniques in other areas of tailings design and operation is also provided.

INTRODUCTION

When assessing the slope stability of tailings storage facility (TSF) embankments, it is the industry standard to use limit equilibrium (LE) methods for a range of loading conditions, namely long-term drained, short-term undrained and post-seismic (ANCOLD Inc, 2012). The shear strength of tailings and embankment materials are a key input to the stability models, and the correct estimation of strength envelopes is therefore paramount.

Throughout the authors' experience in geotechnical and tailings engineering, it is clear that most consultants define shear strength envelopes based on data in one of two ways:

- Best estimate: this generally involves fitting a curve to the databased on the least square method ('trendlines' in Microsoft Excel).

- Low bound estimate: this generally involves defining a curve that generates lower strength than a certain percentage of the data (ie 85th percentile).

Depending on the type of material and the problem analysed, one of the above is selected. For compacted embankments, most designers use the best estimate approach, while the opposite is true for tailings, where a more conservative approach is the norm. These considerations make perfect sense as the triggering mechanisms for liquefaction are not well understood and the consequences of overestimating residual strength could be catastrophic, while there is more confidence on the behaviour of compacted embankments.

The outcome from both methods are deterministic strength parameters for input into LE software, and a deterministic factor of safety (FoS) is obtained for the slope loading condition analysed. Throughout the paper these methods will be referred as the conventional approach.

While this approach has been used successfully for years, it does not consider the uncertainty of the shear strength estimation. Although generally not required by regulators or TSF owners, ANCOLD Inc (2012) does recommend the use of reliability analyses (section 6.1.9), as it '_provides a means of evaluating the combined effects of uncertainties and a means of distinguishing between conditions_ *where uncertainties are particularly high or low'*. Bayesian statistics can be used to calculate probability distributions and correlation factors of the shear strength parameters that enable reliability calculations as recommended by ANCOLD.

This paper describes the Bayesian approach for inference applied to shear strength parameters and uses real project data to compare the Bayesian and conventional methods, including a discussion on the merits of both. Finally, the results of the inference are used in a stability analysis including reliability results.

It is also worth noting that the paper does not cover the interpretation of laboratory or in situ tests. It is assumed that the interpretation has already been done and the starting point is a database comprised of shear strength results (τ_f) and the corresponding effective normal stress (σ'_n).

BAYESIAN APPROACH

The advantages of the Bayesian approach for inference of parameters are described in detail by Contreras, Brown and Ruest (2018) within the context of intact rock strength and by Contreras, Linero-Molina and Dixon (2022) on the characterisation of a waste rock material. The Bayesian analysis enables the estimation of the more likely parameter values given three elements:

1. A model.

2. The data to feed the model.

3. Any prior information on the parameter values.

The model corresponds to a mathematical function that represents the performance of a system of interest. The data correspond to measurements of the system's actual performance that are compared with the model predictions to define errors. The prior information typically consists of valid ranges of credible values of the uncertain parameters to be inferred. These three elements (model, data, and prior information) are combined in a probabilistic function that contains the set of parameters for inference. This function corresponds to a posterior probability distribution within the Bayesian concept and gives probability values for particular sets of parameters. The objective of the analysis was to define the sets of parameters that produce the largest probability values (minimum errors). Generally, the parameters are defined by evaluating the posterior function using a Markov Chain Monte Carlo (MCMC) (Kruschke, 2015).

The methodology is illustrated in the diagram of Figure 1 adapted from Contreras, Brown and Ruest (2018) for the evaluation of the shear strength of a soil material. In this case, the model corresponds to the Mohr–Coulomb (M-C) strength model represented by Equation 1.

$$\tau_f = c + \sigma'_n \, tan(\phi) \tag{1}$$

Where:

τ_f	=	shear strength at failure
c	=	cohesion
σ'_n	=	effective normal stress on the failure plane
ϕ	=	friction angle

In the context of the present analysis, τ_f is the predicted variable to be compared with the data values, σ'_n is the predictor variable that points to the predicted variable within the model, and c and ϕ are the target uncertain parameters to be inferred with the Bayesian approach to minimise the errors, ie the differences between the model predictions and the data measurements. The data consist of results of consolidated undrained (CU) triaxial strength tests. The data include measurements of τ_f for particular values of σ'_n. The prior information consists of the known ranges of credible values of the uncertain parameters in the M-C equation.

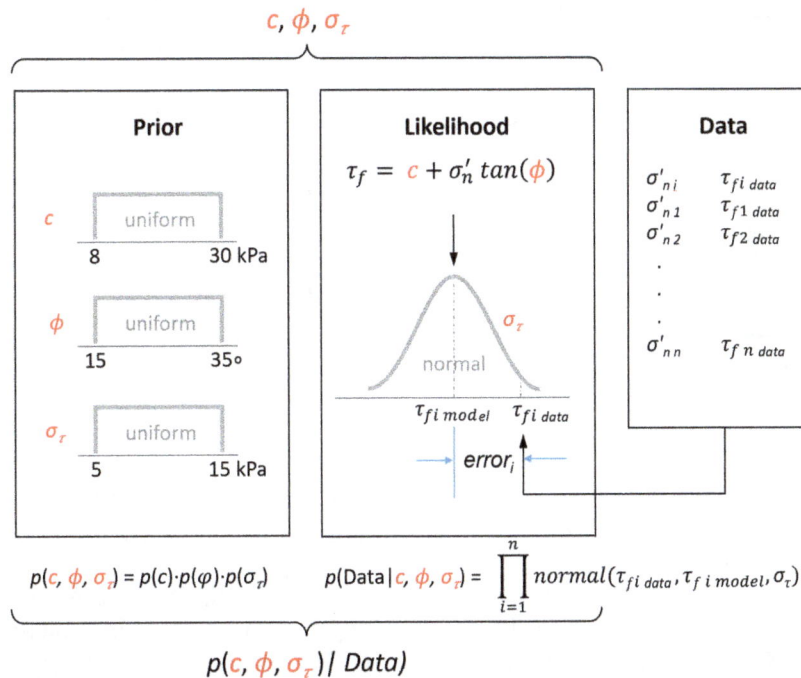

c, ϕ, σ_τ

Prior	Likelihood	Data
	$\tau_f = c + \sigma'_n \, tan(\phi)$	

Prior

c uniform
8   30 kPa

ϕ uniform
15   35°

σ_τ uniform
5   15 kPa

Likelihood

σ_τ

normal

$\tau_{fi\ model}$ $\tau_{fi\ data}$

$error_i$

Data

$\sigma'_{n\,i}$ $\tau_{fi\ data}$
$\sigma'_{n\,1}$ $\tau_{f1\ data}$
$\sigma'_{n\,2}$ $\tau_{f2\ data}$
.
.
$\sigma'_{n\,n}$ $\tau_{f\,n\ data}$

$p(c, \phi, \sigma_\tau) = p(c) \cdot p(\varphi) \cdot p(\sigma_\tau)$

$p(Data \,|\, c, \phi, \sigma_\tau) = \prod_{i=1}^{n} normal(\tau_{fi\ data}, \tau_{f\,i\ model}, \sigma_\tau)$

$p(c, \phi, \sigma_\tau) \,|\, Data)$

FIG 1 – Layout of the method for the Bayesian estimation of the M-C strength parameters of a soil.

The diagram in Figure 1 shows the main elements of the Bayesian approach for the inference of the M-C parameters, c and ϕ, from the triaxial test data. The method combines the prior and the likelihood parts to define the posterior function according to the Bayesian concept. The errors in τ_f are evaluated with a normal distribution to construct the likelihood function. The normal distribution is centred in the model prediction and is used to quantify the probability of each data point. In this way, small errors would have larger probability values and vice versa. The product of the probabilities of the data points gives the probability of the data set for a particular selection of parameters. The objective of the method is to find the set of parameters that result in large probability values for the data set ie small errors. For this search, the method should try normal distributions with different widths and, for this reason, the standard deviation of the normal distribution used to quantify the errors (σ_τ) is an additional parameter that needs to be inferred.

The method considers prior distributions on the three target parameters, c, ϕ, and σ_τ. The specification of the prior distributions was based on the characteristics of the data sets and consists of setting up values sufficiently vague to avoid constraining the result. The vague priors for the case presented in Figure 1 are defined with uniform distributions with ranges from 8 to 30 kPa for c, 15° to 35° for ϕ, and 5 to 15 kPa for σ_τ. The posterior probability distribution is evaluated by sampling the parameters with an MCMC algorithm. The methodology outlined in Figure 1 was implemented in the Python programming language (Python Software Foundation, 2001). The analysis was carried out with the affine-invariant ensemble sampler algorithm implemented in the emcee Python package developed by Foreman-Mackey *et al* (2013).

The evaluation of errors in the Bayesian analysis can be done by considering absolute or relative residuals as indicated in Figure 2. Relative residuals are calculated by normalising the shear stress at the data points relative to the shear stress given by the model. In this way, all the errors are measured on a similar scale regardless of the level of effective normal stress. The advantage of using relative residuals for the evaluation of errors is that it reduces the bias caused by the relatively larger errors occurring at higher levels of stress.

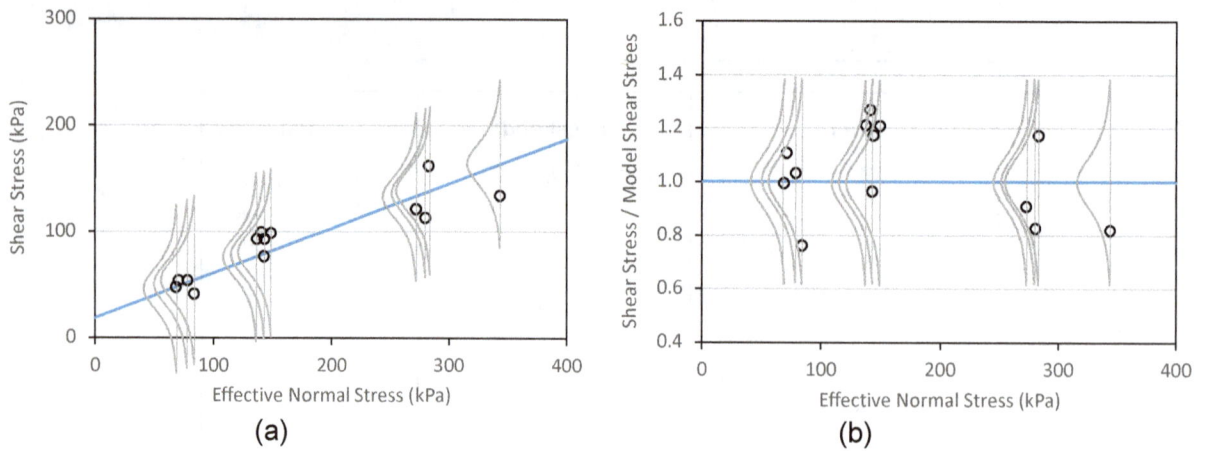

FIG 2 – Measurement of errors with a normal distribution centred on the model prediction: (a) with absolute residuals, (b) with relative residuals.

The main aspects of a typical Bayesian analysis are illustrated with the inference of the parameters c, ϕ, and σ_τ used to define the more likely shear strength of a soil material given a data set of 64 triaxial test results. A simplified representation of the Bayesian model implemented is shown in Figure 3. In this case, the three parameters are fine-tuned within the limits prescribed with the prior information to get the best agreement (minimum errors) between the measurements of τ_f with the triaxial tests and the respective predictions with the M-C strength model, considering relative residuals. The search for the parameter values that produce the maximum probabilities with the posterior function is carried out with the MCMC sampling procedure. During the sampling process, a full probability distribution of τ_f values is also constructed at selected σ'_n locations. These distributions are used to define prediction intervals (PIs) of τ_f with the desired level of confidence. The PIs constructed in this way account for the variability of the mean and standard deviation of the parameters.

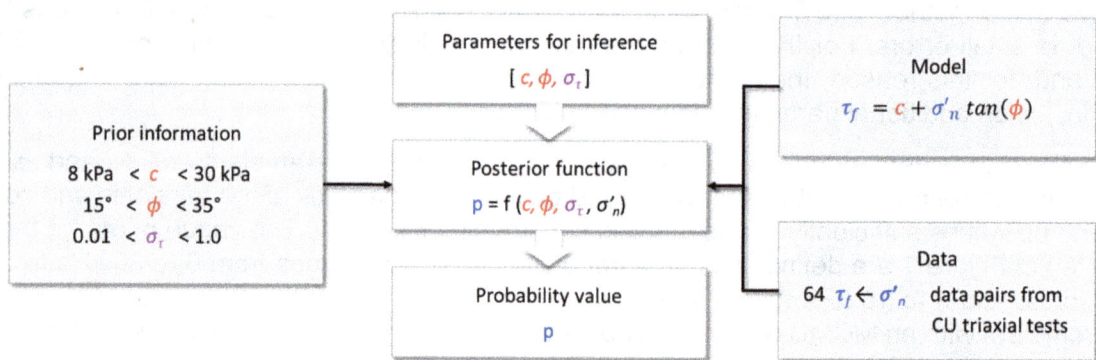

FIG 3 – Conceptual Bayesian model for the inference of the M-C parameters of a soil material using triaxial test results.

The results of the analysis are summarised in the corner plot of Figure 4. The scatter plots in this figure show the correlation between the inferred parameters, and the histograms on the diagonal define the ranges of likely values. These results suggest a strong negative correlation between c and ϕ. The sampled values of the parameters represented in the scatter plots produce a spread of the M-C envelopes that causes the variability of the estimates of τ_f as indicated in the graph of τ_f versus σ'_n shown in Figure 5.

The graph in Figure 5 includes the data points, the more likely fitting envelope, and the band of the 95 per cent highest density interval (HDI) of the τ_f values calculated with the more likely parameter values inferred. This band reflects the uncertainty of the parameters c, ϕ and σ_τ. The graph also includes the band defining the 5 per cent and 95 per cent prediction intervals (PI) of τ_f. The 95 per cent PI is highlighted in red and corresponds to a low bound strength defined by conservative values of c and ϕ that are recommended for the deterministic analysis of stability of critical structures.

Mine Waste and Tailings Conference 2023 | Brisbane, Australia | 13–14 July 2023

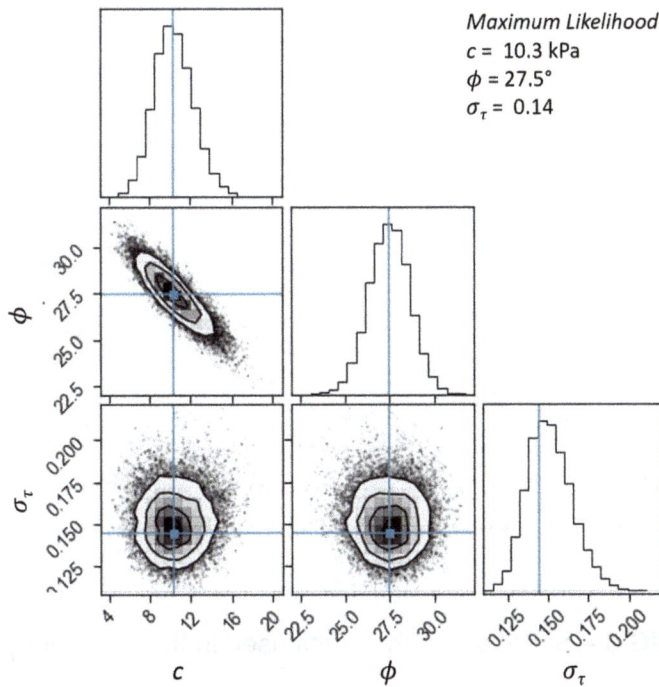

FIG 4 – Corner plot with scatter plots and histograms of the M-C strength parameters inferred with Bayesian analysis using relative residuals based on the 64 triaxial test results.

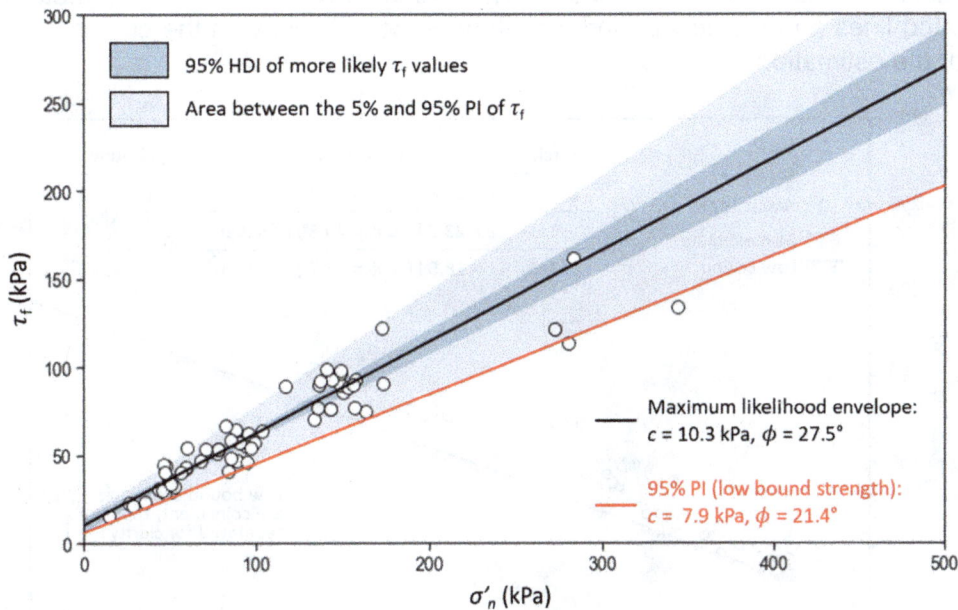

FIG 5 – Data, fitted envelope with maximum likelihood, band of 95 per cent HDI of the more likely τ_f values, and band between the 5 per cent and 95 per cent PI of τ_f.

CASE STUDY

The data in Figure 6 shows a data set of CU tests undertaken in three successive geotechnical investigations from a real project in Australia (batch 1, batch 2 and batch 3). The points represent the tangents to Mohr circles of stress at failure for a sandy clay used to build a compacted embankment, and were compacted to 98 per cent maximum dry density (MDD), saturated and tested as per AS1289.6.4.2 (Standards Australia, 2016).

The objective of this case study was to use the data to infer the shear strength parameters of the compacted embankment for input into a LE slope stability model using both the conventional and Bayesian methods. 'Best estimate' and 'low bound' strength envelopes were developed for both methods and their merits were discussed in detail.

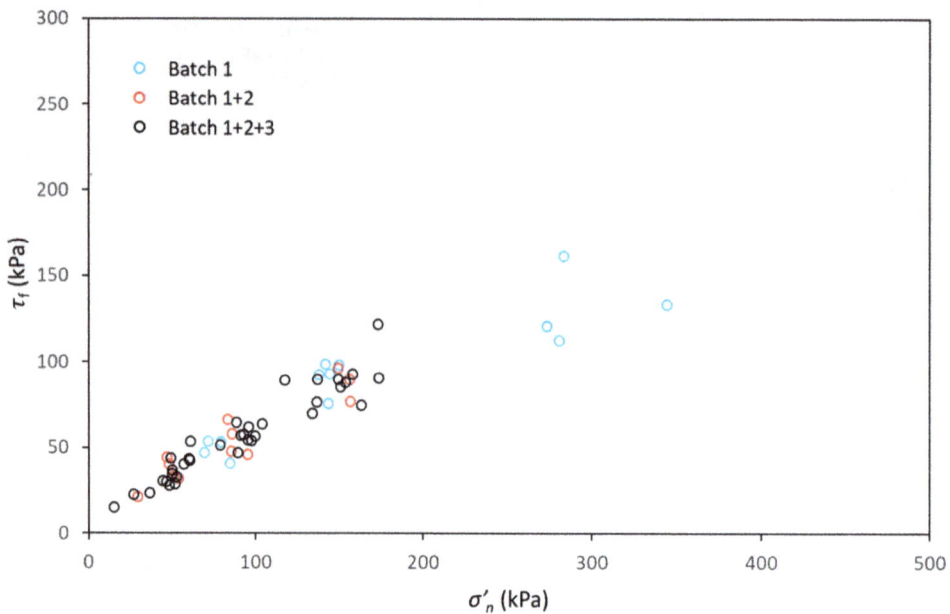

FIG 6 – Shear strength results used in the case study.

Conventional inference

Figure 7 shows the CU data with the M-C envelopes estimated with conventional linear regression methods for batch 1, batch 1+2, and batch 1+2+3 respectively. The 'best estimate' envelope is shown in dotted lines and the 'low bound' envelope in solid lines, with the colours referring to the data used for the estimation.

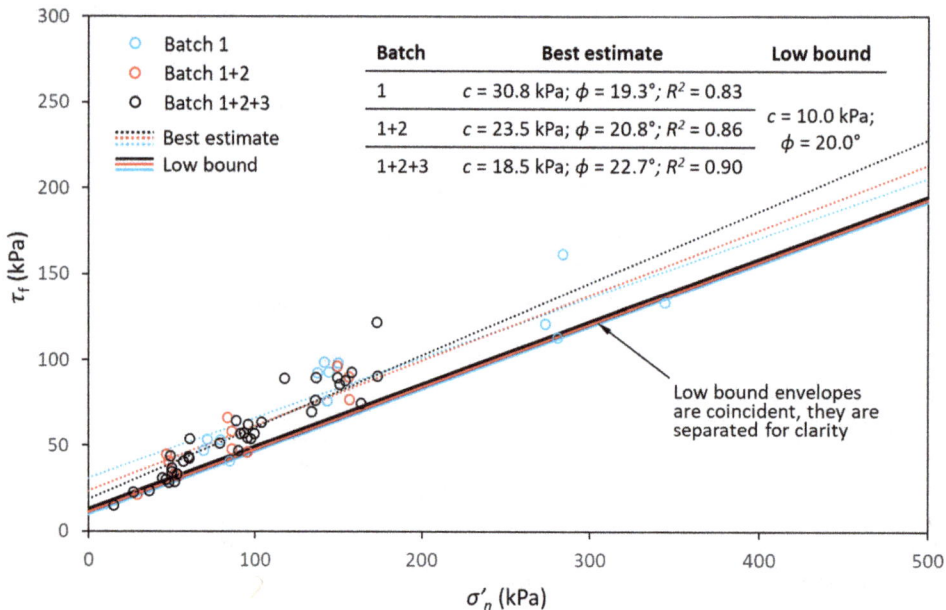

FIG 7 – Shear strength estimations using a conventional linear regression analysis.

The 'best estimate' lines are the envelopes that best fit the data points, defined with the least squares regression method using absolute residuals. The results are the M-C parameters c, ϕ and the coefficient of determination R^2 that provides a measure of the goodness of fit. The added batches of data increase the quality of the fit, however, there is no measure of the uncertainty of the parameters.

The low-bound envelopes were selected using judgement to get a conservative estimate of the strength. However, this procedure is subjective and does not account for the information contained within the structure of the data set. Therefore, the interpreted envelope has no measure of confidence attached to it. In this particular case, the low-bound envelope was not affected by the addition of new data.

Bayesian inference

The inference of the strength parameters using the Bayesian approach was carried out as described in the previous section. The evaluation of the posterior function included the selection of 50 000 samples of the parameters with the MCMC procedure.

The results of the Bayesian analysis are shown in Figures 8, 9 and 10 for batches 1, 1+2 and 1+2+3, respectively. Each figure includes the data and envelopes with the measure of uncertainty of the strength on the left and the corner plot with the results of the MCMC sampling of c and ϕ on the right. The explanation of the information contained in these figures was presented in the previous section and Figures 4 and 5. The width of the more likely (blue) and PI (lavender) strength bands provide a visual indication of the uncertainty of the estimation. Therefore, the results show that the width of the bands of uncertainty reduces as more data is available. This is also reflected in the reduction of the width of the distributions of the M-C parameters, although this is not visually evident in the corner plots due to the re-scaling of the histograms. The corner plots also include the coefficient of correlation between c and ϕ, which is useful to reproduce the more likely strength envelopes in a probabilistic analysis of stability.

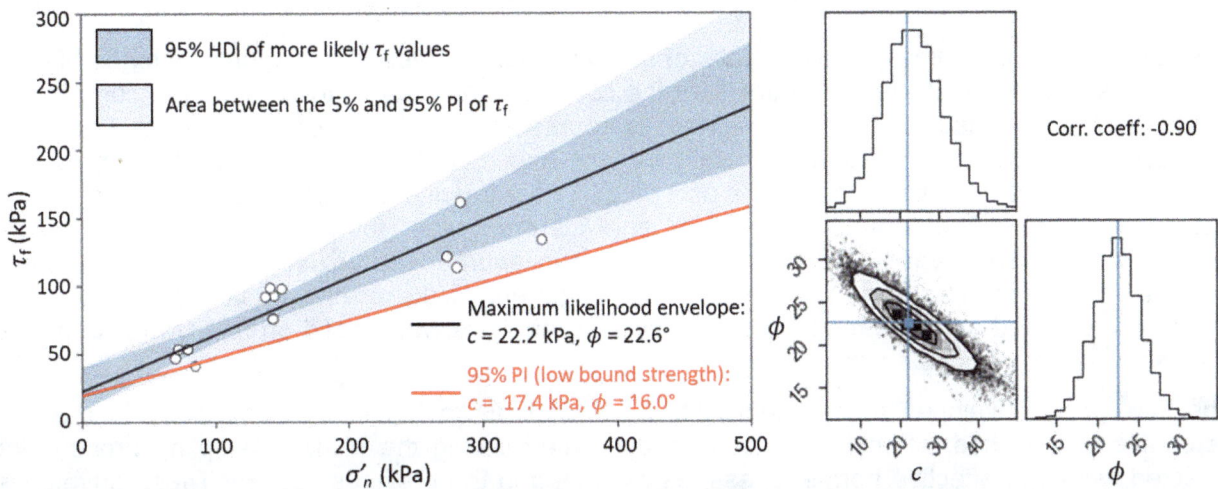

FIG 8 – Data and envelopes with the measure of uncertainty (left) and corner plot (right) for batch 1 of triaxial data.

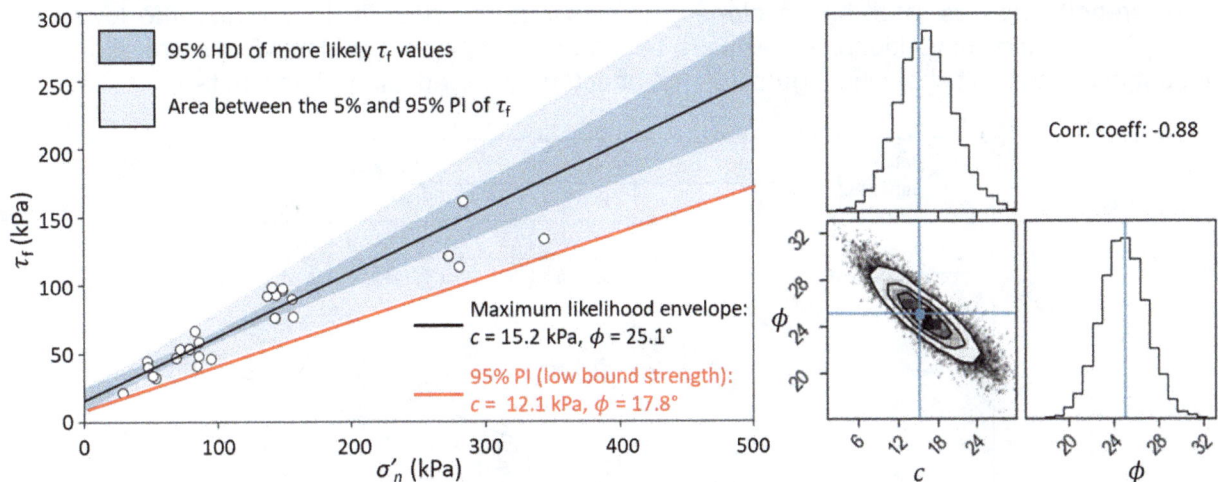

FIG 9 – Data and envelopes with the measure of uncertainty (left) and corner plot (right) for batches 1+2 of triaxial data.

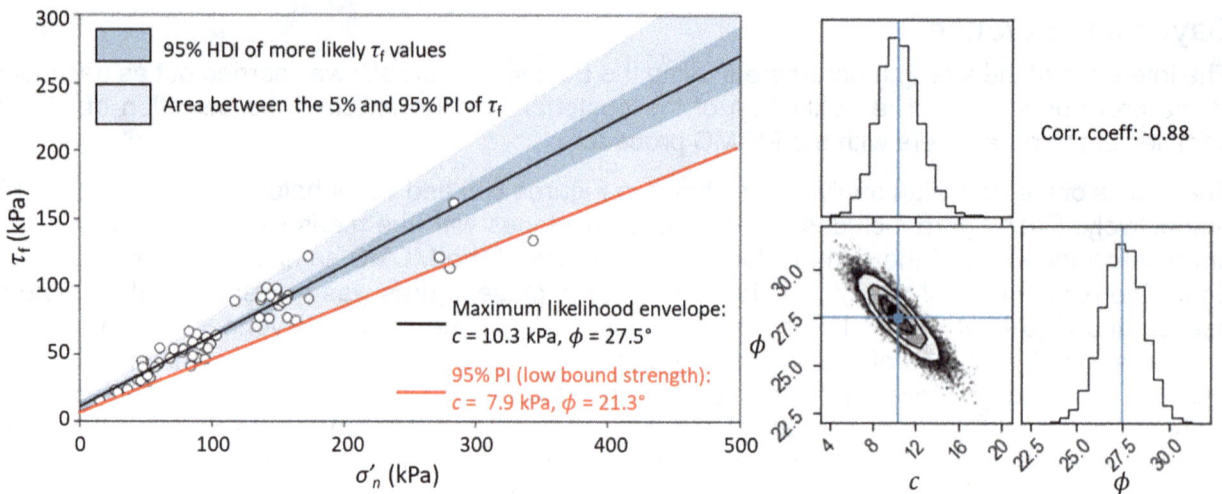

FIG 10 – Data and envelopes with the measure of uncertainty (left) and corner plot (right) for batches 1+2+3 of triaxial data.

Figure 11 shows a summary of the results of the Bayesian analyses for the three batches of data using the same layout utilised in Figure 7 for the conventional analysis to facilitate the comparison of the two approaches. The best-fit envelopes represented with the dotted lines in Figure 11 correspond to the set of inferred parameters with the maximum likelihood (ie maximum probability of occurrence given the prior knowledge and data). These results are comparable with the best-fit envelopes from the least squares regression analysis in Figure 7 for the conventional approach. Generally, the best-fit envelopes defined with the two approaches are very close. In this case, however, the differences are mainly due to the treatment of the errors. While the conventional method considered absolute residuals, the Bayesian analysis was based on relative residuals as discussed in the previous section.

The low-bound envelopes represented with the solid lines in Figure 11 correspond to the fifth percentile of the shear strength distributions generated during the MCMC sampling process for selected values of effective normal stress, as explained in the previous section. These envelopes have a 95 per cent confidence and are referred to as the 95 per cent PI of the shear strength. They depend on the data sets and account for the variability of the inferred parameters. The low-bound envelopes in Figure 11 show that as the size of the data set increases, the uncertainty of the inferred shear strength reduces, and the envelope gets closer to the best fit. In contrast, the low-bound envelopes from the conventional analysis are based on a simple and subjective interpretation of the data set and the result shown in Figure 7 is not affected by the addition of new data.

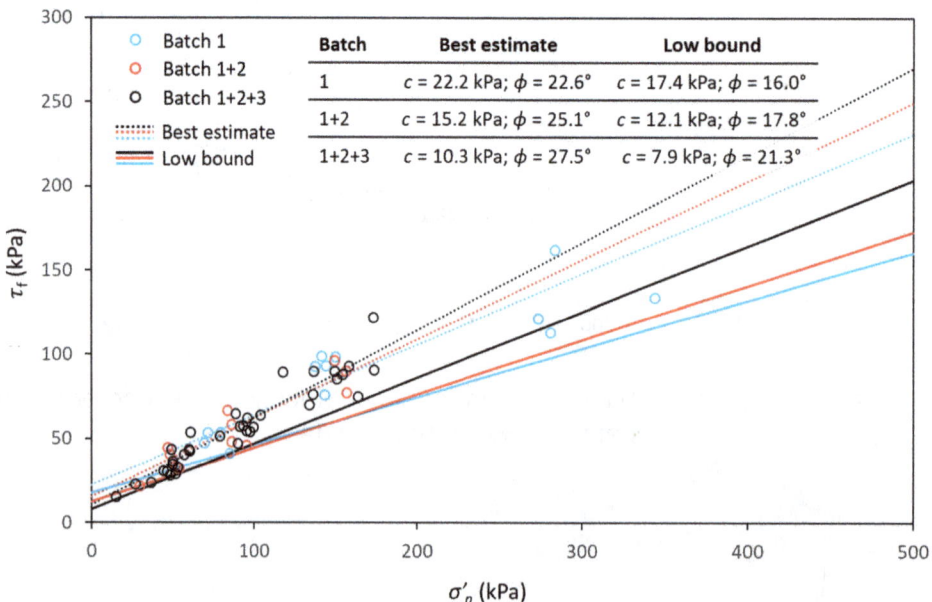

FIG 11 – Shear strength estimations using Bayesian methods.

Rate of 'conservatism'

When analysing the residual or post-peak strength of tailings, it is common to use an overall low-bound strength, which consists in assuming a shear strength envelope that is equal to or lower than all the data points, as it is the case of the solid lines in Figure 7. The general consensus is that the design is safe because it is based on a strength lower than all the tested samples, which appears to be a reasonable assumption, although it does not account for the number of test results. There is always the possibility of getting a lower test result in the future, particularly if the number of data points is insufficient. A conservative shear strength with a specified level of confidence can be defined with the construction of prediction intervals using conventional statistics or during the MCMC sampling of the Bayesian approach as described in the previous section.

STABILITY ANALYSIS

The shear strength parameters inferred using both the conventional and Bayesian methods were used in a slope stability analysis of a typical TSF embankment cross-section as shown in Figure 12. The model comprised an 8 m high starter embankment and two 4 m high upstream raises. The analysis considered non-circular surfaces with Cuckoo search (500 iterations) and used the GLE/Morgenstern Price method. The foundation was modelled as a 2 m layer of residual soil (γ = 21 kN/m^3, c = 10 kPa; ϕ = 35°) overlain by bedrock (infinite strength). The tailings were modelled with γ = 15 kN/m^3, and peak strength using SHANSEP with A = 0; S = 0.3; OCR = 1. The embankment materials were modelled using γ = 20 kN/m^3 and the shear strength parameters as per Table 1.

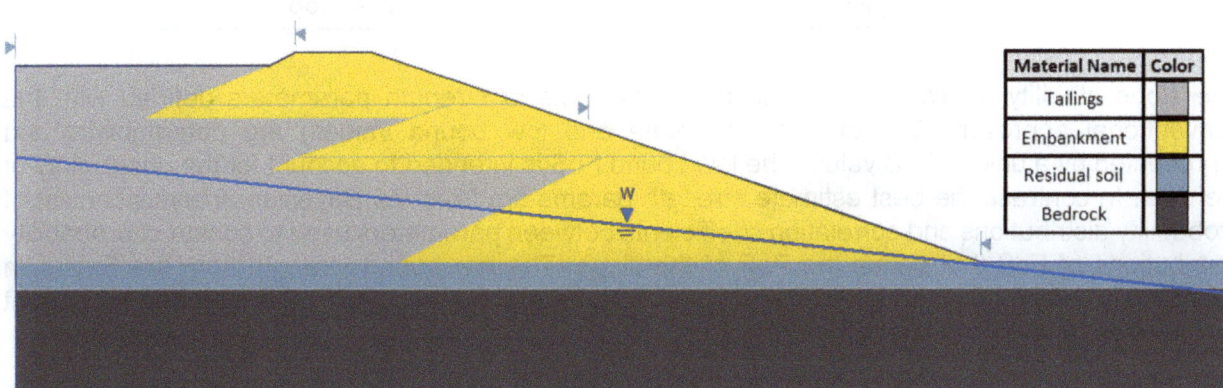

Material Name	Color
Tailings	
Embankment	
Residual soil	
Bedrock	

FIG 12 – Stability analysis cross-section.

TABLE 1

Embankment strength parameters used in stability analysis.

Case	Data batch	Conventional approach	Bayesian approach
Best estimate strength	1	c = 30.8 kPa ϕ = 19.3°	c = normal (μ = 23.7 kPa; σ = 7.81 kPa) ϕ = normal (μ = 22.4°; σ = 2.82°) $c - \phi$ correlation: -0.90
	1+2	c = 23.5 kPa ϕ = 20.8°	c = normal (μ = 16.2 kPa; σ = 4.21 kPa) ϕ = normal (μ = 24.9°; σ = 2.06°) $c - \phi$ correlation: -0.88
	1+2+3	c = 18.5 kPa ϕ = 22.7°	c = normal (μ = 10.4 kPa; σ = 1.95 kPa) ϕ = normal (μ = 27.5°; σ = 1.15°) $c - \phi$ correlation: -0.88
Low bound strength	1	c = 10.0 kPa ϕ = 20.0°	c = 17.4 kPa; ϕ = 16.0°
	1+2		c = 12.1 kPa; ϕ = 17.8°
	1+2+3		c = 7.9 kPa; ϕ = 21.4°

A total of ten slope stability scenarios were analysed, one for each set of shear strength parameters of the compacted embankment listed in Table 1. The analyses are described as conventional/ Bayesian and best estimate/low bound referring to the estimation approach and type of shear strength parameters used, respectively. The results are presented in Table 2 and consist of factors of safety (FoS) and probabilities of failure (PoF) for the deterministic and probabilistic analysis, respectively.

TABLE 2

Stability analysis results.

Case	Data batch	Conventional approach	Bayesian approach
Best estimate strength	1	FoS = 2.04	Mean FoS = 2.04; SD = 0.088 FoS range = 1.68–2.42; PoF = 0%
	1+2	FoS = 1.96	Mean FoS = 1.98; SD = 0.060 FoS range = 1.77–2.23; PoF = 0%
	1+2+3	FoS = 1.92	Mean FoS = 1.95; SD = 0.031 FoS range = 1.82–2.09; PoF = 0%
Low bound strength	1		FoS = 1.57
	1+2	FoS = 1.52	FoS = 1.50
	1+2+3		FoS = 1.56

The slope stability analyses based on the single sets of strength parameters defined with the conventional approach (for both best estimate and low bound values) are deterministic and represented by a unique FoS value. The low bound FoS is intended to account for the uncertainty of the data. In contrast, the best estimate strength parameters from the Bayesian analysis consist of probability distributions and correlation coefficients between parameters used to construct probability distributions of FoS that define the PoF of the slope. The low bound strength from the Bayesian approach is used for conservative analysis of stability where the resulting FoS is intended to account for the uncertainty of the strength parameters.

The results of the six stability analyses based on the strength parameters determined with the Bayesian approach are summarised in Figure 13. The histograms of FoS for the best estimate strength parameters derived from the data set batches 1, 1+2 and 1+2+3 indicate very stable slope conditions with PoF of 0 per cent in all cases. Nevertheless, the results show that the increase of the amount of data causes a gradual decrease of the mean FoS and the reduction of the spread of the FoS distributions, which is a manifestation of the reduction of the uncertainty of the strength parameters. Although the FoS values from the low bound strength parameters are noticeably smaller than those from the analysis with best estimate inputs, they exceed the generally accepted criterion of stability of 1.5. As expected, this result is also consistent with the PoF of 0 per cent from the best estimate analysis.

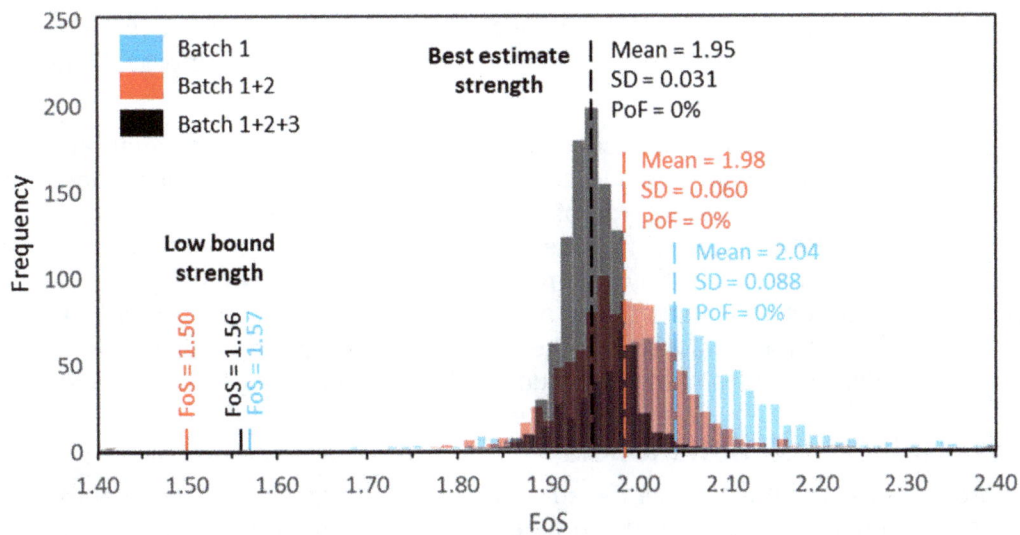

FIG 13 – Results of the stability analyses with the shear strength from the Bayesian approach.

The summary of the results of the four stability analyses based on the strength parameters defined with the conventional approach is shown in Figure 14. The results consist of unique FoS values, one for the low bound strength case that is unaffected by the size of the data set and three values for the best estimate strength case with the three data sets considered. The histograms in the background in Figure 14 correspond to the Bayesian result and are included only for reference. The FoS results of the best estimate strength case are consistent with the results of the Bayesian case, but they do not carry any information on the uncertainty of the estimation as it is the case with the FoS distributions from the probabilistic analysis.

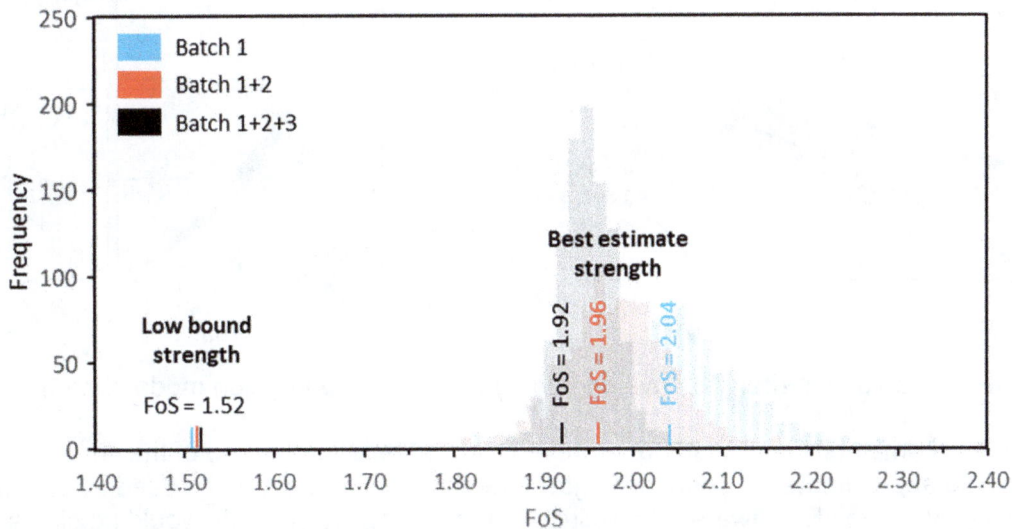

FIG 14 – Results of the stability analyses with the shear strength from the conventional approach.

DISCUSSION

The Bayesian approach presented in this paper is a probabilistic method of analysis of triaxial data that provides a quantification of the uncertainty of the strength parameters, enabling the evaluation of the reliability of TSF embankments. The approach includes the definition of maximum likelihood strength envelopes with their associated band of uncertainty and low bound envelopes with a specified level of confidence used for conservative analysis of stability. In contrast, the conventional approach for the estimation of the strength parameters for TSF analysis is partly based on the classical frequentist approach of statistical analysis for the calculation of the envelopes that best fit the data, with no measure of uncertainty and with a strong subjective component for the selection of conservative strength properties to account for the uncertainties in the design. The conventional approach does not consider other available frequentist tools of analysis such as the construction of confidence and prediction intervals. There are many benefits of the Bayesian approach for the inference of parameters when compared with the frequentist approach, however, a full discussion on this topic is outside the scope of this paper and can be consulted in Contreras, Brown and Ruest (2018).

The commissioning of quantitative risk assessments (QRAs) by TSF owners is currently a common practice and the use of a probabilistic approach for the analysis of data as described in this paper can provide the required inputs for this type of analysis. The QRAs generally consider estimates of PoF derived from calculated values of FoS using generic empirical correlations between the two parameters such as that proposed by Silva, Lambe and Marr (2008). However, the PoF values based on the strength parameters determined with a Bayesian analysis as described in the paper are specific to the TSF analysed. Moreover, the methodology can be applied for the strength characterisation of other materials of the TSF such as foundation, tailings, rock fill and so forth.

Use of other failure criterion models

The cases presented in the paper were based on a shear strength model represented by the M-C failure criterion, although the approach could be easily adapted to consider other strength models. To illustrate this point, a non-linear potential strength model represented by Equation 2 was used to run the analysis.

$$\tau_f = A \cdot {\sigma'_n}^B \tag{2}$$

In this case, A and B are model parameters with no physical meaning. Figure 15 show the results of the analysis with the complete data set for batches 1+2+3.

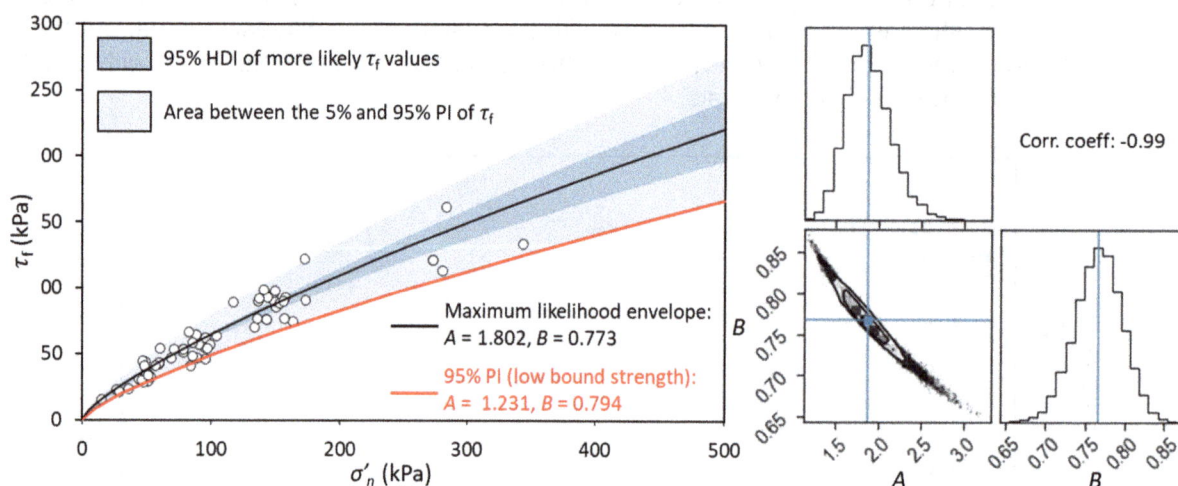

FIG 15 – Bayesian shear strength envelope using a potential regression model (batches 1+2+3).

Although the parameters A and B do not represent any material properties of the soil, they still allow the designer to approximate the shear strength values to a curve that can be used as an input in any commercially available LE software. The results from the stability analysis would be similar to those presented in Table 2, including a histogram of FoS and the estimation of the PoF. The Bayesian model used for the analyses presented in this paper could be extended further with minor adjustments to consider more elaborated models with many more parameters for inference.

Factors of safety and probability of failure

In Australia, it is common to use the FoS values from Table 8 of ANCOLD Inc (2012) guideline as the minimum requirements for a TSF to be considered safe. It is noted, however, that the text reads the following:

> There are no 'rules' for acceptable factors of safety, as they need to account for the consequences of failure and the uncertainty in material properties and subsurface conditions. Table 8 shows the ANCOLD recommended factors of safety for tailings dams under various loading conditions.

Note 2 of ANCOLD's Table 8 indicates that the FoS for the post seismic loading condition depends on the confidence of the residual shear strength value used.

The FoS is a stability indicator calculated for a fixed set of input conditions and does not account for the uncertainty in the material properties in itself. The uncertainty is accounted for indirectly by the level of conservatism used in the selection of the strength inputs and the criterion used to select an acceptable FoS for design. This is particularly important when the FoS values obtained are slightly below the ANCOLD recommendations. The implications of the lack of information on the uncertainty of the parameters with stability analysis based on FoS values is illustrated in the sketch of Figure 16 from Steffen *et al* (2008). This figure shows the definition of PoF and its relationship with FoS according the magnitude of the uncertainty and indicates that a larger FoS not necessarily represent a safer slope, as the magnitude of the implicit uncertainties is not captured by the FoS value.

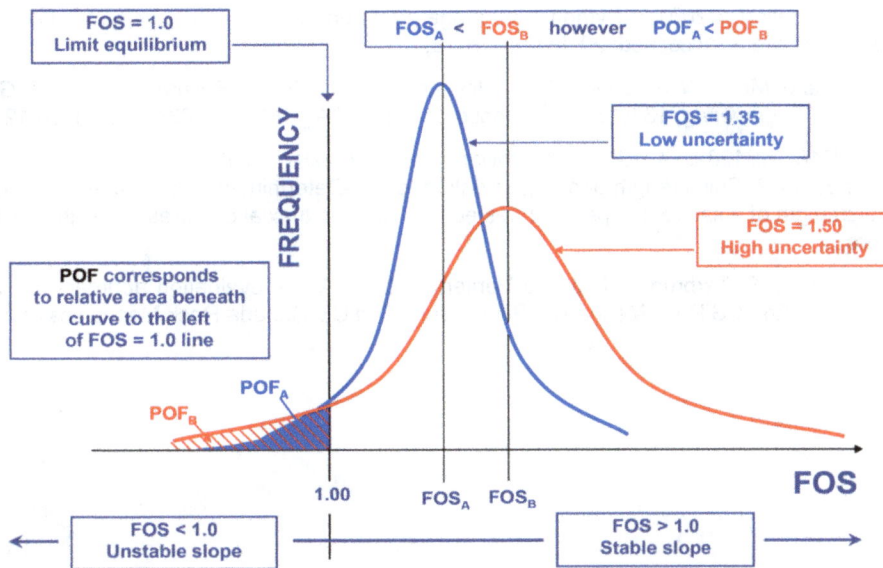

FIG 16 – Definition of PoF and relationship with FoS according to uncertainty magnitude (Steffen *et al*, 2008).

There are no recommended values of PoF for design of TSFs, as they need to be determined depending on the severity of the consequences of the failure, as dictated by ANCOLD Inc (2022) or a similar guideline. Nevertheless, it is acknowledged that owners generally have their own requirements for FoS and that some of them go above and beyond the ANCOLD recommendations.

CONCLUSIONS

This paper compares the conventional industry practice of selecting a single value for shear strength parameters when analysing the slope stability of tailings storage facility (TSF) embankments to the use of Bayesian statistics for obtaining a probabilistic evaluation of the strength envelope based on available data and prior knowledge of the parameters. The paper shows the advantages of the Bayesian approach, including quantification of uncertainty, traceability of results, incorporation of prior knowledge, and update with new information. The Bayesian method presented in this paper is an intuitive and powerful way to undertake parameter inference and can be readily implemented with minimal computational power and open-source tools. The study uses a data set from a real project in Australia and performs shear strength parameter inference with both conventional and Bayesian methods, showing how the results change as additional data is included.

The authors believe that the proposed Bayesian approach is robust and systematic and aligns with the risk approach to tailings management proposed by ANCOLD and the Global Industry Standard on Tailings Management (GISTM). They conclude that using statistical methods that quantify uncertainty for the estimation of shear strength parameters is the desirable approach, as it can be used to estimate the variability of the factor of safety and the probability of failure. The use of the proposed Bayesian approach provides the required inputs to estimate the probability of failure for an embankment, which is considered a preferred outcome when compared with a factor of safety, which is usually the practice in the industry.

REFERENCES

Australian National Committee on Large Dams Incorporated (ANCOLD Inc), 2012. Guidelines on Tailings Dams – Planning, Design, Construction, Operation and Closure.

Australian National Committee on Large Dams Incorporated (ANCOLD Inc), 2022. Guidelines on Risk Assessment.

Contreras, L F, Brown, E T and Ruest, M, 2018. Bayesian data analysis to quantify the uncertainty of intact rock strength, *Journal of Rock Mechanics and Geotechnical Engineering*, 10(1):11–31.

Contreras, L F, Linero-Molina, S and Dixon, J, 2022. Bayesian approach to improve the confidence of the estimation of the shear strength of coarse mine waste using Barton's empirical criterion, *Australian Geomechanics*, 57(2):163–174.

Foreman-Mackey, D, Hogg, D W, Lang, D and Goodman, J, 2013. emcee: The MCMC Hammer, *Publications of the Astronomical Society of Pacific*, 125(925):306–312.

Kruschke, J K, 2015. Doing Bayesian Data Analysis: A Tutorial with R, JAGS and STAN (Academic Press/Elsevier).

Python Software Foundation, 2001. Python language reference, version 3.4 [online]. Available from <https://www.python.org/> [Accessed: 20 January 2015].

Silva, F, Lambe, T W and Marr, W A, 2008. Probability and Risk of Slope Failure, *Journal of Geotechnical and Geoenvironmental Engineering*, 134(12). https://doi.org/10.1061/(ASCE)1090–0241(2008)134:12(1691)

Standards Australia, 2016. AS1289.6.4.2:2016. Australian Standard 1289. Methods of testing soils for engineering purposes, Method 6.4.2: Soil strength and consolidation tests—Determination of compressive strength of a soil—Compressive strength of a saturated specimen tested in undrained triaxial compression with measurement of pore water pressure.

Steffen, O K H, Contreras, L F, Terbrugge, P J and Venter, J, 2008. A risk evaluation approach for pit slope design, Proceedings of the 42nd US Rock Mechanics Symposium, 2nd US-Canada Rock Mechanics Symposium, ARMA.

On the selection of appropriate tailings geotechnical characteristic values for design – common pitfalls and recommended approach

B Wentzinger[1] and A Arenas[2]

1. Principal geotechnical engineer, Red Earth Engineering – a Geosyntec Company, Brisbane Qld 4000. Email: benoit@redearthengineering.com.au
2. Principal geotechnical engineer, Red Earth Engineering – a Geosyntec Company, Perth WA 6000. Email: alfredo@redearthengineering.com.au

ABSTRACT

Ten geotechnical engineers will generally provide ten different estimates of tailings characteristic values, based on the same testing data set. Depending on which method is used to derive characteristic values the estimates may be vastly different. This paper explores selected current guidelines and recommendations from relevant standards, codes of practice and tailings literature and highlights the lack of consistency in the matter. Illustrative examples are provided to demonstrate how data distribution, which is rarely a textbook normal distribution, and interpretation methods may lead to overly conservative, or unsafe characteristic values for geotechnical design and analysis of tailings dams. Recommendations are provided, based on the authors' experience, for a consistent and robust approach in data clean up, distribution review and development of characteristic values. The proposed workflow considers the material's variability, its impact on the design, and the type of analysis carried out such as limit equilibrium or finite element/finite difference analysis.

INTRODUCTION

In deterministic design, a minimum factor of safety of 1.5 against slope instability is often used to design earthen embankments in water dams and tailings storage facilities (TSFs) under normal operating conditions. In limit equilibrium analyses, and without using a partial factor of safety approach, this provides for a 33 per cent allowance for uncertainty across loading conditions and resisting forces. However, it is also known that the undrained shear strength of soils may exhibit very high variability: Duncan (2000) cites a coefficient of variation between 13 per cent and 40 per cent for the undrained shear strength. The coefficient of variation is defined as the standard deviation of the parameter divided by its mean or expected value. It is therefore clear that using the mean undrained strength of a highly variable soil in slope stability analyses would not provide an acceptable level of safety. This is why geotechnical engineers have always, with more or less rigour, used conservative 'design values' that account for the material's measured or expected variability.

Uncertainty about shear strength is usually the largest uncertainty involved in slope stability analyses (Duncan, Wright and Brandon, 2014). This is all the more relevant in the stability analysis of TSFs that include saturated, contractive, and potentially brittle tailings. The following discussion therefore focuses mostly on the methods available to derive appropriate design values for the undrained strength of tailings, while avoiding some common errors made in their application.

METHODS TO ASSESS GEOTECHNICAL CHARACTERISTIC VALUES

Many methods are available in the literature to assess field and laboratory soil data for a given soil and derive parameters that may be used for geotechnical design or analysis. For simplification this paper excludes the limit state approach, and therefore characteristic value and design value will herein be used interchangeably. Practitioners should however be aware that the selection of a design value also depends on the type of analysis being carried out. For instance, reliability-based design, factored loads and resistance design, and working stress design may each require different kinds of soil parameters. This matter is beyond the scope of this paper.

The notations in this paper follow these conventions:

- x_m is the mean value of the material parameter considered

- x_k is the characteristic value (same as the design value here)

- σ is the standard deviation

- V_x is the coefficient of variation of x, which is equal to σ divided by x_m
- S_u ratio is the undrained strength ratio, equal to S_u divided by the initial effective vertical stress.

Tailings and water dams standard and guidelines

A review of relevant tailings and water dams guidelines and standards was completed to find guidance related to the development of geotechnical characteristic values. While many publications (not all referenced here) discuss in detail deterministic (factor of safety) and probabilistic (risk-based) approaches to design, unfortunately no clear directions were identified regarding the selection of parameters required to inform these assessments, although the need to consider material variability was often mentioned.

The Global Industry Standard on Tailings Management (ICMM, UNEP and PRI, 2020) requires that a *'robust design'* be developed for a TSF (Requirement 5.2). Robust design is further defined in the standard as being associated with various aspects, including the degree of brittleness and variability of the materials. The degree of robustness is related to the facility maintaining its overall integrity despite less than ideal performance of one or more of its components.

The Guidelines on Tailings Dams (ANCOLD, 2019) discuss in detail the appropriate selection of testing methods and material models to assess and represent the strength of tailings for stability analysis. However, the guidelines do not provide guidance regarding the selection of geotechnical characteristic values. The effects of undrained strength uncertainty are indirectly mentioned in a note related to the selection of a minimum factor of safety for post-seismic loading conditions, which is recommended *'to be related to the confidence in selection of residual strength'*.

The CDA's technical bulletin on geotechnical considerations for dam safety (CDA, 2007) refers to the Canadian Geotechnical Society's Foundation Engineering Manual (CGS, 2006), which states that *'the selection of nominal or characteristic strength for design varies with local state-of-practice and with the training, intuition, background, and experience of the individual geotechnical engineer'*.

The USBR's design standard 13 on embankment dams (USBR, 2011) notes that determination of the shear strength parameters is the most important phase of a stability analysis, yet the most difficult, especially for undrained strengths, and that judgment and experience play an extremely important role in the evaluation of test results. The USBR's best practice guide for embankment slope instability (USBR, 2019) warns that the selection of shear strength for evaluation of levees and dams has to address uncertainty in strength properties, and notes their possible large coefficients of variation due to small-scale variation in soil properties, measurement errors, model bias, improper interpretation, etc.

Intuition, judgment, and experience alone do not form a reliable basis for consistent design practices and outcomes. The methods presented in the following sections seek to provide a more consistent approach to the development of geotechnical characteristic values for design.

Methods based on sample population

Geotechnical engineers often have to heavily rely on their engineering judgement and experience to make a reasonably conservative estimate of the characteristic value: the key drivers for this being the limited number of tests available in relation to the volume of soil considered (natural foundation, embankment, and stored tailings) and the spatial variability of the soil properties. The selected value is typically 'eyeballed' of a test results chart, implicitly accounting for:

- Uncertainty related to the test method.
- Number of data points in relation to the soil volume.
- Screening of outliers.
- Scatter and distribution of the data.
- Design application (stability analysis, settlement estimation, foundation design etc).

It is easy to understand why the method above is likely to yield different outcomes based on the engineer's skill, experience and personal biases.

In their landmark book *Geotechnical Engineering of Dams*, Fell *et al* (2018) provide some guidance in the selection of design parameters for design of dam embankments and landslide stabilising works. They recommend a *bias towards the conservative by selecting a line with, say, 75 per cent of the test points above* [the selected design line] *and 25 per cent below*. This is equivalent to the 25 per cent fractile.

In the context of assessing the characteristic strength of soils for liquefaction problems, Jefferies and Been (2015) conclude that a 20 per cent fractile value offers a high level of confidence in the selected value.

The Eurocode 7 uses different definitions of the characteristic values depending on the soil volume involved: it can be either a cautious mean with a 95 per cent confidence level, or a local low value estimated at the 5 per cent fractile. Frank *et al* (2004) however note that if brittle failure or strain softening of an embankment is possible the 5 per cent fractile should be used.

Statistical methods

In order to provide a more consistent and predictable approach to parameter selection, statistical methods have been applied to geotechnical data. The standard deviation is a simple mathematical tool to describe the scatter of data around the mean. Most statistical methods applied to this problem are of a similar form:

$$x_k = x_m - f\sigma \tag{1}$$

Where f is a statistical coefficient related to the data distribution, the size of the data set and level of confidence. The other terms are as defined previously.

Schneider (1997) proposed 0.5 as an approximation for f, providing characteristic values within reasonable accuracy and valid for several distributions typical for soils.

In Eurocode 7 a 95 per cent reliable mean implies f = 1.645 (for a normal distribution). This corresponds to the 5 per cent fractile mentioned previously.

This statistical approach may be refined if the coefficient of variation of the material is known, or estimated using one of the many available references, for example as compiled by Duncan (2000). Equation 1 is then refactored as:

$$x_k = x_m(1 - fV_x) \tag{2}$$

If reliable data is not available to calculate the standard deviation, Duncan (2000) also proposes using the '*three-sigma rule*', which relies on the fact that 99.73 per cent of all values of a normally distributed parameter fall within three standard deviations of the mean. This allows an estimation of the standard deviation as:

$$\sigma = \frac{HCV - LCV}{6} \tag{3}$$

Where HCV is the highest conceivable value of the parameter and LCV is the lowest conceivable value of the parameter. The three-sigma rule should however be used with great care, as Duncan (2000) points out that engineers have been shown to poorly estimate appropriate values of HCV and LCV.

EXAMPLE APPLICATION ON TAILINGS UNDRAINED STRENGTH

The next sections detail common errors that the authors have observed in the assessment of the characteristic value for tailings geotechnical parameters.

The first example data set is introduced in Figure 1, which shows the distribution of undrained strength (S_u) ratio of a tailings material, interpreted from high quality cone penetration testing data. The data was captured across five test locations and represents a five metre thick layer of saturated tailings that were deposited sub-aqueously.

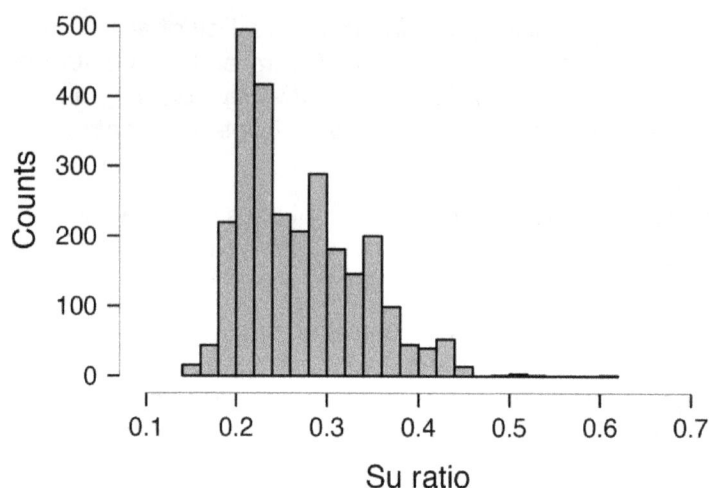

FIG 1 – Distribution of tailings undrained strength ratio – example 1.

Basic statistics for the data set are presented in Table 1. Selected percentiles aligned with the references above are included. It can be noted that the coefficient of variation of 0.237 is relatively low compared to the typical range of 0.3 to 0.5 suggested by Schneider (1997) for soil cohesion.

TABLE 1

Tailings undrained strength ratio – example 1 – basic statistics.

Statistic	Value
No. of samples	2703
Mode	0.220
Median	0.260
Mean	0.274
Std. Deviation	0.065
Coefficient of variation	0.237
Minimum	0.150
Maximum	0.610
5th percentile	0.200
20th percentile	0.220
25th percentile	0.220*

* For this specific data set, calculation of the 20th and 25th percentiles returns the same value.

COMMON ERRORS

The normal distribution issue

Design values are references to a normal distribution as this is what is applied to steel and concrete design, and many codes apply this normality concept also to soil and rock (Look, 2014). This is the case in the Eurode 7 approach discussed previously. Look (2015) used test results for California Bearing Ratio (CBR), rock point load index ($Is_{(50)}$) and pile compression capacity to show that the normal distribution cannot be applied to certain real-life data: in the case studies reviewed, the interpreted characteristic value based on a 5 per cent fractile or $x_m - 1.645\sigma$ would provide unrealistically low or negative values. This is explained by the large coefficients of variation and non-normal distributions of the data reviewed by Look.

The data presented in Figure 1 is visibly positively skewed and may be better fitted to another type of probability distribution function. In the authors' experience, the undrained strength of tailings often follows a log-normal distribution. This may be checked by calculating the logarithm of the data and assessing if the resulting data is normally distributed. This was applied to the data in Figure 1 and resulted in calculated $\log(S_u \text{ ratio})_m$ = -1.32 and σ = 0.23. The calculated log-normal cumulative distribution (bold line) is shown over the data set in Figure 2 and provides a reasonable fit.

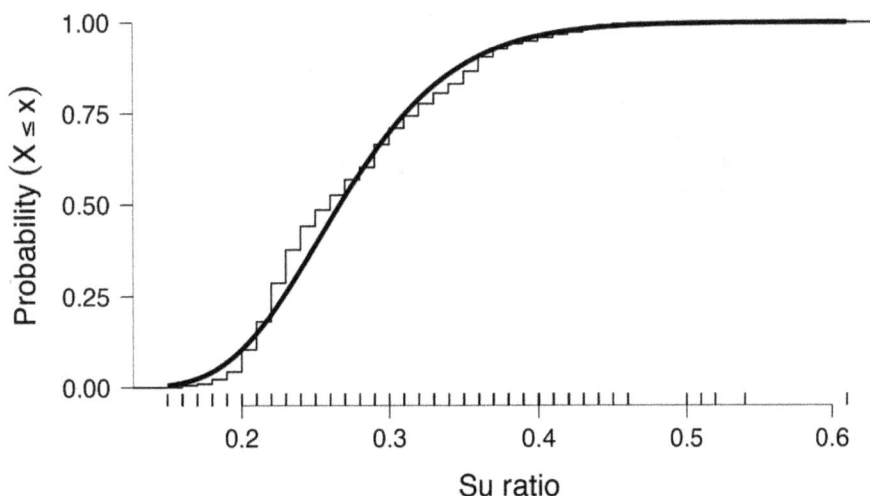

FIG 2 – Log-normal distribution fitted to tailings undrained strength ratio data – example 1.

Frank *et al* (2004) recommend that if a log-normal distribution is used, statistical methods for the estimation of characteristic values must be applied to the logarithm of the parameter. This was applied to the data above and is summarised in Table 2 where the various methods presented have been applied.

TABLE 2

Comparison of characteristic S_u ratio estimates – example 1.

Characteristic S_u ratio	Normal distribution	Log-normal distribution	'Three-Sigma rule'
5th percentile		0.200	
20th percentile		0.220	
25th percentile		0.220	
Schneider (f = 0.5)	0.242	0.238	0.236
Eurocode 7 (f = 1.645)	0.167	0.183	0.148

Methods based on sample population (percentiles) are insensitive to the actual distribution of the data and its standard deviation. In example 1, a range of percentiles from 5 per cent to 25 per cent only changed the estimated changed the characteristic value by 10 per cent. This relative robustness may explain why many practitioners prefer relying to this form of characteristic value estimation.

The Schneider method using the recommended 'typical' statistical coefficient of 0.5 provided the highest values of $(S_u \text{ ratio})_k$. This coefficient may not be suited for use with tailings undrained strength data.

The Eurocode 7 method using f = 1.645 provided the lowest estimate of $(S_u \text{ ratio})_k$ when applied to the untransformed data. Applying the same method to the logarithm of S_u increased the $(S_u \text{ ratio})_k$ value by 10 per cent and brought it closer to the 5 per cent fractile, which is expected as this method is designed for use with normally distributed data. As discussed previously, in the case of soft, saturated, and potentially brittle tailings this level of conservatism may be appropriate.

The 'three-sigma' rule was applied to the data set by simply using the maximum and minimum values measured as HCV and LCV. In practice, engineering judgement would be required to assess whether these represent reasonable estimates of HCV and LCV. For the values used here the resulting standard deviation of 0.077 is higher than the calculated value, resulting in lower estimates of $(S_u \text{ ratio})_k$ than with the Schneider and Eurocode 7 methods.

Outliers

Example 2 in Figure 3 shows the distribution of measured tailings S_u ratio at the same site as example 1, but in a different location of the TSF. Basic statistics for the data set are presented in Table 3. The distribution of S_u ratio is again positively skewed, however the mean strength measured is lower than in example 1 and the tail is much longer (higher maximum value). The calculated coefficient of variation of 0.44 is much higher than in example 1, and in the upper range of published typical values for undrained strength.

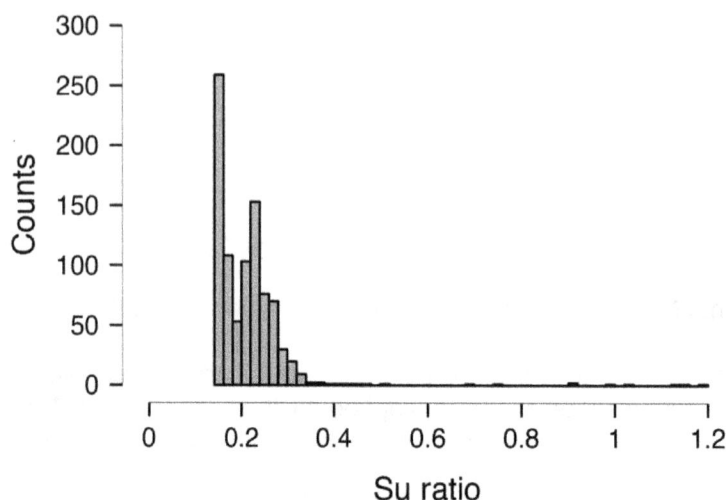

FIG 3 – Distribution of tailings undrained strength ratio (not filtered) – example 2.

TABLE 3

Tailings undrained strength ratio – example 2 – basic statistics (not filtered).

Statistic	Value
No. of samples	900
Median	0.207
Mean	0.214
Std. Deviation	0.094
Coefficient of variation	0.438
Minimum	0.141
Maximum	1.194
5th percentile	0.146
20th percentile	0.154
25th percentile	0.158
99th percentile	0.502

Looking more closely at the data, it is observed that 99 per cent of the values are lower than about 0.5. It is however not expected that subaqueously deposited slurry tailings would reach S_u ratio values as high as observed in this data set (up to 1.2). The high values of S_u ratio values measured at this location are possibly explained by changes in tailings deposition that would have exposed the tailings surface and led to consolidation by desiccation, or by changes in the tailings properties leading to coarser particles being deposited at certain times. Therefore, the 1 per cent values exceeding a value of 0.5 are not considered representative of the tailings strength profile and must be removed from the data set before attempting to derive a characteristic value. The coefficient of variation of the filtered data set is 0.25, which is very close to the value calculated in example 1.

The characteristic values estimation methods presented above were applied to the data from example 2, using the filtered and unfiltered data sets. Results are presented in Table 4.

TABLE 4

Comparison of characteristic S_u ratio estimates – example 2.

Characteristic S_u ratio	Unfiltered data	Filtered data (<0.50)	'3σ rule' (unfiltered data)	'3σ rule' (filtered data)
5th percentile			0.146	
20th percentile			0.154	
25th percentile			0.158	
Schneider (f = 0.5)	0.167	0.181	0.126	0.178
Eurocode 7 (f = 1.645)	0.060	0.121	-0.075	0.115

Methods based on sample population (percentiles) are insensitive to the removal of the 1 per cent outliers and method of calculating the standard deviation. A variation of less than 10 per cent is observed for the calculated $(S_u$ ratio$)_k$ across the range of percentiles considered. This may however not be the case if a larger portion of the raw data were to be removed.

The Schneider method yields slightly higher values than the percentile methods for the filtered and unfiltered cases. The effect is more pronounced in the filtered case due to the reduction in calculated standard variation. However, if the three-sigma rule is used to calculate the standard deviation based on unfiltered minimum and maximum values, the calculated $(S_u$ ratio$)_k$ is greatly reduced.

The Eurocode 7 method provides the lowest estimates of $(S_u$ ratio$)_k$, and the cases where unfiltered data were used provided unrealistically low or negative values.

Log-normal transformation was also trialled for the unfiltered and filtered data. The calculated $(S_u$ ratio$)_k$ are presented in Table 5. The difference of $(S_u$ ratio$)_k$ between unfiltered and filtered data is relatively small, indicating that log-normal transformation can minimise the impact of outliers for a positively skewed data set.

TABLE 5

Log-normal transformation applied to data set with outliers – example 2.

Characteristic S_u ratio	Log-normal distribution unfiltered data	Log-normal distribution filtered data (<0.50)
Schneider (f = 0.5)	0.176	0.178
Eurocode 7 (f = 1.645)	0.127	0.135

Variability of tailings density and pore pressures in calculating S_u ratio

The undrained strength ratio approach to modelling tailings strength relies not only on the accurate measurement of the undrained strength, but also on the appropriate estimation of vertical effective stresses. Vertical effective stress depends on two variables: the tailings density and porewater pressures.

The density of natural soils is generally much less variable than the undrained strength. Duncan (2000) cites a range of published typical coefficients of variation from 3 per cent to 7 per cent. However, in the case of mine tailings it is likely to be a much wider range: firstly, because the source materials may be coming from vastly different orebodies (eg pit location, seam depth etc) and in time accumulate in the same TSF. Secondly, the accurate measurement of in situ tailings density can be extremely challenging, especially with very low density, low plasticity tailings that are notoriously difficult to sample. Advanced sampling techniques such as mini-block samplers have made progress in the past few years but remain limited in use and achievable sampling depth. It is the authors' opinion that the true upper-bound variability of measured tailings density may be greater than 10 per cent, considering the material's intrinsic variability and measurement challenges discussed above.

The tailings pore pressures must be deducted from the total vertical stress to calculate the effective vertical stress. In paddock-style, unlined TSFs, it is common that the tailings water pressures do not follow a hydrostatic profile: if the foundation and embankments are significantly more permeable than the tailings, sub-hydrostatic conditions are likely to develop. This has been observed by the authors across many Australian TSFs of this type. Artesian layers are also common, both in the foundation and within the tailings storage. Lastly, excess pore pressures may be generated due to undrained loading, for example during upstream construction over low permeability tailings.

The implications of the above in the calculation of the undrained strength ratio are important and must not be overlooked, as shown in the following simple, but realistic example: take a 15 m thick body of tailings with an average bulk density of 17 kN/m^3 and an observed standing water level at the tailings' surface. The undrained strength measured at 15 m depth using cone penetration or field vane testing was 25 kPa. The effects of a 50 per cent sub-hydrostatic porewater profile and 10 per cent variability of the material's density are shown in Table 6.

TABLE 6
Effect of pore pressure and density on undrained strength ratio.

Effective stress conditions	Calculated S_u ratio	Difference
Hydrostatic conditions, average tailings density	0.232	-
50% sub-hydrostatic conditions, average tailings density	0.138	-41%
Hydrostatic conditions, 10% error in density	0.187 to 0.304	-19% to +31%

It is therefore recommended obtaining suitably accurate measurements of tailings densities and porewater pressures before attempting any determination of $(S_u$ ratio$)_k$. To this effect, multiple sampling techniques may be trialled (eg thin-walled piston sampling, block sampling, indirect density measurements using small samples of saturated tailings that are also tested for specific gravity), and multiple piezocone dissipation tests spaced regularly from top to bottom of the tailings layer.

Bias towards large number of data points

While assessing large data sets of tailings strength data, cone penetration test data is often over-represented in terms of number of values: this is simply because modern electric cones capture data every 10 mm to 20 mm of penetration, resulting in thousands of readings for each test location. This data richness, combined with the CPT repeatability, provides unmatched soil profiling capabilities. For a typical TSF site, while thousands of CPT readings may be available, only a few dozen field vane test results may be available, due to the much increased time and effort required in performing this test. Given the tailings sampling difficulties discussed above, it is also common that only a few usable undisturbed samples are available for laboratory testing.

The danger here is combining the thousands CPT data points with a limited number of other test results to create a single data set for assessment of a characteristic value. It is recommended using appropriate weighting of each of the techniques utilised to avoid over-representing one or the other. For example, all the cone-interpretated S_u ratio values for a tailings layer may be averaged to a single value, which is then measured against the field vane/flat-plate dilatometer/laboratory test results obtained in the same layer. Other ways of combining/comparing the data are possible, but the key here is not falling to a sense of increased confidence due to the large number of data points available.

Lastly, it is worth noting that the interpretation of undrained strength from cone penetration relies on the cone factor N_{kt}. Robertson and Cabal (2022) cite a typical range for the cone factor from 10 to 18, with an average of 14. The undrained strength is estimated by dividing the net cone resistance by the cone factor. This means that the estimated undrained strength may vary by ±28.6 per cent across this range. Some practitioners use the results of field vane tests to 'calibrate' the cone factor. This may lead to the opposite problem to the discussion above, as in this case greater confidence is given to the field vane test, influencing the results of all CPT-interpreted strength values. The field vane test is a useful test for the direct estimation of undrained strength, however normal stresses and porewater conditions during vane testing are unknown. In case of partial drainage during testing the undrained strength may be overestimated (Wentzinger and Keulemans, 2021; Reid, 2016). It is therefore advised using vane testing for 'calibration' of N_{kt} with care. It is the author's experience that field vane test results are generally more variable and less repeatable than the CPT. It is recommended that as many field vane tests as possible be carried out and the data be critically reviewed. Invalid test results (eg too high, too low, peak not reached etc) are common.

Impact on design outcomes

The last common error often observed by the authors in the development of tailings characteristic values is the belief that 'lower is always more conservative'. While this is often true for strength values when stability or bearing capacity are considered, there are specific cases and parameters for which a characteristic value higher than the mean would be more conservative.

One typical example is material density: in dam stability analysis, the soil mass in the driving side of a potential slip surface will generate higher disturbing forces with a higher density, while the soil mass in the resisting side of the same slip surface will provide lower resisting forces with a lower density.

Another example is linked to the previous discussion about porewater pressures in tailings: if higher porewater pressures than actual are used in stability analysis this will likely result in a lower (more conservative) factor of safety due to the reduction in vertical effective stresses. However, as shown previously, mistakenly using hydrostatic conditions for the interpretation of the S_u ratio may lead to unconservative results.

It is therefore recommended carefully assessing the impact of each design parameter before deciding whether a characteristic value should be selected above, or below the mean measured value.

NUMERICAL MODELLING CONSIDERATIONS

A common practice used when numerically analysing TSFs is assuming a worst state condition for the entire tailings impounded. This is not only a poor use of the numerical analysis tool, but also it will lead to overconservative designs. This is especially true in the case of upstream raised facilities.

When good quality field data is available, for example CPT readings, it is recommended using that information for defining the material state distribution throughout the numerical model.

Many techniques for interpolation of the material state are readily available. Most of them use the distance between the point of interest and the closest data point as a weighted measurement of the final value. Care needs to be considered when using CPT data. As previously mentioned, CPT information can be very dense in the vertical direction but with tens of metres of spacing in the other two directions. If non-tailored schemes are used, the resulting distribution will have local bias given by the high amount of information in the CPT data in the vertical direction. This is especially important in tailings deposit where the shared tailings characteristic are horizontally predominant. A solution for this bias is using a quadrant approach, whereby not all data blindly used but instead it is first arranged in the quadrant defined by the point of interest positions and then the preferred interpolation scheme is used.

Another consideration related to numerical modelling is somehow contrasting with what has been discussed in the preceding. Given the numerical modelling mesh describes a continuum of information that can vary spatially, there is no need to obtain a characteristic value for all elements in the mesh. Rather, it is more accurate to use a local representative value that changes with changes in the data set.

Finally, as previously mentioned the CPT data can be measured as densely as 10 mm to 20 mm in the vertical direction. In the other hand, TSF numerical models are seldom using an element height less than 0.5 m to 1 m. Therefore, for every numerical element in the vertical direction, there are at least 25 to 50 measured points. If some form of weighted interpolation scheme is used, care must be taken to where the element's centroid is in relation to the data set. For example, if the state parameter is interpolated and the CPT results show a thin contractive layer within a dilative layer, that contractive layer could be easily missed if the element centroids are located too far from that contractive layer. An immediate solution would be using a much finer mesh, but that is not always possible given the associated computational cost.

CONCLUSIONS

This paper presented a number of methods that may be considered for the development of geotechnical characteristic values, and commons errors that should be avoided in their application. The recommendations provided are summarised as follows:

1. When investigating tailings materials and TSF foundation, multiple field and laboratory test methods should to utilised to provide the data required for parameter interpretation. Care should be taken to not over-represent a test method, or imparting greater confidence in a test method unless is it proven to be more accurate.

2. If the undrained strength ratio is utilised to model tailings strength, the tailings density and porewater pressures must be accurately measured during the investigation to provide reliable values of vertical effective stresses.

3. Use histogram views of the data sets and asses the best suited statistical distribution for the parameter; undrained strength is often better fitted to a log-normal distribution, which has implications in the development of characteristic values based on statistical methods.

4. Assess the presence of outliers and remove them prior to using statistical methods. The 99 per cent fractile, or 3σ, are good starting points. If too many 'outliers' appear, question the test results' reliability or the need for sub-dividing the layer.

5. Compare the outcomes of multiple methods to derive characteristic values and select best-suited:

   o If the data is showing much scatter and does not seem to follow a typical distribution function: use a percentile-based method.

   o If the data can be fitted to a distribution function: use a statistical-based method with an appropriate statistical coefficient. The typical value of 0.5 is generally not suited to application with tailings undrained strength.

   o Consider the level of risk related to the material. For tailings with a risk of brittle failure use the 5 per cent fractile or statistical method with f = 1.645.

6. For numerical analyses use the actual distribution of the values instead of a single characteristic value. If interpolation of CPT-derived data is used:

   o Use the quadrant approach to avoid bias from the inherent CPT spatial data distribution.

   o Use statistical analysis to determine the interpolated value. This prevents the bias produced by the elements' centroid position relative to the data. The authors recommend using a bin size between half and one element height and using the 50th percentile.

REFERENCES

Australian National Committee on Large Dams (ANCOLD), 2019. *Guidelines on tailings dam: planning, design, construction, operation and closure,* Australian National Committee On Large Dams.

Canadian Dam Association (CDA), 2007. *Dam Safety Guidelines Technical Bulletin: Geotechnical Considerations for Dam Safety,* Canadian Dam Association.

Canadian Geotechnical Society (CGS), 2006. *Canadian Foundation Engineering Manual,* Canadian Geotechnical Society.

Duncan, J M, 2000. Factors of Safety and Reliability in Geotechnical Engineering, *Journal of Geotechnical and Geoenvironmental Engineering*, 126(4):307–316.

Duncan, J M, Wright, S G and Brandon, T L, 2014. *Soil strength and slope stability* (Wiley New: Jersey).

European Committee for Standardization, 2004. EN 1997–1, *Eurocode 7: Geotechnical design – Part 1: General rules*, 168 p (European Committee for Standardization: Brussels).

Fell, R, Macgregor, P, Stapledon, D, Chartered, B and Foster, M, 2018. *Geotechnical engineering of dams*, 2nd edition (CRC Press: Boca Raton).

Frank, R, Bauduin, C, Driscoll, M, Kavvadas, M, Krebs Ovesen, N, Orr, T and Schuppener, B, 2004. *Designer's guide to EN 1997–1 Eurocode 7: Geotechnical design general rules,* 216 p (Thomas Telford Publishing: Bodmin).

International Council on Mining and Metals (ICMM), United Nations Environment Programme (UNEP) and Principles for Responsible Investment (PRI), 2020. *Global Industry Standard on Tailings Management.*

Jefferies, M and Been, K, 2015. *Soil liquefaction: a critical state approach* (CRC Press: Boca Raton).

Look, B G, 2014. *Handbook of geotechnical investigation and design tables* (CRC Press: Boca Raton).

Look, B G, 2015. Appropriate probability distribution functions for geotechnical data, in 12th Australia – New Zealand Conference on Geomechanics, Wellington.

Reid, D, 2016. Effect of rotation rate on shear vane results in silty tailings, in *Proceedings of Geotechnical and Geophysical Site Characterization 5* (Australian Geomechanics Society: Sydney).

Robertson, P K and Cabal K, 2022. *Guide to cone penetration testing* (Gregg Drilling LLC: Signal Hill).

Schneider, H R, 1997. Panel discussion: Definition and determination of characteristic soil properties, in *14th International Conference on Soil Mechanics and Foundation Engineering*, pp 2271–2274.

US Department of the Interior Bureau of Reclamation, 2011. *Design Standards No. 13, Embankment Dams, Chapter 4: Static Stability Analysis*, USBR.

US Department of the Interior Bureau of Reclamation, 2019. *Best Practice, Chapter D-5: Embankment Slope Instability*, USBR.

Wentzinger, B and Keulemans, Y, 2021. The effect of vane shear rotation speed on the estimation of tailings undrained strength, in *Proceedings of the Mine Waste and Tailings Conference 2021*, pp 184–190 (The Australasian Institute of Mining and Metallurgy: Melbourne).

Tailings dam breach and runout assessment

Adoption of non-Newtonian fluid routing in dam breach analysis for a closed tailings storage facility

Y Chen[1], Y Wang[2] and T Matuschka[3]

1. Senior Geotechnical Engineer, Engineering Geology Ltd, Auckland 0632, New Zealand.
   Email: yuanzhi.chen@egl.co.nz
2. Geotechnical Engineer, Engineering Geology Ltd, Auckland 0632, New Zealand.
   Email: yuri.wang@egl.co.nz
3. Director, Engineering Geology Ltd, Auckland 0632, New Zealand.
   Email: trevor.matuschka@egl.co.nz

ABSTRACT

In a simple tailings dam breach analysis (TDBA) for an active tailings storage facility (TSF), the mixture of liquefiable tailings and supernatant water is often conservatively assumed as Newtonian fluids in the runout assessment, similar to a dam breach analysis for water dams. For a closed or inactive TSF with dry capping, there is no permanent ponded water to be released. The contractive saturated tailings susceptible to flow-type liquefaction is often limited to the lower portion of the impoundment in a closed TSF, and the breach flow may exhibit different behaviours depending on the site-specific liquefaction potential. In such cases, numerical modelling using non-Newtonian fluid is required to provide a more realistic estimation of the runout extent. This paper presents a case study of a hypothetical tailings dam breach analysis using non-Newtonian flow, in which the selection of critical parameters was mainly based on site-specific geotechnical investigations. The geotechnical investigations comprise field tests to determine the phreatic surface within tailings, density, water content, particle size distribution, critical state parameters and residual shear strength. The judgements applied to overcome the limitations are discussed, including the common issue of no available rheological testing data for deposited tailings, the limitations of the selected commercial computer program, and using a set of yield stress and viscosity to simulate a mudflow formed by tailings with distinct characteristics. Sensitivity analyses are undertaken to evaluate the effect on the estimated runout potential due to the extent of saturated tailings in the impoundment, the run-off due to extreme rainfall event, the assumptions, and uncertainties of the rheological properties. The gaps of knowledge in the current literatures of tailings with high solids concentrations are highlighted. The results indicate that the estimated runout extent is not that sensitive to the assumption of viscosity, despite its range of uncertainty being more than ± two order of magnitudes. The estimated runout extent can be significantly affected by the assumption of yield stress. The study highlights the importance of understanding of the tailings undrained residual strengths at various solids concentrations, and the estimation of solids concentration in a Rainy-day breach that could be affected by significant volume of run-off water.

INTRODUCTION

The design and operational requirements of a tailings storage facility (TSF) are based on the consequence of dam failure (eg ANCOLD, 2019; GISTM – ICMM, 2020). The consequences are evaluated by undertaking a dam breach analysis with consideration of the credible potential failure modes. The methodology used in the tailings dam breach analysis (TDBA) in Australia are described in ANCOLD Consequence Guidelines (2012), with appropriate consideration of the potential increase in damage and safety consequence associated with mine process tailings. This requires considering a hypothetical dam breach simulation under various conditions of flooding, including 'Sunny-day' (no flooding) and extreme flood events. The methodology for TDBA can involve complex hydrological studies, or in some cases simple empirical or qualitative methods, eg in the early stage option study. For tailings dams the simulation often assumes that tailings are replaced with water, ie Newtonian fluids, or use more sophisticated methods to model mudflow, ie non-Newtonian fluids. Modelling of the flow of mixed tailings and water is complex. Considerable judgement is required to determine a realistic mudflow scenario. Simulation of a dam breach scenario of a closed or inactive TSF with dry capping is a typical example. In such cases, there is no permanent ponded water to be released. The contractive saturated tailings susceptible to flow-type liquefaction is often limited to the lower portion of the impoundment in a closed TSF, and the breach flow may exhibit different behaviours depending on the site-specific liquefaction potential.

Numerical modelling using non-Newtonian fluid is required to provide a more realistic estimation of the runout extent due to tailings dam breach. Most of the commercial computing programs require assumptions of a set of rheological parameters to represent the properties of a mudflow. Tailings in a closed TSF located above the phreatic surface are unsaturated with a degree of saturation less than the critical level that could trigger liquefaction. The tailings below the phreatic surface are often closed to fully saturated. As the liquefaction resistance of soils could increase sharply with the decrease in the saturation degree (Mele *et al*, 2019), tailings above the phreatic surface could assumed as non-liquefiable, and tailings below the phreatic surface have a flow potential. Selection of rheological properties requires careful judgement for a closed TSF, as the mudflow comprises tailings with distinct characteristics. This paper presents case studies of hypothetical tailings dam breach analyses. The study adopts the recommendations in the Canadian Dam Association Technical Bulletin on Tailings Dam Breach (CDA, 2021). It is considered to represent the current best practice and is a referenced document in the Global Industry Standard on Tailings Management (GISTM). A comprehensive TDBA usually requires the identification and consideration of potential failure modes, assessment of tailings characteristics and rheology, estimation of outflow volume, dam breach flood routing, mapping of the extent of flood inundation, and evaluation of the peak flood depth, flow velocity, time of flood arrival, time of flood peak and inundation duration at key locations. The discussion in this paper will focus on the assessment of tailings characteristics, estimation of outflow volume, and determination of rheological properties adopted in flood routing simulation, and their impacts on the estimation of runout extent. The results indicate that the estimated runout extent is not sensitive to the assumption of viscosity but can be significantly affected by the assumption of yield stress.

DESCRIPTION OF HYPOTHETICAL DAM BREACH SCENARIOS

Case study background

The tailings dam in this case study is of upstream construction. The upstream constructed section of the dam comprises eight upstream lifts typically of 3 m height. The maximum height of the dam is approximately 31 m. A typical cross-section is shown in Figure 1. The dam is a zoned earth fill embankment. Earlier stages were constructed from materials excavated from within the pond footprint. Later stages have been constructed mostly from a stockpile of historical Mine Open Pit overburden material. During operation water is impounded on the tailings surface and the edge of the decant pond is at approximately 500 m away from the upstream edge of embankment crest. After the TSF is closed, the tailings surface is recontoured and capped, so no water is impounded.

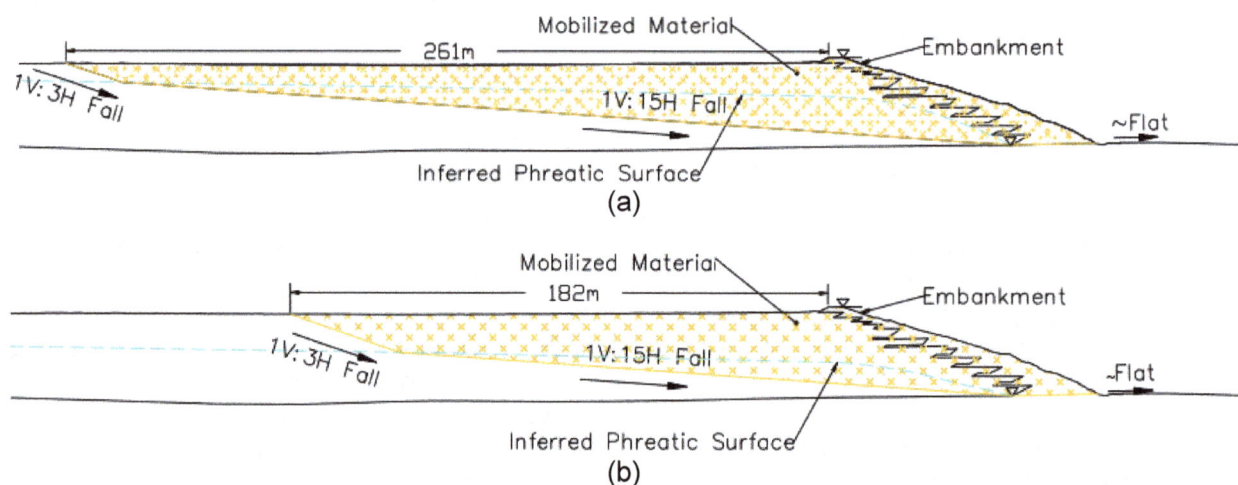

FIG 1 – Dam breach cross-sections (a) during operation and (b) post-closure.

Potential failure modes and dam failure scenarios

A potential failure mode is a hypothetical failure mechanism which is developed considering the physical characteristics of the dam structure and its foundation, subsurface drainage conditions, surface water management, operation, maintenance, and surveillance, and natural hazards (eg floods, earthquakes). Potential failure modes can and do vary during the life cycle of the TSF. Different failure modes each have different scenarios which lead to failure and different scenarios that may occur during the breach. The failure modes and description of failure scenarios considered in this case study are summarised in Table 1.

TABLE 1

Summary of potential failure modes and failure scenarios in the case study.

No.	TSF life cycle	Potential failure mode	Description of failure scenario
1	Operation	An extreme earthquake event causes slope instability of the embankment.	The failure is triggered by an extreme earthquake during normal hydrologic conditions in the downstream rivers and streams, ie a Sunny-day scenario. Tailings are released due to slope instability. There is a potential of releasing the decant pond water, which depends on the distance to the pond at the dam breach location.
2	Post-closure	An extreme earthquake event causes slope instability of the embankment.	The failure is triggered by an extreme earthquake during normal hydrologic conditions in the downstream rivers and streams, ie a Sunny-day scenario. Tailings are released due to slope instability.
3	Post-closure	Infiltration from extreme rainfall through defects within embankment fill causes saturation of the embankment and tailings at low elevation. Static liquefaction is triggered and results in slope instability.	The failure is triggered by an extreme rainfall event, ie a Rainy-day scenario. Tailings are released due to slope instability.

Tailings dam breach analysis case

The CDA Technical Bulletin (CDA, 2021) suggests that the TDBA should be based on a selection of appropriate TDBA cases. There are two main factors that are expected to have an important impact on the characteristics and volume of the outflow from a TSF during a breach event:

1. The presence of fluids on the surface of the impoundment near the dam.

2. The potential of liquefaction induced flowability of the tailings material, which may be due to various trigger mechanisms, including the breach itself.

These factors are used to define four types of tailings dam breaches. The supernatant pond is located a considerable distance away from the embankment during normal operation. It is unlikely that pond water will be discharged in the Sunny-day scenario. No supernatant water is present after closure. Therefore, according to the CDA Technical Bulletin (2021), the dam failure scenarios are classified as 'Case 2A – liquefiable tailings without a supernatant pond' or 'Case 2B – non-liquefiable tailings without a supernatant pond' type of breach depending on whether the tailings are liquefiable or non-liquefiable, respectively. Tailings above and below the phreatic surface are assumed to be non-liquefiable or liquefiable, respectively. The breach in this case study is considered as a combination of these two TBDA cases. The runout analysis assumes mudflow as a mixture of liquefied flowable tailings below the phreatic surface and slumped tailings above the phreatic surface. It is less fluid compared to water and consequently deposited immediately downstream of the breach location.

ASSESSMENT ON TAILINGS PROPERTIES

Investigations of tailings properties in this case study are focused on its geotechnical characteristics. It comprises field geotechnical investigations, eg cone penetration tests (CPTs), and laboratory tests to enable assessment of the degree of saturation, density, particle size distribution, shear strengths.

CPTs have been undertaken on the tailings in the TSF with pore pressure dissipation tests undertaken at constant intervals to investigate the piezometric levels within the tailings and to infer the phreatic surface which is compared with the piezometric levels measured in nearby piezometers.

Tailings in this case study are non-plastic sandy silt with an average fines content (<75 μm) of about 60 per cent. Laboratory tests were undertaken to determine the critical state parameters of typical tailings samples. Estimation of *in situ* state parameters was based on the CPT data and only considers the portion of saturated tailings, which is below the depth of the inferred phreatic surface. The interpretation indicates that the tailings have contractive behaviour, and the saturated tailings are potentially liquefiable.

No laboratory tests have been undertaken to study the rheological characteristics of the deposited tailings. It is also a common limitation for some TSFs. In the cases without site specific testing, the judgement on rheological properties can be initially based on literatures on materials with similar properties. If the assumption be found critical to the results of analysis and affecting decision-making, laboratory tests for the rheological properties should be scheduled.

BREACH ANALYSIS

Breach parameters

Breach parameters describe the shape of the breach, the breach development time, and how the breach develops over time. Dam breach model tools (eg Froehlich, 2008; Von Thun and Gillette, 1990; Xu and Zhang, 2009) have been developed for breaches of water storage embankments, which are either homogeneous or zoned earth fill, or rock fill dams. No such models are currently available for tailings dams. Some are under development. Similar breach parameters have been adopted for the three dam failure scenarios considered in this study; bottom width of 130 m, side slopes of 1V:0.5H, and development time of 15 minutes. These are based on the typical ranges of breach parameters of water retaining dams with additional considerations of historical dam failure cases (eg the 2015 Fundão and 2019 Feijão dam failures). Earthquake induced sudden dam failure could have a much shorter development time. Runout distance will be further with assumption of a shorter duration. Sensitivity analysis of this case study indicates that the estimated maximum runout will increase by 10 per cent if a development time of 5 minutes is assumed.

Estimation of breach outflow volume

The breach outflow volume is the total volume of materials that could be mobilised in a breach. A cone of depression analysis of a full height breach was conducted for different breach scenarios. The breach geometries of the breach analysis are shown on the cross-sections in Figure 1. The phreatic surface for the case during operation was inferred based on CPTs and piezometer monitoring data. The phreatic surface for the post-closure case was estimated from a seepage analysis considering the conditions after closure. The phreatic surface in the tailings will drop down following removal of the supernatant pond.

The geometric surfaces of the tailings remaining within the TSF, ie the angle of tailings coming to rest, are based on the shear strength of the tailings. The residual undrained shear strength ratio of the tailings based on CPTs has a minimum value of 0.12. Such material would have a natural angle of repose of about 1V:8H. The slope at rest was estimated in a range of 1V:10H to 1V:14H based on the method of Lucia, Duncan and Seed (1981). In this method, the behaviour of tailings was assumed undrained, and its strength could be represented as an equivalent shear strength. The residual shear strength of tailings may vary through the mass due to different degrees of saturation, variations in particle size or void ratio, age of tailings, etc. This results in the uncertainty in the estimation. Considering the active pressure at the back of a liquefied wedge would result in a flatter slope at rest, 1V:15H was adopted for the liquefiable tailings in the TBDA based on engineering judgements. It is slightly more conservative than the estimated range using the method of Lucia, Duncan and Seed (1981).

Laboratory tests show that the tailings above the inferred phreatic surface are generally with degree of saturation less than 80 per cent and considered as non-liquefiable. For the non-liquefiable tailings, the at rest slope is estimated as 1V:1.6H which is based on the critical state friction angle of 32° derived from the laboratory tests. Considering the effect of void redistribution that may occur at the plane of failure during breach, a flatter slope of 1V:3H was adopted for the non-liquefiable tailings in the current TBDA, which is in the typical range of the angle of repose for wet silt of 15° to 25° (Mitchell and Soga, 2005).

The breach outflow volumes adopted in the analysis and the range due to the uncertainty in estimation of angles coming to rest are summarised in Table 2. In the rainy-day dam breach scenario after closure, the breach outflow volume includes the volume of rainfall run-off due to the extreme rainfall event. The breach outflow volume is more sensitive to the assumption of the at rest angle of liquefiable tailings than the non-liquefiable tailings. As shown in Table 2, the difference between the minimum and maximum breach outflow volume, due to the steepest and flattest angles coming to rest, is about 60 per cent. The difference between the minimum and maximum peak breach outflow is about 30 per cent.

TABLE 2

Summary of Dam Breach Outflow Volumes for Different Scenarios

Scenario	Breach outflow volumes (Mm³)		Peak breach outflow (m³/s)	
	Adopted	Range[1]	Adopted	Range[1]
1. Sunny-day under operation	2.16	1.35–2.16	5920	4510–5920
2. Sunny-day after closure	1.84	1.13–1.85	5430	4150–5430
3. Rainy-day after closure	1.99	1.28–2.00	5630	4380–5630

Note: [1] The range was estimated based on different assumptions of angles coming to rest. For tailings below the phreatic surface, the range considered in the sensitivity analysis is 1V:10H to 1V:15H. For the tailings above the phreatic surface, the range considered in the sensitivity analysis is 1V:1.6H to 1V:3.7H.

The dam breach outflow hydrographs used in the TDBA are shown in Figure 2. They were derived by using HEC-RAS, with the consideration of the volume change from the breach bottom elevation to the impoundment elevation at failure, the breach geometry and development time, and the type of materials used in the construction of dam.

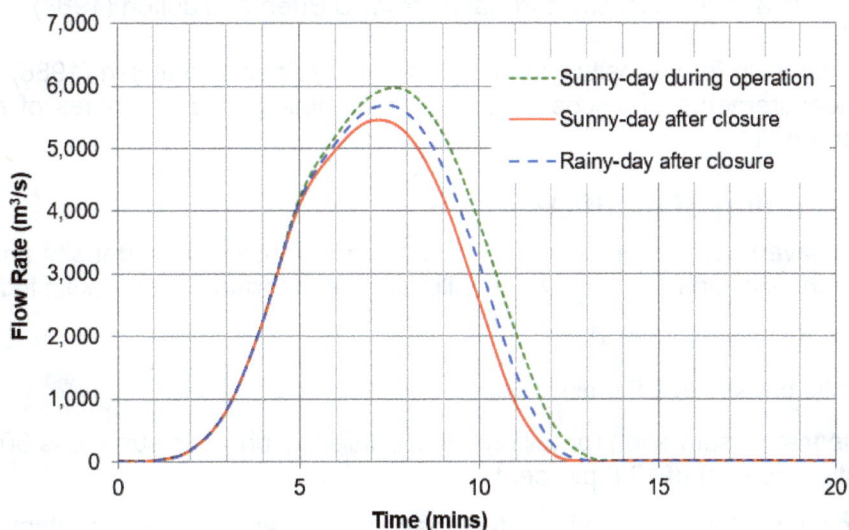

FIG 2 – Adopted outflow hydrographs for different dam failure scenarios in the TDBA.

RUNOUT ANALYSIS

The topography of the area downstream of the dam in this study is flat. The runout analysis of the mudflow formed by the liquefied and slumped tailings was modelled as a breach of non-Newtonian fluids. The runout analysis in this case study uses the computer program HEC-RAS (Gibson and Sanchez, 2020). However, HEC-RAS does not account for changing solids concentration in routing, ie no dilution effect. Therefore, in a Rainy-day scenario the volume of run-off from an extreme rainfall event will need to be added to the breach outflow to include the effect of dilution. This limitation may increase the estimated runout extent. Some other commercial computing program can model a time varying solids concentration, eg FLO-2D. A comparison between HEC-RAS with FLO-2D as tools in a TDBA was conducted by Scholtz and Chetty (2021), and the results were found similar. The stormwater may not be blended uniformly with the tailings mudflow. If the results from HEC-RAS or FLO-2D analysis affect the decision-making and require further improvement on accuracy, runout analysis could be undertaken using a model simulate the fluid and solid fractions as a quasi-two-phase flow, eg D-Claw (George and Iverson, 2014).

Rheological properties of non-Newtonian fluid

Yield stress and viscosity are generally functions of sediment concentration and can vary significantly for tailings. As shown in Figure 3, both the yield stress and viscosity increase with the increase of solids concentration. Since there is no site-specific rheological test data available, the residual shear strength obtained in the geotechnical lab tests for shear strength was used to derive the yield stress. Figure 3(a) shows the relationship between the residual shear strength and the solids concentration by volume based on the results of direct simple shear tests on representative tailings samples.

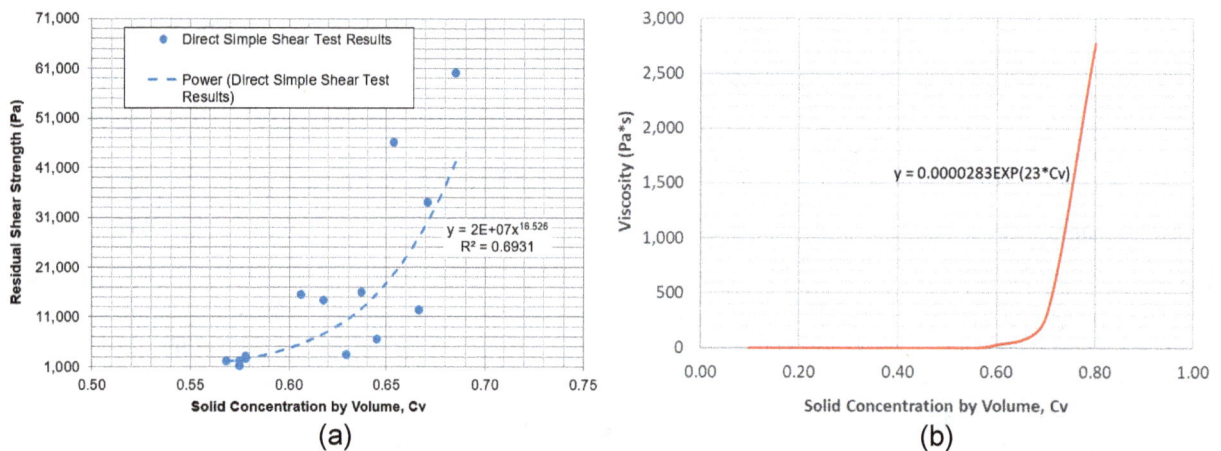

(a)   (b)

FIG 3 – Correlation adopted to determine yield stress and viscosity based on solids concentration by volume: (a) residual shear strength correlation based on direct simple shear tests on tailings on-site and (b) viscosity correlation from O'Brien and Julien (1988).

The adopted relationship for viscosity was adopted from O'Brien and Julien (1988). It was obtained by laboratory measurements of viscosity of non-plastic fine-grained samples of natural mudflow deposits in Colorado.

Solids concentration of mudflow

Laboratory tests have been undertaken to obtain the dry density, solid density and water content from high-quality thin-walled tube samples of tailings. The laboratory test results indicate:

- average dry density of 1.8 t/m³

- average solid density of 2.9 t/m³

- average degree of saturation for tailings above inferred phreatic surface is 60 per cent, equal to a moisture content of 12.6 per cent.

The tailings below the inferred phreatic surface have an average moisture content of 21.1 per cent and are fully saturated or close to fully saturated.

The solids concentrations of the mudflow in different dam failure scenarios were estimated based on the volume of partially saturated soils, which includes the embankment fill and tailings above the phreatic surface, volume of saturated tailings and volume of water. The estimated average solids concentrations for the three scenarios are in a range of 82.1 per cent to 87.6 per cent by weight, equivalent to 61.2 per cent to 70.9 per cent by volume. They are summarised in Table 3.

TABLE 3

Average solids concentration, yield stress and viscosity of mudflow in different scenarios.

Scenario	Average Solids Concentration (%)		Yield Stress (kPa)		Viscosity (kPa*s)	
	By Weight	By Volume	Adopted	Range	Adopted	Range
1. Sunny-day under operation	86.2	68.3	32.3	0.280–46.7	1.88	0.013–143
2. Sunny-day after closure	87.6	70.9	58.6	0.438–95.3	3.41	0.019–342
3. Rainy-day after closure	82.1	61.2	5.2	0.087–10.1	0.37	0.005–13.2

Adopted values and uncertainties

Since the solids concentrations of the mudflow in this study are all greater than 60 per cent and the tailings mainly comprise silt and fine sand, the mudflow is similar to the hyper-concentrated flow (Gibson and Sánchez, 2020; O'Brien and Julien, 1984). The Bingham model was selected in the runout analysis.

Relationships for yield stress and viscosity in the literature are mostly derived based on testing undertaken on samples with solids concentration by volume less than 60 per cent by volume (eg O'Brien and Julien, 1984, 1988; Major and Pierson, 1992; Martin, Adria and Wong, 2022). Extrapolation is required to obtain estimates for the mudflow in this study. The upper and lower bound estimates of yield stress and viscosity based on published literature are calculated and summarised in Table 3. The lower bound estimate of yield stress is more than two orders of magnitude less than the adopted value. While selecting the yield stress in analysis, more weighting was given to the results laboratory tests on site-specific tailings samples with high solids concentration than the values based on extrapolation. The possible range of viscosity can be ± three orders of magnitude from the adopted value. The medium value of viscosity is adopted in the analysis.

RESULTS AND DISCUSSION

Comparison between dam breach scenarios

The results of the runout analyses are shown in Figure 4. The differences between dam breach scenarios can be seen by comparing Figures 4(a), (b) and (h). The estimated runout extent is not governed by the volume of the dam breach outflow, but more dependent on the solids concentration. The Rainy-day post-closure scenario is estimated to have the furthest runout distance. Although its breach outflow volume is less than the Sunny-day scenario under operation, the runout distance is longer due to the lower estimated solids concentration due to dilution from run-off in the analysis.

The results indicate that careful consideration should be given to the effect of dilution for a closed TSF with significant catchment area in the Rainy-day scenario. It also applies for the case of a TSF close to a major natural watercourse in both the Sunny-day and Rainy-day scenarios.

FIG 4 – Results of runout analyses for different dam failure scenarios and sensitivity assessment of yield stress and viscosity assumptions.

Sensitivity analysis on viscosity and yield stress

Yield stress and viscosity increase exponentially with solids concentration and become very sensitive to the solids concentration when the solids concentration exceeds a threshold, as shown in Figure 3. For a TSF with a large proportion of the tailings in the impoundment could be partially saturated, the estimated solids concentration of the dam breach mudflow is likely to exceed the threshold and becomes sensitive in the estimation of rheological parameters. The relationship of rheological parameters with solids concentrations varies between different mineral tailings, particle size distribution and Atterberg limits. It is possible to have a large variation in rheological parameters in the same TSF. At present, a sensitivity analysis to cover a wide range of uncertainty in rheological parameters is necessary to understand the effects on the results from a TDBA.

Sensitivity analyses were undertaken for the Sunny-day and Rainy-day scenarios after closure. The sensitivity of viscosity assumptions is shown in Figures 4(b)-(d) for the Sunny-day scenario and Figures 4(h)-(j) for the Rainy-day scenario. In this analysis, the maximum value is approximately three orders of magnitude greater than the minimum value. The results indicate that the runout distance increases with an increase of viscosity. However, the amount of change is relatively minor.

The runout distance could be increased by up to 17 per cent due to per order of magnitude increase of viscosity. The result is contrary to the authors' anticipation. It could be due to the limitation of computing method with assumption of a constant yield stress.

The sensitivity of yield stress assumptions is shown in Figures 4(b), (e) and (f) for the Sunny-day scenario. The runout distance has a notable increase with a reduction of yield stress as can be seen comparing Figures 4(b), (e), and (f). In this analysis, the minimum value of yield stress is approximately an order of magnitude less than the adopted value (5 kPa versus 58.6 kPa), and the runout distance due to a lower yield stress is approximately double of the case with the adopted assumption (430 m versus 190 m). Should there be items of interest in the close vicinity of the dam breach location, eg houses, roads and critical infrastructures, that may affect the dam failure consequence classification, additional testing should be considered to provide higher confidence in the estimates of yield stress.

Method to verify results of analysis

Methodology to confirm the results of TDBA usually involves comparison with case histories. However, there are no existing case histories for breaches of a TSF with significant volume of mobilised tailings being non-liquefiable. A tailings deposition analysis using three-dimensional civil modelling program can be undertaken. The runout extent could be estimated by considering the volume slumped from the breach of the embankment equal to the volume of tailings deposited adjacent to the breach location and forming a stable flatter slope. The method of Lucia, Duncan and Seed (1981) could be adopted to estimate the final angle at rest. However, the accuracy of this method will be limited by the downstream terrain. The downstream terrain generally needs to be less than 4° to be consistent with the case histories, from which the slope at rest was derived.

CONCLUSIONS

The paper provides a case study of a tailings dam breach analysis using non-Newtonian fluid runout analyses for the scenario of a closed TSF with dry capping (ie no permanent pond water) where there is a notable portion of non-liquefiable tailings. The scenario is typical for TSFs in dry climates. The determination of rheological parameters and dam breach outflow volumes are mainly based on geotechnical investigations, including field and laboratory testings. The conclusions from the results and limitations of a TDBA for such scenarios are summarised below.

1. The solids concentration of a mudflow from a breach of a TSF with notable non-liquefiable tailings and no supernatant pond is likely greater than 60 per cent by volume, which exceeds most of the rheological test data available in the literature. Extrapolation based on models in literatures potentially leads to variations in estimates with multiple orders of magnitude.

2. The runout distance is more dependent on the estimated solids concentration of the breach mudflow than the dam breach outflow volume.

3. The sensitivity analysis indicates that the runout extent increases with the increase of viscosity, but it is not that sensitive. The result is contrary to the authors' anticipation and could be due to the limitation of model assuming a constant yield stress.

4. The runout distance is more sensitive to the assumption of yield stress. An order of magnitude less for the yield stress can result in double the estimated runout distance.

5. Considerable judgements on the effect of dilution should be given, for the case of a TDBA for a closed TSF with significant catchment area in the Rainy-day scenario, and for the case of TSF close to a major natural watercourse in both Sunny-day and Rainy-day scenarios.

6. Should there be items of interest in the close vicinity of the dam breach location, eg houses, roads and critical infrastructures, that may affect the dam failure consequence classification, additional testing should be considered to provide higher confidence in the estimation of yield stress.

REFERENCES

Australian National Committee on Large Dams (ANCOLD), 2012. Guidelines on the Consequence Categories for dams.

Australian National Committee on Large Dams (ANCOLD), 2019. Guidelines On Tailings Dams: Planning, Design, Construction, Operation and Closure.

Canadian Dam Association (CDA), 2021. Technical Bulletin: Tailings Dam Breach Analysis, 59 p.

Froehlich, D C, 2008. Embankment dam breach parameters and their uncertainties, *Journal of Hydraulic Engineering*, 134(12):1708–1721.

George, D L and Iverson, R M, 2014. A depth-averaged debris-flow model that includes the effects of evolving dilatancy, II, Numerical predictions and experimental tests, *Proceedings of the Royal Society A: Mathematical, Physical and Engineering Sciences*, 470(2170):20130820. https://doi.org/10.1098/rspa.2013.0820.

Gibson, S and Sánchez, A, 2020. HEC-RAS 6.0 mud and debris manuals, US Army Corps of Engineers, Hydrologic Engineering Center.

International Council on Mining & Metals (ICMM), 2020. Global Industry Standard on Tailings Management (GISTM).

Lucia, P C, Duncan, J M and Seed, H B, 1981. Summary of research on case histories of flow failures of mine tailings impoundments, Proceedings Bureau of Mines Technology Transfer Workshop.

Major, J J and Pierson, T C, 1992. Debris flow rheology: Experimental analysis of fine-grained slurries, *Water Resources Research*, 28(3):841–857.

Martin, V, Adria, D and Wong, H, 2022. Inundation modelling of non-Newtonian tailings dam breach outflows, in the Congress of the International Commission on Large Dams (ICOLD), pp 501–520 (ICOLD: Marseille).

Mele, L, Tian, J T, Lirer, S, Flora, A and Koseki, J, 2019. Liquefaction resistance of unsaturated sands: Experimental evidence and theoretical interpretation, *Géotechnique*, 69(6):541–553.

Mitchell, J K and Soga, K, 2005. *Fundamentals of soil behavior* (John Wiley & Sons: New York).

O'Brien, J S and Julien, P Y, 1984. Physical properties and mechanics of hyperconcentrated sediment flows, Proceedings ASCE HD Delineation of Landslides, Flash Flood and Debris Flow Hazards.

O'Brien, J S and Julien, P Y, 1988. Laboratory analysis of mudflow properties, *Journal of Hydraulic Engineering*, 114(8):877–887.

Scholtz, A and Chetty, S, 2021. A comparison between HEC-RAS 6.0 and FLO-2D as tools used in tailings dam breach analyses, in *Global Tailings Standards and Opportunities 2021*, pp 101–111.

Von Thun, J L and Gillette, D R, 1990. Guidance on Breach Parameters, U.S. Bureau of Reclamation, Denver, Colorado, 17 p.

Xu, Y and Zhang, L M, 2009. Breaching parameters for earth and rockfill dams, *Journal of Geotechnical and Geoenvironmental Engineering*, 135(12):1957–1970.

Flood failure sub-categories for tailings dam break failure impact and consequence category assessment

Z Lu[1] and B Pirouz[2]

1. Senior Engineer, ATC Williams, Mordialloc Vic 3195. Email: thomasl@atcwilliams.com.au
2. Senior Principal, ATC Williams, Mordialloc Vic 3195. Email: behnamp@atcwilliams.com.au

ABSTRACT

With several catastrophic tailings dam failures happening in the recent decade, claiming human lives and causing environmental damage, it has become crucial that the tailings dam break analysis (TDBA) be properly conducted to represent the credible failure modes and scenarios, and to assist with the assessment of the critical failure impact. This also urged the mining industry and regulatory authorities such as ANCOLD, GISTM and Dam Safety NSW etc to either develop new guidelines or revise the existing guidelines and regulations to more accurately define the realistic failure modes and Failure Impact Assessment (FIA) for the tailings storage facilities (TSF).

Most of the current guidelines and regulations categorise the dam failure and consequence category scenarios based on the initial hydrologic conditions, which include sunny day failure and flood failure. Nonetheless, the flood failure scenario among different guidelines is generally only defined as the storm-related failure events with concurrent downstream flood condition, without much details about the failure mode, and is usually only assessed for the incremental dam break impact over the natural flood in an embankment overtopping scenario. This may not necessarily represent the worst-case failure impact and may consequently underestimate the consequence category for the TSF.

Based on the authors' experience in tailings dam FIA, there should generally be two sub-categories of the flood failure for the TSF, namely the incremental flood failure (IFF) and the post-flood failure (PFF), with the former assessing the incremental failure impact over the natural flood event, while the latter one assesses the direct dam break impact once the downstream flood has subsided. These two scenarios should both be assessed following the potential credible failure modes, and the more severe impact from these two scenarios should then be considered for the flood failure consequence category assessment.

Key differences between the IFF and PFF for FIA are discussed in the paper, with several case studies presented to demonstrate the validity and necessity to consider both scenarios in assessing the credible, worst-case failure impact.

INTRODUCTION

Although tailings dams are generally formed by well-engineered containment structures, unfortunately, failure of these structures and loss of containment are still reported worldwide over the recent years, releasing both tailings and contaminated water. As summarised in Table 1 and shown in Figure 1, these failures can lead to significant environmental, social and economic impacts and mostly involve the loss of human lives.

TABLE 1

Summary of the recent major tailings dam failures (Morrison, 2020; Morrison and Gomide, 2019; WISE Uranium Project, 2023; Rivet, 2023).

TSF and Location	Date	Failure impact
Mount Polley, Canada	August, 2014	~10 Mm^3 of water and ~7 Mm^3 of tailings released
Fundão Dam, Brazil	November, 2015	~43 Mm^3 tailings released, 19 fatalities
Córrego do Feijão Dam 1, Brazil	January, 2019	~12 Mm^3 tailings released, 270 fatalities
Jagersfontein Tailings Dam, South Africa	September, 2022	~5 Mm^3 tailings released, multiple injuries and deaths

FIG 1 – Jagersfontein tailings dam failure, South Africa (Planet, 2022).

Understanding the potential impact of the hypothetical dam break scenarios for all tailings storage facilities (TSFs) has become crucial for all parties involved in the planning, permitting, operation and management of TSFs, including mine owners, design engineers, and regulatory authorities (Morrison, 2022). Therefore, tailings dam break analysis (TDBA) and failure impact assessment (FIA) are required to assist with quantifying the hazard and assigning appropriate consequence category to the TSFs, as well as the development of the emergency response plan for the overall safety management of the TSF.

As a response to this, both the mining industry and regulatory authorities have developed and revised guidelines and regulations to more accurately define the realistic failure modes and FIA.

A summary of the key requirements from different guidelines is provided below:

- Global Industry Standard on Tailings Management (GISTM) (ICMM, UNEP and PRI, 2020) specifies that the *'(tailings facility) breach analyses must be based on credible failure modes, and it is primarily used to inform emergency preparedness and response planning and the consequence of failure classification. The classification is then used to inform the external loading component of the design criteria.'*

- ANCOLD Guidelines on Tailings Dams (ANCOLD, 2019) states that *'the critical input to determining the Dam Failure Consequence Category is the assessment of the consequences of failure, which involves considering a dam break simulation under various conditions of flooding, including 'sunny day' (no flooding) and extreme flood events.'*

- Dam Safety NSW Guideline – Emergency Plans (Dam Safety NSW, 2020) specifies that a dam break study shall examine dam break failure scenarios associated with 'credible mode of failure', which *'include sunny day and flood dam breaks, from acceptable flood capacity up to probably maximum flood.'*

- CDA Technical Bulletin: Tailings Dam Breach Analysis (CDA, 2021) states that *'the failure modes, in conjunction with applicable hydrologic conditions, constitute the dam failure scenarios selected for tailings dam breach analysis. The two initial hydrologic conditions include: fair weather or sunny day scenario, and flood induced or rainy day scenario.'*

FAILURE MODES AND SCENARIOS

When performing any detailed or quantitative tailings dam break modelling and FIA, comprehensive numerical modelling and flood routing are generally required by the guidelines, and shall incorporate the following considerations:

- Credible failure modes and critical failure scenarios.

- Realistic, site-specific tailings physical parameters.

- Non-Newtonian flood routing based on tailings rheological parameters, should the tailings be identified as liquefiable and flowable.

Notwithstanding the detailed aspects of adopting site-specific parameters for the numerical dam break modelling, which are described by the authors in a separate study (Lu, Tan and Pirouz, 2022), the approach to identify credible failure modes and critical failure scenarios itself forms a crucial component in assessing the tailings dam break FIA.

There are various potential mechanisms that may trigger the failure of the impounding structure of a TSF, but generally these mechanisms can be grouped into three different categories of failure modes.

- <u>Embankment instability</u> – failure due to loss of structural integrity (failure mechanism such as foundation failure) which results in uncontrolled mobilisation and release of storage material.

- <u>Overtopping</u> – overflow from the embankment crest (often in a severe flood event) due to loss of available freeboard or exceedance of spillway design capacity which results in excessive scouring of the embankment material, loss of containment and outflow discharges.

- <u>Piping</u> – internal erosion of the embankment material due to hydraulic gradient and insufficient filtering, resulting in progressive embankment failure and loss of containment.

Care is then exercised to determine whether certain failure modes are credible, or physically plausible, for the TSF under assessment. Failure scenarios are then only assessed based on the identified credible failure modes, to find out the critical or worst-case failure scenarios.

The failure scenarios are commonly categorised into two types among the majority of the current guidelines and regulations based on the initial hydrologic conditions, namely the 'sunny day failure' and 'flood failure'.

- The <u>'sunny day'</u> or non-flood induced failure scenario (SDF) is commonly accepted as the failure event occurring during normal operating conditions of the TSF, without any influence from any storm event. Consequently, there will generally be no antecedent flood or adequate advanced warning of failure.

- The <u>'flood failure'</u> or flood-induced failure scenario (FF) is commonly accepted as the failure event triggered by extreme precipitation, and/or associated with various concurrent major flooding conditions downstream of the TSF. This scenario is generally associated with some degree of warning and emergency planning system.

Regardless of the description of the flood failure scenario, generally it is only required by the guidelines to assess the inundation impact based on *'the difference between the consequence of a natural rainfall and flood event without dam break and the consequences of the superimposed dam break flood resulting from the failure of the dam during that event'* (ANCOLD, 2012), to assess the incremental dam break failure impact.

From the authors' perspective, this may not necessarily represent the worst-case flood failure impact due to a series of common misunderstandings and simplified assumptions, and may consequently underestimate the consequence category for the TSF. Hence, some discussions and insights have been provided in this study regarding different possible types of flood-induced failure scenarios and the potential sub-categories that should be considered in the TDBA that might affect the critical failure impact assessment.

FLOOD FAILURE SUB-CATEGORIES

The flood-induced failure scenarios can be correlated to a variety of failure modes, and a range of variables related to the overall failure conditions. For example, the more common failure modes during a storm event include overtopping due to exceeded spillway capacity, or piping due to long-term poor seepage management. However, there are also less common failure modes that are post-storm yet still flood-induced, like piping failure after a prolonged period of excessive stormwater pond being kept within the TSF, or upstream-raised embankment instability triggered by static liquefaction of tailings after re-saturation due to extreme storm storage.

These less common post-storm failure events are still credible failure modes that can potentially lead to catastrophic failures, but can easily be overlooked as they do not fit into the standard 'sunny day' or 'concurrent flood' scenarios.

Furthermore, the following common misunderstandings about the flood failure scenario assessment can be summarised that may prevent a thorough consideration of all different potential failure impacts:

1. Simplifying the tailings dam break modelling by assuming that the failure impact with extreme flood storage into a 'non-flooded' downstream environment is always more critical. This approach arbitrarily adopts no downstream concurrent flood, so the absolute inundation impact becomes the 'maximum incremental impact'.

2. Assessing incremental impact only during the extreme flood (eg Probable Maximum Flood, PMF), and hence concluding that the incremental impact is negligible due to the significant catchment size differences.

3. Assuming that the concurrent downstream flood event has to align with the event that triggers the dam failure, regardless of the catchment size difference.

4. Assuming that the dam break impact is more critical only with the concurrent downstream flood, as the failure event should be happening during a flood event.

As such, to avoid overlooking the critical dam break failure impact from the credible failure modes associated with different flood-induced failure scenarios, it is proposed here that two sub-categories of the flood failure shall be incorporated into the TDBA, namely the <u>incremental flood failure</u> (IFF) and the <u>post-flood failure</u> (PFF) scenarios, to expand the current definition in the guidelines, and to capture the more severe impact.

Adopting this approach provides confidence that the appropriate flood failure consequence category will be assigned to the TSF under study.

Incremental flood failure

The definition of the IFF scenarios follows the conventional description of the 'flood failure' listed in the guidelines and regulations, which assesses the dam failures that are triggered during an extreme storm event and the corresponding downstream impact with the antecedent flood. This sub-category of the flood-induced failure scenario is particularly important when the downstream receiving environment is sensitive to the incremental flood impact caused by the release from the tailings dam failure. For example, the inundation impact from a certain flood event within a river system (downstream of a dam) for a specific 'area of interest' may be exacerbated by the additional inflow from the dam break failure release.

In general, *'a (downstream) flood that is equal or larger than the inflow design flood for the TSF is typically used for this scenario; however, the downstream incremental impacts can also be analysed using a range of floods for both the failure of the dam and for the downstream receivers'* (CDA, 2021).

Essentially, for similarly sized TSF and downstream catchments, the incremental impact of a dam failure shall be assessed via dynamic, real-time hydraulic modelling where the dam failure is triggered while the corresponding downstream flood condition is established. On the contrary, for significantly different catchment sizes between the TSF and the downstream environment (eg major river course), downstream flood events from average conditions to PMF should be considered to arrive at the most significant incremental dam break impact (Dam Safety NSW, 2020).

Post-flood failure

The PFF scenario, as the name suggests, assesses the flood-induced dam failures after a major storm event, when the downstream flood has receded or reverted to a normalised condition. This failure scenario is not equivalent to a 'sunny day' failure, as it only shares the same downstream condition, but the initial water stored within the TSF and also the tailings storage conditions can be completely different from an SDF. Furthermore, this scenario typically would not be associated with the same risk level and emergency planning as the SDF.

Unlike traditional water dams where the initial condition prior to a dam failure is entirely focused on the water storage volume, the change in the tailings storage conditions due to water pond variation may facilitate post-flood credible failure modes. Consequently, depending on the TSF operation and management, certain PFF scenarios can potentially generate a more severe impact towards the downstream receivers than the IFF scenario.

There are generally two conditions under which the PFF scenario should be assessed:

1. When a dam failure is unlikely to be physically initiated during an extreme storm event, eg overtopping failure of a TSF that has been designed with enough freeboard and spillway capacity to safely pass the PMF.

2. When a dam failure is initiated by the change in tailings conditions due to keeping excessive stormwater storage within the TSF for a prolonged period, eg tailings become potentially liquefiable after re-saturation, or excessive seepage forms due to high hydraulic gradient.

CASE STUDIES

Two case studies are presented in this section, to demonstrate the distinctive characteristics of each of the discussed flood failure sub-categories (IFF and PFF), and to highlight the potential significance of the different consequence category assessments from these scenarios.

It should be noted that although the presented case studies are non-existent projects and do not represent any actual tailings dam break FIA, the TSF storage conditions, tailings parameters, downstream antecedent flood assessment etc adopted in the following case studies are all based on the authors' actual experience with similar type of projects.

Case study 1 – max incremental impact

A cross-valley TSF is constructed approximately 2 km west of a major river channel and is situated along the contributing sub-catchment of the river, as shown in Figure 2. The embankment is constructed out of earthfill material with no raises and is equipped with an emergency spillway to the north which is designed to safely pass 1:1000 Annual Exceedance Probability (AEP) flood. The decant pond normally forms away from the embankment, over the potentially liquefiable tailings. The river to the east of the TSF can be typically described as an incised, trapezoidal shaped flow channel, with a nominal gradient from the south to the north at 0.06 per cent. The river channel is approximately 7 m deep and some 60 m wide at the riverbed, and approximately 90 m wide at the bank top. The river extends some 40 km in length and encompasses an overall catchment area of approximately 1800 km^2. In comparison, the catchment that drains into the TSF is approximately 1 km^2. There is no critical infrastructure between the TSF and the surrounding drainage paths, and there is only one low-lying road across the river 2.5 km north-east of the TSF which may be subject to flooding.

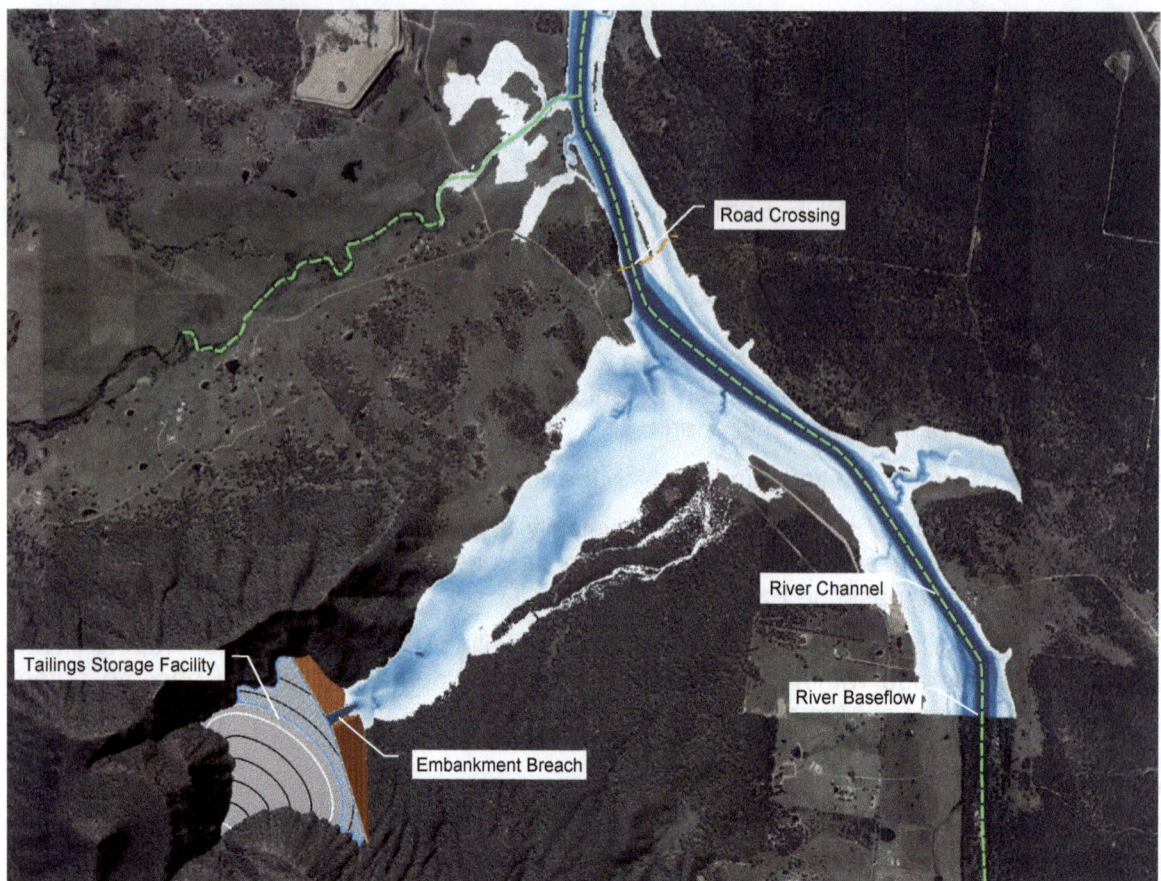

FIG 2 – Case study 1 site layout.

Traditionally, it is common for this type of projects to be assessed only for the SDF or PFF failure scenario, as due to the significant catchment size difference, the incremental impact from the dam break failure release over the downstream flood is 'expected' to be minimal. Consequently, as the dam break failure outflow enters the river with just the baseflow (ie 'sunny day' condition), the inundation is assessed to be fully contained within the river channel with no direct impact over the road crossing level. As a result, a consequence category of 'Low' is assigned.

However, due to the unknown timing of the dam failure, the concurrent flood within the river channel does not necessarily have to align with the flood that triggers the dam failure. As indicated in Figure 3, based on the water level time series extracted from the river flood study, the downstream road can be frequently flooded with different arrival times and durations. As such, to assess the maximum incremental impact at this low-lying road across the river, the IFF scenario should assume that the TSF is overtopped due to an extreme storm event exceeding the spillway capacity, at the time when the downstream flood is just reaching the surface level of the road. The TSF overtopping storm event could generate more severe flooding at the road crossing, but by the time the dam failure release enters the river channel, the flood level might not be at its peak and could potentially be just aligning with the road level. This assumption ensures that all the liquefied tailings and water released from the TSF failure contribute directly to the impact on the road and its potential users. Consequently, based on the numerical modelling output, it is assessed that the maximum incremental impact due to the dam failure would result in a PAR of 1 to 10 and the consequence category would increase to 'Significant'.

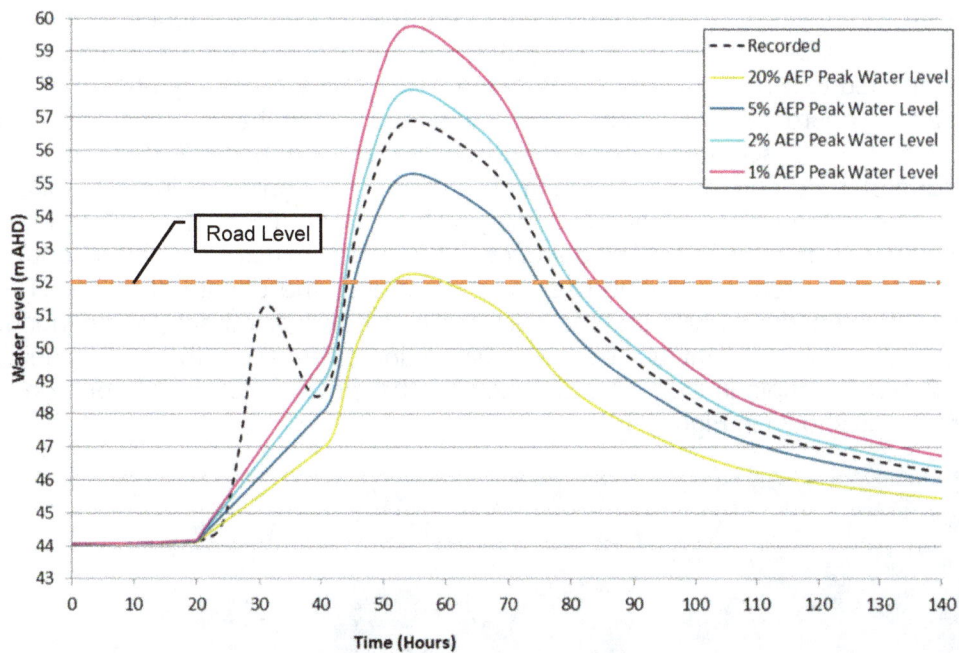

FIG 3 – Water level time series from the river flood study in case study 1.

Case study 2 – post-flood tailings failure

Another paddock style TSF with perimeter discharged tailings and a central decant structure is presented in Figure 4. Due to the centralised decant pond location and the distinctive dry/wet seasons (tropical monsoon climate) at this TSF site, the majority of the tailings around the perimeter area of the TSF are drained with a low degree of saturation. The earthfill embankment is formed by one starter and three consecutive upstream raises. The TSF is not fitted with an emergency spillway and hence stormwater would be stored within the facility until pumped out. However, the only extreme flood event that can overtop this TSF is identified to be the PMF due to the catchment being limited to its impoundment area. Only the process plant to the north of this TSF is identified as the potential impacted site, and the surrounding catchment is just slightly bigger than the TSF impoundment area.

FIG 4 – Case study 2 site layout.

Comprehensive liquefaction assessment based on the *in situ* geotechnical investigation has confirmed that the tailings in this TSF are not potentially liquefiable at their existing condition, unless they get re-saturated to a high degree of saturation. Based on the non-liquefiable assessment, the commonly adopted approach is to assess an SDF scenario with embankment collapsing and tailings slumping only, and an IFF scenario based on overtopping failure during a PMF event and the release of internal storm storage only (ie no tailings mobilisation). No significant impact is identified in the SDF scenario due to the limited tailings slumping distance. The IFF scenario is then assessed via dynamic precipitation modelling within both the TSF and downstream catchment, allowing for overtopping to happen from the northern embankment near the decant structure. The modelling results for the IFF scenario indicate that the maximum incremental impact (failure inundation minus natural flood) at the process plant includes a maximum incremental depth of 0.8 m and a maximum velocity of 0.2 m/s. Example inundation map showing the incremental depth from the modelling of the IFF scenarios is shown in Figure 5.

FIG 5 – Case study 2 IFF inundation map – incremental depth only.

However, based on experience from the operation of this TSF, the decant pond has the potential to rise and reach to the perimeter embankment during the wet season, which takes months to be pumped out of the facility. The hydraulic gradient over the tailings beach surface when this extreme stormwater pond is formed can facilitate constant infiltration of water into the drained tailings and re-saturate the tailings. Hence, a transient seepage model has been developed by the authors using Seep/w (Geostudio 2021.4, GEOSLOPE International Ltd, 2021), to simulate the response and re-saturation of the tailings due to the extreme storm storage along with the pumping. The tailings' unsaturated soil behaviour in relation to the changes in the negative pore water pressure has been defined in the modelling process by using laboratory test results. The technical details of the re-saturation simulation are beyond the scope of this study and are therefore not discussed herein, however, the simulation confirms that tailings against and beneath the north embankment have the potential to re-saturate after the prolonged period of keeping the extreme water pond in the storage. Consequently, tailings may then undergo static liquefaction, resulting in the potential embankment failure and uncontrolled release of liquefied tailings and the water pond towards the process plant.

Since this identified credible failure mode does not match with a typical 'sunny day' condition, and is still flood-induced, this failure scenario should be assessed separately as a PFF and hence distinguished from the existing predefined SDF and IFF scenarios, which do not involve any liquefied tailings routing as non-Newtonian fluid from the dam failure.

Due to the high solids concentration of the re-saturated, liquefied tailings flow, high rheological properties have been measured and adopted for the modelling of this PFF scenario. As a result, slow mobilisation of the tailings flow generates a limited inundation footprint, but with a more severe inundation impact. A maximum inundation depth of 1.5 m with a maximum velocity of 1.5 m/s is then identified at the process plant area, which is considerably more critical than that assessed from the IFF scenario. An example inundation map from the modelling of this PFF scenario is shown in Figure 6.

FIG 6 – Case study 2 PFF scenario inundation depth map.

CONCLUSIONS

Most of the recent guidelines and regulations require that TDBA should consider credible failure modes and the associated failure scenarios, which are generally categorised into two categories: flood induced and non-flood induced failures. However, due to the lack of details in describing the failure modes and misunderstandings for the flood-induced failure scenarios, this study proposes that two sub-categories of the flood-induced failure for the TDBA should be considered, namely the incremental flood failure (IFF) and the post-flood failure (PFF).

The IFF focuses on assessing the maximum incremental dam break failure impact over the natural flood event in the downstream environment, while the PFF assesses the direct dam break impact once the downstream flood has subsided. These two scenarios should both be assessed as potential credible failure modes, and the more severe impact from the two analyses should then be considered for the flood failure consequence category assessment.

As demonstrated via the case studies presented in this paper, the IFF scenario can be the most critical when the maximum incremental impact is properly assessed. In the meantime, the PFF scenario can also generate a more severe impact should the tailings condition deteriorates due to the storage of extreme flood.

In summary, the authors would like to promote more realistic, site-specific TDBA, not just via representative tailings parameters, but also by incorporating the comprehensive range of failure scenarios, to assess the credible, worst-case failure impact.

REFERENCES

Australian National Committee on Large Dams (ANCOLD), 2012. Guidelines on the Consequence Categories for Dams, Australian National Committee on Large Dams.

Australian National Committee on Large Dams (ANCOLD), 2019. Guidelines on Tailings Dams – Planning, Design, Construction, Operation and Closure, Australian National Committee on Large Dams.

Canadian Dam Association (CDA), 2021. Technical Bulletin: Guidelines for Tailings Dam Breach Analysis, Canadian Dam Association.

Dam Safety NSW, 2020. Guideline – Emergency Plans, NSW Department of Planning, Industry and Environment.

GEOSLOPE International Ltd, 2021. GeoStudio Seep/w 2021.4, version 11.3.2.23783.

International Council on Mining and Metals (ICMM), United Nations Environmental Programme (UNEP) and Principles for Responsible Investment (PRI), 2020. Global Industry Standard on Tailings Management. Available from: <https://globaltailingsreview.org/> [Accessed: 15 September 2020].

Lu, Z, Tan, Y and Pirouz, B, 2021. Non-Newtonian Tailings Dam Break Analysis and Released Volume Estimation Using Site-Specific Parameters, in Tailings 2022: 8th International Conference on Tailings Management (Gecamin Publications).

Morrison, K F, 2020. Tailings dam failures and regulatory or social response: A reflection of lack of trust?, in *Mine Tailings—Perspectives for a Changing World* (eds: R Furey and J F Lupo), pp 97–104 (SME: Englewood).

Morrison, K F (ed), 2022. *Tailings Management Handbook: A Life-Cycle Approach* (Society for Mining, Metallurgy & Exploration).

Morrison, K F and Gomide, P C R, 2019. Post-Córrego do Feijão: Continued evolution of tailings dam regulations in Brazil, in *Tailings and Mine Waste 2019: Proceedings of the International Conference on Tailings and Mine Waste*, pp 391–400 (University of British Columbia: Vancouver).

Planet, 2022. Tailings Pond Failure, Planet Labs Skysat, released under Creative Commons license, used with permission, Available from: <https://www.planet.com/gallery/#!/post/tailings-pond-failure/> [Accessed: March 22, 2023].

Rivet, S, 2023. Preliminary Survey Report for Jagersfontein, Orthophotos, Topographic Surveys (DEM), and Reconstructed Surfaces Produced from Archive Satellite Imagery [online]. PhotoSat Information Ltd. Available from: <https://storage.pardot.com/840113/1674671854iRP0gLZV/Preliminary_Survey_Report_for_Jagersfontein.pdf> [Accessed: 24 February 2023].

World Information Service on Energy (WISE) Uranium Project, 2023. Chronology of major tailings dam failures. Available from: <https://wise-uranium.org/mdaf.html> [Accessed: 24 February 2023].

Comprehensive tailings dam breach analyses – a framework for overcoming new challenges of adhering to competing industry requirements

N C Machado[1], H A Duarte[2] and N Moon[3]

1. Senior Resources Engineer, WSP, Brisbane Qld 4000. Email: nathalia.machado@wsp.com
2. Senior Water Resources Engineer, WSP, Perth WA 6000. Email: helvecio.duarte@wsp.com
3. Principal Water Resources Engineer, WSP, Brisbane Qld 4000. Email: nigel.moon@wsp.com

ABSTRACT

Tailings storage facilities (TSF) are an essential part of mining. Understanding their risks is vitally important for regulators as well as for nearby communities, lenders and shareholders. The tailings dam breach analyses (DBA) are an important input to understanding these risks. They provide information for the consequence classification assessments, risk assessments, emergency response plans, etc. These assessments are the driving force behind safely managing existing TSF and safely designing new TSF.

Identifying the hazards associated with a TSF is a vitally important step in this process. For this, there are currently several industry guidelines around the world such as the Global Industry Standard on Tailings Management (GISTM, 2020), Australian National Committee on Large Dams (ANCOLD, 2019), and the Canadian Dam Association (CDA, 2021), along with industry-specific regulatory requirements. In recent years, detailed company requirements have also been introduced. Together these create high standards for the DBA, and now, in most cases, they supersede the previous minimum requirements. Today, as industry pushes for increased consideration of credibility in the DBA's input assumptions, a high degree of technical detail must be assessed. This poses challenges for consultants completing the work. To undertake a reliable and credible DBA, consultants now must consider a range of requirements: the TSFs technical aspects, inputs from stakeholders and the Engineer of Record, while balancing all this with data gaps and the inherent uncertainties of our current methodologies, eg model uncertainties.

This paper aims to outline the main challenges encountered in studies carried out in Brazil, Canada and Australia. It also presents a framework, which includes suggested steps to overcoming these challenges. The key phases of this approach are a thorough desktop review and gap analysis; a workshop on credible failure modes with stakeholders; outflow volume definition considering embankment and tailings characteristic inputs; a rheology assessment on deposited tailings; definition of the downstream hydrological setting; breach and runout modelling; and sensitivity analysis on inherent uncertainties associated with the DBA.

INTRODUCTION

Dam breach analyses (DBA) are critical tools to assessing risks and informing management decisions during all stages of a tailings storage facility's (TSF) life cycle. According to the Global Industry Standard on Tailings Management (GISTM, 2020), conducting human rights due diligence is necessary to address the risks associated with a TSFs a credible failure scenario – noting that the understanding of the risks are significantly more challenging to address than for water dams.

Evaluating a TSF is complex due to several factors such as the tailings characterisation, liquefaction, rheology, development management and construction method, geographic and hydrologic setting, all of which are unique to each facility. Therefore, DBA is a crucial component at each stage of a study and may require different levels of detail. For example, during a feasibility study the screening method may be sufficient to meet the requirements. However, for facilities with greater consequence classifications, a comprehensive assessment is required as per Australian National Committee on Large Dams (ANCOLD, 2019). The comprehensive assessment should offer a quantitative precise definition of credible failure modes, which will lead to more reliable results for failure depth, velocity, inundation extents, failure travel times and other essential data. These data support consultants as they define the population at risk (PAR), potential loss of life (PLL), and other consequences that will eventually be fed into consequence category assessments (CCA) and emergency response plans (ERP).

Following recent dam failures at Mount Polley in 2014, Fundão dam in 2015 and Brumadinho in 2019, new guidance on tailings management and on DBA for tailings facilities were published by the International Council on Mining and Metals (GISTM, 2020) and the Canadian Dam Association (CDA, 2021), respectively. Similarly, mining companies also have their own guidance documents that must be considered. Despite the industry's acceptance of water-based models due to inherent limitations associated with DBA of tailings facilities (ie empirical breach parameters), best practices continue to evolve as demand continues for more realistic assessments.

Another challenging aspect of completing a DBA is also related to the different sources of guidance. This paper's objective is to outline approaches that can be applied in this scenario. The proposed framework is divided in two phases: phase 1 being the desktop review and gap assessment and phase 2 the DBA development. The framework is based on the following industry guidelines:

- ANCOLD's current *Guidelines on Tailings Dam* (ANCOLD, 2012a) and its addendum (ANCOLD, 2019).

- *Guidelines on the Consequence Categories for Dams* (ANCOLD, 2012b).

- GISTM (2020).

- *Tailings Dam Breach Bulletin* (CDA, 2021).

PHASE 1 – DESKTOP REVIEW AND GAP ANALYSIS

This gap analysis phase is proposed for two main reasons: 1) most sites have a previous DBA which can be built upon; and 2) there is a higher technical expectation on this type of study which is quickly evolving and a gap analysis mitigates the need for rework later in the project.

The main objective of the gap analysis phase is to inform stakeholders of the gaps encountered in the previous DBA prior commencing the study's next stage, which is the development of a new DBA. The gap analysis should:

- Identify gaps by comparing existing DBAs with relevant requirements outlined in the respective industry guidelines and best practices references, such as CDA, ANCOLD and GISTM.

- Present the desktop review findings, including recommendations on how to conform to the relevant requirements.

- Outline what is needed to update the existing assessments and/or undertake new assessments with consideration of the latest designs and future TSF developments.

The main objective of the gap analysis is to identify where the existing DBA and failure impact assessment (FIA) are not in accordance with best-practice references and respective industry standards, if existent. The gap analysis is also an appropriate point to evaluate if the most recently collected data is sufficient to complete an updated DBA and fill the identified gaps. This is particularly important if there is a transition to a new stage of the TSF life cycle, or if major changes have occurred at the TSF. If the new data is sufficient, the recommendation is to go ahead with the study and complete a new DBA. Otherwise, consider filling the information gaps with new information prior to starting another study that will likely be considered non-compliant according to the requirements. Figure 1 illustrates the proposed workflow.

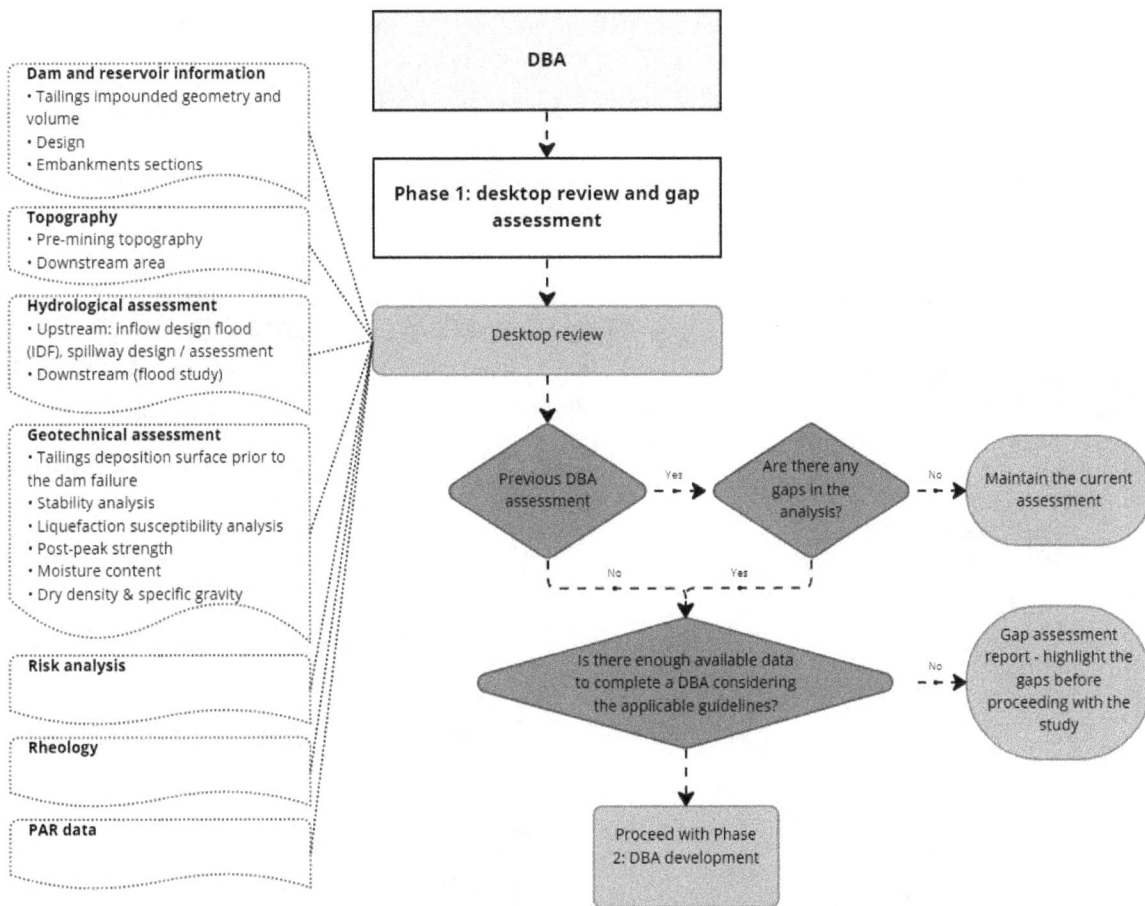

FIG 1 – Phase 1 framework.

PHASE 2 – DBA DEVELOPMENT

Phase 2 involves developing a dam breach study, which comprises the following stages:

- Credible failure mode determination
- Definition of the failure scenarios:
  - TSF configuration
  - Breach location
  - Initial hydrological condition
  - Downstream hydrological conditions.
- Tailings characteristics and rheology
- Outflow volume
- Runout modelling and mapping.

Credible failure mode

The latest DBA guidelines are undergoing a shift from a conservative and unfeasible approach to a more pragmatic and evidence-based approach. This can be observed in Requirement 2.3 of the GISTM (2020), and also in CDA (2021) and some industry guidelines. These guidelines states that breach analyses should be based on the credible and physically possible failures modes where loss of containment is possible. The recommendation is to always conduct a DBA scenario for the worst credible case, regardless of the probability of the failure scenario.

To help identify and justify the failures modes prior to starting the DBA, an event-based and/or risk-based approach should be used such as failure mode and effects analysis (FMEA), potential failure mode analysis (PFMA), fault tree analysis (FTA), event tree analysis (ETA). From such a risk analysis, the credible failure mechanism and trigger event will be identified for each structure.

In addition, and in accordance with GISTM Requirement 2.1 (GISTM, 2020), it is expected that all the credible failure modes and site conditions be considered. However, not all failure mechanisms will lead to a dam breach (ICOLD, 2022). If a risk analysis determines there is no possible failure that would lead to an uncontrolled discharge of tailings, then the analysis may be documented in the report – making it unnecessary to conduct the study.

To ensure compliance, it is essential to verify industry guidelines since there are certain guidelines that require the assessment of susceptibility to flow liquefaction, regardless of credibility.

Figure 2 illustrates the proposed decision tree for the credible failure mode stage. Notably, Figure 2's correlation with Figure 1 is predicated on the availability of a risk analysis as it facilitates informed decision-making regarding the best study approach.

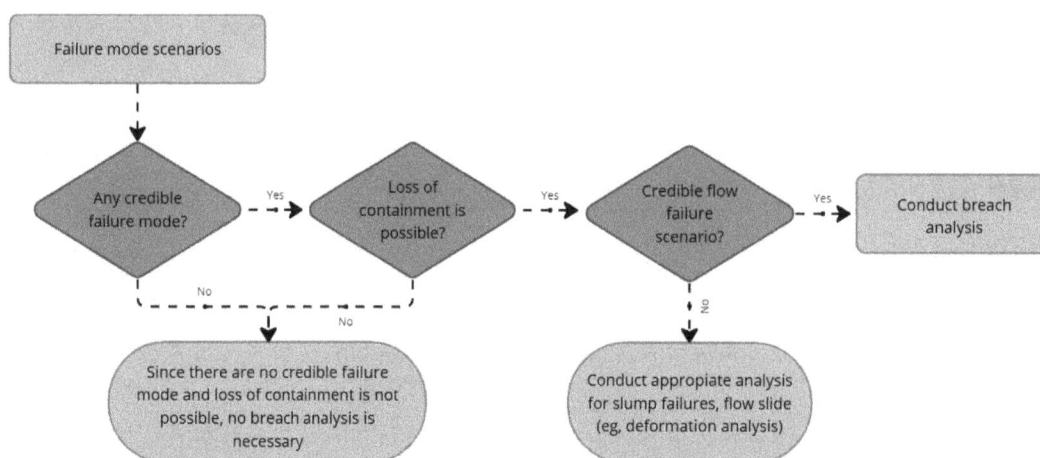

FIG 2 – Decision tree for credible failure mode.

Definition of scenarios

The selection of an appropriate failure scenario is based on a combined set of definitions for each of the following: TSF phase, breach location, hydrological condition, failure mechanism and trigger event. Defining each aspect is necessary as they can have a great impact on each other and consequently on subsequent DBA steps such as runout modelling.

TSF configuration

The risks of loss and damage associated with a TSFs failure may change significantly during its life cycle. Risk is affected by factors such as deposition capacity increases, operation of a supernatant pond and freeboard management. Based on this and the GISTM (2020) recommendation, a DBA should be completed at all phases in the TSF life cycle, including at closure and post-closure. During project conception or design level, a DBA should be completed to provide estimates on the consequence for multi-criteria analysis, which eventually may have an impact on TSF site selection. At this stage, depending on the maturity level, it may be considered that a TSFs future scenario could represent a worst-case scenario, which is usually associated with the TSF being at full capacity when operations end. Subsequent DBA updates should be carried out to inform risks associated with failure during operation. It should be noted that if there is a planned stage transition or any change to the facility in the near future (eg within the next five years), it may be preferable to evaluate a construction stage or a near-future phase, instead of the current condition. Updates to the DBA may also be undertaken if the TSF is under standby care, if closure is planned, or if there is an expected change in PAR or downstream environmental conditions.

Breach location

The assessment should consider potential breach locations at sections of confining embankments along the facility that could influence downstream effects. This is usually included in the risk assessment, and it is closely associated with the failure mechanism. A ring-dyke type of TSF will likely require selection of several breach locations, unlike cross-valley facilities. The beach location definition is a necessary part of the risk assessment (ANCOLD, 2012b) and therefore should be aligned with an existing FMEA, PFMA or similar (CDA, 2021).

Initial hydrological conditions

TSF failures should be assessed under non-hydrologic (ie sunny day) and hydrologic (ie flood day) conditions to represent the facility under normal operation conditions and during an extreme rainfall event, respectively.

The surface water level at the facility under normal operating conditions is usually defined as the normal operating level (NOL). Conservatively, ANCOLD (2012b) and CDA (2021) recommend adopting the maximum normal operating level (MOL) despite the persistent lower water levels at the reservoir. This can also be called the full supply level (FSL) (ANCOLD, 2000). For TSF that are not designed to allow discharge from the reservoir, the MOL should preserve the water surface levels up the required storage allowance and freeboard in case of an extreme event. In these cases, generally there are no emergency spillways constructed at the facility. Eventually, the water surface can be assumed to be at the TSF crest if the supernatant pond levels have been close to the top of the dam. However, this is not a credible scenario for a TSF that is operating with storage allowance and adequate freeboard (CDA, 2021). For a TSF that does not have a spillage restriction, the MOL may correspond to the spillway invert, especially if discharges have been registered in the past.

Consideration should be given to supernatant pond management for a facility that controls the supernatant pond at levels near the minimum operating level. For example, continuous supply of return water may be necessary to sustain process plant operations and operational trigger levels, particularly when tailings deposition cannot cease. The historic water levels at a facility and the water balance outcomes may be important to support the definition of the MOL. Particularly when considering facilities without a spillway, with considerable storage allowance above the MOL, and which are under routine surveillance.

CDA (2021) suggests the flood day scenario be associated with an inflow design flood (IDF), in addition to the MOL considered for the sunny day scenario. CDA recommends the adoption of a flood equal or larger than the IDF used in the TSF design, while ANCOLD (2012b) recommends the use of the dam crest flood (DCF) and the probable maximum flood (PMF). For facilities without a spillway, there should have sufficient capacity above the MOL to store the PMF without leading to overtopping. Typically, the 24-hour duration storm events are used. However, longer duration storms (eg up to 72 hours) may be considered as they are critical in terms of the volume of mobilised tailings and pond water (CDA, 2021). If the facility contains an emergency spillway, the critical duration of the IDF should be determined in terms of peak discharge and maximum water surface elevation at the reservoir.

Table 1 summarises the proposed water surface levels for failure scenarios and Figure 3 presents a framework based on assessed guidelines. It is important to note that GISTM (2020) does not specify the initial water condition for each scenario, focusing instead on the credibility aspects of the failure.

TABLE 1

Proposed water level for sunny day and flood day failure scenarios.

Scenarios	CDA (2021)	ANCOLD (2019)
Sunny Day	MOL, which can be one of the following depending on operation: <ul><li>No discharge structure: below storage allowance and freeboard *or*</li><li>TSF with spillway: elevation of the spillway invert *or*</li><li>Elevation of the top of the dam</li></ul>	MOL (also known as FSL)
Flood Day	Flood equal or larger than the IDF	DCF PMF

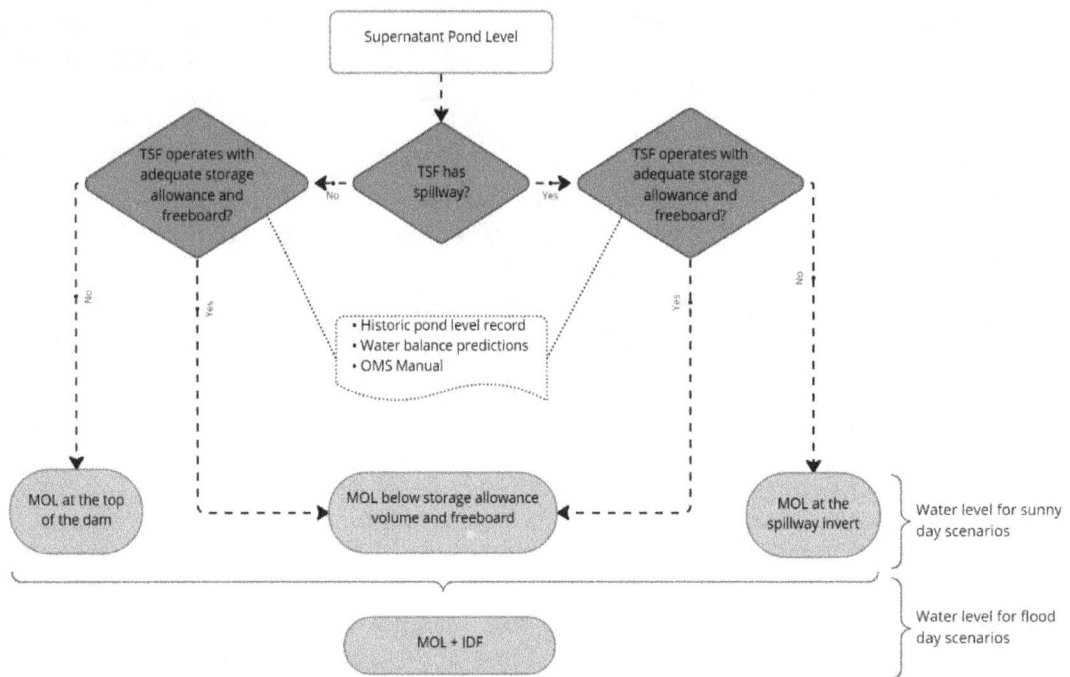

FIG 3 – Framework to define the initial water pond level for Sunny Day and Rainy Day Scenarios.

The IDF duration can have great influence on the failure mechanism and outflow volume. As mentioned before, the initial water level, usually the MOL, should be a plausible condition that can occur in the facility. If not, the additional volume required to overtop the TSF may be greater than the PMF, which is not plausible (Martin, Fontaine and Cathcart, 2015). In this case, flood-induced failure mechanisms besides overtopping should be assessed.

The initial hydrologic condition also plays a relevant role in defining the failure mechanisms, the released supernatant pond volume, and the estimate of PAR as occupancy at the mining site may vary under different hydrologic conditions (eg site may be evacuated under extreme rainfall events).

Downstream hydrological condition

The GISTM (2020) requires the TSFs failure consequence on the downstream drainage network be determined based on incremental environment, social, cultural, and economic losses. This means understanding the magnitude of the flow in adjacent catchments is important to identifying the incremental impact of a dam failure during sunny day and flood day scenarios.

For sunny day scenarios, CDA (2021) recommends setting the drainage network to the mean annual discharge of rivers and creeks. Eventually, if the flow downstream of the facility is expected to be at least two time less than the failure flow, it can be disregarded. Since the extent of the affected zone may also be influenced by topography (ANCOLD, 2012b), judgement should be made in selecting a natural flood event when only topography (without detail on river and creek bathymetry) is available. The addition of the mean annual discharge or a 2-year annual exceedance probability (AEP) flood event will likely represent a worse flooding condition and occupation of the floodplain.

The downstream drainage network should be taken at flood levels expected to occur concurrently with the IDF selected for the TSF catchment. Ball *et al* (2019) propose a concurrent tributary flow method to identify a flow in an adjacent catchment that is likely to coincide with design floods that can be applicable, which is recommended by CDA (2021). For small drainage network catchments, the flood frequency may be similar to that of the TSF. For large drainage network catchments, the flood frequency may be smaller than that of the TSF. If the concurrent flood method proves difficult to apply for large catchments, it may be acceptable to evaluate several AEPs as background floods to determine the largest incremental impact following the breach. Generally, the smaller the background flood applied to the downstream network drainage, the greater the incremental effects of the failure. By contrast, the greater the background flood of the downstream network drainage, the greater the overall impacted area and the smaller the incremental effects.

Tailings characteristic and rheology

In accordance with GISTM Requirement 2.1 (GISTM, 2020), the DBA should use a methodology that considers slurry properties. It is important to note that tailings characteristics tested for pipeline or thickener designs may not be suitable for DBA. Ideally, characteristics of consolidated tailings deposited in the facility should be considered, along with the inherent spatial variability, the heterogeneity of *in situ* properties, and rheological parameters.

In the absence of *in situ* and laboratory testing data, information on tailings composition and characteristics can be obtained from literature. Material types and their applicability should be considered as tailings behaviour differs from hard (eg copper, nickel) and soft (eg coal) rock mine materials (Martin, Adria and Wong, 2022). The percentage of fine material, particle solids distribution and plasticity index (ie Atterberg limits) can also influence the viscosity and yield stress of tailings, and such information can be used to predict the rheology when no other data are available (Fitton and Seddon, 2012).

Sensitivity analysis of the rheology parameters should be carried out even when there is data available since the inundation extent can vary significantly when varying the rheology (CDA, 2021). Extrapolation of *in situ* data or literature is often required to represent the solids concentrations expected in the DBA's released volume and in the eventual mixing with the downstream background flood, especially for flood day scenarios.

Outflow volume

The outflow volume that is mobilised during a tailings dam failure depends on several factors. These include the geometry of the structure and its surrounds, the failure mechanism and previous weather conditions. It is often a challenging assessment to represent the physical process of a TSF failure and consequently precisely predict the tailings released volume. Because of that a simplified approach to assume a geometrical method such as a conical failure surface to estimate the released tailings volume (Moon *et al*, 2019) has been widely adopted. This type of method is based on the assumption that the remaining tailings in the reservoir present a stable post-failure slope.

Due to the lack of guidance on this subject, tailings characteristics should be based on the existing geotechnical assessment and site investigations data from the facility. A framework is proposed in Figure 4, which suggests the failure slope may be defined by the strength of deposited tailings. The peak or residual strength of tailings may be considered depending on the failure mechanics associated with the TSF. For example, post peak strength are used for saturated and liquefiable tailings. Otherwise, Adams *et al* (2022) suggest adopting the peak strength ratio if the failure mechanism is an embankment slump and the tailings is not likely to liquefy. According to CDA (2021), the boundary between liquefiable and non-liquefiable material may not be well defined since both contractive and dilative tailings may coexist within the TSF. It is necessary to evaluate the TSFs physical condition, the characteristics of the impounded material, and the overall understanding of the physical process that could occur during a failure process.

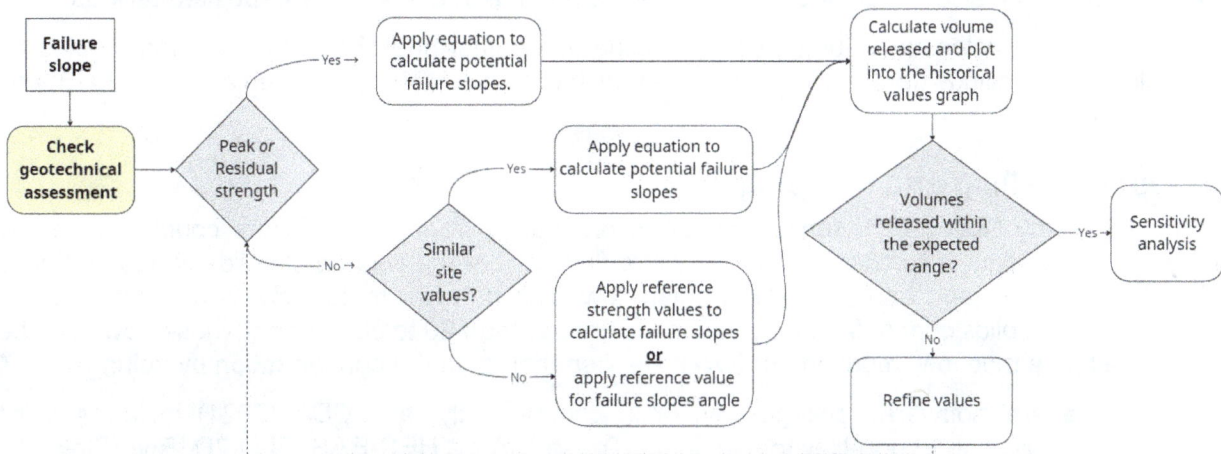

FIG 4 – Framework to determine the outflow volume.

The tailings slope can be estimated by simplified methods that calculate the infinite slope angle. Barrera and Riveros (2006) suggest a method to determine the infinite slope, assuming that the sliding mass's density and shear-strength parameters are constant along the slope and that the slope is saturated. Errazuriz (2018) proposes a method to estimate the maximum stable beach slope considering an infinite slope limit equilibrium for both saturated and unsaturated conditions.

Technical judgement should be made to these methods as the deposited tailings present inherent spatial and vertical variability in tailings properties that often could not be captured in the current site investigation (eg peak and residual strength, degree of saturation, dry density, moisture content).

To understand the impact of these uncertainties on the study, one of the key aspects, the post-peak strength should be tested and also varied. The peak and post-peak strength can be used to estimate the material mobilised in a failure case, which is expected to vary according to the density and plasticity of the tailings. Adams *et al* (2022) mentioned the range of the expected values based on their experience (Table 2).

TABLE 2
Residual strength used to define the stable slope angle (Adams *et al*, 2022).

Material	Peak undrained shear-strength ratio	Residual strength ratio
Loose saturated coarse tailings sands	0.1	0.04 to 0.05 – susceptible to liquefaction
Denser saturated coarse tailings sands or fine tailings with some plasticity	0.2 to 0.3	0.15 to 0.2

The methodology proposed to calculate the failure slope and volume released, including the sensitivity analysis, is presented in Figure 4.

Obtained failure slopes can be compared to available historic data from studies by Blight and Fourie (2003), which indicate that failure slopes typically vary between 3 and 7 per cent and can be up to between 17 and 33 per cent around the perimeter of the cone. Observations from historical tailings dam breaches by Rourke and Luppnow (2015) indicate that post-failure tailings slopes are typically within the range of 5 to 18 per cent.

Engineer of Record (EoR) involvement is also crucial in this stage of the study since their knowledge of the structure is broad enough to fulfil all the functions described in the GISTM (2020), such as responsibility for the design basis report, design report, performance reviews, etc.

Additional tailings volume may be released by hydraulic erosion due to fluid discharge from the facility and downcutting through the embankment, as shown by Fontaine and Martin (2015). Nevertheless, Chen (2022) shows that not only the supernatant volume may influence the eroded tailings volume, but also the geotechnical properties and the integrity of the deposited tailings.

Finally, calculated released volumes can be validated against published research and studies on available historic tailings dam failures, as presented in Rico *et al* (2008), Larrauri and Lall (2018) and Rana *et al* (2021).

Runout modelling and mapping

As mentioned by CDA (2021) runout modelling originating from a TSF failure could involve two different types of flow, a Newtonian flow (or water flow, where the volume of solids concentration is below 20 to 30 per cent) and a non-Newtonian flow. The non-Newtonian flow is characterised by fluids where the solids concentration by volume is greater than 20 to 30 per cent; these flows can be categorised as a mudflow, mudflood or flow slide, depending on the concentration by volume.

There are several software packages available commercially, and CDA (2021) includes a list selected for application to non-Newtonian fluids. These include HEC-RAS, FLO-2D, RiverFlow-2D, TUFLOW and FLOW-3D. However, it is important to note the software should not only be capable of running the non-Newtonian flow simulation, but also of estimating the required output for the case study. A clear example of this is the Requirement 2.1 of the GISTM (2020). It states that results should include estimates of the physical area impacted by a potential failure (a common output), but also flow arrival times, depth and velocities, and the depth of material deposition. Not every software is able to calculate the material deposition. The guideline also highlights that all these results are essential for the development of an ERP, turning this into one of the main criteria when choosing which software to use for the simulation.

Other aspects to consider when choosing the software are the type of failure, the available data, along with the outputs required. Another common criterion found in the industry guidelines is the two-phase flow modelling.

Regarding the available data, it is important to note that rheology is one of the most recent inputs added to the DBA. Since it is a recent requirement, many sites do not have the rheological characteristics of their tailings measured yet. Figure 5 suggests a framework showing the runout modelling decisions needed to produce the end result of inundation maps.

As mentioned, if a TSF failure's outflow volume is calculated and the solids volume concentration is bigger than 20 to 30 per cent, it will be characterised as a non-Newtonian flow. If there is a lack of rheological testing, it is suggested that some additional modelling is undertaken. A Newtonian simulation and non-Newtonian simulation using rheological theoretical values can be used as 'upper' and 'lower' inundation extent. Due to the inherent uncertainties of a study that lacks rheological data, it is suggested that the results are presented in terms of ranges, with maximum and minimum expected extents, and not with a unique and deterministic value.

It is essential to present the outcomes in the form of maps with appropriate cross-referencing to the corresponding reports containing the design basis, methodology, assumptions, and textual findings. The report should also indicate the main uncertainties and specific limitations of the study.

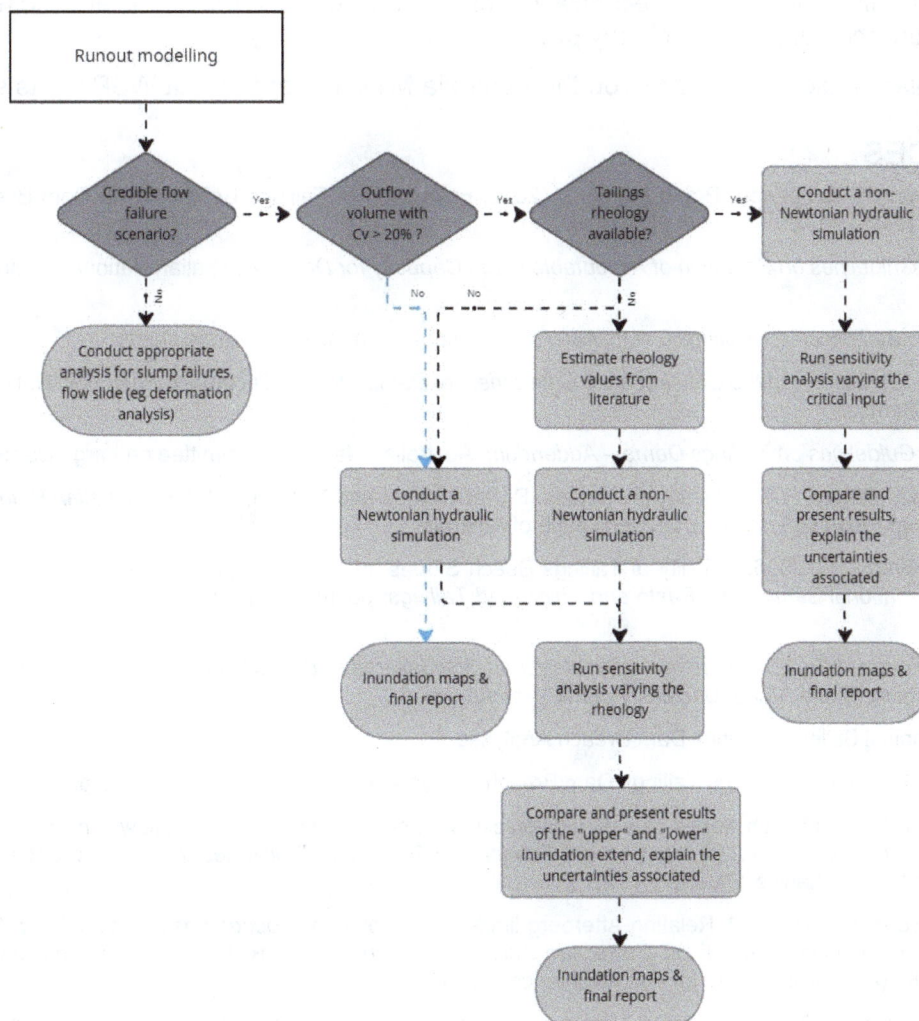

FIG 5 – Runout modelling framework.

CONCLUSIONS

Dam breach modelling and flood wave routing is a complex process. Recognising the several uncertainties inherent in each step of the DBA study and running sensitivity analyses is recommended to provide an overall understanding on how the adopted assumptions may impact the results, which will consequently feed into the CCA, risk assessment or ERP.

It has been emphasized the relevance of the desktop review and gap analysis stage of a DBA. The outcomes of this process may indicate the need to plan for additional studies or field data to meet requirements or may clearly state the uncertainties associated to data availability to all stakeholders if decision is to carry the assessment forward. Other stages require attention as available methods are simplification of complex physical processes, such as estimates of mobilised volume, breach parameters and rheology behaviour. Sensitivity analysis should be extensively used to determine the most appropriate parameters and assumptions that represents a physical and realistic outcome.

Even though the dam breach studies carry many uncertainties, when combined with professional experience and best practices defined by guidance, they can be a powerful tool to inform the risks associated with TSF. Complying with existing guidelines takes effort, as the available guides themselves may present different recommendations that may impact DBA completion. The present paper aimed to shed light on the existing challenges in adhering to competing requirements and present a suggested approach to deal with it.

ACKNOWLEDGEMENTS

The authors acknowledge that this paper was prepared on First Peoples land. We recognise Aboriginal and Torres Strait Islander Peoples as the first scientists and engineers and pay our respects to Elders past and present.

The authors greatly appreciate our technical editor, Kelly Stock, who provided valuable technical text review to ensure the content was clearly presented.

The authors also would like to thank you the Australia Mine Water team at WSP for its support.

REFERENCES

Adams, A, Brouwer, K, Robertson, P and Martin, V, 2022. Evaluation of Tailings Behaviour for Dam Breach Analyses. *Mine and Waste 2022.*

ANCOLD, 2000. *Guidelines on Selection of Acceptable Flood Capacity for Dams,* Australian National Committee on Large Dams.

ANCOLD, 2012a. *Guidelines on Tailings Dams,* Australian National Committee on Large Dams.

ANCOLD, 2012b. *Guidelines on the Consequence Categories for Dams,* Hobart, Tasmania: Australian National Committee on Large Dams.

ANCOLD, 2019. *Guidelines on Tailings Dams – Addendum,* Australian National Committee on Large Dams.

Ball, J, Babister, M, Nathan, R, Weeks, W, Weinmann, P, Retallick, M and Testoni, I, 2019. *Australian Rainfall and Runoff – A Guide to Flood Estimation,* Commonwealth of Australia.

Barrera, S and Riveros, C, 2006. Stability of Tailings Beach Slopes, in *Proceedings of Paste 2006: Proceedings of the Ninth International Seminar on Paste and Thickened Tailings,* pp 169–179 (Australian Centre for Geomechanics, Perth).

Blight, G E and Fourie, A B, 2003. *A review of catastrophic flow failures of deposits of mine waste and municipal refuse,* Johannesburg, South Africa: University of the Witwatersrand.

CDA, 2021. Technical Bulletin: Tailings Dam Breach Analyses.

Chen, J H, 2022. Hydraulic Erosion in Tailings Dam Breach Analysis, *Tailings and Mine Waste,* 9 p.

Errazuriz, T, 2018. Tailings beach slopes as a dimensionless parameter of non-Newtonian flows, in *Proceedings of Paste 2018: Proceedings of the 21st International Seminar on Paste and Thickened Tailings,* pp 515–530 (Australian Centre for Geomechanics, Perth).

Fitton, T D and Seddon, K D, 2012. Relating Atterberg limits to rheology. in *Proceedings of Paste 2012: Proceedings of the 15th International Seminar on Paste and Thickened Tailings* (eds: R Jewell, A B Fourie and A Paterson), pp 273–284 (Australian Centre for Geomechanics, Perth).

Fontaine, M and Martin, V, 2015. Tailings mobilization estimates for dam breach studies, *Tailings and Mine Waste,* 14 p.

GISTM, 2020. *Global Industry Standard on Tailings Management,* International Council of Mining and Metals.

ICOLD, 2022. Tailings Dam Safety Bulleting No 194, version 1.0.

Larrauri, P and Lall, U, 2018. Tailings dams failures: updated statistical model for discharge volume and runout, *Environments Journal,* 5(28).

Martin, V, Adria, D and Wong, H, 2022. Inundation Modelling of Non-Newtonian Tailings Dam Breach Outflows, in *Proceedings of the 27th International Congress on Large Dams World Congress,* pp 501–520.

Martin, V, Fontaine, D and Cathcart, J, 2015. Challenges with conducting tailings dam breach studies, *Tailings and Mine Waste*, 15 p.

Moon, N, Parker, M, Boshoff, H and Clohan, D, 2019. Advances in non-Newtonian dam break studies, in *Proceedings of the 22nd International Conference on Paste, Thickened and Filtered Tailings* (eds: A J C Paterson, A B Fourie and D Reid), pp 165–172 (Australian Centre for Geomechanics, Perth).

Rana, N M, Ghahramani, N, Evans, S G, McDougall, S, Small, A and Take, W A, 2021. Catastrophic mass flows resulting from tailings impoundment failures, *Engineering Geology,* vol 292.

Rico, M T, Benito, G and Díez-Herrero, A, 2008. Floods from tailings dam failures, *Journal of Hazardous Materials,* 154:79–87.

Rourke, H and Luppnow, D, 2015. The Risks of Excess Water on Tailings Facilities and Its Application to Dam-break Studies, *Tailings and Mine Waste Management for the 21st Century*.

Tailings dewatering
and placement

Impact of filter feed solids on filter sizing and equipment cost

C Braun[1] and K Rahal[2]

1. Global Process and Study Manager – Tailings, FLSmidth, Salt Lake City UT 84047, USA.
   Email: chris.braun@flsmidth.com
2. Global Director of Tailings, FLSmidth, Salt Lake City UT 84047, USA.
   Email: ken.rahal@flsmidth.com

ABSTRACT

More mining companies are investigating filtered tailings as part of their long-term sustainability plans and to reduce risks associated with tailings storage. FLSmidth (FLS) recommends that miners conduct tailings technology trade-off studies which includes, analysing different tailings solutions options, including different flow sheets and equipment to determine the optimal dewatering process. This analysis includes determining the most economic approach for each tailings dewatering option. The primary factors that affect filtration are the settling rate of the solids for thickening, feed density to the filters, target filter cake moisture, particle size distribution (PSD), mineralogy and particle shape. The impact of feed density on filtration will be the focus of this paper.

Thickeners are frequently used to increase feed solids concentration to filters, reducing the quantity of water that filters need to remove from the tailings slurry. Less water in the filter feed, increases filtration rates and can reduce the quantity of filters required decreasing CAPEX and OPEX. Surface disposal of tailings requires a filter to remove enough water to provide the geotechnical stability to stack. There are many types of filters used on mine tailings such as, rotary vacuum disc filters, horizontal vacuum belt filters and filter presses. Gold mines tend to have a finer grind than other metal tailings and therefore filter presses are usually specified for surface stacking.

This paper presents a case study for a greenfield gold site looking at filtering tailings for increased water recovery. The tailings dewater options considered in this project was: 1) direct filtration of the gold tailings using a filter press; and 2) thickening followed by filtration with a filter press(es). A comparison of the filtration rates and filter costs with and without thickening prior to filtration will be presented in this paper.

INTRODUCTION

The demand for minerals and metals, especially copper is growing. As the World Bank's 2017 report titled 'The Growing Role of Minerals and Metals for a Low-Carbon Future' shows, demand for metals, including copper, could rise tenfold by 2050 if the world moves towards a low-carbon energy future. This growth in demand is coinciding with continued declines in orebody quality. Mudd (2009) showed that orebody grades for copper, lead, zinc, and other minerals have been declining globally. This has forced companies to process larger tonnages to achieve economy of scale. These larger throughputs require larger amounts of water which can cause conflict with local communities due to increasing water scarcity. UNICEF predicts water demand to increase by 20 to 30 per cent by 2050 due to population growth and higher living standards as well as food and energy demands (UNICEF, 2021).

This increase in water consumption and tailings generation are also coinciding with investor pressure for more sustainable mining practices (Global Sustainable Investment Alliance, 2018; RBC Global Asset Management, 2018). More mining companies are investigating alternate dewatering solutions for their tailings as part of their long-term sustainability plans, to decrease the amount of fresh water needed, and to reduce the risks associated with tailings storage.

Tailings dewatering should always begin with a holistic flow sheet approach. Minor changes to upstream unit operations can have significant impact downstream with tailings dewatering. As the high-grade, easy to recover, mineral deposits become depleted, lower grade ore is required to maintain output for mining companies. These lower grade deposits need to be ground finer and finer to liberate the valuable minerals in the current deposits to maintain production. This causes the problem of increasing throughputs of tailings for the same valuable recovery as well as, ever-smaller particle size distribution that needs to be dewatered for disposal.

As the grind size of tailings becomes finer, a greater level of pressure differential is required to dewater tailings to a solids concentration that can be disposed of on the surface (paste backfill is not considered in this paper). Thickening is the cheapest form of water recovery in a tailings stream. The driving force is gravity which does not cost money, and the advancements of polymer flocculants has allowed settling rates of tailings to increase dramatically. But thickening can only produce a liquid tailings stream. Therefore, these liquid streams from a direct disposal method, up into the paste regime, still behave like fluids, requiring some retention of the slurry such as a dam (in the case of non-paste thickened tailings) or a retention wall (paste regime tailings such as a central thickened discharge). To produce tailings that behave like a plastic or solid, filtration techniques are required.

Filtering tailings can be performed using different methods which employ different driving forces. The driving force for filtration can be as low as gravity drainage (belt press filter) to as high as several g's (centrifuges). A belt press relies on gravity drainage dewatering of tailings prior to squeezing in-between two belts at up to 500 psi (very short high-pressure dewatering time). A centrifuge uses a rotation to induce g-forces to separate the solids from the liquid as in a solid bowl centrifuge. Both machines work very well in niche applications typically at lower throughput mines. For larger throughput mines that require low cake moistures, the main filter market is vacuum and pressure filters with the maximum pressure for vacuum filters at 101.4 kPa_v and 1550 kPa for pressure filters. Due to the higher driving force, pressure filters are usually employed on finer tailings such as gold tailings for tailings requiring stacking.

When developing a tailing dewatering strategy, it is advantageous to look at several different methods. During the conceptual stage of the project development, test work along with equipment CAPEX and OPEX can be a low-cost option to narrow down the options. Equipment suppliers have decades of data correlation between lab-scale testing and full-scale operation that lab testing can provide pretty accurate simulations of dewatering tailings at the full-scale operation. While each filter vendor may have different testing equipment and scale up methodologies, they all use a filter device that relies on a chamber with filter media as the separation medium and a filtration driving force such as pumping slurry or air. For this study, the laboratory testing was done with a lab-scale recessed chamber filter (filter media on two sides) with slurry pumped into the chamber by a pump at pressures up to 1550 kPa. An image of the filtration apparatus is shown in Figure 1.

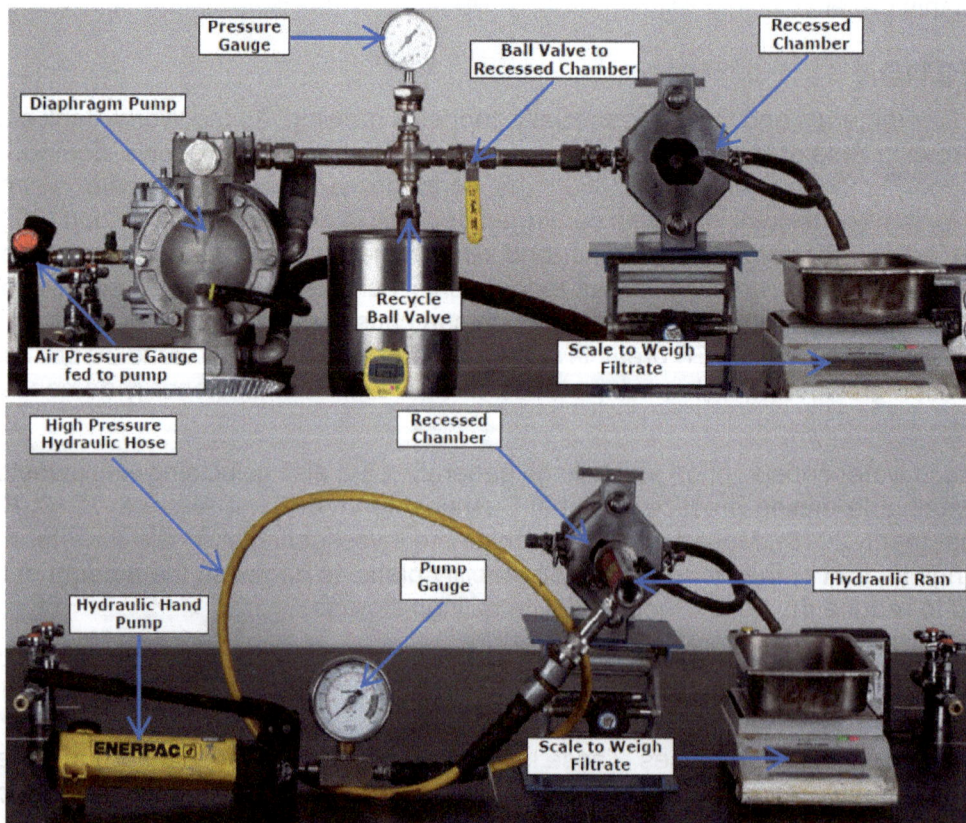

FIG 1 – Pressure filtration test apparatus.

As the focus of this paper is on filtration, it must be noted that similar test equipment and methodologies are used for sizing and selection of thickeners as well. Once the lab testing is completed, the proper equipment is selected with the size based upon the throughput and whether the site would like any offline spare filters. If offline spare filters are excluded, 15 per cent excess capacity should be sized for downtime for maintenance and repairs. For a CAPEX evaluation the quantity of filters along with ancillaries (sized for the process) is calculated. Operating expenses such as consumables, labour, and utility costs are used in the calculation of the annual operating expense for the process. In the conceptual phase of the project, only the major equipment with ancillaries is used for the CAPEX and OPEX comparison. It is assumed that most of the smaller equipment will change factorially with each option and therefore excluded for time and cost reasons.

Once the CAPEX and OPEX calculations are completed, the option(s) selected based upon the success criteria of the project, are used in the next phase of the project.

CASE STUDY

The basis for this study was a prefeasibility project for a greenfield gold mine in North America. Due to the site's commitment to minimising the usage of water for the project, filtered tailings was a preferred method of tailings disposal. The site was specified with typical flotation recovery followed by cyanide destruct and thickening of the tailings to 45 wt%. The incoming battery limit for this study was the tailings at 45 wt%. The two options for this case study were:

1. Direct filtration of the tailings at 45 wt% solids.

2. Thickening the tailings prior to filtration.

The equipment was size based upon lab-scale test work with CAPEX and OPEX calculations. The project was capex sensitive and therefore the lowest CAPEX option was preferred.

The tailing stream feed to the dewatering plant was set at 500 tonnes per hour (tph) at 45 wt%. The target filter cake solids concentration was 86 wt% solids. The tailings feed characteristics are shown in Table 1.

TABLE 1

Gold tailings characteristics.

Parameter	Units	Value
Suspended Solids	Wt%	45
Solids Specific Gravity	-	2.80
Liquor Specific Gravity	-	1.00
Slurry pH	-	8.5
PSD – P_{80}	µm	153
PSD – P_{20}	µm	12
PSD – P_{10}	µm	4

Lab-scale testing results

For the non-thickened filtered tailings options, the tailings were taken directly at 45 wt% and filtered using a lab-scale pressure filter (simulating a filter press). For the thickened option, the results obtained from standard lab-scale sedimentation testing are shown in Table 2.

TABLE 2

Sedimentation testing results.

Test variable	Units	Thickened
Feed Slurry	wt%	45
Optimum Thickener Feed Solids	Wt%	10
Solids Retention Time	Hr.	<1
Underflow Density	wt%	70
Underflow Yield Stress	Pa	<30
Flocculant Dosage	g/t	25
Thickener Size (diameter)	m	20

The results from the sedimentation testing indicate the 500 t/h throughput can be dewatered the tailings to a solids concentration of 70 wt% using a 20 m high-rate thickener.

The pressure used in the filter press testing was 1000 Pa for cake formation/consolidation and 689 Pa for cake air blow. The results for the non-thickened and thickened filter testing are shown in Table 3.

TABLE 3

Filter testing results.

Test variable	Units	Non-thickened	Thickened
Feed Slurry	wt%	45	70
Chamber Thickness	mm	50	50
Cycle Time	min.	16	7.6
Final Cake Solids	wt%	86	86
Dry Cake Density	kg/m^3	1692	1850
Filter size (plate size)	m × m	2.5 × 2.5	2 × 2
Number of Chambers	-	107	120
Number of Filters	-	3	2

The lab-scale testing indicated that the non-thickened tailings option requires three larger sized filter presses relative to the thickened tailings option which requires two smaller filter presses and one thickener. Aside from the equipment benefits, an additional benefit of the thickened option is feed slurry solids concentration variations. Without a thickener prior to filtration, any variations in the feed will have a much higher impact on filtration than if otherwise. Specifically, dilute feed will increase the cake formation time and overall cycle time. With a thickener in front of the filter, a dilute feed will settle and thicken in the same capacity as the design feed solids.

CAPEX and OPEX analysis

After the lab-scale testing and equipment sizing was complete, a comparison based upon major equipment CAPEX and annual operating (OPEX) costs was performed. The costs provided by the site and used in the determination of OPEX are shown in Table 4. All values in this report are in United States Dollars ($USD) and have a conceptual level accuracy of ±30 per cent.

TABLE 4

Values used in OPEX analysis.

OPEX variable	Units	Value
Normalised Labour Cost	$/hr	$50.00
Normalised Power Rate	$/kWh	$0.10
Flocculant Cost	$/ton	$3500

Water consumption is not included in the OPEX analysis. The best practice for filters that include a thickener prior to the filter is to return any filtrate, cloth wash, and core wash water back to the thickener feed. The solids are removed in the thickener and the process water is recovered from the thickener overflow.

Equipment cost analysis

The lab-scale results were used as inputs into the equipment selection and sizing. Along with process, utility and site assumptions provided by the client, a CAPEX and OPEX analysis was performed. The CAPEX was calculated for the major equipment and does not account for construction and installation. Pumps, tanks, filters and ancillaries, thickener and ancillaries were used in the calculation of CAPEX. Tables 5 and 6 show the equipment List for the Non-thickened and thickened flow sheets respectively.

TABLE 5

Equipment list for non-thickened flow sheet.

Equipment name	Quantity
Filter Feed Tanks	2
Filter Feed Pump (10 Bar)	3
Filter Press (AFP 2500)	3
Filter Press Ancillary Pump Package	1
Air Compressor and Receiver	2
Filtrate Tank	1
Filter Cake Feeder	3
Total System Price ($MUSD)	**$10.9**

TABLE 6

Equipment list for thickened flow sheet.

Equipment name	Quantity
High-Rate Thickener (20 metre)	1
Thickener Underflow Pump	2
Thickener Overflow Tank	1
Flocculant Dosing System	1
Filter Feed Tank	2
Filter Feed Pump	2
Filter Press	2
Filter Press Ancillary Pump Package	1
Air Compressor and Receiver	2
Filtrate Tank	1
Filter Cake Feeder	2
Total System Price ($MUSD)	**$7.0**

Although the thickened option has additional unit operations in the flow sheet, the CAPEX was reduced as compared to the non-thickened option. With the addition of the thickener, there was approximately 400 m³ less of process water to filter. Less water in the filter feed means that the cycle times are faster, and more cake can be filtered per hour.

Annual operating costs were calculated based upon site inputs, historical operation data and factorial methods. The categories included in the analysis included, labour, power usage, maintenance and spares and major consumables (flocculant and filter media). The annual operating estimate is shown in Table 7.

TABLE 7

Annual operation cost.

Category	Units	Non-thickened	Thickened
Annual OPEX	$MUSD/yr	$4.52	$2.60

The annual operating for the thickened was lower than for the non-thickened option. The major additional operating expense found in the thickened option was flocculant, in which the cost is minimal compared to the extra power requirements to run an additional filter in the noon-thickened option. This study indicated that a thickened filter feed will reduce both CAPEX and OPEX for the same amount of water recovered. A comparison of the CAPEX and OPEX can be found in Figure 2.

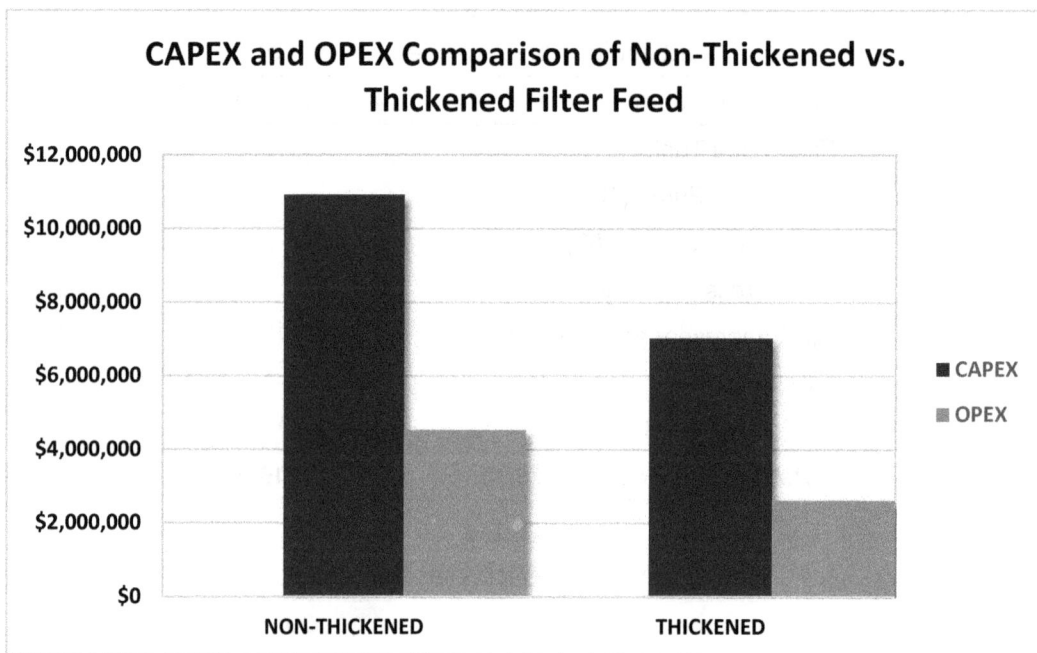

FIG 2 – CAPEX and OPEX comparison.

Discussion

The lab-scale test results were used as inputs into the CAPEX and OPEX analysis for the project. Having lab-scale test results for equipment sizing is one of the best ways to compare flow sheets for a specific project.

Another advantage of performing lab-scale test work is that the equipment sizing from the testing is more accurate than databases and comparable processes as each site's tailings are unique and the properties make it difficult to use results from another site for equipment selection and sizing. Having more accurate equipment sizing leads to more accurate CAPEX and OPEX which allows for a better flow sheet comparison analysis. Conceptual trade-off studies allow for many flow sheet options be compared quickly and cost less that performing at a later stage of the project.

For this case study, there was a 36 per cent reduction in CAPEX and a reduction of 42 per cent OPEX by adding a thickening the filter feed.

CONCLUSIONS

More mining companies are investigating filtered tailings as part of their long-term sustainability plans and to reduce risks associated with tailings storage. FLSmidth recommends that miners conduct tailings solutions technology trade-off studies which includes analysing different tailings solution options, including different flow sheets and equipment. This analysis includes determining the most economic approach for each solution.

However, it can be difficult to justify the extra capital expense associated with filtered tailings during studies. One option that is often investigated to see if it will reduce the cost is to evaluate removing the thickener from the flow sheet and sending a lower density feed to the filters. This paper presented a case study that included thickened and non-thickened feed to the filters. This case study showed that:

- Thickening the filter feed reduced the CAPEX by 35 per cent.
  - Addition of the thickener reduces the number of filters required from three to two.
  - The cost savings from removing one filter and its ancillaries is greater than the cost of the thickener.
- Thickening the filter feed reduced the OPEX by 42.5 per cent.
  - The main driver of OPEX for the thickener is the cost of flocculant as its power requirement for operation is quite low.
  - The main drivers of OPEX for pressure filters is power for pumps and compressors and filter media.
  - The cost impact of the flocculant is offset by the reduction in filter media and pumping power.

REFERENCES

World Bank Group, 2017. The Growing Role of Minerals and Metals for a Low Carbon Future, www.worldbank.org.

Mudd, G M, 2009. The Sustainability of Mining in Australia: Key Production Trends and Environmental Implications; Research Report No. RP5; Department of Civil Engineering, Monash University and Mineral Policy Institute: Melbourne, Australia.

UNICEF, 2021. Water Security for All, www.unicef.org.

Global Sustainable Investment Alliance, 2018. Global Sustainable Investment Review, https://www.gsi-alliance.org/wp-content/uploads/2019/03/GSIR_Review2018.3.28.pdf.

RBC Global Asset Management, 2018. Responsible Investing Survey: Executive Summary, https//us/rbcgam/com/resources/docs/pdf/HTML-files/files/ESF/Responsible_Investing_Executive_Summary_2018.pdf.

Investigation on the efficient selection of flocculants for dewatering of coal tailings

A Khazaie[1], B Samali[2] and S Gilmour[3]

1. Tailings and Dams Engineer, GHD Pty Ltd, Sydney NSW 2065.
   Email: atousa.khazaie@ghd.com
2. Professor, Centre for Infrastructure Engineering, Western Sydney University, Sydney NSW 2753. Email: B.Samali@westernsydney.edu.au
3. Geochemist, GHD Pty Ltd, Hobart Tas 7000. Email: sarah.gilmour@ghd.com

ABSTRACT

Coal processing plants generate a large volume of tailings containing fine and colloidal particles. The most conventional method for coal waste disposal is using tailing storage facilities (TSFs) (Khazaie et al, 2022). Tailings dam failures have always been a serious environmental threat to the contamination of surface and groundwater due to seepage, blockages, and insufficient capacity of spillway systems which leads to overtopping (Fourie, 2009; Spain and Tibbett, 2012). Moreover, mine tailings disposal imposes additional cost to mining companies. According to NSW Environment Protection Authority (EPA), the waste levy rate for Coal Washery Reject (CWR) in 2019–2020 is $15 per tonne (NSW Environmental Protection Authority (NSW EPA), 2020). Sustainable alternative proposals for coal tailings disposal in tailings dams include mechanical dewatering methods such as coagulation, flocculation, and sedimentation, followed by filtration. Efficient selection of flocculants is an important factor in this process.

In this study, mono flocculation experiments were conducted using various polyacrylamide (PAM) based flocculants with different molecular weight and charge type to evaluate the effect of molecular weight on settling rate and turbidity removal of coal tailings. Microstructural analysis of coal tailings including the X-Ray Diffraction (XRD) and Scanning Electron Microscopy and Energy Dispersive X-ray Spectroscopy (SEM-EDS) results, demonstrated that coal tailings were mostly composed of quartz, magnesium calcite, dawsonite, muscovite, and different clay minerals, including kaolinite, illite, and montmorillonite.

INTRODUCTION

Coal processing plants produce large volumes of tailings annually. Tailing storage facilities (TSFs) are the most commonly used method for tailings disposal. Tailing dams have always been a serious threat to the surrounding environment (Fourie, 2009; Spain and Tibbett, 2012). Moreover, they impose additional risk and cost to mining industry. According to NSW Environment Protection Authority (EPA), the waste levy rate for coal tailings disposal in 2019–2020 is $15 per tonne (NSW EPA, 2020).

Sustainable alternative proposals for TSFs include mechanical dewatering through coagulation, flocculation, and sedimentation processes, followed by filtration or centrifugation. Polymeric flocculants are normally used in solid-liquid separation of tailings before disposal which can result in saving huge volumes of fresh water consumed in mineral processing plants (Ahmari, Chen and Zhang, 2012; Ahmari and Zhang, 2012).

Polyacrylamide (PAM) flocculants are the most commonly used in coal processing plants that can be classified based on charge type into four categories including cationic, anionic, non-ionic and amphoteric. The polymers can vary in charge density, molecular weight and polymeric structure. The performance of flocculation is evaluated by two different parameters, ie the turbidity rate of supernatant water and the settling rate of flocculated particles (Sabah and Cengiz, 2004). Various factors involved in the flocculation process such as mineral composition, polymer type and dosage, mixing speed, pH etc (Hogg, 2000; Hogg, Bunnaul and Suharyono, 1993).

This research studies the effect of molecular weight on the flocculation efficiency of cationic PAM (CPAM), anionic PAM (APAM), and non-ionic PAM (NPAM) for dewatering of coal tailings collected from a coalmine in New South Wales (NSW), Australia. The settling rate and turbidity removal have been considered as criteria for optimisation of the flocculation conditions.

MATERIALS AND METHOD

Coal tailings were collected from a coalmine in New South Wales, Australia. The electrokinetic potential in coal tailings at its natural pH was measured using a zeta-meter. The pH was also obtained using pH metre. The elemental composition of coal tailings was also obtained using Scanning Electron Microscopy (SEM) and Energy Dispersive Spectroscopy (EDS) technique. The mineral composition of coal tailings was determined after oven drying by X-ray diffraction (XRD) using a Bruker D8 Advance Powder Diffractometer. Six polyacrylamide-based (PAM) flocculants with different charge type and molecular weight have been used in this study purchased from Henan Runquan Purification Material Co., Ltd, China. The characteristics of these polymers are shown in Table 1.

TABLE 1

Summary of flocculant characteristics.

Name	Charge type	Charge density %	Molecular weight
CPAM-C5	Cationic	40	~5 million
CPAM-C5	Cationic	40	~10 million
APAM-A10	Anionic	40	~10 million
APAM-A14	Anionic	40	~14 million
NPAM-N10	Non-ionic	-	~10 million
NPAM-N14	Non-ionic	-	~14 million

The flocculation experiments were conducted using a Velp JLT6 jar test with speed control. A stock solution of 0.1 g/L of each of the polymers was prepared using tap water. For each flocculation test, 200 ml slurry was added into a 1000 ml glass beaker and stirred for 1 minute at 150 rev/min to obtain a homogeneous suspension. Then, the specified volume of polymer solution was added continuously in a way that the total volume of suspension became 1000 ml and stirred for an additional 1 minute. The water interface and the height of the sediment bed were recorded at selected times for the calculation of the settling rate. The turbidity rate was measured after 15 min of settling using turbidity metre in accordance with ASTM D2035–19 and ASTM D7315–17.

RESULTS AND DISCUSSION

Characterisation of coal tailings

The general characteristics of coal tailings and their elemental composition are presented in Tables 2 and 3 respectively. The Zeta potential value for coal tailings at its natural pH showed that suspended particles carry a negative charge in an aqueous environment. The comparison between the conductivity value of coal tailings and distilled water indicates that the coal tailings are very saline; as a result, they affect the zeta potential significantly. The EDS analysis results specified that coal tailings are mostly composed of Si, Al, and O elements. Mineralogical analysis results show that the coal tailings are mainly composed of quartz, magnesium, calcite, dawsonite, muscovite, and different clay minerals including kaolinite, illite, and montmorillonite.

TABLE 2

Summary of flocculant characteristics.

Natural pH	Carbon content %	Conductivity (mS/cm)	Zeta potential at natural pH	Original solid content %
8.1	19.75	4.06	− 32	12

TABLE 3

Elemental composition of coal tailings.

Si	Al	O	Na	Mg	Ca	Fe	K	S	Ti
26.7	14.1	48	0.6	1.2	1.1	3.9	2.5	1.1	0.7

Effect of flocculant type and dosage on sedimentation

Sedimentation of coal tailings using CPAM

The turbidity measurements and settling rate of coal tailings using CPAMs with different molecular weights are plotted as a function of flocculant dosage in Figure 1. Increasing the molecular weight of CPAM significantly improves the settling rate and turbidity removal. The reason can be attributed to the increased micro-ion length, which is in favour of bridge mechanism. The maximum settling rate is obtained by CPAM-C10 at lower dosage compared to the CPAM-C5. In other words, using CPAM-C10 requires less polymer to get the maximum settling rate. In terms of turbidity removal, the optimum dosage is 250 g/t solid for both polymers and after that, increasing the dosage does not have any effect on the turbidity rate while it has an adverse effect on the settling rate. Experimental results show that CPAM-C10 can improve turbidity removal significantly, up to 97.3 per cent, which results in a clear water. The interaction mechanism between CPAM molecules and coal tailing particles is mostly composed of charge neutralisation. Through this mechanism, negatively charged coal tailing particles absorb on positively charged polyacrylamide molecules as a result of electrostatic attraction force (Sabah and Erkan, 2006).

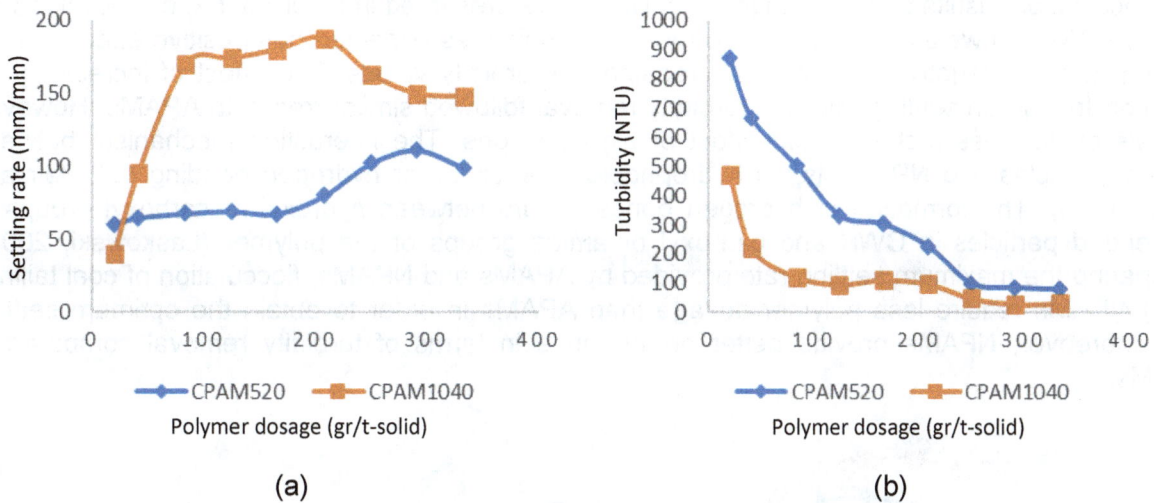

(a)

(b)

FIG 1 – Effect of CPAM on settling rate (a) and turbidity removal (b).

Sedimentation of coal tailings using APAM

The effect of APAMs on settling rate and turbidity removal of coal tailings is presented in Figure 2. Increasing the molecular weight of APAM does not have a specific effect on settling rate and turbidity removal as follows similar trends. In the flocculation of coal tailings using APAM, the bridging mechanism is the main mechanism for the formation of flocs. In this circumstance, the electrostatic repulsion force between functional groups of PAMs and negatively charged coal particles leads to stretching the hydrocarbon chain of polymers. As a result, suspended particles are adsorbed on the polymer chain through the hydrogen and/or chemical bonding (Sabah and Erkan, 2006; Duong et al, 2000). For both polymers, increasing the polymer dosage positively contributes to increasing the settling rate up to a particular dosage. After the optimum dosage, increasing the polymer dosage adversely affects the flocculation rate due to the increased electrostatic repulsion force between functional groups of PAMs and the negatively charges coal particles which compress the electrical double layer (Sabah, Yüzer and Celik, 2004).

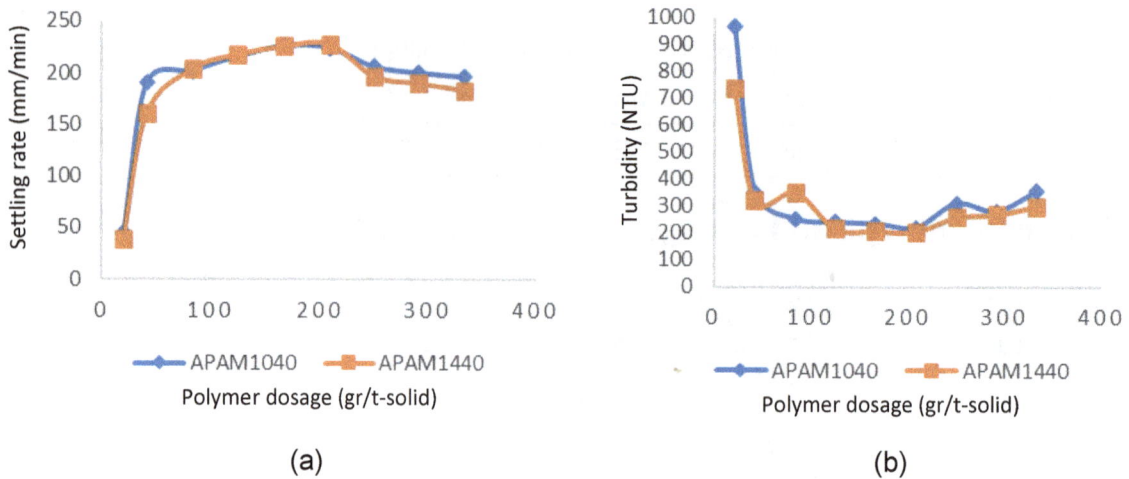

FIG 2 – Effect of APAM on settling rate (a) and turbidity removal (b).

Comparing the maximum settling rate and turbidity removal obtained by the APAM and the CPAM, anionic polymers shows a better performance in terms of settling rate, while cationic polymers are more favourable for turbidity removal purposes. Increasing the polymer dosage provides more surfaces for the adsorption of suspended particles and improves the settling rate.

Sedimentation of coal tailings using NPAM

The flocculation results of coal tailings using NPAM are presented in Figure 3. Experimental results using NPAMs shows that increasing the molecular weight as expected has positive effects on the settling rate but negative impact on decreasing the turbidity values. The effect of increasing the polymer dosage on settling rate and turbidity removal followed similar trends to APAMs. However, NPAMs exhibit less restabilisation effect at high dosages. The interaction mechanism between mineral particles and NPAMs is via hydrophobic interactions or hydrogen bonding (Littlefair and Lowe, 1986). The formation of hydrogen bonds occurs between hydroxyl or carboxyl groups of suspended particles in CWR and carboxyl or amide groups or the polymer (Laskowski, 2001). Comparing the maximum settling rate provided by APAMs and NPAMs, flocculation of coal tailings using NPAMs require less polymer dosage than APAMs in order to obtain the optimum settling rate. Moreover, NPAMs provide better performance in terms of turbidity removal compared to APAMs.

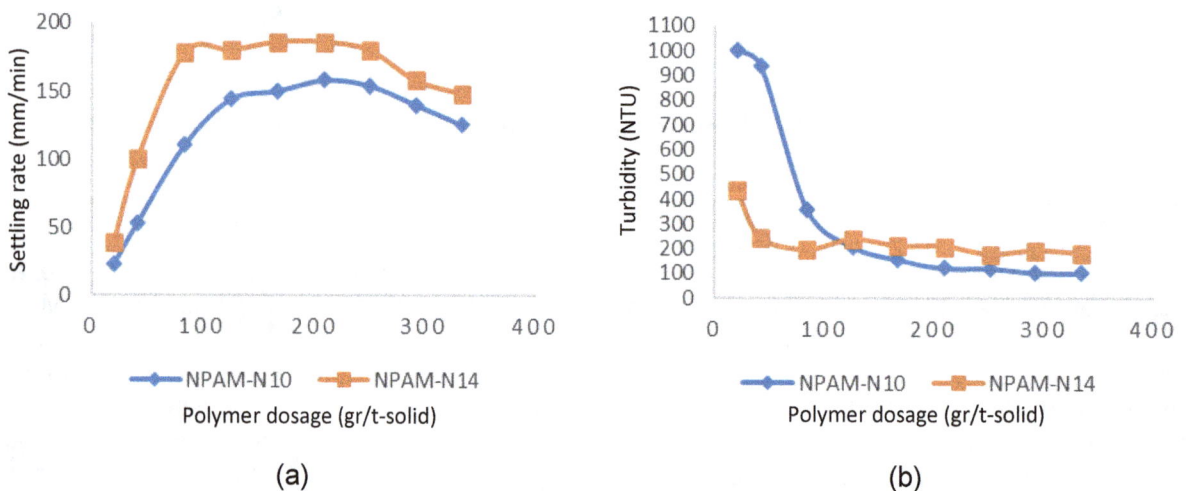

FIG 3 – Effect of NPAM on settling rate (a) and turbidity removal (b).

CONCLUSIONS

XRD and SEM-EDS analysis results demonstrate that coal tailings are mostly composed of quartz, magnesium calcite, dawsonite, muscovite, and different clay minerals, including kaolinite, illite, and montmorillonite. Six commercial polymeric flocculants with different charge type and molecular weight have been tested to identify the effect of molecular weight on flocculation performance at different dosages in terms of settling rate and turbidity value. According to sedimentation results, APAMs provide a high settling rate whereas CPAMs and NPAMs are more effective for turbidity removal. The results also show that increasing the molecular weight of cationic PAM improves the settling rate and turbidity removal while it does not have a significant effect on results obtained from anionic PAM. Although the non-ionic polymer with higher molecular weight achieves a higher settling rate, it is less effective in decreasing the turbidity rate at high dosages.

REFERENCES

Ahmari, S and Zhang, L, 2012. Production of eco-friendly bricks from copper mine tailings through geopolymerization, *Construction and Building Materials,* 29:323–331.

Ahmari, S, Chen, R and Zhang, L, 2012. Utilization of mine tailings as road base material, GeoCongress 2012: State of the Art and Practice in Geotechnical Engineering.

Duong, C, Choung, J, Xu, Z and Szymanski, J, 2000. A novel process for recovering clean coal and water from coal tailings, *Minerals Engineering,* 13(2):173–181.

Fourie, A, 2009. Preventing catastrophic failures and mitigating environmental impacts of tailings storage facilities, *Procedia Earth and Planetary Science,* 1:1067–1071.

Hogg, R, 2000. Flocculation and dewatering, *International Journal of Mineral Processing,* 58:223–236.

Hogg, R, Bunnaul, P and Suharyono, H, 1993. Chemical and physical variables in polymer-induced flocculation, *Mining, Metallurgy & Exploration,* 10:81–85.

Khazaie, A, Mazarji, M, Samali, B, Osborne, D, Minkina, T, Sushkova, S, Mandzhieva, S and Soldatov, A, 2022. A review on coagulation/flocculation in dewatering of coal slurry, *Water,* 14:918.

Laskowski, J, 2001. *Coal flotation and fine coal utilization* (Elsevier).

Littlefair, M and Lowe, N, 1986. On the selective flocculation of coal using polystyrene latex, *International Journal of Mineral Processing,* 17:187–203.

NSW Environmental Protection Authority (NSW EPA), 2020. Levy regulated area and levy rates [online]. Available: https://www.epa.nsw.gov.au/your-environment/waste/waste-levy/levy-regulated-area-and-levy-rates

Sabah, E and Cengiz, I, 2004. An evaluation procedure for flocculation of coal preparation plant tailings, *Water Res,* 38:1542–9.

Sabah, E and Erkan, Z, 2006. Interaction mechanism of flocculants with coal waste slurry, *Fuel,* 85:350–359.

Sabah, E, Yüzer, H and Celik, M, 2004. Characterization and dewatering of fine coal tailings by dual-flocculant systems, *International Journal of Mineral Processing,* 74:303–315.

Spain, A and Tibbett, M, 2012. Coal mine tailings: development after revegetation with salt-tolerant tree species, in *Mine Closure 2012, Proceedings of the Seventh International Conference on Mine Closure,* pp 583–594 (Australian Centre for Geomechanics: Perth).

Farming method and assessment of its application in bauxite residue using a scaled laboratory model

A Ranabhat[1], C Zhang[2], S Quintero Olaya[3] and D J Williams[4]

1. PhD candidate, School of Civil Engineering, Brisbane Qld 4072.
   Email: a.ranabhat@uqconnect.edu.au
2. Research Fellow, Geotechnical Engineering Centre School of Civil Engineering, Brisbane Qld 4072. Email: chenming.zhang@uq.edu.au
3. Senior Research Technologist, School of Civil Engineering, Brisbane Qld 4072.
   Email: s.quintero@uq.edu.au
4. FAusIMM, Professor of Geotechnical Engineering. Director, Geotechnical Engineering Centre, School of Civil Engineering, Brisbane Qld 4072. Email: d.williams@uq.edu.au

ABSTRACT

Dewatering is an important process in red-mud tailings management due to the high amount of water remaining after the extraction of aluminium. Excess water will not only reduce the storage capacity of the facility but also increases the liquefaction potential and seepage of the toxic chemicals into the groundwater or vegetation nearby. 'Farming' or ploughing/rolling of the surface with special amphibious tractors, known as Amphirols or 'scrollers' is usually carried out on settled bauxite residue, to increase the surface area and break the sealed surface, hence accelerating the dewatering process due to evaporation. However, it is unclear how much farming can enhance dewatering, and how the farming should be scheduled to maximise the dewatering outcomes. Laboratory experiments on the dewatering of bauxite residue were conducted to quantify the enhancement of dewatering due to farming. Two tanks were initially filled with settled bauxite residue at the same solids content. The ambient condition was controlled, and the drying of the residues was enhanced by an electrical fan. The residue in one of the tanks was subjected to farming using a downscaled farming scroller while the other tank was not farmed. Moisture sensors, load cells, and cameras were used to quantify the water loss during dewatering. Differences in the dewatering process in the two tanks were analysed, including:

- evaporation rates

- cumulative evaporation

- degree of saturation

- dry density

- crack formation.

This study finds some significant improvement in some of the parameters such as evaporation rate, cumulative evaporation, degree of saturation, dry density, and shrinkage which establishes that the farming method is worthy of further investigation.

INTRODUCTION

The disposal of mine tailings is an ongoing challenge to the mining industry, exacerbated by the increase in demand for the minerals, the ability to process lower grade ores which leads to the production of higher volume of residue, and the limited availability of convenient disposal sites because of land constraints and environmental regulations (Somogyi, 1976).

Bauxite residue (commonly known as red-mud) is a type of tailings which is a concern to the mining industry due to all the problems mentioned above. Red-mud is produced during the Bayer process of alumina production as an insoluble by-product after bauxite is digested in the presence of sodium hydroxide (NaOH) at high pressure and temperature. It consists of Hematite (Fe_2O_3), Aluminium oxide (Al_2O_3), Quartz (SIO_2), Sodium oxide (Na_2O), calcium oxide (CaO), and rutile (TIO_2) (Khairul, Zanganeh and Moghtaderi, 2019). Two to three tonne of bauxite ore is required to produce one tonne of alumina mud and depending on the geography 1–2.5 t of red-mud is produced as a by-product (Patil and Thorat, 2022).

The tailings produced are disposed in various ways including as dry stacking or solid cake disposal; however, the most prevalent method of tailings storage is to deposit it in a subaerial environment in the form of large pond called a tailings storage facility (TSF), often supported by dams using hydraulic pumping. This is a simple approach and reasonably cost-effective compared to other storage options like dry stacking (Williams, 2015).

Red-mud sample obtained for this test is deposited with solid content of around 27 per cent with high residual water content. Consequently, the tailings at that stage has low strength which leads to a less stable storage facility, potentially vulnerable to failure. The storage capacity of the storage facility will also reduce significantly as most of the space will be consumed by water. In addition to this, red-mud is highly alkaline with pH ranging from 11 to 13 with high concentrations of salt and potential harmful elements (PHEs). The status of tailings with high water content combined with their chemical content makes it a serious environmental concern if there is seepage to groundwater or vegetation (Wang *et al*, 2019).

Hence, the significant amount of residue produced raising downstream effects and the environmental concerns underscore the pressing need to minimise the effects of excess water present in the TSF.

There are numerous approaches to managing water both before and after TSF deposition. The method selected will rely on several variables, including the water retention capacity of tailings, environmental concerns, the amount of investment required, and the local water supply. The methods selected may also be influenced by local regulations, which may limit the amount of water that may be utilised for mining.

Prior to deposition in the TSF, water can be decreased by utilising thickeners or filtration of the tailings. Deep cone thickeners can be used to produce solid content of 30–50 per cent and can be disposed in the TSF as thickened tailings. Vacuum filtration can also be used for the processing which produces semi-dry cake with 50 to 60 per cent solid content and is 'dry stacked' at the disposal site. The automated filter-press is the most common water separation method because of its ability to maximise soda recovery and produce a high solid content of 70 to 75 per cent in the form of cake and can be disposed in dry form. The most advanced semi-continuous type of pressure filtration technology is HiBar steam pressure filtration in which disc filters are installed in a pressure vessel for filtration with steam cabins which are used for dewatering, washing, and cake drying. It can produce overflows with solid content of up to 75 to 80 per cent (Patil and Thorat, 2022).

Flocculation is also another method to manage water before deposition using flocculants like starch or synthetic polymers. In this process, fine particles are destabilised to eliminate repulsion and aggregated to larger units through interparticle collision and adhesion. This process is completed through sedimentation, resulting in the dewatering of fine particles in suspension (Patil and Thorat, 2022).

Finally, the dewatering of tailings can be done in the facility itself. This process can use external equipment for removal/collection of water or/and expose the tailings for evaporation.

Mud farming is one of the common methods of TSF dewatering that is employed by Queensland's alumina businesses. This is a mechanical technique that combines bulldozers with a piece of screw-propelled machinery known as an amphirol. The tailings are placed in the dam in a sloping manner forming a beach which allows the removal of surface run-off and the water that bleeds into the trenches formed by farming. The machinery is utilised after the tailings have developed a bearing pressure of 3–5 kPa sufficient to support an amphirol. Amphirols use an Archimedes screw mechanism to create trenches in the soil layer where water is accumulated from the surrounding red-mud with high potential energy. The water can then be collected in decant ponds and transferred. A thick crust develops on the surface of red-mud sometime after deposition. In addition to this, salts accumulate on the surface which combined with the fine tailing of low permeability leads to blockage of the evaporation path. The farming technique will break the crust exposing the tailings underneath and increasing the surface area for evaporation and thus has the potential to accelerate desiccation and expedite the rate of consolidation. It is repeated until the tailings gain enough strength (about 35 kPa) to support a bulldozer, which can continue the compaction process by trafficking.

Red-mud is a challenging material to densify due to its low permeability. The dry density that can be achieved without any treatment is about 0.7 t/m^3 and the effect is limited to 300 mm from the surface.

However, with the use of amphirols, density can reach up to 0.9 t/m^3, and subsequent dozing can develop it up to 1.4 t/m^3 (Williams, 2021).

Although mud farming in tailings for water management is gaining popularity there are not many studies that quantify its effectiveness. There is only a limited understanding of the improved rate of dewatering compared to the natural method in numerical terms. The change in the degree of saturation and dry density over time compared to the natural method is not yet well addressed.

This study evaluates the effect of mud farming on the bauxite residue. The main objective of the study is to understand and quantify the dewatering process of red-mud under amphiroling. To answer this question, this study will observe the rate of evaporation, degree of saturation, shrinkage, and dry density of samples after mud farming and compare properties with a control sample where natural dewatering is allowed under as similar conditions as possible.

MATERIAL AND METHOD

Tailings sample

The red-mud sample for this experiment was obtained from Queensland Alumina Limited (QAL), Gladstone, Queensland. The red-mud is neutralised with sea water before being disposed of as an aqueous slurry that has been thickened to a nominal 26 per cent solids by mass (Olaya *et al*, 2017). The major characteristics of the red-mud sample is illustrated in Table 1.

TABLE 1

Characteristics of QAL red-mud.

Characteristic		Unit	Value
Specific gravity			3
Liquid limit (LL)	Casagrande method	%	47
	Fall cone method	%	51
Particle size distribution	Clay	%	15
	Silt	%	58
	Sand	%	27
Plastic limit		%	32
Plasticity index		%	15
Linear shrinkage			6.2
pH			9.25
Average *in situ* density (compacted)		t/m^3	1.462
In situ gravimetric moisture content (compacted)		%	34.6
In situ hydraulic conductivity		m/s	$1*10^{-9}$

Experimental set-up and procedure

The current study builds upon the foundation established by (Herrera Cabrera, 2020) in his research thesis 'Assessment of a farming method applied to bauxite residue using a laboratory model'. The scope of this experiment was to study the optimum duration for drainage and desiccation of red-mud following farming and the optimum use of a scroller to facilitate subsequent dozing. This experiment serves as a logical extension by introducing novel variables and methodologies. By leveraging the insights gained from Herrera Cabrera's work, I aim to further study the complex relationship between mud farming and the dewatering of red-mud.

For this study, two heavy-duty tanks with dimensions 355 mm, 510 mm, and 770 mm representing height, width, and length are employed.

Capacitive moisture sensors and load cells are used for data collection from the experiment. A total of 24 capacitive moisture sensors, 12 in each tank were used to record the real-time water content of the red-mud. These sensors have a measurement range of 260 to 530, with 260 to 350 representing saturated soil, 350 to 430 representing unsaturated soil, and 430 to 520 representing dry soil. These are electrical values sent by sensors which are later calibrated to soil properties. The capacitive moisture sensor used for the study is illustrated in Figure 1.

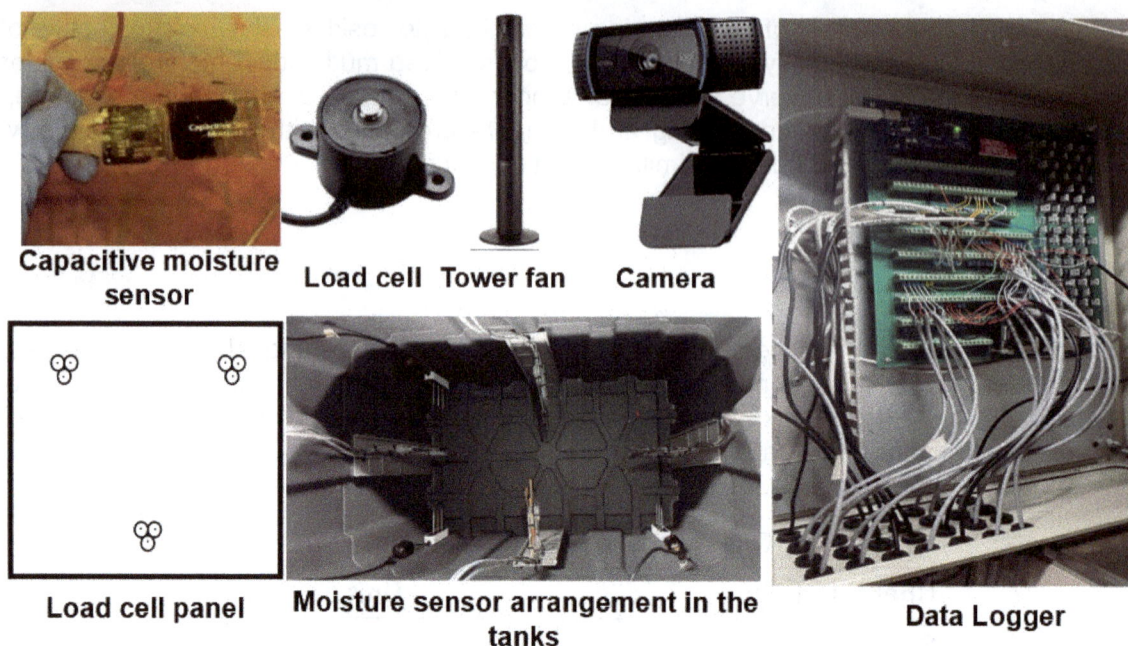

FIG 1 – Sensor assembly (Olaya *et al*, 2017).

This experiment is conducted indoors wherein significant efforts were undertaken to create a comparable environment for both tanks. The test tanks were initially marked for the installation of the capacitive moisture sensors. Each tank has three levels of moisture sensors positioned at 50, 100, and 200 mm below the top of the tank. The capacitive moisture sensors are positioned nearer the surface since it is anticipated that the greatest amount of change will take place there. Four moisture sensors in total are positioned on each level and affixed in the middle of each wall, with sensors pointing towards the centre of the tank.

The weight of the red-mud decreases because of the loss of water due to evaporation. Compression load cells, each with a capacity of 22 kg, are utilised to record the real-time change in weight. A total of nine load cells were employed on each load cell panel. These load cells are calibrated against the known weight and an equation is generated to convert the electrical values from the load cells to the actual weight. The configuration of the load cells on the load cell panel is shown in Figure 1.

All the sensors are connected to the data logger system and Python code is prepared to periodically instruct as well as retrieve data from the datalogger and upload it to the network.

The solid content of the sample was measured to be 30 per cent outdoors with natural settlement already taking place. The sample was taken out from the tank and calculated water was removed to prepare the sample with 50 per cent solid content. Prior to placing red-mud in two tanks, it was meticulously mixed in small 16 L batches and distributed to both the control and experiment tanks each time to enhance consistency. Scrolling over the red-mud in the experiment tank was only conducted after the red-mud was able to support the electric scroller we prepared for the experiment. Straight trenches were formed by the scroller on the surface of the red-mud. Both tanks were allowed to desiccate after the initial scrolling. A tower fan has been running continuously throughout this time, blowing a gentle breeze across both tanks. After 29 days, a crust had formed on the top surface of tank B (farmed tank). In addition to that, salt build-up on the surface was also observed. 29 days after the first scrolling, a second scrolling was conducted. Firstly, this process broke the crust surface and exposed the tailings underneath and secondly, it increased the surface area allowing for more evaporation. Figure 2 shows the condition of the red-mud samples on the first day and 29th day of farming.

FIG 2 – Before and after images of the 1st and 29th days.

Cumulative evaporation

Cumulative evaporation is an important component of the water balance which would add to the knowledge of water and energy partitioning at the interface between the soil surface and atmosphere (Xiao *et al*, 2011). In this study, the scale reading from load cells is conducted every 30 mins to measure the loss of weight considered as evaporation of water.

The total number of days taken for the control sample to lose 50 mm of water by evaporation was 25 days whereas it was only 19 days for farmed sample saving 6 days. Similarly, it took 53 and 40 days for the control and farmed samples respectively to evaporate 100 mm of water. The total amount of water loss at the end of the experiment is slightly different. Farmed tank lost 200 mm of water whereas the unfarmed lost 180 mm. tanks as they both had the same amount of water. Figure 3 compares the cumulative evaporation and evaporation rate between the two samples.

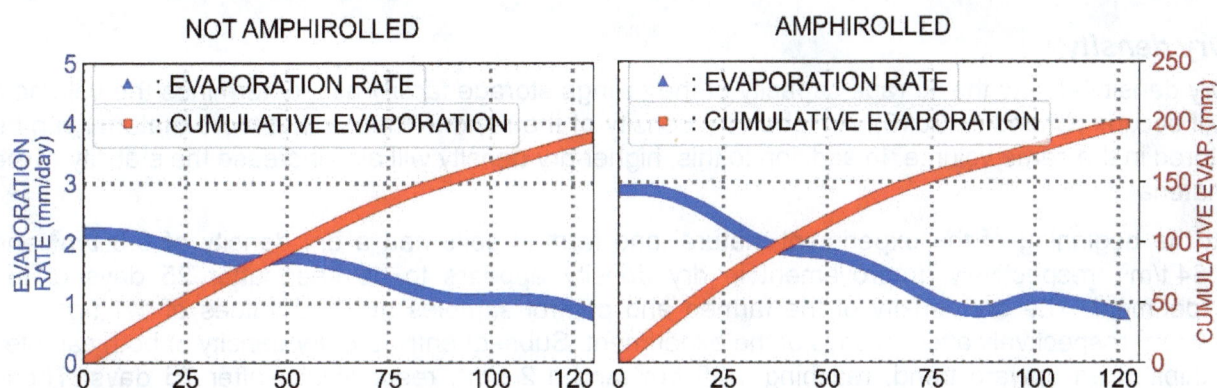

FIG 3 – Comparison of cumulative evaporation and evaporation rate between farmed and non-farmed red-mud.

Evaporation rate

This parameter holds significant importance in the understanding of the potential impact of farming on the dewatering process of red mud. It will shed light on the degree to which farming will enhance the rate of evaporation. In this study, data readings are done every 30 mins and the weight is measured in grams.

The evaporation rate was 3 mm per day during the first 12 days of the experiment in the farmed sample which is 50 per cent higher than the control sample with the evaporation rate of 2 mm/day. The rate of evaporation became similar to each other after 35 days with slight variation due to the weather conditions. As this experiment was conducted indoors, the rate of evaporation was found to be significantly less than what would be in the natural environment.

Degree of saturation

This experiment was allowed to run for two years and was re-wetted after 7 months from the start of the experiment. For calibration of moisture sensors, the tank is assumed to be completely desaturated at the end of seven months. The electrical values recorded at this time and at the beginning of the experiment when red-mud was saturated are used for the calibration of sensors to approximate the degree of saturation. Figure 4 illustrates the comparison of the degree of saturation between the control and farmed residue.

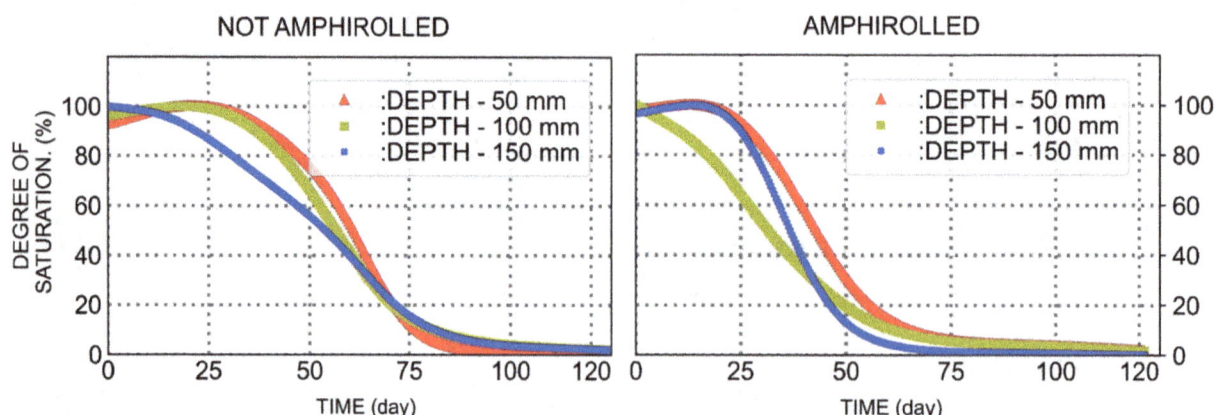

FIG 4 – Degree of saturation between farmed and non-farmed red-mud.

The degree of saturation is near 100 per cent for the first 20 days of the experiment. Both the samples went through consolidation instead of desaturation at this phase. Desaturation began after 20 days in the farmed sample; however, it was only after 30 days in the control sample. The upper layer of both tanks are desaturated later than the layers below. This changes after the degree of saturation become closer to 30 per cent. The degree of saturation of the farmed sample decreases by 80 per cent within a span of 30 days from 20 to 50 days of the experiment, whereas the control sample takes 45 days for a similar reduction, occurring between 30 to 75 days.

Dry density

Dry density affects the storage capacity of the tailings storage facility as it determines the volume it will occupy (Chandler, 2016). A higher dry density of the residue means that more material can be stored in the same volume. In addition to this, higher dry density will also increase the stability of the material.

At the beginning of the experiment, natural and farmed soils had a dry density of 0.8 t/m³ and 0.84 t/m³, respectively. Improvement in dry density appears to be seen after 25 days of the experiment. The dry density of the farmed and control samples attained values of 1.1 t/m³ and 0.9 t/m³ respectively after 25 days of the experiment. Subsequently, the dry density of both samples exhibited an upward trend, reaching 1.35 t/m³ and 1.2 t/m³, respectively, after 75 days. These findings indicate a significant increase in the dry density of both samples over time, with the farmed sample exhibiting a higher increase in dry density compared to the control sample. The comparison of dry density gained over time is depicted in Figure 5.

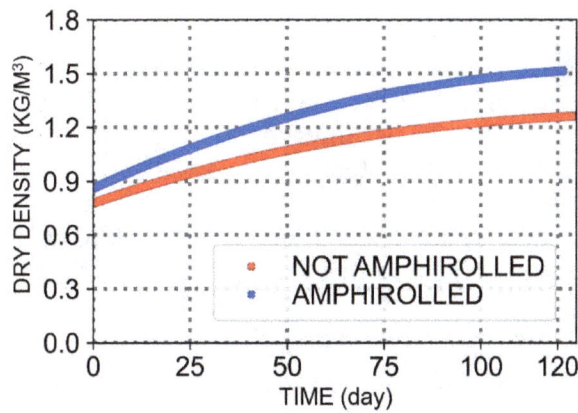

FIG 5 – Comparison of dry density between farmed and control sample.

Shrinkage

Dewatering will cause major volume shrinkage of the red-mud (Chandler, 2016). To capture this, a total of two cameras are employed, with one camera positioned to capture a top view and the other to capture a front view of each tank. Results are presented in Figure 6.

The vertical and horizontal shrinkage in the control sample was found to be similar throughout the dewatering process with total shrinkage reaching 35 per cent in 75 days after which there wasn't a significant increment in shrinkage. However, the vertical shrinkage was double horizontal in the farmed sample. In addition, this sample shrunk by 40 per cent within 50 days of the experiment which is 35 days sooner than the control sample to reach similar shrinkage.

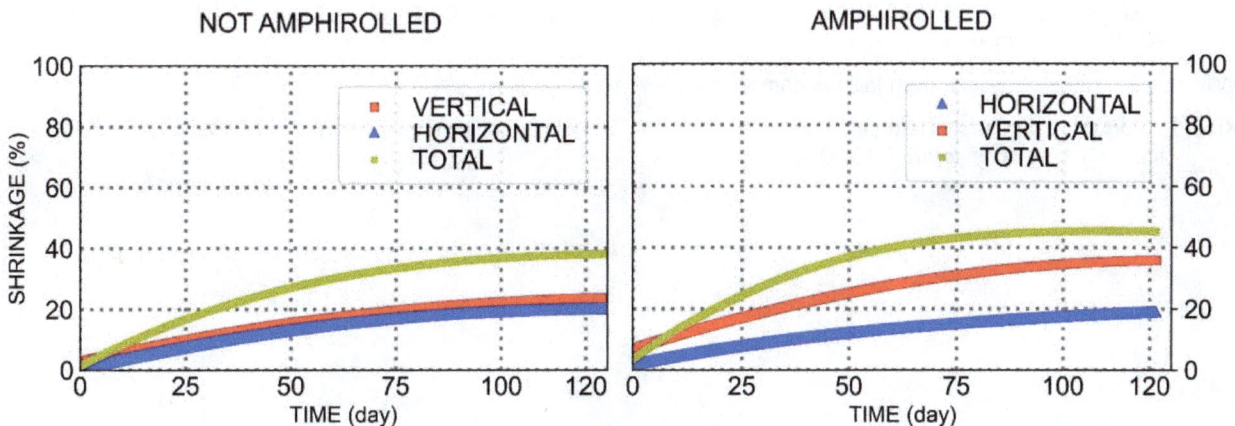

FIG 6 – Comparison of shrinkage between farmed and control sample.

CONCLUSION

According to the results of the experiment, significant improvement in dewatering can be realised by mud farming. Although dewatering did not begin in the early stage of the experiment, the farmed sample lost 50 mm of water six days earlier than the control sample. The rate of evaporation is up to 50 per cent higher than the natural method of dewatering immediately after farming. Similarly, 80 per cent of desaturation occurs in 30 days in farmed red-mud whereas it takes 45 days for the control sample. In addition to this, the farmed sample increased the dry density by 50 per cent in 37 days whereas it only improved by 25 per cent in the control sample. The shrinkage was also larger by 10 per cent in the farmed sample in 120 days, but 80 per cent of the shrinkage occurred in this sample in 50 days of the experiment whereas it took 85 days to attain the same shrinkage for the control sample.

Hence, it appears that farming can be useful in dewatering red mud in wet surface storage to contribute towards the safety of the facility as well as increase the storage capacity. This study indicates that the benefit of red-mud farming can be realised in the early stages of deposition. Targeted further research should be conducted to support the findings.

ACKNOWLEDGEMENTS

We would like to express my sincerest gratitude and appreciation to Lieven Tan, Zicheng Zhao, Wenqiang Zhang, Bolin Wang, and Shusen Chen for their invaluable contributions towards the completion of this research paper. Their unwavering support, guidance, and feedback were instrumental in shaping the direction of this study and in achieving its goals. Their assistance in data collection, analysis, and interpretation has been immensely valuable, and their dedication to excellence has been an inspiration. I am grateful for the privilege of working with such talented and supportive colleagues, and I extend my heartfelt thanks to each of them.

REFERENCES

Chandler, J L, 2016. Solar Drying of Red Mud, *Essential Readings in Light Metals: Volume 1 Alumina and Bauxite*, pp 938–943.

Herrera Cabrera, F, 2020. Assessment of farming method applied to bauxite residue using a laboratory model.

Khairul, M, Zanganeh, J and Moghtaderi, B, 2019. The composition, recycling and utilisation of Bayer red mud, *Resources, Conservation and Recycling*, 141:483–498.

Olaya, Q, Williams, D J, Pilar, M, Ingunza, D and Serati, M, 2017. Geotechnical characterisation of neutralised red mud for use in civil engineering construction and manufacturing, Proceedings of the 32nd International Conference on Solid Waste Technology and Management. Available from: <http://solid-waste.org/past-conferences-and-proceedings/proceedings/2017-2/>

Patil, S V and Thorat, B N, 2022. Mechanical dewatering of red mud, *Separation and Purification Technology*, 121157.

Somogyi, F, 1976. Dewatering and drainage of red mud tailings, University of Michigan.

Wang, L, Sun, N, Tang, H and Sun, W, 2019. A Review on Comprehensive Utilization of Red Mud and Prospect Analysis, *Minerals*, 9:362.

Williams, D J, 2015. Towards the elimination of conventional surface slurried tailings storage facilities, in *Proceedings Tailings and Mine Waste Management for the 21st Century*, pp 11–24 (The Australasian Institute of Mining and Metallurgy: Melbourne),.

Williams, D J, 2021. Lessons from tailings dam failures—where to go from here?, *Minerals*, 11:853.

Xiao, X, Horton, R, Sauer, T, Heitman, J and Ren, T, 2011. Cumulative soil water evaporation as a function of depth and time, *Vadose Zone Journal*, 10:1016–1022.

Tailings liquefaction and seismic response

Evaluation of the triggering of static liquefaction of tailings dams considering the SWCC

F L Rivarola[1,2], N Tasso[3,7], K Bernardo[4,7], M Sottile[5,7] and A Sfriso[6,7]

1. SRK Consulting, Chile 300, Buenos Aires C1098AAF, Argentina. Email: flopez@srk.com.ar
2. Universidad de Buenos Aires – Facultad de Ingeniería, Buenos Aires, C1063ACV, Argentina. Email: flopez@fi.uba.ar
3. SRK Consulting, Chile 300, Buenos Aires C1098AAF, Argentina. Email: ntasso@srk.com.ar
4. SRK Consulting, Chile 300, Buenos Aires C1098AAF, Argentina. Email: kbernardo@srk.com.ar
5. SRK Consulting, Chile 300, Buenos Aires C1098AAF, Argentina. Email: msottile@srk.com.ar
6. SRK Consulting, Chile 300, Buenos Aires C1098AAF, Argentina. Email: asfriso@srk.com.ar
7. Universidad de Buenos Aires – Facultad de Ingeniería, Buenos Aires, C1063ACV, Argentina.

ABSTRACT

Standard practice to evaluate global stability of tailings storage facilities (TSFs) entails the use of limit equilibrium analyses, which considers either peak or residual undrained shear strength of the saturated area and drained shear strength for the unsaturated soils and tailings. This practice might still be non-conservative, as nearly saturated materials can develop excess pore pressures and behave in an undrained mode. Moreover, limit equilibrium analyses do not account for the work input required to drive the stress state past the peak and into the softening process. Numerical deformation models, on the other hand, can readily model the transition from saturated to unsaturated conditions and from peak to residual strength, and therefore are able to provide further insight of the static liquefaction vulnerability of a given dam.

This paper describes the application of finite element modelling in Plaxis 2D to the analysis of the vulnerability of a TSF to static liquefaction. The procedure employs off-the-shelf features like the Hardening Soil Model with Small Strain Stiffness (HSS) constitutive model and Van Genuchten Soil Water Characteristic Curve (SWCC) to simulate the shear-induced generation of pore pressure in the saturated and nearly saturated materials while providing a smooth and physics-friendly transition from the fully undrained to the fully drained behaviours of the various portions of the dam. It is demonstrated that, by providing a step-like SWCC, the procedure produces nearly identical results to those obtained by ignoring suction altogether and zoning the dam by hand. This, by itself, is useful, because the position of the phreatic surface can be varied within the model without the need of remeshing. On the other hand, if a realistic SWCC is employed, the vulnerability of the facility can be better understood and reported.

INTRODUCTION

Conventional tailings, deposited hydraulically in tailings storage facilities (TSF), result in a loose and saturated soil-like material that may exhibit a sudden loss of strength in undrained shear and be prone to static liquefaction. Analysis of tailings static liquefaction has become a very relevant topic in mine waste engineering since recent upstream-raised TSFs failures such as those of Cadia Valley, Fundâo and Corrego deo Feijâo in Brumadinho. Guidelines with international acceptance (eg ANCOLD, 2019) recommend that the stability of the dam be carried out by limit equilibrium methods, assuming residual undrained shear strength for all contractive saturated tailings. This approach is, in general terms, conservative and safe, but provides little insight about the vulnerability of a given dam.

Numerical deformation analyses, on the other hand, are useful for assessing the robustness of a given dam by imposing external perturbations to its current or future configuration. One simple approach, taken from structural engineering, is the so-called 'pushover analysis' where the system – in this case the dam – is loaded monotonically until failure occurs. When strain-softening materials are involved, the external action usually grows until it reaches a point where progressive failure is triggered. For this progressive failure to occur, the loss of potential energy due to deformation, plus the release of stored elastic energy of the non-strain softening portions of the dam, must provide enough work input to keep the process going until the fully softened state is reached in a portion of the dam or a localisation zone. Whether this progressive failure happens or not depends on several factors including the brittleness of the tailings, the geometry and zoning of the dam etc. Typical triggers are loads suddenly applied at various places, imposed deformations at the toe of the dam, raises in the phreatic surface, surface water inflow etc (Ledesma, Sfriso and Manzanal, 2022).

Several critical-state constitutive models exist that can reproduce the undrained strain-softening behaviour of the tailings and can therefore be used in these analyses. For purely undrained shear, however, it has been shown in Sottile, Cueto and Sfriso (2020) that the Hardening Soil Model with Small Strain Stiffness (HSS), available in most commercial software packages, suffices for practical applications.

The state-of-the-practice of trigger analyses assumes drained behaviour in the unsaturated tailings and undrained behaviour in the saturated region, with no transition zone. Apart from being non-physics-friendly, this procedure is also not practical as it creates two regions in the finite element (or finite difference) mesh. This implies, for instance, that a change in the position of the phreatic surface requires remeshing.

In this paper, we show that the transition from the fully drained to the fully undrained behaviours can be readily accomplished by resourcing to the now conventional unsaturated soil mechanics and the Soil Water Characteristic Curve (SWCC). A complete trigger analysis is performed to a hypothetical TSF using both approaches: zoning by hand and incorporating a smooth transition with the SWCC. The off-the-shelf HSS constitutive model and Van Genuchten SWCC formulations implemented in PLAXIS 2D® are used. Apart from the introduction of the SWCC, the modelling strategy is the same as in previous publications (see Sottile, Cueto and Sfriso, 2020; Sottile *et al*, 2021; Rivarola *et al*, 2022).

CALIBRATION OF HSS FOR UNDRAINED SHEAR

HSS is an effective stress, nonlinear elastic, isotropic hardening plasticity constitutive model, able to represent the behaviour of materials undergoing plastic compression, consolidation, and monotonic shear. HSS is not a critical state model, ie void ratio is not a state variable, so a critical state cannot be achieved in drained shear. In undrained shear, however, a critical state model is not required because void ratio does not change, so it doesn't matter whether it is a state variable or not. All that is required is a stress-dilatancy formulation able to capture the evolution of plastic volumetric strain with shear with enough accuracy. To this end, HSS employs the Li–Dafalias stress-dilatancy formulation (Li and Dafalias, 2000), so it essentially inherits the simulation capabilities of the SANISAND family of constitutive models. Mathematically, the total volumetric strain increment ε_v is always zero in undrained shear. Decomposition into the elastic and plastic components yields $\dot{\varepsilon}_v^e + \dot{\varepsilon}_v^p = 0$. $\dot{\varepsilon}_v^p$ is controlled by shear hardening and the stress-dilatancy equation, and hence depends on stiffness parameters and on the shear stress ratio, the friction angle ϕ' and the dilatancy angle ψ, while $\dot{\varepsilon}_v^e$ depends on the elastic parameters in HSS. If $\phi' = \phi_{cv}$ and $\psi = 0$, the ultimate state becomes the critical state, because the stress-dilatancy formulation yields $\dot{\varepsilon}_v^p = 0$ at that state, ie yielding at constant stress and constant volume (for more details see Sottile, Cueto and Sfriso, 2020; Rivarola *et al*, 2022).

Once $\phi' = \phi_{cv}$ and $\psi = 0$ are adopted, the peak undrained shear strength s_u^{peak} and the residual one s_u^{res} depend on the combination of elastic and plastic hardening stiffness parameters only. Calibration is then performed to achieve both a target peak ratio $\left(s_u^{peak}/\sigma_{1,0}'\right)$ and a target residual ratio $\left(s_u^{res}/\sigma_{1,0}'\right)$ using a unique set of parameters. An example of such calibration is shown in Figure 1 for a Direct Simple Shear test (DSS) (Sottile, Cueto and Sfriso, 2020).

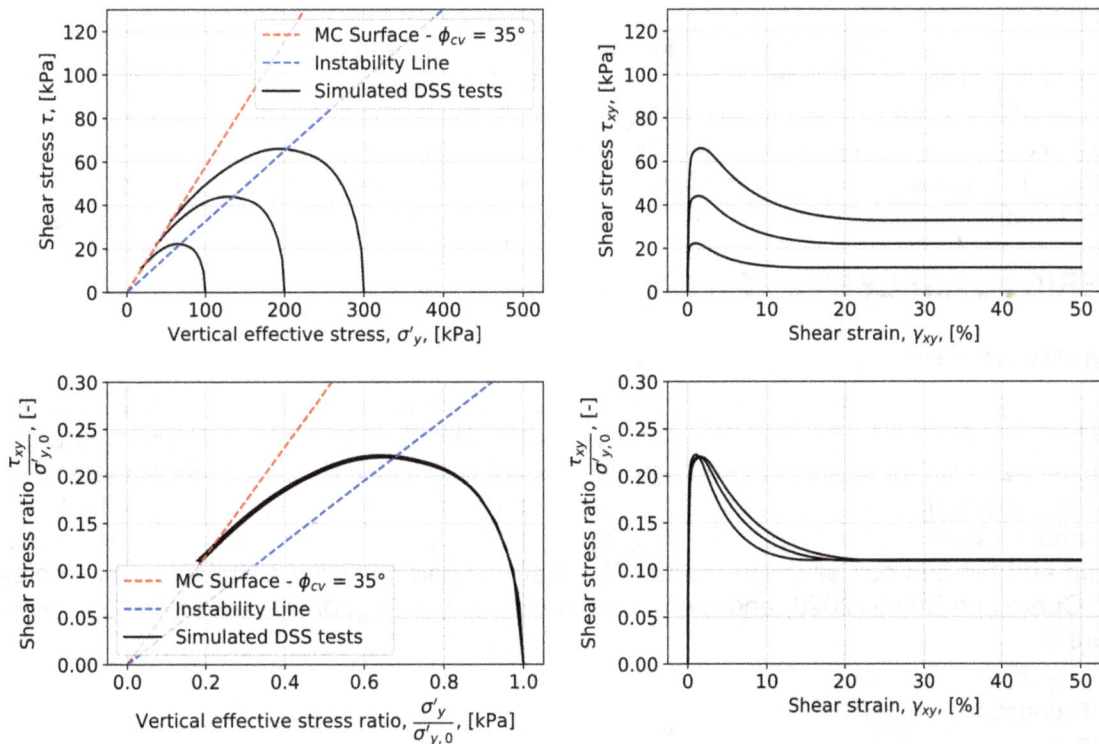

FIG 1 – Direct Simple Shear test (DSS) simulations with HSS (Sottile, Cueto and Sfriso, 2020).

SWCC – VAN GENUCHTEN MODEL

The Soil Water Characteristic Curve (SWCC) determines the relationship between suction (water in tension, $u_w > 0$) and degree of saturation S_w. One of the most popular SWCCs is the Van Genuchten model (van Genuchten, 1980):

$$S_{eff}[u_w] = [1 + (g_a \cdot u_w)^{g_n}]^{g_m}$$

where g_a, g_n and g_m are fit parameters and $S_{eff} = (S_w - S_{res})/(S_{sat} - S_{res})$, S_{res} and S_{sat} are the residual and saturated degree of saturation. Bishop's effective stress:

$$\sigma' = \sigma - S_{eff} \cdot u_w \mathbf{1}$$

is employed, where σ is the total Cauchy stress (negative in compression) and u_w is pore pressure (negative in compression). As the degree saturation is reduced, suction increases monotonically, but the term $S_{eff} \cdot u_w$ reaches a maximum value and then decreases to 0. This provides a practical transition from the saturated state ($S_{eff} = 1$, the classical definition of effective stress is recovered) to dry state ($S_{eff} = 0$ and therefore $\sigma' = \sigma$, see Figure 2).

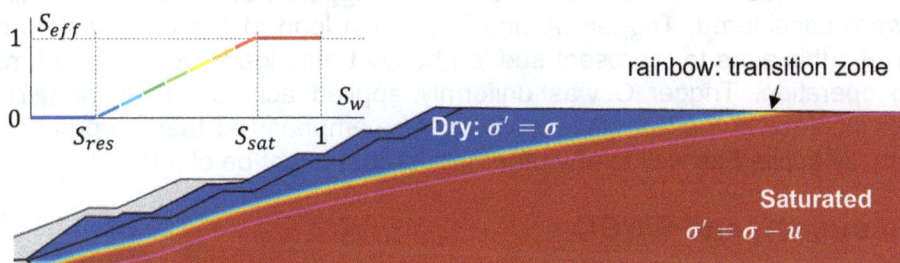

FIG 2 – Use of the SWCC in defining the saturated and dry behaviours.

The resulting saturation S_w (for cases 1 to 3 that ignore suction) or effective saturation S_{eff} (for cases 4 to 9) are shown in the following section. It can be seen that the phreatic surface, shown by the pink curve, is almost the same in all cases.

To facilitate the selection of Van Genuchten parameters in PLAXIS 2D, data sets for common types of soil are available. These data sets can be accessed based on standardised soil classification systems. One such system is the HYdraulic PRoperties of European Soils (HYPRES) database, which is an internationally recognised soil classification system (Wösten *et al*, 1999). The classification distinguishes between Topsoil and Subsoil, with subsoil referring to soils located more than 1 m below the ground surface. The Type drop-down menu in the HYPRES data set includes options such as Coarse, Medium, Medium fine, Fine, Very fine, and Organic soils.

NUMERICAL ASSESSMENT OF A TSF

Geometry and mesh

As an example, a hypothetical upstream-raised TSF was modelled in PLAXIS 2D and its vulnerability against static liquefaction was evaluated. The dam is 45 m high and has a 1V:3.5H average slope, including a berm. The numerical model has 9145 triangular 15-node elements and entails four geotechnical units: tailings, embankment raises, foundation soil and bedrock. The embankment raises and tailings were modelled using HSS, the upper foundation using Mohr–Coulomb and the bedrock as linear elastic. The material parameters are the same as employed in Sottile, Cueto and Sfriso (2020) and are not repeated here. The geometry and mesh are presented in Figure 3.

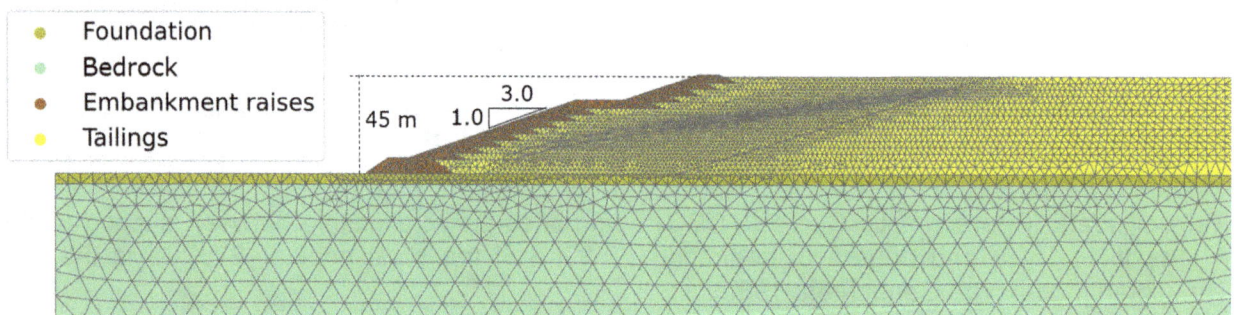

FIG 3 – Geometry and mesh of the model.

Modelling strategy

The TSF was raised in several stages until its final configuration using an average embankment raise height of 3.0 m and consolidation phases with an average rate of rise of 2.0 m/a. The aim is to reasonably capture the non-linearities associated with the staged construction process, which play a critical role in determining the *in situ* stress field and pore pressure distribution. A steady state groundwater flow was computed at each deposition phase. The position of the phreatic surface, therefore, is an outcome of the model, defining the following regions: submerged, below the phreatic surface, saturated, in the capillary zone, partially saturated above.

Once the construction was completed, flow liquefaction triggers were analysed. In this paper, three load triggers were considered. Triggers A and B apply a load at the current dam crest and the berm, respectively; this aims to represent sudden heavy traffic loads or stockpiled material during regular mining operation. Trigger C was uniformly applied across the entire tailings beach to simulate a rapid increase in the tailings level. It must be emphasized that the procedure presented here can be equally applied as well to any other trigger that might be of interest.

The effect of suction and SWCC

To demonstrate the value added by incorporating the SWCC to the analysis, nine cases were analysed. The first three ignore suction as follows:

Case 1: Suction is ignored. Undrained behaviour is enforced both above and below the phreatic surface. This is an over-conservative scenario, included for reference only.

Case 2: Suction is ignored. Undrained behaviour is enforced below the water table only.

<u>Case 3</u>: Suction is ignored. Undrained behaviour is enforced below the capillary fringe only (approximately 3 m above the water table).

It must be noted that these first three cases require that the mesh be partitioned to accommodate the different geotechnical units.

The remaining six cases include a SWCC and suction, so there is no partitioning of the tailings into sub-subunits. Different SWCCs are employed for the tailings as follows:

<u>Case 4</u>: A very high air entry value (AEV) is adopted ($S_{res} = 10\%$, $g_a = 0.01/m$, $g_n = 10.00$, $g_m = -0.90$). As the AEV is high, the capillary rise is also high, and all tailings are almost fully saturated above the water table, which is conservative and unrealistic. However, unlike case 1, the saturated zone above the water table does have suction, so the tailings body has a higher stiffness and strength than in case 1.

<u>Case 5</u>: Parameters are calibrated to yield a step-like SWCC with a capillary rise of $\approx 3\ m$ followed by a rapid desaturation ($S_{res} = 10.0\%$, $g_a = 0.45/m$, $g_n = 10.00$, $g_m = -0.90$). A very narrow unsaturated zone remains, so the practical effects of suction are negligible. Hence, similar results to case 2 or 3 are expected.

<u>Case 6</u>: HYPRES Subsoil Very Fine type is adopted ($S_{res} = 1.3\%$, $g_a = 1.30/m$, $g_n = 1.20$, $g_m = -0.17$).

<u>Case 7</u>: HYPRES Subsoil Coarse type is adopted ($S_{res} = 6.8\%$, $g_a = 4.30/m$, $g_n = 1.52$, $g_m = -0.34$).

<u>Case 8</u>: SWCC is calibrated from real tests on gold tailings ($S_{res} = 5.0\%$, $g_a = 0.43/m$, $g_n = 4.33$, $g_m = -0.77$).

<u>Case 9</u>: SWCC is calibrated from real tests on iron ore tailings ($S_{res} = 8.7\%$, $g_a = 231/m$, $g_n = 1.42$, $g_m = -0.30$).

The six SWCC are shown in Figure 4.

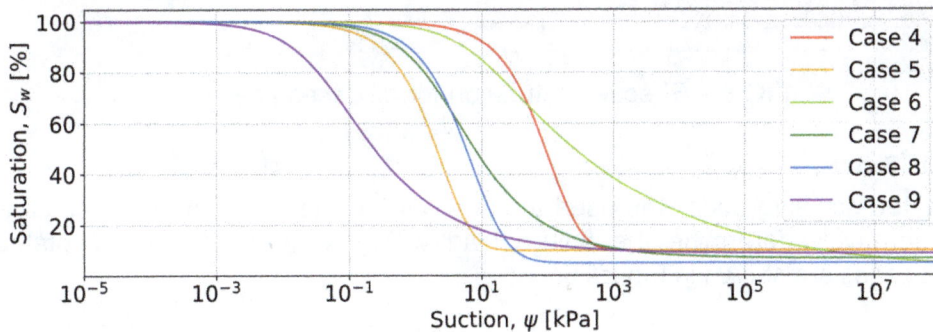

FIG 4 – SWCCs used in the six cases analysed that included suction.

The resulting saturation S_w (for cases 1 to 3 that ignore suction) or effective saturation S_{eff} (for cases 4 to 9) are shown in Figure 5. It can be seen that the phreatic surface, shown by the pink curve, is almost the same in all cases.

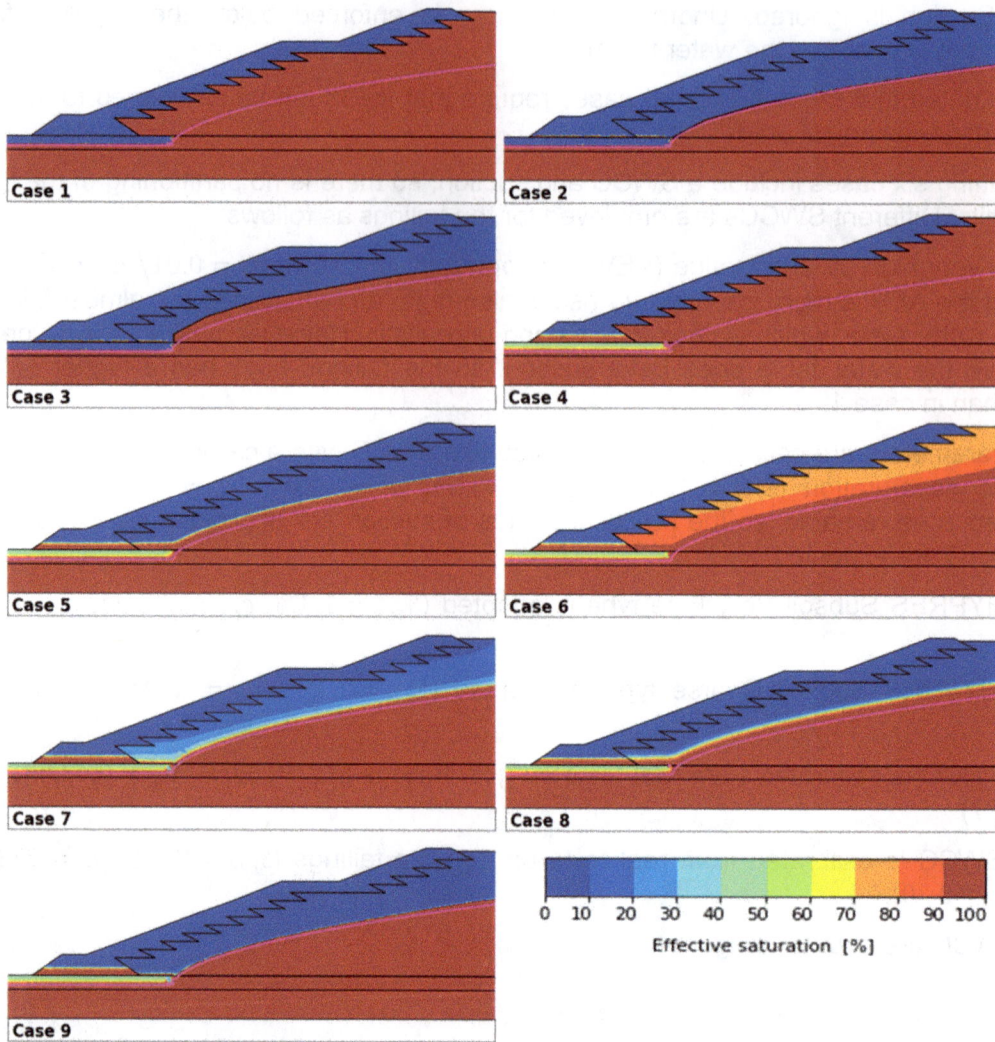

FIG 5 – Effective saturation for all cases analysed.

Trigger loads

A fully automated scripting code was used to carry out the trigger analyses. It employs numerical methods to iteratively determine the failure load within a user-defined acceptable tolerance. Results for all cases are shown in Figure 6.

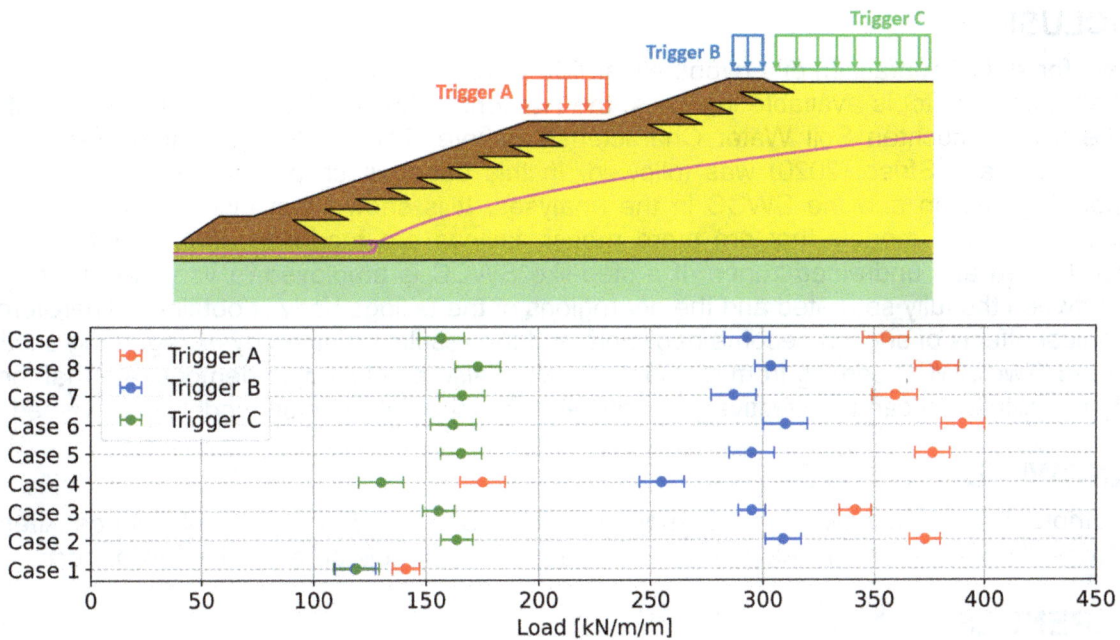

FIG 6 – Trigger values. The trigger load is bracketed as shown, the midpoint is reported.

As expected, case 1 gives an absolute lower bound for all three triggers. Cases 2 and 3 give larger values than case 1, slightly lower for case 3 as the saturated zone is ≈ 3 m higher. In this example the different is small; however, the effect of capillary rise might be significant if the water table is closer to the slope.

Case 4 provides a lower bound for the cases considering suction, as all tailings are saturated and excess pore pressure is generated even well above the phreatic surface. Trigger values obtained for case 5 (a step-like SWCC) are in the same range as to those found for cases 2 and 3, where suction was ignored, proving that incorporating the SWCC does not necessarily mean a lower degree of conservatism but does mean a more robust and practical modelling strategy. Cases 6 to 9 show how the SWCC affects the trigger values, demonstrating that the real effect of unsaturation can be effectively incorporated into the deformation models if so wished. Figure 7 shows some stress paths along one of the failure surfaces.

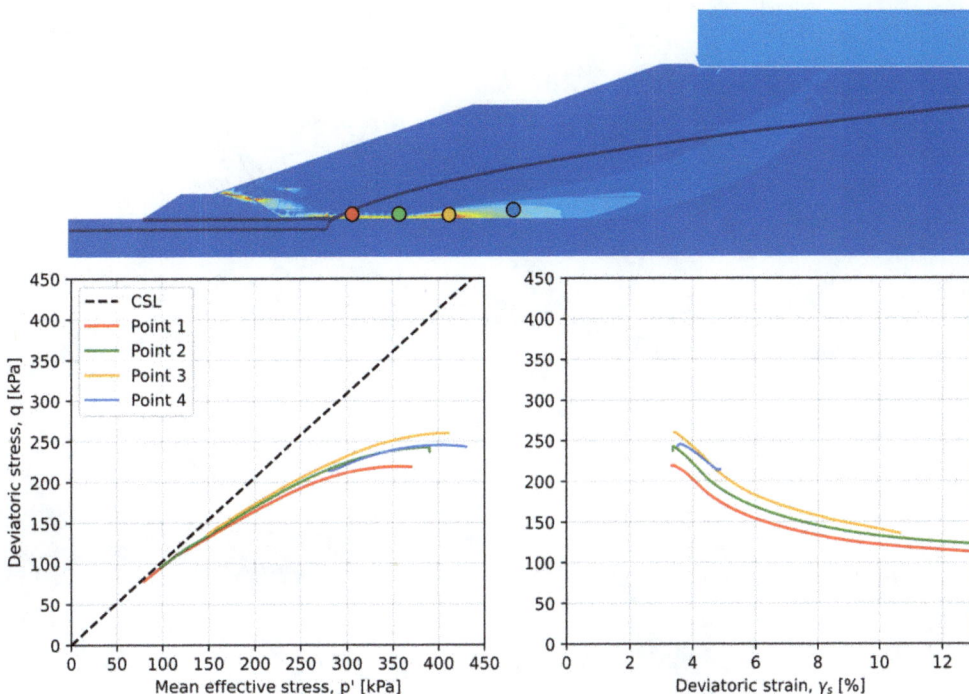

FIG 7 – Failure surface for Trigger C of case 5 (step-like SWCC) and associated stress paths.

CONCLUSIONS

Triggers for static liquefaction of a hypothetical TSF were determined through the application of off-the-shelf numerical tools available in Plaxis: the Hardening Soil Model with Small Strain Stiffness and the Van Genuchten Soil Water Characteristic Curve. The modelling strategy proposed by Sottile, Cueto and Sfriso (2020) was followed. In this paper, such procedure was extended by incorporating suction and the SWCC in the analyses. It is shown that incorporating the SWCC produces numerical models that are more robust: tailings are modelled as one unit, instead of distinct drained and undrained zones. If a step-like SWCC is employed, a very narrow transition zone between the fully saturated and the dry regions of the tailings body is obtained. Therefore, the mechanical effects of suction become negligible and the results obtained by disregarding suction is recovered. Two SWCC coming from real projects were also employed, to demonstrate that the real effect of unsaturation can be effectively incorporated into the deformation models if so wished.

ACKNOWLEDGEMENTS

The authors wish to acknowledge the support of the team at the University of Buenos Aires and SRK Consulting during the development and testing of the models presented in this paper.

REFERENCES

ANCOLD, 2019. Guidelines on tailings dams – Planning, design, construction, operation and closure.

Ledesma, O, Sfriso, A and Manzanal, D, 2022. Procedure for assessing the liquefaction vulnerability of tailings dams, *Computers and Geotechnics*, 144:104–632.

Li, X S and Dafalias, Y F, 2000. Dilatancy for cohesionless soils, *Geotechnique*, 50(4):449–460.

Rivarola, F, Tasso, N, Bernardo, K and Sfriso, A O, 2022. Numerical aspects in the evaluation of triggering of static liquefaction using the HSS model, *Mecánica Computacional*, XXXIX:983–992.

Sottile, M G, Cueto, I A and Sfriso, A O, 2020. A simplified procedure to numerically evaluate triggering of static liquefaction in upstream-raised tailings storage facilities, Campinhas, COBRAMSEG.

Sottile, M G, Cueto, I A, Sfriso, A O, Ledesma, O N and Lizcano, A, 2021. Flow liquefaction triggering analyses of a tailings storage facility by means of a simplified numerical procedure, Proceedings of the 20th International Conference on Soil Mechanics and Geotechnical Engineering.

van Genuchten, M, 1980. A closed-form equation for predicting the hydraulic conductivity of unsaturated soils, *Soil Science Society of America Journal*, 44(5):892–898.

Wösten, J H M, Lilly, A, Nemes, A and Bas, C L, 1999. Development and use of a database of hydraulic properties of European soils, *Geoderma*, 90(3):169–185.

Characterisation of in situ saturation profile through tube measurements and SCPTu

E Coyle[1], B Otieno[2] and D Reid[3]

1. Senior Geotechnical Engineer, SLR Consulting, Denver CO. Email: ecoyle@slrconsulting.com
2. Senior Geotechnical Engineer, KCB, Brisbane Qld 4101. Email: botieno@klohn.com
3. Research Fellow, UWA, Perth WA 6009. Email: david.reid@uwa.edu.au

ABSTRACT

As more scrutiny is given to upstream tailings facility's liquefaction potential and resulting strengths around the world, saturation of the tailings becomes an important question for tailings practitioners, especially in conditions of near, but not fully saturated tailings. The difference in seismic behaviour between near, but not fully saturated and fully saturated tailings can have significant implications in the stability outcomes and design configurations for tailings storage facilities. Near fully saturated tailings have beneficial effects that include suction that can contribute to increased resistance against liquefaction. This may not be fully understood under the influence of cyclic loading and a complex tailings column comprising interbedded saturated and near saturated layers. Current *in situ* techniques for measuring Saturation Ratios (S_r) are still in development, leaving designers and owners with uncertainty around stability outcomes and conservative costly remediation in some instances.

This paper presents a unique and extensive site investigation performed at an Australian gold mine upstream tailings facility, in predominantly near (but not fully) saturated tailings. The paper presents the method and results obtained by direct measurement of S_r through the collection of 82 piston tube samples to full depth of the tailings column that were used to determine void ratio and density. The tube samples were supplemented with adjacent SCPTu dynamic Pore Water Pressures (PwP), and profiles of Compression wave velocity (V_p) and shear wave velocity (V_s) that will be presented for comparison and correlations. This data has been used to support and validate the instrumented columns along the tailings facility and improve the interpretation of the tailings saturation regime and resulting assessment of the susceptibility against liquefaction.

INTRODUCTION

Studies present the use of *in situ* methods, in particular geophysical methods that output compression wave velocity (V_p) and shear wave velocity (V_s) to define the degree of saturation, a key parameter in the liquefaction resistance and strength of soils (Jamiolkowski, 2012). The use of *in situ* testing and seismic cone penetration techniques provides a valuable tool for the evaluation of difficult saturation conditions that may exist in many upstream tailing dams that experience partial drainage and segregation of fines resulting in perched water tables (Jamiolkowski and Masella, 2014). In geotechnical stability evaluation, the range of near to fully saturated presents a unique challenge when determining the extents of a tailings column that are liquefiable and will exhibit liquefied strengths. An opportunity was presented to expand the utility of *in situ* testing beyond the typical screening level evaluation of the presence of a fully saturated or partially saturated layer and an estimate of the corresponding water table at depth. Studies have demonstrated how the relative cyclic resistance ratio can increase considerably as a function of the degree of saturation. This paper presents a unique approach and extensive site investigation that aimed to improve confidence in the estimates of saturation level, water table and allow comparison with *in situ* methods to develop relationships that support characterisation for the tailings column at an Australian gold mine tailings storage facility (TSF). The objective targeted an improved definition of the tailings saturation regime and to define key parameters evaluating liquefaction susceptibility to support the evaluation of geotechnical stability.

FIELD PROGRAM

A site investigation in 2020 involving the collection of 82 piston tube samples was performed at the upstream tailings' facility, in predominantly nearly (but not fully) saturated tailings. The investigation targeted a direct measurement of S_r for the tailings column, to determine its susceptibility to liquefaction should potential triggering occur. Complimentary piezocone penetration tests with seismic measurement (SCPTu) were performed adjacent to the sampling locations for comparison of seismic measurements with the direct S_r measurements. The samples were then transported to the University of Western Australia (UWA) Soils Laboratory where a program comprising index testing and advanced strength tests was completed. The scope of advanced testing is not covered in this paper.

The site investigation followed previous SCPTu campaigns that were completed two years earlier in 2018 in nearby locations. These investigations have been included in the comparison to support the interpretation of the saturation levels.

The sections below detail the procedures followed to collect the field tube samples, handle, transport, and complete measurements in the laboratory.

Field tube sampling procedure

A total of 82 thin-walled 86 mm ID, 600 mm long tube samples were retrieved from eight locations along the internal beach perimeter of the TSF using a custom piston sampler in general accordance with the procedures outlined in ASTM D1587 (2016). The samples were retrieved at depths up to 43 m. The sampler was deployed from a Cone Penetration Testing (CPT) rig, to ensure no introduction of water associated with traditional drilling. The custom piston tube sampling apparatus shown in Figure 1 uses a CO_2 gas injection system that 'cuts' the retrieved sample just above the level of the tube's bottom cutting edge and breaks the suction force generated at the base of the sample. Previous attempts to recover samples of smaller diameter without gas cutting had generally been unsuccessful.

NARROW·GAS· CYLINDER¶

FIG 1 – Tube sampler showing gas cylinder (left) and tube cutting face (right) showing CO_2 injection hole and squared off sample.

In total 97 samples were attempted, giving a retrieval rate of approximately 85 per cent.

After tube retrieval, the sample was squared-off at both ends until a uniform, planar surface was achieved. This allowed for a neater fit of the sealing plug and greater accuracy when estimating sample length and associated in situ sample volume. Once a uniform, planar surface was achieved, the distance from the sample surface to the top of the tube was measured at three points using vernier (at both ends). This measurement, along with the tube's total length and ID, was accepted as the best estimate of the sample's total volume (V_t), representing the closest approximation to in situ void ratio.

The samples were packed in boxes with padding and transported vertically to the UWA Soils Laboratory. Despite efforts made to reduce transportation-induced densification, some densification of low Plasticity Index (PI) samples was expected and observed. This was noted and taken into corrections when measurements were taken in the laboratory.

At four of the eight sampling locations, complimentary shear wave measurements were taken from the CPT rig at downhole depth intervals of 1 m, prior to performing the piston tube sampling (2020 Campaign). The previous CPT seismic campaign (2018 Campaign) measuring shear and compression waves presented for comparison included soundings located at six of the eight tube locations, at an average distance away of 27 m.

Laboratory procedure

Laboratory measurements taken on the recovered tubes were carried out as follows:

1. Upon arrival in the laboratory the specimens were re-weighed, with the masses measured in the laboratory consistently matching the site measurements, confirming that the sample sealing procedures had been successful at avoiding moisture loss.

2. Specimen dimensions were taken again in the laboratory in two ways:

   o Distance to the J-plug seals at each end, along with thickness of each J-plug.

   o Direct measurement to the sample after removal of the J-plug. This measurement consistently indicated greater distances between the specimen and the end of the tubes (ie a smaller sample length) than that taken in the field, consistent with transport-induced densification.

3. The entire specimen was then extruded into a large drying tray, then allowed to dry at room temperature for about one week. The mass of tray, wet specimen and 'dry' specimen after air drying were recorded in this process.

4. The specimen was then transferred to a sample bag and mixed thoroughly. A subsample was taken for Gravimetric Water Content (GWC) measurement in a hot (~105°C oven). Only a small portion of the sample was exposed to hot oven drying to avoid potentially affecting the mechanical behaviour and mineralogy of the majority of the recovered material which was needed for subsequent reconstituted triaxial and direct simple shear testing. The process of measuring initial moisture loss of the entire sample through air drying, and the GWC of a sub sample, enabled the original mass of water and solids in the piston sample to be calculated.

5. In some cases the process of subsampling and hot oven drying was carried out in a beaker, facilitating measurement of the salt concentration of the pore fluid. This additional step involved:

   o Addition of a known mass of deionised water to the specimen in the beaker, after drying.

   o Thoroughly mixing the specimen and Deionised water, then leaving the soil to settle.

   o Removing clear water from the top of the beaker after settling, then measuring the salt concentration of the mixture. This enabled calculation of the original salt concentration of the water within the piston sample. If this salt concentration was sufficiently high, it can affect the density of the fluid. This can result in either underestimating or overestimating your sample and/or fluid density.

RESULTS

Tailings properties

The properties of the piston tube samples were characterised through index tests that include particle size distributions (PSDs) through hydrometer tests, water contents and salt content tests.

The results of the PSDs are shown in Figure 2 indicating a predominantly fine-grained low plasticity silt. The hydrometer tests conducted reported soil particle densities ranging from 2.73 to 2.85 Mg/m^3.

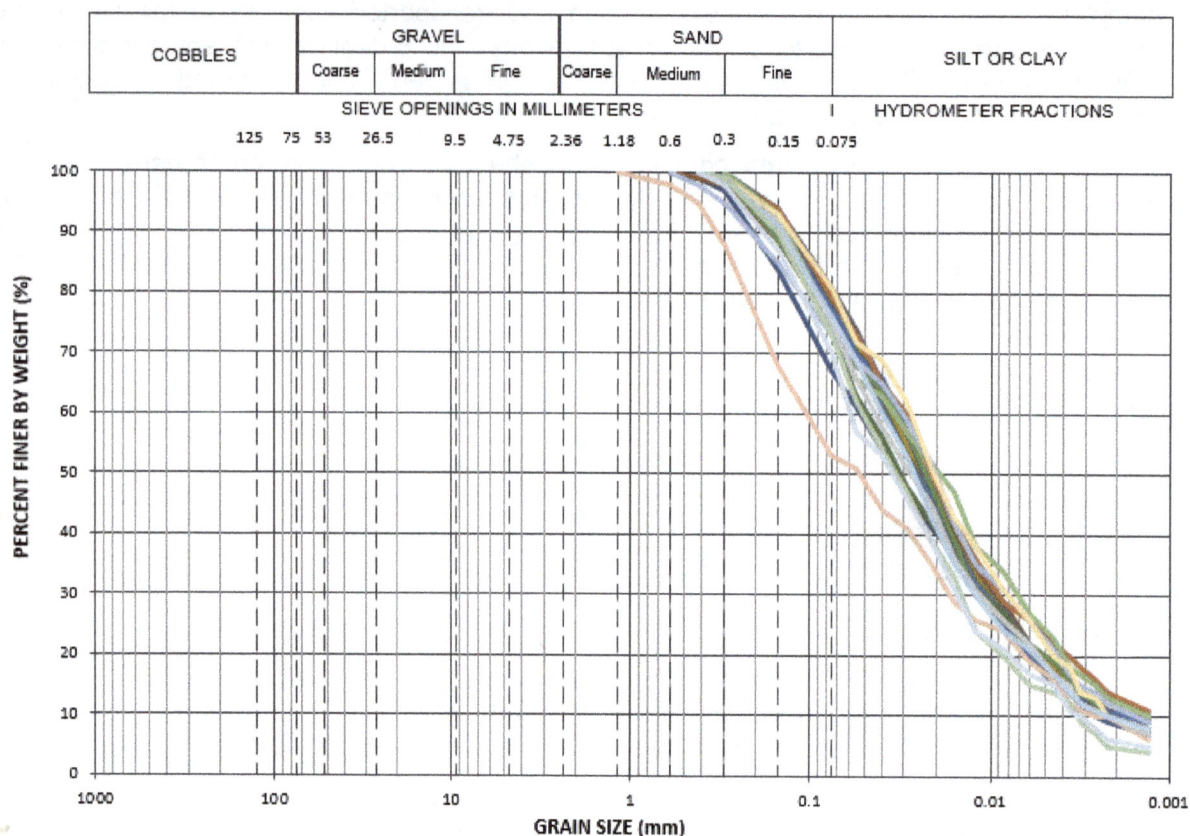

FIG 2 – Particle size distribution test results.

The salt content tests were carried out to determine the effect of dissolved salts on the saturation level. An initial base case assumption of 40 g/L was used as a sensitivity resulting in a slight to negligible difference on the overall saturation level in comparison to actual results and a zero-salt case, as shown in Table 1.

TABLE 1

Salt content results.

Tube sample	Salt content (g/L)	Saturation (actual)	Saturation (base case)	Saturation (0 g/L)	Difference
PT01 01	11.3	78%	80%	78%	-2%
PT01 02	15.3	90%	91%	89%	-1%
PT01 06	27.3	79%	79%	77%	0%
PT01 07	19.2	130%	131%	129%	-1%
PT01 08	15.5	79%	81%	78%	-2%
PT01 09	6.2	78%	79%	77%	-1%
PT01 010	14.2	87%	88%	86%	-1%
PT01 012	11.8	97%	99%	97%	-2%

Seismic testing

The results of the 2020 campaign normalised shear wave V_{s1} profiles with adjacent tube S_r measurements are presented in Figure 3. In most instances the correlation between V_{s1} and measured S_r is not evident to infer a relationship in the near to partially saturated range.

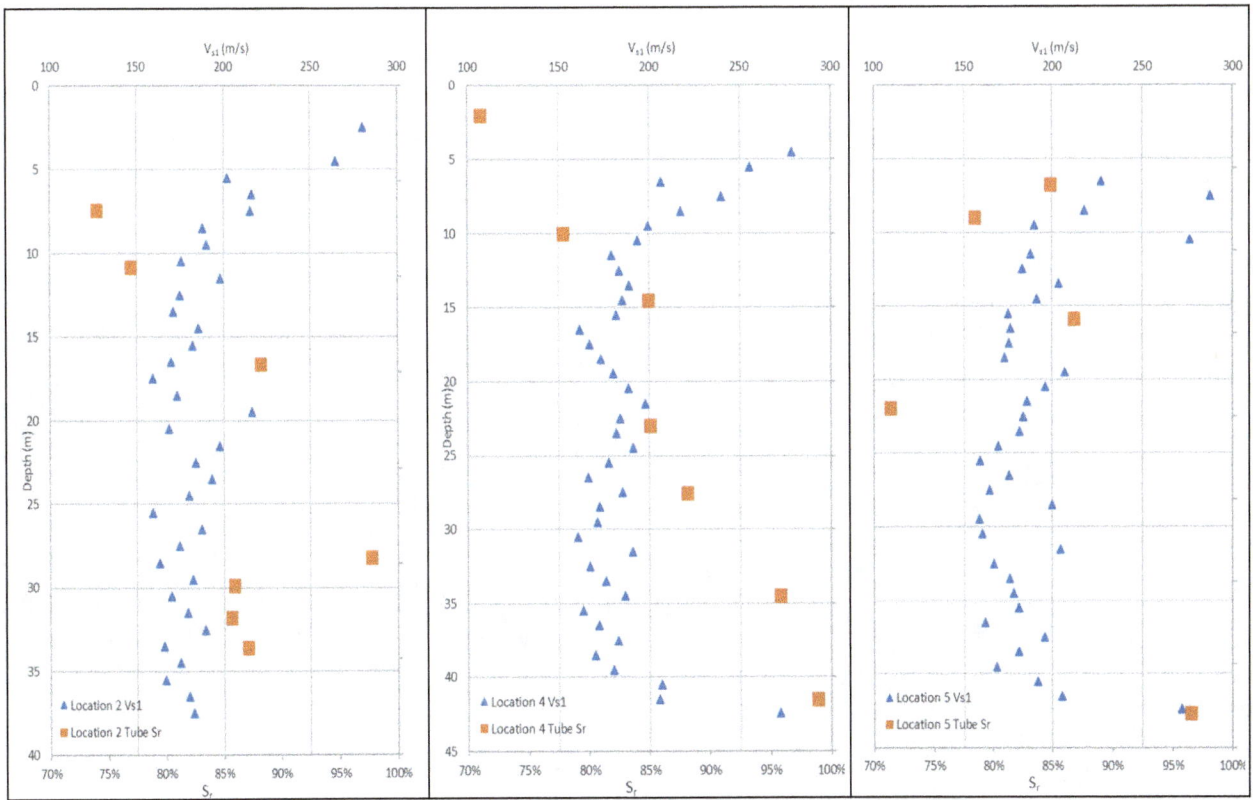

FIG 3 – 2020 V_{s1} versus 2020 tube S_r.

A subset of the data from the 2018 campaign V_{s1} profiles with adjacent tube S_r measurements similarly did not present any relationship with the saturation level.

The campaign compression wave velocity (V_p) profiles completed in 2018 with the adjacent tube S_r measurements are presented in Figure 4. In most instances, a correlation between V_p and the measured S_r is observed and can be inferred in comparison to the V_{s1} profile.

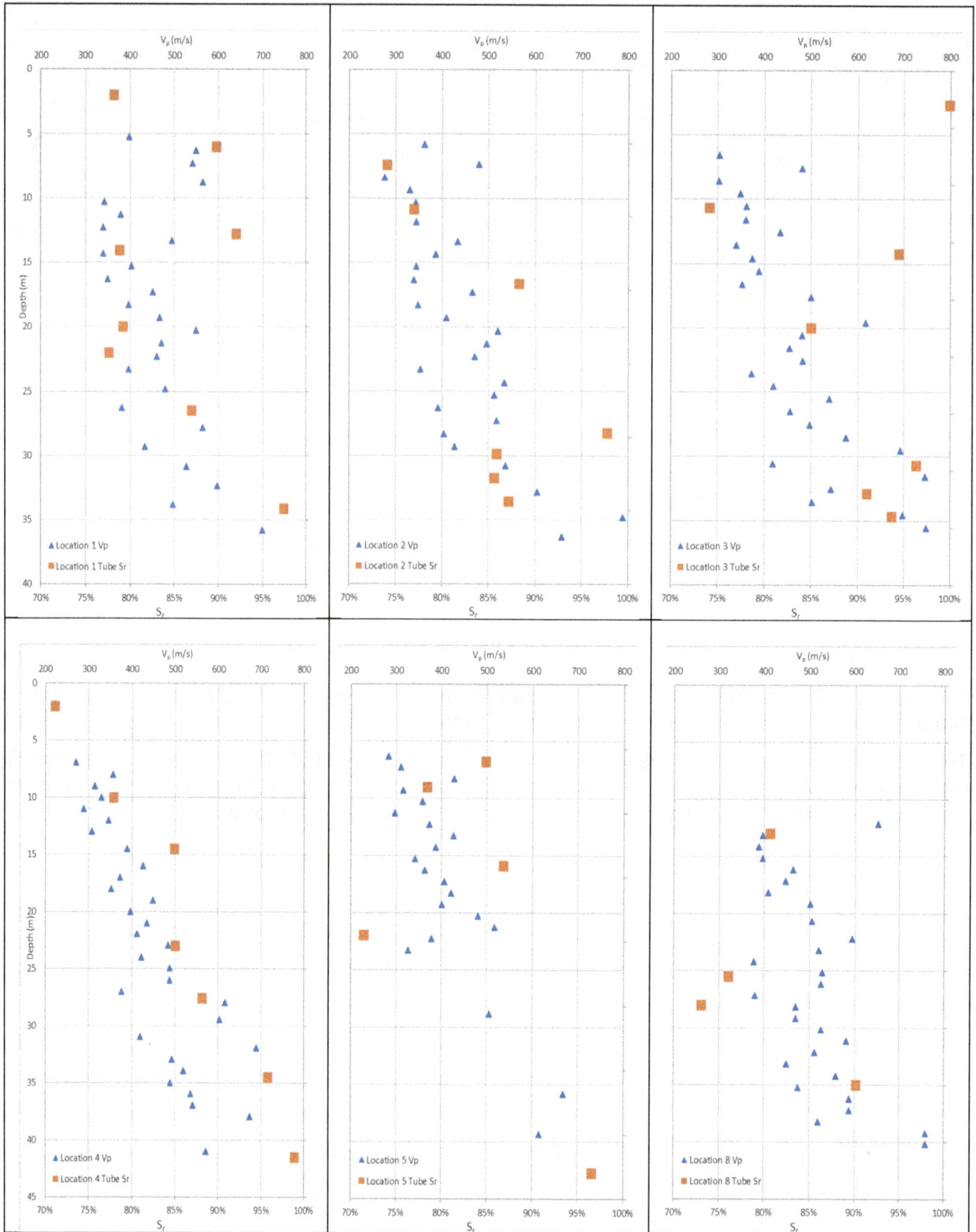

FIG 4 – 2018 V_p versus 2020 tube S_r.

The final profiles compared are the pore pressure ratio B_q with adjacent tube S_r measurements presented in Figure 5. The correlation between B_q and measured S_r can be inferred on a few plots but in most instances is not as evident without further deduction and adjustments of the horizontal axes in comparison to the V_p profile.

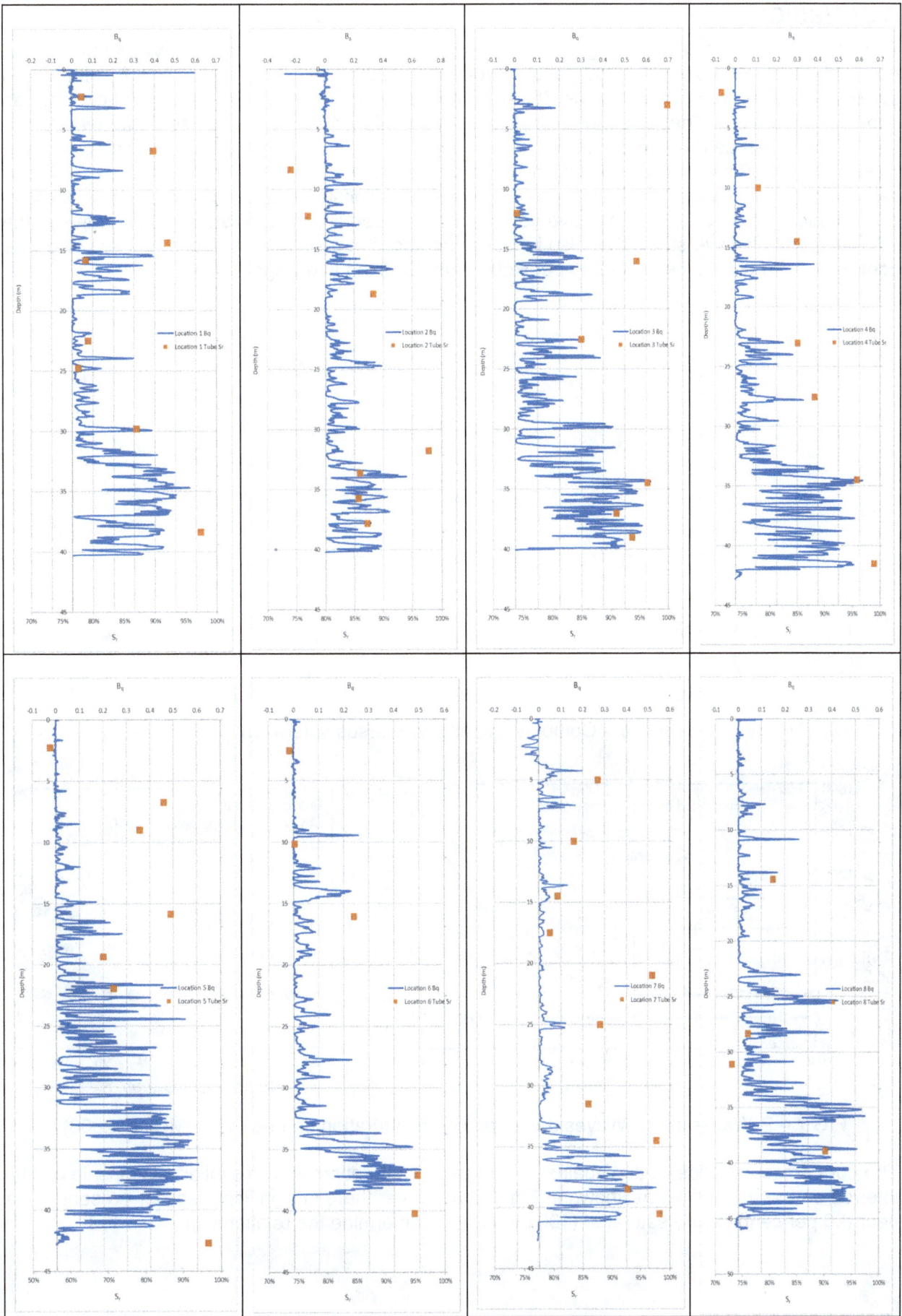

FIG 5 – 2020 B_q versus 2020 tube S_r.

DISCUSSION

The collated data displayed in Figure 6 suggests a similar correlation between V_p and S_r that has been made by Kiyoto *et al* (2006) and Valle-Moline (2006) as shown in Figure 7, with useful correlations of S_r from V_p as low as 85 per cent observed. A simple site-specific relationship as shown in Figure 6 outlines thresholds of V_p at approximately 400 m/s to represent the near to fully saturated (90 per cent) regime with full saturation defined at a V_p of 1000 m/s. In comparison, a threshold of 1700 m/s defined full saturation as per Kiyoto *et al* (2006) and Valle-Molina (2006) for sands. The presented V_{s1} did not provide any evident relationship to infer or support the evaluation of saturation, however the shear wave velocities have been linked with relationships between the void ratio and effective stress that also influence liquefaction resistance that can be used to define contractive and dilative behaviour as per Robertson, Cunning and Sego (1995).

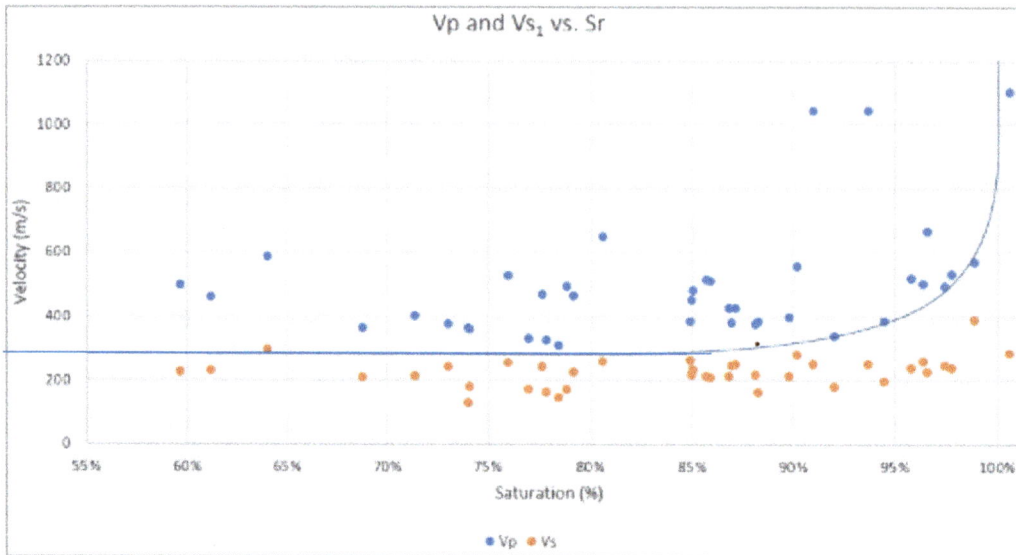

FIG 6 – Combined V_p and V_{s1} versus saturation.

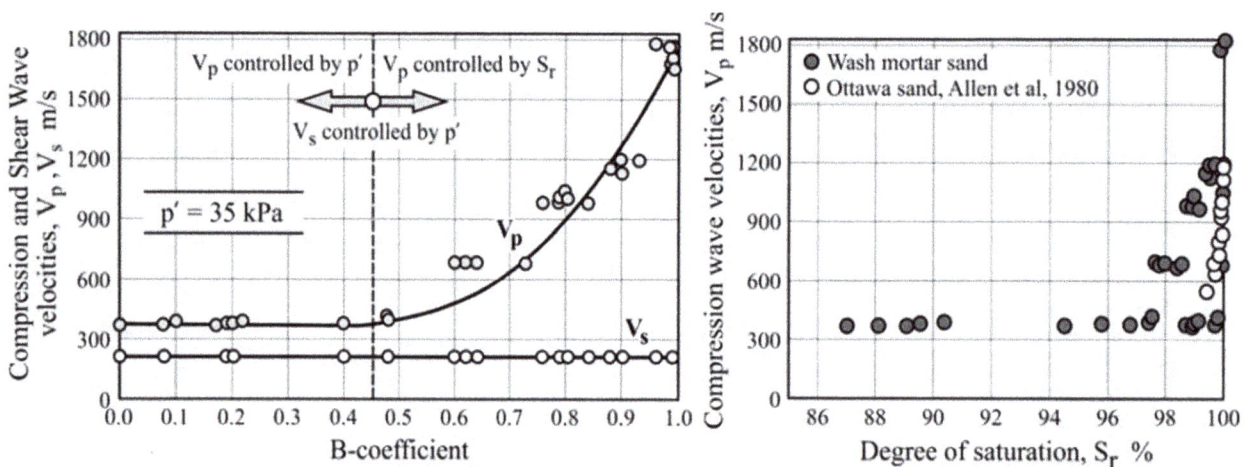

FIG 7 – P-Waves and S-Waves dependence on saturation degree (Valle-Molina, 2006).

The pore pressure ratios plotted against the degree of saturation show a similar relationship can be inferred as the V_{s1} profiles as shown in Figure 8 however there is significant scatter present in the near (90 per cent) to fully saturated (100 per cent) to determine the relationship.

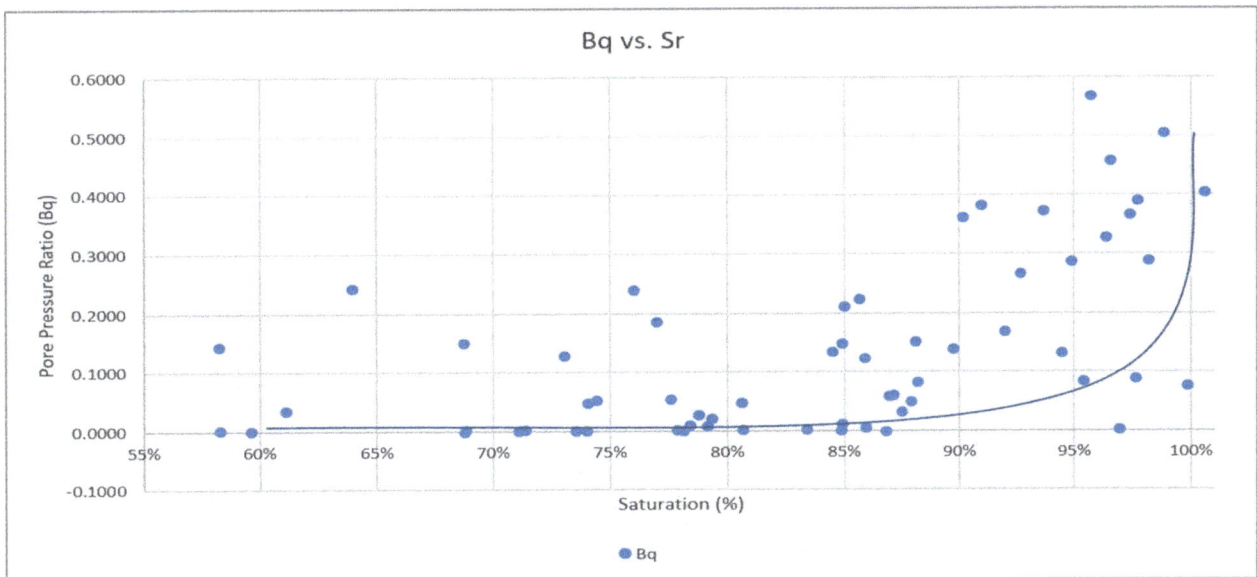

FIG 8 – Combined B_q versus saturation.

While the ability to infer a relationship can be further developed by validating the *in situ* data with bender element tests on laboratory samples, the collected piston tube samples were successful in providing good quality data that were critical to improving the soil characterisation. In addition to the saturation ratio, the void ratio, density, and water contents were obtained. These were extended into comparisons against estimated unit weights, and void ratios based on CPT correlations and allowed to validate state parameter following advanced tests. The 85 per cent retrieval rate demonstrates the ability to successfully collect reliable data and validate *in situ* data through advancements in tube sampling that preserve *in situ* moisture condition and density, that would otherwise be estimated.

CONCLUSIONS

A program of testing involving SCPTu, in combination with tube sampling undertaken for the tailings provided data to infer the degree of saturation of an upstream storage facility with a complex pore pressure regime. This program demonstrated the value of a well-executed site investigation and current capabilities of obtaining high quality data through a combination of *in situ* testing and tube sampling to sufficiently characterise the *in situ* state of tailings to enable evaluation of the expected behaviour. The ability to infer the degree of saturation from SCPTu can be further improved with a suite of laboratory tests using bender element tests to eliminate noise that was present in the *in situ* evaluations of compression and shear wave velocity. This work is presently in development.

REFERENCES

ASTM, 2016. ASTM D1587 Standard Practice for Thin-Walled Tube Sampling of Fine-Grained Soils for Geotechnical Purposes.

Jamiolkowski, M, 2012. Role of Geophysical Testing in Geotechnical Site Characterisation, *Soils and Rocks*, 35(2):May–August, 2012, 21 p. doi: 10.28927/SR.352117

Jamiolkowski, M and Masella, A, 2014. Geotechnical Characterization of Copper Tailings at Zelazny Most Site, *Geology*, 18 p.

Kiyoto, T, Sato, T, Koseki, J and Tsutsumi, Y, 2006. Comparison of liquefaction properties of in-situ frozen and reconstituted sandy sails, *Geomechanics and Geotechnics of Particulate Media, Yamaguchi, Japan*, pp 113–119.

Robertson, P K, Cunning, J C and Sego, D C, 1995. Shear Wave Velocity to Evaluate In Situ State of Cohesionless Soils, *Canadian Geotechnical Journal*, pp 848–858.

Valle-Molina, C, 2006. Measurements of Vp and Vs in Dry, Unsaturated and Saturated Sand Specimens with Piezoelectric Transducers, PhD Dissertation, University of Texas, Austin.

Why is -0.05 state parameter the boundary between contractive and dilative behaviour?

D Reid[1], R Fanni[2] and A B Fourie[3]

1. Research Fellow, The University of Western Australia, Crawley WA 6159.
   Email: david.reid@uwa.edu.au
2. Principal Tailings Engineer, Golder WSP, Perth WA 6159. Email: iccardo.fanni@wsp.com
3. MAusIMM, Professor, The University of Western Australia, Crawley WA 6159.
   Email: andy.fourie@uwa.edu.au

ABSTRACT

The use of critical state soil mechanics (CSSM) in tailings engineering has become well established. While CSSM in engineering practice includes many different procedures and analysis frameworks, the selection of an *in situ* state at which contractive or dilative behaviour is expected is one of the most important choices to be made. Over the past few decades, an evolving understanding of full scale flow liquefaction behaviour and state parameter Ψ interpretation methods has led to a value of Ψ = -0.05 being commonly applied as the transition value. However, while justifiable from case history data, the rationale for, and implications of, this contractive/dilative boundary are often not appreciated. For example, viewing CSSM in a 'steady state' interpretation framework, where shearing soil tends towards the critical state line (CSL) at large strains, is quite inconsistent in outlook from a contractive/dilative boundary of Ψ = -0.05. To illustrate the rationale for this boundary, a review is first carried out of relevant case history data, then of element test behaviour when sheared from initial states near to the CSL, with a particular emphasis on localisation and fabric anisotropy. Finally, the wisdom and implications of extending the Ψ = -0.05 boundary far from where it was originally justified are outlined.

INTRODUCTION

The series of major tailings storage facility (TSF) failures over the past decade has led to several changes and improvements to engineering practice. One of these is an increased application of critical state soil mechanics (CSSM) to the characterisation of tailings. This generally includes the determination of the critical state line (CSL) for some gradations of the tailings under consideration, and estimation of the *in situ* state parameter Ψ (Been and Jefferies, 1985), representing the distance, in void ratio terms, between the current void ratio of an element of soil and the CSL at the same mean effective stress p'. Application of CSSM and Ψ in engineering practice also involves the use of – universally, as far as the authors are aware – a criterion of liquefaction susceptibility (and, more generally, contractive undrained response) of Ψ = -0.05 (Shuttle and Cunning, 2008; Jefferies and Been, 2015).

Despite how commonly a Ψ = -0.05 criterion is applied, it is the authors' view that little critical appraisal of the reason for this boundary (rather than simply using Ψ = 0), and the history of its development, are carried out by practitioners. Further, although the Ψ = -0.05 criterion was initially developed based on the element behaviour of sands and the field scale response of flow liquefaction case histories made up primarily of sandy soils, it seems to now be widely applied to all soil types. This paper has been prepared to address what the authors see as a general lack of awareness around the development of the Ψ = -0.05 criterion in much of current engineering practice, the limited discussion on its rationale, and perhaps the even lesser discussion on how applicable this criterion may be for finer-grained, more compressible materials.

BACKGROUND

Historical development

The state parameter was first proposed in 1985, as part of works carried out to better characterise sands deposited in the Beaufort Sea for the oil industry (Been and Jefferies, 1985). A key initial use was to develop trends of soil behaviour, for example dilatancy and peak undrained shear strengths, against Ψ. In parallel, the Nerlerk Berm failure in the Beaufort Sea was under investigation (and a variety of hypotheses put forward as to the cause of failure – not dissimilar to the current debate over the proximate trigger of the Feijão failure). Two general hypotheses were initially suggested for Nerlerk: (i) liquefaction of the sand deposit (Sladen *et al*, 1985), (ii) significant contribution of the clay foundation to the failure, owing to the inferred dilative condition of the sand (Been *et al*, 1987).

In the initial debates it was first suggested that based on triaxial compression laboratory element tests, $\Psi > +0.02$ would be required for a strain softening response of the sand (Been *et al*, 1987). However, further consideration of laboratory data trends led to the insight that the initial contraction of specimens at, or even slightly dense of, the CSL was such that a loss of strength could occur for soils in such a state (Been *et al*, 1988; Jefferies *et al*, 1988). In other words, although specimens with a Ψ in the range of -0.05 to 0.0 would tend to be dilative at large strains, the initial contraction and loss of strength is such that liquefaction and flow cannot be excluded. This process of initial contraction of such 'medium dense' sands (ie Ψ = -0.05 to 0), and its potential implications for slope stability, are discussed subsequently.

Quasi steady state and idealised soil behaviour

It is first useful to present three idealised soil responses and define the current terminology used to describe these common behaviours. This is presented in Figure 1 as deviator stress versus strain and q versus p' plots for tests carried out under undrained conditions. Three idealised samples are presented, loose (ie $\Psi > +0.05$), medium dense ($\Psi \sim 0$) and dense ($\Psi < -0.05$). Loose specimens reach a peak undrained shear strength followed by a monotonic decrease to a residual value, which is the critical state strength – in other words, the specimen exhibits a monotonic decrease in mean effective stress p' to the CSL. It seems safe to say that this is the most commonly understood behaviour of flow liquefaction of loose soils in the context of CSSM. Similarly, the dense specimen, despite a slight initial contraction, dilates near-continuously towards the CSL, resulting in a continuous increase in undrained shear strength. The point at which the mean effective stress begins to increase being referred to as the Phase Transformation (PT) point.

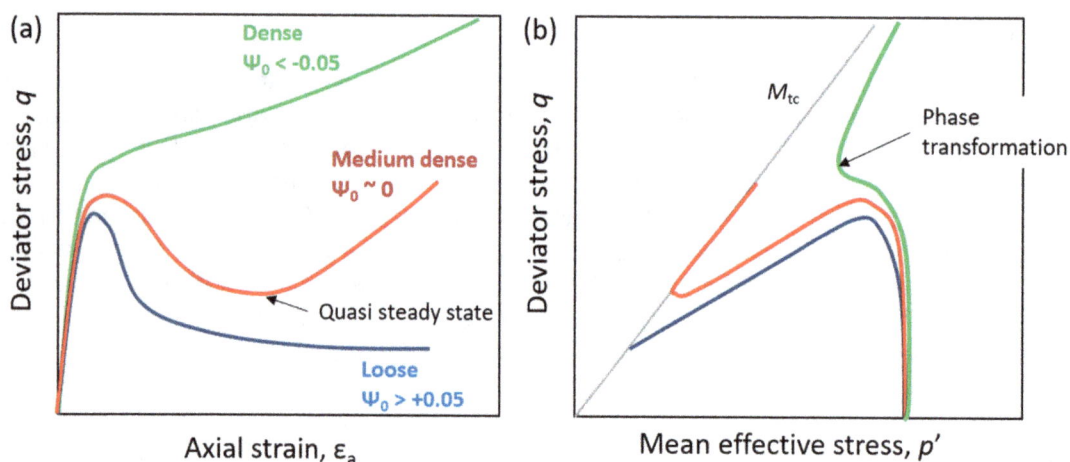

FIG 1 – Idealised soil response of loose, medium dense, and dense specimens in undrained shearing: (a) axial strain – q plot, (b) p' – q plot.

While the loose and dense specimen behaviour is well understood and conceptualised by practitioners in the context of CSSM, it is the medium dense, ie specimen near to the CSL (but looser than Ψ -0.05) that is our primary concern. This specimen first exhibits significant contraction, including a post-peak reduction in deviator stress q, reaches a 'local minimum' strength, then subsequently dilates towards the CSL. This local minimum of strength, and the condition at which it occurs, being referred to as the quasi-steady state (QSS). Such behaviour is ubiquitous in medium dense sands (Alarcon-Guzman, Leonards and Chameau, 1988; Konrad, 1990). At first inspection, while this behaviour may seem only of academic interest, or useful in describing qualitative soil response, the important implications of this response to considerations of slope stability are now addressed.

Relevance of the QSS to liquefied strengths

Consider the medium dense ($\Psi \sim 0$) element of soil discussed previously and imagine such an element of soil is located below a slope. Then consider an increment of undrained loading occurs resulting in an exceedance of the initial peak strength, leading then to further contraction and post-peak weakening to the QSS. In an element test, where loading is strain controlled at a typically slow rate, subsequent dilation will occur after the QSS. However, conditions in a real slope differ significantly. First, a slope is stress, rather than strain, controlled. If elements of below-slope material have sheared to the QSS – typically at least 3 per cent strain, this means the slope has begun a meaningful amount of movement. Vibrations are occurring – the situation is now dynamic, with momentum effects playing a role. Finally, the post-QSS dilation seen in element tests requires a reduction in pore water pressure. While possible in an element test that is globally undrained (sealed with valves), *in situ* this dilation may be attempting to occur in what is a relatively thin layer. The presence of nearby layers of other saturated material may lead to the 'donation' of pore water into the soil attempting to dilate, thus suppressing such dilation through partial drainage. These explanations summarise why many (most?) current workers in this area emphasise the relevance of the QSS strength to field-scale flow liquefaction behaviour.

Current engineering practice

Jefferies and Been (2015) carried out a detailed review of the flow liquefaction case histories which included estimating the characteristic (~80th percentile) *in situ* state parameter Ψ_k and slope of the critical state line (λ_{10}) for each case history. This built on previous interrogation of the same data reviewed by Jefferies (1998) and (Been *et al*, 2012), and the insights from the Nerlerk case history. The results of this review and synthesis are presented in Figure 2 as liquefied strength ratio against Ψ. The following items are outlined in the figure:

- The 'steady state' prediction of liquefied strength against Ψ. This line is the result of an analytical calculation with inputs of Ψ, slope of the CSL, and the critical state friction ratio M. These calculated values assume that elements of soil are sheared in a globally undrained manner and tend to the CSL. As such, the predicted strengths become far higher than those observed in flow liquefaction case histories less than about Ψ +0.05. This aligns well with when QSS-type behaviour commences in many soils, and thus where the assumptions of the steady state prediction approach are likely to become inapplicable.

- A trendline representing a best-fit to the flow liquefaction case history data – much of which, at least for relatively stiff soils (ie sands, silty sands) is located in the range of Ψ = -0.05 to +0.05, a range in which the steady state predictions do not produce reasonable estimates of liquefied shear strength. The leading hypothesis of which the authors are aware as to the origin of this discrepancy is the QSS behaviour of soil sheared from this range of Ψ. This therefore would suggest that mapping the extent of QSS behaviour for a particular soil will provide useful guidance as to its range of liquefaction susceptibility.

- It can also be seen in Figure 2 that different ranges of compressibility result in different Ψ – liquefied strength ratio trends. For the steady state predictions, this is simply a result of the analytical calculation – the same Ψ for a compressible soil will result in less effective stress reduction when tending to the CSL, and thus less loss in strength, than an incompressible/stiff soil. However, the case history trends, where the steady state predictions are no longer applicable, are informed by observed behaviour and back analyses. This then implies that QSS behaviour in this range may be less pronounced in compressible soils compared to stiffer sands and silty sands. This is investigated subsequently.

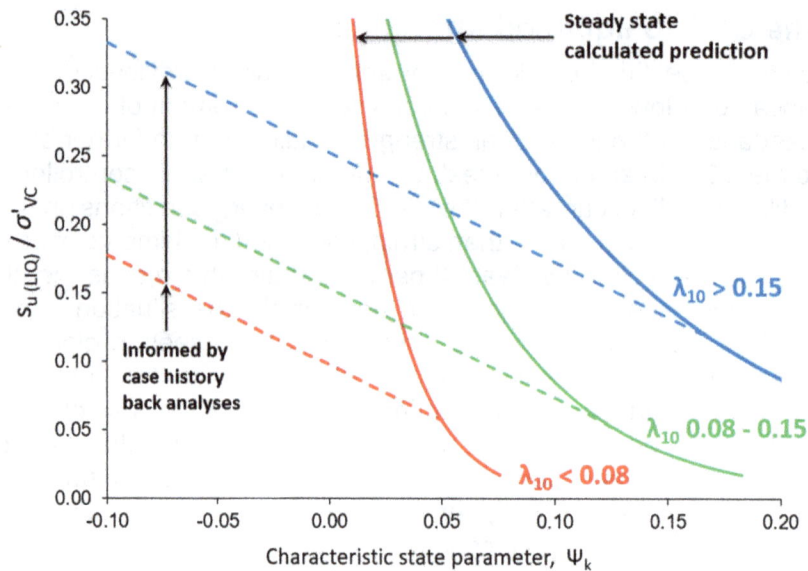

FIG 2 – Synthesis of the Jefferies and Been (2015) framework for comparing the inferred state parameter and back-analysed liquefied strength ratios.

Ψ uncertainty

Before proceeding further to real element test data, it is worth noting that the authors have previously heard speculation that the Ψ = -0.05 flow liquefaction cut-off is largely, or even solely, the result of uncertainty around the estimation of *in situ* Ψ. Indeed, it should certainly be acknowledged that the industry is currently unable to identify the elevation of the CSL at a better accuracy than about +/-0.02 Ψ (eg Reid *et al*, 2021), which in some analyses may play a role in the estimated *in situ* Ψ, and thus (were such a deposit to undergo flow liquefaction) the development of trends such as Figure 2. Further, refinements around the estimation of Ψ from cone penetration tests are ongoing (eg Ayala, Fourie and Reid, 2022). However, it is emphasised that the liquefaction criterion of Ψ = -0.05 is distinct from this uncertainty: no amount of plausible uncertainty will 'push' the estimated *in situ* states plotted in Figure 2 sufficiently that the steady state predicted liquefied strength could explain the field-scale behaviour seen. Further, the Nerlerk berm case history benefits from detailed calibration chamber testing on similar sands and many soundings, further reducing uncertainty in this case. It is therefore argued that while uncertainty in the estimation of Ψ must be acknowledged, this is not the primary cause of the Ψ = -0.05 flow liquefaction criterion.

FURTHER INVESTIGATION AND ILLUSTRATION OF THE QSS

Rationale

Accepting the previous discussion on the importance of the QSS as it relates to the behaviour of deposits with an *in situ* Ψ from about -0.05 to +0.05, and how the limit of QSS behaviour appears to be a plausible explanation for the limit of flow liquefaction, further investigation of factors affecting the QSS is useful. This can provide insight into how element test studies can better inform estimates of liquefied strength, and, if the lack of QSS behaviour seen in some more fine-grained clayey soils is such that a different liquefaction criterion may be warranted in such cases.

Effect of fabric/sample preparation method

An important aspect of the QSS is that it occurs at a relatively small strain, far before critical state conditions are achieved. As such, it is affected by fabric – that is, the initial orientation and position of soil grains, rather than simply Ψ. This then adds difficulties in measuring a QSS in laboratory element tests that can be reliably related to *in situ* conditions, given the difficulties of successfully replicating *in situ* soil fabric in the laboratory (Høeg, Dyvik and Sandbækken, 2000; Chang, Heymann and Clayton, 2011; Reid and Fanni 2022; Reid, Fourie and Fanni, 2022).

The effect of fabric on QSS strength is demonstrated through a synthesis of the series of isotropically consolidated (CIU) triaxial compression tests carried out by Murthy *et al* (2007) on Ottawa sand.

Specimens with a Ψ ranging from -0.05 to +0.05 were created by means of slurry deposition (SD) and moist tamping (MT), enabling comparison of the QSS strengths for the two sample preparation types (and, thus, two distinct fabrics). These are summarised in Figure 3 as QSS strength against Ψ_0. The MT specimens are seen to give higher QSS strengths than SD specimens across the same range of Ψ_0. This is most likely a result of some combination of tamping-induced overconsolidation (OC) in the MT specimens resulting from what is a relatively high density to achieve using MT-based approaches (Reid, Fanni and DiDonna, 2021; Reid, Fanni and Fourie, 2022a) and/or differences in fabric between MT and SD. This may then point to some important implications for *in situ* liquefaction susceptibility of different fabrics, when one considers that SD is likely to produce a fabric more similar to slurry-deposited tailings, while MT more closely resembles filtered tailings placement and compaction.

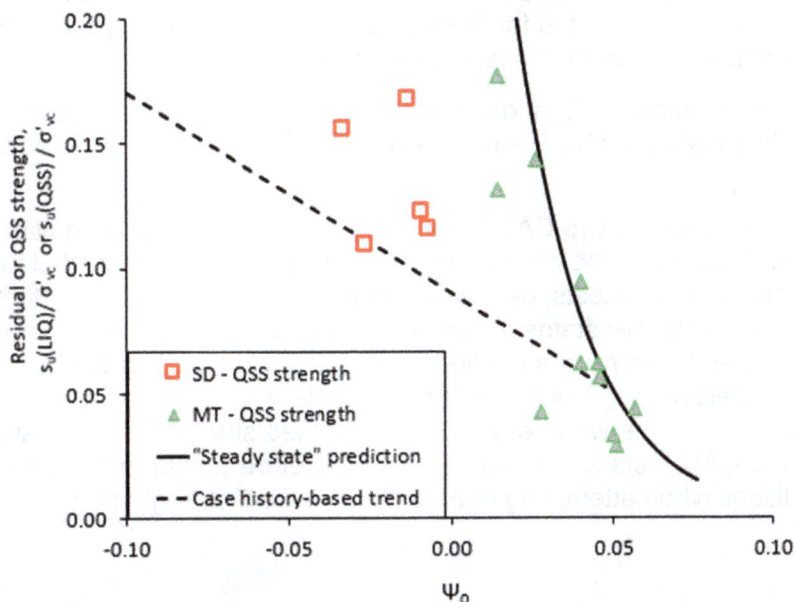

FIG 3 – Examination of QSS strength versus Ψ trends for the Ottawa sand tests carried out by Murthy *et al* (2007).

Effect of consolidated stress conditions

As noted previously, the tests carried out by Murthy *et al* (2007) were isotropically consolidated – ie CIU tests. While a useful consolidation condition for tests to provide much of the required characterisation data for soil such as the CSL, isotropically consolidated tests do not represent the stress conditions *in situ* – which in most cases are anistotropic. As the QSS is a small – to medium – strain phenomenon, examining the effects of consolidation conditions is also important to obtain reliable estimates. However, there appears to be relatively limited data in the literature to allow comparison of the QSS obtained from isotropic and anisotropic consolidation (CAU) conditions, at least in soils where the CSL has also been determined, therefore allowing interpretation of the results in a CSSM framework. The limited testing available also being generally inconclusive as to the effects (eg Kato, Ishihara and Towhata, 2001).

To further examine the effect of consolidated stress conditions, a set of tests was carried out on a soil mixture developed to support this paper as follows:

- A 'synthetic' blend of 80 per cent silica fine sand (SFS) and 20 per cent Eckalite (EK) kaolin was developed that allows QSS behaviour to be produced from samples deposited as a slurry.

- The CSL was determined through conventional test procedures.

- A pair of CIU and CAU tests were carried out to enable comparison of QSS response under these different consolidated stress conditions.

The results of the two CAU tests are summarised in Figure 4 as a $p' - q$ plot, q versus axial strain, and as a state diagram to show the test responses in the context of the CSL measured for the 80SFS-20EK soil. The specimens are seen to exhibit a peak strength at extremely small strains of

0.01–0.02 per cent, followed by brittle post-peak strength loss and contraction to a QSS condition between 4–5 per cent strain. Following the QSS an increase in strength is seen from the specimen beginning to dilate back towards the CSL. As indicated by the state diagram, both specimens were sheared from essentially 'on' the CSL. Several important points regarding *in situ* soil behaviour are provided by these tests:

- From an anisotropic consolidation condition such as exists *in situ*, very small strains and increments of undrained loading are required to initiate brittle post-peak strength loss.

- While strength gain after the QSS is seen, this requires about 5 per cent of axial strain to manifest. Although in a strain-controlled triaxial device this may be plausible, in a real slope if this process had occurred within material relevant to stability (ie a likely slip plane or zone), significant slope movement would have occurred by the time the was QSS reached. For the reasons outlined previously, it is far from clear whether the post-QSS dilation seen in strain-controlled laboratory element tests would manifest *in situ*.

- The strengths seen at the QSS, and the post-peak brittleness of the specimens, are consistent with expected behaviour for back analyses of flow liquefaction case histories with $\Psi \sim 0$ as will be seen.

The QSS strengths seen for the two CAU tests presented in Figure 4 along with two CIU tests are compared in Figure 5, again in the framework developed by Jefferies and Been (2015) as used throughout this paper. The CAU tests, despite adopting the same preparation technique as the CIU tests are seen to achieve denser states – this likely being a result of shear densification processes that occur when a higher stress ratio is applied during consolidation in a drained manner. However, despite the slightly different range of Ψ for the two tests the isotropic tests, when normalised by vertical effective stress as done when examining undrained strengths, clearly would exhibit higher QSS values than the CAU tests at a given Ψ. This therefore points to the importance of realistic consolidation conditions when attempting to measure the QSS strength that may apply *in situ*.

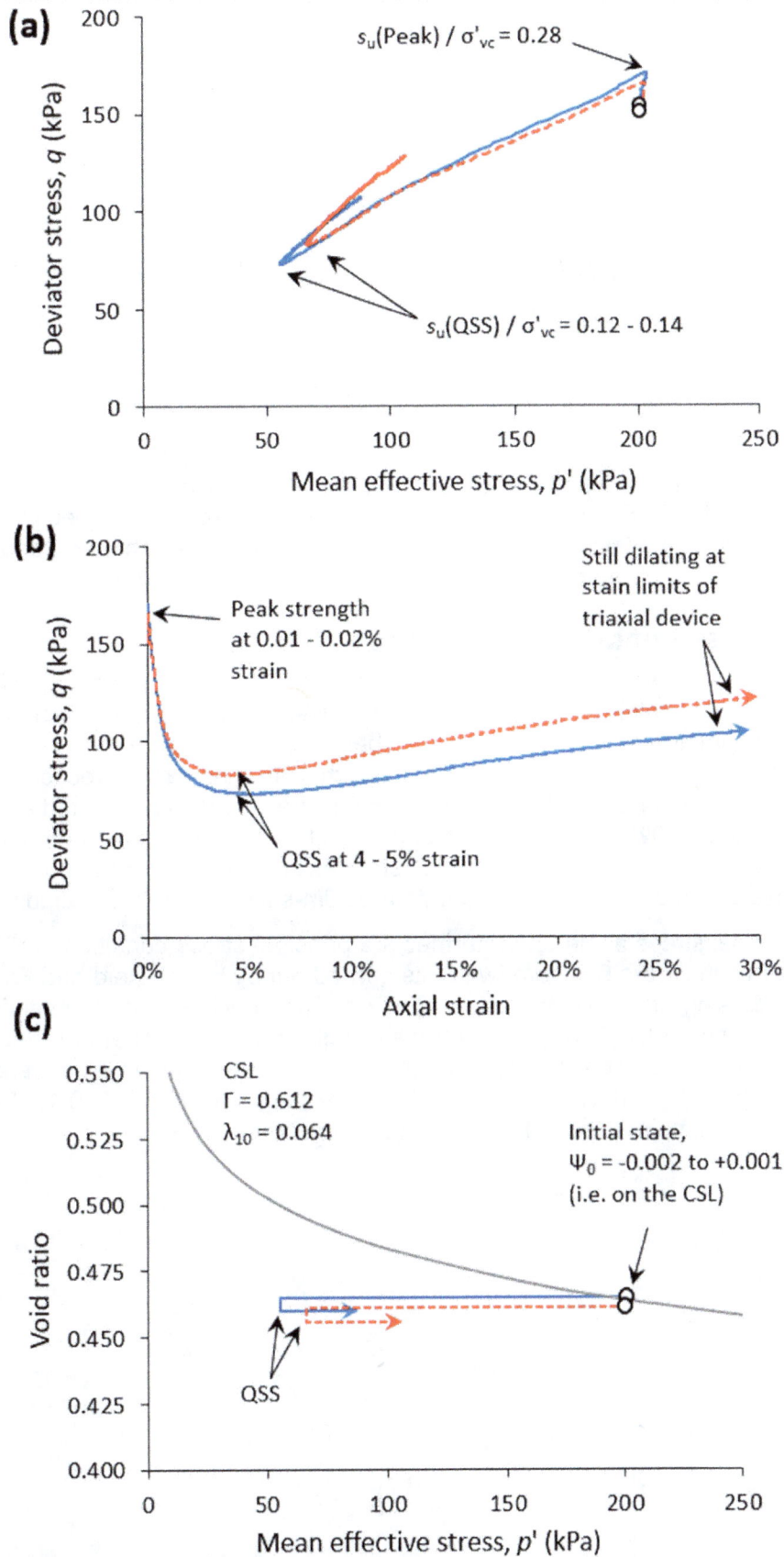

FIG 4 – Results of two CAU triaxial compression tests on slurry consolidometer-prepared specimens of 80SFS-20EK soil: (a) q versus p', (b) q versus axial strain, and (c) state diagram.

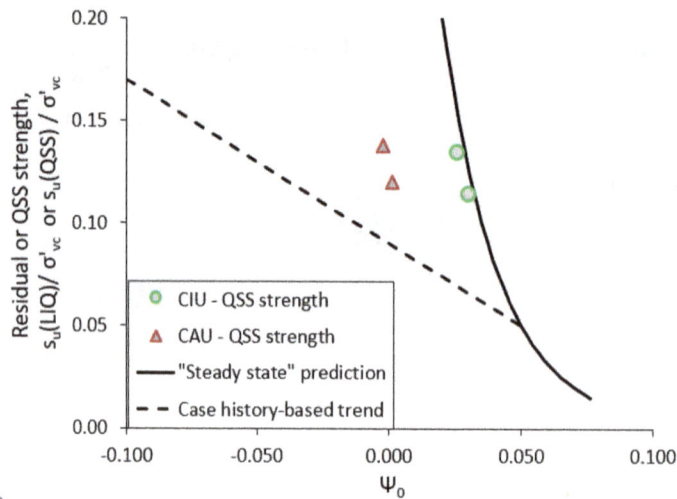

FIG 5 – Summary of the CIU and CAU tests on 80SFS-20EK slurry consolidometer-prepared samples that exhibit a QSS response, presented in the framework developed by Jefferies and Been (2015) and including the 'steady state' predicted soil response and the trend derived from case history data for soils with similar compressibility.

Below slope stress conditions/shearing modes

The laboratory data presented thus far has all been triaxial compression testing, either isotropically or anisotropically consolidated. While undoubtedly a useful test, the stress conditions within the triaxial differ substantially from those present below slopes. For example, within a triaxial compression test the principal stress angle is vertical – that is, in the same direction as the deposition condition for most soils, whereas below slopes principal stress angle has rotated away from vertical (Reid, 2020; Reid *et al*, 2022b). Further, the triaxial test is axisymmetric, whereas below a slope plane strain conditions generally prevail. Both differences between the triaxial compression test and below slope stress conditions can affect small- to medium-strain behaviour, including the QSS.

The effect of principal stress angle and intermediate principal stress conditions on QSS behaviour for SFS is presented in Figure 6, where two tests carried out by Fanni, Reid and Fourie (2022b) on samples prepared using dry pluviation are compared. The specimens had essentially identical Ψ (-0.03 and -0.04). The test with $\alpha = 0$ undergoes negligible loss of strength, rather exhibiting a monotonic increase in shear strength throughout. In contrast, the specimen sheared with $\alpha = 45°$ exhibits a significant strength decrease to a QSS representing $s_u(QSS)/\sigma'_{vc} = 0.12$. Clearly, principal stress angle can play a major role in QSS development in some cases.

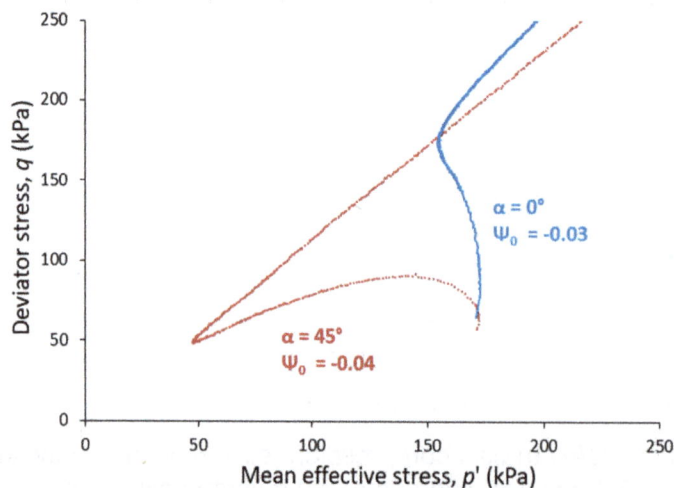

FIG 6 – Comparison of two undrained HCTS tests on SFS prepared using dry pluviation, after Fanni, Reid and Fourie (2022).

Extending the data presented in Figure 6, Figure 7 presents a summary of a series of HCTS tests on Toyoura (Yoshimine, Ishihara and Vargas, 1998), Ottawa 20–30 (Alarcon-Guzman, Leonards and Chameau, 1988), and Fontainebleu (Georgiannou, Konstadinou and Triantafyllos, 2018) sands, building on previous synthesis of such data (Reid *et al*, 2022a; Reid, Fanni and Fourie, 2022c). The previous SFS test carried out with a fixed principal stress angle is shown, whereas all the other data are from simple shear tests carried out using a HCTS – with the simple shear loading mode a useful analogue to below-slope stress conditions for TSFs (eg Reid, 2018) and the HCTS providing a fuller picture of stress conditions than a conventional direct simple shear for such tests (eg Fanni, Reid and Fourie, 2022a). The strengths presented from each test are those measured at the QSS. Also included, as with previous discussions in the paper, are the steady state calculated values and the trend derived from case history back analyses. The alignment of the laboratory test QSS strengths and the field-scale behaviour is striking. It is also noted that for each of the sands presented, triaxial compression tests across a similar range of Ψ exhibit far higher strengths. This comparison highlights the clear importance of shearing specimens in a loading mode consistent with below-slope conditions if attempting to directly extract relevant strengths from the test.

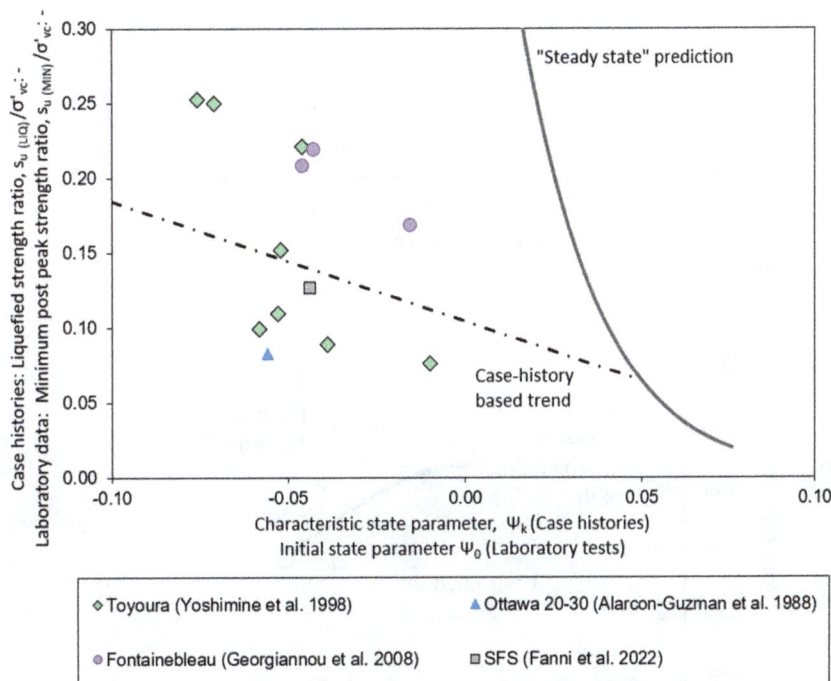

FIG 7 – Summary of QSS strengths for HCTS tests carried out by Yoshimine, Ishihara and Vargas (1998), Alarcon-Guzman, Leonards and Chameau (1988), Georgiannou, Konstadinou and Triantafyllos (2018) in the Jefferies and Been (2015)-developed framework, including 'steady state' predicted values and the trend developed from case histories with similar compressibility.

QSS behaviour in more compressible silts and clays

The previous discussions were focused primarily on sands, with the finest-grained soil presented thus far comprising only 20 per cent fines. This selection was based on the commonly observed QSS behaviour and the greater prevalence of relevant test data for such soils. However, many tailings are of much higher fines content and plasticity. Is QSS behaviour seen in such cases? To attempt to answer this question and extend data for QSS behaviour to finer-grained soils with different sample preparation techniques, the authors have recently carried out testing on an iron ore tailings and a gold tailings under simple shear loading within the HCTS at specimens intentionally prepared to a relevant Ψ range for QSS behaviour.

The iron ore tailings has a liquid limit of 39, plastic limit of 21, and plasticity index (PI) of 18, with 100 per cent < 75 μm and λ_{10} = 0.244 – thus being more finer-grained and clayey than any of the predominately sandy soils discussed thus far. A specimen was prepared by slurry deposition into a column, vertically loading to 80 kPa, extruded, trimmed to length, and finally the inner bore required for HCTS testing was cored out. Given the plasticity and general consistency of the iron ore tailings, this rather involved sample preparation process had been seen in early work to slightly densify this sample, thus fortuitously leading to the desired consolidated Ψ values necessary to investigate potential QSS behaviour.

The results of this HCTS simple shear test are presented in Figure 8, with the CSL for this material shown from parallel studies (Reid, Fanni and Fourie, 2023a). An initial contraction and strength loss is seen, to a QSS representing an undrained strength ratio of $s_u(QSS)/\sigma'_{vc} = 0.24$. In summary, QSS behaviour was seen, yet the magnitude of the QSS strength is fairly high in the context of typically observed liquefied strengths in cases of flow liquefaction.

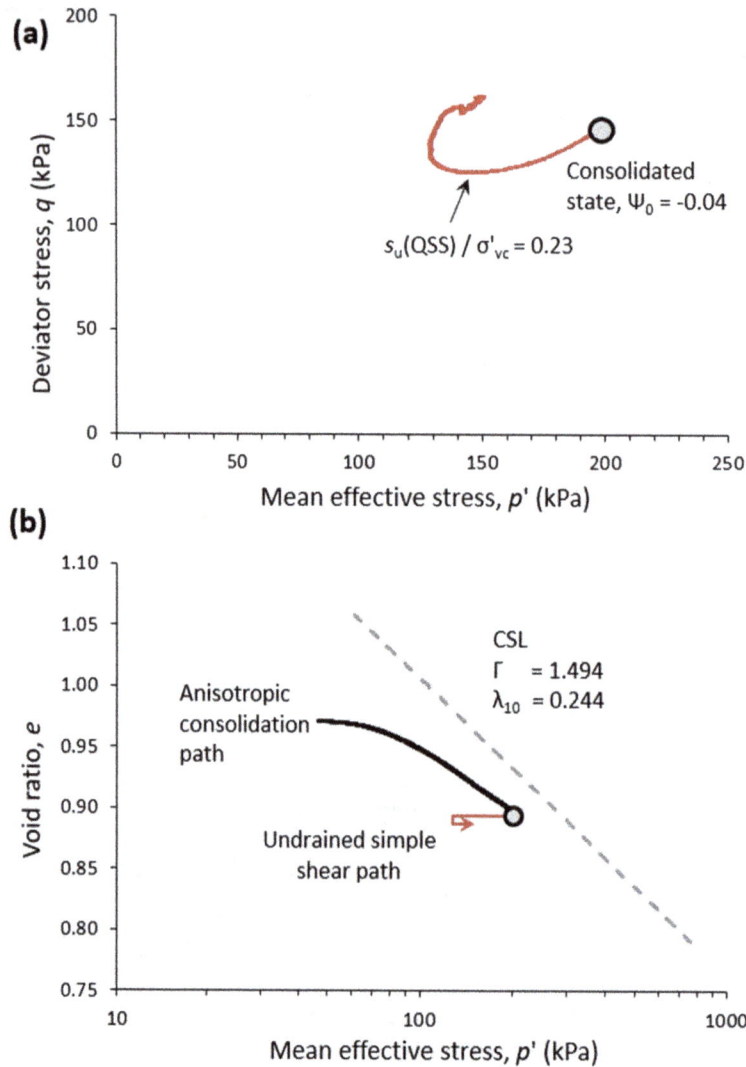

FIG 8 – Iron ore tailings HCTS simple shear testing results: (a) state diagram, and (b) p' – q plot. CSL location after Reid, Fanni and Fourie (2023a).

The gold tailings was predominately silt-sized (59 per cent < 75 μm), is non-plastic, has $\lambda_{10} = 0.063$, and has been thoroughly characterised in other recent studies (Ayala, Fourie and Reid, 2020; Reid, Fanni and Fourie, 2022a). In the interests of further investigating QSS behaviour of soils of greater fines content/compressibility than sands, Reid, Fanni and Fourie (2023b) carried out simple shear tests on the material within a HCTS. To produce specimens of the desired density while avoiding potential tamping-induced overconsolidation, the 'air dried' preparation technique (Reid *et al*, 2022c) was adopted.

The results of the gold tailings simple shear tests are presented in Figure 9 as a p' – q plot and state diagram. An initial contraction of both specimens is seen, despite them being dense of the CSL (Ψ -0.012 to -0.015). However, negligible post-peak loss of strength is apparent, with both specimens rather appearing to begin dilating, with an associated increase in strength near the end of shearing. It is noted in this case that the air dried specimens tended to undergo clear localisation at higher strains, a common difficulty in HCTS specimens of certain fabric and densities.

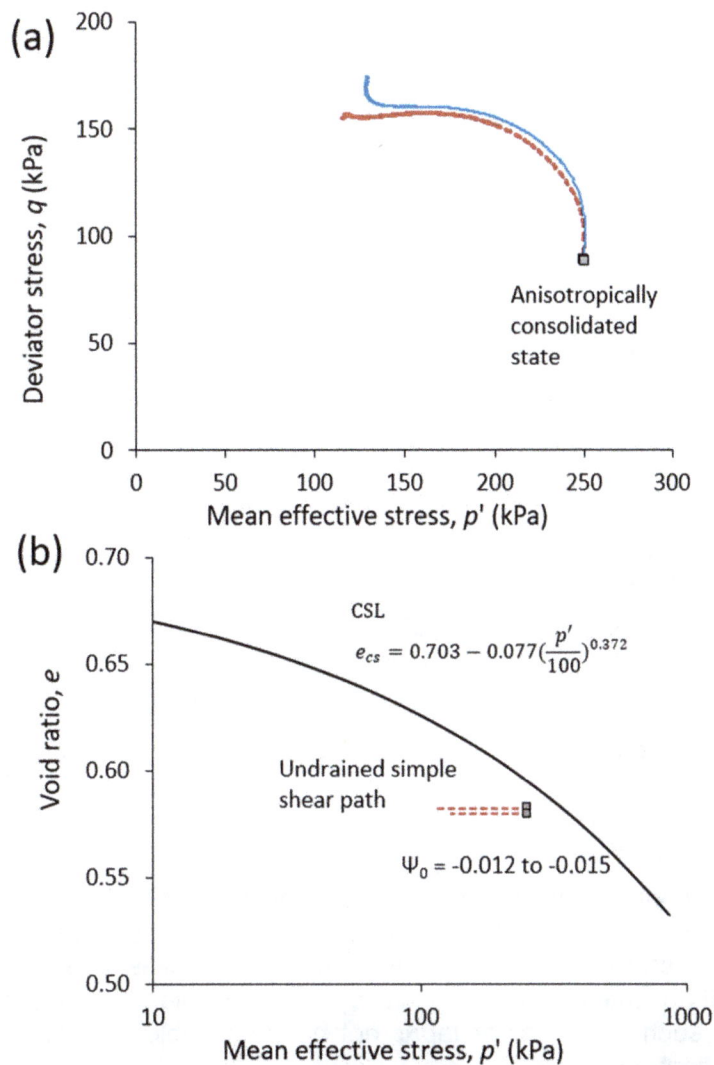

FIG 9 – Gold tailings HCTS simple shear testing results: (a) state diagram, and (b) $p' - q$ plot. CSL location after Reid, Fanni and Fourie (2023b).

The results of the iron ore and gold tailings HCTS simple shear tests are summarised in Figure 10 in the Ψ versus liquefied strength format plot for comparison. Although very subtle, both gold tailings tests were seen to exhibit a very slight reduction in strength at moderate strains that have been identified as the QSS for the purpose of plotting the data. The iron ore tailings QSS strength plots in a somewhat reasonable manner compared to the trend line developed by Jefferies and Been (2015) for similarly compressible soils – indeed, the iron ore tailings QSS strength is somewhat lower than the trend line would predict for Ψ = -0.04. However, while some post-peak strength loss was seen in the iron ore tailings, it is again emphasised that a 'liquefied' strength ratio of 0.23 is of a quite different magnitude, and would have dramatically different implications, than much lower values often seen in looser and/or stiffer soils.

The gold tailings results do not align well with the trend developed by Jefferies and Been (2015) from liquefaction case histories with similar stiffness (λ_{10} <0.08). This may be a result of the preparation method used for these specimens – homogenous air dried specimens prepared from a non-segregating slurry are likely to result in a more isotropic fabric than *in situ* deposition more likely to comprise segregation across layers. Indeed, block samples of the gold tailings obtained from a subaerially-deposited beach, which clearly included layering/segregation, did indicate cross-anisotropic undrained strengths (Reid, Fanni and Fourie, 2022b). It could then be speculated that thickened tailings deposits, much less likely to comprise the thin layering/segregation of subaerial conventional deposition, are less susceptible to QSS behaviour in the Ψ = -0.05 to 0 range that is the focus of the current paper – this being a particularly important range of densities, given the propensity for subaerial and thickened tailings to achieve such states (Reid and Jefferies, 2017,

2018). However, much further confirmatory evidence would be required before extending this tentative hypothesis too far.

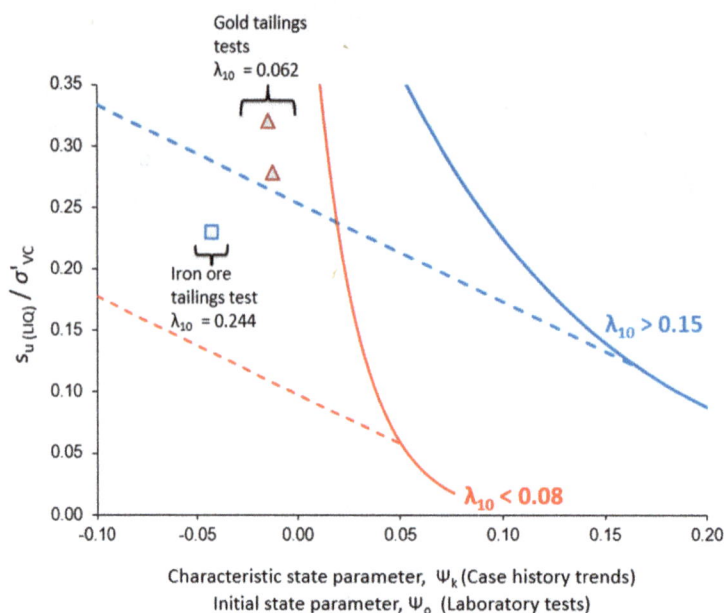

FIG 10 – Synthesis of iron ore and gold tailings data in the context of the Jefferies and Been (2015) liquefied strength–state parameter framework.

Practical considerations

At first inspection it may appear that whether QSS behaviour, and thus, implicitly, a liquefaction cut-off of Ψ -0.05, applies to more clayey soils may be merely academic, normally consolidated clayey soils and tailings being usually far looser. However, there are at least two practical situations where the use of a Ψ = -0.05 criterion in more clayey/compressible tailings may lead to excessive conservatism, should such a soil and/or fabric not be susceptible to post-peak strength loss and resulting flow liquefaction:

1. Design of the structural zone of filtered tailings for tailings that undergo compaction to a specified density. The application of CSSM, including a Ψ -0.05 liquefaction criterion; see current use in this area (Rola, 2021). In addition to compressibility, the observations of a reduced QSS response from MT specimens, with a fabric plausibly similar to filtered tailings placement, may also be important in this context (refer to Figure 3 and related discussion).

2. Analysis of the expected behaviour of tailings having undergone desiccation, which can result in overconsolidated conditions at low overburden pressure but will tend towards a more contractive state with increasing effective stress (Reid *et al*, 2018; Reid, Fourie and Russell, 2021). Further, testing presented here on air dried specimens may tentatively suggest that fabric created through the drying process is less susceptible to a QSS response (refer to Figure 9 and 10 and related discussion).

In both the cases indicated, analyses will likely include assessing at which point in the future the tailings under consideration may transition to a contractive and/or liquefiable condition. While a cut-off of Ψ -0.05 would be used in current practice, it seems possible that a looser cut-off may be appropriate for more compressible filtered tailings and/or desiccated specimens. Future experimental evidence of this behaviour, extending the initial testing attempts shown here, would be useful to help refine realistic limits of liquefaction susceptibility for these commonly encountered situations.

CONCLUSIONS

Current tailings engineering practice involves the frequent use of the CSSM framework with a criterion for liquefaction of Ψ > -0.05 being ubiquitous. This is based on historical development since the 1980s, with strong field-scale evidence supporting the potential for deposits with a characteristic Ψ in the range of -0.05 to 0 to undergo flow liquefaction, despite having a dilative response at high strains in element tests. The link between element tests and the field-scale behaviour being the QSS, which plausibly represents the post-peak strength that can be relied on *in situ*.

The current paper outlined the historical development of these ideas, while presenting a synthesis of previously published data and some new experimental results to demonstrate these issues and highlight the factors that can affect the QSS strength measured in laboratory element tests. The favourable alignment between the liquefied strengths estimated through back analyses and that measured at the QSS in appropriate laboratory element tests with below-slope shearing mode were highlighted. Finally, the paucity of available experimental evidence for QSS behaviour in finer-grained soils and tailings was outlined, including specific (unsuccessful) efforts by the authors to produce such behaviour, which appears to align with the field-scale evidence of *in situ* states for more-compressible case histories. The authors advocated further investigation of the QSS in finer-grained soils with a variety of fabrics to potentially reduce what may be excessive conservatism in assigning a liquefaction/contractive cut-off of Ψ -0.05 for more-compressible soils having undergone certain deposition methods.

REFERENCES

Alarcon-Guzman, A, Leonards, G A and Chameau, J L, 1988. Undrained Monotonic and Cyclic Strength of Sands, *Journal of, Geotechnical, Engineering*, 114:1089–1109.

Ayala, J, Fourie, A and Reid, D, 2020. Cone penetration testing on silty tailings using a new small calibration chamber, *Geotechnique, Letters*, 10:492–497.

Ayala, J, Fourie, A and Reid, D, 2022. Improved cone penetration test predictions of the state parameter of loose mine tailings, *Canadian Geotechnical Journal*, 59:1969–1980.

Been, K and Jefferies, M G, 1985. A state parameter for sands, *Géotechnique*, 35:99–112.

Been, K, Conlin, B H, Crooks, J H A, Fitzpatrick, S W, Jefferies, M G, Rogers, B T and Shinde, S, 1987. Back analysis of the Nerlerk berm liquefaction slides: Discussion, *Canadian Geotechnical Journal*, 24:170–179.

Been, K, Crooks, J H A, Conlin, B H and Horsfield, D, 1988. Liquefaction of hydraulically placed sand fills, Hydraulic Fill Structures: A Specialty Conference Sponsored by the Geotechnical Engineering Division of the American Society of Civil Engineers, Co-sponsored by the Society of the Mining Engineers of AIME, pp 573–590.

Been, K, Obermeyer, J, Parks, J and Quinonez, A, 2012. Post-liquefaction undrained shear strength of sandy silt and silty sand tailings, in *Proceedings of the 16th International Conference on Tailings and Mine Waste*, pp 325–336 (University of British Columbia: Vancouver).

Chang, N, Heymann, G and Clayton, C, 2011. The effect of fabric on the behaviour of gold tailings, *Géotechnique*, 61:187–197.

Fanni, R, Reid, D and Fourie, A B, 2022a. On reliability of inferring liquefied shear strengths from simple shear testing, *Soils and Foundations*, 62:101151.

Fanni, R, Reid, D and Fourie, A B, 2022b. Effect of principal stress direction on the instability of sand under the constant shear drained stress path, *Geotechnique*, 17 p. https://doi.org/10.1680/jgeot.22.00062

Georgiannou, V N, Konstadinou, M and Triantafyllos, P, 2018. Sand Behavior under Stress States Involving Principal Stress Rotation, *Journal of Geotechnical and Geoenvironmental Engineering*, 144:04018028.

Høeg, K, Dyvik, R and Sandbækken, G, 2000. Strength of Undisturbed versus Reconstituted Silt and Silty Sand Specimens, *Journal of Geotechnical and Geoenvironmental Engineering*, 126:606–617.

Jefferies, M and Been, K, 2015. *Soil Liquefaction: A Critical State Approach*, second edition (CRC Press).

Jefferies, M G, 1998. A critical state view of liquefaction, *Physics and mechanics of soil liquefaction*, pp 221–236.

Jefferies, M G, Rogers, B T, Stewart, H R, Shinde, S, James, D and Williams-Fitzpatrick, S, 1988. Island construction in the Canadian Beaufort Sea, Hydraulic Fill Structures: A Specialty Conference Sponsored by the Geotechnical Engineering Division of the American Society of Civil Engineers, Co-sponsored by the Society of the Mining Engineers of AIME, pp 816–884.

Kato, S, Ishihara, K and Towhata, I, 2001. Undrained Shear Characteristics of Saturated Sand Under Anisotropic Consolidation, *Soils and Foundations*, 41:1–11.

Konrad, J-M, 1990. Minimum Undrained Strength of Two Sands, *Journal of Geotechnical Engineering*, 116.

Murthy, T G, Loukidis, D, Carraro, J A H, Prezzi, M and Salgado, R, 2007. Undrained monotonic response of clean and silty sands, *Géotechnique*, 57:273–288.

Reid, D and Jefferies, M, 2017. State parameter as a geological principle in tailings, in *Proceedings of the 21st International Conference on Tailings and Mine Waste 2017*, pp 305–315 (University of Alberta Geotechnical Centre and Oil Sands Tailings Research Facility (OSTRF)).

Reid, D and Jefferies, M, 2018. A geological principle for the density of thickened tailings, in *Proceedings of Paste 2018, the 21st International Seminar on Paste and Thickened Tailings*, pp 117–126 (Australian Centre for Geomechanics, Perth).

Reid, D, Fourie, A B, Castro, J and Lupo, J, 2018. Undrained shear strength evolution with loading on an undisturbed block sample of desiccated gold tailings, in *Proceedings Tailings and Mine Waste 2018*, pp 365–376 (Colorado State University).

Reid, D, 2018. Importance of the simple shear loading direction to the stability of tailings storage facilities, in *Proceedings Mine Waste and Tailings Stewardship Conference 2018*, pp 274–284 (The Australasian Institute of Mining and Metallurgy: Melbourne).

Reid, D, 2020. On the effect of anisotropy on drained state liquefaction triggering, *Géotechnique Letters*, 10:393–397.

Reid, D, Fanni, R and DiDonna, P, 2021. The effect of tamping conditions and sample preparation on undrained shear strengths of a non-plastic sandy silt tailings, *Canadian Geotechnical Journal*, 59.

Reid, D, Fourie, A B, Ayala, J L, Dickinson, S, Ochoa-Cornejo, F, Fanni, R, Garfias, J, Viana da Fonseca, A, Ghafghazi, M, Ovalle, C, Riemer, M, Rismanchian, A, Olivera, R and Suazo, G, 2021. Results of a critical state line test round robin program, *Géotechnique*, 71:616–630.

Reid, D, Fourie, A B and Russell, A, 2021. Effects of desiccation on shear strength of tailings – comparison of clayey and sandy tailings, in *Proceedings of the 25th International Conference on Tailings and Mine Waste 2021*, pp 612–620 (University of Alberta Geotechnical Centre and Oil Sands Tailings Research Facility (OSTRF)).

Reid, D and Fanni, R, 2022. A comparison of intact and reconstituted samples of a silt tailings, *Géotechnique*, 72:176–188.

Reid, D, Chiaro, G, Fanni, R and Fourie, A B, 2022a. Arguments for increasing the use of hollow cylinder testing in the characterization of tailings, Tailings and Mine Waste 2022, 12 p (Colorado State University).

Reid, D, Dickinson, S, Mital, U, Fanni, R and Fourie, A, 2022b. On some uncertainties related to static liquefaction triggering assessments, *Proceedings of the Institution of Civil Engineers – Geotechnical Engineering*, 175:181–199.

Reid, D, Fanni, R and Fourie, A, 2022a. Effect of Tamping Conditions on the Shear Strength of Tailings, *International Journal of Geomechanics*, 22:04021288.

Reid, D, Fanni, R and Fourie, A, 2022b. Assessing the undrained strength cross-anisotropy of three tailings types, *Géotechnique Letters*, 12:1–7.

Reid, D, Fanni, R and Fourie, A B, 2022c. Discussion of 'Evaluation of flow liquefaction and liquefied strength using the cone penetration test: an update', *Canadian Geotechnical Journal e-first*.

Reid, D, Fourie, A and Fanni, R, 2022. Layering – the missing factor in fabric studies?, presented at the 20th International Conference on Soil Mechanics and Geotechnical Engineering (ICSMGE).

Reid, D, Simms, P, Sivathayalan, S, Fanni, R and Fourie, A, 2022c. The 'air dried' specimen reconstitution technique – review, refinements, and applications, presented at the *26th International Conference on Tailings and Mine Waste 2022* (Colorado State University).

Reid, D, Fanni, F and Fourie, A, 2023a. A SHANSEP approach to quantifying the behaviour of clayey soils on a constant shear drained stress path, *Canadian Geotechnical Journal*, under review.

Reid, D, Fanni, R and Fourie, A, 2023b. Linking laboratory quasi-steady state strengths to field scale performance, IS-Porto 2023, 8th International Symposium on Deformation Characteristics of Geomechanics.

Rola, J, 2021. Critical state soil mechanics in filtered TSF design, in *Proceedings of Mine Waste and Tailings Conference 2021*, pp 380–388 (The Australasian Institute of Mining and Metallurgy: Melbourne).

Shuttle, D A and Cunning, J, 2008. Reply to the discussion by Robertson on 'Liquefaction potential of silts from CPTu', *Canadian Geotechnical Journal*, 45:140–145.

Sladen, J A, D'Hollander, R D, Krahn, J and Mitchell, D E, 1985. Back analysis of the Nerlerk berm liquefaction slides, *Canadian Geotechnical Journal*, 22:579–588.

Yoshimine, M, Ishihara, K and Vargas, W, 1998. Effects of principal stress direction and intermediate principal stress on undrained shear behaviour of sand, *Soils and Foundations*, 38:179–188.

Water and AMD management

Multiphase gas flow model of a geomembrane covered buttress to inform hydrogeochemical TSF design

J Fourie[1], J Zheng[2], B Usher[3] and J Durocher[4]

1. Senior Hydrogeochemist, KCB Australia, Brisbane Qld 4101. Email: jfourie@klohn.com
2. Geochemist, KCB Australia, Brisbane Qld 4101. Email: jzheng@klohn.com
3. Principal, KCB Australia, Brisbane Qld 4101. Email: busher@klohn.com
4. Geochemist, Associate, Klohn Crippen Berger, Sudbury Ontario P3E 3B6, Canada. Email: jdurocher@klohn.com

ABSTRACT

Gas modelling was completed to inform stability upgrade designs for a tailings storage facility (TSF). Significant iron-rich precipitates have been accumulating at the toe of a 40+ year-old tailings storage facility (TSF). From a geotechnical perspective, the concern is that precipitation at the toe result in increased piezometric levels in the dam, which might compromise the long-term dam stability. A buttress with an integrated High-Density Polyethylene (HDPE) geomembrane liner above the drain elevation is being designed, which would limit the ingress of oxygen. The aim of this design is that reduced mineral precipitation will prevent the further rise of piezometric levels in the dam.

A 2D gas model was constructed to model the performance of the liner to limit oxygen ingress with cognisance of the buttress geometry. The following critical aspects were required for the modelling of the gas migration into the proposed sealed buttress:

- The Oxygen Consumption Rate (OCR) of the buttress waste rock.

- The Oxygen Diffusion Coefficient (ODC) of the HDPE liner.

- The degree of saturation in the layers overlying the HDPE.

The gas modelling indicated that below the liner, the air phase reaches sub-oxic conditions within a couple of years. Saturated parts of the buttress reach sub-oxic conditions within a couple of months as gas diffusion into the water phase is too slow to keep up with the consumption of oxygen.

It is anticipated that the sub-oxic conditions will create a geochemical environment that would not favour mineral precipitation and ideally, may even promote mineral dissolution. The results of the gas modelling were further used in a follow up study in a Reactive Transport Model, to calculate the potential for precipitation in the buttress.

INTRODUCTION

Significant iron-rich precipitates have been accumulating at the toe of a 40+ year old tailings storage facility (TSF). From a geotechnical perspective, the concern is that precipitation at the toe result in increased piezometric levels in the dam, which might compromise the long-term dam stability. The ferric iron precipitates are a direct result of the Eh-change experienced by the anoxic, ferrous iron-rich tailings water when it flows out of the tailings pore space.

An updated stability design for the TSF includes a (rock-filled) buttress with an integrated high-density polyethylene (HDPE) geomembrane liner above the drain elevation to limit the ingress of oxygen and reduce the rate of precipitate development.

Site observations have shown that flooded and sealed buttress designs in existing areas, using liner and/or clay cover encapsulation, provide a means of reducing the rate of precipitate development.

This paper discusses the gas modelling that was performed to assess whether sub-oxic conditions could be achieved in the buttress below the liner included in the updated design and, if so, the expected time for this condition to be reached.

CONCEPTUALISATION

Pore-water quality from the saturated TSF has been measured from pumped monitoring wells in the TSF. The deeper pore water is circumneutral (±pH 6–7), is generally anoxic (±Eh <0.3 V) and contains elevated total iron (±50–1000 mg/l).

When the tailings water emerges as seepage, conditions become oxic and the hydrolysis of the ferric iron (following the oxidation of the ferrous iron) results in a drop in the pH at the buttress (±pH 2.5–3.5). Precipitation of mostly iron-(oxy)-hydroxides and -sulfates occurs at the toe in drainpipes and seepage collection ponds. Crystalline precipitates observed from field and laboratory observations on-site include goethite ($FeOOH$), ferrihydrite ($Fe(OH)_3$), jarosite ($KFe_3(SO_4)_2(OH)_6$) and gypsum ($CaSO_4.2H_2O$).

A photo of toe seepage and associated precipitation are depicted in Figure 1.

FIG 1 – Precipitation of ferric iron precipitates at the toe seepage point.

The construction of a (waste rock fill) buttress with the potential inclusion of an HDPE liner is under consideration to provide stability support. A further objective of the proposed buttress is to limit oxygen ingress, ferrous iron oxidation, and subsequent precipitation of secondary minerals. A conceptual design is provided in Figure 2.

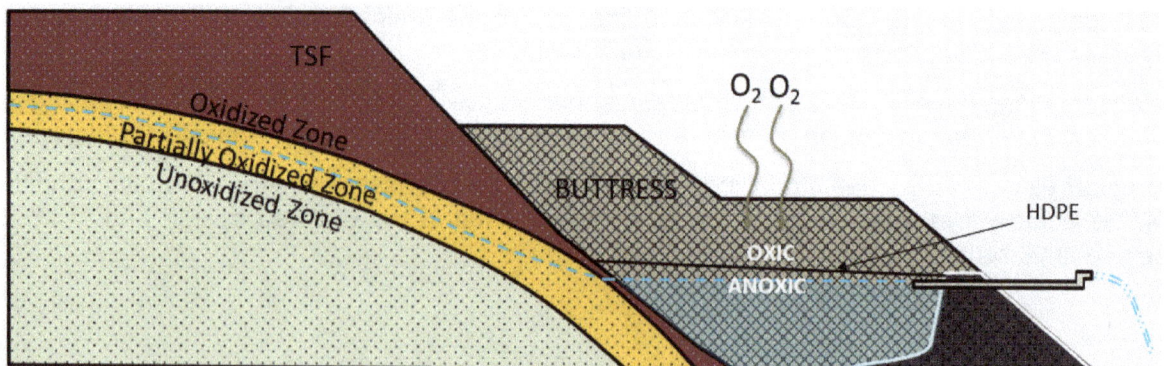

FIG 2 – Precipitation of ferric iron precipitates at the toe seepage point.

The buttress will only be partially flooded below the liner and will therefore include both an unsaturated and saturated zone. The objective of the airflow and geochemical models is to predict the potential effectiveness of the liner in limiting oxygen infiltration into the buttress, which would allow the water to pass to the outlet with a reduced likelihood of mineral precipitation.

METHODOLOGY

The methodology used for the modelling included:

- Calculation of the Oxygen Consumption Rate (OCR) of pyrite oxidation in the buttress material.
- Analytical calculation of the decrease of oxygen in the buttress as a first-order estimation.
- 2D numerical gas flow model to assess the changing oxygen concentrations across the facility.

OXYGEN CONSUMPTION RATE (OCR) OF BUTTRESS MATERIAL

Introduction

The oxygen consumption rate (OCR) was used to estimate the amount of oxygen that is consumed in the buttress. The OCR of typical waste rock considered for the buttress was calculated from six kinetic columns performed on norite, gabbro and granitic gneiss waste rock. The waste rock material had %S ranging between 0.3–3 per cent. However, waste rock with %S <1 per cent will most likely be used for the buttress fill.

Methodology

The following methodology was used:

- The late-sulfate production rate of the columns (sulfate load after it reached an approximate constant concentration in leachate) was calculated from the column test results.
- The OCR was estimated using the Geochemist Workbench (GWB) software as a spreadsheet calculation cannot consider the potential effect of secondary mineral precipitation and dissolution.
- The pyrite effective surface area was calibrated to obtain the late-sulfate production rate observed. The OCR was calculated from the modelled pyrite oxidation rate. It is expressed as the mass of oxygen consumed per mass of rock over time.

Results

An example of the simulation of one of the columns is shown in Figures 3 and 4. The following comments relate:

- The simulation in GWB indicated dissolution of secondary sulfates (eg gypsum), resulted in the high calcium and sulfate in the initial column leachate. Iron and sulfate will initially precipitate as schwertmannite which will redissolve in favour of $Fe(OH)_3(s)$ to form.
- The waste rock has a Sulfide S% of 0.13–0.66 (25–75 percentile). The selected oxygen consumption rate used for modelling was 5.8×10^{-11} kg/kg/s, which is in the range for the waste rock with <1 per cent Sulfide S.
- The rate may be slightly lower under field conditions because of the lower temperature expected in the field (±10°C in monitoring wells). Using the Arrhenius equation, the calculated rate for field conditions is about a factor of 0.44 less at 2.6×10^{-11} kg/kg/s.

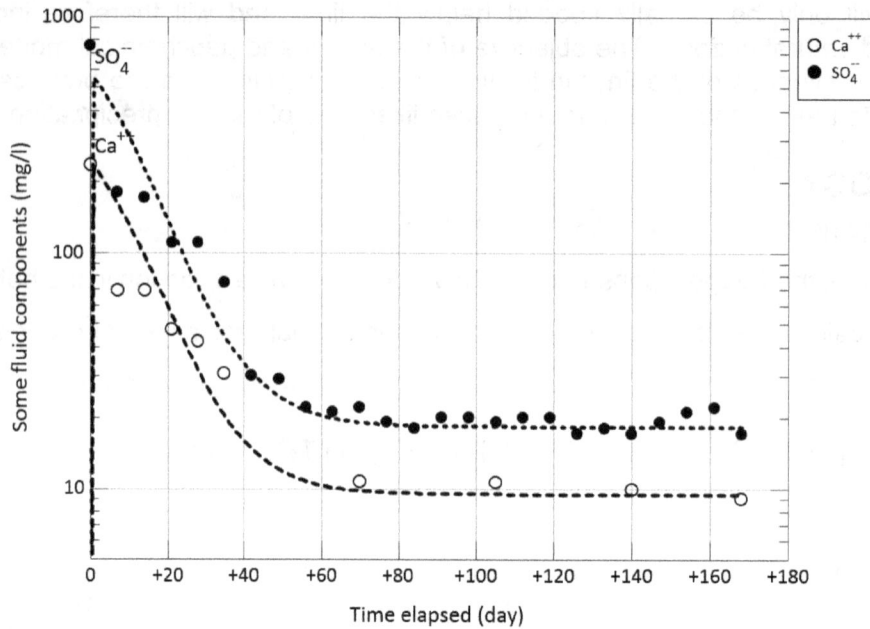

FIG 3 – Estimating the produced sulfate and calcium release using GWB.

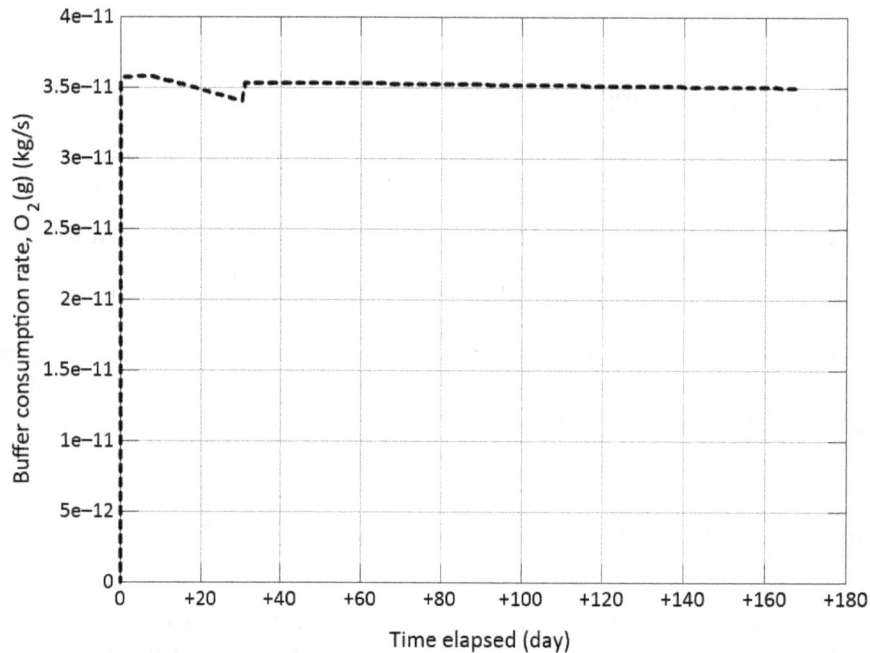

FIG 4 – Estimating the oxygen consumption rate using GWB.

FIRST-ORDER ANALYTICAL CALCULATION

The buttress will only be partially flooded below the liner, therefore reactions in both the unsaturated and saturated zones were considered. It was assumed that immediately after the construction of the buttress, the oxygen in the unsaturated waste rock pores would be at atmospheric concentration (~280 g/m³). An analytical calculation was performed to calculate the decrease of oxygen in the buttress below the liner, assuming no further migration of oxygen through the liner (as a reference for the best case).

The rate of change in the oxygen concentration in the buttress could be calculated using the first-order reaction rate equation for a chemical reaction (Chemistry Libretext, 2013):

$$dC/dt = KC_0$$

where C is the oxygen (mass) concentration, C_0 is the initial concentration, and K is the rate constant. From the column leach test results, the oxygen consumption rate for the buttress waste rock is estimated at 5.8 × 10⁻¹¹ kg/kg/s at 20°C or 2.6 × 10⁻¹¹ kg/kg/s at 10°C. Using these two rates, the decrease in oxygen was calculated for the buttress material, assuming that no further oxygen ingress occurs, as shown in Figure 5. For the two rates, the oxygen concentration decreases 50 per cent within 108 and 245 days (the half-life), and 90 per cent within 353 and 788 days.

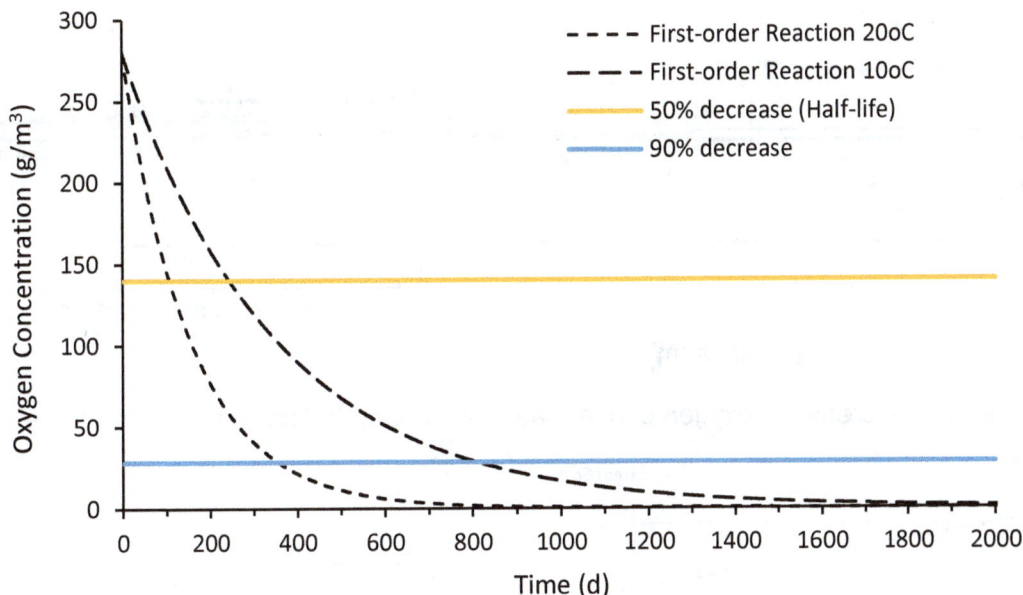

FIG 5 – Decrease in oxygen concentration according to first-order reaction rate.

2D NUMERICAL GAS FLOW MODEL

Introduction

The objective of the 2D model was to determine whether sub-oxic conditions could be maintained within the buttress to limit ferrous iron oxidation and mineral precipitation. The 2D model considers the performance of the liner with cognisance of the buttress geometry.

Methodology

Multiphase modelling was performed using the Geostudio 2020 suite of software. SEEP/W and CTRAN/W were used, which are finite element software packages for modelling water flow, solute and gas transfer in porous media.

A single representative cross-section through the TSF and buttress is presented in this paper. A base scenario (No Liner) was compared to three model scenarios, which included different liner setups. The model scenarios are summarised in Table 1.

TABLE 1

Multiphase model scenarios.

Model	Scenario	Description
No Liner Scenario	No Liner	Buttress with no liner. This was used as the initial condition for all the other scenarios as it was assumed that there would be some oxygen in the buttress just after construction.
Liner Scenarios	Scenario A	Buttress liner extent to the top of the buttress.
	Scenario B	Buttress liner 20 m up on the slope of the TSF.
	Scenario C	Liner only intersects the TSF. No liner on slope.

The model extent and discretisation are depicted in Figure 6. In Figure 7 the extent of the liner up the slope of the TSF is shown for the various scenarios. The model extent used for the gas model included the pond on the left with the buttress on the right of the tailings dam. The liner is on the flat surface in the batter and the 1:5 slope towards its right. A finite element grid was assigned to the model with approximate 2–5 m element size in the saturated tailings and aquifer. The element size in the unsaturated tailings and buttress was further refined to be at 0.5–1 m.

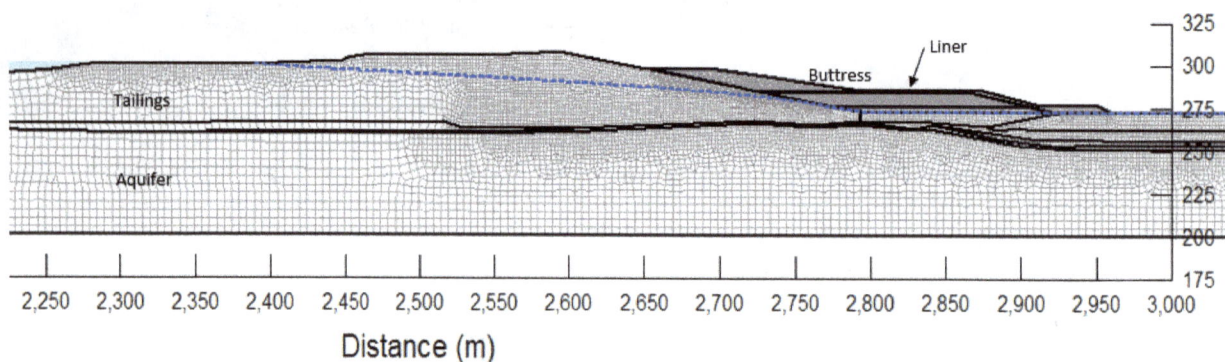

FIG 6 – Decrease in oxygen concentration according to first-order reaction rate.

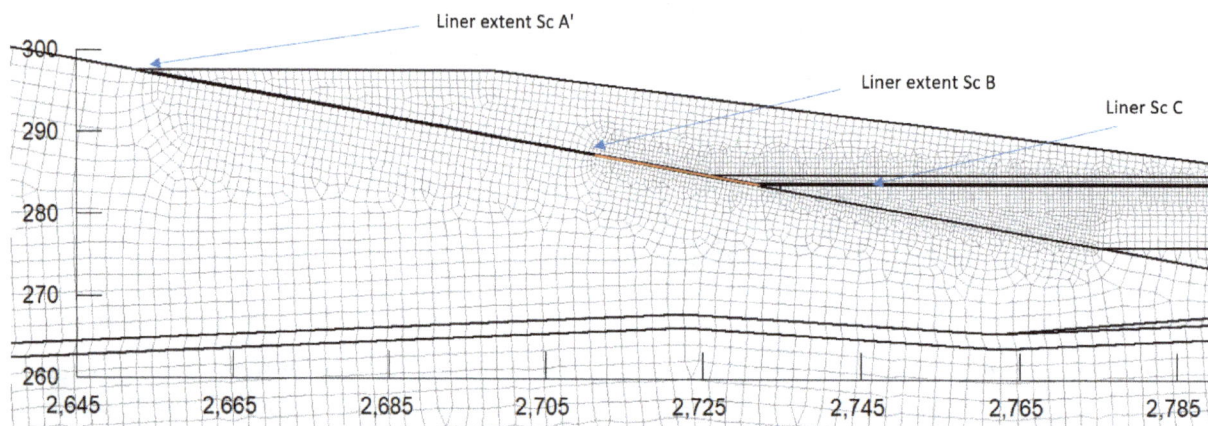

FIG 7 – Extent of liner in Scenario A, B and C (zoomed in).

Material properties and boundary conditions related to heat generation, airflow and oxygen transport were based on site-specific data, data from analogous sites, and/or typical literature values. The material properties are summarised in Table 2 and the boundary properties in Table 3.

TABLE 2

Multiphase model layer properties.

Material property	Symbol	Value	Units	Source/reference
Porosity – Waste Rock and Tailings	N	0.32 (WR) 0.44 (tails)	Dimensionless	Laboratory testing
Air Permeability (dry) – Waste Rock and Tailings	k	1×10^{-8} (WR) 1×10^{-9} (tails)	m^2	From literature (eg Lefebvre *et al*, 2001; Steinpress, 2017; Wang *et al*, 2016)
Oxygen Consumption Rate	-	2.6×10^{-11}	kg/kg/s	Kinetic tests (unit is mass of gas consumed per bulk mass of material per second).
Oxygen Diffusivity in Air	Da	1.8×10^{-5}	m^2/s	Theoretical Physical Value (eg Lefebvre *et al*, 2001)
Oxygen Diffusivity in Water	Dw	2.5×10^{-9}	m^2/s	Theoretical Physical Value (eg Lefebvre *et al*, 2001)
Oxygen Diffusion Coefficient (Liner)	De	1.157×10^{-12}	m^2/s	As calculated from the liner design specification.
Solubility Coefficient of Oxygen in Water	-	0.031805	Dimensionless	Theoretical Physical Value (eg Lefebvre *et al*, 2001)

TABLE 3

Boundary and initial conditions assigned to the model.

Boundary	Value	Units	Source/reference
Oxygen – Concentration at Surface (atmospheric)	280	g/m^3	Literature
Temperature – Surface energy Balance	Modelled	°C	Calculated from site climatology
Temperature – Initial Waste Rock and Tailings	±10°C	°C	Field Data
Temperature – Bottom of foundation	±10°C	°C	Estimate from field data
Temperature – Heat Generation of Sulfide	1.258	kJ/Mg O$_2$	Calculation

2D gas flow results

For each of the scenarios, an initial condition was established by modelling the system with a buttress in place but without a liner (No Liner Scenario). Initial conditions are reported as Year 0 results in this section, as depicted in Figure 8. Year 0 results were used as the initial conditions for all the other scenarios, as it was assumed that there would be some oxygen in the buttress just after construction. From the results, the following observations could be made:

- After construction, the oxygen concentration in the air phase below the liner was mostly oxic (Figure 8), varying roughly between 210–280 g/m^3 (15–21 per cent O$_2$ in air). This is the oxygen that will be present in the buttress if no liner is present (No Liner Scenario).

- After four years, the oxygen concentration in the buttress is below sub-oxic conditions (10–30 g/m^3, 0.8–2 per cent atm) in Scenario A and B (see Figures 9 and 10).

- With the liner extended up the slope of the tailings dam to the top of the buttress in Scenario A, oxygen migrating through the unsaturated tailings will be below sub-oxic conditions (<10 g/m^3, <0.8 per cent) before even reaching the buttress (the 10 g/m^3 contour is towards the left of the 'window' area between the horizontal liner and the phreatic zone; about 20 m closer to the tailings than in Scenario C (Figure 11).

- With the liner extended 20 m up the slope of the tailings dam in Scenario B, oxygen migrating through the unsaturated tailings will reach the buttress at, or very close to, sub-oxic conditions (10 g/m^3, <0.8 per cent) (the 10 g/m^3 contour is in the 'wedge' between the horizontal liner and the phreatic zone; about 5 m closer to the tailings than in Scenario C (Figure 11)).

- In Scenario C (no liner up the slope), a very small part close to the unsaturated tailings 'window' (see Figure 10) is above these concentrations. In Scenario C, some oxygen is bypassing the liner and is being transported through the unsaturated tailings. The rate of transport is, however, very slow as the tailings are typically at least 30–40 per cent water saturated; and as a result, the oxygen bypassing the liner does not influence the development of sub-oxic conditions in the bulk of the buttress, and especially in the zone that is below the phreatic surface in the buttress, oxygen concentration remains very low – eventually reaching zero.

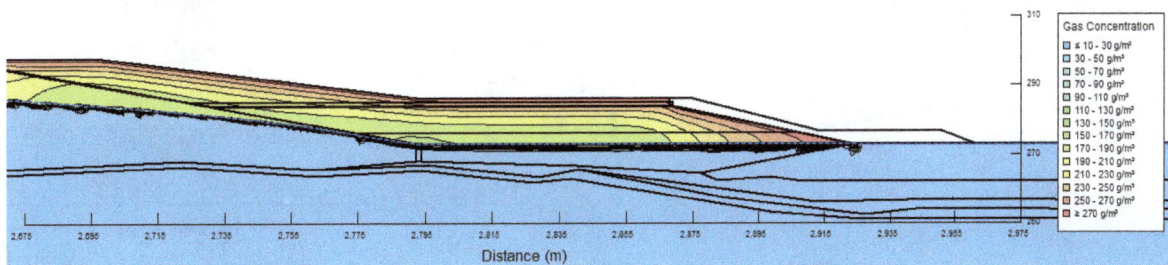

FIG 8 – No Liner: Initial Oxygen Concentration (Year 0).

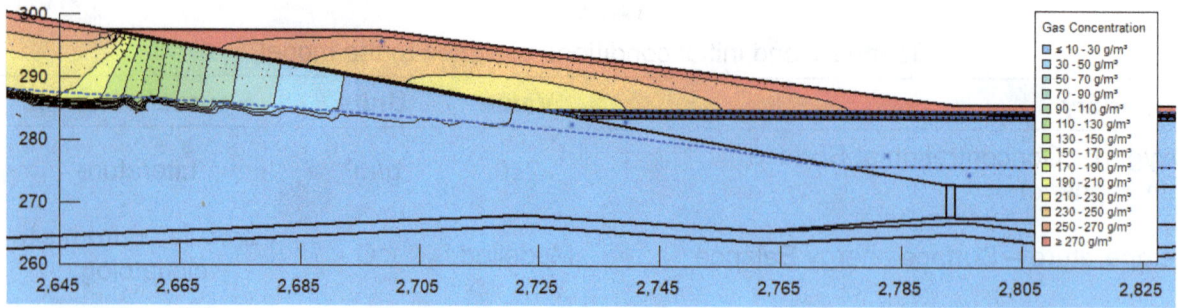

FIG 9 – Scenario A: Oxygen Concentration after four years with liner up to the top of the buttress (zoomed-in; with gas mass flux arrows).

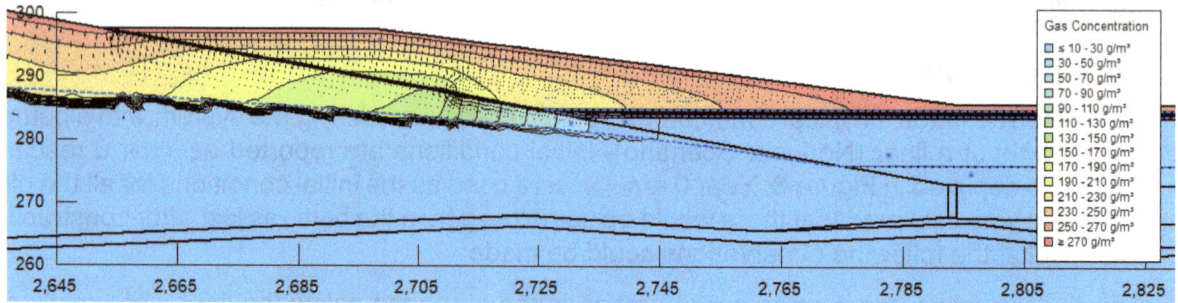

FIG 10 – Scenario B: Oxygen Concentration after four years with liner 20 m up the slope (zoomed-in; with gas mass flux arrows).

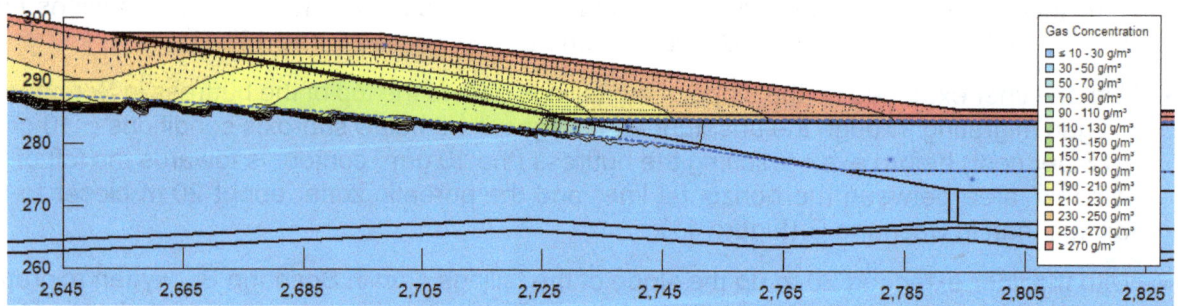

FIG 11 – Scenario C: Oxygen Concentration after four years (zoomed-in; with gas mass flux arrows).

Figure 12 indicates the position of the oxygen concentration observation points in the model. Figure 13 graphs the oxygen concentration at these five points in the buttress over time. The oxygen concentration above the liner (#1) is close to that in the atmosphere, while below the liner (#4), the air phase reaches sub-oxic conditions within three to four years, as discussed above. The oxygen in the saturated part of the buttress (#5) decreases to zero within less than six months because oxygen is consumed by both waste rock material in the buttress and (the ferrous iron in the) water phase; gas diffusion into the water phase is too slow to keep up with the consumption of the oxygen.

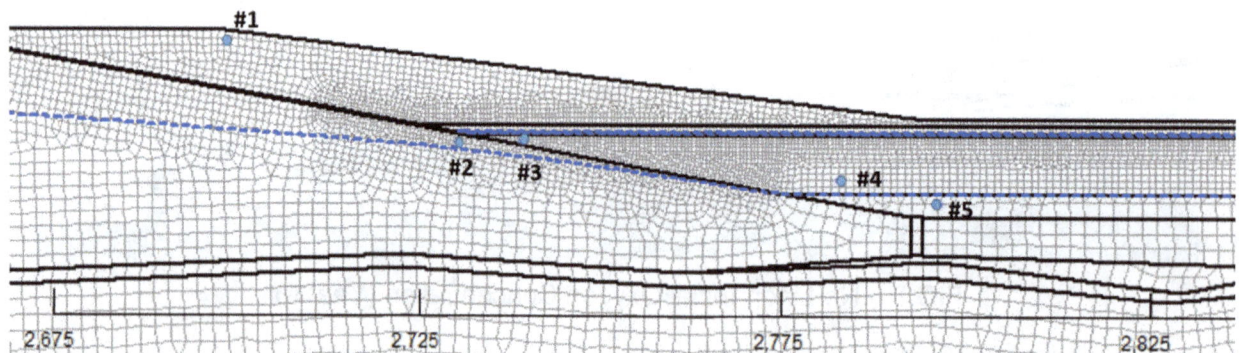

FIG 12 – Scenario C: Oxygen concentration model observation points.

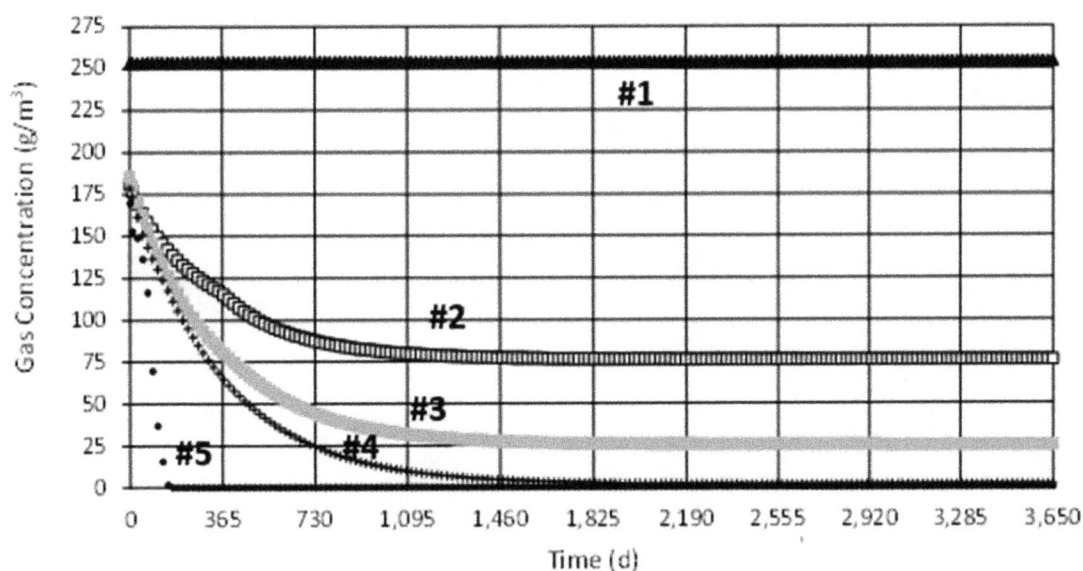

FIG 13 – Oxygen concentration logs in various parts of the buttress.

CONCLUSION

The objective of the 2D multiphase model was to determine whether sub-oxic conditions will be obtained within a buttress to promote sub-oxic conditions and limit ferrous iron oxidation. The gas modelling indicated that the oxygen concentration above the liner is close to that in the atmosphere, while below the liner, the air phase reaches sub-oxic conditions within three to four years. The oxygen in the saturated part of the buttress decreases to zero within less than six to nine months. Oxygen bypassing the liner and being transported through the unsaturated tailings is also insignificant in changing the sub-oxic conditions below the liner, suggesting that the proposed design should be suitable for limiting secondary mineral precipitates from forming in the drainage zone of the TSF.

From a buttress design perspective, it will be important to keep the oxygen in the buttress at a minimum by ensuring the integrity of the liner (ie good installation to prevent holes). The liner should be embedded in medium to fine-grained material to serve as a protection layer against sharp rocks from the buttress.

REFERENCES

Chemistry Libretext, 2013. First-Order Reactions [online], Chemistry LibreTexts. https://chem.libretexts.org/Bookshelves/ Physical_and_Theoretical_Chemistry_Textbook_Maps/Supplemental_Modules_(Physical_and_Theoretical_Che mistry)/Kinetics/02%3A_Reaction_Rates/2.03%3A_First-Order_Reactions [Accessed 30 August 2021].

Lefebvre, R, Hockley, D, Smolensky, J and Gélinas, P, 2001. Multiphase transfer processes in waste rock piles producing acid mine drainage: 1: Conceptual model and system characterization, *Journal of Contaminant Hydrology,* 52:137–164.

Steinpress, M, 2017. Investigation of Gas Transport Rates Through a Covered Waste Rock Pile and Synchrotron Studies on the Sulfide Oxidation Reaction, unpublished MSc Thesis, University of Waterloo.

Wang, T, Huang, Y, Chen, X and Chen, X, 2016. Using Grain-Size Distribution Methods for Estimation of Air Permeability, *Groundwater,* 54:131–142.

Accurately forecasting the extra treatment cost associated with the removal of lesser regulated PFAS species in water

M Kearney[1] and J Reardon[2]

1. Technical Manager, SciDev, Newcastle NSW 2300. Email: mkearney@scidev.com.au
2. Technical Director, SciDev, Sydney NSW 2000. Email: jreardon@scidev.com.au

ABSTRACT

Per-and poly-fluoroalkyl substances (PFAS), a well-known group of molecules worldwide, are still considered emerging contaminants. These molecules are poorly understood, and little data exists on the toxicology and transformation/degradation pathways. As a result, regulators perceive the risk of contamination differently, and human health and environmental guidelines can differ significantly across local, state, and national borders. Multinational firms in the mining sector operate across multiple regulatory environments and are often held to different regulatory standards for discharging PFAS waters.

This paper reviews the data from two case studies using adsorptive and ion exchange filtration techniques to quantify the added cost of treating all PFAS molecules versus treating only regulated compounds. The first case study was from a site with hotspot contamination, with surface water concentrations as high as 600 µg/L for sum of PFAS. Comparing the cost of treatment for 1000 ML (megalitres) of PFAS-contaminated surface water, the total cost was $0.027/L for treating the full suite of PFAS and the Total Oxidisation Precursor Assay (TOPA) for PFAS down to 0.002 µg/L. The cost to treat only the regulated molecules was $0.0257/L. This is an increase of only $0.0013/L to protect the surrounding environment and reduce the risk to human health.

The second case study was from a typical mine site application with high flow rates and low PFAS concentrations of 0.57 µg/L of sum of PFAS. Reviewing the cost of treatment for 1000 ML of PFAS-contaminated surface water, the total cost per litre was $0.0078/L to treat to below 0.0002 µg/L for the sum of PFAS, or $0.007/L to treat the regulated molecules for the region. This is an increase of $0.0008/L to treat the sum of PFAS down to the lowest detectable concentration.

INTRODUCTION

Per-and poly-fluoroalkyl substances (PFAS) are a group of recalcitrant, man-made, fluorinated molecules. Due to their unique surfactant properties and resistance to decomposition, PFAS substances have been heavily used in manufacturing since their invention in the 1930s. They are most commonly associated with their use in Aqueous Film Forming Foams (AFFFs) and in surface coating and protectant manufacturing.

Concerns about the toxic nature of PFAS substances were raised in the early 2000s. Since then, connections have been made between elevated PFAS levels in the blood and adverse health effects such as heart disease, liver damage and cancer. Since PFAS contamination was first discovered in water supplies, regulation worldwide has been growing and constantly changing. Historically, the primary molecules of concern were Perfluorooctane sulfonic acid (PFOS) and perfluorooctanoic acid (PFOA), with perfluorohexane sulfonic acid (PFHxS) garnering more attention shortly after. Although there are over 9000 known PFAS compounds, most research around the toxicology, degradation pathways and behaviour of PFAS has been about these three molecules.

Due to the mobility of PFAS in water, their abundant use and their recalcitrant nature, the number of sites contaminated with PFAS across the world is estimated to be in the high tens of thousands. Most contaminated sites are airports, defence bases, mines and Fire and Rescue facilities. In light of the toxic nature of these substances and the widespread contamination, the Stockholm Convention restricted the use of PFOS in 2009 and prohibited the use of PFOA in 2019. They are also currently investigating the need to include PFHxS and several long-chain carboxylic acids.

Since these restrictions came into place, several countries have regulated the PFAS levels acceptable in waterways and drinking water. However, the response to the threat of PFAS consumption varies worldwide, with many federal agencies opting to provide screening values instead of solid legislation. Australia is often considered at the forefront of regulation with a national environmental management plan for PFAS, which provides standards for PFAS concentrations in drinking water, recreational water, freshwater, and marine water bodies. The most strict of these regulations is the 99 per cent species protection criteria requiring PFOS concentrations to be under 0.2 parts per billion (PPB) or 0.0002 µg/L. However, even Australian regulation is still focused on the three molecules highlighted by the Stockholm Convention.

The global PFAS remediation market is estimated to be worth over $80 billion over the next five years, with the Australian Defence Force spending over $505 million on their PFAS contamination response since 2015. For the mining and infrastructure sectors, PFAS remediation is often triggered by new developments requiring environmental impact assessments or investigation into water reuse, discharge and dewatering activities. These sectors operate across multiple regulatory environments and, as a result, must manage legacy PFAS contamination on different sites in significantly different ways.

The most common technologies for ex-situ remediation of PFAS-contaminated water is the use of both adsorptive filter media, such as Granular Activated Carbon (GAC) and ion exchange (IX) media in a conventional water treatment plant (WTP). This research investigates long-term breakthrough trends for PFAS families and individual PFAS molecules at different stages of treatment within PFAS WTPs. TOPA and carboxylic acids families are the focus, with data sets gathered from two different industrial scale, temporary WTPs designed, built, and operating in Australia. Despite both WTPs utilising adsorptive media and IX media as the primary treatment technology, each WTP is significantly different (low versus high flow rate, heavy versus trace influent contamination). However, similar trends of breakthroughs and, therefore, cost increases to manage lesser/unregulated PFAS molecules can be seen.

CASE STUDY ONE

The first case study is a historical fire training ground operational from 1973 to 2015 when it was closed for remediation. This site was typical of hotspot/source contamination, with surface water concentration levels ranging up to 600 µg/L for sum of PFAS and 685 µg/L for TOPA. The historical use of old-age AFFFs resulted in the majority of the PFAS compounds being in the sulfonic acid family, with 50.88 per cent being (n:2) fluorotelomer sulfonic acids, 32.54 per cent being sulfonic acids, 16.57 per cent being in the Carboxylic acid family, and only 0.01 per cent being Sulfonamides. Despite the national guidelines only regulating PFOS, PFOA and PFHxS, the Environmental Protection Licence (EPL) issued by the Environmental Protection Authority (EPA) required the treated water to be below 0.002 µg/L for all PFAS molecules and TOPA that can be determined in Australian National Association of Testing Authorities (NATA) accredited laboratories. The WTP for this project operated at a flow rate of 5.8 L/s with a maximum instantaneous flow rate of 8 L/s and treated around 80 ML of contaminated surface water on-site over a 12-month period. A block flow diagram outlining the WTP process flow can be seen in Figure 1.

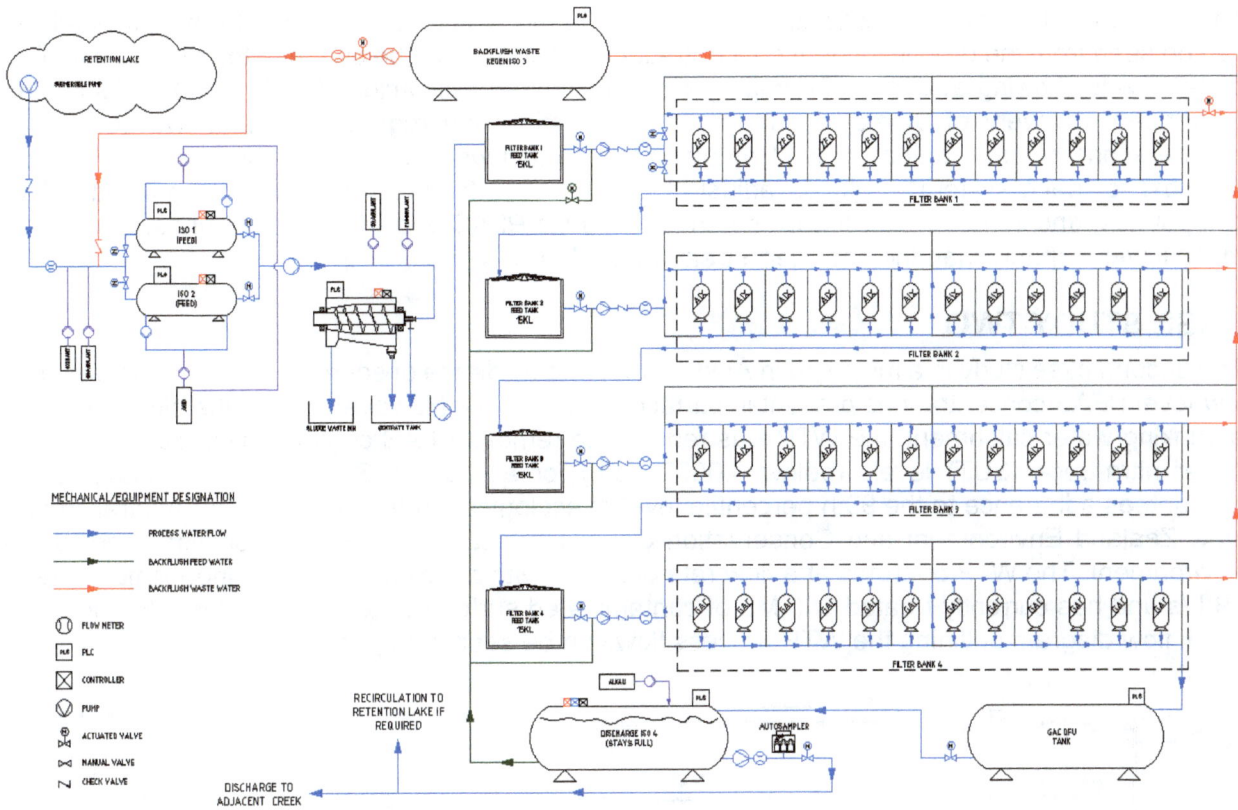

FIG 1 – Block flow diagram of the case study 01 water treatment plant.

As seen in Figure 2, the regulated PFAS molecules (PFOA, PFHxS and PFOS) at this site were removed entirely from the process water after filter bank 02. If this site was regulated in accordance with the national guidelines, we could remove every treatment stage after filter bank 02. We also could replace four of the AIX filters in filter bank 02 with Granular Activated Carbon (GAC), which is a significantly cheaper adsorptive media. This simplification of the process flow would save over A\$ 470 000 in initial procurement alone for this project. To compare the cost of design, procurement, installation, commissioning, operations, maintenance and then demobilisation post-completion, the cost of this project was \$0.0372/L of water treated. Treating only the regulated PFAS compounds would cost \$0.0312/L of water treated.

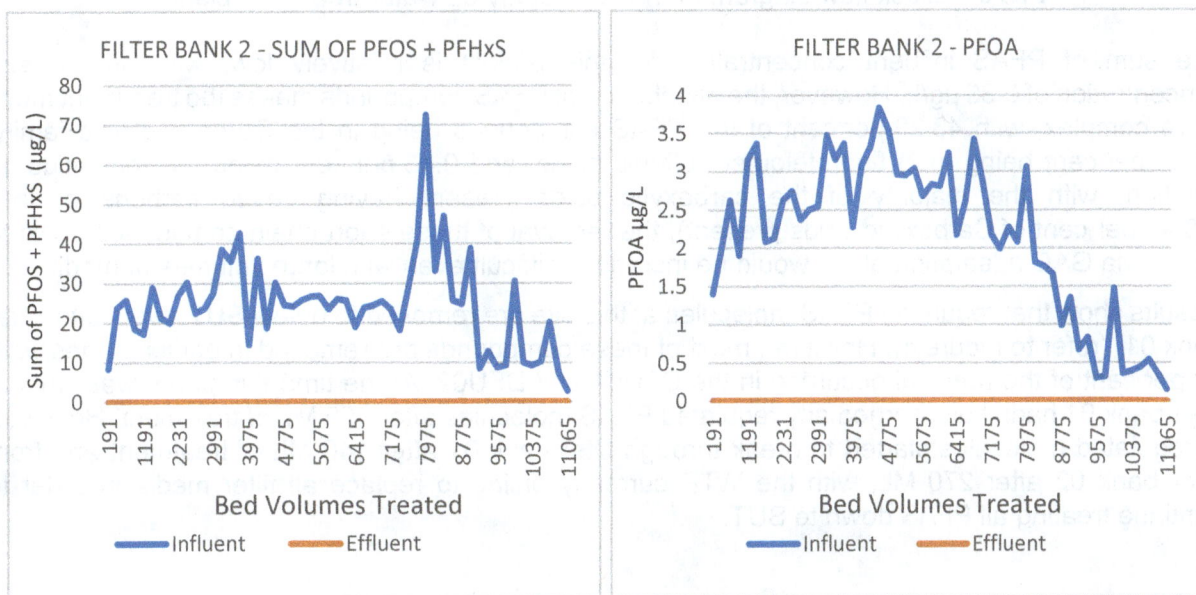

FIG 2 – Case study one: sum of PFOS and PFHxS, and PFOA from filter bank 02.

The GAC from filter bank 01 was replaced after 8000 bed volumes (BV) of water was treated (as can be seen from the drop in influent concentration to filter bank 02 below), and the IX media in filter bank 02 was showing evidence of dumping short chain carboxylic acids after 11 000 BV. Assuming 65 per cent recovery of IX media capacity after caustic brine regeneration and two successful regenerations before requiring new IX media, if this project were to continue long-term (1000 ML), the cost for each system would be adjusted to $0.0257/L for the treatment of regulated PFAS compounds, and $0.0270/L for the treatment of full-suite PFAS and TOPA compounds. Showing a much lower disparity than the short-term treatment costs.

CASE STUDY TWO

The second case study is a mine site in Australia with no evidence of source contamination and only low-level PFAS concentrations present in surface water run-off. However, due to the site residing in a drinking water catchment, the WTP was required to remove PFAS contamination down to below super ultra-trace (SUT) levels (below 0.0002 µg/L) for all 28 PFAS compounds. This criterion guarantees adherence to the 99th percentile level of ecological protection under the Australian and New Zealand Environment and Conservation Council (ANZECC) guidelines for discharge to the environment. The WTP operates at a flow rate of 20 L/s with a maximum instantaneous flow rate of 29 L/s and has currently treated 355 ML of contaminated surface water over the last 16 months. A block flow diagram outlining the WTP process flow can be seen in Figure 3.

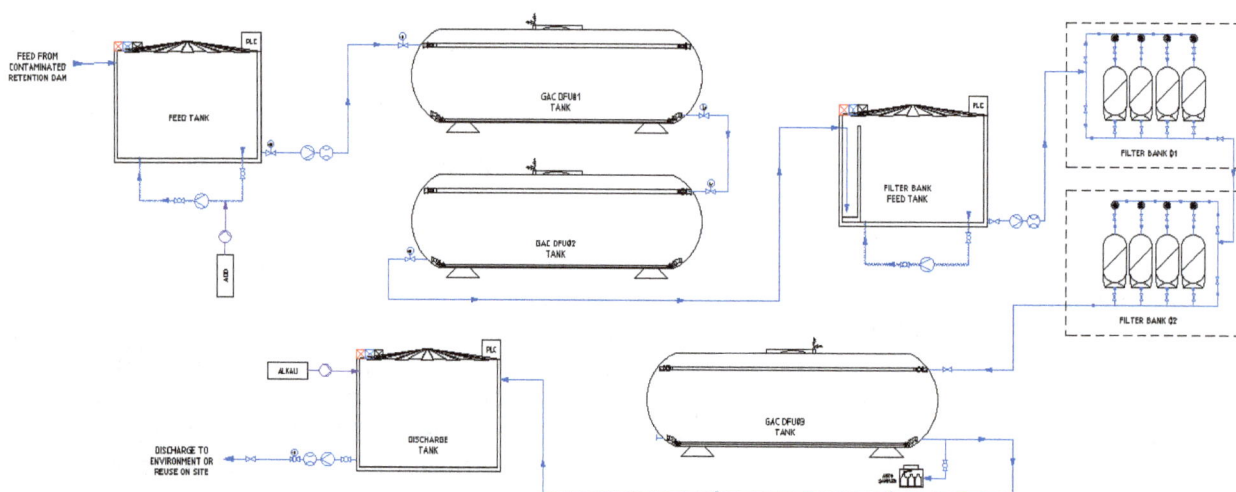

FIG 3 – Block flow diagram of the case study 02 water treatment plant.

The sum of PFAS influent concentration for this project is relatively low, with an average concentration of 0.36 µg/L. However, the distribution of PFAS compounds makes the treatment much more complex, with 43.20 per cent of the PFAS compounds being in the Carboxylic acid family, 56.37 per cent being (n:2) fluorotelomer sulfonic acids, and 0.43 per cent being sulfonic acids. In addition, with the majority of the carboxylic acids present having seven carbons or less (98.49 per cent of Carboxylic acids present), the removal of these short-chain compounds down to a SUT via GAC adsorption alone would be incredibly difficult even with large volumes of media.

Results show that regulated PFAS molecules at this site are removed to below SUT levels after filter bank 01 (refer to Figure 4). However, most of these compounds are removed in earlier stages, with 88 per cent of the removal occurring in the DFU01 and DFU02. At the time this paper was written, filter bank 01 hadn't discharged any regulated PFAS molecules after 355 ML of treatment. However, unregulated molecules started to break through filter bank 01 after 140 ML of treatment and from filter bank 02 after 270 ML, with the WTP currently offline to replace all filter media in order to continue treating all PFAS down to SUT.

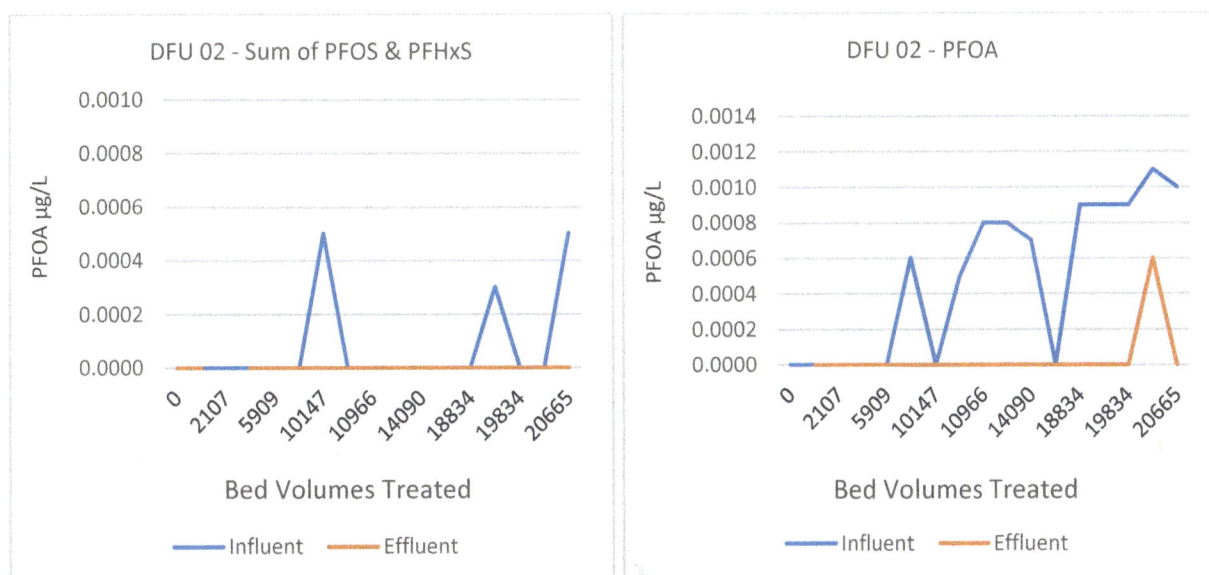

FIG 4 – Case study two: sum of PFOS and PFHxS, and PFOA from DFU 02.

If this project only required treatment of regulated compounds to super ultra-trace levels, not only could we remove every treatment stage after filter bank 01, but we wouldn't need to replace any media, or at least not the AIX resin, for roughly another 50–70 ML. Simplifying the process alone would result in a cost-saving of over $327 000 in initial procurement. So far, the overall cost of this project has been $0.008/L of water treated compared to $0.007/L to treat only the regulated compounds down to SUT. Comparing treatment models with long-term treatment (1000 ML), the costs would be adjusted to $0.0078/L for treating full-suite PFAS and $0.0070/L for treating regulated compounds.

CONCLUSIONS

Although both of these case studies look at sites with different source water contamination and very different flow rates, they come to the same conclusion. The cost increase in treating PFAS-contaminated water for all NATA testable PFAS molecules and TOPA PFAS is minimal compared to only treating regulated compounds. Although regulation is still adapting across the world, mining firms can acknowledge the potential risk to the environment and human health now and opt to upgrade their WTPs for a relatively small increase in cost.

ACKNOWLEDGEMENTS

I want to express my deepest appreciation to Jake Reardon for not only his help with this work, but also for mentoring me through my time at SciDev. I wouldn't have half the knowledge or the passion for the work we do without his help.

I would also like to thank SciDev for allowing Jake and I to share the knowledge and the data we obtained from these two case studies.

Calibration of the water balance model for an inactive tailings storage facility using Sentinel-2 satellite imagery

M Liu¹ and W Ludlow²

1. Senior Water Engineer, Red Earth Engineering a Geosyntec Company, Brisbane Qld 4000. Email: mike.liu@redearthengineering.con.au
2. Senior Principal | Director, Red Earth Engineering a Geosyntec Company, Brisbane Qld 4000. Email: wade.ludlow@redearthengineering.con.au

ABSTRACT

Calibration of a water balance model usually requires measured pond water levels of sufficient duration and frequency. However, such data is not always available for tailings storage facilities, especially for inactive facilities. Satellite remote sensing products may be used as the alternative data set in water balance model calibration when in situ monitoring data is not available. This paper presents the water balance model calibration case study for an inactive tailings storage facility located in Northern Australia. The calibration used the Sentinel-2 open-access remote sensing data which provides high-resolution historical satellite imageries captured every five days for the study site. The historical satellite imageries were used to develop a long-term historical supernatant pond level record for the facility. The process involves using the Normalized Difference Water Index (NDWI) to calculate the supernatant pond size and shape. The identified pond boundary was then converted to a pond water level using the available topographic survey of the tailings beach. This approach developed a three year historical pond water level record which was then used for calibration of the water balance model. The calibrated water balance model was used to conduct a probabilistic water balance assessment which established a more reliable facility flood risk profile for both the dam owner and the local authority. This case study reveals the potential application of the Sentinel-2 Satellite remote sensing data set for the calibration of tailings storage facility water balance models. The paper also discussed the precondition and limitations of this application.

INTRODUCTION

One key aspect of developing a water balance calibration for a tailings storage facility (TSF) is to have a well-established decant pool water level as this is a key verification of parameters used in the water balance. A water balance was required for an inactive TSF which had limited water level records but did have a survey of the embankments and the internal tailings area. The author approached this issue by using Sentinel-2 imagery to build a database of historical water level records for the facility and to assist in the calibration of the water balance model.

This paper primarily focuses on process of the calibration approach applied to the inactive TSF. However, similar approaches and principles have also been applied to active TSFs.

WATER BALANCE MODEL

A water balance model has been developed for an inactive TSF in Northern Australia to simulate the yield and loss of water on a daily time step and forecast long-term water levels within the decant pond. The schematisation of the water balance model is illustrated in Figure 1.

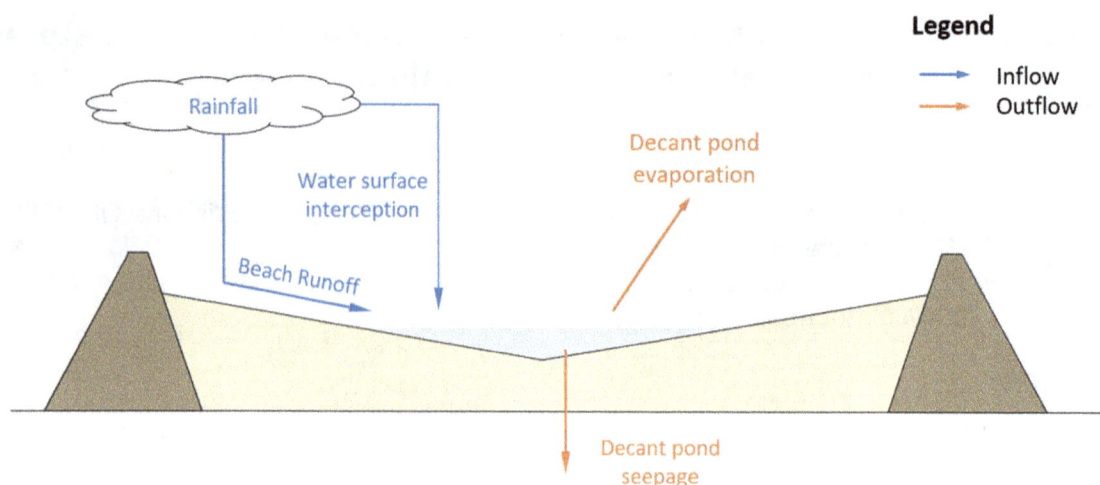

FIG 1 – Water Balance Schematic.

The storage characteristics of the decant pond are derived by the water balance model from an elevation versus plan area relationship, determined from the available topographic surveys of the TSF. These relationships are also used by the water balance model for calculating the decant pond water surface areas.

The calibration of the water balance model will span a historical period of four years. In order to accommodate this, the model used two elevation versus plan area relationships, each derived from surveys conducted on different dates, as shown in Figure 2. This ensures that the changes in storage characteristics, which occur due to the excavation around the decant pond area and consolidation of tailings, are accurately represented within the water balance.

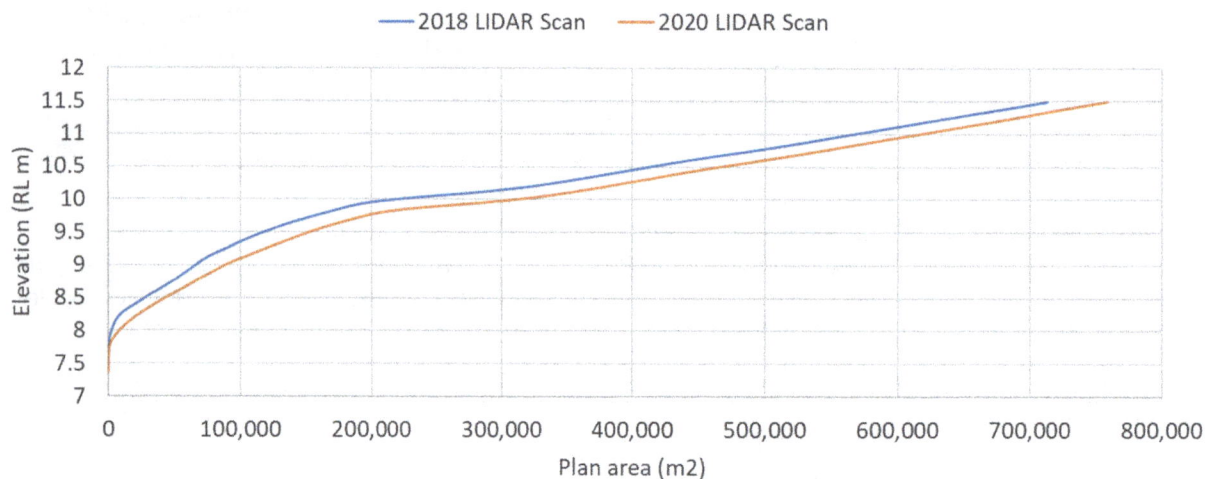

FIG 2 – Elevation versus Plan Area relationship determined from available Lidar survey.

Historical pond level estimation

Calibration of the water balance model requires relatively small effort because it was an inactive TSF that does not involve tailings deposition and process water extraction. The challenge was to develop a historical decant pond level as this was not recorded for the TSF. The historical pond water levels were estimated using the freely available remote sensing product – Sentinel-2 imagery. Sentinel-2 is an earth observation mission that systematically acquires optical imagery at high spatial resolution (10 m to 60 m) over land and coastal waters. The mission is a constellation with two twin satellites, Sentinel-2A and Sentinel-2B, which have five days of re-visit time. Whilst other satellite imagery products are available, the benefits of using Sentinel-2 imagery is that it is easily accessible and readily downloaded into GIS programs and has a good spatial resolution (10 m) when compared with other non-commercial satellite imagery product such as Landsat 8 which has approximately 30 m spatial resolution.

Sentinel-2 imagery for the TSF area was available from October 2017. All available Sentinel-2 satellite images were acquired and used to calculate the historical water inventory. The satellite images were enhanced using the Normalized Difference Water index (NDWI) to help identify the water surface in the decant pond. NDWI is used to monitor changes related to water content in water bodies, using green and near-infrared (NIR) wavelengths, defined by McFeeters (1996).

The water surface area in the Sentinel-2 imagery was compared to existing survey contours of the tailings beach to calculate the historical decant pond water levels. This process was applied to all available satellite images with identifiable water bodies within the TSF. An example is shown in Figure 3. It should be noted that although Sentinel-2 images were produced on average every five days, some images were not able to be utilised due to high cloud coverage (this is the main drawback of using satellite imagery).

FIG 3 – Sentinel-2 Images – True Colour (upper image) and NDWI (lower image, blue is water surface estimate) used to calculate water levels.

Historical climate data

As there was no available at-site gauged rainfall and evaporation data, the Long Paddock Scientific Information for Land Owners (SILO) patched point data was used to acquire historical climate data for the site. SILO climate data set for the TSF location is available from 1889. A review of nearby

Bureau of Meteorology (BOM) climate stations shows their earliest rainfall record is from 1914, which indicates the SILO climate data before 1914 is likely sourced from other climate stations and may not be representative of site conditions. Therefore, SILO climate data from 1 January 1914 to 31 December 2020 was adopted for the water balance assessment. Annual values of the adopted site rainfall and Morton's lake evaporation data are presented in Figure 4.

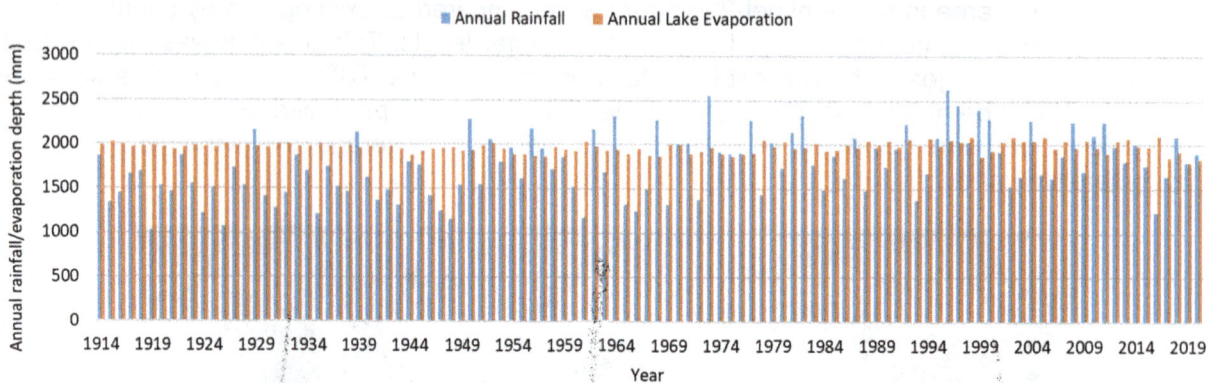

FIG 4 – SILO annual rainfall and lake evaporation.

The water balance calibration used the SILO climate data from 1 January 2017 to 31 December 2020 which covers the date range of the established pond level record. The historical decant pond water levels calculated using the above approach were plotted against the SILO daily rainfall depths as shown in Figure 5.

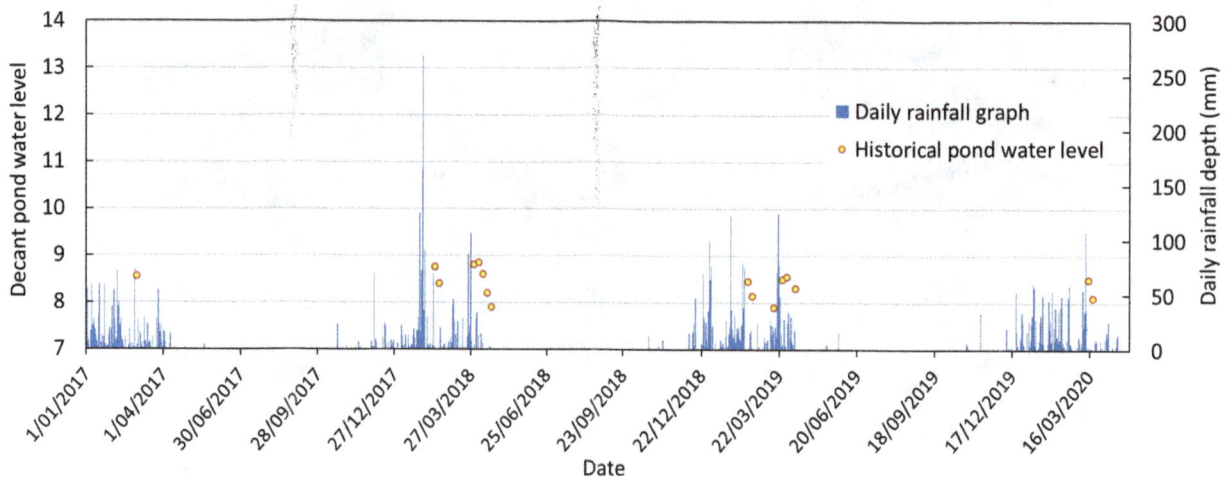

FIG 5 – Historical pond water levels against SILO daily rainfall graph.

Pond seepage loss estimate

The TSF was expected to have significant seepage loss from the decant pond area, based on the noticeable pond size changes following storm events as observed by the mine site staff. The high seepage loss rate may be attributed to the coarse nature of the tailings. However, without a pre-existing hydrogeological assessment to guide the seepage loss rate for the water balance assessment, the seepage loss volume from the decant pond was calculated using an average seepage loss rate estimated from the historical pond water levels shown in Figure 5.

The average seepage loss rate was estimated using water level changes during a dry period, with minimal or no rainfall, thus reducing the impact of rainfall run-off. Review of SILO daily rainfall data identified four dry periods that were accompanied by well-defined decant ponds as captured by Sentinel-2 satellite images: from 14/02/2018 to 19/02/2018, 10/04/2018 to 15/04/2018, 15/04/2018 to 20/04/2018, and 15/03/2020 to 20/03/2020.

Assuming that the water level changes during these dry periods were caused only by seepage and pond evaporation, an average hourly seepage loss rate could be derived for each dry period using the following formula:

$$\text{Average seepage rate} = \frac{\text{Water level drop} - \text{Total evaporation depth during the period}}{\text{Duration of the period}}$$

The total evaporation was calculated using SILO daily Morton's lake evaporation data. The daily Morton's evaporation is a result of matching pan evaporation rates to average monthly lake evaporation rates calculated as per Morton (1983). The evaporation is taken across the area of the decant pond water surface; the exposed tailings beach is not considered.

The estimated average hourly seepage rate for each dry period is summarised in Table 1.

TABLE 1
Seepage estimate using period of decant pond change between satellite images.

Selected dry period	Water level drop (m)	Duration (h)	Total lake evaporation (mm)	Estimated seepage rate (mm/h)
14/02/2018 to 19/02/2018	0.35	120	31.9	2.65
10/04/2018 to 15/04/2018	0.40	120	27.3	3.11
15/04/2018 to 20/04/2018	0.30	120	29.8	2.25
15/03/2020 to 20/03/2020	0.40	120	27.2	3.11

Table 1 shows the estimated seepage loss rate ranges from 2.25 mm/h to 3.11 mm/h. For conservatives, the minimum infiltration rate of 2.25 mm/h was adopted for the water balance assessment.

Calibration of volumetric run-off coefficient

The rainfall run-off inflows are calculated using a direct rainfall method. This method uses volumetric run-off coefficients to convert rainfall depth to run-off volumes based on catchment areas. Different run-off coefficients were assigned to the water surface and exposed tailings beach, which was calculated for each time-step based on decant pond inventory.

For the decant pond water surface, a volumetric run-off coefficient of C_1 = 100 per cent was used to simulate 100 per cent of rainfall becoming inflow into the decant pond.

For the exposed tailings beach, a single volumetric run-off coefficient, C_2, was adopted for the sake of simplifying the model calibration. It's acknowledged that using a single run-off coefficient may not accurately represent the physical process of run-off generation on tailings beach. Hence, a conservative value for C_2 was adopted through calibration, which will be discussed later. This approach was deemed suitable for the purpose of this assessment. It is recommended that for operating TSFs and facilities that often maintain a large pond, a more comprehensive rainfall run-off model, such as the Australian Water Balance Model (Boughton, 2004), should be used.

The model calibration was conducted using a GoldSim daily timestep water balance model using the SILO daily climate data input and a decant pond seepage loss rate of 2.25 mm/h. For calibration, the volumetric run-off coefficient C_2 was adjusted to match the simulated water level curve to historical water levels. As the water balance model will be used for a spill risk assessment, to avoid underestimation of run-off volume, a conservative volumetric run-off coefficient of 10 per cent was adopted. This run-off coefficient ensures all historical water levels are lower than the simulated water level curve, as shown in Figure 6. It should be noted that the daily rainfall depth in early 2018 has an AEP between 2 per cent and 1 per cent. This indicated the 10 per cent run-off coefficient is likely to be conservative for both frequent and rare rainfall events.

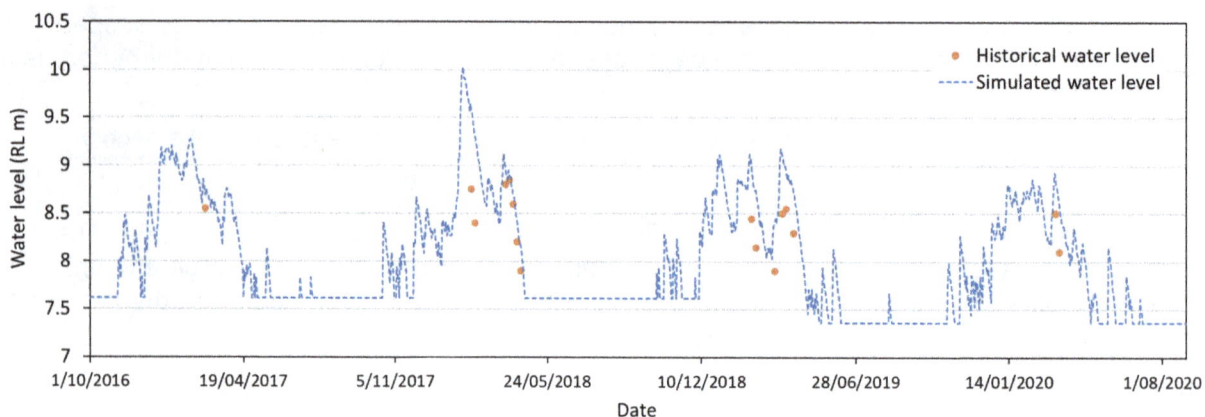

FIG 6 – Comparison of simulated water levels to historical water levels.

It is worth noting that in Figure 6, there is a noticeable shift in the lower bound water level during early 2019. This can be attributed to the use of the two Elevation versus Plan Area curves shown in Figure 2 to represent the changes in storage characteristics due to the excavation around the decant pond area and the consolidation of tailings over time.

SIMPLIFIED DAM SPILL PROBABILITY ASSESSMENT

A simplified dam spillway probability assessment was conducted using the calibrated water balance model. The adopted model parameters from the model calibration area are listed below:

- SILO daily rainfall and Morton's lake evaporation data.

- Elevation versus Plan Area relationship determined from the 2020 Lidar survey.

- Average seepage rate for decant pond area = 2.25 mm/h.

- Volumetric run-off coefficient for decant pond water surface, C_1 = 100 per cent.

- Volumetric run-off coefficient for exposed tailings beach, C_2 = 10 per cent.

A long-term water balance simulation was conducted using the SILO climatic data from 1914 to 2020. Model outputs were extracted and post-processed to identify the simulated maximum historical pond water levels occurring every wet season. The wet season peak water levels were then fitted to a Log-Person III distribution to develop the maximum water level probability curve. The fitted maximum pond water level versus Annual Exceedance Probability (AEP) curve is shown in Figure 7.

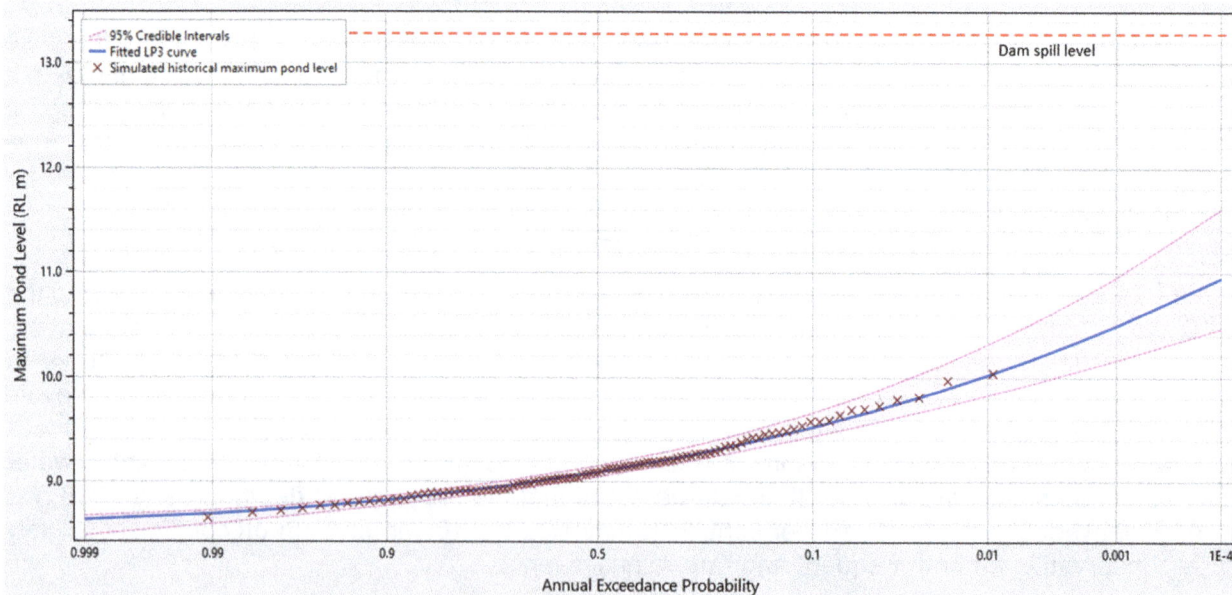

FIG 7 – Maximum pond level probability curve.

The maximum pond level probability curve suggests that the pond level with a 1:10 000 AEP remains more than 2 m below the elevation of the existing spillway. It's important to recognise the uncertainty related to the 1:10 000 AEP pond level, which is extrapolated from the Log-Pearson Type III curve that was fitted using a data set of 107 years of peak pond levels. Despite this uncertainty, the results clearly illustrate that the probability of the water level reaching the current spillway level is extremely low.

Even though the TSF is currently inactive, an upgrade to the spillway is necessitated due to an update in the TSF's consequence category. With a comprehensive understanding of the TSF's spill risks via the water balance analysis, it has been possible to select a spillway design capacity that is lower than initially expected. The adopted spillway capacity has been endorsed by the regulator and other relevant stakeholders, potentially leading to significant cost savings and the possibility of reallocating resources to other areas.'

PRECONDITION AND LIMITATIONS

This paper has presented a method to develop historical decant pond water levels for TSFs using freely available Sentinel-2 imagery. However, it is important to note that several preconditions and limitations may impact the application of the method.

One precondition is that topographic data of the tailings beach is available for the period to develop a historical pond level. Additionally, the approach is best suited for Turkey Nest type TSFs, which have a relatively flat surface and a decant pond managed away from the embankment. This allows for changes in pond size and shape to be identified from satellite imagery. For TSFs with steeper beaches, a reduced accuracy is anticipated due to the limitation imposed by the resolution of satellite imagery in determining the horizontal movement of the decant pond boundary. Correspondingly, for a given water level change, a steeper beach slope results in a shorter horizontal displacement of the pond boundary.

Furthermore, a limitation of the method is that the availability of satellite imagery is affected by cloud coverage. This could create difficulties in estimating decant pond water levels during wet seasons for facilities located in tropical areas.

CONCLUSION

This paper presents a case study that demonstrates the application of satellite remote sensing products in calibrating a water balance model for a TSF located in Northern Australia. Whilst the example presented has been used for an inactive TSF that was not receiving tailings during the period, the author has applied a similar analysis for active TSFs where the process water flows are metered.

Further improvements can be made to the water balance model by additional calibration runs for more recent data and by improving the rainfall run-off model with more sophistication, such as using the Australian Water Balance Model (Boughton, 2004) rather than a direct run-off coefficient model.

Furthermore, satellite remote sensing products could be used as part of the regular TSF monitoring program to identify pond size, tailings moisture levels (an index is available from various satellite remote sensing products to estimate this parameter) and management of change via review of satellite imagery. The Sentinel program is planned to continue until at least 2030 with the next satellites (Sentinel 2C and 2D) planned to be launched in the coming years and the author recommends that all practitioners take advantage of this type of technology in TSF management.

REFERENCES

Boughton, W J, 2004. The Australian water balance model, *Environmental Modelling & Software*, 19:943–956.

McFeeters, S K., 1996. The use of normalized difference water index (NDWI) in the delineation of open water features, *International Journal of Remote Sensing*, 17:1425–1432.

Morton, F I, 1983. Operational Estimates of Areal Evapotranspiration and their Significance to the Science and Practise of Hydrology, *Journal of Hydrology*, 66(1983):1–76.

Alternative treatment and management of mine influenced water

S Mancini[1] and R James[2]

1. Senior Principal, Geosyntec Consultants International Inc, Toronto ON M8X 1Y9, Canada.
   Email: smancini@geosyntec.com
2. Senior Principal, Geosyntec Consultants International Inc, Toronto ON M8X 1Y9, Canada.
   Email: rjames@geosyntec.com:

ABSTRACT

Mining companies are increasingly challenged with water treatment obligations that may endure in-perpetuity. Methods for conventional treatment of mine effluent is costly and, in some cases, ineffective at meeting environmental water quality objectives. Due to these challenges, mining companies are undergoing a transformation from providing conventional to innovative environmental management solutions; implemented during initial mine planning through source control and water management, to operational and post closure water treatment and reclamation. Technologies that manage and treat mine influenced water through stimulation of biogeochemical processes support this transformation, and may minimise compliance issues and operational requirements, and assist in meeting sustainability goals. This presentation will present case studies to exhibit the development, design and implementation of alternative technologies that have been successfully applied at mine sites for the treatment and management of mine influenced water. Applications of technologies including in situ and ex situ treatment reactors such as Gravel Bed Reactors™ and bioreactors, phytotechnologies, constructed and engineered wetlands, pit lake in-pit treatment, and permeable reactive barriers will be presented. The deployment of mobile treatment systems to mine sites, such as containerised treatment pilot systems, will be discussed as an important stage to facilitate treatability studies, regulatory approval, and advancement of technology application to full-scale. Advantages of using alternative water treatment and management tools will be presented along with new challenges and opportunities to apply these technologies at Australian mine sites.

INTRODUCTION

The implementation of innovative technologies at mine sites has the potential to treat and manage MIW effectively and reduce operational requirements while assisting in meeting sustainability goals. MIW can carry elevated concentrations of metal cations (eg cadmium, lead, zinc), transition metals (iron, manganese, copper, chromium, mercury), non-metals (sulfur, nitrogen, selenium), metalloids (arsenic, antimony), and actinides (uranium) and represent one of the greatest challenges at mine sites (USEPA, 2014). MIW typically contain high concentrations of sulfate from iron-sulfide oxidation, which in-turn generates acidity, reduces pH, and can further influence metal leaching potential of mine waste and generate deleterious water quality. *In situ* passive and semi-passive treatment of MIW can treat a variety of constituents using biological and abiotic treatment mechanisms to reduce metal mobility and improve water quality (Simon, Meggyes and McDonald, 2002; Duncan *et al*, 2004; Sobolewski, 2005; Zhang and Frankenberger, 2005; Zagury and Neculita, 2007; Martin, Jones and Buckwalter-Davis, 2009; Herbert *et al*, 2014; USDIBR, 2011; USEPA, 2014; Higgins *et al*, 2017; Seervi *et al*, 2017; Skousen *et al*, 2017; Rezadehbashi and Baldwin, 2018; Nielsen *et al*, 2018).

Biological treatment systems typically rely on shifting the biogeochemical conditions of a treatment system to stimulate microbial populations that can either directly or indirectly reduce, adsorb, assimilate and/or precipitate inorganic constituents. Microbial populations are diverse, found in a variety of pristine and contaminated waters (Nancharaiah and Lens, 2015; Nielsen *et al*, 2018; Rezadehbashi and Baldwin, 2018), and can harness energy by catalysing oxidation and reduction (redox) reactions while simultaneously assimilating carbon and nutrients (eg nitrogen) to sustain their growth. The chemical transformations performed by these organisms are dependent on the energy the redox reaction yields (ie electron acceptor and donor pairing), and types of micro-organisms present, and can include partitioning between aqueous and solid phases. Constituents influenced by biogeochemical reactions include major (carbon, nitrogen, sulfur, iron, manganese) and trace elements (arsenic, chromium, selenium, uranium). Additionally, non-redox sensitive element mobility, such as metal cations (eg copper, zinc, cadmium, cobalt), may be indirectly impacted by coprecipitation with sulfide minerals or adsorption onto organic matter and iron and manganese (oxyhydr)oxide minerals.

Non-biological or abiotic treatment relies on physical-chemical treatment mechanisms that can include precipitation, cation exchange, adsorption, and reduction. Abiotic adsorption systems typically utilise a fixed bed or barrier to house adsorptive media. Abiotic reduction of inorganic constituents can be accomplished using ferrous hydroxide, electrocoagulation, ZVI, ferrous hydroxides, and others.

Passive treatment systems rely on stimulating and optimising natural biological and/or geochemical processes to remove inorganic constituents from impacted water. Passive water treatment systems generally require little to no maintenance once installed, whereas active tank-based water treatment systems use mechanical devices under specialised operation and require continual flow control, reagent addition, solids/residual management, and power. Semi-passive treatment systems combine concepts from passive treatment, relying on natural biological and geochemical treatment, and active treatment, such as chemical amendment and flow control. Due to their low maintenance requirements and lower operating costs, the incorporation of passive and semi-passive treatment systems can be ideal for long-term water management for mine sites with predicted long-term water treatment requirements.

STATE OF APPLICATIONS

Biological treatment

Biotreatment solutions are gaining acceptance at mine sites as several full-scale operations and pilot-scale tests have proven that the approach is reliable in the treatment of metals, metalloids, sulfate, and nitrate (USEPA, 2014). In Pennsylvania, a passive bioreactor treated ARD for eight years during which some maintenance (mixing and addition of reactants) was required to restore the system's permeability (Skousen *et al*, 2017). At another mine site in Utah, an active biotreatment system was implemented to reduce and immobilise dissolved selenium (USEPA, 2014) at flow rates from 28 m³/d to 7600 m³/d while meeting effluent discharge levels of 5 µg/L of selenium. Sulfate reduction bioreactors have been used at the United Keno Hill Mine in the Yukon (pilot-scale; Ness, Janin and Stewart, 2014) and Leviathan Mine in California (full-scale; USEPA, 2006) to treat metals and have effectively removed between 80 per cent and 100 per cent of antimony, manganese, and zinc and aluminium, copper, iron, nickel, and zinc in their respective two and five year tenures. Anaerobic sulfate bioreactors have also been combined with wetlands and limestone drains to remove metals at a high efficiency at Teck Metals Smelter in British Columbia (> five years; Duncan *et al*, 2004) in Indiana County, Pennsylvania (seven years).

Passive and semi-passive anaerobic bioreactors range widely in design specifications, and utilise the high concentrations of nitrogen, sulfur, and iron associated with MIW. Examples include gravel bed reactors (GBRs), *in situ* mine pit treatment, engineered wetlands, flow-through bioreactors, and permeable reactive barriers (PRBs) (Higgins *et al*, 2017; ITRC, 2013; USEPA, 2014). A semi-passive treatment system can effectively reduce and immobilise constituents of concern in MIW while typically providing a higher control of treatment of water and a smaller footprint compared to an engineered wetland. Soluble electron donor (eg ethanol or methanol)-dosed reactors such as a GBR, in contrast to compost or organo-rich substrate reactors, provide several advantages including higher porosity (rock matrix versus compost), controlled reduction rate, and higher rates of reduction and microbial activity (ITRC, 2013). Semi-passive treatment requires less intensive infrastructure and lower operation and maintenance costs compared to active bioreactor treatment options (INAP, 2009; USEPA, 2014), which can include moving bed bioreactor (MBBR), packed bed bioreactor (PBR), fluidised bed bioreactor (FBR), upflow anaerobic sludge bed (UASB) reactors, and biofilters (ABMet) (ITRC, 2013).

Abiotic treatment

Abiotic treatment of MIW may comprise of abiotic reduction, adsorption, and ion-exchange. Abiotic reduction of metal(loid)s is generally accomplished using ferrous hydroxide, electrocoagulation or zero-valent iron (ZVI). Abiotic reduction using ferrous hydroxides also promotes transformation of metal(loid)s and the subsequent adsorption of metal(loid)s onto iron oxyhydroxides that are formed in the previous reduction step. In contrast, electrocoagulation uses sacrificial iron anodes to form a high-density iron – metal(loid)s solid sludge (USEPA, 2014; NAMC and CH2MHill, 2010).

Abiotic adsorption systems utilise a fixed bed or barrier to house an adsorptive media (ie activated alumina, activated carbon, clays etc). Water is passed through the bed or barrier to maximise contact time to encourage optimal adsorption capacity (USEPA, 2014; NAMC and CH2MHill, 2010). Adsorption may be limited by poor metal(loid) adsorption onto mineral surfaces, requiring pre-treatment stages.

Abiotic ion-exchange systems perform bypassing water through an ion-exchange media in which metal(loid)s are exchanged for non-hazardous ions. Abiotic ion-exchange may be paired with electrocoagulation for the removal of metal(loid)s (Littlejohn, Mohamm and Kratochvil, 2017). Ion-exchange has similar shortcomings to electrocoagulation, with the added cost of regenerating ion-exchange media (USEPA, 2014; NAMC and CH2MHill, 2010).

These systems can lead to a high degree of operational sophistication and increased chemical dosing and O&M costs, specifically in remote locations. In addition, these processes may not achieve low target treatment requirements. Therefore, innovative biotreatment or biogeochemical technologies can offer beneficial opportunities for operational and closed mines.

PROSPECTIVE TECHNOLOGIES

The following sections provide summaries and case studies of highlighted passive and semi-passive technologies that have been demonstrated to treat MIW including GBRs, constructed wetlands, in-pit lake treatment, and PRBs; and to manage MIW including TreeWell® technology and organic cover systems

Gravel bed reactors

A GBR is a fixed-film bioreactor, whereby an engineered bed of gravel or crushed rock media is placed in a lined cell or container in which the treatment conditions are controlled to support the growth and activity of microbes and biofilms on the media surface. Fixed-film bioreactors have been heavily applied across numerous industrial water treatment processes to biodegrade or immobilise contaminants that may be present in impacted water streams. GBRs are a unique configuration of fixed-film technology which have been developed to address low porosity and resultant high back pressures of organic bed systems, more controlled reduction rate and increased rates of removal. GBRs can be installed either below ground surface in an excavation, on-surface within a bermed or constructed cell, or as a hybrid of both (some portion below grade surrounded by berms above grade). Figure 1 presents a schematic for a GBR. A backfilled pit in situ water treatment system is another example that has been used to create conditions for biological semi-passive treatment of MIW, whereby an existing backfilled mine pit is dosed with carbon to support biological treatment mechanisms in groundwater (Heffernan, 2019). A GBR can be considered a purpose-built backfilled pit that allows more control and hydraulic isolation within the treatment system.

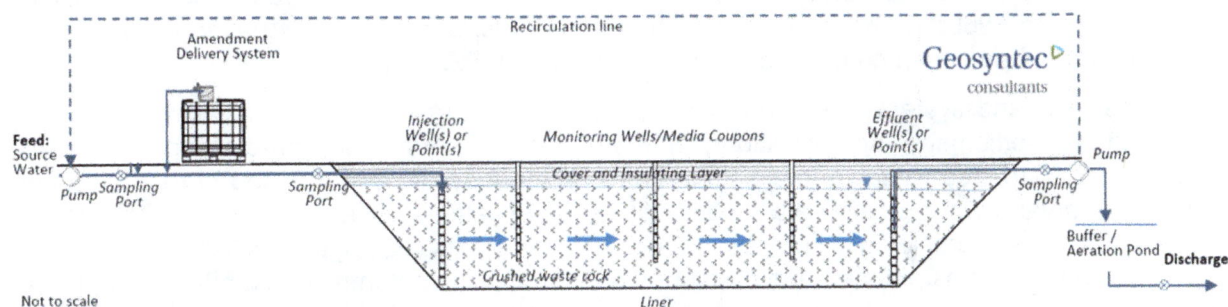

FIG 1 – Schematic of a Gravel Bed Reactor (GBR™).

Impacted industrial water, surface water and/or groundwater (the influent) is typically pumped into the GBR, and the treated effluent is discharged to surface water or recharged to the subsurface. In some cases, post-GBR polishing is needed to adjust oxygen levels, pH or other chemistry to match influent conditions or meet specific regulatory criteria. Depending on the constituents to be treated, chemical amendments such as electron donors (typically carbon substrates) or electron acceptors (typically oxygen or anionic substrates such as nitrate or sulfate) are dosed into the influent to promote the desired operating conditions and resulting biological reactions. GBRs are conceptually simple and very flexible reactors that can be engineered in varying ways to treat a wide variety of metals, inorganics, and organics. GBRs have successfully treated selenium-impacted water at several sites to concentrations below applicable water quality guidelines.

One example GBR case study is that of a water treatment project conducted on behalf of a coal mining company in West Virginia to mitigate selenium impacts to surface water emanating from seeps (Mancini *et al*, 2019). The GBR was installed beneath a parking lot at the facility to accommodate limited available space. The field application was designed to be capable of treating flows up to 550 m³/day, although during operations it averaged a flow rate of approximately 270 m³/day. The field application ran from March to May 2012 and was intended to continue operating, however the coal company went into bankruptcy and GBR operations, although successful, were suspended.

The seep source water had an aerobic ORP averaging 55 mV, DO averaging 4.8 mg/L, near neutral pH (7.03), a temperature ranging from 13 to 18°C, selenium ranging from 15 to 25 µg/L, and nitrate averaging 6 mg/L as N. A field application of the GBR technology was initiated in March 2012. The GBR was designed as a lined excavated trench underneath a parking lot at the facility. The gravel media bed was constructed inside a plastic-lined trench of approximately 28 m length × 8 m width × 1.5 m depth (approximately 360 m³). The media of the reactor was 2-cm sandstone gravel. The GBR had a treatment capacity varying from 165 to 550 m³/day. Citric acid and later acetic acid were used as electron donors. Main equipment included:

- The transfer pump to transfer water from the sump to the GBR.

- An electron donor dosing system to add citric acid or acetic acid to the influent of the GBR.

- An ORP probe at the outlet of the GBR to monitor effluent ORP to provide feedback to guide the electron donor dosing rate (adjustments to the dosing rate were conducted manually).

- Associated controls and monitoring equipment for the GBR.

The performance of the GBR was assessed from March through May of 2012, after which the GBR operations were suspended as the company entered bankruptcy. During operations, the GBR effectively reduced selenium concentrations in the effluent to levels below the treatment goal of <5 µg/L; nitrate concentrations were treated to below detection (<6.0 mg/L as N). The system operated at a flow rate of approximately 270 m³/day.

Constructed wetlands

Constructed (ie engineered) wetlands have a long and successful history in treating a variety of MIW. Over 40 years of research and practical full-scale applications of constructed wetlands throughout the world support the technical feasibility of the approach (ITRC, 2003). Research on the use of constructed wetlands for wastewater treatment began in Europe in the 1950s and in the United States in the late 1960s (USEPA, 2000). By the late 1990s, over 650 constructed wetlands sites were reported to have been constructed in North America (ITRC, 2003).

Constructed wetland systems incorporate biological (ie plant uptake, microbial activity), physical (ie adsorption, settlement, and filtration), and chemical (ie reduction, precipitation) treatment processes. Constructed wetlands are engineered treatment systems comprised of treatment cells planted with a variety of typically native wetland plants (eg cattails, rushes, reeds etc). The primary treatment involves microbiological processes mediated by plants (eg phytovolatilisation etc). Plants uptake nutrients such as nitrogen and phosphorus in addition to potential contaminants like metals and metalloids.

In general, there are two main types of constructed wetlands:

1. Surface-flow wetlands, which mimic natural wetland systems.

2. Subsurface-flow wetlands, where the water table remains below ground surface (USEPA, 2000; ITRC, 2003).

Within the subsurface-flow wetlands, there are two sub-categories: horizontal-flow and vertical-flow wetlands. Vertical-flow wetlands have the advantages of increased contact time, decreased potential for clogging, and increased oxygen delivery to the subsurface as compared to horizontal-flow wetlands and surface-flow wetlands. This results in more efficient treatment on a smaller footprint. The system is operational throughout the year and is able to mitigate problems with odours and mosquitoes (when properly maintained). However, a surface-flow wetland component may still be useful for flow equalisation and pre-treatment of higher-strength MIW. Figure 2 presents a schematic of a vertical sub-surface flow wetland.

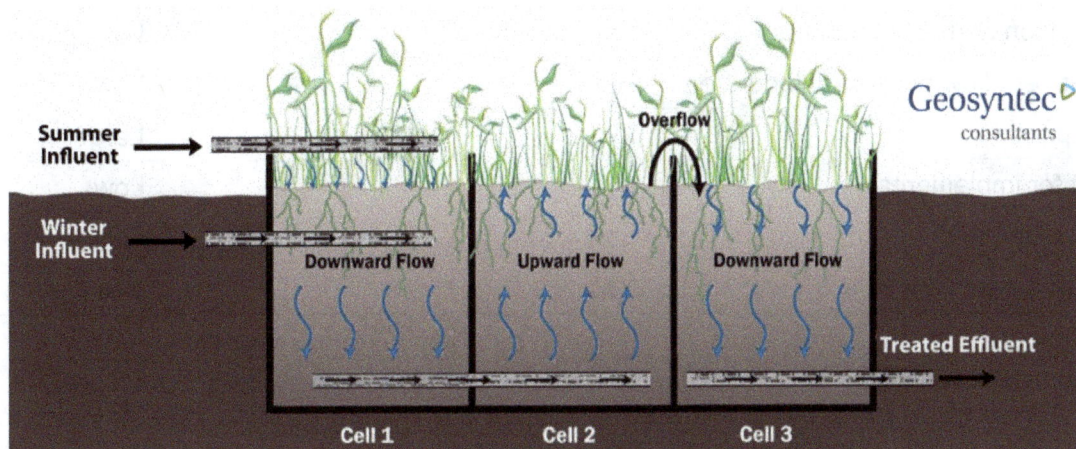

FIG 2 – Schematic of a three-cell subsurface vertical flow wetland system.

Much of the treatment success of a wetland system is dependent on the ability to affect redox conditions. Metal and metalloid treatment in wetlands can occur at faster rates when sub-oxic and anoxic conditions are achieved. A subsurface vertical-flow wetland system allows for the manipulation of redox conditions within individual treatment cells (ie alteration between aerobic and anaerobic/anoxic cells). This is especially important for the treatment of complex MIW, which may require aerobic treatment of various compounds including ammonia nitrogen, biological oxygen demand, chemical oxygen demand, and volatile organic compounds as well as anaerobic/anoxic conditions to accomplish treatment of metals, metalloids, and inorganics. The manipulation of redox conditions is accomplished by adjusting water levels within treatment cells coupled with addition of organic carbon (eg mulch, wood chips) as an electron donor source for micro-organisms in the anoxic treatment cell(s) to facilitate target treatment mechanisms.

The operation of constructed wetlands in cold weather climates can be a challenge; however, operation of subsurface wetlands extends the zones in which cold-climate operation of these systems can be operated. Depending on the location and climate, these wetlands can be cycled between winter and summer operation to effectively treat impacted water year-round.

Vertical flow wetlands, such as the example hybrid vertical subsurface flow wetland system (Figure 2) have a significant advantage over surface or horizontal-flow wetlands most notably due to the design of isolated treatment cells to combine aerobic with anaerobic processes which can target specific treatment mechanisms as required by site specific conditions. The depth of this type of design also provides advantages for cold climate treatment. Other advantages of this hybrid vertical design wetland system are summarised in Table 1.

TABLE 1

Summary of treatment efficacy comparing a hybrid vertical subsurface flow wetland system and surface flow wetland.

Efficacy category	Hybrid vertical subsurface Flow wetland system	Surface flow wetland
Cold climate/seasonal treatment	Yes	Limited
Long-term stability	Adaptable design; isolated treatment	Limited
Operation and maintenance	Low; passive	Low; passive
Treatment flow per unit of land area	High	Low
Footprint	Mid	Large
Cost for implementation	Mid	Low
Power demand	Low	Low
Isolated treatment cells	Yes	Yes

One example case study for a constructed wetland is that of an ongoing hybrid passive water treatment system, which is being operated to support the closure of silver mine in Western USA through water quality management to meet discharge permit limits. Bench-scale and pilot-scale treatability testing was conducted to design a passive water treatment system to treat approximately 6500 cubic metre per day (m^3/d) of MIW emanating from a mine portal prior to discharge to the receiving environment. Target treatment constituents included cadmium, copper, iron, lead, manganese, nickel, silver, and zinc. Bench-scale testing included: batch mesocosm testing for coagulation, flocculation and settling; manganese treatment with potassium permanganate; and adsorptive media tests using biochar, zeolite, and granular activated carbon. Pilot-scale testing involved construction of an approximate 130 m^3/d passive treatment system that was operated for 27 weeks. The pilot-scale system demonstrated that an in-mine pilot passive water treatment system could achieve metals removal in a remote, high-altitude environment which includes cold temperatures, high flow variability, and variability in metals concentrations. Construction of the ultimate treatment system was a sulfate reducing GBR followed by constructed wetlands as a polishing system. The passive treatment system provided a successful closure option for the mine with an approximate 75 per cent cost reduction compared to active treatment. The hybrid treatment system is currently operating successfully at full-scale.

In-pit lake treatment

In-pit lake treatment strategies are an example of a semi-passive water treatment systems which require little process equipment and minimal operational supervision (except for chemical dosing equipment). Such systems could include adding chemical amendments directly to a MIW collection pond or pit lake and allowing treatment to take place *in situ* or installing a passive treatment system as part of a naturally flowing system (eg backfilling the collection pond or pit lake, or installing reactive barriers). Either of these options would require occasional monitoring and maintenance to ensure proper operation. Modified in-pit lake treatment through backfilling a collection pond or pit lake with inert substrate is one method to increase surface area and biological film density to achieve sub-oxic conditions more readily in shallow water.

The microbial community within the collection pond or pit lake tend to use alternative electron acceptors if the process is energetically favourable under specific conditions such as redox potential, availability of carbon (organic matter), geochemical parameters (hardness, pH, alkalinity) and temperature. In the absence of DO, nitrate is usually consumed first, passing through a reduction pathway to nitrogen gas, next are selenite and selenate which can be reduced to elemental Se, while the sulfate is reduced to form sulfides.

For in-pit lake treatment systems where naturally occurring organic carbon is low, an external carbon source is provided through either a simple or complex media to serve as a source of electron donor for the microbial reduction of inorganic constituents. The carbon serves two purposes:

1. It increases the oxygen demand of the water and depletes DO resulting in anoxic conditions conducive for constituent reduction.

2. It supplies carbon, an essential energy source for the growth and sustenance of the microbial community capable of degrading the constituent.

Table 2 presents a summary of applications of in-pit lake treatment that have been documented in literature.

TABLE 2

Summary of applications of in-pit lake treatment for mine water treatment (Gammons and Icopini, 2020).

Lake name	Distribution method
Berkeley Pit-Lake	Shore-based plant, lime applied using pipeline
Meirama Mine Pit-Lake	Liming of influent stream
Rävlidmyran Pit-Lake	Lime injected through a pipeline from trucks
Lake Senftenberg	Lime applied using hopper barges without mixing
Lake Geierswalde	Resuspension of lime from bottom of pit-lake and applied using sprinklers
Lake Bockwitz	Soda applied into the lake using air/solid mixture injection
Lake Hain	Shore-based plant, slurry applied using sprinklers and distributed via lake convective currents
Lake Bernstein	Limed using a commercial barge and underwater pipeline
Lake Scheibe	Shore-based plant for application of lime slurry and CO_2 for buffering
Lake Nero	Shore-based plant used to create slurry and applied via pipeline

The effectiveness of in-pit lake treatment systems is contingent upon successful implementation under site-specific conditions. Some of the essential site-specific conditions from an implementation perspective include geographic location and access to the pit lake with suitable geomorphology, resource availability within the area, shape and limnology of the pit lake being treated, seasonal access constraints and climatic conditions. Engineered cover systems and other similar systems to maintain sub-oxic conditions can improve the effectiveness of the treatment technologies.

Permeable reactive barriers (PRB)

A PRB is a wall or trench of reactive material that is constructed in the ground, perpendicular to groundwater flow which allows impacted water to contact the reactive zone for treatment. A PRB can utilise biological or physical-chemical treatment mechanisms. In biological PRB systems treatment of dissolved metals requires an initial reduction of sulfate via microbial processes. Bacteria use sulfate as the terminal electron acceptor in the oxidation of the labile carbon to carbon dioxide. If any aqueous metals (eg iron, zinc, cobalt, cadmium, and others) are present, these metal cations will precipitate as metal sulfides. Carbon substrate based PRBs containing a wide range of solid-phase organic carbon (eg plant mulch, compost, wood chips, saw dust etc) have been successfully implemented for the treatment of dissolved metals in groundwater. A biological treatment zone can also be established (sometime called a permeable reactive zone or bio barrier) through injection of liquid carbon media directly into groundwater, for example emulsified vegetable oil or similar substrate can be injected into permanent wells installed along the remedial alignment. In physical-chemical PRB systems abiotic adsorptive media (eg zeolite, phyllosilicate minerals, activated alumina, activated carbon, ZVI, and others) or material that can stimulate chemical fixation (eg phosphate-based minerals) or pH adjustment (eg calcium carbonate, lime, caustic magnesia, steel slag, and others) is housed within a fixed bed or barrier. In some cases, amendments such as elemental sulfur, phosphate minerals, reductants such as ZVI and others can provide system performance benefits for both biological and abiotic treatment technologies. Table 3 presents an example of the treatment amendments and processes that have been applied to treat water impacted with zinc, cadmium, cobalt, and nickel.

TABLE 3

Example amendments and water treatment processes that have been used in a permeable reactive barrier deployment for treatment of metals-impacted water.

Amendments and processes	Zinc	Cadmium	Cobalt	Nickel
Biosequestration, microbial reduction and/or precipitation	X	X	X	X
ZVI-based approaches – reductive precipitation of dissolved metals and sorption on surface adsorption sites of ZVI	X	X	X	X
Phosphate based materials, such as Apatite, for chemical fixation as insoluble phosphate mineral phases	X	X	X	X
Zeolite, phyllosilicate minerals or activated alumina	X	X	X	X
Iron mineral sorbents – added as a treatment wall or generated *in situ* via biostimulation or hydrolysis of injected iron salts	X	X	X	X
Activated carbon and other sorptive solid carbon materials	X	X	X	X
Alkaline materials, such as calcium carbonate, lime, caustic magnesia, steel slag for acidic pH buffering and precipitation	X	X	X	X

One example PRB case study is that of a ZVI-based PRB that was designed, constructed, and operated to treat hexavalent chromium [Cr (VI)] in groundwater at an industrial site. A pre-design laboratory column study was conducted to determine the Cr (VI) removal capacity by a commercial ZVI source, including assessment of the influence of background inorganic chemistry on the removal kinetics and the rate of ZVI passivation under a continuous exposure to site water containing approximately 100 mg/L of Cr (VI). Using the obtained Cr (VI) treatment rate, the anticipated maximum Cr (VI) concentration and the site groundwater velocity, and the PRB design lifetime of 20 years, the dimensions of ZVI required for treatment was calculated. Based on the lateral and vertical distribution of the Cr (VI) plume at the PRB location, the design PRB length was 78 m and the PRB depth ranged from 1 m below ground surface (bgs) to approximately 12 m bgs in saturated silty sediments underlain by a heavy clay unit. The laboratory study demonstrated the tested commercial ZVI product could treat Cr (VI) of up to 100 mg/L in site groundwater and the ZVI permeability loss due to precipitate formation was minimal and would not impact PRB performance. The PRB was successfully installed at the site using biopolymer-supported excavation and backfill.

Innovative water management technologies

The implementation of water management at mine sites is crucial to limiting water contact with point sources and is typically implemented via water collection infrastructure to collect mine waste seepage and run-off water; and conveyance and diversion infrastructure to isolate clean water from mine waste solid and liquid storage facilities. Innovative water management using phytotechnologies such as TreeWell® technology and innovative source control technologies such as organic cover systems can limit mobilisation of MIW at its source thereby minimising the volume of impacted water that requires management and treatment.

Phytoremediation can be an appropriate treatment strategy for mine sites. Plant-based remediation technologies can be effective as either a sequestration or containment strategy or for accumulation and recovery of metal(loids) from soil and/or groundwater. Plants can effectively remove inorganic constituents from impacted mine sites via root uptake. Once absorbed by roots, plants have the capacity to (i) sequester the metal in plant tissue and/or (ii) phytovolatilise the contaminant in a volatile and non-toxic form. The TreeWell® technology is an innovative type of phytoremediation that provides hydraulic control and contaminant sequestration from surface water and groundwater. The TreeWell® system uses a patented design to focus groundwater extraction from a targeted depth interval using a specialised planting unit; optimum planting media that promotes downward root growth and focuses groundwater uptake from a targeted depth interval. One example application of a TreeWell® system includes treatment of mercury-impacted groundwater at an industrial site. Approximately 200 TreeWell® planting units were installed over a one-acre area of a former water collection pond to prevent the flux of mercury impacted groundwater from discharging to the receiving environment. The magnitude of an inward hydraulic gradient to the TreeWell® plot grew each growing season for several years. The water treatment objectives for the project were achieved via hydraulic control, plant uptake and transpiration of target groundwater concomitant to metal sequestration in the root zone around each planting.

Innovative source control technologies can provide cost savings to mine waste management and provide improved containment of mine waste in storage areas. One example of source control is the use of organic covers systems to prevent air and water influx to mine waste storage piles. Potential benefits of organic covers include use of site and local materials as substrate for vegetation, waste reuse thereby supporting circular economy, and competitive cost compared to other types of cover systems, such as geosynthetic clay liners and subaqueous disposal. Organic covers are designed utilising carbon layers from mulch or municipal compost. The organic material can mitigate mobilisation of contaminants from mine waste through oxygen consumption through carbon degradation, preventing oxygen ingress and subsequent mineral oxidation; and reduction of water infiltration thereby mitigating seepage water contact. Table 4 provides a summary of mine sites that have implemented organic cover systems using a variety of organic substrates.

TABLE 4

Summary of organic cover systems that have been applied at mine sites for source control of metals-impacted water.

Organic material	Site name/location	Commodity mined
Pulp and paper biosolids	Copper Cliff Tailings[1]	Ni, Cu, Au, Ag
Mixed municipal compost and biosolids fertiliser	Strathcona Tailings[2,5]	Ni, Cu, Au, Ag
	Delnite Tailings[3]	Au
Sawdust and municipal biosolids	East Sullivan Mine[4]	Au

Notes: (1) Beauchemin *et al* (2018); (2) Asemaninejad *et al* (2021); (3) Paktunc (2013); (4) Germain, Tassé and Cyr (2009); (5) McAlary *et al* (2019).

SCALE UP AND PERMITTING OF INNOVATIVE TECHNOLOGIES

Successful implementation of an innovative technology requires proof of concept and demonstration to gain buy in from stakeholders, regulators and rightsholders. This is achieved through an assessment of the technology along the scale of technology readiness and can involve laboratory treatability testing to field pilots and demonstrations. Treatability testing conducted in batch mesocosm and flow-through vertical and horizontal bench scale columns provides the capability of testing all stages of technology development. To increase level of certainty on performance, costing and design requirements it is important to identify any technical or design uncertainties prior to field demonstration and full-scale design.

Figure 3 presents a summary of the stages of innovative technology scale up, from technology screening and conceptualisation to full-scale technology deployment. The Level of Certainty of the engineering design, treatment performance, and long-term reliability of the technology increases with through each stage of technology testing.

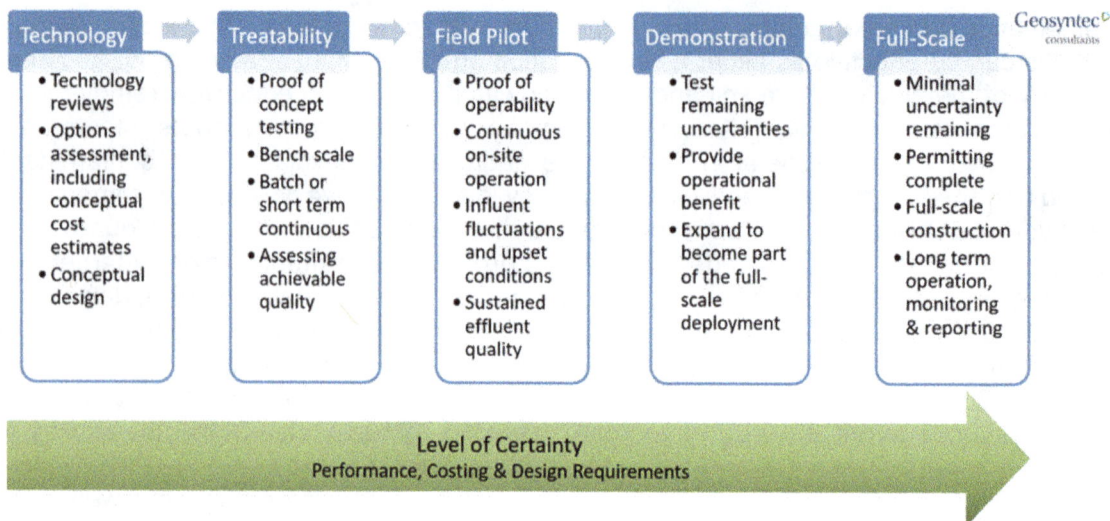

FIG 3 – The stages of innovative technology development to full-scale implementation.

Example field-scale mobile treatment pilot systems are presented in Figure 4. In Figure 4a, a constructed wetlands approach was piloted to achieve pre-treatment of site leachate using mobile wetlands pilot system. In Figure 4b, site water is treated in five horizontal large-scale columns within a temperature-controlled container. The system provides a relatively low discharge water volume that allows site specific treatment uncertainties such a low temperature, carbon dose substrate type and mechanism, and other operational variables to be tested to evaluate potential worst case effluent water quality for various operational scenarios. This information is used to inform engineering design for field pilot, demonstration, and full-scale water treatment systems. Results from pilot testing provide confidence to environmental regulators, stakeholders and rightsholders on the upper and lower bound scenarios that may be anticipated using the technology thereby facilitating permit approvals. Furthermore, robust system and water quality monitoring programs and mitigation procedures can be developed to manage off specification water should the tested scenarios be implemented.

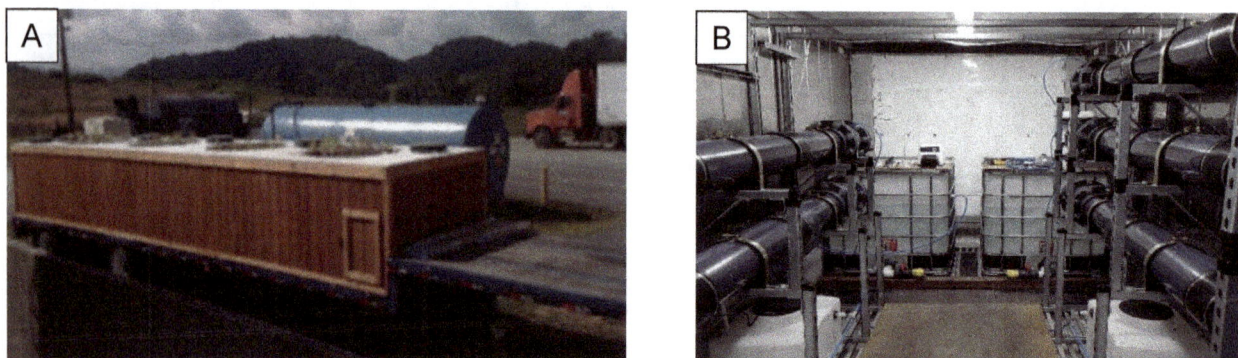

FIG 4 – Example field-scale mobile treatment pilot treatment systems.

CONCLUSIONS

Passive and semi-passive treatment of inorganic constituents is gaining acceptance at many mine sites. The use of GBR, constructed wetland, pit lake in-pit treatment, and PRB treatment technologies can provide flexible MIW treatment options with lower operational costs when compared to active treatment systems. Passive and semi-passive approaches to treatment can be a potential option to treat MIW to meet regulatory requirements for operational and closed mines and should be further investigated to validate potential site-specific risks, challenges, and opportunities. The selection of an appropriate passive and semi-passive treatment system can depend on several factors including impacted water flow rate, footprint, climate, available local carbon, and media material, etc. Alternative water management technologies including phytotechnologies like TreeWells® and organic covers systems can provide hydraulic and source control to limit deleterious constituents from leaching from waste rock and tailings and thus prevent mobilisation of MIW at its source. Passive and semi-passive treatment and management of MIW can lead to many advantages when compared to active treatment; primarily, the ability to operate these systems with lower operating cost and oversite which is a major advantage in longer term water quality management for closure planning of mines.

ACKNOWLEDGEMENTS

The authors would like to acknowledge Kevin Dufresne, Evan Cox, Ron Gestler, Jason Kersteins, Brent Miller, Brian Petty, and Andrzej Przepiora at Geosyntec Consultants Inc for their contributions of technical findings from technology development and pilot programs included herein.

REFERENCES

Asemaninejad, A, Langley, S, Mackinnon, T, Spiers, G, Beckett, P, Mykytczuk, N and Basiliko, N, 2021. Blended municipal compost and biosolids materials for mine reclamation: Long-term field studies to explore metal mobility, soil fertility and microbial communities, *Science of the Total Environment*, 760(15):143393.

Beauchemin, S, Clemente, J S, Thibault, Y, Langley, S, Gregorich, E G and Tisch, B, 2018. Geochemical stability of acid-generating pyrrhotite tailings 4 to 5 years after addition of oxygen-consuming organic covers, *Science of the Total Environment*, 645:1643–1655.

Duncan, W F A, Mattes, A G, Gould, W D and Goodazi, F, 2004. Multi-stage biological treatment system for removal of heavy metal contaminants, in *Proceedings of the Fifth International Symposium on Waste Processing and Recycling in Mineral and Metallurgical Industries,* pp 469–483.

Gammons, C H and Icopini, G A, 2020. Improvements to the Water Quality of the Acidic Berkeley Pit Lake due to Copper Recovery and Sludge Disposal, *Mine Water and the Environment*, 39(3):427–439.

Germain, D, Tassé, N and Cyr, J, 2009. The East-Sullivan Mine Site: Merging Prevention and Treatment of Acid Mine Drainage, in Proceedings of the 14th British Columbia MEND Metal Leaching/Acid Rock Drainage Workshop. https://bc–mlard.ca/files/supporting-documents/2009–16a-GERMAIN-ETAL-east-sullivan-mine-prevention-treatment-amd.pdf

Heffernan, V, 2019. Genomics applications to wastewater start proving their worth, *Canadian Mining Journal*, April 2019, pp 15–17.

Herbert, R B, Winbjörk, H, Hellman, M and Hallin, S, 2014. Nitrogen removal and spatial distribution of denitrifier and anammox communities in a bioreactor for mine drainage treatment, *Water Research*, 66:350–360.

Higgins, J, Mattes, A, Stiebel, W and Wootton, B, 2017. Eco-Engineered Bioreactors Advanced Natural Wastewater Treatment, report prepared for Teck Coal Limited (CRC Press).

International Network for Acid Prevention (INAP), 2009. Global Acid Rock Drainage Guide (GARD Guide).

Interstate Technology and Regulatory Council (ITRC), 2003. Technical and Regulatory Guidance Document for Constructed Treatment Wetlands, Interstate Technology and Regulatory Council, December 2003.

Interstate Technology and Regulatory Council (ITRC), 2013. Biochemical Reactors for Mining Influenced Waters, Interstate Technology and Regulatory Council.

Littlejohn, P, Mohamm, F and Kratochvil, D, 2017. Advancement in non-biological selenium removal treatment systems–Results of continuous pilot scale operations, in *Proceedings of the Water Environment Federation Technical Exhibition and Conference 2017*, 1:369–393.

Mancini, S, Cox, E, deVlaming, L, Bechard, K, James, R E, Przepiora, A and Risacher, F, 2019. Gravel Bed Reactors: A Water Treatment Technology for Industrial and Mining Applications, *The British Columbia Environment Industry Guide*, British Columbia Environment Industry Association (BCEIA), May 2019, pp 19–20.

Martin, A J, Jones, R and Buckwalter-Davis, M, 2009. Passive and semi-passive treatment alternatives for the bioremediation of selenium from mine waters, BC's 33rd Annual Miner Reclamation Symposium, Cranbrook, British Columbia.

McAlary, M, Ptacek, C, Bain, J, Blowes, D, Fyfe, J and Marshall, S, 2019. Effectiveness of Organic Carbon Cover Systems on Sulfide-rich Tailings, in Proceedings of the 24th British Columbia MEND Metal Leaching/Acid Rock Drainage Workshop. https://bc–mlard.ca/files/presentations/2019–8-MCALARY-ETAL-effectiveness-organic-carbon-cover.pdf

North American Metals Council (NAMC) and CH2MHill, 2010. *Review of Available Technologies for the Removal of Selenium from Water*, Charlotte, NC.

Nancharaiah, Y V and Lens, P N L, 2015. Ecology and Biotechnology of Selenium-Respiring Bacteria, *Microbiology and Molecular Biology Reviews*, 79(1):61–80.

Ness, I, Janin, A and Stewart, K, 2014. Passive Treatment of Mine Impacted Water In Cold Climates: A review, Yukon Research Centre, Yukon College.

Nielsen, G, Hatam, I, Abuan, K A, Janin, A, Coudert, L, Blais, J F, Mercier, G and Baldwin, S A, 2018. Semi-passive in-situ pilot scale bioreactor successfully removed sulfate and metals from mine impacted water under subarctic climatic conditions, *Water Research*, 140:268–279.

Paktunc, D, 2013. Mobilization of arsenic from mine tailings through reductive dissolution of goethite influenced by organic cover, *Applied Geochemistry*, 36:49–56.

Rezadehbashi, M and Baldwin, S A, 2018. Core Sulphate-Reducing Microorganisms in Metal-Removing Semi-Passive Biochemical Reactors and the Co-Occurrence of Methanogens, *Microorganisms*, 6(1):16.

Seervi, V, Yadav, H L, Srivastav, S K and Jamal, A, 2017. Overview of Active and Passive Systems for Treating Acid Mine Drainage, *IARJSET*, 4(5):31–137.

Simon, F G, Meggyes, T and McDonald, C, 2002. *Advanced Groundwater Remediation: Active and Passive Technologies* (Thomas Telford Publishing).

Skousen, J, Zipper, C E, Rose, A, Ziemkiewicz, P F, Nairn, R, McDonald, L M and Kleinmann, R L, 2017. Review of Passive Systems for Acid Mine Drainage Treatment, *Mine Water and the Environment* (Springer: Heidelberg).

Sobolewski, A, 2005. Evaluation of Treatment Options to Reduce Water-Borne Selenium at Coal Mines in West-Central Alberta, Report prepared for Alberta Environment Water Research Users Group, Edmonton, https://open.alberta.ca/dataset/456eee9c-86d5–46e6-bc2e-e605c6599eba/resource/b96ac61e-78db-4b2c-b0d5–40d9c0bf772e/download/3877073–2006-evaluation-treatment-options-reduce-water-borne-selenium.pdf

US Department of the Interior Bureau of Reclamation (USDIBR), 2011. Reclamation Managing Water in the West: Western Water and Power Solution Bulletin, Research and Development Office – Denver, Colorado, Bioreactors Remove Selenium from Water at Reduced Cost, Bulletin No. 30, September 2011.

US Environmental Protection Agency (USEPA), 2000. Constructed Wetlands Treatment of Municipal Wastewaters Manual, EPA/625/R-99/010.

US Environmental Protection Agency (USEPA), 2006. Leviathan Mine Compost Free sulfate reducing bioreactor treatment of Aspen Seep, EPA Hardrock Mining Innovative Technology Case Study.

US Environmental Protection Agency (USEPA), 2014. Reference Guide to Treatment Technologies for Mining-Influenced Water.

Zagury, G J and Neculita, C M, 2007. Passive treatment of acid mine drainage in bioreactors using sulfate-reducing bacteria: critical review and research needs, *Ottawa Geo,* pp 1439–1446.

Zhang, Y and Frankenberger, W T, 2005. Removal of selenium from river water by a microbial community enhanced with Enterobacter taylorae in organic carbon coated sand columns, *Science of the Total Environment*, 346:280–285.

AUTHOR INDEX

www.ingramcontent.com/pod-product-compliance
Lightning Source LLC
Chambersburg PA
CBHW061103210326
41597CB00021B/3960